Energy Systems

Series editor

Panos M. Pardalos, Gainesville, USA

For further volumes:
http://www.springer.com/series/8368

Panos M. Pardalos · Steffen Rebennack
Mario V. F. Pereira · Niko A. Iliadis
Vijay Pappu
Editors

Handbook of Wind Power Systems

Springer

Editors
Panos M. Pardalos
Vijay Pappu
Industrial and Systems Engineering
University of Florida
Gainesville
USA

Steffen Rebennack
Division of Economics and Business
Engineering Hall
Colorado School of Mines
Golden
USA

Mario V. F. Pereira
Centro Empresarial Rio Praia de Botafogo
Rio de Janeiro
Brazil

Niko A. Iliadis
EnerCoRD—Energy Consulting, Research and Development
Athens
Greece

ISSN 1867-8998 ISSN 1867-9005 (electronic)
ISBN 978-3-642-41079-6 ISBN 978-3-642-41080-2 (eBook)
DOI 10.1007/978-3-642-41080-2
Springer Heidelberg New York Dordrecht London

Library of Congress Control Number: 2013957555

Printed on acid-free paper

Springer is part of Springer Science+Business Media (www.springer.com)

Preface

World wind generation capacity has more than quadrupled between 2000 and 2006, doubling every 3 years. At the end of 2012, worldwide wind capacity was 282 GW, growing by 44 GW over the preceding year. Due to this reason, it is considered as the fastest growing energy resource in the world currently. Several factors have led to this transformation in the last decade. In particular, advances in technologies have certainly played a major role in the wind energy conversion. In addition, in countries like United States, wind energy production receives several subsidies to increase its attractiveness, and also to compensate for subsidies received by other forms of production.

The *Handbook on Wind Power Systems* is divided into four parts: optimization problems in wind power generation, grid integration of wind power systems, modeling, control and maintenance of wind facilities, and innovative wind energy generation.

Several optimization problems arising in wind power systems are addressed in the book. Reliability assessment unit commitment with uncertain wind power is addressed by Wang et al. while a wind farm layout optimization problem is discussed by Samorani et al. Additionally, several risk management tools for wind power trades are given by Yamada et al. Another chapter on innovative wind energy models and prediction methodologies by Sen et al. is also included.

Grid integration of wind power systems is an important problem and has been addressed by several authors in the book. Vespucci et. al. talk about stochastic models for integration of wind power production in a conventional power production system, while Santoso et al. discuss modeling of wind power plants. Deterministic and probabilistic approaches for steady-state analysis of distribution systems with wind farms is discussed by Carpinelli et al. Additionally, advanced control functionalities for grid integration of large-scale wind generation is addressed by Resende et al. Network stability issues arising under high wind power penetration is discussed by Tsikalakis et al. while power system operations with large penetrations of wind power is discussed by Denny et al. Rosa et al. assess the operational reserves under wind power fluctuations in power systems.

Several chapters in the handbook are dedicated to modeling, control, and maintenance of wind facilities. Namak et al. give a comprehensive review of floating wind turbine controllers, while Ramìrez et al. discuss in detail the modeling aspects of wind turbines and wind parks. Another chapter on grid support capabilities of wind turbines is discussed by Michalke et al. Coordination between wind farms and storage devices is discussed by Castronuovo et al. while a hybrid offshore wind and tidal turbine generation system is discussed by Rahman et al. Wind energy facility maintenance and monitoring are discussed by Ding et al. and Milan et al.

Satellite remote sensing in offshore wind energy is discussed by Hasager et al, while Optimization of AC electric power systems of offshore wind farms in addressed by Ramos et al. Low power wind energy conversion systems are discussed by Bratcu et al. and small wind-driven devices are discussed by Ahmed et al.

The chapters of the handbook are contributed by experts working on different aspects of wind energy generation and conversion. We would like to thank all the authors for their contribution, all reviewers for their constructive comments, and the Springer team for making this project possible.

Panos M. Pardalos
Steffen Rebennack
Mario V. F. Pereira
Niko A. Iliadis
Vijay Pappu

Contents

Part I
Optimization in Wind Power Generation

Reliability Assessment Unit Commitment with Uncertain Wind Power

Jianhui Wang, Jorge Valenzuela, Audun Botterud, Hrvoje Keko, Ricardo Bessa and Vladimiro Miranda

Abstract This book chapter reports a study on the importance of modeling wind power uncertainty in the reliability assessment commitment procedure. The study compares, in terms of economic and reliability benefits, the deterministic and stochastic approaches to modeling wind power. The report describes the mathematical formulation of both approaches and gives numerical results on a 10-unit test system. It is found that scenario representation of wind power uncertainty in conjunction with a proper reserve margin to accommodate for wind power uncertainty may provide higher benefits to market participants.

Nomenclature

Indices

i	Index for wind unit, $i = 1.. I$
j	Index for thermal unit, $j = 1.. J$
k	Index for time period, $k = 1.. 24$
l	Index for generation block, thermal units, $l = 1.. L$
m	Index for reserve demand block, m = 1..M
s	Index for scenario, $s = 1.. S$

J. Wang (✉) · J. Valenzuela · A. Botterud
Decision and Information Sciences Division, Argonne National Laboratory, Argonne, IL 60439, USA
e-mail: jianhui.wang@anl.gov

J. Valenzuela
Department of Industrial Systems Engineering, Auburn University, Alabama, USA

H. Keko · R. Bessa · V. Miranda
INESC Porto, Institute for Systems and Computer Engineering of Porto, Porto, Portugal

H. Keko · R. Bessa · V. Miranda
Faculty of Engineering, University of Porto, Porto, Portugal

P. M. Pardalos et al. (eds.), *Handbook of Wind Power Systems*, Energy Systems,
DOI: 10.1007/978-3-642-41080-2_1,

Constants

a, b, c	Unit production cost function coefficients
$\alpha(s)$	Operating reserve percentage, scenario s
$WR(k)$	Additional wind reserve, period k
$D(k)$	Load, period k
C_{ens}	Cost of energy not served
$CR_{rns,m}$	Cost of reserve not served, block m
A_j	Operating cost at minimum load, thermal unit j
$MC_{l,j}$	Marginal cost (or bid), block l, thermal unit j
$\overline{PT}_j$	Capacity, thermal unit j
$\underline{PT}_j$	Minimum output, thermal unit j
$\overline{\Delta}_{l,j}$	Capacity, block l, thermal unit j
CC_j	Cold start cost, thermal unit j
HC_j	Hot start cost, thermal unit j
$\mathbf{G}(\cdot)$	Generalized network constraints
T_j^{cold}	Time for cold start cost (in addition to minimum downtime), thermal unit j
T_j^{up}	Minimum up-time, thermal unit j
$T_j^{up,0}$	Minimum up-time, initial time step, thermal unit j
T_j^{dn}	Minimum down-time, thermal unit j
$T_j^{dn,0}$	Minimum down-time, initial time step, thermal unit j
SU_j	Start-up ramp limit, thermal unit j
SD_j	Shut-down ramp limit, thermal unit j
RL_j	Ramping limit (up/down), thermal unit j
$W_i(k)$	Actual maximum wind generation, wind unit i, period k
$PW_i^{f,s}(k)$	Forecasted maximum generation, wind unit i, period k, scenario s
$prob_s$	Probability of occurrence, wind scenario s

Variables

$c_j^p(k)$	Production cost, thermal unit j, period k
$c_j^u(k)$	Start-up cost, thermal unit j, period k
$pt_j(k)$	Generation, thermal unit j, period k
$\delta_{l,j}(k)$	Generation, block l, thermal unit j, period k
$\overline{pt}_j(k)$	Maximum feasible generation, thermal unit j, period k
$v_j(k)$	Binary on/off variable, thermal unit j, period k
$pw_i^s(k)$	Generation, wind unit i, period k, scenario s
$cw_i^s(k)$	Curtailed wind generation, wind unit i, period k, scenario s
$ens^s(k)$	Energy not served, period k, scenario s
$rns_m^s(k)$	Reserve curtailed, period k, scenario s
$r^s(k)$	Reserve requirement (spinning), scenario s, period k

1 Introduction

In many electricity markets, power producers and buyers submit bids to an Independent System Operator (ISO/RTO) for an amount of energy and the price they are willing to offer or pay [1]. The electricity market is usually structured into day-ahead (DA) and real-time (RT) markets. In the DA, power producers submit generation offers, and consumers submit demand bids. The ISO/RTO computes market clearing prices for the next 24 h by using a least-cost security-constrained unit commitment (SCUC), and a security-constrained economic-dispatch (SCED) optimization model. The objective of ISO/RTO is to meet the demand at minimum cost while maintaining the reliability of the system.

However, high penetrating of wind power has caused great challenges to the ability of power system operators to reliably operating the system due to the uncertainty and variability from wind power, especially in the unit commitment stage. How to commit the generating units optimally to address the fluctuations of wind power becomes very critical. To account for load fluctuations, outages, and wind power output uncertainties, the unit commitment algorithms co-optimize energy and ancillary services. Most of current research is focused on how to integrate the wind power forecasts into the day-ahead SCUC only without simulating a complete market procedure. Barth et al. [2] presented the early stage of the Wind Power Integration in the Liberalised Electricity Markets (WILMAR) model in [3]. More recently, a more comprehensive UC algorithm based on Mixed Integer Linear Programming (MILP) has been introduced in WILMAR. However, the model is mainly a planning tool. Tuohy et al. [4] extended their previous studies in [5] and [6] to examine the effects of stochastic wind and load on the unit commitment and dispatch of power systems with high levels of wind power by using the WILMAR model. The model builds on the assumptions needed for the hours-ahead or day-ahead system scheduling. The analysis compares only the scheduling alternatives at the scheduling stage. The effectiveness of the methods should be examined further by analyzing the operational impact in the real-time market, where the realized wind generation is likely to deviate from the forecast. Ummels et al. [7] analyzed the impacts of wind power on thermal generation unit commitment and dispatch in the Dutch system, which has a significant share of combined heat and power (CHP) units. Bouffard and Galiana [8] proposed a stochastic unit commitment model to integrate significant wind power generation while maintaining the security of the system. Rather than being pre-defined, the reserve requirements are determined by simulating the wind power realization in the scenarios. Ruiz et al. [9] proposed a stochastic formulation to manage uncertainty in the unit commitment problem. The stochastic alternative to the traditional deterministic approach can capture several sources of uncertainty, and system reserve requirements can be determined for each scenario. In a related paper [10], the authors consider uncertainty and variability in wind power in the

UC problem by using the same stochastic framework. Wang et al. [11] presented a SCUC algorithm that takes into account the intermittency and variability of wind power generation. The wind power uncertainty is represented by scenarios, and Bender's decomposition is used to solve the problem.

In most of the existing markets, after the clearing of the DA market, the ISO/RTO performs a revised commitment procedure focusing on the reliability of the power system. This procedure is called Reliability Assessment Commitment (RAC). In this procedure, the demand bids are replaced with the forecasted load for the next day. Since this procedure is performed several hours after the DA market is cleared, the ISO/RTO may change the commitment schedule from the DA market clearing because of reliability issues. After the RAC procedure, the SCED is executed again every 5 min to economically dispatch the units. The 5-minute prices are integrated over the hour to obtain the real-time hourly prices. As in the DA market clearing, the RAC generates a new commitment schedule. In the real time, the SCED determines the hourly dispatch results and energy prices. The market settlement is based on real-time deviations from the DA transactions over the hour. A RT demand that exceeds its DA quantity pays a RT price for the shortage and a demand that is below its DA quantity is paid the RT price for the surplus. In the market procedure, wind power forecasting can play a role in several places as shown in Fig. 1. As can be seen in Fig. 1, the wind power forecast can help determine the operating reserve requirements, as a part of the required reserves may be used to accommodate the uncertainty and variability from wind power. Wind power bidding strategies should also be based on wind power forecasts to predict the actual wind power output in the real-time market. Furthermore, updated wind power forecasts can be used by the ISO/RTO in the RAC process to provide more accurate information. Moreover, wind power forecasts can be used in the operating day to guide the real-time system operations. The use of wind power forecasting in U.S. electricity markets is further discussed in [12].

Given that the day-ahead market is generally cleared as a financial market, we investigate the role of wind power forecasting in the reliability unit commitment by simulating the market procedure through a market simulation model in this book chapter. The model is used to investigate the effects of wind power uncertainty on unit commitment and dispatch decisions and to analyze its impact on reserve requirements for system operations. The purpose of this book chapter is to demonstrate that modeling wind power uncertainty properly can efficiently deal with the uncertainty and variability associated with high penetration of wind power generation in current markets. Also, a stochastic approach that utilizes a set of scenarios to represent the wind power uncertainty is developed and described in this book chapter. A deterministic approach that uses a wind power point forecast is implemented with the purpose to compare it to the stochastic approach.

The following sections give the mathematical formulation of both approaches and preliminary results are given for a 10-unit power system.

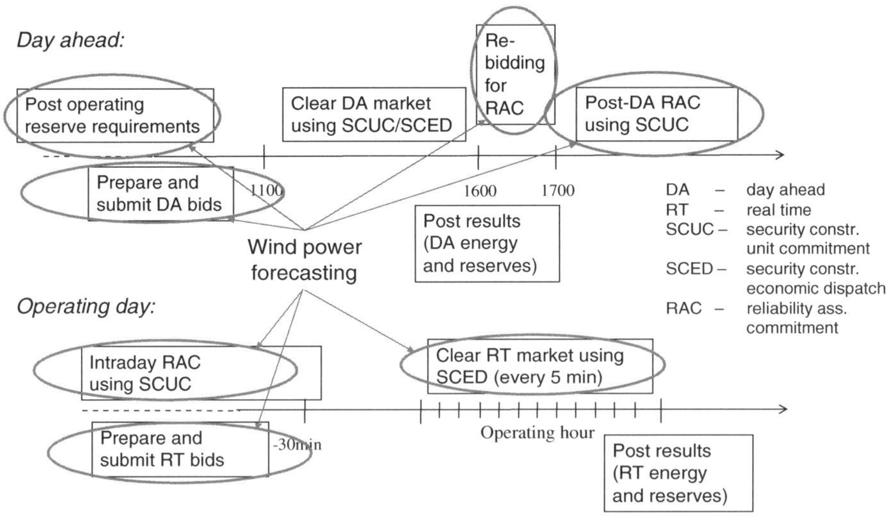

Fig. 1 Market operations timeline for midwest ISO, indicating where wind power forecasting could play an important role[1]

2 Unit Commitment and Dispatch Formulations

The general UC constraints follow the deterministic model in [13]. However, we make adjustments in this stochastic version based on the introduction of wind power and wind power forecasting uncertainty, which is represented in the form of scenarios. Our stochastic unit commitment model was first presented in [14]. We include the formulation here for completeness.

1. *Objective Function*

The objective is to minimize the sum of expected production costs, the expected cost of unserved energy and reserve curtailment, and start-up costs, as shown in (1). Constraints on load and operating reserves are represented in (2)–(3). We use a step-wise reserve demand curve to mimic the reserve requirement practiced by some system operators, such as MISO, as shown in (4). This formulation allows the reserve requirement, represented by a percentage of wind power, to be reduced at the reserve curtailment cost in some cases to avoid load curtailment. The idea is that this wind power reserve helps accommodate the uncertainty and variability from wind at the day-ahead UC stage since we do not simulate load forecasting errors and contingencies. Wind units may also be curtailed if necessary, as shown in (5). It is of note that the thermal dispatch, and therefore the production cost and the curtailed energy and reserve costs, all vary by wind scenario. The constraints

[1] Reprinted from [12], with permission from Elsevier.

for load, operating reserves, and wind curtailment must be met in all wind scenarios. In contrast, the start-up costs are independent of wind scenarios. This is because we assume that the commitment of thermal units has to be fixed at the day-ahead stage.

We assume that each thermal unit is offered into the market as a step-wise price-quantity offer function, and that the offers can be derived by linearizing a standard quadratic production cost function. Hence, we can express the operating cost for one thermal unit with the equations in (6)–(9). The coefficients for the generation blocks are derived from the quadratic production cost function. The last part of the objective function is the start-up cost. This part is modeled by assuming that there is a cold start-up cost and a warm start-up cost, depending on the length of time that the unit has been down. The mathematical formulation is shown in (10)–(12).

2. *Thermal Unit Constraints*

The constraints for the operation of thermal units include generation limits, ramping-up limits, ramping-down limits, minimum-up time, and minimum-down time. The upper and lower generation limits for the thermal plants are shown in (13). The maximum power output of a unit, $\overline{pt}_j^s(k)$, is constrained by the generation limit of a unit in (14), limitations on start-up and ramp-up rates in (15) shut-down ramp rates in (16), and ramp-down limits in (17). The availability of spinning reserves is equal to the difference between the maximum potential generation and the actual generation, that is, $\overline{pt}_j^s(k) - pt_j^s$. Hence, the reserve requirement in (5) takes into account the constraints imposed by (13)–(17). The reserve requirement is maintained for each individual wind scenario.

The final constraints included are the minimum-up and -down time constraints. Minimum-up times are represented by (18)–(20), which represent the initial status, the intermediate time periods, and the final time steps of the planning period, respectively. The minimum-down time constraints are represented analogously by (21)–(23). It is of note that the equations for generation and ramping limits, (13)–(17), must be included for all wind scenarios, because thermal dispatch depends on the wind generation. In contrast, the minimum-up and minimum-down time constraints, (18)–(23), are functions of commitment only and do not vary with wind scenarios. Generalized network constraints are represented in (24).

$$\begin{aligned} Min \sum_{s=1}^{S} prob_s \cdot & \left\{ \sum_{k=1}^{K} \sum_{j=1}^{J} c_j^{p,s}(k) + \sum_{k=1}^{K} C_{ens} \times ens^s(k) \right. \\ & \left. + \sum_{k=1}^{K} \sum_{m=1}^{M} CR_{rns,m} \times rns_m^s(k) \right\} + \sum_{k=1}^{K} \sum_{j=1}^{J} c_j^u(k) \end{aligned} \quad (1)$$

s.t.

$$\sum_{i=1}^{I} pw_i^s(k) + \sum_{j=1}^{J} pt_j^s(k) = D(k) - ens^s(k), \quad \forall k, \forall s \tag{2}$$

$$\sum_{j=1}^{J} \left[\overline{pt}_j^s(k) - pt_j^s(k) \right] \geq r^s(k), \quad \forall k, \forall s \tag{3}$$

$$r^s(k) = WR(k) - \sum_{m=1}^{M} rns_m^s(k), \quad \forall k, \forall s \tag{4}$$

$$pw_i^s(k) + cw_i^s(k) = PW_i^{f,s}(k), \quad \forall i, \forall k, \forall s \tag{5}$$

$$c_j^{p,s}(k) = A_j v_j(k) + \sum_{l=1}^{L} MC_{l,j}(k) \cdot \delta_{l,j}^s(k), \quad \forall j, \forall k, \forall s \tag{6}$$

$$pt_j^s(k) = \underline{PT}_j \cdot v_j(k) + \sum_{l=1}^{L} \delta_{l,j}^s(k), \quad \forall j, \forall k, \forall s \tag{7}$$

$$\delta_{l,j}^s(k) \leq \overline{\Delta}_{\mathrm{l,j}}, \quad \forall l, \forall j, \forall k, \forall s \tag{8}$$

$$\delta_{l,j}^s(k) \geq 0, \quad \forall l, \forall j, \forall k, \forall s \tag{9}$$

$$c_j^u(k) \geq CC_j \cdot \left[v_j(k) - \sum_{n=1}^{N} v_j(k-n) \right], \quad \forall j, \forall k \quad \text{where } N = T_j^{dn} + T_j^{cold}. \tag{10}$$

$$c_j^u(k) \geq HC_j \cdot \left[v_j(k) - v_j(k-1) \right], \quad \forall j, \forall k \tag{11}$$

$$c_j^u(k) \geq 0, \quad \forall j, \forall k \tag{12}$$

$$\underline{PT}_j \cdot v_j(k) \leq pt_j^s(k) \leq \overline{pt}_j^s(k), \quad \forall j, \forall k, \forall s \tag{13}$$

$$0 \leq \overline{pt}_j^s\ (k) \leq \overline{PT}_j \cdot v_j(k), \quad \forall j, \forall k, \forall s \tag{14}$$

$$\begin{aligned} \overline{pt}_j^s(k) \leq\ & pt_j^s(k-1) + RL_j \cdot v_j(k-1) \\ & + SU_j \cdot \left[v_j(k) - v_j(k-1) \right] \\ & + \overline{PT}_j \cdot \left[1 - v_j(k) \right], \quad \forall j, \forall k, \forall s \end{aligned} \tag{15}$$

$$\overline{pt}_j^s(k) \leq \overline{PT}_j \cdot v_j(k+1) + SD_j \cdot \left[v_j(k) - v_j(k+1) \right], \quad \forall j, \forall k = 1\ldots23, \forall s \tag{16}$$

$$pt_j^s(k-1) - pt_j^s(k) \le RL_j \cdot v_j(k) + SD_j \cdot \left[v_j(k-1) - v_j(k)\right] + \overline{PT}_j \cdot \left[1 - v_j(k-1)\right], \quad \forall j, \forall k, \forall s \tag{17}$$

$$\sum_{k=1}^{T_j^{up,0}} \left[1 - v_j(k)\right] = 0, \quad \forall j \tag{18}$$

$$\sum_{n=k}^{k+T_j^{up}-1} v_j(n) \ge T_j^{up} \cdot \left[v_j(k) - v_j(k-1)\right], \quad \forall j, \forall k = T_j^{up,0}+1, \ldots, T - T_j^{up}+1 \tag{19}$$

$$\sum_{n=k}^{T} \left\{v_j(n) - \left[v_j(k) - v_j(k-1)\right]\right\} \ge 0, \quad \forall j, \forall k = T - T_j^{up}+2, \ldots, T \tag{20}$$

$$\sum_{k=1}^{T_j^{dn,0}} v_j(k) = 0, \quad \forall j \tag{21}$$

$$\sum_{n=k}^{k+T_j^{dn}-1} \left[1 - v_j(n)\right] \ge T_j^{dn} \cdot \left[v_j(k-1) - v_j(k)\right], \quad \forall j, \forall k = T_j^{dn,0}+1, \ldots, T - T_j^{dn}+1 \tag{22}$$

$$\sum_{n=k}^{T} \left\{1 - v_j(n) - \left[v_j(k-1) - v_j(k)\right]\right\} \ge 0, \quad \forall j, \forall k = T - T_j^{dn}+2, \ldots, T \tag{23}$$

$$\boldsymbol{G}\left(pw_i^s(k), pt_j^s(k)\right) \le 0 \tag{24}$$

3. *Deterministic Formulation*

In a simplified representation, the formulation above would consider only one scenario for forecasted wind generation. In this case, the formulation is equivalent to a deterministic version of the UC problem. The selected scenario could be the expected wind power generation or point forecast or could also represent a certain quantile in the forecasting probability distribution.

4. *Economic Dispatch*

In order to assess the dispatch cost in real-time, we also develop an economic dispatch formulation. The commitment variables are now assumed to be fixed from the UC run. The representation of wind power generation by scenarios is replaced by the realized wind power output (without considering potential wind power curtailment). Hence, we formulate a deterministic economic dispatch problem consisting of Eqs. (1)–(9) and (13)–(17) with only one wind power scenario and fixed values for the thermal commitment variables, $v_j(k)$. The start-up cost and minimum-up and -down time constraints are not considered because of the fixed commitment. Ramping constraints are in (13)–(17), and the 24 h problem is solved simultaneously. It is of note that the operating reserve requirement in (3) is also imposed in the ED formulation.

3 Market Simulation

A market simulation set up has been put in place, which first clears the day-ahead market. This is done by first running UC and then ED on the basis of a day-ahead wind power forecast. Next, the RAC is performed with a new forecast, which could be a deterministic point forecast or a set of scenarios, as explained below. Finally, the real-time ED is run on the basis of the realized wind conditions. This action is performed in sequence for multiple days. An updated wind power forecast along with the unit status and generation output for the thermal units from the previous day are taken as initial conditions for the UC problem for the next day. The main results for the day-ahead and real-time market operations (UC, dispatch, available reserves, unserved load, curtailed reserve, prices, etc.) are calculated and stored after each simulation day.

Because the focus is the impact of wind power forecasts on system operation, the only uncertainty we consider is from wind generation. Other uncertainties, such as load or forced outage, are not directly considered. Hence, the additional amount of "wind power reserve" becomes critical to address the impact of uncertainty from use of wind power. With the stochastic UC formulation, the need for additional operating reserves is arguably already addressed because we include a representative set of wind power outcomes in the scenarios. However, because of the accuracy of the scenarios in capturing all of the possible wind power outputs, additional reserve may be needed even in the stochastic formulation to cover the extra uncertainty that the simulated scenarios are not able to capture. In the case study below, we run a number of different cases to investigate the impact of UC strategy and operating reserve policy on the system dispatch.

3.1 Deterministic Approach

In the deterministic approach, the DA market uses a wind power point forecast for the next 24 h of the operation day. Since the objective of the RAC procedure is to assure the reliability of the power system, a conservative estimate of the wind power, e.g. the 20th percentile, may be used in the procedure. After the units are

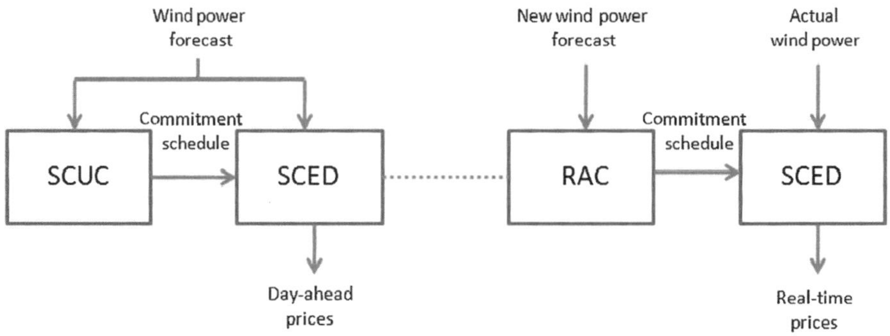

Fig. 2 Wind power forecast in the deterministic approach

committed in the RAC procedure, the units are economically dispatched by the SCED using the actual wind power. It is also assumed that there is no demand side bidding. Figure 1 illustrates the different stages of the market procedure and how the wind power data is used in the deterministic approach.

3.2 Stochastic Approach

In the stochastic approach, the DA market uses the same wind power forecast for the next 24 h of the operation day as in the deterministic approach. However, in comparison with the point forecast used in RAC in the deterministic approach, the RAC uses a set of wind power scenarios in the stochastic case. The scenarios are derived by using a combination of quantile regression and Monte-Carlo simulation [15, 16]. A comprehensive review of the state-of-the art in wind power forecasting, including uncertainty forecasts is provided in [17]. As in the deterministic approach, the dispatch of the units in the real-time market is completed by the SCED using the actual wind power. Again, we assume no uncertainty on the load and no generator outages. Figure 2 illustrates the use of the wind power data. It is also assumed that there is no demand side bidding in the stochastic approach.

4 Case Study

A test 10-unit power system is used to study the reliability unit commitment problem. Since we focus on the impact of different UC strategies and reserve requirement with wind power, we do not consider transmission constraints in this case study. Hence, the SCUC, SCED and RAC become UC and ED without transmission constraints and are solved in sequence as described in Figs. 2 and 3. The technical characteristics of the thermal units are given in Table 1. The values

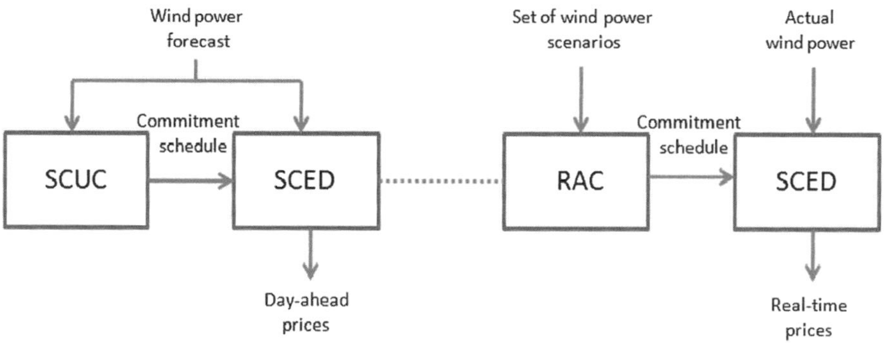

Fig. 3 Wind power forecast in the stochastic approach

Table 1 Generator data

Unit	$\overline{PT}_j$ (MW)	$\underline{PT}_j$ (MW)	RL_j (MW/h)	T_j^{up} [h]	T_j^{dn} [h]	In. state (h)
1	455	150	200	8	8	8
2	455	150	200	8	8	8
3	130	20	100	5	5	−5
4	130	20	100	5	5	−5
5	162	25	100	6	6	−6
6	80	20	80	3	3	−3
7	85	25	85	3	3	−3
8	55	10	55	1	1	−1
9	55	10	55	1	1	−1
10	55	10	55	1	1	−1
Unit	a_j (\$/h)	b_j (\$/MWh)	c_j (\$/MW^{2}h)	CC_j (\$/h)	HC_j (\$/h)	T_j^{cold} (h)
1	1,000	16	0.00048	9,000	4,500	5
2	970	17	0.00031	10,000	5,000	5
3	700	30	0.002	1,100	550	4
4	680	31	0.0021	1,120	560	4
5	450	32	0.004	1,800	900	4
6	370	40	0.0071	340	170	2
7	480	42	0.00079	520	260	2
8	660	60	0.0041	60	30	0
9	665	65	0.0022	60	30	0
10	670	70	0.0017	60	30	0

(*) Start-up and shut-down ramps, SU_j and SD_j, are equal to ramp rate RL_j

in this table are based on the case studies presented in [13, 18]. Ramp rates have been added to the table and the cost coefficients have been slightly modified. Each unit is assumed to have four blocks of equal size. The bid price of each block is based on the quadratic cost function. The production cost increases from unit 1 to unit 10, with units 1 and 2 being the baseload plants.

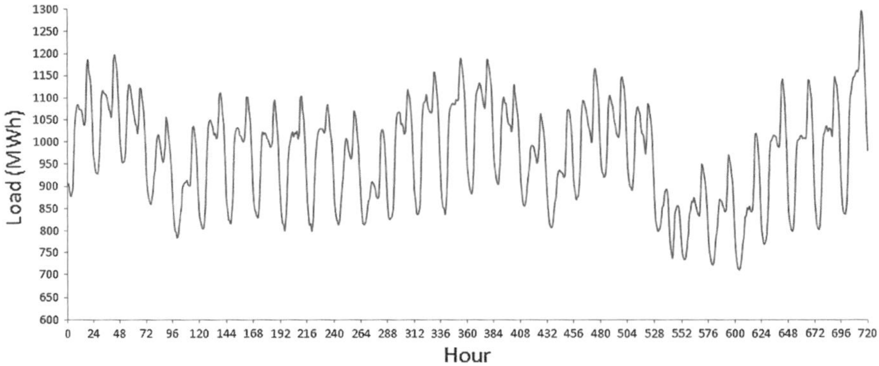

Fig. 4 Hourly load of November, 2006

The power system is simulated for 91 days. The hourly load profile corresponds to the historical data from two utilities in the state of Illinois for the months of October to December of 2006. The load has been scaled down to match the generation capacity of the test power system. Figure 4 shows the hourly load for the month of November. It is assumed perfect information about the load, i.e. the forecasted load is assumed to be equal to the actual load. Again, the reason for this assumption is to isolate the effects of wind power uncertainty from load uncertainty. Outages of thermal plants and wind farms are not simulated either. Thus, the results of the simulated cases show the effects of wind power uncertainty only. The cost of reserve curtailment is 1,100 ($/MWh) and the cost of unserved energy is 3,500 ($/MWh). The wind power plants do not provide reserves and therefore the operating reserve requirement is met by the thermal power plants .

The wind power data corresponds to wind power forecasts and realized wind power generation for 15 hypothetical locations in the state of Illinois for 2006. Time series of wind power generation for the 15 sites were obtained from NREL's EWITS study [19]. This data was produced by combining a weather model with a composite power curve for a number of potential sites for wind power farms. The forecasts were generated based on observed forecast errors from four real wind power plants. The wind power data for the 15 sites were aggregated into one time series. The accuracy of the day-ahead wind power forecast varies from day to day. For the forecast, the normalized mean average errors (NMAE) over a 91-day simulation period vary between 8.4 and 12.4 % for different hours of the day, with the highest forecast errors occurring in the afternoon between noon and 6 pm.

The total installed capacity of wind power is assumed to be 500 MW, and for simplicity it is modeled as one large wind power plant. For the simulated 91-day period, the wind power capacity factor is 40 %, and the wind power meets 20 % of the load (with no wind curtailment). Wind power and load are uncorrelated with a correlation coefficient of 0.01. With these assumptions, the total installed capacity of the thermal units is 10.8 % higher than the peak load. If we assign a capacity value of 20 % to the wind power capacity, the system reserve margin increases to 17.4 %.

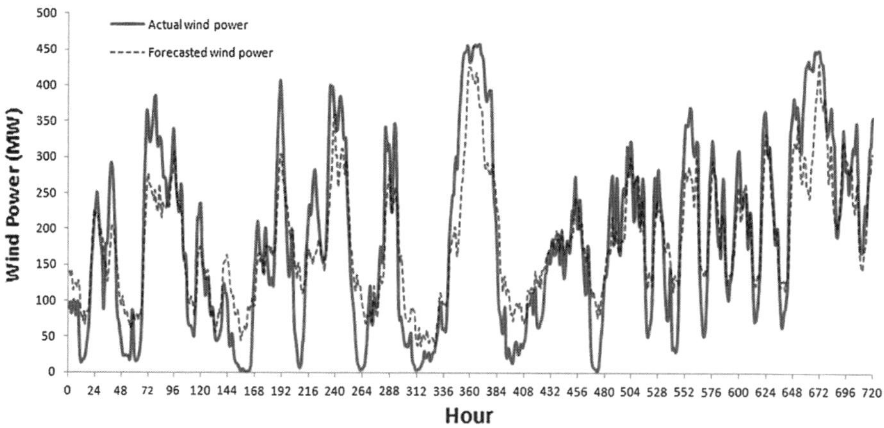

Fig. 5 Actual and forecasted hourly wind power of November, 2006

Table 2 Description of deterministic cases (10-unit system)

Case	UC and ED Reserve margin α (%)	RAC and ED Reserve margin α (%)
D0[a]	20	20
D1	20	20
D2	20	40
D3	40	20
D4	40	40
D5	No reserve	No reserve

[a] Case D0 uses a perfect forecast

Figure 5 shows the actual and forecasted hourly wind power for the month of November.

In the following sections, the results for the deterministic and stochastic approaches are compared. The purpose is to study how the use of wind power forecasts, reserve requirements, and unit commitment strategy influence cost, prices, and reliability in system operations.

1. *Deterministic Approach*

To study the impact of different reserve margins, five cases are developed. Their characteristics are shown in Table 2. All cases solve the UC and ED in the DA market using the same wind power forecast. The reason is to clear the DA market with the same amount of information on wind power in all cases. The RAC formulation uses an estimate of the 20th percentile of the wind power as a forecast. The 20th percentile is a conservative forecast of the actual wind power, which means that the actual wind power is expected to be higher than this forecast. In this way, the difference between the actual wind power and the 20th percentile acts as a natural reserve for the system. In order to clarify the structure of the cases, let us consider case D2 in Table 2. In case D2, the UC and ED use a forecast of wind power to commit and dispatch the thermal power plants in the day-ahead market. The reserve margin is set to 20 % of the wind power forecast to accommodate wind power variations. The RAC, on the other hand, commits the thermal units in the real-time market using a 20th percentile estimate of the wind power and a reserve margin equal to 40 % of the 20th percentile estimate. The thermal units are dispatched in the real-time market by ED using the actual wind power and with the same reserve margin as in the RAC procedure. Notice that all reserve requirements are functions of the wind power forecast. In addition, a hypothetical case D0 consisting of using a perfect wind power forecast (the forecast for the UC, ED, and RAC is the actual wind power) is also simulated. The fact that the forecast is a perfect prediction of the actual wind power is assumed to be unknown to the operator and a reserve of 20 % of the forecast is used to accommodate for the possible wind power variations.

A summary of the results for the six cases are provided in this section. The cases are solved using the mathematical programming formulation given in Sects. 2.1–2.4 and the commercial solver Lingo 12.0. A Lingo tolerance of 0.01 (1 % gap) is used in all cases for solving the integer programs. Microsoft Excel 2007 is used as the input and output interface. On average, it takes 12 s to simulate one day using a personal computer with 8 GB RAM and Intel(R) Core™2 Duo processor 3.33 GHz. Table 3 summarizes the overall performance during the real-time dispatch over the 91-day simulation.

As expected, D0 gives the best performance. It serves the load at all hours, and it has the lowest startup and production costs while maintaining the lowest percentage of units on-line. However, this case is unrealistic since the wind power cannot be predicted with absolute certainty. Case D5, on the other hand, is the worst case in terms of serving the load and total costs. 0.072 % (1,556.21 MWh over a total of 2161,989 MWh) of the total load is curtailed. The loss of load probability is equal to 0.0238 (52 h over a total of 2,184 h). It has also the higher total cost (startup + production + unserved load costs), a 16 % higher than the best case. Case D2 is the next best case in regard to serving the load. Under this case, 0.006 % of the total load is curtailed representing a 91 % improvement from the worst case D5. The loss of load probability is equal to 0.004, which represents

Table 3 10-unit system overall performance (91-day simulation of deterministic approach)

Case	Unserved load (MWh)	Unserved reserve (MWh)	Startup cost (M$)	Production cost (M$)	Unserved load cost (M$)
D0	0.00	5.5	100.70	34,357.24	0.00
D1	465.87	927.1	110.22	34,538.52	1,630.53
D2	137.37	1,678.6	110.70	34,676.18	480.80
D3	465.87	927.1	109.73	34,537.52	1,630.55
D4	137.37	1,678.6	110.49	34,676.46	480.80
D5	1,556.21	0.0	141.16	34,448.38	5,446.73

Table 4 10-unit system market prices (91-day simulation of deterministic approach)

Case	Average DA energy price ($/MWh)	Average RT energy price ($/MWh)	Load-weighted DA energy price ($/MWh)	Load-weighted RT energy price ($/MWh)
D0	26.91	26.91	27.60	27.60
D1	22.84	72.88	23.66	77.83
D2	22.76	64.64	23.56	68.88
D3	22.32	72.88	22.90	77.84
D4	22.76	64.64	23.50	68.89
D5	20.58	103.58	21.04	105.32

an 83 % improvement. This gain is at the cost of maintaining 5.4 % more units on-line than in the worst case. In addition to these improvements, case D2 reduces the total cost by 12.5 %. Case D4 performs similarly to case D2, except that the production cost is slightly higher. Cases D1 and D3 have similar performance but they do not provide better results than cases D2 and D4.

The six cases are also compared in terms of market prices. Table 4 summarizes the day-ahead and real-time prices over the 91-day simulation of each case.

In Table 4, Case D0 gives the lowest RT prices. There is no difference between the DA and RT prices because the actual wind power is used to clear both the DA and RT markets. The prices in the DA market are similar in all cases because the UC and ED in the DA market are solved using the same wind power forecast. However, the market prices in the RT market are much higher than those of the DA market. The reason is that the 20th percentile estimate of the wind power overestimates (27 % of the time) the actual wind power, which in turn increases the reserve requirements in the real-time market. In some hours, the reserve requirement causes the dispatch of more expensive thermal plants and sometimes the curtailment of the reserve or the load. For example, in case D2 the reserve is curtailed in 68 h and the load in 9 h. Case 5 produces the highest average RT price. The reason is that having no reserve requirements causes the ramp up of more expensive power plants to accommodate variations of the wind power. At the same time, the price goes up to the cost of unserved energy (3500 $/MWh) as soon as load curtailment occurs. The average DA price is the lowest because there is no cost attributed to reserve curtailment.

The results in Tables 3 and 4 demonstrate the importance of modeling the uncertainty of wind power and choosing an appropriate reserve margin to account for the wind power uncertainty.

2. *Stochastic Approach*

Three cases are developed to test the stochastic approach. Their characteristics are shown in Table 5. All cases solve the UC and ED in the DA market using the same wind power forecast as in the deterministic approach. In all cases, the RAC formulation uses a set of scenarios for representing the uncertainty of the wind power. For each scenario, a reserve margin is set to accommodate wind power uncertainty not captured by the scenarios. Within a case, the same percentage of reserve is

Table 5 10-unit system description of stochastic cases

Case	UC and ED Reserve margin α (%)	RAC and ED Reserve margin α(s)[a] (%)
S1	20	20
S2	40	20
S3	No reserve	No reserve

[a] The same percentage of reserve is used in all scenarios

Table 6 10-unit system overall performance (91-day simulation of stochastic approach)

Case	Unserved load (MWh)	Unserved reserve (MWh)	Startup cost (M$)	Production cost (M$)	Unserved load cost (M$)
D0	0.00	5.5	100.70	34,357.2	0.00
D2	137.37	1,678.6	110.70	34,676.2	480.8
S1	0.00	0.0	158.84	35,501.0	0.00
S2	0.00	0.0	160.92	35,511.5	0.00
S3	0.00	0.0	158.43	35,317.3	0.00

Table 7 10-unit system market prices (91-day simulation of stochastic approach)

Case	Average DA energy price ($/MWh)	Average RT energy price ($/MWh)	Load-weighted DA energy price ($/MWh)	Load-weighted RT energy price ($/MWh)
D0	26.91	26.91	27.60	27.60
D2	22.76	64.64	23.56	68.88
S1	22.78	19.97	23.58	20.33
S2	22.79	19.94	23.53	20.30
S3	20.42	20.03	20.90	20.39

used in all the scenarios. To produce the wind power scenarios for the months of October, November and December, data (forecasts and realized generation) for the period from January to July were used to train a quantile regression [20] and to estimate the co-variance matrix for the Monte-Carlo simulations. The months from August to December were used as a test dataset.

In this section, detailed results are provided for the three cases in Table 5. The cases are solved using the mathematical programming formulation given in Sects. 3.1–3.4 and the commercial solver Lingo 12.0. A Lingo tolerance of 0.01 (1 % gap) is used in all cases for solving the integer programs. On average, it takes 3 min to simulate one day. Table 6 summarizes the overall real-time performance over the 91-day simulation of each case. The results of cases D0 and D2 are shown again for comparison.

The cases are also compared in terms of market prices. Table 7 summarizes the day-ahead and real-time prices over the 91-day simulation of each case.

Except for case D0, all cases produce comparable total costs (startup + production + unserved load costs). However, they differ on the percentage of units that are kept online. The small dissimilarity observed in the DA average prices is due to the different reserve margins established in each case. As for the RT prices, a large difference between the stochastic approach cases and the deterministic case D2 exists. The three stochastic cases produce RT prices much lower than case D2. Moreover, the prices are lower than the prices produced by D0, which uses a perfect forecast of wind power. The reason is that more thermal plants are kept online in the stochastic cases than in the case D0, which in turn provides more flexibility in economically dispatching the thermal plants. Notice that case S3, which uses no reserve requirements, provides similar performance when it is compared to the other two stochastic cases. This fact implies that the wind power scenarios may be sufficient for dealing with the uncertainty of wind power.

5 Conclusion

In this book chapter, a market simulation model with reliability unit commitment is used to compare various unit commitment strategies to address the uncertainty from wind power. It is shown that the deterministic approach may not be suitable for coping with the complicating aspects of operations planning with large penetration of wind power. The deterministic approach intends to manage the uncertainty in wind power forecast through determining the commitments of thermal plants and scheduling sufficient levels of operating reserves to follow the rapid variations of the wind power. The stochastic approach, on the other hand, uses scenarios in modeling the uncertainty of wind power and determines an appropriate unit commitment by considering the dispatch of thermal plants across all the scenarios. Based on the simulation results, we conclude that the stochastic formulation is a promising alternative for coping with the uncertainty and variability of wind power and establishing an adequate reserve margin in the reliability unit commitment.

Acknowledgments The authors acknowledge the US Department of Energy, Office of Energy Efficiency and Renewable Energy through its Wind and Hydropower Technologies Program for funding the research presented in this paper. The submitted manuscript has been created by UChicago Argonne, LLC, Operator of Argonne National Laboratory (Argonne). Argonne, a US Department of Energy Office of Science laboratory, is operated under Contract No. DE-AC02-06CH11357.

References

1. Midwest ISO Energy and Operating Reserve Markets Business Practices Manual. Available at http://www.midwestiso.org/home
2. Barth R, Brand R, Meibom P, Weber C (2006) A stochastic unit commitment model for the evaluation of the impacts of the integration of large amounts of wind power. In: 9th international conference probabilistic methods applied to power systems, Stockholm, Sweden
3. Wind Power Integration in Liberalised Electricity Markets (Wilmar) Project. Available at: http://www.wilmar.risoe.dk

4. Tuohy A, Meibom P, Denny E, O'Malley M (2009) Unit commitment for systems with significant wind penetration. IEEE Trans Power Syst 24(2):592–601
5. Tuohy A, Denny E, Malley MO' (2007) Rolling unit commitment for systems with significant installed wind capacity. IEEE Lausanne Power Tech, pp 1380–1385, 1–5 July, 2007
6. Tuohy A, Meibom P, Malley MO (2008) Benefits of stochastic scheduling for power systems with significant installed wind power. In: Proceedings 10th international conference probabilistic methods applied to power systems (PMAPS), Mayagüez, Puerto Rico
7. Ummels BC, Gibescu M, Pelgrum E, Kling WL, Brand AJ (2007) Impacts of wind power on thermal generation unit commitment and dispatch. IEEE Trans Energy Convers 22(1):44–51
8. Bouffard F, Galiana F (2008) Stochastic security for operations planning with significant wind power generation. IEEE Trans Power Syst 23(2):306–316
9. Ruiz PA, Philbrick CR, Zak E, Cheung KW, Sauer PW (2009) Uncertainty management in the unit commitment problem. IEEE Trans Power Syst 24(2):642–651
10. Ruiz PA, Philbrick CR, Sauer PW (2009) Wind power day-ahead uncertainty management through stochastic unit commitment policies. In: Proceedings power systems conference and exhibition, Seattle, WA, Mar
11. Wang J, Shahidehpour J, Li Z (2008) Security-constrained unit commitment with volatile wind power generation. IEEE Trans Power Syst 23(3), Aug 2008
12. Botterud A, Wang J, Miranda V, Bessa RJ (2010) Wind power forecasting in U.S. electricity markets. Electr J 23(3):71–82
13. Carrion M, Arroyo JM (2006) A computationally efficient mixed-integer linear formulation for the thermal unit commitment problem. IEEE Trans Power Syst 21(3):1371–1378
14. Wang J, Botterud A, Bessa R, Keko H, Carvalho L, Issicaba D, Sumaili J, Miranda V (2011) Wind power forecasting uncertainty and unit commitment. Appl Energy 88(11):4014–4023
15. Nielsen HA, Madsen H, Nielsen TS (2006) Using quantile regression to extend an existing wind power forecasting system with probabilistic forecasts. Wind Energy 9(1–2):95–108
16. Pinson P, Papaefthymiou G, Klockl B, Nielsen HA, Madsen H (2009) From probabilistic forecasts to statistical scenarios of short-term wind power production. Wind Energy 12(1):51–62
17. Monteiro, C., R. Bessa, V. Miranda, A. Botterud, J. Wang, G. Conzelmann. Wind Power Forecasting: State-of-the-art 2009. Technical report ANL/DIS-10-1. Argonne National Laboratory, 2009. Available online at: http://www.dis.anl.gov/projects/windpowerforecasting.html
18. Kazarlis SA, Bakirtzis AG, Petridis V (1996) A genetic algorithm solution to the unit commitment problem. IEEE Trans Power Syst 11(1):83–92
19. Eastern Wind Integration and Transmission Study (EWITS), National Renewable Energy Laboratory (NREL). Information at: http://www.nrel.gov/wind/systemsintegration/ewits.html
20. Nielsen HA, Madsen H, Nielsen TS (2006) Using quantile regression to extend an existing wind power forecasting system with probabilistic forecasts. Wind Energy 9(1–2):95–108

The Wind Farm Layout Optimization Problem

Michele Samorani

Abstract An important phase of a wind farm design is solving the Wind Farm Layout Optimization Problem (WFLOP), which consists in optimally positioning the turbines within the wind farm so that the wake effects are minimized and therefore the expected power production maximized. Although this problem has been receiving increasing attention from the scientific community, the existing approaches do not completely respond to the needs of a wind farm developer, mainly because they do not address construction and logistical issues. This chapter describes the WFLOP, gives an overview on the existing work, and discusses the challenges that may be overcome by future research.

1 Introduction

Wind energy is the fastest growing source of renewable energy, as the worldwide production has doubled between 2005 and 2008, reaching 121.2 GW of total installed capacity. The transformation of wind power into electrical power is performed by wind turbines, which are usually grouped into wind farms in order to exploit considerations relative to economies of scale, such as lower installation and maintenance costs. But as costs decrease, grouping turbines leads to a reduction in the power produced because of the presence of wake effects within the wind farm. When a turbine extracts power from the wind, it generates a "wake" of turbulence that propagates downwind, so that the wind speed and therefore the power extracted by the turbines affected are reduced. In large wind farms wake effects lead to considerable power loss [1], and thus it is desirable to minimize them in order to maximize the expected power output. The wind farm layout optimization

M. Samorani (✉)
Alberta School of Business, University of Alberta, Edmonton, AB T6G 2R6, Canada
e-mail: samorani@ualberta.ca

P. M. Pardalos et al. (eds.), *Handbook of Wind Power Systems*, Energy Systems,
DOI: 10.1007/978-3-642-41080-2_2,

problem (WFLOP) consists of finding the turbine positioning (wind farm layout) that maximizes the expected power production. Finding high quality solutions may ultimately lead to high profits for the wind farm developers. Currently, this problem is usually solved using very simple rules that lead to rectilinear layouts, where turbines are often organized in identical rows that are separated by a conveniently large distance (http://www.offshorewindenergy.org). Recently, a few papers showed that irregular layouts result in a higher expected energy production than regular grids [2–8].

The existing work that has tackled this problem is very limited and has been carried out by the wind engineering and wind energy communities, whereas no effort has been done by the optimization community. Existing algorithms include only genetic algorithms and simulated annealing. There is therefore potential for improvement by using other optimization techniques, such as mixed-integer programming, dynamic programming, stochastic programming, etc... As it will be clear in the following, the main reasons why this problem has been largely disregarded by the operations research community are its nonlinearity and the difficulty in obtaining data about the problem instances.

In this chapter we present the process of building a wind farm, discuss the problems caused by the wake effects and how they impact energy production and maintenance costs, review the existing literature on the WFLOP, comment on the advantages and shortcomings of the existing methods, and lay out the lines of research that can be developed.

2 Construction of a Wind Farm

This section briefly describes the phases of a wind farm development project. The first step is finding a windy site to ensure the economic profitability of the project. Sites are usually classified in 7 different wind power classes (http://www.awea.org) that correspond to 7 different average wind speed intervals. Usually, sites with wind power class 4 or higher are considered potentially profitable for large size projects; nevertheless, not all sites belonging to high wind power classes are feasible sites for a project. In fact, some sites can be very far from the electrical grid or reachable only by roads that are not wide enough to transport very long trucks. In the first case, the site is not profitable because of the high cost to connect it to the electrical grid, in the second case, because of the high cost of the necessary road work.

Once a suitable site is found, land owners are contacted and asked if they are interested in hosting wind turbines on their land. Land owners that participate in the project usually receive a percentage of the profit generated by the turbines on their land, and extra money if roads or other infrastructure are built on their terrain. The process of contacting land owners and agreeing on the lease terms is not straightforward though, and usually takes a few months. In the meantime, the wind developer installs measurement towers in order to assess the wind distribution (or wind rose) of all the parts of the site. This measurement may take as short as one month for sites where the wind is known to blow from the same direction all year around, or as long as 2 years for sites where seasonal winds blow from different directions. The accuracy of these measurements is critical for the project because they are used to find the optimal layout and to assess the expected annual profit of the wind farm. Nevertheless, the measurements may be taken at significantly lower heights than the hub height of the wind turbines because of the high cost of installing tall towers. In these cases, atmospheric models such as the "power law model" [9] use the wind speed at measurement height to extrapolate the wind speed at hub height. Alternative measurement methods are Doppler SODARs and Doppler LIDARs. As described in [10], Doppler SODARs "measure the wind from acoustic energy that is backscattered from small-scale turbulent fluctuations of temperature (density)," whereas Doppler LIDARs measure the wind "using the light energy backscattered from microscopic particulates or aerosols being transported by the wind". Both technologies have the advantage of accurately computing the wind distribution in all parts of the site with no need to install measurement towers. Since it is so accurate [11], LIDAR is becoming a more and more popular tool used by wind farm developers. When the land that can be used for the project is known and the wind distribution has been obtained, the WFLOP is solved.

How to choose the number and the model of the turbines to install depends on a variety of factors. First, it is important to note that a more powerful turbine is usually preferred to a less powerful one since both the cost of a turbine and the energy it generates are usually proportional to its nominal power. Thus, the net profit generated by the turbine is also proportional to its nominal power. Obviously this property does not hold for extremely powerful (and therefore state-of-the-art) turbines because spare parts may be very expensive and maintenance costs very high. Since the trend is to build turbines that are more and more powerful, the cost of unused smaller turbines dramatically decreases. Therefore turbine manufacturers may offer advantageous discounts on small turbines in order to reduce their inventory. Other times, large wind farm development companies have their own turbine inventory. In this case, they may have extra turbines in stock that need to be used before they become obsolete. Given these considerations, the existing works on the WFLOP assume that the type and number of the turbines to install is predetermined.

3 Wind Turbines and Wake Effects

The characteristics of a wind turbine that are related to wind farm layout optimization are the following:

- Cut-in speed c_i
- Cut-out speed c_o

- Nominal speed
- Nominal power
- Power curve
- Thrust coefficient curve
- Rotor diameter d
- Hub height z

When the wind speed is greater than c_i, the blades of the turbine start spinning and therefore the turbine generates power. The power produced roughly increases with the cube of the wind speed until the *wind speed* reaches the nominal speed, at which point the control system of the turbine modifies the pitch of the blades so that the power produced is constant and equal to the *nominal power*. When the speed reaches c_o, it is considered too high, and the turbine is shut down to avoid damaging the blades.

Other important characteristics of interest are the *power curve* and the *thrust coefficient curve*. They report respectively the power produced and the value of the thrust coefficient (C_t) at every wind speed included between c_i and c_o. Roughly speaking, the thrust coefficient measures the proportion of energy captured when the wind passes through the blades of the turbine [12]. For both power curve and thrust coefficient curve, manufacturers usually provide a few data points, which need to be interpolated to obtain the intermediate points. For example, Fig. 1 shows the two curves of the Vestas turbine V63 ($c_i = 5$ m/s, $c_o = 25$ m/s, nominal power = 1.5 MW, nominal speed = 16 m/s). The data points, which are provided only for 1, 2, … 25 m/s, have been linearly interpolated.

By extracting energy from the wind, a turbine creates a cone *(wake)* of slower and more turbulent air behind it. This phenomenon, which is referred to as *wake effect* (see Fig. 2), has been studied by several authors in the fluid-aerodynamic field. In their survey of these studies, Vermeer et al. [13] focus on the experiments aimed at identifying mathematical models that accurately describe the wake effect, both in terms of wind speed reduction and turbulence intensity. Some of these models are only valid near the turbine that generates the wake (near wake models), and others are only valid far from the turbine that generates the wake (far wake models). Beyond 3 rotor diameters distance, Vermeer et al. [13] suggest to use the far wake models. In the near wake, the turbulence intensity is so large that the affected turbines must be shut down in order to avoid blade damage. Although more accurate computational methods have been proposed (in particular CFD—see [14]), the existing works on wind farm layout optimization use the model proposed by [15, 16] because its simplicity makes it more practical to embed within optimization procedures and in computer programs, such as PARK [17]. Although simple, this wake model has been shown to accurately compute the wind speed reduction in the far wake case [18, 19].

Let us explain the Jensen model through the example in Fig. 2. The wind blows from left to right at speed u_0 and hits a turbine (represented as a black rectangle on the left) whose rotor radius is r_r. At a distance x downwind, the wind speed is u and the wake radius (initially equal to r_r) becomes $r_1 = \alpha x + r_r$. The a-dimensional

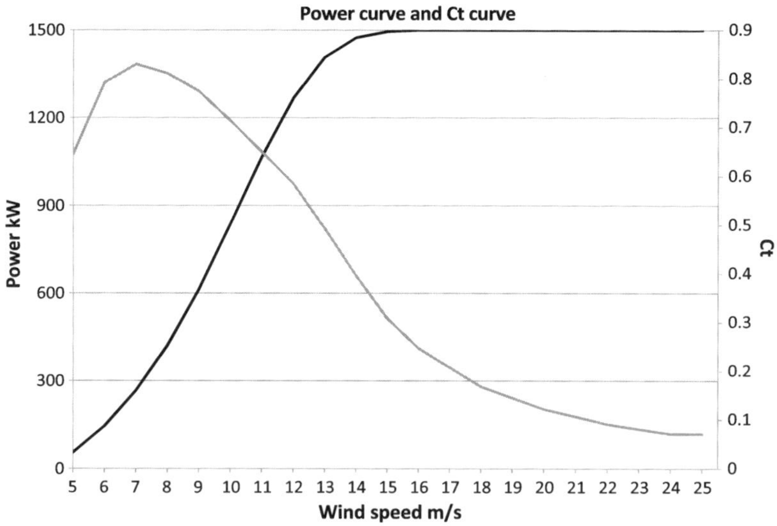

Fig. 1 Power curve (*black line*) and thrust coefficient curve (*grey line*) of the turbine Vestas V63

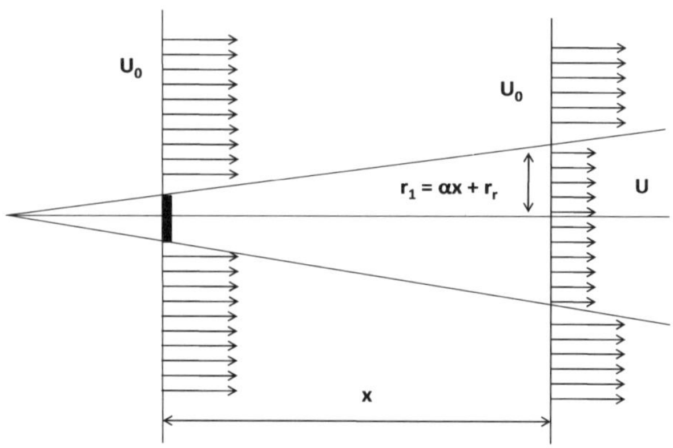

Fig. 2 Schematic representation of the wake effect [6]

scalar α determines how quickly the wake expands with distance and it is defined as:

$$\alpha = \frac{0.5}{\ln \frac{z}{z_0}} \tag{1}$$

where z is the hub height of the turbine generating the wake and z_0 is a constant called surface roughness, which depends on the terrain characteristics.

Let i be the position of the turbine that generates the wake, j the position affected by it, u_0 the ambient wind speed, and u_j the wind speed at j. Then:

$$u_j = u_0(1 - vd_{ij}) \tag{2}$$

where vd_{ij} is the *velocity deficit* induced on position j by the wake generated by i. vd_{ij} is computed as follows:

$$vd_{ij} = \frac{2a}{1 + \alpha\left(\frac{x_{ij}}{r_d}\right)^2} \tag{3}$$

The term a that appears in the numerator is called *axial induction factor* and is computed by the following expression:

$$a = 0.5\left(1 - \sqrt{1 - C_T}\right) \tag{4}$$

The term r_d that appears in the denominator is called *downstream rotor radius* and is equal to:

$$r_d = r_r\sqrt{\frac{1-a}{1-2a}} \tag{5}$$

The term x_{ij} is the distance between positions i and j. The notation we propose is coherent to the ones used in [2–4, 6, 8], and it is equivalent to the one proposed by [20].

Since many turbines are installed in a wind farm, wakes can intersect and affect turbines downwind at the same time. In the Jensen model, the total velocity deficit $v_{def}(j)$ at a location j that is affected by more wakes is obtained as follows:

$$v_{def}(j) = \sqrt{\sum_{i \in W(j)} vd_{ij}^2} \tag{6}$$

where $W(j)$ is the set of turbines affecting position j with a wake. $v_{def}(j)$ is then used in (2) in place of vd_{ij} to compute u_j. Let us illustrate this with the example reported in Fig. 3.

In this example the wind blows from left to right at a speed equal to $U_0 = 12$ m/s and the wakes generated by turbines A and B affect position C. We are interested in computing the wind speed U_C. The data of the problem are:

- $x_{AC} = 500$ m
- $x_{BC} = 200$ m
- $z = 60$ m
- $z_0 = 0.3$ m
- $r_r = 20$ m (i.e. $D = 40$ m)
- $C_T = 0.88$

In order to solve the problem, we use (1), (3)–(5) to compute vd_{AC} and vd_{BC} and obtain:

- $vd_{Ac} = 0.0208$
- $vd_{Bc} = 0.1116$

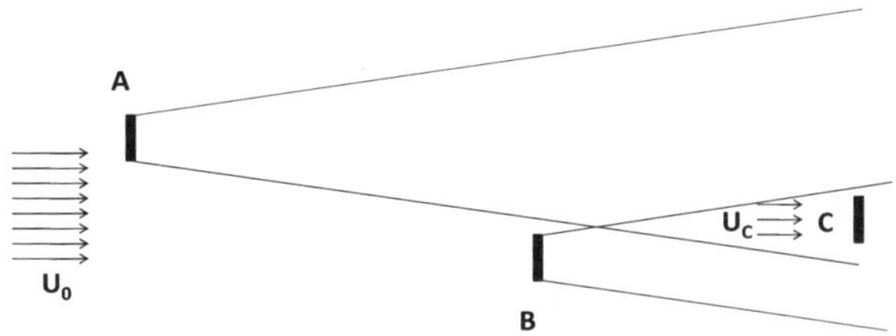

Fig. 3 Example of multiple wakes affecting a position

These partial results can be interpreted as follows:

- if C was affected only by A, then the wind speed at C would be reduced by 2.08 % from the ambient wind speed
- if C was affected only by B, then the wind speed at C would be reduced by 11.16 % from the ambient wind speed

Then we use (6) to compute the total velocity deficit $v_{def}(C) = 0.1135$ (wind speed at C reduced by 11.35 %) and (2) to compute $U_C = U_0\big(1 - v_{def}(C)\big) = 10.64\,\mathrm{m/s}$. Note that the computations would be the same even if B was inside the wake created by A.

This example highlights a very important property of multiple wake combinations: the total velocity deficit mostly depends on the closest turbine that generates a wake. In the example, the total velocity deficit $v_{def}(C)$ is very close to the velocity deficit vd_{BC} caused by turbine B. In other words, the presence of turbine A does not substantially affect the wind speed in C since U_C would be equal to 10.66 m/s if A was not present. However, the interaction of multiple wakes is not fully understood and is subject of many studies in the aerodynamics field. Vermeer et al. [13] report that recent studies highlighted significant discrepancies between the real and the estimated energy production in large wind farms, where this type of interaction has a big impact.

Under scenario s, characterized by a wind direction and an ambient wind speed, the power produced by the wind farm can be obtained by computing the wind speed v_j^s at each turbine location $j \in L$. Let $P(v)$ be the function, which we assume known, that computes the power generated by one turbine if the wind speed is v at the turbine location. The total power produced is obtained by summing up the contribution of all the turbines. If more scenarios are present, we are interested in the expected power produced, which is calculated by summing up the power produced in each scenario s weighted by the probability of its realization r_s. In the existing literature, the objective is to maximize the total power (T) function:

$$T = \sum_{s\in S} r_s \sum_{j\in L} P\left(v_j^s\right) = \sum_{s\in S} r_s \sum_{j\in L} P\left[U_s \cdot \left(1 - \sqrt{\sum_{i\in W^s(j)} vd_{ij}^2}\right)\right] \tag{7}$$

where $W^s(j)$ is the set of turbines affecting position j with a wake under scenario s. Although the decision variables (i.e., the locations of the turbines) do not explicitly appear in (7), the sets $W^s(j)$ directly depend on them. Obviously, a mathematical model should also include the constraints on the $W^s(j)$, in order to guarantee the correctness of the model. For example, if under scenario s position a affects position b and b affects position c, then a must affect also c. Since this is outside the scope of this survey, we leave the definition of a complete mathematical model to future work.

4 Work on Wind Farm Layout Optimization

The WFLOP has been largely neglected by the operations research and operations management communities. To date, most of the published works that address this problem appear in journals whose topics are related to energy and wind engineering. These articles apply existing optimization methodologies to solve different versions of this problem without focusing on the solution method itself. However, this topic has been receiving increasing attention, as reported in Fig. 4, which shows the number of papers published from 1992 to 2009 retrieved through a google scholar search with the following keywords: *"wind farm", wake, turbine, position, optimization*. Note that some of the retrieved papers focus on wake effect modeling and not on wind farm layout optimization. Nevertheless, the increasing presence of this type of papers indicates an increasing interest in accurately assessing the energy production of a wind farm so that it can be designed more carefully.

In the following discussion we identify and review the most prominent published works on this topic, with a particular emphasis on both their shortcomings and on the research opportunities that can be addressed by the optimization community. Mosetti et al. [6] were the first to take into consideration the WFLOP. They model the wind farm site as a 10×10 square grid, where the centers of the 100 squares are the possible positions of the turbines. To ensure the validity of the Jensen model, the side of each cell is $5D$—although $3D$ would be enough [13]. The turbine used has a hub height $z = 60$ m, diameter $D = 40$ m, and a constant thrust coefficient $C_T = 0.88$. The power curve, depicted in 5, is expressed by the following:

$$P(U) = \begin{cases} 0 & \text{for } U < 2 \\ 0.3U^3 & \text{for } 2 \leq U < 12.8 \\ 629.1 & \text{for } 12.8 \leq U < 18 \\ 0 & \text{for } 18 \leq U \end{cases}$$

where the wind speed U is expressed in m/s and the power in kW (Fig. 5).

Mosetti et al. [6], who do not assume a predefined number of turbines to install, define their goal as maximizing the power produced P_{tot} while minimizing the installation cost $cost_{tot}$. The power produced is derived as explained in the previous section, whereas the installation cost is defined as:

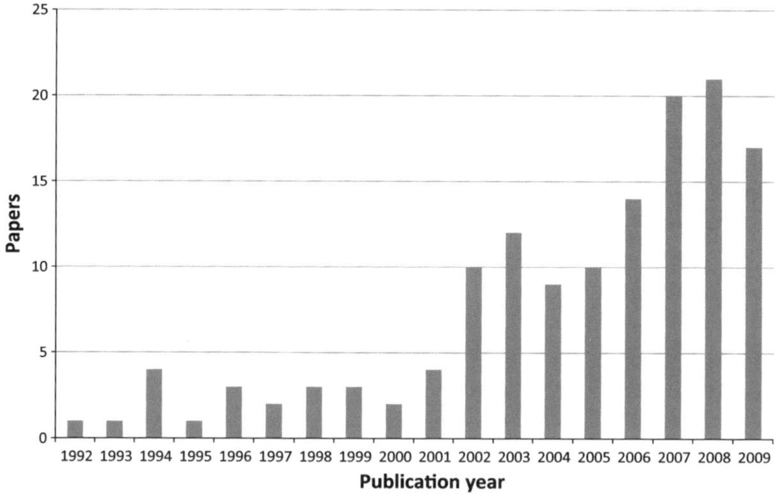

Fig. 4 Number of papers published from 1992 to 2009, retrieved through a google scholar search with the following keywords: "wind farm", wake, turbine, position, optimization

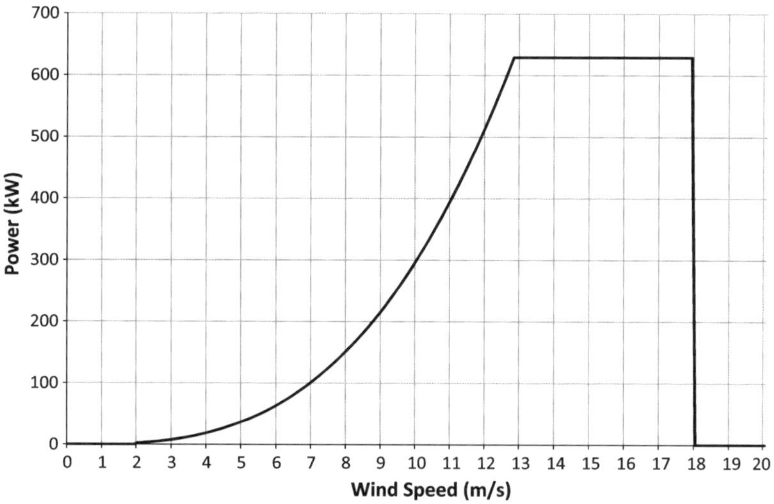

Fig. 5 Power curve used in [6]

$$cost_{tot} = N_t\left(\frac{2}{3} + \frac{1}{3}e^{-0.00174N_t^2}\right) \tag{8}$$

where N_t is the number of turbines installed. The cost of a turbine, which is the expression in parenthesis in the formula above, decreases as N_t increases, which reflects the economies of scale considerations reported in the introduction. The adopted objective function is:

$$Obj_{MOS} = \frac{1}{P_{tot}} w_1 + \frac{cost_{tot}}{P_{tot}} w_2 \tag{9}$$

where the weight w_1 has been kept small compared to w_2.

The solution method they use is based on Genetic Algorithms (GAs). GAs keep a population of solutions which iteratively evolves through combinations and selections. At each iteration, solutions are combined and a new solution is obtained whose components are inherited from one of two parents. A solution is represented by a vector of 100 binary variables x_i (with $i = 1,\ldots 100$), each indicating the presence of a turbine in position i. Therefore, combining two solutions effectively consists of generating a new solution that has some turbines in the same positions of the first parent and the others in the same positions of the second parent. After generating a new solution, some of its components may be changed in order to introduce diversity in the population. This mechanism is called mutation, given its resemblance to the genetic changes that are involved in the evolution of the species. The reader should consult [21] for an overview of GAs. Mosetti et al. [6] let a population of 200 individuals evolve for 400 iterations.

They introduce 3 problem instances: A, B, and C. In A the wind constantly blows from North at 12 m/s; in B the wind speed is 12 m/s but the direction is uniformly distributed across 36 angles having the same angular sectors width 10° in C the wind blows at 3 possible different speeds (8, 12, 17 m/s) from one the 36 directions described above. The probability distribution that describes the occurrence of each wind speed and direction is reported in Fig. 6.

A computational study shows that the solutions obtained by GAs have a higher objective function value and a higher efficiency than solutions obtained by installing turbines in random positions. The efficiency, which is a very common way of evaluating and comparing solutions, is defined in [6] as

> the ratio between the total energy extracted by the windfarm having N_t turbines and N_t times the energy extracted by an isolated turbine with the same undisturbed wind

Grady et al. [3] replicated the experiments presented in [6] by modifying the settings of the GA. In particular, they show that by letting 20 subpopulations evolve for 3,000 iterations one can achieve better solutions. More recently, Hou-Sheng [4] improved upon these results using a distributed GA where a small fraction of the highest quality individuals of each subpopulation periodically migrates to another sub- population. Sisbot et al. [8] proposed a multi-objective genetic algorithm, where two objectives are the one of maximizing the power produced and the one of minimizing the cost. Although interesting, their method has only been tested on the case where wind blows from a constant direction at a constant speed. They claim that this assumption makes it possible to have rectangular cells instead of square ones, so that the minimum distance between two turbines is 8D in the prevailing wind direction and 2D in the crosswind direction. This type of consideration is also the basis of the following rule of thumb that can be used to design a wind farm layout (http://guidedtour.windpower.org/en/tour/wres/park.htm, last access on 01/06/2010):

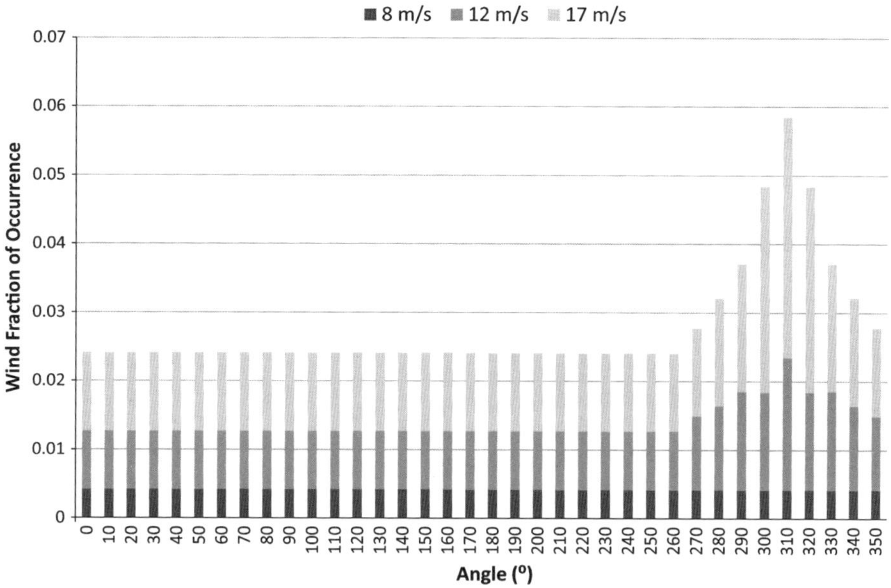

Fig. 6 Wind distribution in C [6]

> turbines in wind parks are usually spaced somewhere between 5 and 9 rotor diameters apart in the prevailing wind direction, and between 3 and 5 diameters apart in the direction perpendicular to the prevailing winds (see Fig. 7)

This simple approach ignores all wind directions except the prevailing one, and it is therefore likely to lead suboptimal wind farm layouts. Furthermore, it fails to describe how to compute the power produced when the wind blows from a crosswind direction, in which case the Jensen model is not valid because the distance between the turbines that generate a wake and the turbines affected by it is too short.

All these GA-based approaches share an evident shortcoming: the solution space is discrete. In other words, there is a predefined set of possible positions—the centers of the cells—of which a subset must be chosen as installation positions of the turbines. Since consecutive positions are spaced by a distance of several rotor diameters and it is not possible to choose intermediate positions, it follows that potentially better positions are not even considered. To solve this problem one may consider a finer grid, i.e., a grid whose cell sides are shorter, as long as proximity constraints are introduced to avoid infeasible solutions. These constraints forbid turbines to be installed too close to each other (at less than 3D distance). Unfortunately GAs do not offer a natural way to embed constraints, which would have to be enforced by introducing a feasibility check in the objective function evaluation routine, making the search significantly slower. Nevertheless, the computational impact of introducing proximity constraints is unknown because it has not been tried yet.

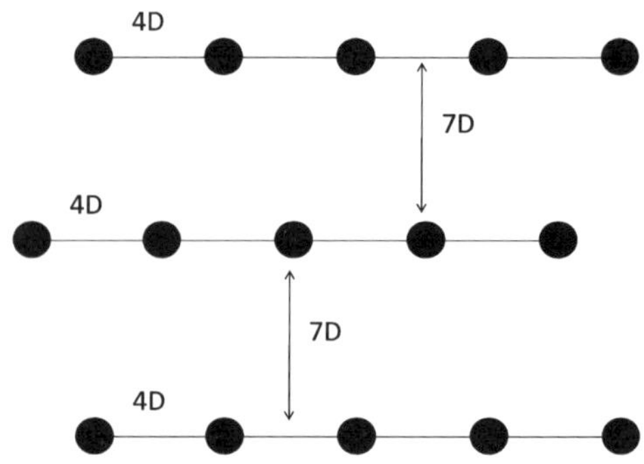

Fig. 7 Three rows of five turbines each, installed according to the rule of thumb

The first attempt to address the limitation of a discrete space was made by Aytun and Norman [2], who proposed a local search that iteratively considers the operations of adding, removing, or moving turbines in an attempt of improving the objective function value. The add operation randomly generates a set of locations, which are individually considered as potential installation positions of new turbines; the remove operation considers the removal of each existing turbine; the move operation attempts to move each existing turbine by 4D along a set of predefined directions. Whenever the add operation is considered, a new set of candidate locations is randomly generated and evaluated, and therefore turbines can be potentially placed in any position of the site. Nevertheless, it would be incorrect to say that this approach considers a continuous solution space; rather, it considers a discrete space where the possible positions are randomly generated during the search instead of being predefined.

A similar work, developed by Rivas et al. [7], consists of a simulated annealing procedure that uses the same set of moves (add, remove, move). Since simulated annealing is a neighborhood search that may accept non-improving moves [22], it overcomes the limitation of the search proposed in [2], which was purely local. Besides improving the solution method, Rivas et al. [7] performed a relevant computational study to assess the difference between the quality of solutions obtained by their approach and the quality of solutions obtained by the rule of thumb. They consider two different problems: one where we must install 106 3 MW turbines and one where we must install 64 5 MW turbines. As they note, the total power installed is similar for the two problems. Each problem is solved by imposing a predefined geographical extension (or site area) of the wind farm, which is equivalent to imposing a predefined density of installed power (the smaller the area, the higher the density). For all site areas considered, Rivas et al. [7] solved the two problems by either using the proposed method or the rule of thumb. Figure 8 summarizes their findings by reporting the efficiency obtained for both problems and for the considered site areas. The light dots represent the results obtained by the rule of thumb, while the dark ones the results obtained by simulating annealing. Obviously, when using the rule of thumb to solve either problem (106 or 64 turbines), if the site area increases, so does the efficiency of the wind

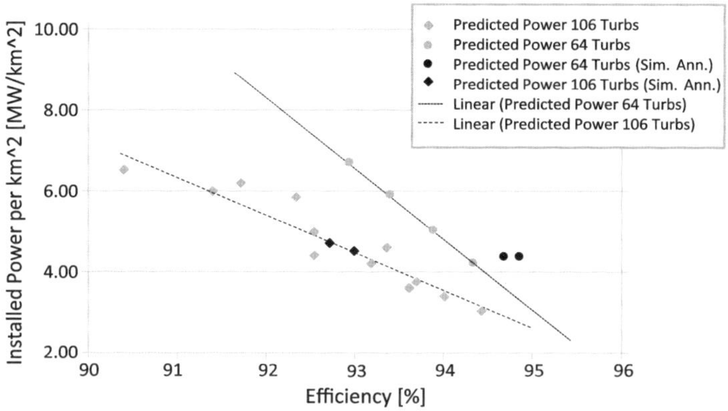

Fig. 8 Study by Rivas et al. [7] on the impact that site area and number of turbines have on efficiency

farm. This trend is highlighted by the two lines in Fig. 8. The relevant finding that is worthwhile noting here is that by using their method one finds higher quality solutions for the 64 turbine problem (they are on the upper half-plane of the line) but not for the 106 turbine problem (they lie on the line).

In other words, the potential improvement due to their method over the rule of thumb is more evident if the turbines are few and big (e.g. 64 5 MW turbines), whereas it tends to disappear if they are many and small (e.g. 106 3 MW turbines). Although this property may hold only for their method, it may hide a more general property, which is valid for any method: the chosen layout strongly impacts the quality of the solution if we install few turbines or, equivalently, the chosen layout does not strongly impact the quality of the solution if we install many turbines. The argument supporting this idea is based on the property of multiple wake superimposition, according to which *the total velocity deficit mostly depends on the closest turbine that generates a wake* (Sect. 3). If we are installing many turbines, we expect that most of them are impacted by at least one wake, regardless to how they are positioned; conversely, if we are installing few turbines, it is possible that only few or none of them are affected by a wake, for example if they are aligned in one row that is perpendicular to the wind direction. In the first case (many turbines) optimizing the layout may reduce the average number of wakes that affect the turbines at the same time, but this will have a little impact on the objective function value; in the second case (few turbines), optimizing the layout may prevent some turbines to be affected by a wake, which will have a great impact on the objective function. Nevertheless, further studies need to be carried out in order to validate this idea.

Let us now consider the first approach that actually considers a continuous solution space: the one proposed by [20]. They aim at finding the optimal wind farm layout for an offshore site where the number of turbines is predefined. An offshore scenario differs from an onshore one in that both costs and energy production strongly depend on the turbine positions. Installation and maintenance costs increase with the depth of the water, and so does the energy production because the wind speed becomes higher as we go farther from shore. Hence, there is an optimal trade-off between these two contrasting effects that determines how far from shore to install the turbines. The objective used is the one of minimizing the levelized cost of energy (LCOE), defined in [23] as:

$$LCOE = \frac{C_C \times FCR + C_{O\&M}}{AEP} \tag{10}$$

where C_c is the total *installed capital cost* of the wind farms (turbine, infrastructure, and transmission costs), FCR is the *fixed charge rate,* a present value factor that considers debt and equity costs, taxes, and insurance, $C_{O\&M}$ is the annual operations and maintenance cost, and AEP is the Annual Energy Production. The objective function is similar to OBJ_{MOS}, but it also includes also operations and maintenance costs, which make it more complete. The AEP is computed as explained in Sect. 3, except that the wind distribution is described through a continuous probability distribution instead of a set of scenarios. In particular, in agreement with [20, 24] fit the wind data with a parametric Weibull distribution that depends on the wind direction, the wind speed, and the position. The variables of the optimization problem are the coordinates of the turbines to install, whose number is fixed, and the optimization is performed through a gradient search that proceeds towards the steepest ascent direction of the objective function. They test their procedure by solving a two-turbine positioning problem in a real world site, obtaining a final solution where one turbine is as close as possible to the connection point of the grid (as to minimize the connection costs to the grid) and the other one is in the position that minimizes the mutual wake effects and is as close as possible to the first turbine (separated by 3.5D) so as to minimize turbine interconnection costs. Clearly, their example is too small to show the validity of their method on real world problems, which usually involve tens of turbines. Nevertheless, their purpose is to provide a framework on top of which more effective optimization procedures can be executed. They achieve their goal by considering an objective function that is realistic and complete, by accurately modeling the wind characteristics through a Weibull distribution, and by considering aspects that are ignored by other works, particularly the connection costs to the grid and the interconnection costs among the turbines. Their objective function can be readily embedded in heuristic solution methods, but on the other hand it is very complex and nonlinear; therefore, it is not suitable to be embedded within exact solution methods.

5 Conclusions

In this chapter we describe the WFLOP, which is a crucial problem that needs to be solved during the design of a wind farm. Being able to find a better layout leads to higher energy production and profit. This problem has only recently been given attention by the scientific community, even if it has been neglected by the operations research area. We have described the mathematical model used to compute the impact of wake effects on energy production, noting that despite its simplicity it has been proven effective and precise.

The main works that have been carried out to date are certainly a good starting point for further research on more effective solution methods, but they cannot be considered satisfactory for several reasons. First, none of the solution methods proposed is able to assess the quality of the solution found. In other words, none of the existing works computes an upper bound on the power produced—with the exception of the power produced if no wake effect is present. The algorithms proposed in these works find a possibly good solution, but none of them can indicate how far it is from optimality. A wind farm developer needs to know this to decide if it is worth spending more time looking for a higher quality layout. Second, all the proposed algorithms are heuristic. An exact solution method, on the other hand, would allow one to find the global optimum or, possibly, to obtain tighter upper bounds. The only attempt in this direction was made by Donovan [25], who formulates the WFLOP as an integer program but does not take into account the wake effects, which are in fact forbidden in his model.

An aspect that should be considered is the topography of the territory for the computation of the wind speed and the wake effects. The existing works always consider a flat area and assume, with the exception of [20], that the wind distribution is identical throughout the entire site. The flat area assumption certainly holds for offshore sites, but it is very unrealistic for onshore sites, where the terrain is rarely flat and uniform; the presence of hills, rivers, forests, roads, or buildings significantly impacts the wind distribution and the behavior of the wakes. All these elements have not been considered so far. One of the reasons for ignoring them is that it is hard to implement a routine that takes into account this information when computing the objective function value. This difficulty is not only technical but it is also caused by the lack of published works that describe how to implement it. Nevertheless, there exist software packages, such as WaSP©, that can compute the objective function value by taking into account the onshore topographical elements. WaSP© is a computer aided design program used to manually design wind farms. A WaSP© user defines a wind distribution, the turbine type to use, the details of the site (which include the presence of natural elements), and the turbine positions. Then, the program computes the expected AEP by taking into account all these inputs. Although WaSP© is not free, its implementation details are available at http://www.wasp.dk/Products/wat/WAThelp/. To the best of our knowledge, this website provides one of the most exhaustive descriptions of how to compute the AEP.

Even if wake effects cause a decrease in wind speed, this is not their only negative consequence. Besides being slower, the air in the wake is also more turbulent, which, in the long run, may lead to blade damage and high maintenance costs. The existing approaches ignore this aspect, although turbulence intensity has been the subject of several studies [13, 26]. The impact of turbulence on maintenance costs has been disregarded because it cannot be described accurately.

Although some effort has been done [27], there is no method to measure the cost of the turbulence. Nonetheless, multi-objective optimization techniques could include the additional objective of minimizing the turbulence intensity (besides the traditional objective of maximizing the power produced). Or, alternatively, a constraint may be included to prevent too large values of turbulence intensity.

The last and most important aspect that has been ignored to date is the installation phase. Particularly for onshore projects, even if one finds the optimal solution to the WFLOP as defined in this chapter, that solution is not necessarily a feasible one or the one that minimizes the construction costs. In fact, there are three aspects that should be considered: landowners, road construction, and setback constraints. Some people may own the land under consideration, and, as mentioned in the introduction, they must be actively involved in the project. Some landowners can be easily convinced to host turbines on their land, while others may be noncooperative. In the latter case, the wind farm developer may offer them more money if their terrain is considered particularly important or strategic. The importance of an area depends on the strength of the wind that blows there and on other considerations related to the second aspect we discuss, the road construction. One fundamental constraint that determines many decisions during the project is that each turbine location must be connected to a road; otherwise, it would be impossible to transport the necessary construction material for its installation. If a road is not present (which is usually the case), the wind farm developer needs to build one. Oftentimes, land owners, who are usually farmers, allow roads to be built only along crop boundaries, so as to minimize the impact on their activity, although in other cases this constraint is not present (for example if an area is dedicated to livestock holdings). Here the purpose of the wind farm developer is to minimize the construction of roads, but also to have a road network that is completely connected, i.e., such that from each point one can reach any other point without passing through public roads. In this way, the cranes used to install the turbines can be easily moved throughout the road network without being disassembled. If, on the other hand, the road network is composed by subnetworks that are separated by traits of public road, the cranes need to be moved through the public road to reach the next location. Unfortunately, an assembled crane cannot be transported on public roads, and therefore it must be disassembled, transported through the public road to another subnetwork, and reassembled. Although it is hard to obtain the cost of this operation (wind farm developers do not release information on costs), it is estimated to be tens of thousands of dollars. Finally, there are other setback constraints that usually impose restrictions on the turbine locations. For example, turbines cannot be installed too close to houses, military facilities, airports, or boundaries of a noncooperative landowner. Furthermore, turbines cannot be installed along the migration path of birds, in locations that may visually impact the landscape, and so on. We refer to [28] for a more complete list of setback constraints.

To the best of our knowledge, all these logistical aspects have been completely ignored in the existing formulations of the WFLOP. We believe that the main reason is the unavailability of published material that treats them extensively.

In fact, this type of information is kept secret by wind farm developers, who are not willing to share it. Nevertheless, without taking into account these aspects, the WFLOP risks remaining an abstract mathematical exercise.

Acknowledgments This work has been made possible thanks to the generosity of Mr. John Callies and a IBM Shared University Research (SUR) Award.

References

1. Méchali M, Barthelemie R, Frandsen S, Jensen L, Rethoré P et al (2006) Wake effects at horns rev and their influence on energy production. In: European wind energy conference and exhibition, Athens
2. Aytun Ozturk U, Norman B (2004) Heuristic methods for wind energy conversion system positioning. Electr Power Syst Res 70:179–185
3. Grady SA, Hussaini MY, Abdullah MM (2005) Placement of wind turbines using genetic algorithms. Renew Energy 30:259–270
4. Hou-Sheng H (2007) Distributed Genetic Algorithm for optimization of wind farm annual profits. In: International conference on intelligent systems applications to power systems, Kaohsiung, Taiwan
5. Kusiak A, Song Z (2009) Design of wind farm layout for maximum wind energy capture. Renewable Energy 35:685–694
6. Mosetti G, Poloni C, Diviacco D (1994) Optimization of wind turbine positioning in large wind farms by means of a Genetic algorithm. J Wind Eng Ind Aerody 51:105–116
7. Rivas RA, Clausen J, Hansen KS et al (2009) Solving the turbine positioning problem for large offshore wind farms by simulated annealing. Wind Eng 33:287–297
8. Şişbot S, Turgut Ö, Tunç M et al (2010). Optimal positioning of wind turbines on Gökçeada using multi-objective genetic algorithm. Wind Energy 13:297–306
9. Petersen EL, Mortensen NG, Landberg L et al (1998) Wind power meteorology. Part I: climate and turbulence. Wind Energy 1:25–45
10. Kelley ND, Jonkman BJ, Scott GN, et al (2007) Comparing pulsed doppler LIDAR with SODAR and direct measurements for wind assessment. In: American Wind Energy Association wind power 2007 conference and exhibition. Los Angeles, California
11. Frehlich R, Kelley N (2010) Applications of scanning Doppler Lidar for the wind energy industry. The 90th American meteorological society annual meeting. Atlanta, GA
12. Ainslie JF (1988) Calculating the flow field in the wake of wind turbines. J Wind Eng Ind Aerodyn 27:213–224
13. Vermeer LJ, Sørensen JN, Crespo A (2003) Wind turbine wake aerodynamics. Prog Aerosp Sci 39:467–510
14. Wilcox DC (1998) Turbulence Modeling for CFD. La Canada, DCW Industries, CA
15. Jensen NO (1983) A note on wind generator interaction. Risø DTU national laboratory for sustainable energy
16. Katic I, Højstrup J, Jensen NO (1986) A simple model for cluster efficiency. In: Europe and Wind Energy Association conference and exhibition, Rome, Italy
17. Katic I (1993) Program PARK, calculation of wind turbine park performance. Release 1.3 ++, Risø National Laboratory, Rosklide
18. Barthelmie R, Larsen G, Pryor H et al (2004) ENDOW (efficient development of offshore wind farms): modelling wake and boundary layer interactions. Wind Energy 7:225–245
19. Barthelmie R, Folkerts L, Larsen GC et al (2006) Comparison of wake model simulations with offshore wind turbine wake profiles measured by Sodar. J Atmos Oceanic Technol 23:888–901

20. Lackner MA, Elkinton CN (2007) An analytical framework for offshore wind farm layout optimization. Wind Eng 31:17–31
21. Goldberg DE (1989) Genetic algorithms in search optimization and machine learning. Addison-Wesley Longman Publishing Co. Inc, Boston
22. Brooks SP, Morgan BJ (1995) Optimization using simulated annealing. J R Stat Soc D (The Statistician) 44:241–257
23. Manwell JF, McGowan JG, Rogers AL (2002) Wind Energy Explained. Wiley, West Sussex
24. Garcia A, Torres JL, Prieto E et al (1998) Fitting wind speed distributions: a case study. Sol Energy 62:139–144
25. Donovan S (2005) Wind farm optimization. University of Auckland, New Zealand
26. Crespo A, Hernandez J (1996) Turbulence characteristics in wind-turbine wakes. J Wind Eng Ind Aerody 61:71–85
27. Kelley, ND, Sutherland HJ (1997) Damage estimates from long-term structural analysis of a wind turbine in a U.S. wind farm environment. In: Prepared for the 1997 ASME Wind Energy Symposium, Reno, Nevada. NREL/CP-440-21672, Golden, CO: National Renewable Energy Laboratory. pp 12
28. Burton T, Sharpe D, Jenkins N et al (2001) Wind energy handbook. Wiley, New York

Risk Management Tools for Wind Power Trades: Weather Derivatives on Forecast Errors

Yuji Yamada

Abstract One of the most difficult issues for using wind power in practice is that the power output largely depends on the wind condition, and as a result, the future output may be volatile or uncertain. Therefore, the forecast of power output is considered important and is key to electric power generating industries making the wind power electricity market work properly. However, the use of forecasts may cause other problems due to "forecast errors". The objective of this chapter is to summerize conventional tools to manage such risks in the wind power electricity market. In particular, we focus on possible insurance claims or the so-called weather derivatives, which are contracts written on weather indices whose values are constructed from weather data.

In this chapter, we introduce weather derivatives based on wind conditions combined with their forecast information. In other words, we consider "wind derivatives" whose payoffs are determined by the forecast errors of wind conditions. In contrast to the standard weather derivatives in which the underlying index is given by observed weather data only (say, temperature), the wind derivatives discussed here take advantage of forecast data and the payoff depends on the difference between the actual and forecast data. Such a derivative contract is expected to be useful for hedging the possible loss (or risk) caused by forecast errors of power outputs associated with the forecast errors of wind conditions in wind energy businesses. We also demonstrate the hedge effect of wind derivatives using empirical data for a wind farm located in Japan.

Y. Yamada (✉)
Graduate School of Business Sciences, University of Tsukuba, 3-29-1 Otsuka, Bunkyo-ku, Tokyo 112-0012, Japan
e-mail: yuji@gssm.otsuka.tsukuba.ac.jp

P. M. Pardalos et al. (eds.), *Handbook of Wind Power Systems*, Energy Systems, DOI: 10.1007/978-3-642-41080-2_3,

1 Introduction

Motivated by a worldwide growing trend towards the production of wind energy, new problems associated with risk management and operation have appeared on the new energy market. Because the wind generated output largely depends on the wind condition, one of the most difficult issues in practice is that the wind generated output may be volatile and is uncertain in advance. This undesirable property may significantly reflect on the schedule of generation and control of frequency in the electricity grid, which leads to increased demand of accurate wind power forecast in various countries.

The forecast of wind condition is considered important and is key to electric power generating industries making the wind power electricity market work properly. However, the use of forecasts may cause other problems due to "forecast errors" of power output. In this chapter, we summerize conventional tools to manage such risks in the wind power electricity market. In particular, we focus on possible insurance claims or the so-called "weather derivatives," which is potentially useful for wind power trades in the wind energy market.

Weather derivatives are contracts written on weather indices, which in turn are variables whose values are constructed from weather data. The payoffs of these contracts are based on weather indices (e.g., temperature, rain, snow, wind, etc.) at a specific site (e.g., Tokyo, Japan) over a prespecified period. Although the underlying variables describing weather dynamics are manifold, a large portion of contracts are currently written on a temperature-based index, such as monthly average temperature, or heating/cooling degree days (HDDs/CDDs). Here HDDs and CDDs are defined relative to a base temperature to represent winter/summer energy demands concerning temperature. For example, an HDD of a day may be defined as the maximum between the base minus daily average temperature and zero [1], and the monthly contracts are based on the cumulative HDDs in a specific calender month.

The difference of weather derivatives from financial derivatives is that the underlying index (i.e., weather data) has no direct cash value unlike stocks or bonds. Therefore, the weather derivatives are usually traded for insurance purpose (not for investment purpose), and may be used by organizations/individuals to reduce risk associated with adverse or unexpected weather conditions. Note that the difference between insurances and derivatives is that, the payoff for insurance is determined by the loss associated with the underlying event, whereas for derivatives, the payoff is determined based on the value of the underlying index, i.e., weather data in the case of weather derivatives; it does not matter if the actual loss was caused due to the change of weather condition.

It should be mentioned that the total amount of transactions has been increasing worldwide. According to the Weather Risk Management Association,[1] the total limit of weather transactions executed amounted to $4.7 billion in the twelve

[1] See http://www.wrma.org/.

months 2003–2004, but in the period 2005–2006, this number jumped nearly tenfold to $45.2 billion. There are a number of ways in which a weather derivative trade can take place. Primary market trades are usually over the counter (OTC), meaning that they are traded privately between the two counterparties. A growing part of the secondary market is traded on the Chicago Mercantile Exchange (CME), which lists, as of June 2010, weather derivatives on temperature for 24 cities in the US, 11 cities in Europe, 6 cities in Canada, and 6 cities in Japan and Australia.[2]

In this chapter, we introduce weather derivatives based on wind conditions combined with their forecast information. In other words, we consider "wind derivatives" whose payoffs are determined by the forecast errors of wind conditions. Here we mainly explain the result of [2] where weather derivative contracts based on forecast errors are first demonstrated. It is fair to say that the most literature on weather derivatives discusses temperature related issues [3–9] in which the underlying index is defined by the observation data (of temperature) only. On the other hand, the weather derivatives considered here take the forecast data into account and the payoff depends on the difference between the actual and forecast data. Such a weather derivative contract is expected to be useful for hedging the possible loss (or risk) caused by forecast errors of power outputs associated with the forecast errors of wind conditions in wind energy businesses.

We use the following notation: For a sequence of observations of a variable, x_n, $n = 1, \ldots, N$, the sample mean and the sample variance are denoted by $\mathrm{Mean}(x_n)$ and $\mathrm{Var}(x_n)$, respectively. $\mathrm{Cov}(x_n, y_n)$ and $\mathrm{Corr}(x_n, y_n)$ represent the sample covariance and the sample correlation, respectively, where y_n, $n = 1, \ldots, N$ is a sequence of observations for another variable. The set of real number is denoted by $\Re$, and an $n \times m$ matrix with real entries is denoted by $A \in \Re^{n \times m}$.

The rest of this chapter is organized as follows: In Sect. 2, we explain the basic structure of derivatives and show some foundations of pricing problems. In particular, we formulate a general pricing problem using a payoff function that satisfies a zero expected value condition. In Sect. 3, we demonstrate wind derivatives on forecast errors of wind speed, and provide hedging problems by introducing a loss function for a wind farm (WF). Then, we define four types of problems based on combinations of payoff and loss functions. An empirical analysis is provided in Sect. 4, where we estimate the hedge effect of wind derivatives using empirical data for a WF located in Japan. In Sect. 5, we make a further discussion concerning the risk management on forecast errors and provide a future research direction. Section 6 offers a summary of this chapter.

[2] See http://www.cmegroup.com/.

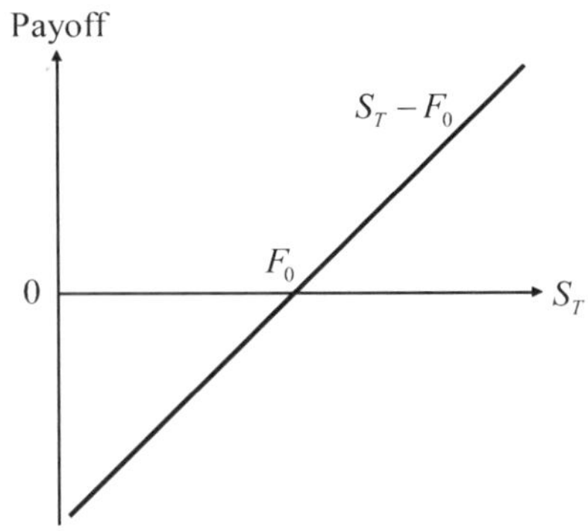

Fig. 1 Payoff function of a forward contract

2 Basic Structure of Derivatives

Before discussing wind derivatives, we need to know some basic terminology and payoff structure which are common with financial derivatives.

2.1 Foundations of Derivative Pricing

A forward contract is probably the simplest example among all types of derivatives, which is a contract made at time 0 (without initial cost) to purchase or sell the underlying asset, say one stock, at a specific price (called the "forward price") at a prespecified future time $T > 0$. If the stock price at time T is above the forward price, the buyer makes a profit by the amount of the difference between the actual (realized) price of stock and the forward price. On the other hand, the buyer makes a loss if the stock price drops below the forward price. In either case, the profit (or loss) for the buyer is given as

$$S_T - F_0 \tag{1}$$

where S_T is the price of stock at time T and F_0 is the forward price determined at time 0. We can say that the forward contract has a "payoff" of $S_T - F_0$ at time T by considering the cash settlement. Figure 1 shows a payoff function of forward contract, where the x-axis refers to S_T and y-axis to the payoff. We see that the payoff function is linear in the case of forward contract.

On the other hand, the payoff function of call option has a nonlinear structure as depicted in Fig. 2, where a call option is a right (not obligation) to buy a stock at a future time $T > 0$ (i.e., the maturity) at a specific price K (called the "strike price"). In contrast with forward contracts, the option holder does not have to buy the underlying stock if $S_T \leq K$ at time T, and in this case, the option becomes worthless. In other words, the value of option at time T is zero when $S_T \leq K$. Note that, if $S_T > K$ holds, the option holder may be considered to earn a profit of

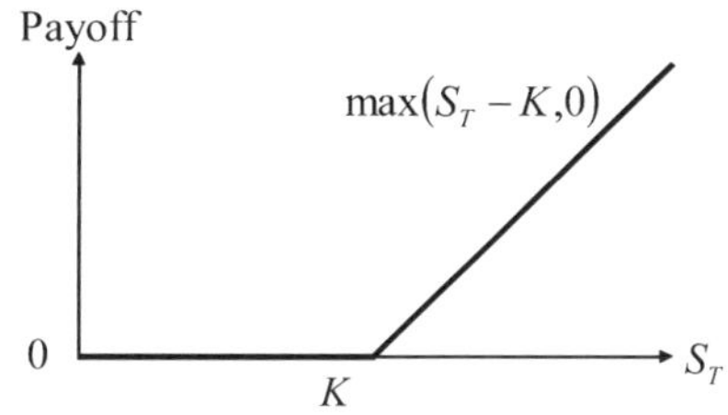

Fig. 2 Payoff function of a call

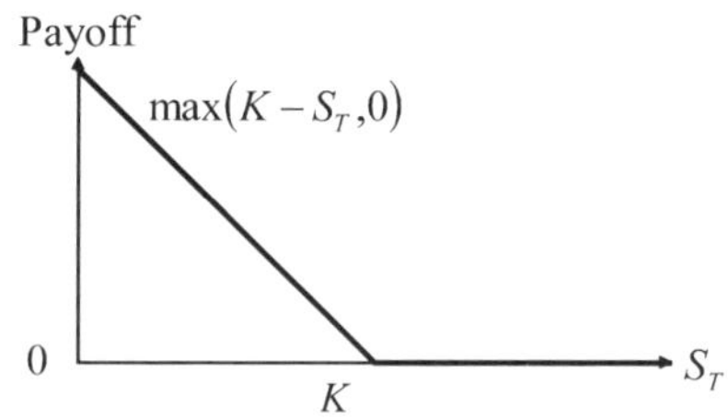

Fig. 3 Payoff function of a put

$S_T - K$ similar to the forward contract. Taking both cases into account, the payoff of call option at time T is given by

$$\max(S_T - K,\ 0), \tag{2}$$

which provides a payoff function of S_T as shown in Fig. 2.

A put option is a right to sell a stock with a strike price K at a specified future time T. With the similar argument, the payoff function of S_T is given by

$$\max(K - S_T,\ 0) \tag{3}$$

and is drawn in Fig. 3.

Although one can introduce an option contract having a flexibility with respect to the choice of timing to sell or buy the underlying stock, here we focus on options and other derivatives whose payoffs occur at a given future time $T > 0$ only. Then the pricing problem at the initial time 0 may be formulated as follows:

> Find the value of option at time 0 whose payoff at time T is defined by a given payoff function, e.g., the one in Fig. 2 for a call option or the one in Fig. 3 for a put option.

Let V_0 be the value of option at time 0. Then V_0 may represent an initial cost to carry out such an option contract. On the other hand, the initial cost for a forward contract is always zero, which is an essential difference between forward and option contracts.

Instead of paying initial costs in option contracts, we can assume that the option holder pays a fixed amount at time T, i.e., at the same time as the payoff is made. The fixed payment may correspond to the initial cost V_0, but we need to take a time value into consideration, where a risk free interest rate $r > 0$ is compounded between time 0 and time T. Then, the fixed payment at time T is computed to be $e^{rT} V_0$, which is known at time 0. We can regard that this type of contract is a

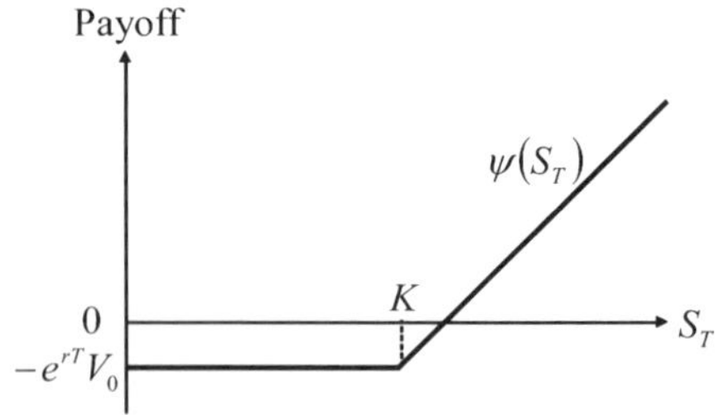

Fig. 4 Payoff function of a swap for call option with strike K

"swap" between a fixed payment, $e^{rT}V_0$, and an uncertain payoff, e.g., $\max(S_T - K,\ 0)$ in the case of call option. In this case, the payoff function of S_T, denoted by $\psi(S_T)$, for a call option is modified to

$$\psi\ (S_T) := \max\ (S_T - K,\ 0) - e^{rT}V_0 \tag{4}$$

as shown in Fig. 4.

Note that the value denoted by Eq. (4) may be thought of a cash flow for the option holder at time T. On the other hand, the cash flow for the counterparty (i.e., the seller for the option) is given as

$$-\psi(S_T). \tag{5}$$

Therefore, a fair contract may satisfy that the expected value of $\psi(S_T)$ is zero, which may be written as follows:

$$\mathbb{E}[\psi(S_T)] = 0 \tag{6}$$

where $\mathbb{E}$ is the expected value operator. Then V_0 is obtained as

$$V_0 = e^{-rT}\mathbb{E}[\max(S_T - K, 0)]. \tag{7}$$

Similarly, in the case of forward contracts, we have

$$\mathbb{E}[S_T - F_0] = 0, \tag{8}$$

which provides the forward price as follows:

$$F_0\ =\ \mathbb{E}[S_T]. \tag{9}$$

According to the fundamental theorem of asset pricing [10], the pricing formula in (7) is valid if $\mathbb{E}$ is defined under a risk neutral probability measure, that coincides with the original results of [11, 12] known as the "Black–Scholes–Merton model" for option pricing. On the other hand, in empirical data analysis, we usually work on the real probability measure so that the sample statistics (such as sample mean and variance) of observed data may provide proxies of the underlying probability distribution. Although one can usually transform the real probability measure to a risk neutral probability measure using a suitable change of measure technique, here we assume that, for simplicity, the real probability measure provides a risk neutral probability measure, and evaluate the value of derivatives under the real probability measure. Note that a further discussion of the relation between the real and risk neutral probability measures is beyond the scope of this article and that the interested reader may refer to [13].

2.2 General Pricing Problem with Payoff Functions

As discussed in the end of Sect. 2.1, the swap contract for a call option has a payoff function of (4) satisfying condition (6). Moreover, the forward price also satisfies (8). These conditions are generally said that swap and forward contracts have zero expected values with respect to payoff functions of S_T, i.e.,

$$\mathbb{E}[\psi(S_T)] = 0 \tag{10}$$

holds with a suitable choice of payoff function ψ. For example, ψ is given as

$$\psi(x) = x - F_0$$

for a forward contract, or

$$\psi(x) = \max(x - K,\ 0) - e^{rT} V_0$$

for a call option.

Then, the problem can be reformulated as follows:

$$\text{Find } \psi \text{ satisfying } \mathbb{E}[\psi(S_T)] = 0 \text{ with } \psi \in \Psi,$$

where Ψ is a set of functions defined as

$$\Psi := \{\psi(x) = x - F_0 | F_0 \in \Re\} \tag{11}$$

in the case of forward contracts, or

$$\Psi := \left\{\psi(x) = \max(x - K, 0) - e^{rT} V_0 | V_0 \in \Re\right\} \tag{12}$$

in the case of call options. In any cases, we see that a given contract is fair as far as condition (10) is satisfied with a specific payoff function in Ψ, so that we can search for a suitable payoff function by taking $\psi \in \Psi$ as a variable. This is a basic idea for constructing payoff functions for wind derivatives in the following sections.

3 Wind Derivatives on Forecast Errors and Hedging Problems

In this section, we demonstrate wind derivatives with respect to forecast errors and define associated hedging problems.

3.1 Loss Functions for WFs and Problem Settings

We will consider two basic positions, a seller and a buyer, for trading wind generated electricity, in which the seller is assumed to be a WF having a responsibility to quote the promised power output in advance using forecast information. A possible sales contract for the power output may be described as follows, which specifies a loss function for a WF on forecast errors of power output:

In general, the value of electricity generated by wind power is considered to be low due to the uncertainty of the tradable volume. Here we assume that the electricity price without forecast is estimated to be 3 yen per 1 kWh. On the other hand, the value of the electricity would be estimated to be higher, if the tradable volume were quoted in advance by forecast, but the seller has to guarantee the quoted volume or has to pay the penalty in case of shortages. Suppose that the value of electricity with forecast is given as 7 yen per 1 kWh and that the penalty of the shortage is 10 yen per 1 kWh. These assumptions are not so far from the current situation discussed in the forecast business [14]. In this case, the loss function caused by forecast errors is depicted in Fig. 5, which shows the relation between the forecast error for the power output $P - \hat{P}$ (the actual power output minus its forecast) and the loss caused by the forecast error. Note that, even if the forecast error is positive, we can also think of this situation as an opportunity loss to sell the output with a suitable price.

Taking the above situation into consideration, we formulate the problem more precisely. Let $n = 1, \ldots, N$ be the time index (say, hourly index) and define the following variables:

P_n : Total power output at time n

$\hat{P}_n$: Forecast of P_n (which is computed, e.g., 1 day in advance).

The buyer is willing to trade the power output by using the reference $\hat{P}_n$, and may require a penalty if the forecast error exceeds a certain level.

Let $\varepsilon_{p,n}(n = 1, \ldots, N)$ be the forecast error of the power output at time n, which causes a loss for the seller due to the penalty or opportunity loss to sell the output. Suppose that the loss associated with the forecast error of the power output is defined using a loss function as $\phi(\varepsilon_{p,n})$. For instance, the loss function may be given as the one shown in Fig. 5 if the seller is a WF owner. Also, there is a case in which the forecast is sufficiently accurate or the forecast error is less than a certain (small) level. In this case, the seller can be thought of getting a bonus because of a higher price of power output with forecast, which results in a profit for the seller and makes the loss negative, i.e., $\phi(\varepsilon_{p,n}) < 0$. We assume that

$$\text{Mean}\left(\phi\left(\varepsilon_{p,n}\right)\right) = 0 \tag{13}$$

so that the sum of profit/loss is zero on average. Note that we take the sample mean instead of using the expected value operator $\mathbb{E}$, because we will work on empirical data in the next section.

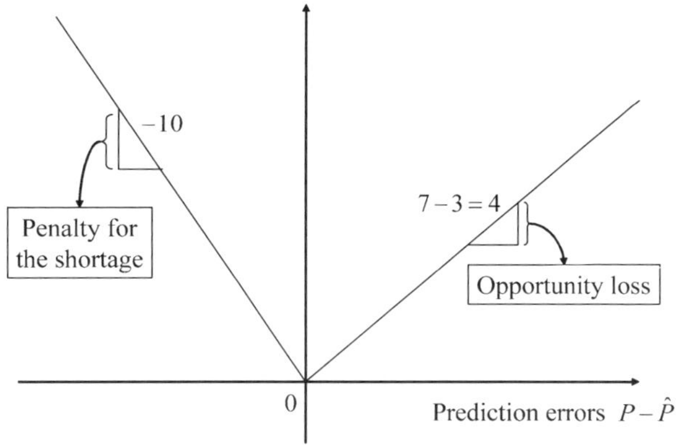

Fig. 5 An example of loss function

Consider a situation in which the seller with $\phi(\cdot)$ would like to compensate their loss on $\varepsilon_{p,n}$ using a weather derivative on the forecast error of the wind speed. To this end, define the following variables:

W_n : Wind speed at time n
$\hat{W}_n$: Forecast of W_n (which is computed, e.g., 1 day in advance).

Let $\varepsilon_{w,n}$ be the forecast error of the wind speed. Without loss of generality, assume that $\text{Mean}(\varepsilon_{w,n}) = 0$. Let ψ be a payoff function such that

$$\text{Mean}\left(\psi\left(\varepsilon_{w,n}\right)\right) = 0. \tag{14}$$

Note that, in the case of simple forward contracts, $\psi\left(\varepsilon_{w,n}\right)$ may be given as a linear function, e.g.,

$$\psi\left(\varepsilon_{w,n}\right) = \varepsilon_{w,n}. \tag{15}$$

Based on the above settings, we will first consider the following problems:

(P1) Given the loss function and the payoff function of wind derivatives, find the optimal volume of wind derivative.
(P2) Given the loss function, find the optimal payoff function of wind derivatives.

We will investigate the hedge effect of wind derivatives and show that using wind derivatives on forecast error of wind speed is highly effective to hedge the loss caused by forecast errors of power output.

Then we will consider a situation in which there already exists a standardized derivative contract with a certain payoff function, but there is some room for improvement on the loss function, e.g., for a WF owner. The problem can be thought of as a reverse problem of (P2), which is given as follows:

(P3) Given the payoff function of wind derivatives, find the optimal loss function against forecast errors of power output.

Finally, we will formulate a simultaneous optimization problem of payoff and loss functions as (P4) below:

(P4) Optimize the payoff function of wind derivatives and the loss function simultaneously.

3.2 Standard Minimum Variance Hedging Problem

With the notation and definitions introduced in the previous subsection, the first optimization problem, (P1), is formulated as follows:

Contract volume optimization problem:

$$\min_{\Delta \in \Re} \operatorname{Var}\big(\phi(\varepsilon_{p,n}) + \Delta\psi(\varepsilon_{w,n})\big). \tag{16}$$

The contract volume optimization problem may be considered as the standard "minimum variance hedge," and the optimal volume Δ^* may be computed analytically as

$$\Delta^* = -\frac{\operatorname{Cov}(\phi(\varepsilon_{p,n}), \psi(\varepsilon_{w,n}))}{\operatorname{Var}(\psi(\varepsilon_{w,n}))}. \tag{17}$$

To estimate the hedge effect, we define the variance reduction rate (VRR) as follows:

$$\mathrm{VRR} := \frac{\operatorname{Var}(\phi(\varepsilon_{p,n}) + \Delta^*\psi(\varepsilon_{w.n}))}{\operatorname{Var}(\phi(\varepsilon_{p,n}))}. \tag{18}$$

Because the minimum variance can be computed as

$$\begin{aligned} &\operatorname{Var}\big(\phi(\varepsilon_{p,n}) + \Delta^*\psi(\varepsilon_{w,n})\big) \\ &= \operatorname{Var}\big(\phi(\varepsilon_{p,n})\big)\Big(1 - \big[\operatorname{Corr}\big(\phi(\varepsilon_{p,n}), \psi(\varepsilon_{w,n})\big)\big]^2\Big), \end{aligned} \tag{19}$$

we obtain

$$\mathrm{VRR} = 1 - \big[\operatorname{Corr}\big(\phi(\varepsilon_{p,n}), (\varepsilon_{w,n})\big)\big]^2. \tag{20}$$

Note that VRR satisfies

$$0 \leq \mathrm{VRR} \leq 1 \tag{21}$$

and that a smaller VRR provides a better hedge effect in terms of minimum variance.

In the case of standard minimum variance hedge, the optimal volume is also found by solving a linear regression problem, where $\phi(\varepsilon_{p,n})$ is regressed with respect to $\psi(\varepsilon_{w,n})$, and the regression coefficient gives the optimal volume for fixed loss and payoff functions. On the other hand, we can expect to obtain a better hedge effect if we could optimize the payoff function of the weather derivative directly. This can be done by applying non-parametric regression techniques introduced in the next section, and we will find that using a non-parametric regression corresponds to optimizing the derivative contract directly by choosing a suitable payoff function.

3.3 Minimum Variance Hedging Using Non-Parametric Regression

In this section, we first introduce a non-parametric regression technique, and then formulate the second optimization problem, (P2).

In the previous section, we showed that the contract volume optimization problem is formulated as standard minimum variance hedging and can be solved by applying linear regression. A similar idea may be employed to solve the payoff function optimization problem of (P2) [or the loss function optimization problem of (P3)] by introducing a non-parametric regression technique. Since we will apply a non-parametric regression to find a payoff function (or loss function) by assuming that a loss function (or payoff function) is fixed, it may be useful to specify which function is given explicitly. To this end, we use overlines as

$$\phi(\cdot) = \overline{\phi}(\cdot) \text{ (or } \psi(\cdot) = \overline{\psi}(\cdot))$$

to indicate that a loss function (or a payoff function) is fixed.

3.3.1 Generalized Additive Models

The non-parametric regression technique introduced here is to find a (cubic) smoothing spline that minimizes the so-called penalized residual sum of squares (PRSS) among all regression spline functions with two continuous derivatives. Let y_n and x_n be dependent and independent variables, respectively, and express y_n as

$$\mathrm{y}_n = h(X_n) + \varepsilon_n, \ \mathrm{Mean}(\varepsilon_n) = 0 \tag{22}$$

using a smooth function $h(\cdot)$ and residuals ε_n. Here the function $h(\cdot)$ is a (cubic) smoothing spline that minimizes the following PRSS,

$$\mathrm{PRSS} = \sum_{n=1}^{N} (y_n - h(x_n))^2 + \lambda \int_{-\infty}^{\infty} (h''(x))^2 \mathrm{d}x \tag{23}$$

among all functions $h(\cdot)$ with two continuous derivatives, where λ is a given parameter. In (23), the first term measures closeness to the data while the second term penalizes curvature in the function. Note that, if $\lambda = 0$ and $h(\cdot)$ is given by a polynomial function, the problem is reduced to the standard regression polynomial and is solved by the least squares method. It is shown that (23) has an explicit and unique minimizer and that a candidate of optimal λ may be found by using the so-called generalized cross validation criteria (See [15]). Note that regression splines can be extended to the multivariable case with additive sums of smoothing splines, known as generalized additive models (GAMs; see e.g., [16]). Also note that GAMs can be computed using free software "R (http://cran.r-project.org/)," and we will refer to the class of smoothing splines for non-parametric regression as GAMs in this chapter. We will apply GAMs to solve (P2)–(P4) and estimate the hedge effect of wind derivatives.

Note that, instead of writing the problem as an unconstrained optimization problem, we can reformulate it as an optimization problem constrained on $h(\cdot)$ as follows:

$$\begin{aligned} &\min_{h(\cdot)} \sum_{n=1}^{N} (y_n - h(x_n))^2 \\ &\text{s.t.} \int_{-\infty}^{\infty} (h''(x))^2 \mathrm{d}x \leq \alpha \end{aligned} \tag{24}$$

where α is a given parameter. Based on the similar argument to that in [15], we can verify that the objective function of problem (24) is quadratic subject to a convex constraint and that the minimization problem (24) is equivalent to the following problem,

$$\max_{\lambda > 0}\left\{\min_{h(\cdot)}\left\{\sum_{n=1}^{N}\{y_n - h(x_n)\}^2 + \lambda\left(\int \{h''(x)\}^2 \mathrm{d}x - \alpha\right)\right\}\right\}, \tag{25}$$

using a Lagrange multiplier $\lambda > 0$. Therefore, we see that fixing λ in (23) corresponds to fixing α in (24) and that the non-parametric regression problem using GAM may be recast as a minimization problem of the sample variance with a smooth constraint.

3.3.2 Optimization of Derivative Contracts

It is in a position to formulate the the second optimization problem, i.e., the payoff function optimization problem, in the context of minimum variance hedge using non-parametric regression as follows:

Payoff function optimization problem:

$$\begin{aligned} &\min_{\psi(\cdot)} \mathrm{Var}\left(\bar{\phi}(\varepsilon_{p,n}) + \psi(\varepsilon_{w,n})\right) \\ &\text{s.t.} \int_{-\infty}^{\infty} (\psi''(x))^2 \mathrm{d}x \le \alpha. \end{aligned} \tag{26}$$

The minimization problem (26) may be recast as (24) by taking $y_n := \overline{\phi}(\varepsilon_{p,n})$, $x_n := \varepsilon_{w,n}$, and $h(\cdot) := -\psi(\cdot)$, and therefore, can be solved by applying GAM. Let $\psi^*(\cdot)$ be the optimal payoff function. Then VRR may be defined as

$$\mathrm{VRR} := \frac{\mathrm{Var}\left(\bar{\phi}(\varepsilon_{p,n}) + \psi^*(\varepsilon_{w,n})\right)}{\mathrm{Var}\left(\bar{\phi}(\varepsilon_{p,n})\right)}. \tag{27}$$

Although it is possible to find the optimal payoff function by solving GAM once, it may be worthwhile to mention that we have a slight improvement by applying a linear regression after finding the optimal payoff function $\psi^*(\cdot)$ as

$$\min_{a \in \Re} \mathrm{Var}\left(\bar{\phi}(\varepsilon_{p,n}) + a\psi^*(\varepsilon_{w,n})\right). \tag{28}$$

In this case, VRR may be given as

$$\mathrm{VRR} = \frac{\mathrm{Var}\left(\bar{\phi}(\varepsilon_{p,n}) + a^*\psi^*(\varepsilon_{w,n})\right)}{\mathrm{Var}\left(\phi(\varepsilon_{p,n})\right)}. \tag{29}$$

or equivalently,

$$\mathrm{VRR} = 1 - \left[\mathrm{Corr}\left(\bar{\phi}(\varepsilon_{p,n}), \psi^*(\varepsilon_{w,n})\right)\right]^2. \tag{30}$$

where $a^* \in \Re$ is the regression coefficient to solve (28). Note that (30) is independent of a^*, or any scaling parameter to $\psi^*(\varepsilon_{w,n})$, and that it can be computed if $\psi^*(\cdot)$ is specified. Therefore, we use the right hand side of (30) as a proxy of VRR. It is readily confirmed that VRR in (27) is actually an upper bound of (30). However, as indicated in the end of Sect. 4.2, the gap between (27) and (30) is very small from our numerical experience.

3.4 *Optimization with Loss Functions and Simultaneous Optimization*

3.4.1 Optimal Loss Function

Next, we will consider a case in which a payoff function of wind derivative is given but we would like to find a loss function that is desirable for using the wind derivative, i.e., in a case where there already exists a standardized derivative contract with a certain payoff function, but there is some room for improvement on the loss function ϕ for a WF owner.

Assume that the loss function $\phi(\varepsilon_{p,n})$ satisfies

$$\mathrm{Mean}(\phi(\varepsilon_{p,n})) = 0, \tag{31}$$

$$\mathrm{Var}(\phi(\varepsilon_{p,n})) = c. \tag{32}$$

We will compute an optimal loss function satisfying the above constraints for a given payoff function $\psi = \bar{\psi}$ such that

$$\mathrm{Mean}(\bar{\psi}(\varepsilon_{w,n})) = 0. \tag{33}$$

The problem is then formulated as follows:

Loss function optimization problem:

$$\begin{aligned} &\min_{\phi(\cdot)} \ \mathrm{Var}(\phi(\varepsilon_{p,n}) + \bar{\psi}(\varepsilon_{w,n})) \\ &\text{s.t.} \int_{-\infty}^{\infty} (\phi''(x))^2 \mathrm{d}x \leq \alpha, \\ &\mathrm{Var}(\phi(\varepsilon_{p,n})) \ = c. \end{aligned} \tag{34}$$

Note that the constraint $\mathrm{Var}(\phi(\varepsilon_{p,n})) = c$ is also quadratic if ϕ is given by a cubic natural spline function, and hence, the problem might be reformulated as an unconstrained optimization problem by introducing another Lagrangian term for the variance constraint. On the other hand, we can still apply GAM directly to solve the problem without the variance constraint (i.e., $\mathrm{Var}(\phi(\varepsilon_{p,n})) = c$), similar to the payoff function optimization problem (26). Then we can scale the minimizing function so that it satisfies the variance constraint (32). In this case, condition (31) is also satisfied.

Let $\hat{\phi}(\cdot)$ be the optimizer of problem (34) without the variance constraint (i.e., $\mathrm{Var}(\phi(\varepsilon_{p,n})) = c$), which can be computed by applying GAM. By scaling $\hat{\phi}(\cdot)$ to satisfy (32), we obtain the optimal loss function $\phi^*(\cdot)$ as follows:

$$\phi^*(\cdot) = \frac{c}{\mathrm{Var}(\phi(\varepsilon_{p,n}))}\phi(\cdot). \tag{35}$$

Note that the optimal volume of wind derivative with the given payoff and loss functions, $\bar{\psi}(\cdot)$ and $\phi^*(\cdot)$, will be found by solving the standard minimum variance hedging problem as in Sect. 3.2, and VRR may be computed as

$$\mathrm{VRR} = 1 - \left[\mathrm{Corr}(\phi^*(\varepsilon_{p,n}), \bar{\psi}(\varepsilon_{w,n}))\right]^2. \tag{36}$$

3.4.2 Simultaneous Optimization

It may be interesting to consider a simultaneous optimization of the payoff and loss functions, $\psi(\varepsilon_{w,n})$ and $\phi(\varepsilon_{p,n})$. Recall that VRR can be computed using the correlation between the payoff function and the loss function as

$$1 - \left[\mathrm{Corr}\left(\phi(\varepsilon_{p,n}), \psi(\varepsilon_{w,n})\right)\right]^2.$$

Since the larger correlation the smaller VRR, the minimization of VRR boils down to the maximization of correlation between $\phi(\varepsilon_{p,n})$ and $\psi(\varepsilon_{w,n})$. Therefore, the simultaneous optimization of the payoff and the loss functions may be formulated as follows:

Simultaneous optimization problem:

$$\begin{aligned} &\max \mathrm{Corr}\left(\phi(\varepsilon_{p,n}), \psi(\varepsilon_{w,n})\right) \\ &\text{s.t.} \int_{-\infty}^{\infty} (\phi''(x))^2 \mathrm{d}x \le \alpha_\phi, \\ &\int_{-\infty}^{\infty} (\psi''(x))^2 \mathrm{d}x \le \alpha_\psi, \\ &\mathrm{Var}\left(\phi(\varepsilon_{p,n})\right) = c. \end{aligned} \tag{37}$$

The simultaneous optimization problem may be solved using an iterative algorithm by solving the payoff function optimization problem with $\phi(\cdot) = \overline{\phi}(\cdot)$ fixed, or the loss function optimization problem with $\psi(\cdot) = \overline{\psi}(\cdot)$ fixed, at each step. The following is the iterative algorithm:

Iterative algorithm:

1. Given $\phi(\cdot) = \overline{\phi}(\cdot)$, find $\psi(\cdot)$ to solve the payoff function optimization problem. Let $\psi^*(\cdot)$ be the optimal function, and let $\overline{\psi}(\cdot) = \psi^*(\cdot)$.
2. Given $\psi(\cdot) = \overline{\psi}(\cdot)$, find $\phi(\cdot)$ to solve the loss function optimization problem. Let $\phi^*(\cdot)$ be the optimal loss function and let $\overline{\phi}(\cdot) = \phi^*(\cdot)$.
3. Repeat steps 2 and 3 until the objective function in (37) does not change.

Note that the optimal loss function obtained from the above iterative algorithm satisfies (32) and that we can consider additional constraints to take more realistic situations into account for the loss and payoff functions. Although we may need to specify α_0 and α_ϕ for solving the iterative algorithm, an optimal selection of smoothing parameters for $\phi(\cdot)$ and $\psi(\cdot)$ may be applicable at each step by using GAMs in stead of fixing these parameters a priori in the algorithm.

Remark 1 The above iterative algorithm is formally in the class of so-called "Alternating Conditional Expectations (ACE) algorithm (see, e.g., Chap. 7 of [16])." The ACE algorithm seeks optimal transformations of $\theta(Y)$ and $f(X)$ for two random variables X and Y so that the squared error loss

$$\mathbb{E}\left[(\theta(Y) - f(X))^2\right]$$

is minimized. Since the zero functions trivially minimize the square error, ACE has a constraint so that $\theta(Y)$ has unit variance at each step, which is exactly the same as our variance constraint (32). Note that the convergence of ACE algorithm is also discussed in [16], although we omit the details for brevity.

4 Empirical Analysis and Numerical Experiment

In this section, we demonstrate the solutions to (P1)–(P4) and estimate their hedge effect using empirical data for the power output, wind speed, and their forecasts.

4.1 Data Description and Preliminary Analysis

Here we consider the power output from a WF located in Japan, where the power output from the WF is predicted based on the numerical weather forecast and the power generating properties for turbines. The numerical weather forecast consists of the following two steps:

- Japan Meteorological Agency announces the hourly data of regional spectral models for the next 51 h twice a day (9 am and 9 pm).
- Using them as initial and boundary values, a public weather forecasting company computes more sophisticated values for the next day's hourly data by 12 pm.

In this chapter, we use the forecast data obtained from the Local Circulation Assessment and Forecast System (LOCALS) developed by ITOCHU Techno-Solutions Corporation for the wind speed and the power output of a WF in Japan [17, 18]. The data set is given as follows[3]:

Data specifications:

Realized and predicted values of total power output for the WF, and those of wind speed for the observation tower in the WF.

[3] All the data used in this chapter were provided by ITOCHU Techno-Solutions Corporation.

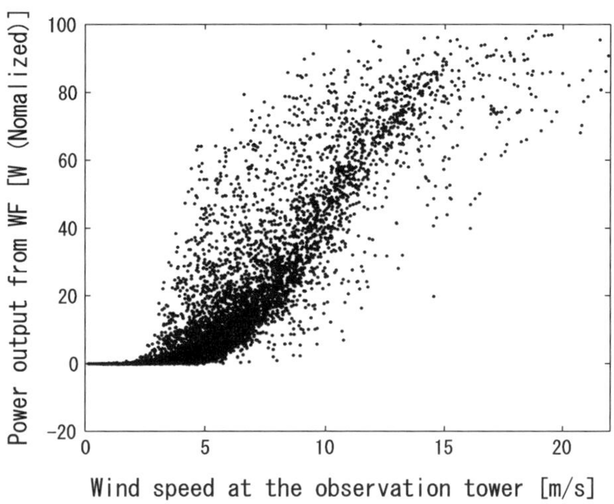

Fig. 6 Wind speed W_n [m/s] versus Power output P_n [W]

Data period:

2002–2003 (1 year), hourly data, everyday.

Total number of data:

8,000 for each variable excluding missing values.

Let $n = 1, \ldots, N$ be the time index (where $N \simeq 8,000$), and assume that the actual power output and the wind speed at time n are, respectively, denoted by P_n and W_n. Also, let $\hat{P}_n$ and $\hat{W}_n$ be the forecasts of the corresponding power output and the wind speed obtained from LOCALS, which are computed by noon one day before the actual data is observed. Figure 6 shows a scatter diagram for the wind speed W_n and the power output P_n, where the power output P_n is normalized so that its maximum equals 100. From Fig. 6, we can see that:

- The generator starts providing the power output when the wind speed exceeds around 2 [m/s].
- The power output increases with the wind speed between 5–15 [m/s].

Also note that, because each electricity generator is controlled so that the maximum output does not exceed a certain value, the total output is also bounded as shown in Fig. 6.

Figure 7 shows a partial residual plot for

$$W_n = a_w \widehat{W}_n + b_w + \varepsilon_{w,n}, \qquad \mathrm{n} = 0, \ldots, N, \ \mathrm{Mean}\left(\varepsilon_{w,n}\right) = 0 \qquad (38)$$

i.e., the scatter diagram of $\left(\hat{W}_n, W_n - b_w\right)$, where a_w and b_w are a regression coefficient and an intercept, respectively, and $\varepsilon_{w,n}$ is a residual satisfying $\left(\varepsilon_{p,n}\right) = 0$. The partial regression line is depicted using a solid straight line as shown in Fig. 7. In this case, the sample variance of residuals is found to be

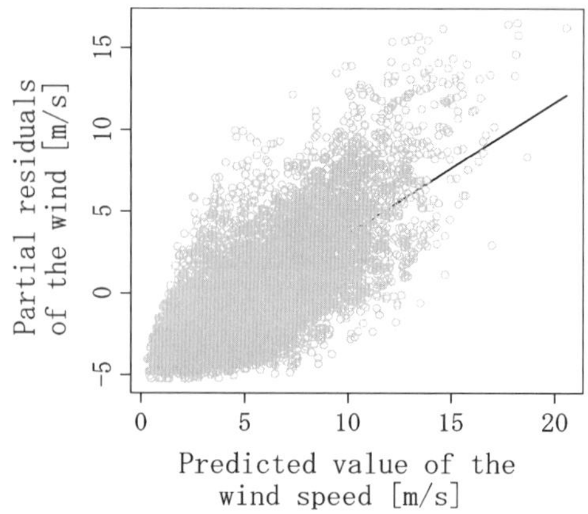

Fig. 7 Predicted versus Measured values for the wind speed

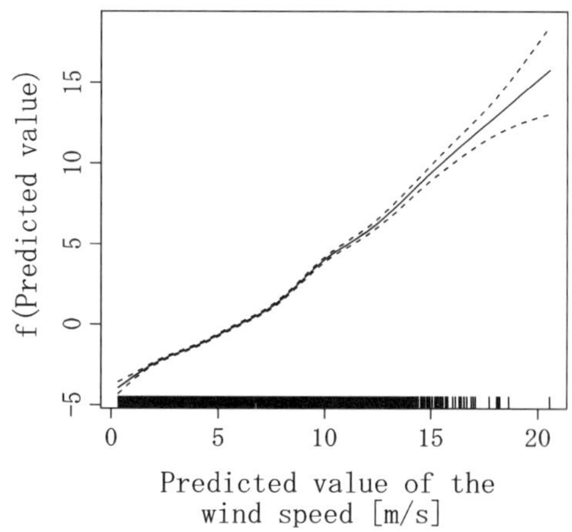

Fig. 8 Spline regression function for the wind speed

$$\mathrm{Var}\left(\varepsilon_{w,n}\right) \simeq 5.12. \tag{39}$$

On the other hand, the regression spline $f(\cdot)$ to fit the same data of Fig. 7 using GAM is shown as a solid line in Fig. 8, where $f(\cdot)$ satisfies

$$W_n = f\left(\widehat{W}_n\right) + \varepsilon_{w,n}. \tag{40}$$

In this case, the sample variance of the residuals is

$$\mathrm{Var}\left(\varepsilon_{w,n}\right) \simeq 4.95. \tag{41}$$

Noting that the sample variance of the measured values is computed as "11.0," we can say that the variance of the wind speed is reduced by 50 % (from "11.0" to "5.12") using the predicted value and the linear regression, and it is improved a little using GAM, i.e., from "5.12" to "4.95." In this section, we define the forecast error of the wind speed as the one given by GAMs, i.e., $\varepsilon_{w,n}$ in (40).

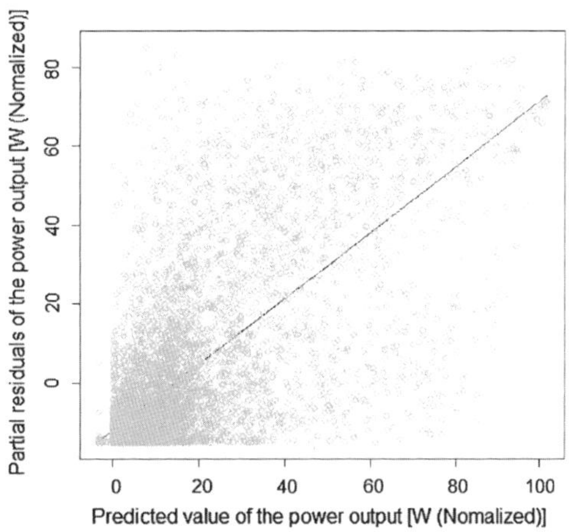

Fig. 9 Predicted versus Measured values for the power output

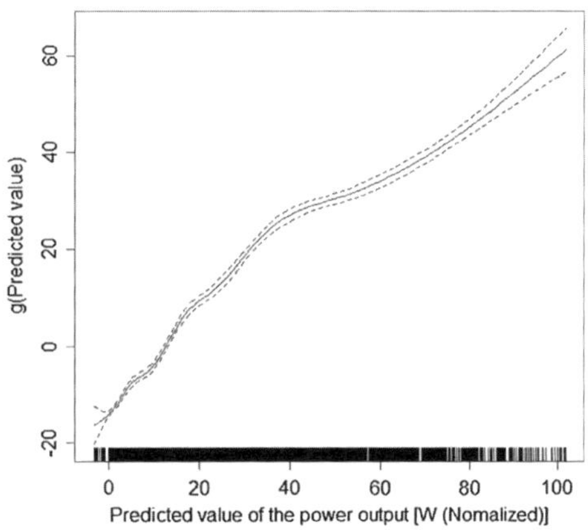

Fig. 10 Spline regression function for the power output

Similarly, we can draw a partial residual plot for the power output P_n with respect to the predicted value $\hat{P}_n$ as shown in Fig. 9, where the solid line is obtained from a linear regression for partial residuals. In this case, the sample variance of the residuals is found to be "249." The solid line in Fig. 10 refers to the regression spline function $g(\cdot)$ satisfying

$$P_n = g\left(\widehat{P}_n\right) + \varepsilon_{p,n}, \quad n = 0, \ldots, N \tag{42}$$

using GAM. Note that the sample variance of residuals in this case is given as "239", whereas the sample variance of the measured value of the power output is "504". Similar to the case of wind speed, we can say that the variance of power output is reduced to less than half (from "504" to "249") using the predicted value and the linear regression, and it is improved a little using GAM, i.e., "249" to "239".

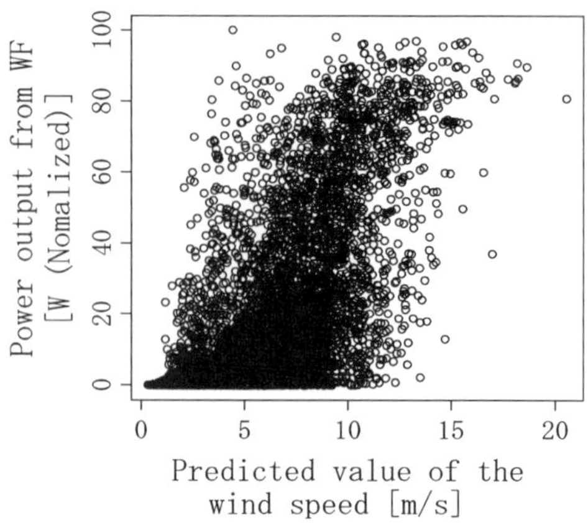

Fig. 11 Predicted value of the wind speed versus Measured values for the power output

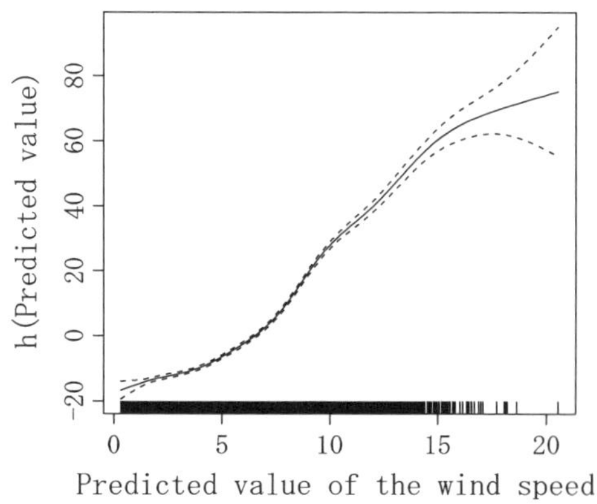

Fig. 12 Spline regression function for the power output using the wind speed forecast

Although we should be able to define the forecast error of the power output using the residual in (42), it might be worthwhile to mention that there is another way to define the forecast error of the power output. As stated in the beginning of this section, the power output is predicted using numerical weather forecast, and therefore, we can define a regression model such that the power output P_n is a dependent variable and the wind speed forecast $\hat{W}_n$ is an independent variable, i.e.,

$$P_n = h\left(\widehat{W}_n\right) + \varepsilon_{p,n}, \tag{43}$$

where $h(\cdot)$ is a regression spline that minimizes PRSS.

Figure 11 shows the relation between the predicted values of the wind speed and the measured values for the power output, where the solid line in Fig. 12 is the regression spline $h(\cdot)$. In this case, the sample variance of the residuals is computed as

$$\mathrm{Var}\left(\varepsilon_{p,n}\right) \simeq 254 \tag{44}$$

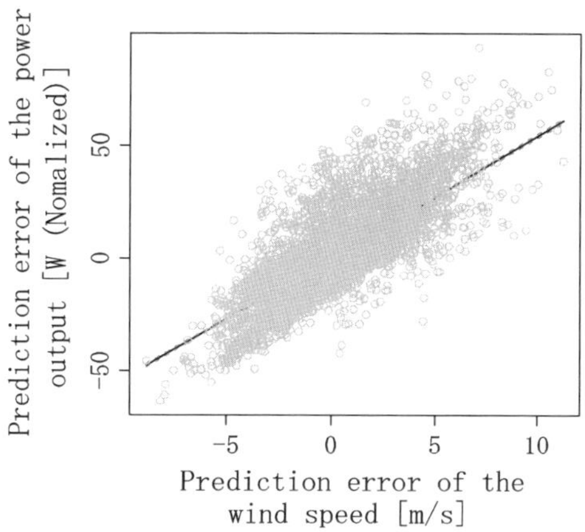

Fig. 13 Wind speed forecast error $\varepsilon_{w,n}$ versus power output forecast error $\varepsilon_{p,n}$

which is, in fact, higher than the one given by (42). However, it will turn out that using the forecast error in (43) provides not only a better hedge effect but also a smaller variance of the hedged loss when combining with the optimal wind derivative. Therefore, we will use the residual $\varepsilon_{p,n}$ in (43) to define the forecast error of the power output. An empirical analysis using the forecast error defined by the residual in (42) may be found in [19].

4.2 Construction of Wind Derivatives

Next, we will construct wind derivatives and demonstrate their hedge effect on wind power energy businesses.

We first solve the minimum variance hedging problem for the simplest case where the loss and the payoff functions are both linear. Let

$$\phi\left(\varepsilon_{p,n}\right) = \varepsilon_{p,n}, \psi\left(\varepsilon_{w,n}\right) = \varepsilon_{w,n} \tag{45}$$

without loss of generality. In this case, the problem is reduced to solving a linear regression for the following regression function:

$$\varepsilon_{p,n} = a_w \, \varepsilon_{w,n} + \eta_n, \tag{46}$$

where η_n is a residual. Since the linear regression computes a_w that minimizes variance of $\eta_n = \varepsilon_{p,n} - a_w \varepsilon_{w,n}$, the regression coefficient provides the optimal volume as

$$\Delta^* = -a_w \tag{47}$$

in the problem (16) under condition (45), where

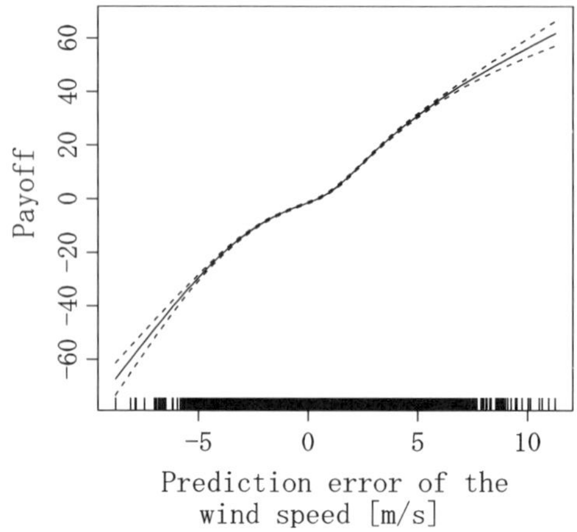

Fig. 14 Optimal payoff function on $\varepsilon_{w,n}$ obtained from GAM

$$a_w = \frac{\mathrm{Cov}\left(\varepsilon_{p,n}, \varepsilon_{w,n}\right)}{\mathrm{Var}\left(\varepsilon_{w,n}\right)}. \tag{48}$$

Figure 13 shows a scatter plot of $\varepsilon_{w,n}$ versus $\varepsilon_{p,n}$ with a linear regression line. The sample correlation is computed as

$$\mathrm{Corr}\left(\varepsilon_{p,n}, \varepsilon_{w,n}\right) \simeq 0.76. \tag{49}$$

and VRR as

$$\mathrm{VRR} = 1 - \mathrm{Corr}(\varepsilon_{p,n}, \varepsilon_{w,n})^2 \simeq 0.43. \tag{50}$$

We see that the forecast errors of the wind speed and the power output, $\varepsilon_{w,n}$ and $\varepsilon_{p,n}$, are highly correlated and that the sample variance is reduced to 43 % from the original one using the wind derivative in the case where the loss and the payoff functions are both linear.

Now, we apply GAMs to compute an optimal payoff function. The solid line in Fig. 14 shows the optimal payoff curve obtained by solving the optimization problem (26) when $\phi(\cdot)$ is linear. In this case, the VRR is computed as

$$\mathrm{VRR} = \frac{\mathrm{Var}\left(\varepsilon_{p,n} + \psi^*\right)\left(\varepsilon_{w,n}\right)}{\mathrm{Var}\left(\varepsilon_{p,n}\right)} \simeq 0.407 \tag{51}$$

where $\psi^*(\cdot)$ is the optimal payoff function. Moreover, the variance of the hedged loss $\varepsilon_{p,n} + \psi^*\left(\varepsilon_{w,n}\right)$ is computed as

$$\mathrm{Var}\left(\varepsilon_{p,n} + \psi^*\left(\varepsilon_{w,n}\right)\right) \simeq 103. \tag{52}$$

The above variance is actually lower than that of the hedged loss using (42) with the optimal wind derivative, which is computed as "119". Therefore, we see that, even though the variance of the original loss might be larger, it can be reduced

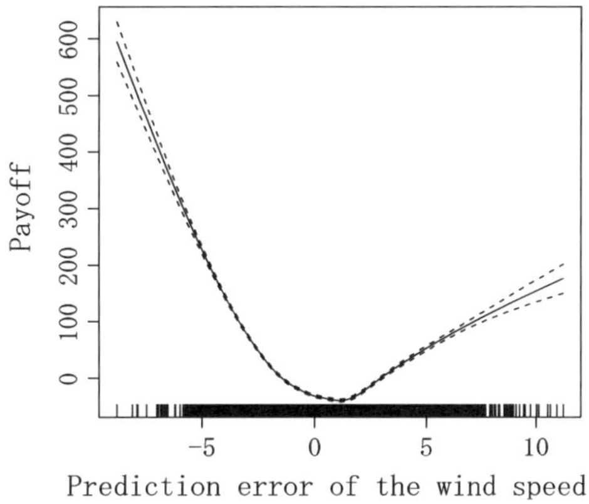

Fig. 15 Optimal payoff function on the wind speed forecast error $\varepsilon_{w,n}$

more effectively by combining it with the wind derivative if we define the forecast error by (43) instead of (42).

Next, we will consider the case in which the loss function $\phi(\cdot) = \overline{\phi}(\cdot)$ is given as shown in Fig. 5 with zero mean constraint (13), i.e.,

$$\overline{\phi}(\varepsilon_{p,n}) := 4|\varepsilon_{p,n}|^{+} + 10|\varepsilon_{p,n}|^{-} - \mu \tag{53}$$

where

$$\mu := \text{Mean}\left(4|\varepsilon_{p,n}|^{+} + 10|\varepsilon_{p,n}|^{-}\right)$$

and $|\cdot|^{+}$ and $|\cdot|^{-}$ are defined as

$$|x|^{+} := \max(x, 0), \ |x|^{-} := \min(x, 0)$$

for $x \in \Re$. The solid line in Fig. 15 shows the optimal payoff function to solve the problem (26). In this case, VRR in (27) is computed as

$$\text{VRR} = 0.5461946\ldots \tag{54}$$

whereas the right hand side of (30) is found to be

$$1 - \left[\text{Corr}\left(\phi(\varepsilon_{p,n}), \psi^{*}(\varepsilon_{w,n})\right)\right]^{2} = 0.5461927\ldots \tag{55}$$

From this example, we see that VRR can be approximated by (30) with high accuracy.

4.2.1 Optimal Loss Function and Simultaneous Optimization

In this subsection, we first provide an illustrative example for solving (P3) to compute an optimal loss function, and then solve the simultaneous optimization problem of (P4).

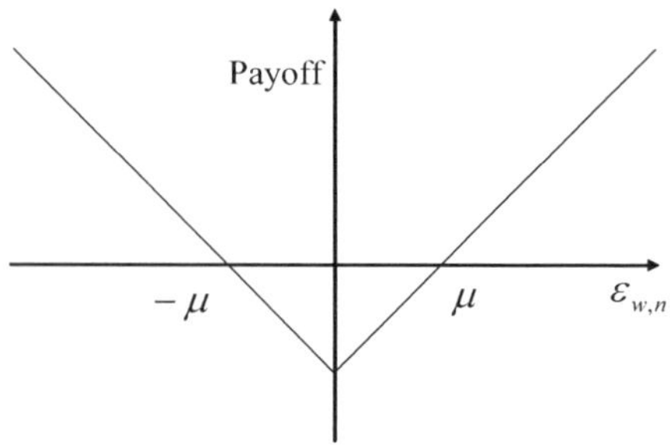

Fig. 16 Payoff function with respect to the wind speed forecast error $\varepsilon_{w,n}$ for the loss function optimization problem

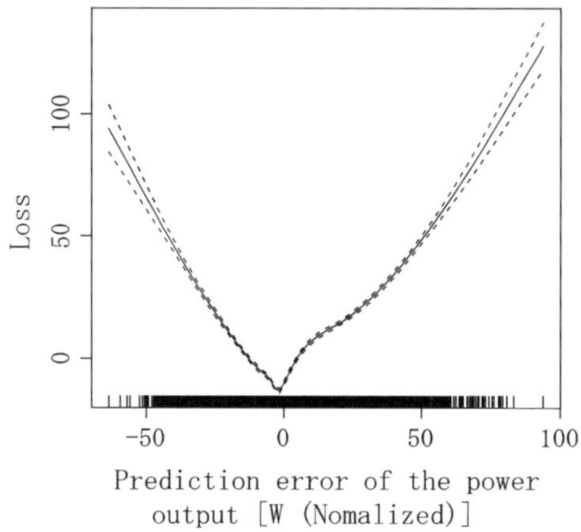

Fig. 17 Optimal loss function on the power output forecast error $\varepsilon_{p,n}$

Since the linear correlation between $\varepsilon_{p,n}$ and $\varepsilon_{w,n}$ is high in this example, it would be more interesting to consider the case where a payoff function is non-linear with respect to $\varepsilon_{w,n}$. Therefore, we assume that there already exists a derivative contract with the payoff being proportional to the size of the wind speed forecast error, $|\varepsilon_{w,n}|$. Noting that $\psi(\varepsilon_{w,n})$ satisfies (14), such a payoff function may be given as

$$\psi(\varepsilon_{w,n}) = \overline{\psi}(\varepsilon_{w,n}) := |\varepsilon_{w,n}| - \text{Mean}(|\varepsilon_{w,n}|), \tag{56}$$

Fig. 16 shows the payoff function with respect to $\varepsilon_{w,n}$.

Now we will solve (P3) with the given payoff function in (56). Assume that the sample variance of the loss, $\phi(\varepsilon_{p,n})$, satisfies

$$\text{Var}\ (\phi\ (\varepsilon_{p,n})) = \text{Var}\ (\varepsilon_{p,n}) \tag{57}$$

and we solve the problem (34) with the assumption that the optimal loss function satisfies the above variance constraint. The solid line in Fig. 17 shows the optimal loss function, which is obtained by applying GAM and scaling the minimizing function as in (35). In this case, VRR is found to be

$$\text{VRR} \simeq 0.56. \tag{58}$$

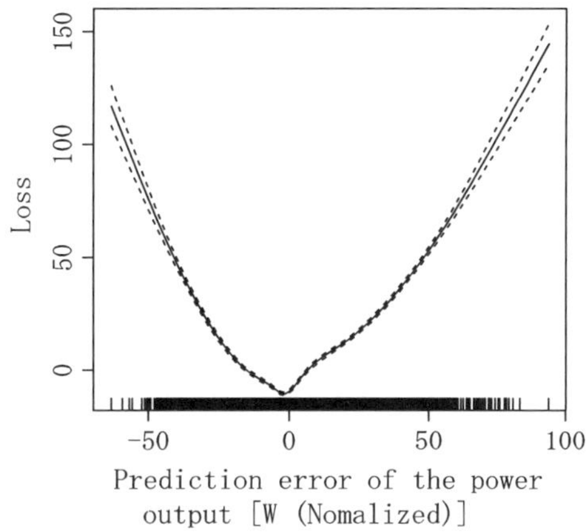

Fig. 18 Optimal loss function after the 8th iteration

Next, we demonstrate the simultaneous optimization of P4). Here we also introduce a nonlinearity using the absolute value of $\varepsilon_{w,n}$. Assume that the payoff of the wind derivative is a function of $|\varepsilon_{w,n}|$, and consider a maximization problem of

$$\operatorname{Corr}\left(\phi\left(\varepsilon_{p,n}\right), \psi\left(|\varepsilon_{w,n}|\right)\right). \tag{59}$$

We apply the iterative algorithm for a fixed loss function $\phi(\cdot)$ or a fixed payoff function $\psi(\cdot)$ at each step to maximize (59). Assume that the payoff function is initially set to the one given in (56) and we solve the loss function optimization problem. The initial loss function in this case is given by the one shown in Fig. 17. We repeatedly apply steps 1 and 2 in the iterative algorithm until the objective function does not change or the relative change of the values of the objective function is less than a sufficiently small number. In this example, we obtained

$$\text{VRR} = 0.53, \tag{60}$$

after the 8th iteration. Figure 18 shows the optimal loss function after the 8th iteration, where the loss function is scaled to satisfy the variance constraint (57). We see that the loss function became smoother compared to the one given in Fig. 17.

5 Further Discussion

The main contribution of this study is summarized as follows:

- Constructed a type of weather derivative contracts based on the forecast errors, which might be applicable for other situations (or businesses) and/or other indices such as temperature, rain falls, and so on.
- Provided an application of non-parametric regression techniques in the context of minimum variance hedge using smooth functions, which can be thought of a generalization of the standard minimum variance hedge based on linear regression.

Here we assumed that the payoff functions are just smooth. Therefore, the approximation of these functions using the standard payoff functions for puts or calls may be required in practice when the standardized derivative contracts are only available. Also, the convergence of the iterative algorithm for simultaneous optimization is an important issue. These are interesting topics to be discussed further in the future work.

It should be mentioned that there is another important issue concerning the risk management on forecast errors such as usage of storage battery system, i.e., a storage battery may be installed to compensate the shortage of the scheduled power output of a WF. Note that the advantage of installing the storage battery is not only to achieve the scheduled power output in certain period, but also to reduce the variability of the wind generated electricity that may significantly effect on the frequency of electricity grid. However, the installation of storage battery requires an additional cost [20], say, as much as the total capacity of the WF in the worst case scenario that the WF quoted the maximum power output but the actual one was zero. As the capacity of storage battery is reduced, the installation cost becomes lower, but in this case, the possibility of shortage may be increased. Therefore, in such a situation, we may still need wind derivatives to hedge the loss caused by forecast errors. For the wind power energy business to be profitable, it should be important to study an optimal balance of the capacity of storage battery and the possible introduction of wind derivatives.

6 Summary

In this chapter, we have introduced weather derivatives based on wind conditions combined with their forecast information, namely ‘‘wind derivatives.’’ The payoffs of wind derivatives were determined by the forecast errors in contrast to the standard weather derivatives in which the underlying index is given by observed weather data only (say, temperature). In other words, the wind derivatives discussed here take advantage of forecast data and the payoff depends on the difference between the actual and forecast data.

Here we began by explaining the basic structure which are common with financial derivatives and showed some foundations of pricing problems. In particular, we demonstrated a general pricing problem using a payoff function that satisfies a zero expected value condition. Then, we introduced wind derivatives on forecast errors, and derived a loss function for a WF based on the possible sales contract using forecasts. The following four types of optimization problems are formulated using combinations of payoff and loss functions: The first is a contract volume optimization problem that computes an optimal volume of wind derivatives for given loss and payoff functions. The second is a payoff function optimization problem that constructs optimal payoff function using a non-parametric regression technique called GAM. The third is a loss function optimization problem to find an optimal loss function that is desirable for a given payoff function of wind derivative, i.e., in a case where there already exists a standardized derivative contract with a certain payoff function, but there is some room for improvement on the loss function for a WF owner. The forth is a simultaneous optimization problem of loss and payoff functions, which may be solved using an iterative algorithm for a given payoff function or a loss function. To estimate the hedge effect, we defined the VRR as the ratio of variances of the losses with and without the hedge.

An empirical analysis was provided using the total power output data of a WF located in Japan, the wind speed data at the observation tower in the WF, and their forecasts. The power output from the WF was predicted based on the numerical weather forecast and the power generating properties for turbines. In particular, we used the forecast data of the power output and the wind speed obtained from LOCALS developed by ITOCHU Techno-Solutions Corporation. Based on the empirical data, we first solved the contract volume optimization problem in the case where the loss and the payoff functions are both linear. In this case, we saw from the VRR that the sample variance was reduced to 43 % from the original one using the wind derivative. Then, we applied GAMs and computed an optimal payoff function, where the VRR was obtained as VRR $\simeq 0.407$, showing the slight improvement by using GAMs. Also, we considered the case in which the loss function is given as the one used in the example of sales contract to compare the original VRR with its approximation formula given by one minus squared correlation coefficient. In this case, we were able to obtain a high accuracy of approximation.

Next, we solved the loss function optimization problem by introducing a nonlinearity using the absolute value of wind speed forecast error. We assumed that the payoff function is proportional to the size of wind speed forecast error and obtained VRR $\simeq 0.56$ by solving the problem. Then, we applied the iterative algorithm to solve the simultaneous optimization problem, where the initial payoff function was set to the same as in the above loss function optimization problem. We repeated the iterative procedure until the relative change of objective function is less than a sufficiently small number. By solving the simultaneous optimization, we concluded that VRR was improved from VRR $\simeq 0.56$ to VRR $\simeq 0.53$ compared to the case of loss function optimization.

Acknowledgments The author would like to thank H. Fukuda, R. Tanikawa, and N. Hayashi from ITOCHU Techno-Solutions Corporation for their helpful comments and discussions.

References

1. Jewson S, Brix A and Ziehmann C (2005) Weather derivative valuation—the meteorological statistical financial and mathematical foundations. Cambridge University Press, Cambridge
2. Yamada Y (2008) Optimal hedging of prediction errors using prediction errors. Asia-Pacific Finan Mark 15(1):67–95
3. Brody DC, Syroka J, Zervos M (2002) Dynamical pricing of weather derivatives. Quant Financ 2:189–198
4. Cao M, Wei J (2004) Weather derivatives valuation and market price of weather risk. J Futures Mark 24(11):1065–1089

5. Davis M (2001) Pricing weather derivatives by marginal value. Quant Financ 1:305–308
6. Kariya T (2003) Weather risk swap valuation Working Paper Institute of Economic Research. Kyoto University, Japan
7. Platen E, West J (2004) Fair pricing of weather derivatives. Asia-Pacific Finan Mark 11(1):23–53
8. Yamada Y (2007) Valuation and hedging of weather derivatives on monthly average temperature. J Risk 10(1):101–125
9. Yamada Y, Iida M, and Tsubaki H (2006) Pricing of weather derivatives based on trend prediction and their hedge effect on business risks. Proc inst stat math 54(1):57–78 (in Japanese)
10. Harrison JM, Pliska SR (1981) Martingales and stochastic integrals in the theory of continuous trading. Stoch Process Appl 11(3):215–260
11. Black F, Scholes M (1973) The pricing of options and corporate liabilities. J Polit Econ 81:637–654
12. Merton RC (1973) Theory of rational option pricing. Bell J Econ Manage Sci 4(1):141–183
13. Shreve SE (2004) Stochastic calculus for finance (2): continuous-time models. Springer, New York
14. Takano T (2006) Natural Energy Power and Energy Storing Technology, trans Inst Electri Eng Jpn (B) 126(9):857–860 (in Japanese)
15. Wood SN (2006) Generalized additive models: an introduction with R. Chapman and Hall, London
16. Hastie T, Tibshirani R (2005) Generalized additive models. Cambridge University Press, Cambridge
17. Enomoto S, Inomata N, Yamada T, Chiba H, Tanikawa R, Oota T and Fukuda H (2001)Prediction of power output from wind farm using local meteorological analysis. Proceedings of European Wind Energy Conference, Copenhagen, Denmark, p 749–752
18. Tanikawa R (2001) Development of the wind simulation model by LOCALS and examination of some studies, Nagare, p.405–415 (in Japanese)
19. Yamada Y (2008) Optimal design of wind derivatives based on prediction errors. JAFEE J 7:152–181 (in Japanese)
20. Tanabe T, Sato T, Tanikawa R, Aoki I, Funabashi T, and Yokoyama R (2008) Generation scheduling for wind power generation by storage battery system and meteorological forecast. IEEE Power and Energy Society General Meeting—Conversion and Delivery of Electrical Energy in the 21st Century, pp 1–7
21. Geman H (1999) Insurance and Weather Derivatives, Risk Books
22. Takezawa K (2006) Introduction to nonparametric regression. Wiley, New Jersey

Innovative Wind Energy Models and Prediction Methodologies

Zekâi Şen

Abstract Energy sources are among the major driving forces almost all the societal activities and its rather easiest way of extraction is from the fossil fuels, especially, coal and petroleum. However, their exploitations have direct and side effects on the most essential substances, air and water, because of greenhouse gas emissions into the atmosphere. Recent climate change effects are all related to fossil fuel exploitation, and therefore, the trend in the world now is towards excessive use, if possible, of clean and hence environmentally friendly energy sources among which apart from other alternatives wind power is present day attraction. In open literature there are many classical wind power calculation methods. In this chapter innovative ones are explained with applications. Among these stochastic temporal, cumulative semivariogram spatial, statistical perturbation, and innovative wind energy formulation and its Betz limit comparisons are presented.

1 Introduction

Presently, the global energy challenge is to tackle the threat of climate change, meet the rising demand for energy and to safeguard security of energy supplies. Renewable energy and especially wind energy technology are effective energy tools that are ready for global deployment today on a scale that can help to tackle climate change and global environment problems. Renewable energy use increment reduces CO_2 emissions, cuts down local air pollution levels, creates high-value jobs and curbs growing dependence on imports of fossil energy (which often come from politically unstable regions).

Z. Şen (✉)
Civil Engineering Faculty, Hydraulics and Water Resources Division, İstanbul Technical University, 34469 Maslak, İstanbul, Turkey
e-mail: zsen@itu.edu.tr

P. M. Pardalos et al. (eds.), *Handbook of Wind Power Systems*, Energy Systems,
DOI: 10.1007/978-3-642-41080-2_4,

Long years' experience gave rise to identification of various energy types and their actual use in daily life throughout the human history. Today, the energy types are categorized into two extensive groups as non-renewable (fossil) energy sources and renewable energy sources, which are clean and environmentally friendly. Among the renewable sources are the solar, wind, wave, and tide, geothermal and solar-hydrogen energy alternatives. It is the main purpose of this chapter to present the bases and modeling alternatives of the wind energy sources.

2 Global Circulation

The solar system and the earth's atmosphere define one very large dynamic system that creates horizontal motion in the atmosphere due to the differential heating of the air which gives rise to mostly horizontal pressure gradients. The words differential and gradient imply local, regional or local variability in the atmosphere. Big temperature differences between the equator and the polar region cause the global scale motions of the air. In addition to the main global wind system there are also local wind patterns as sea breezes and mountain valley winds. However, local wind speeds such as lake breezes are caused by local temperature differences. There are also gradient winds that are associated with curved isobars. Within the planetary boundary layer there are wind speed retardations which are functions of the terrain surface roughness in addition to and the pressure and temperature gradients in the lower atmosphere. Wind energy is equal to the kinetic energy of these air mass movements.

In general, wind energy is affected by two sets of variables. These are meteorological variables within the boundary layer of the atmosphere and surface features due to the natural or human activities such as urbanization. Among the primary meteorological variables are the temperature and pressure which change even in micro-scale instantaneously within the lower layers of the atmosphere and consequently force air masses to move giving rise to wind speed. In the mean time, simultaneously there are changes also in the wind speed direction which is a vector quantity rather than a scalar variable. Furthermore, the changes in the wind speed are also shown to be chaotic, and therefore, its prediction presents scientific difficulties.

On the other hand, geomorphologic (surficial) features such as the surface roughness, troughs, extended valleys, mountain ranges, high plateaus, existence of boundary between the free surface body (oceans, lakes) and land masses still modify general atmospheric affects on the wind speed depending on the location. However, the changes from geomorphologic affects have spatial characteristics only.

Although the atmosphere is on the average only 10 km thick all of the meteorological events occur within this conservation layer, which is referred to as the troposphere. Convective movements also occur at global scales. The seasonal cycle of the Earth orbiting around the sun creates global differences in the rate at

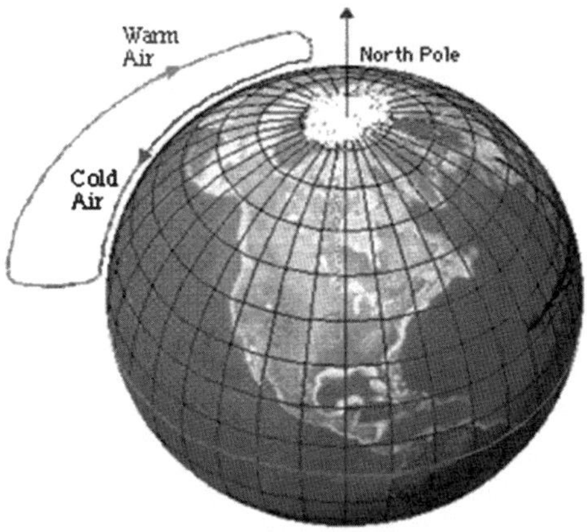

Fig. 1 Atmospheric circulation in northern hemisphere without Earth rotation

which it is heated. The equator is heated most directly due to perpendicular solar irradiation on the surface and higher latitudes (both north and south pole) experience seasonal changes depending on the angle of the hemisphere in relation to the sun. These global and seasonal differences give rise to an unequal distribution of heat across the globe and an imbalance in the density of air at any given latitude. Due to extra heat at the equator and cooler places at the poles, convection cells originate that circulate air through the atmosphere from equator towards both poles (Fig. 1).

The two main driving factors in global circulation are the solar radiation and the atmosphere and the Earth rotations. The seasonal variation is due to the tilt of the Earth's axis to the plane of the Earth's movement around the sun. Since the solar radiation is greater per unit area when the sun is directly overhead, there is a transport of heat from the regions near the equator toward the poles (Fig. 1). Since, the atmosphere is not attached to the Earth it can spin independently of the atmosphere. These convection cells of air rotate from the equator to the poles with the Earth spinning beneath it. The air moves eastward as the Earth moves eastward at about a rate of 28 km/min along the equator. However, the Earth's spin rate is not the same at the equator as it is at the poles diameter of the Earth at higher latitudes is less than the diameter of the Earth at lower latitudes. The Earth moves half as slow at 60°N as it does at the equator. The net result is that the air is moving eastward faster than the Earth and thus it appears to be deflected to the right (left) in the northern (southern). The predominate surface winds are due to the earth's rotation on its axis and conservation of angular momentum. The wind is shifted in the northern hemisphere as it moves along a longitudinal direction. Superimposed on this circulation is the migration of cyclones and anti-cyclones across the mid-latitudes, which disrupt the general flow. Also the jet streams influence the surface winds as the fast core of the general westerlies at upper levels.

Any movement in the universe at various scales indicates the existence of energy, which is the ability of temporal and/or spatial motions. Accordingly, wind

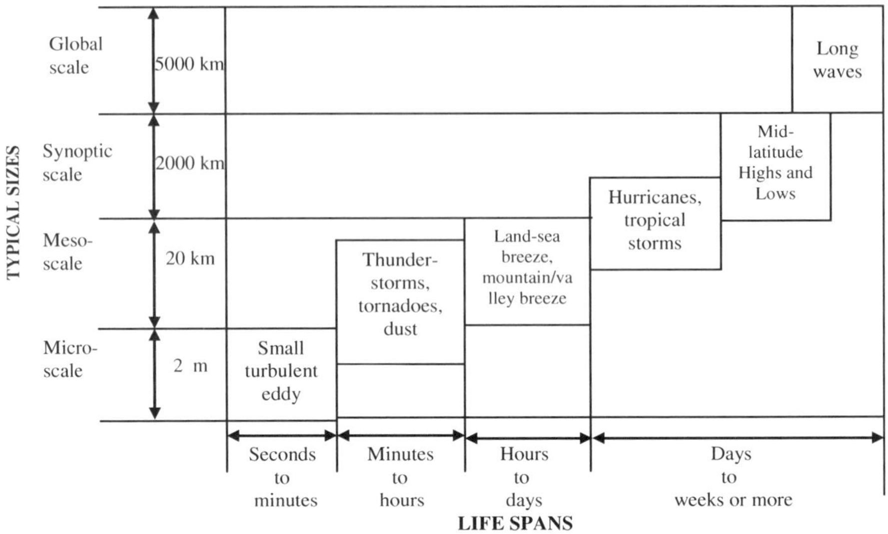

Fig. 2 Atmospheric motion scales

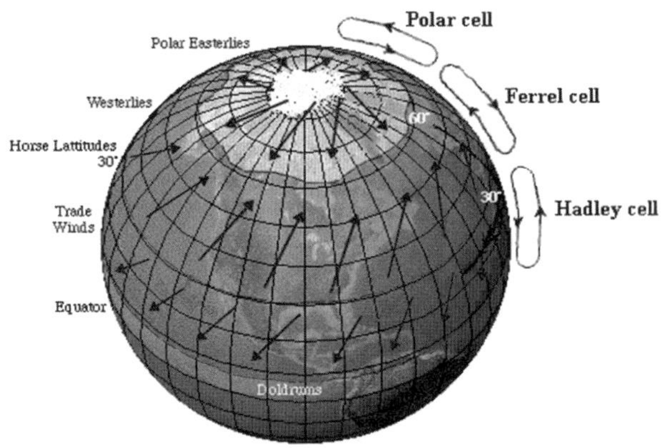

Fig. 3 General atmospheric circulations in northern hemisphere after Earth rotation

energy has also temporal and spatial types that are interrelated with each other. As in Fig. 2 atmospheric motion occurs in scales between 1 mm and 1,000 km.

The motion of the atmosphere can vary in distance and time from the very small to the very large scales. There is an interaction between each of these scales and the flow of air (wind patter) is complex. The global circulation encloses eddies which enclose smaller eddies until finally the micro-scale is reached.

As a result of the Coriolis force rather than one giant convection cell from the equator to the poles (Fig. 1), there are three convection cells as in Fig. 3. As air moves towards the north from the equator, it is deflected to the right (in the northern hemisphere). The end result is that at about 30°N, the wind is "turned around" such that it moves back towards the equator. The same thing happens to air moving equator-ward from the poles; it deflects to the right and takes a turn at about 60°N back towards the poles. This results in the formation of three giant atmospheric circulation cells in both hemispheres.

The equatorial circulation cells between equator and 30°N latitude are known as Hadley cells. The atmospheric circulation cell between 30 and 60°N is the Ferrel cell, and finally, the poleward circulation cell between 60°N and the pole is known as polar cell.

Wind blows not only air of different temperature or humidity into a region, but also air of different momentum, which per unit mass is referred to as the wind and it can change due to advection (horizontal movement). On the other hand, pressure gradient force acts always perpendicular to the isobars (equal pressure lines) from high to low pressure. It exists regardless of wind speed, starts the horizontal winds and can accelerate, decelerate or change the direction of existing winds. This gradient force is also the only force that can drive the horizontal winds in the atmosphere. The other forces such as Coriolis, drag, centrifugal and even advection disappear for zero wind speed. Hence, they can change the direction and speed of an existing wind, but they can not create a wind out of calm conditions.

3 Meteorological and Forces

Physically energy can be regarded as the work, which is force multiplied by distance and hence the question is what is the force and how can one describe the distance so that energy calculations can be achieved. This is rather straightforward for wind energy where the force is the wind pressure (force per area) and the distance is equal to velocity multiplication by time (Sect. 7). It is also known from physical principles that the pressure exists where there is wind velocity and there is a direct relationship between the pressure and the wind velocity and hence wind energy. It is, therefore, logical to end up with saying that wind energy is a function of wind velocity and direction, and therefore, velocity information is enough for simple wind energy calculations. The source of wind velocity and direction are the meteorological conditions in the lower atmosphere (troposphere) over the Earth surface features.

Some of the meteorological characteristics as renewable energy sources are wind, solar irradiation, water power and wave. If meteorological characteristics of these renewable energy sources are not well know, then there will be important gaps for energy investments [43]. Wind velocity changes with time, area and elevation (see Fig. 2). The most obvious relationship among these factors is between the wind velocity and height (elevation) from the earth surface because as the elevation increases the wind velocity increases in a directly proportional and non-linear manner but the rate of increase decreases with height (Sect. 4).

Since, the surface of the Earth is neither flat nor homogeneous, the amount of absorbed heat energy varies spatially as well as temporally. This results in temperature, pressure and density (specific mass) differences, and in turn creates forces that cause air to move from place to another. It is evident that depending on the surface features of the Earth, some areas would be preferable to others for extracting kinetic energy from the wind in the atmospheric boundary layer. This situation gives a new tendency for wind speed modeling and searching. Wind takes main role in meteorological trajectory movement. Petersen et al. [37] took into account wind power meteorology in their paper and searched for relations between meteorology and wind power in detail. They said that wind power meteorology is an applied science, which is based on boundary layer meteorology depicting high relations with climatology and geography. Wind power meteorology is not a standard meteorological term and it must be thought as a composition of meteorology, applied climatology and fluid physics.

Among the most effective meteorological variables temperature, pressure and moisture take important role in wind occurrence (Sect. 6). Generally, in wind engineering, moisture changeability is negligible and air is assumed at dry conditions. This situation can give important errors in calculations and energy plans [43].

Spatio-temporal variation of meteorological variables such as pressure, temperature, density and moisture generates dynamism, and hence movement in the lower atmosphere (troposphere). It is possible to quantify this variability through mass, momentum, energy conservations in addition to Newton's mechanical force balance and state equation of gases. Atmospheric movements such as geostrophic, gradient, height and surface winds around low air pressure and atmospheric waves can be explained on the basis of relevant equations, which represent atmospheric spatial variables with convenient numerical models that depend on initial and boundary conditions. The solution of these simultaneous equations is rather difficult and they can be solved only by ratios of time variables and finite element with variation intervals [22].

3.1 Wind Types

Geostrophic wind has two main features as no friction and blows parallel to isobars. Winds in nature are good approximations to geostrophic winds especially in the upper troposphere. This is because winds are considered truly geostrophic only when the isobars are straight and there are no other forces acting on it, and these conditions are not found too often near the Earth surface. The geostrophic assumption implies that there is a balance between the Coriolis and pressure gradient forces.

Gradient winds do not have straight flow and can represent real wind due to adding of the centrifugal force to other forces that are considered for geostrophic wind. All the winds that occur at low and high pressure systems with consideration of centrifugal force give a real approach to wind [37, 39].

Topographical effects are very affective for wind occurrence and hence another wind type is topographic wind. For instance, it is easy to observe that in a narrow mountain the air becomes compressed on the windy side of the mountains, and its speed increases considerably, which is known as a tunnel effect. In the very complex terrain wind speed not only has weak power but also high turbulence. Once again, this is due to the fact that the wind becomes compressed on the windy side of the hill, and as the air reaches the ridge, it can expand again as its soars down into the low pressure area on the lee side of the hill. This situation physically is another kind of tunnel effect called as hill effect. The sea and lake surfaces are obviously very smooth, thus the roughness of a seascape is very low. With increasing wind speeds some of the energy in the wind is used to build waves on sea surfaces, i.e. the roughness increases. Roughness classes or lengths are very significant for wind conditions in a landscape. A high roughness refers to landscapes with many trees and buildings, while a sea surface is in the lowest roughness class. These high roughness classes generate high turbulence areas. For wind, laminar flow means that air flows along streamlines without whirling motion mixing of the air in different stream layers. However, in turbulent flow, there is a whirling motion and air mixture.

Winds are very much influenced by the ground surface at altitudes up to 100 m. The wind will be slowed down by the Earth's surface roughness and obstacles. Wind directions near the surface will be slightly different from the direction of the geostrophic wind because of the Earth's surface properties. Among the topographical winds are sea–land and valley breezes together with adiabatic–catiabatic winds with thermal origin. Land masses are heated by the sun quicker than the sea in the daytime and hence the air rises, flows out to the sea, and finally creates a low pressure at ground level, which attracts the cool air from the sea, which is called a sea breeze. At nightfall there is often a period of calm when land and sea temperatures are equal. At night the wind blows in the opposite direction. The land breeze at night generally has lower wind speeds, because the temperature difference between land and sea is smaller at night. For instance, the monsoon originating from South-East Asia is in reality a large-scale form of the sea and land breezes, varying in the direction between seasons, because land masses are heated or cooled more quickly than the sea.

Mountain regions display many interesting weather patterns. One example is the valley wind which originates on south (north) facing slopes in the southern (northern) hemisphere. When the slopes and the neighboring air are heated the density of the air decreases, and the air ascends towards the top following the surface of the slope. At night the wind direction is reversed, and turns into a down slope wind. If the valley floor is sloped, the air may move down or up the valley as a canyon wind. Winds flowing down the leeward sides of mountains can be quite powerful.

Synoptic system is another main factor for windy areas. Generally, land–sea interaction areas are windy with stable conditions for wind speed and direction. High differences between heating and cooling during the day lead to the secondary windy areas in valleys. Astronomical seasonal time is important for heating and cooling depending on Sun position and area location [50]. Although these features provide advantages in wind engineering applications, discontinuities cause to disadvantages in wind energy studies. In practice, topographically stable areas can be found easier than temporally stable areas. In other words, discontinuity is the main problem in wind energy studies.

Wind means motion of air masses in synoptic scale, where these masses have potential and kinetic energies because of their conditions and motions. Winds occur as a result of potential energy transformation to kinetic energy by pressure forces. Horizontal winds move and cover more extensive areas than vertical winds. In wind energy calculations and applications vertical winds are negligible.

Solar energy drives the wind, which is then dissipated due to turbulence and friction at the Earth's surface. A comparison can be made on the basis of the kinetic energy of the winds per unit area of the Earth's surface. Incoming extra-terrestrial solar irradiation is converted only 2 % into wind power and 35 % out of this is dissipated within 1 km of the earth's surface. This is the wind power available for conversion to other forms of energy. Gustavson [21] assumed the extractable limit as 10 % of the available wind power within 1 km of the surface. The extractable amount is dependent on the local climate variation, where there are large uncertainties in determining such criterion. So wind energy represents a very large energy source.

4 Wind Speed Change with Height

The troposphere extends from the ground up to an average altitude of 11 km, but often only the lowest few kilometers are directly modified by the underlying surface features. One can define the atmospheric boundary layer (ABL) as that part of the atmosphere, which is directly influenced by the presence of the Earth's surface, and responds to surface forcing with a timescale of about an hour or less. These forcing include frictional drag, evaporation and transpiration, heat transfer, pollutant emission and terrain induced flow modification. The boundary layer thickness is quite variable in time and space, ranging from hundreds of meters to a few kilometers.

The mixed layer can be subdivided into a cloud and sub cloud layers. Wind speed changes with height at stable boundary layer (SBL) conditions given in Fig. 4. It is seen that wind speed changes exponentially at stable boundary layer (approximately 500 m) with height. This situation indicates that as height from the Earth surface increases so does the wind energy production. The wind increases with height to become equal to the geostrophic wind in the free atmosphere, both in direction and velocity, at the top of the ABL.

In the unstable boundary layer, UBL, the temperature of the surface is greater than the temperature of the air (the surface heats the air). Hence, there is a temperature inversion which caps the ABL. In the capping inversion zone, the turbulence is damped by a strongly stable stratification. The UBL occurs, generally, during the day when the solar heating is important. In the SBL, the temperature of the surface is lower than the temperature of the air (the air heats the surface).

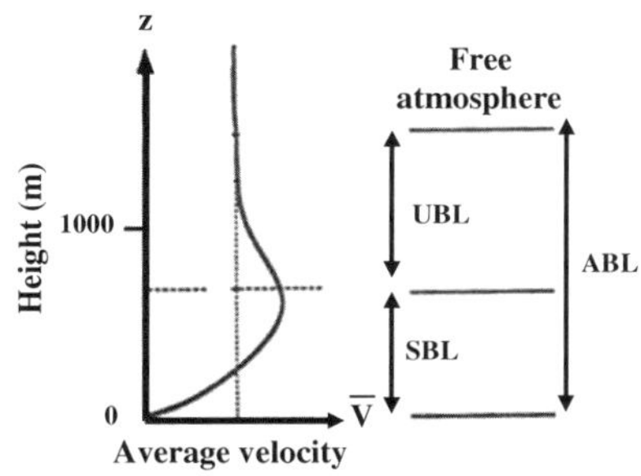

Fig. 4 Average wind speed, $\overline{V}$, profiles

The wind is, generally, weak near the surface and a maximum (low level jet) is often found at the top of the temperature inversion zone. The SBL usually occurs during the night.

In horizontally homogeneous terrains, the turbulence intensity is a function of height and roughness length, in addition to stability, whereas turbulence intensity is not too far from the ground and may be considered as a function of stability only. General properties can be given for air motions, which occur at ABL as,

1. Wind speed increases with height,
2. There are wind speed fluctuations within ABL,
3. Turbulence is distributed to wide frequency area,
4. There are relationships between turbulence at different heights. These relations are stronger at low than high wind speeds.

Wind boundary layer is evaluated by Erasmus [18] for complex topographical conditions by objective analysis methods. In this model, topographical effects are conserved and do not impair wind flows. In any mathematical algorithms, surface roughness, valley slopes, change of wind direction and mass conservation factors are taken into account. Another work concerning the terrain topography for wind energy calculations is presented by Şen [47].

The wind speed usually varies approximately logarithmically with height in the ABL. For non-neutral situations, the wind profile deviates slightly from logarithmic pattern. In SBL the wind profile is concave downward while unstable boundary layers are concave upwards (see Fig. 4). The mean wind profile, $\overline{V}$, averaged over periods up to 1 h, is often described for engineering purposes by a power law approximation. The following relationship relates two velocity values (V_1 and V_2) at two different heights (z_1 and z_2) through such a power law as,

$$\frac{V_1}{V_2} = \left(\frac{z_1}{z_2}\right)^n \tag{1}$$

where n is the power law exponent with a typical value of 1/7 = 0.14 [42]. A serious problem with this approach is that n varies with height, surface roughness and stability, which means that Eq. (1) is of quite limited usefulness. There are various assumptions involved in the suggestion and application of the power law. Among these are the following significant points.

1. The wind speed, v, profile change with elevation, z, is assumed to abide with a power law having zero speed value at the earth surface,
2. An implied assumption in Eq. (1) is that wind speed records at different elevations are independent from each other. This is not quite valid in actual situations as will be documented later in this section. Physically, the closer the two elevations, the greater is the cross-correlation coefficient between them,
3. In Eq. (1) the wind profile is assumed as a smooth curve defined by the arithmetic averages of wind speeds at different elevations. However, there are always fluctuations around these averages and consequently in the definition of wind profile the variances and standard deviations must also be taken into consideration.

A more realistic expression for the mean wind speed at height z, with much more general validity, can be obtained from the so-called logarithmic wind profile with stability correction. This expression, which is well supported by theoretical considerations, can be written as,

$$v = \frac{u_*}{k} \ln\left(\frac{z}{z_0}\right) \tag{2}$$

where u_* is the friction velocity, k is the von Karman constant (0.4), z_0 is the roughness length. The wind speed gradient is diminished in unstable conditions (heating of the surface, increased vertical mixing) and increased during stable conditions (cooling of the surface, suppressed vertical mixing). The term roughness length is really the distance above ground level where the wind speed theoretically should be zero. A simple, yet subjective method represents the method of terrain classification. Summarizing the results stated in the literature, which were attained for z_0 from measurements of wind speed and direction for several heights a table can be compiled where for each surface type a certain roughness length is assigned. From radiosonde measurements and maps (preferably from the standard scales 1:10.000 to 1:50.000) z_0 can be derived by means of the surface types within a radius of some kilometers around the site. Unfortunately in the literature the tables for z_0 differ considerably. The oldest table comes from Davenport [15] as Table 1.

Frictional effects proportional to the surface roughness retard the movement of air near the earth's surface. The surface roughness implies the effects and the nature of terrain, the location of forests, lakes, depressions, hills, valleys and their sizes in addition to the density of trees. Furthermore, the buildings produce different wind and velocity gradients in the vertical direction [53]. Practically, the wind speed profile levels out at approximately 600 m and above this height there are no significant roughness effects and the wind speed becomes equivalent with the gradient wind speed [24].

Table 1 Roughness lengths from terrain classification

Class surface	Landscape description	z_0 (m)
1	Sea open sea, fetch at least 5 km	0.0002
2	Smooth mud flats, snow; little vegetation, no obstacles	0.005
3	Open flat terrain; grass, few isolated obstacles	0.03
4	Roughly open low crops; occasional large obstacles	0.1
5	Rough high crops; scattered obstacles	0.25
6	Very rough orchards, bushes; numerous obstacles	0.5
7	Closed regular large obstacle coverage (suburb, forest)	1.0
8	Chaotic city centre with high- and low rise buildings	>2

Table 2 Stability conditions, n and p values

Stability condition	p	n
Large lapse rate	0.20	0.11
Zero or small lapse rat	0.25	0.14
Moderate inversion	0.33	0.20
Large inversion	0.50	0.33

The ABL thickness decreases with the decrease in the surface roughness. With decreasing roughness the profile is also steeper near the surface. Because of the appreciable change in the wind speed with height in the ABL, any wind speed value must be quoted with respect to the elevation at which it is measured. The internationally accepted standard height is 10 m for surface winds. In many meteorological stations the wind speed measurements are taken at elevations other than the standard height. Many studies have been made to develop suitable analytical expressions for height extrapolations as in Eq. (1). Hence, the atmospheric stability plays significant role in the form of wind speed profile and the depth of ABL. As a first approximation, the exponent n must vary in relation to the stability characteristics of the atmosphere. Sutton [42] has suggested related n to another parameter, p, which specifies the atmospheric stability condition as

$$n = \frac{p}{2 - p}$$

Table 2 shows the values of p for various stability conditions. Furthermore, the same author has related n to temperature differences as shown in Table 3.

In general the exponent value is found as 0.40 suitable for built-up areas; 0.28 for heavily wooded areas, cities and suburbs; and 0.16 for flat open country, lakes and oceans.

Table 3 Temperature difference and n

Temperature difference	n
−20 to −19	0.145
−19 to −17	0.170
−19 to −18	0.250
−18 to −17	0.290
18 to −17	0.320
−17 to −16	0.440
−16 to −14	0.530
−14 to −13	0.630
−13 to −12	0.720
−12 to −11	0.770

5 Power Law Dynamics

In many meteorology stations all around the world, there are towers reaching to 100 m in the atmospheric surface layer. Anemometers located at different heights on such as tower record with time the changes in the wind speed and direction. It is logical that as stated before all the environmental, boundary layer and atmospheric factors play roles of different weights on the power law exponent value. However, all these factors are hidden in a way in the wind records and hence there should be a method for the determination of the exponent value estimation from simultaneous wind speed time series at two fixed heights, z_1 and z_2. Since these elevations are fixed in the field the only variations appear in the velocity records of, v_1 and v_2 in Eq. (1). It is therefore possible to express these speeds in terms of the mean speeds and erratic deviations around them as perturbations,

$$v_1 = \overline{v}_1 + v'_1 \tag{3}$$

and

$$v_2 = \overline{v}_2 + v'_2 \tag{4}$$

where over bars indicate mean values and dashes the perturbation term. It is important to remark that statistical analysis of wind speed data from non-overlapping time intervals have shown that there are different estimates of the mean wind speed data at a given location and height. Substitution of these two expressions into Eq. (1) leads to

$$\left(\frac{\overline{v}_1 + v'_1}{\overline{v}_2 + v'_2}\right) = \left(\frac{z_1}{z_2}\right)^n$$

or after some algebraic manipulations as

$$\left(\frac{\overline{v}_1}{\overline{v}_2}\right)\left(1 + \frac{v'_1}{\overline{v}_1}\right)\left(1 + \frac{v'_2}{\overline{v}_2}\right)^{-1} = \left(\frac{z_1}{z_2}\right)^n \tag{5}$$

The third parenthesis on the left hand side can be expanded into a Binomial expansion which leads to the following expression

$$\left(\frac{\overline{v}_1}{\overline{v}_2}\right)\left(1+\frac{v_1'}{\overline{v}_1}\right)\left[1-\left(\frac{v_2'}{\overline{v}_2}\right)+\left(\frac{v_2'}{\overline{v}_2}\right)^2-\left(\frac{v_2'}{\overline{v}_2}\right)^3+\left(\frac{v_2'}{\overline{v}_2}\right)^4-\cdots\right]=\left(\frac{z_1}{z_2}\right)^n \tag{6}$$

It must be considered that terms $v_1'/\overline{v}_1'$ and $v_2'/\overline{v}_2'$ are significantly smaller than 1, and therefore, they may be ignored in further formulations. After the necessary algebraic calculations on the left hand side of this expression and then by taking the arithmetic averages of both sides one can obtain finally

$$\left(\frac{\overline{v}_1}{\overline{v}_2}\right)\left[1-\frac{\overline{v}_1'\overline{v}_2'}{\overline{v}_1\overline{v}_2}+\left(\frac{\overline{v}_2'}{\overline{v}_2^2}\right)\right]=\left(\frac{z_1}{z_2}\right)^n \tag{7}$$

Although by definition $\overline{v}_1'=\overline{v}_2'=0$ but Eq. (7) is an approximate expression because $\overline{v}_1'\overline{v}_2'^2$ is equal to zero approximately. For symmetrical (i.e. Gaussian) perturbation terms the odd power arithmetic averages such as $\overline{v}_1'\overline{v}_2'^2$ are also equal to zero by definition. From the stochastic process theory [14] the term $\overline{v}_1'\overline{v}_2'^2$ can be written in terms of the standard deviations, s_1 and s_2 and the autocorrelation coefficient, r_{12} between v_1 and v_2 wind speed time series and hence the final expression becomes

$$\left(\frac{\overline{v}_1}{\overline{v}_2}\right)\left(1-\frac{s_1 s_2}{\overline{v}_1\overline{v}_2}r_{12}+\frac{s_2^2}{\overline{v}_2^2}\right)=\left(\frac{z_1}{z_2}\right)^n \tag{8}$$

Given wind speed time series at two elevations all the terms in this expression are known with only one unknown n. Hence, Eq. (8) provides an objective way for determining the value of n. This expression reduces to the classical Eq. (1) only when the two wind speed time series at different elevations are independent from each other. In such a case $r_{12}=0$. Otherwise, the term within the second bracket on the left hand side varies between zero and one. Besides,

$$n=\frac{\ln(\overline{v}_1/\overline{v}_2)+\ln\left(1-C_{v_1}C_{v_2}r_{12}+C_{V_2}^2\right)}{\ln(z_1/z_2)} \tag{9}$$

where C_{v1} and C_{v2} are the coefficient of variations at heights z_1 and z_2, respectively. In order to illustrate the appropriateness of the methodology wind speed data from the Millstone nuclear power plant in Waterford, Connecticut for the year 1975 was used as presented by Kaminsky and Kirchhoff [33]. At this location, 15 min averages with a 96 % recovery rate, were recorded at four different heights above the earth surface as shown in the first column of Table 4.

The cross-correlation coefficients are between wind speed records at specified height and wind speed at 136.25 m. It is obvious from this table that as the height increases all the statistical parameters increase. Especially, the cross-correlations in the last column shows that as the difference between two elevations decreases

Table 4 Wind speed summary statistics

Height (m)	Mean speed (m/s)	Standard deviation (m/s)	Cross-correlation
9.75	4.15	2.43	0.795
43.28	5.45	2.95	0.875
114.00	6.73	3.32	0.975
136.25	6.87	3.55	1.000

Table 5 Exponent calculations

Height (m)	Classical	Extended (Eq. 9)
9.75	0.19	0.129
43.28	0.20	0.158
114.00	0.11	0.174

the correlation coefficient increases, which is an expected result. This also implies that the closer the height to the Earth surface the greater is the instability of the air. The power law exponent, n, calculation between height 136.25 m and others are found from Eq. (9) and presented in Table 5.

Although the amount of wind energy is economically insignificant for the time being in many parts of the world, mankind took the advantage of its utilization since early years whenever he/she found the chance to provide power for various tasks. Among these early utilizations are the hauling of water from a lower to a higher elevation, grinding grains in mills by water and other mechanical power applications for many centuries. It is still possible to see at some parts of the world these types of marginal benefits from the wind speed. Perhaps the wind-mills in Holland exemplify the most publicized application of wind power. All of the previous activities have been technological and the scientific investigation of wind power formulations and accordingly development of modern technology appeared after the turn of twentieth century. In recent decades the significance of wind energy originates from its friendly behavior to the environment so far as the air pollution is concerned although there are, to some extent, noise and appearance pollution in the modern wind-farms. Due to its cleanness, wind power is sought wherever possible for conversion to the electricity with the hope that the air pollution as a result of fossil fuel burning will be reduced [11]. In some parts of USA, up to 20 % of electrical power is generated from wind energy. In fact, after the economic crises in 1973 its importance increased by forcing the economic limitations and today there are wind-farms in many western European countries [1, 19, 52].

Although the technology in converter-turbines for the wind energy is advancing rapidly, there is a need to assess its spatiotemporal behavior scientifically with approaches. It is the purpose of this paper to develop a complete approach for spatiotemporal wind energy formulations and then interpretations based on the available meteorological variables. These formulations provide a basis to understand, control and predict the spatial, height, temporal and meteorological variations in total wind energy calculations.

In open literature, various researchers [7] have treated the problem of wind energy evaluation and assessment by different statistical approaches, but they did not present spatiotemporal and meteorological variability. The methodology presented in the following section has differences from their approaches by being deterministic and analytical; and by considering the time, space and meteorological variability in their most explicit behaviors during any wind energy calculations.

6 Stochastic Wind Energy Calculation Formulation

Almost all the wind power investigations relied mostly upon arithmetic average of the wind speed which represents the simplest central tendency measure of the wind power density, i.e., probability distribution function [34, 46]. On the other hand, many authors based the wind power estimates on elaborated wind speed statistics including the standard deviation, variance, skewness coefficient and kurtosis. In these studies several probability density functions (PDF) such as the log-normal, Gamma (Pearson-type III), Weibull and Rayleigh distributions are tried as possible useful fit to empirical wind speed frequency. Especially, Justus et al. [26, 29] advocated for the use of the two-parameter Weibull distribution in wind velocity applications. Consequently, their suggestions have been taken in granted in many parts of the world during the wind power calculations. Unfortunately, in some of the studies without testing the wind speed data for the suitability to the two-parameter Weibull distributions, the calculations are carried out automatically. In some researches not the genuine wind velocity data but their cubes are assumed to have two-parameter Weibull distribution as noticed by Hennessey [23]. Auwera et al. [4] showed that a three-parameter Weibull distribution fits the wind speed data in a more refined manner than the two-parameter Weibull PDF.

For standard atmosphere where the average air density $\rho = 1.225$ g/cm^3 at sea level and the temperature 15 °C the instantaneous wind energy, E_U, per unit time and per unit area is formulated as,

$$E_U = \frac{1}{2}\rho V^3 \tag{10}$$

where V is the wind speed and the unit of power is W/m^2 provided that V is in m/s. Although ambient air density is not a dominant variable in the wind power integration assumption of sea-level standard density had been questioned. At high altitude stations, the sea level density assumption causes available wind power to be overestimated by nearly 30 % [38]. An air-density correction factor is provided by the same author in order to convert the sea-level wind power estimates to the site location. This density correction factor is dependent on the site elevation and the annual cycle of monthly mean temperatures. The density of air changes with temperature and air pressure, i.e., altitude (Sect. 6). Furthermore, the power curve for any given wind turbine depends on air density. Solar radiation that reaches the earth's atmosphere plays a distinctive role in the creation of winds because of temperature differences and subsequent pressure differences. Wind is defined as the horizontal movement of air due to pressure difference. In this statement the word 'air' needs to be specified whether it is dry or moist. In all the applications so far it is understood as the dry air with the assumption of standard atmosphere which has the following specific points.

1. The air is completely dry,
2. The pressure at the average sea level is 1013.25 hPa,
3. The average temperature at the same level is 15 °C; and finally,
4. The air density at the same level is taken as a constant equal to 1.225 kg/m^3.

Each one of these assumptions led to rather restrictive application in the practice. According to the dependent random variables theory in the stochastic processes [36] if the air density and the wind speed were dependent on each other then taking the expectation of both sides in Eq. (10) should lead to,

$$E(E) = \frac{1}{2}E(\rho V^3) \tag{11}$$

However, the multiplication of two dependent random variables can be written in terms of the covariance and the multiplication of expectations of the individual random variables according to the definition of the covariance expression as,

$$\mathrm{Cov}(\rho, V^3) = E(\rho V^3) - E(\rho)E(V^3) \tag{12}$$

Furthermore if the correlation coefficient between the wind speed and the air density is denoted by r its classical definition in terms of the covariance and standard deviations of the individual random variables can be written as,

$$r = \frac{\mathrm{Cov}(\rho, V^3)}{S_\rho S_{V^3}} \tag{13}$$

Hence, the substitution of $\mathrm{Cov}(\rho,V^3)$ between these two last expressions gives after simple algebraic manipulations,

$$E(\rho V^3) = E(\rho)E(V^3) + rS_\rho S_{V^3} \tag{14}$$

The substitution of this last expression into Eq. (11) yields the most general form of the wind power formulation as,

$$E(E) = \frac{1}{2}\left[E(\rho)E(V^3) + rS_\rho S_{V^3}\right] \tag{15}$$

This expression reduces to some simple approaches that are available in the practical applications.

1. The second term in the brackets makes the major difference from the so far presented formulations in the literature. This term vanishes for constant air density assumption because r = 0, and Eq. (15) reduces to Eq. (10).
   On the other hand, if the air density is constant over the whole time duration, i.e., $E(\rho) = \rho$ then

$$E(E) = \frac{1}{2}\rho E(V^3) \tag{16}$$

2. For instantaneous air density and wind velocity measurements there is no cross correlation and hence Eq. (10) becomes valid. In the general stochastic formulation of Eq. (15), r plays the significant role depending on its actual value between −1 and +1.Logically, the greater the air density the smaller the wind speed. Since, moist air is lighter than dry air, the moisture plays a lubricant role in the air giving rise to higher wind velocities. Hence, the cross correlation is expected to have negative sign between wind speed and air density. This argument indicates that Eq. (11) yields greater value than Eq. (15) which can be rewritten as

$$E(E) = \frac{1}{2}E(\rho)E(V^3)\left[1 + r\frac{S\rho S_V^3}{E(\rho)E(V^3)}\right] \tag{17}$$

Hence, the term within the brackets can be defined as the correction factor, α, and it does not have any dimension.

$$\boldsymbol{\alpha = 1 + r\frac{S\rho S_V^3}{E(\rho)E(V^3)}}$$

It is also possible to rewrite this expression by considering the coefficient of variation, C, definition, in general, as the ratio of the standard deviation to the arithmetic average. Hence, two of such ratios appear in the last expression as the coefficient of variation for air density $C\rho$ and wind speed cube C_{V^3}. Finally, one can write succinctly that

$$\boldsymbol{\alpha = 1 + rC\rho C_{V^3}} \tag{18}$$

This expression indicates that in the cases of small and especially less than one coefficient of variations, the second term on the right hand side becomes negligible. For such situations, the traditional formulation is quite acceptable. Furthermore, although there are so many other uncertainties that the effort involved for a correction does not seem worthwhile. On the other hand, for relatively big coefficient of variations the second hand term may amount to significant level which means that the traditional formulation overestimates the wind energy potential. It is obvious that only when actually r = 0, the relative error becomes equal to zero. The formulations in this paper provide a basis for assessing the size of the correction factor for a given pair of air density and wind speed time series records.

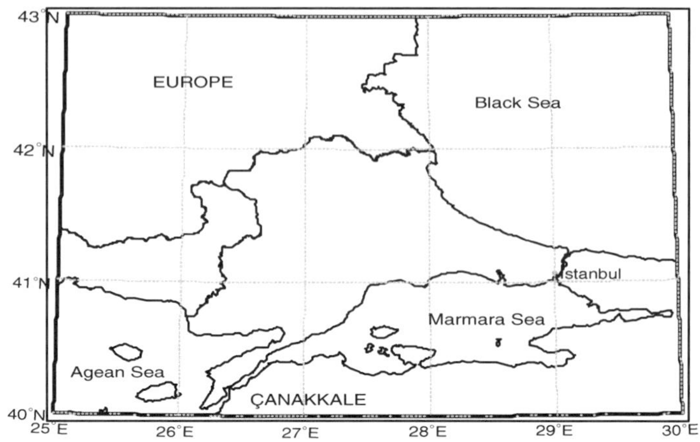

Fig. 5 Çanakkale station location

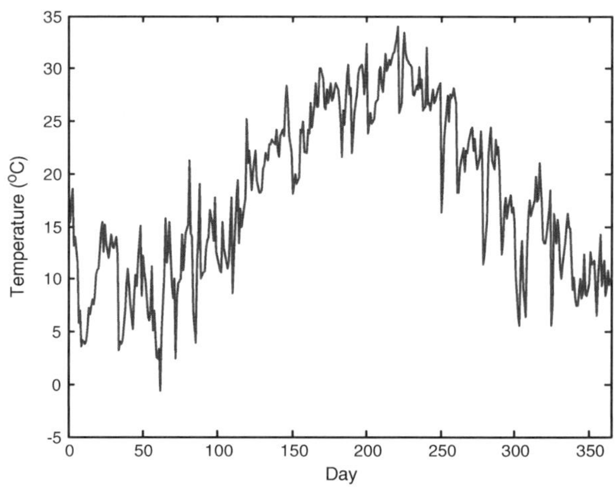

Fig. 6 Daily temperature time series

6.1 Application

In order to show the random variability in the actual situation daily temperature, pressure, air density and wind speed measurement time series as recorded at Çanakkale meteorological station in the northwestern part of Turkey (Fig. 5) are presented for complete year of 1995 in Figs. 6, 7, 8 and 9.

This is one of the most potential wind energy generation area in Turkey located at longitude 40°08′ and latitude 26°24′ with average velocity 4.13 m/s and total annual wind energy production as 93.50 W/m^2 [45]. Especially, temperature, pressure and air density time series exhibit within year seasonal variations rather distinctively, the wind speed fluctuations are quite stationary without explicitly

Fig. 7 Daily pressure time series

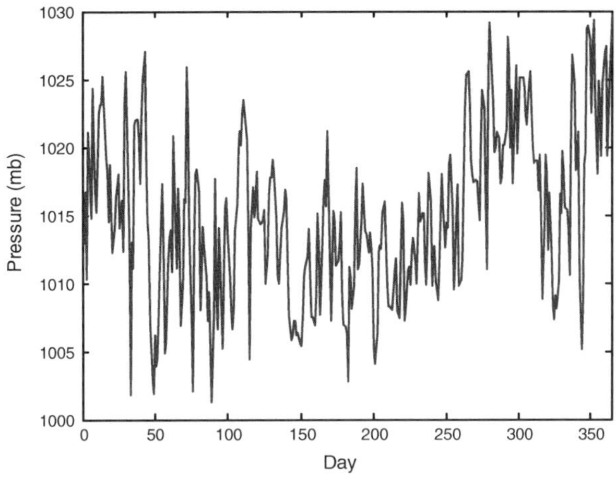

Fig. 8 Daily density time series

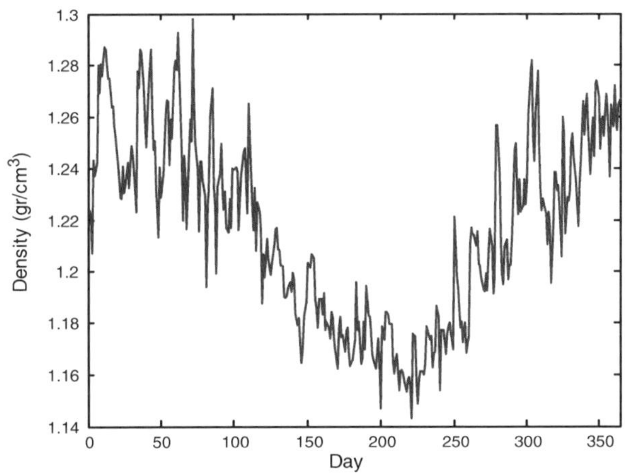

Fig. 9 Daily wind speed time series

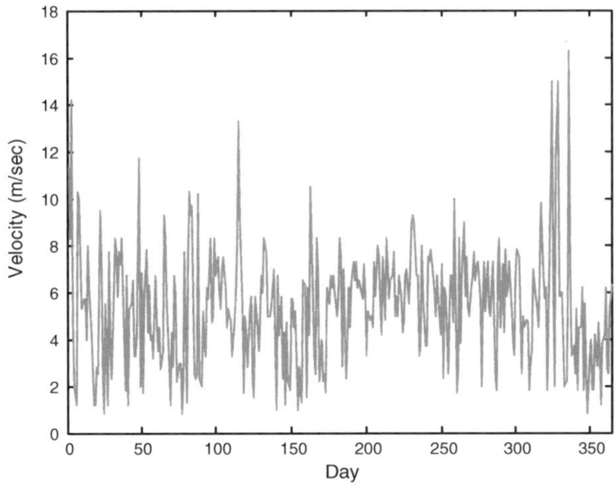

Table 6 Çanakkale station statistics

Parameter	Air density (g/cm^3)	Velocity cube (m/sn)^3
Average	1.216	266.037
Median	1.216	166.375
Mode	1.216	300.763
Standard deviation	0.036	441.203
Maximum	1.298	4330.750
Minimum	1.143	0.512

observable periodicities or trends. This feature of the wind speed supports the use of Eq. (16) as a rough approximation for wind energy calculation.

Table 6 presents the statistical properties of air density and wind velocity. Additionally, it is calculated that the cross-correlation between the air density and cube of the wind velocity time series is equal to −0.52.

The variations in the wind speed are comparatively very much than the air density variations. As a first impression, since the variations in the air density are very small, one might tend to ignore these variations and assume a constant air density equal to the arithmetic mean which is 1.216 g/cm^3 from Table 6. However, although the variations in the air density are small their impact on the wind energy calculations might be significant. In order to confirm this point, let us apply first Eq. (18) to data given in Table 6.

First of all the coefficient of variation for air density and velocity cube series are $C_\rho = 1.216/0.036 = 33.778$ and $C_V^3 = 226.037/441.203 = 0.603$, respectively. Hence, the substitution of these and cross-correlation value into Eq. (18) yields correction factor as $\alpha = 0.975$.

7 Theoretical Formulations of Spatio-Temporal Wind Energy and Dynamics

The total amount of wind energy E_T in a region with turbine surface area, A, perpendicular to the wind direction and during a time interval, t, can be calculated ideally from Eq. (10) as $E_T = AtE_U$. The implied assumptions in this equation are,

1. That the wind speed over the region considered remains almost the same at every point without any turbulence effect,
2. That the average wind speed along the time interval considered does not change significantly. The air density ρ at any point is dependent on the absolute temperature T (K) and pressure P (hPa) according to the universal gas law as $P = R\rho T$ where R is the universal gas constant equal to 2.87 for dry air when P in millibars, T in degrees Kelvin and ρ in kg/cm^3. Elimination of ρ between universal gas law and Eq. (10) leads to [46]

$$E_u = \frac{1}{2R}\left(\frac{P}{T}\right)V^3 \tag{19}$$

This is the most explicit wind energy formulation showing the individual effects of meteorological variables. The application of Eq. (19) is possible at any station and time instant that instantaneous measurement of (T,P,V) triple is available.

7.1 Energy Ratios

Eq. (19) can be expanded to two-point formulation by considering its ratio at two different times or places. Let us define this type of two-point ratio from Eq. (19) between points i and j as,

$$\frac{E_{u_i}}{E_{u_j}} = \left(\frac{P_i}{P_j}\right)\left(\frac{T_j}{T_i}\right)\left(\frac{V_i}{V_j}\right)^3 \tag{20}$$

here i and j indicate the position of two distinct points in the space or along the time axis. The pressure ratio on the right hand side is similar to the sigma-coordinate system in meteorology. As Holton [24] states, the isobaric coordinate systems have many advantages such as that the meteorological data is normally referred to isobaric surfaces and hence sound waves are completely filtered. On the other hand, in the sigma system the vertical coordinate is the pressure normalized with the surface pressure as P_j, hence, $\sigma_{i,j} = (P_i/P_j)$. The value of σ_{ij} assumes 1 at the ground level and 0 at the top of atmosphere where the wind energy is equal to zero.

Similarly, temperature ratio $\tau_{j,i} = (T_j/T_i)$ shows that since $T_j > T_i$ within the troposphere $\tau_{j,i} < 1$, except in the case of inversion layer existence, when $\tau_{j,i} > 1$. Finally, because of the surface roughness effects on vertical direction $V_i > V_j$ and accordingly velocity ratio $\upsilon_{i,j} = V_i/V_j$ is expected as greater than one. The substitutions of these ratios into Eq. (20) lead to a non-dimensional form of energy as

$$e_{i,j} = \sigma_{i,j}\tau_{j,i}\upsilon_{i,j}^3 \tag{21}$$

where $e_{i,j}$ is the energy ratio. It is possible to interpret this expression for various iso-surfaces of different meteorological variables, as follows.

1. **Isobaric Surfaces:** These are equal pressure surfaces along which there is no pressure gradient, and hence as long as points i and j remain on such a surface then $\sigma_{i,j} = 1.0$, and Eq. (21) becomes

$$e_{i,j} = \tau_{j,i}\upsilon_{i,j}^3 \tag{22}$$

This formulation shows that the wind velocity changes due to temperature differences only between any two equal pressure points (Fig. 10). Physically, if $\tau_{j,i} < 1.0$ then $\upsilon_{i,j} > 1.0$ or vice versa.

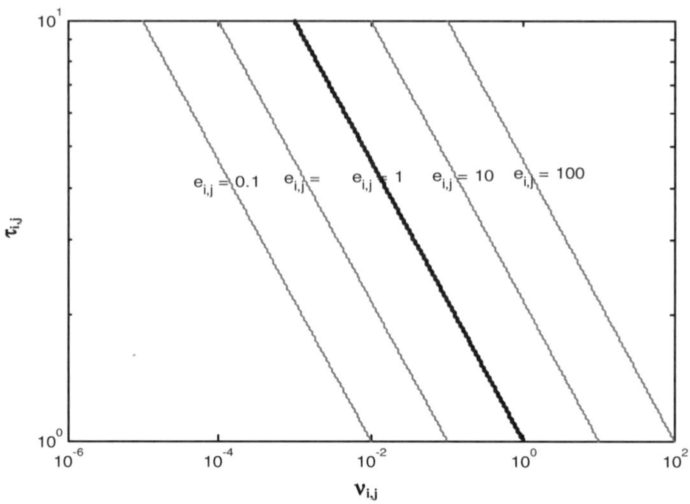

Fig. 10 Wind speed ratio versus temperature ratio variation

2. **Isothermal Surfaces:** Absence of temperature gradient along any surface implies that it is isothermal surface. In this case, the winds are generated due to pressure differences only and hence $\tau_{i,j} = 1.0$ which converts equation to

$$e_{i,j} = \sigma_{i,j} \upsilon_{i,j}^3 \tag{23}$$

It is obvious that similar formulations are valid for $e_{i,j}$ and $\upsilon_{i,j}$ variation provided that $\tau_{j,i}$ is replaced with $\sigma_{i,j}$. It is physically plausible that air flows from high pressure to low pressure zone. Since during such a flow due to the friction losses the wind velocity is expected to decrease toward low pressure centers and consequently it is always $\sigma_{i,j} > 1.0$ for $\upsilon_{i,j} < 1.0$ or vice versa.

7.2 Vertical Wind Energy Variation

At a fixed time instant and location it is possible to calculate the correlation between wind energy variation along the vertical direction and height h from the earth's surface. Bennett et al. [7] have shown the effect of upper air data on the lower level wind estimation. First isothermal atmospheric case will be considered which implies no vertical temperature change. In this case, the remaining effective variables in wind energy generation are the pressure difference and the wind speed. The vertical pressure change is given in many classical textbooks as

$$P(h) = P_0 \exp\left(-\int_0^h \frac{dh}{H} \right) \quad (24)$$

where H = RT/gM is defined as the local scale height of the atmosphere. Herein, M is the average molecular weight of the atmosphere and it can be considered as a constant up to 100 km. Additionally, g is the gravitational acceleration and it depends on h, but its variation within 100 km is negligible at 3 % error level. Consequently, in isothermal atmosphere, Eq. (24) can be rewritten simply as

$$P(h) = P_0 e^{-h/H} \quad (25)$$

where H would indicate the height at which the pressure has decreased by a factor $e^{-1} = 0.37$, i.e., the height within about 2/3 of the atmosphere mass.

On the other hand, assuming an atmosphere with a constant lapse rate, β, temperature at height h can be expressed as $T = T_0 - \beta h$ where T_0 is the temperature corresponding to P_0. In this case the change of pressure with height is given by

$$P(h) = P_0 \left(1 - \frac{\beta}{T_0} h \right)^{3.41/\beta} \quad (26)$$

It is to be noticed that in this formulation β is in °C/100 m. Although the pressure-height expressions have physical basis, the velocity-height models have empirical basis. The first of such empirical expressions is the classical power law model which reads as

$$V(h) = V_m \left(\frac{h}{h_m} \right)^{\gamma} \quad (27)$$

where V_m is the velocity measurement at height h_m and γ is an exponent depending on the surface roughness around the measurement site. Finally, substitution of Eqs. (26) and (27) into Eq. (20) leads after some algebra to

$$\frac{E}{E_m} = \left[\left(1 - \frac{\beta}{T_0} h \right) / \left(1 - \frac{\beta}{T_0} h_m \right)^{-1} \right]^{\frac{3.41}{\beta} - 1} \left(\frac{h}{h_m} \right)^{3\gamma} \quad (28)$$

This expression shows the variation of wind energy with height in its general form. However, specially, the International Commission for Air Navigation (ICAN) specified a standard atmosphere up to 11 km as $P_0 = 1013.2$ millibar; $T_0 = 288$ K and $\beta = 0.65$ °C/100 m. Hence, for these values Eq. (28) becomes

$$\frac{E}{E_m} = \left(\frac{1 - 2.25 \times 10^{-5} h}{1 - 2.25 \times 10^{-5} h_m} \right)^{4.25} \left(\frac{h}{h_m} \right)^{3\gamma} \quad (29)$$

7.3 Meteorological Wind Energy Dynamics

Atmospheric dynamics are governed by fundamental physical principles such as conservation of energy, mass, momentum, state of gas, etc. Similar to basic meteorological variables, namely, temperature, pressure and velocity the wind energy calculations, energy field are also spatiotemporally continuous. Therefore, energy changes can be calculated by considering the concept of continuous media. Two types of energy calculations are commonly used in fluid dynamics. In the Eulerian frame of reference system a fixed point is considered with its relevant meteorological variables. Wind energy field variables are T, P and V which are related to each other succinctly as

$$E = \alpha \frac{P}{T} V^3 \tag{30}$$

where α is a constant equal to 1/2R. The energy change, dE, may be expressed in a Taylor series expansion by assuming the first order approximation as

$$dE = \left(\frac{\partial E}{\partial P}\right) dP + \left(\frac{\partial E}{\partial T}\right) dT + \left(\frac{\partial E}{\partial V}\right) dV + \varepsilon \tag{31}$$

in which ε indicates an error component resulting from higher order terms contribution. Substitution of partial derivatives from Eq. (31) leads approximately to

$$dE = \alpha \left(\frac{V^3}{T} dP - \frac{P}{T^2} V^3 dT + 3 \frac{P}{T} V^2 dV\right) \tag{32}$$

This is a key expression from which it is possible to evaluate the rate of energy change with respect to many significant meteorological, space and time variables as follows:

1. **Change with velocity:** Division of Eq. (32) by very small increments in the wind speed, dV, leads to

$$\frac{dE}{dV} = \alpha \left(\frac{V^3}{T} \frac{dP}{dV} - \frac{P}{T^2} V^3 \frac{dT}{dV} + 3 \frac{P}{T} V^2\right) \tag{33}$$

where dE/dV is the rate of energy change per wind velocity increment. Similarly, the pressure and temperature increments due to wind velocity are dP/dV and dT/dV, respectively. These latter rates can be obtained from meteorological measurements at a given station as will be explained in the application section.

2. **Change with temperature:** On the other hand, division of Eq. (32) by dT leads to rate of energy change with temperature as

$$\frac{dE}{dT} = \alpha \left(\frac{V^3}{T} \frac{dP}{dT} - \frac{P}{T^2} V^3 + 3 \frac{P}{T} V^2 \frac{dV}{dT}\right) \tag{34}$$

3. **Change with pressure:** Furthermore, wind energy rate with respect to pressure can be obtained as

$$\frac{dE}{dP} = \alpha\left(\frac{V^3}{T} - \frac{P}{T^2}V^3\frac{dT}{dP} + 3\frac{P}{T}V^2\frac{dV}{dP}\right) \tag{35}$$

The following general interpretations can be drawn from the last three energy rate expressions.

1. If there are no pressure, temperature and velocity variations at a given measurement point then Eqs. (33)–(35) yield equivalent result to Eq. (19). Such a situation does not rise in any practical study for some time duration (short or long term) during which there is always fluctuations in meteorological variables. Therefore, direct application of Eq. (19) is possible for either instantaneously measured or averaged meteorological variables.
2. If the pressure and temperature changes are neglected and the wind energy calculations are based only on the wind velocity then energy rate with respect to velocity increment becomes from Eq. (32) as

$$\frac{dE}{dV} = 3\alpha\frac{P}{T}V^2 + \varepsilon_{PT} \tag{36}$$

in which the amount of error, ε_{PT}, is

$$\varepsilon_{PT} = \alpha\left(\frac{V^3}{T}\frac{dP}{dV} - \frac{P}{T^2}V^3\frac{dT}{dV}\right) = \alpha\frac{P}{T}V^3\left(\frac{1}{P}\frac{dP}{dV} - \frac{1}{T}\frac{dT}{dV}\right) \tag{37}$$

8 New Wind Energy Formulation and Betz Limit Comparison

The advent and development of the air plane in the first decades of the twentieth century gave rise to intense analysis and design studies of the propeller that could immediately be applied to the wind turbine. Betz [10] evolved in clear form, the magic value of 16/27 as the ideal maximum utilization factor which is called as Betz limit pertaining to the flux of energy available from the wind.

There are two basic physical processes that limit the maximum rotor power coefficient of an inducted wind turbine. First, a rotor increases the upwind static pressure, reducing the mass flow rate through its swept area, and the wind energy available for conversion. Second, a rotor converts some of the linear kinetic energy of the wind to rotational kinetic energy in its wake, which is no longer available for conversion to mechanical energy. The first retardation process limits the rotor power coefficient at all tip-speed ratios to 0.593 (16/27), which is referred to as Betz or, more accurately Lanchester-Betz limit [9]. This limit can be approached when the wake rotation is low.

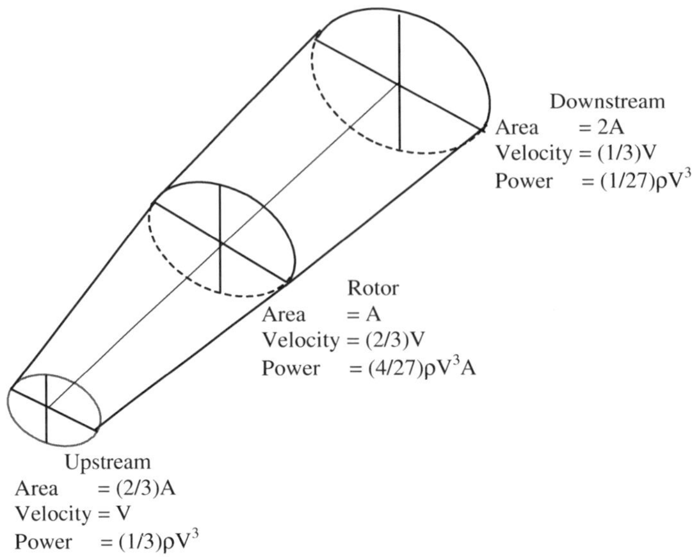

Fig. 11 Betz limit, condition for maximum power

It is not possible to extract all the energy from the air, which passes through the windmill rotor, because the air does not cease to move and pile up behind the rotor. The air flow through the stream tube and rotor disk is shown in Fig. 11 [17].

It has been shown by Betz [10] that under conditions of maximum power extraction, the air velocity at the rotor will have fallen to 2/3 of the value of the upstream wind velocity, V, and further decreases to a final value of 1/3, the initial wind velocity well downstream of the rotor. This means that it is possible to extract only a fraction 24/27 of the initial wind energy. In general, it is more convenient to express the performance of a wind turbine as the power output divided by the power in the wind passing through an area equal to the swept area of the mill. It can be seen from Fig. 11 that this is equivalent to a fraction 16/27.

8.1 Classical Formulation Coupled with Betz Limit

It is well known from any classical textbook on wind energy that the wind power per area per time is given by Eq. (10). On the other hand, for Betz limit approach the wind power formulation can be derived as follows by considerations from momentum rate. If the wind speeds are $\mathbf{V_u}$ and $\mathbf{V_d}$ for upstream (before the rotor) and downstream (after the rotor), respectively, then the power that can be absorbed by the wind turbine is,

$$\mathbf{P_A} = \mathbf{m}(\mathbf{V_u} - \mathbf{V_d})\mathbf{V_m} \tag{38}$$

in which m is the mass of air passing through the rotor in unit time and $\mathbf{V_m} = 0.5(\mathbf{V_u} - \mathbf{V_d})$ represents average wind speed actually passing through the rotor. The rate of kinetic energy change in wind can be expressed as,

$$\mathbf{E_K} = (\mathbf{1/2})\mathbf{m}\left(\mathbf{V_u^2} - \mathbf{V_d^2}\right) \tag{39}$$

It is obvious that these two expressions explain the same phenomenon, and therefore, they should be equal to each other. On the basis of assumptions that the direction of wind velocity through the rotor is axial and that the velocity is uniform over the entire rotor area, **A**, the retardation of the wind $(\mathbf{V_u} - \mathbf{V_m})$ before the rotor is equal to the retardation $(\mathbf{V_u} - \mathbf{V_m})$ behind it. Hence, the power extracted by the rotor can be expressed in terms of air density as

$$\mathbf{P_E} = \boldsymbol{\rho}\mathbf{A}\mathbf{V_m}(\mathbf{V_u} - \mathbf{V_d})\mathbf{V_m} \tag{40}$$

which first by substitution of $\mathbf{V_m} = 0.5(\mathbf{V_u} - \mathbf{V_d})$ and after algebraic arrangements can be written as

$$\mathbf{P_E} = (\mathbf{1/4})\boldsymbol{\rho}\mathbf{A}\mathbf{V_1^2}(\mathbf{1} + \boldsymbol{\alpha})(\mathbf{1} - \boldsymbol{\alpha}^2) \tag{41}$$

where $\alpha = V_d/V_u$ is the velocity ratio, $(0 < \alpha < 1)$. In order to determine the possibility of maximum power extraction, it is necessary to differentiate this last expression with respect to α and then equate the result to zero which yields $\alpha = 1/3$. Thus the maximum power which is referred to herein as the Betz limitation formulation can be found by substitution of is $\alpha = 1/3$ into the last expression which leads to,

$$\mathbf{P_{Betz}} = (8/27)\boldsymbol{\rho}\mathbf{V_u^2} \tag{42}$$

As compared with an ideal wind turbine can extract (Eq. 10), it is possible to rearrange Eq. (42) as follows.

$$\mathbf{P_{Betz}} = (1/2)(16/27)\boldsymbol{\rho}\mathbf{V_u^2} \tag{43}$$

where the Betz limit $16/27 = 0.593$ appears explicitly. Hence, if the power extraction without any losses would be possible, only 59 % of the wind energy could be utilized by a wind turbine. The derivation in Eq. (43) does not take into consideration the change in momentum directly. In the following section the wind power will be derived by taking into consideration the change in the momentum right from the beginning of the evaluations.

8.2 New Derivation

In order to arrive at a more reliable and accurate formulation, it is necessary to consider the fluid property of the air and hence the density prior to the kinetic energy formulation. Physically, the retardation force of the wind passing through a wind turbine occurs in two steps, namely before and after the passage through the rotor (see Fig. 11). This force can be expressed in term of the mass of air that passes through the rotor in unit time and the velocity change. Furthermore, the mass is also function of air density and the volume of air that passes through the rotor. In turn, the air density is a function of air pressure and air temperature which both are the functions of the height above sea level. The power in the wind is the total available energy per time unit. The power in the wind is converted into the mechanical-rotational energy of the wind turbine rotor, which results in a reduced speed of the air mass. The power in the wind cannot be extracted completely by a wind turbine, as the air mass would be stopped completely in the intercepting rotor area. Under the light of the aforementioned statements, it is possible to write the force that affects the wind turbine physically by considering Newton's second law as,

$$F = m \cdot dV/dt \tag{44}$$

Provided that m is the mass of air passing through the rotor in unit time, then the rate of momentum change per unit time (for constant m) can be written as mdv which is equal to the resulting thrust. Herein, dV indicates the difference between the upstream and downstream velocities from the rotor. During a time period, T, the upstream mass covers VT distance and hence the volume of the air passing during this period through the rotor is equal to AVT where A is the rotor area. Hence, the air mass can be expressed as ρAVT the substitution of which into Eq. (44) gives

$$F = \rho ATVdV/dt \tag{45}$$

On the other hand, the energy (or work) is defined physically as the multiplication of force by distance, say dL, as

$$dE = FdL = \rho ATVdVdL/dt \tag{46}$$

where the ratio, dL/dt is equal physically to the definition of velocity, V, and hence, this expression can be rewritten as,

$$dE = \rho ATV^2dV \tag{47}$$

Finally, the total energy can be obtained after the integration of both sides according to

$$E = (1/3)\rho ATV^3 \tag{48}$$

or similar the wind power per area per time becomes,

$$E_W = (1/3)\rho V^3 \tag{49}$$

It is obvious that the only difference between this expression and the classical counterpart in Eq. (43) is the numerical factor of (1/3) instead of (16/27)(1/2). It is suggested herein that the use of Eq. (49) is more accurate theoretically because the fluid property of the air is considered right from the beginning of the physical derivations.

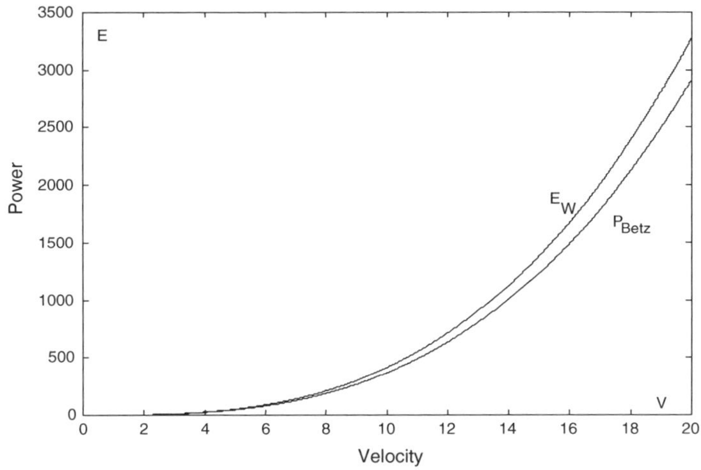

Fig. 12 Betz and the new power formulation comparison

8.3 Comparison of Formulations

As explained above, Eq. (43) takes into consideration the momentum effects after the basic power formulation derivation based on the concept of kinetic energy in Eq. (10). The Betz limit as 16/27 comes as a factor into this expression and yields the Betz wind power formulation as given in Eq. (43). On the other hand, herein derived wind power expression in Eq. (49) is physically based right from the beginning on the rate of momentum change. Eqs. (43) and (49) are shown graphically in Fig. 12 which indicates that practically these expressions are very close to each other with slightly less wind power estimation according to the Betz limitation formulation.

However, the difference within the practically valid range of velocities does not exceed more than 10 %. For small velocities they both yield almost equal results, but the difference increases by the velocity increase. Betz did not consider the impact of unavoidable swirl losses. For turbines with a high tip speed ratio, and optimum blade geometry, these losses are very low.

9 Wind Power Perturbation

Perturbation methodology has been used extensively since Taylor [51] research in many studies concerning turbulent flow in channels. Since the wind also presents a turbulent flow in the atmosphere the instantaneous wind speed, $\mathbf{V}$, can be thought of consisting two components namely, average wind speed, $\overline{\mathrm{V}}$, and the perturbation term $\mathbf{V}'$ as in Eq. (3). The perturbation term has the expected value equal to zero, i.e.,

$\mathbf{E(V')} = \mathbf{0}$ and its variance is equal to the variance of the instantaneous wind speed record, that is $\mathbf{V(V)} = \mathbf{V(V')}$. The substitution of Eq. (3) into Eq. (11) leads after stochastically manipulations to,

$$E(P) = \frac{1}{2}\rho\left[E^3(V) + 3E(V)E(V'^2) + E(V'^3)\right] \tag{50}$$

where P is used instead of E. This expression is common in many papers but the last term in the big bracket is ignored and it is written in an approximate form as

$$E(P) = \frac{1}{2}\rho\left[E^3(V) + 3E(V)E(V'^2)\right] \tag{51}$$

This expression is valid exactly for symmetrical probability distributions such as the Gaussian distribution but the wind speed is never symmetrically distributed in nature. Most often the wind speed distributed according to the Weibull, Gamma, Chi squared, logarithmic-normal distribution etc. In this paper, refined wind power expectation in Eq. (50) will be applied for the Weibull distribution in the following section.

The general stochastic definition of variance of a random variable such as the vend energy is defined as

$$V(P) = E(P^2) - E^2(P) \tag{52}$$

The second term on the right hand side is the square of expectation given in Eq. (50) and it is necessary to calculate $\mathbf{E(P^2)}$ for finding $\mathbf{V(P)}$. Substitution of Eq. (3) into Eq. (11) and then taking the square of both sides leads to the desired expectation as

$$\mathbf{E(P^2)} = \frac{1}{4}\boldsymbol{\rho}^2\mathbf{E}\left[\overline{\mathbf{V}}^6 + 6\overline{\mathbf{V}}^5\mathbf{V}' + 15\overline{\mathbf{V}}^4\mathbf{V}'^2 + 20\overline{\mathbf{V}}^3\mathbf{V}'^3 + 15\overline{\mathbf{V}}^2\mathbf{V}'^4 + 6\overline{\mathbf{V}}\mathbf{V}'^5 + \mathbf{V}'^6\right] \tag{53}$$

Since by definition $E(V') = 0$, this last expression becomes in an explicit form as

$$E(P^2) = \frac{1}{4}\rho^2\left[E^6(V) + 15E^4(V)E(V'^2) + 20E^3(V)E(V'^3) + 15E^2(V)E(V'^4) + 6E(V)E(V'^5) + E(V'^6)\right] \tag{54}$$

However, in the case of symmetric probability distribution this expression will simplify to the following form because the odd order expectations of the perturbation term are equal to zero. Hence,

$$E(P^2) = \frac{1}{4}\rho^2\left[E^6(V) + 15E^4(V)E(V'^2) + 15E^2(V)E(V'^4) + E(V'^6)\right] \tag{55}$$

Substitution of this expression together with Eq. (50) into Eq. (52) leads after algebraic manipulations to

$$\begin{aligned}V(P) = \frac{1}{4}\rho^2\Big[&6E^4(V)E(V'^2) + 15E^2(V)E(V'^4) - 9E^2(V)E^2(V'^2) - E^2(V'^3) - 2E^3(V)E(V'^3)\\&-6E(V)E(V'^2)E(V'^3) + E(V'^6)\Big]\end{aligned} \tag{56}$$

This is the general variance equation after perturbation of the wind speed value that can be applied specifically to any probability distribution function. The application of Eqs. (50) and (56) is performed for the Weibull wind speed distribution in the following sequel.

9.1 Weibull Distribution and Wind Power

All over the world, the two parameter Weibull probability distribution (WPD) is employed in most of the wind speed assessments [26, 28]. It has been shown that WPD of wind speed is a useful practical tool for estimation of future power in many climatic regions. Petersen et al. [37] have shown that the Weibull wind speed probability distribution is a useful practical tool for estimation of future power from windmills in Denmark. A review of the relevant statistical methods for estimation of Weibull parameters is given with emphasis on efficiency [12]. The wind speed is treated as a random variable which abides with a two-parameter Weibull distribution. The density function of such a distribution is given mathematically as,

$$f(V) = \frac{a}{b}\left(\frac{V}{a}\right)^{b-1}\exp\left[-\left(\frac{V}{b}\right)^{b}\right] \tag{57}$$

where **a** is a scale parameter with the same dimension as **V**, and **b** is a dimensionless shape parameter. In many applications of the Weibull distribution, it is necessary to use the statistical moments. In general, k-th order statistical moment of the WPD wind speed, **V**, is given as

$$E(V^k) = a^k\Gamma\left(1 + \frac{k}{b}\right) \tag{58}$$

Hence, the expected value ($k = 1$) and the variance from Eq. (52) of a Weibull distribution can be obtained analytically as

$$E(V) = a\Gamma\left(1 + \frac{1}{b}\right) \tag{59}$$

and

$$V(V) = a^2\left[\Gamma\left(1+\frac{2}{b}\right) - \Gamma\left(1+\frac{1}{b}\right)\right] \quad (60)$$

In order to obtain an analytical expression for the Weibull wind speed probability distribution wind power expectation according to Eq. (50), it is necessary to evaluate various terms in the big brackets. The first term is ready from Eq. (58) but other two perturbation terms are obtained as

$$E(V©^2) = a^2\left[\Gamma\left(1+\frac{2}{b}\right) - \Gamma^2\left(1+\frac{1}{b}\right)\right]$$

and the third moment expectation becomes, in general, as

$$E\left(V'^3\right) = E(V-\overline{V})^2 = E\left(V^3\right) - 3E(V)E\left(V^2\right) + 2E^2(V)$$

Under the light of Eq. (58) this expression takes the following form for the Weibull distribution

$$E(V'^3) = a^3\left[\Gamma\left(1+\frac{3}{b}\right) - 3\Gamma\left(1+\frac{1}{b}\right)\Gamma\left(1+\frac{2}{b}\right) + 2\Gamma^3\left(1+\frac{1}{b}\right)\right]$$

Finally, the substitution of the necessary expectation terms into Eq. (50) leads after some algebra to

$$E(P) = \frac{1}{2}\rho a^3\Gamma\left(1+\frac{3}{b}\right) \quad (61)$$

On the other hand, in order to find an explicit expression for the wind power variance according to Eq. (56), it is necessary to evaluate some other perturbation term higher order moments as follows: The fourth order perturbation term expectation, in general, is

$$\begin{aligned} E\left(V'^4\right) &= E(V-\overline{V})^3 \\ &= E\left(V^4\right) - 4E(V)E\left(V^3\right) + 6E^2(V)E\left(E^2\right) - 4E^3(V)E(E) + E^4(V) \end{aligned}$$

However, for the Weibull distribution after the substitution of the necessary moments from the Weibull distribution (use of Eq. 58), one can obtain

$$E\left(V'^4\right) = a^4\left[\Gamma\left(1+\frac{4}{b}\right) - 4\Gamma\left(1+\frac{3}{b}\right)\Gamma\left(1+\frac{1}{b}\right) + 6\Gamma\left(1+\frac{2}{b}\right)\Gamma^2\left(1+\frac{1}{b}\right) - 3\Gamma^4\left(1+\frac{1}{b}\right)\right]$$

On the other hand, the general fifth moment expression for the perturbation velocity term can be expressed as

$$E\left(V'^5\right) = E(V - \overline{V})^5$$
$$= E\left(V^5\right) - 5E(V)E\left(V^4\right) + 10E^2(V)E\left(E^3\right) - 10E^3(V)E\left(E^2\right)$$
$$+ 5E^4(V)E(V) - E^5(E)$$

Its specific form for the WPD becomes

$$E\left(V'^5\right) = a^5\left[\Gamma\left(1+\frac{5}{b}\right) - 5\Gamma\left(1+\frac{4}{b}\right)\Gamma\left(1+\frac{1}{b}\right) + 10\Gamma\left(1+\frac{3}{b}\right)\Gamma^2\left(1+\frac{1}{b}\right)\right.$$
$$\left. - 10\Gamma\left(1+\frac{2}{b}\right)\Gamma^3\left(1+\frac{1}{b}\right) + 4\Gamma^5\left(1+\frac{1}{b}\right)\right] \tag{62}$$

Furthermore, the sixth perturbation moment by stochastic definition can be expanded into the following form

$$E\left(V'^6\right) = E(V - \overline{V})^6 = E\left(V^6\right) - 6E(V)E\left(V^5\right) + 15E^2(V)E\left(E^4\right) - 20E^3(V)E\left(E^3\right) + 15E^4(V)E\left(V^2\right)$$
$$- 6E^5(E)E(E) + E^6(V)$$

Its specific counterpart for the Weibull distribution is

$$E\left(V'^6\right) = a^6\left[\Gamma\left(1+\frac{6}{b}\right) - 6\Gamma\left(1+\frac{5}{b}\right)\Gamma\left(1+\frac{1}{b}\right) + 15\Gamma\left(1+\frac{4}{b}\right)\Gamma^2\left(1+\frac{1}{b}\right)\right.$$
$$\left. - 20\Gamma\left(1+\frac{3}{b}\right)\Gamma^3\left(1+\frac{1}{b}\right) + 15\Gamma\left(1+\frac{2}{b}\right)\Gamma^4\left(1+\frac{1}{b}\right) - 5\Gamma^6\left(1+\frac{1}{b}\right)\right] \tag{63}$$

The substitution of all these necessary terms into Eq. (56) leads after the necessary manipulations to desired variance of wind power as

$$V(P) = \frac{1}{4}\rho^2 a^6\left[\Gamma\left(1+\frac{6}{b}\right) - \Gamma^2\left(1+\frac{3}{b}\right)\right] \tag{64}$$

The standard deviation of the wind power can be found by taking the square root of this expression which leads to

$$S_P = \frac{1}{2}\rho a^3\sqrt{\Gamma\left(1+\frac{6}{b}\right) - \Gamma^2\left(1+\frac{3}{b}\right)} \tag{65}$$

Finally, it is possible to define the coefficient of variation as the ratio of the standard deviation to the expectation of the wind power which gives

$$C_v = \frac{S_P}{E(P)} = \frac{\sqrt{\Gamma\left(1+\frac{6}{b}\right) - \Gamma^2\left(1+\frac{3}{b}\right)}}{\Gamma\left(1+\frac{3}{b}\right)} \tag{66}$$

or succinctly,

$$C_v = \sqrt{\frac{\Gamma\left(1+\frac{6}{b}\right)}{\Gamma^2\left(1+\frac{3}{b}\right)} - 1} \tag{67}$$

9.2 Vertical Extrapolation Weibull Distribution Parameters

Extrapolation of wind data to standard elevations poses a rather subjective approach based on the mean wind velocity values only. Any unreliability in such extrapolations is reflected in the subsequent wind energy, E, calculations by taking into consideration Eq. (10). Most often the wind speed at a meteorology station is measured along a tower at different elevations and it is desired to be able to find the wind profile at this station for further wind loading or energy calculations. Justus et al. [26, 28], Justus and Mikhail [27] have employed the Weibull probability distribution function (PDF) for empirical wind speed relative frequency distribution, i.e., histogram and a set of formulas are derived for the extrapolation of the Weibull PDF parameters. In general two-parameter Weibull PDF, its mean and variance are given in Eqs. (58)–(60). On the other hand, it is already shown that wind velocities at two different heights from the Earth surface on the same vertical are related to each other through Eq. (9), where the dependence coefficient is denoted by r_{12} in addition to the coefficient of variations and variances. In the case of the two-parameter Weibull PDF, the coefficient of variation is presented in Eq. (67). The value of a can be obtained from Eq. (59) as

$$a = \frac{E(V)}{\Gamma\left(1+\frac{1}{b}\right)} \tag{68}$$

By using approximate expansion of Gamma function as given in standard mathematical textbooks, it is possible to obtain b value after some algebraic approximations from Eq. (67) as,

$$b = \frac{1}{C_v^{1.086}} \tag{69}$$

The substitution of last two expressions for two elevations into Eq. (9) leads to

$$\left(\frac{v_1}{v_2}\right)\left[1 - (b_1 b_2)^{-0.921} r_{12} + b_2^{-1.841}\right] = \left(\frac{z_1}{z_2}\right)^n \tag{70}$$

or the power exponent becomes

$$n = \frac{\mathrm{Ln}(v_1/v_2) + \mathrm{Ln}\left[1 - (b_1 b_2)^{-0.921} r_{12} + k_2^{-1.841}\right]}{\mathrm{Ln}(z_1/z_2)} \tag{71}$$

Table 7 Exponent calculations

Height (m)	Classical	Extended (Eq. 18)
9.75	0.19	0.13
43.28	0.20	0.16
114.00	0.11	0.18

Hence, once the Weibull PDF parameters are determined the power exponent can be calculated provided that the cross-correlation coefficient is found from the available wind speed time series data. It must be noticed that this last expression reduces to the classical counterpart by substitution of $r_{12} = 0$ and $\mathbf{S}_{\mathbf{V}_2} = 0$.

9.3 Application

Wind speed data from the Millstone nuclear power plant in Waterford, Connectitut, for the year 1975 is adopted to illustrate the validity of the methodology developed in the previous section. The data is presented by Kaminsky and Kirchhoff [33]. In fact, it is very difficult to find suitable data for the application of the developed methodology. At the Waterford location, 15 min averages with 96 % recovery rate were recorded at different heights (9.75, 43.28, 114, and 136.25 m). The average wind speed V, the standard deviation S, and the correlation coefficient r are given in Table 3.

The cross-correlation coefficients are between wind speed at specified height and wind speed at 136.25 m. It is obvious from this table that as the height increases all the statistical parameters increase. Especially, the cross-correlations in the last column shows that as the difference between two elevations decreases the correlation coefficient increases which is an expected result. This also implies that the closer the height to the earth surface the greater is the instability of the air. The power law exponent, n, calculation between height 136.25 m and others are found from Eq. (69) and presented in Table 7.

10 Statistical Investigation of Wind Energy Reliability and its Application

The energy, E, in the wind is equivalent to flux of the kinetic energy of a moving air mass, m, with a speed, v (Eq. 10).

Since, in hydro- and aero-dynamics, the measurement of m is almost impossible, it is preferable to use specific mass $\rho = m/V$. The substitution of m between this expression and Eq. (10) gives,

$$E = \frac{1}{2}\rho V v^2 \tag{72}$$

However, the wind is a horizontal air movement and, therefore, the volume, V, can be defined as V = AL in which A is vertically fixed control cross section and L is the horizontal distance. This distance is related to the wind velocity as L = vt. The necessary substitutions lead to,

$$E = \frac{1}{2}\rho A t v^3 \tag{73}$$

For practical purposes, it is preferable to consider the wind energy per vertical unit area per time which will be referred herein onwards as the unit wind energy Eu as in Eq. (10). In general, the variations in absolute temperature, T, pressure, p, and specific mass, ρ are interrelated to each other through the state equation of gases,

$$p = \rho R T \tag{74}$$

where R is the universal gas constant equal to 2.87 for dry air when p is in millibars, T in degrees Kelvin and ρ in kg/cm^3. Finally, the elimination of ρ between Eqs. (73) and (74) gives,

$$E_U = \alpha\left(\frac{p}{T}\right)v^3 \tag{75}$$

where α is a constant and equal to 0.5/R or 0.174 with the aforementioned units. The meteorological variables on the right hand side are conventionally measured at any meteorology station, but Eu cannot be measured directly. Hence, the questions are to find ways of estimating wind energy from these measurements whenever the basic meteorological data are available.

10.1 First Order Statistical Analysis

It is obvious from any record on p, T and v that their measurements fluctuate with time in a random fashion depending on weather patterns. They are regarded as random variables with a persistence to a certain extent. Consequently, wind energy evaluations along time axis in a given station are also random in their characters. This enables one to use the statistical techniques in the wind energy assessments. Eq. (96) is a nonlinear relationship between four variables, namely, Eu, p, T and v. A nonlinear function of random variables can be approximated using the linear terms of a Taylor series expansion [2, 8]. For instance, if Y is a function of several random variables as,

$$Y = f(X_1, X_2, \ldots, X_n) \tag{76}$$

then $f(X_1, X_2,\ldots, X_n)$ can be expanded into a Taylor series about the mean values $\overline{X}_1$ to $\overline{X}_n$. Considering the linear first order terms and over bars as indicators of mean values,

$$Y = f(\overline{X}_1, \overline{X}_2, \ldots, \overline{X}_n) + \sum_{i=1}^{n} (X_i - \overline{X}_i)\frac{\partial f}{\partial X_i} + \varepsilon \tag{77}$$

where the partial derivatives are evaluated at respective mean values and ε represents error term consisting of higher-order terms. The mean of Y, $\overline{Y}$ then becomes,

$$\overline{Y} = f(\overline{X}_1, \overline{X}_2, \ldots, \overline{X}_n) \tag{78}$$

The variance of Y, is $\sigma Y2$ with its implicit expression as

$$\sigma_Y^2 = \sum_{i=1}^{n}\left(\frac{\partial f}{\partial X_i}\right)^2 \sigma_{X_i}^2 + \sum_{i=1}^{n}\sum_{j=i}^{n}\left(\frac{\partial f}{\partial X_i}\right)\left(\frac{\partial f}{\partial X_j}\right)\mathrm{Cov}\left(X_i, X_j\right) \tag{79}$$

in which $\sigma_{X_i}^2$ is the variance of random variable, X_i, and $\mathrm{Cov}(X_i, X_j)$ is the variance of X_i and X_j which is given by definition as

$$\mathrm{Cov}\left(X_i, X_j\right) = \rho_{X_iX_j}\sigma_{X_i}\sigma_{X_j} \tag{80}$$

where $\rho_{X_iX_j}$ is the correlation coefficient between X_i and X_j. However, if all the random variables are independent, the covariance equals zero, and accordingly Eq. (79) reduces to,

$$\sigma_Y^2 = \sum_{i=1}^{n}\left(\frac{\partial f}{\partial X_i}\right)^2 \sigma_{X_i}^2 \tag{81}$$

It is interesting to note that these equations require no assumptions about the form of the probability distributions of the component random variables.

10.2 Statistical Properties of Wind Energy

The main energy considered as a random variable will have mean value and variance according to the expressions mentioned in the previous section. In order to derive their explicit forms, all of the variables on the right hand side (rhs) of Eq. (75) are considered as random variables. Hence, the mean value of wind energy under the light of Eq. (78) becomes,

$$\overline{E}_U = \alpha\frac{\overline{p}}{\overline{T}}\overline{V}^2 \tag{82}$$

On the other hand, the variance of wind energy, σ_E^2 can be written, in general, by considering Eqs. (79) and (80) which leads explicitly to

$$\sigma_E^2 = \left(\frac{\partial E_U}{\partial p}\right)\sigma_p^2\left(\frac{\partial E_U}{\partial T}\right)\sigma_T^2 + \left(\frac{\partial E_U}{\partial V}\right)\sigma_V^2 + 2\left(\frac{\partial E_U}{\partial p}\right)\left(\frac{\partial E_U}{\partial T}\right)\rho_{pP}\sigma_p\sigma_T$$
$$+ 2\left(\frac{\partial E_U}{\partial p}\right)\left(\frac{\partial E_U}{\partial V}\right)\rho_{pV}\sigma_p\sigma_V + 2\left(\frac{\partial E_U}{\partial T}\right)\left(\frac{\partial E_U}{\partial V}\right)\rho_{TV}\sigma_T\sigma_V \quad (83)$$

where σ_i^2 and ρ_{ij} are the variances and linear correlation coefficients of relevant indices. The partial derivations in Eq. (83) are as follows.

$$\frac{\partial E_U}{\partial p} = \frac{\overline{E}_U}{\overline{p}} \quad (84)$$

$$\frac{\partial E_U}{\partial T} = -\frac{\overline{E}_U}{\overline{T}} \quad (85)$$

and

$$\frac{\partial E_U}{\partial V} = 3\frac{\overline{E}_U}{\overline{V}} \quad (86)$$

The substitution of which into Eq. (83) yields after some simple manipulations to

$$\sigma_E^2 = \left(\frac{\sigma_p^2}{\overline{p}} - \frac{\sigma_T^2}{\overline{T}} + 3\frac{\sigma^2}{\overline{V}}\right)\overline{E}_U - 2\overline{E}_U^2\left(\frac{\sigma_p\sigma_T}{\overline{p}\overline{T}}\rho_{pT} - 3\frac{\sigma_p\sigma_V}{\overline{p}\overline{V}}\rho_{pV} + 3\frac{\sigma_T\sigma_V}{\overline{T}\overline{V}}\rho_{TV}\right) \quad (87)$$

If the meteorological variables p, T and v are linearly independent from each others, i.e., $\rho_{pV} = \rho_{pV} = \rho_{TV} = 0$, then Eq. (87) simplifies to,

$$\sigma_E^2 = \left(\frac{\sigma_p^2}{\overline{p}} - \frac{\sigma_T^2}{\overline{T}} + 3\frac{\sigma_V^2}{\overline{V}}\right)\overline{E}_U \quad (88)$$

Given time series of p, T and v at any meteorological station one can calculate their statistical parameters and, subsequently, substitutions into Eqs. (82) and (87) yield the average and variance of wind energy. Eq. (87) assumes simpler forms for isobaric and isothermal atmospheric situations. For instance, if the wind velocity field is measured within an isobaric environmental, i.e., $\sigma_p = 0$ then the variance of wind energy simplifies to,

$$\sigma_E^2 = \left(3\frac{\sigma_V^2}{\overline{V}} - \frac{\sigma_T^2}{\overline{T}}\right)\overline{E}_U - 6\frac{\sigma_T\sigma_V}{\overline{T}\overline{V}}\rho_{TV}\overline{E}_U^2 \quad (89)$$

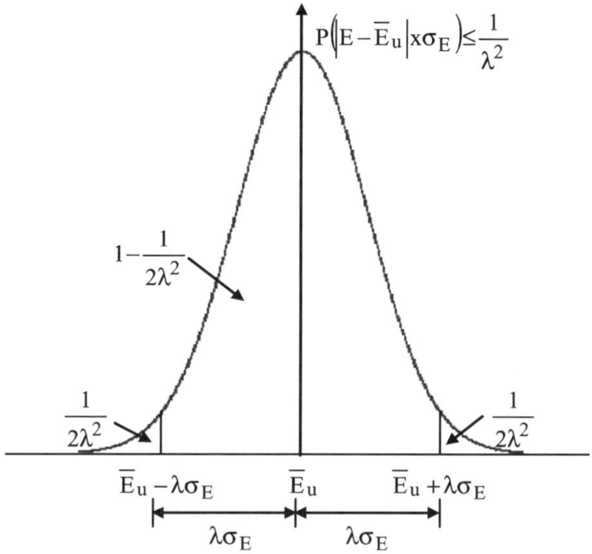

Fig. 13 Graphical representation of Chebyschev's inequality

On the other hand, if the atmospheric events take place within an isothermal environment then $\sigma_T = 0$ and consequently,

$$\sigma_E^2 = \left(\frac{\sigma_P^2}{\overline{P}} + 3\frac{\sigma_V^2}{\overline{V}}\right)\overline{E}_U + 6\frac{\sigma_P\sigma_V}{\overline{p}\overline{V}}\rho_{pV}\overline{E}_U^2 \tag{90}$$

10.3 Wind Energy Risk and Reliability

In general, reliability of an event is defined as the percentage of cases that this event does not cause any undesirable consequences. For instance, there is certain cut-in velocity, v_0, which gives rise to significant wind energy to be greater than any prefixed threshold value, E_0. Hence, the probability, $P(E > E_0)$, is the reliability whereas the complementary event $P(E < E_0) = 1 - P(E > E_0)$, represents the failure, i.e., risk. Unfortunately, the exact probability distribution function of E is not known but the Chebyschev's inequality helps us to find the reliability as in terms of $\overline{E}$ and σ_E. In the absence of exact probability distribution function (PDF) of the wind energy, the mean and standard deviation alone are sufficient to make certain exact statements on the probability of a random variable which lies within a given upper and lower boundaries. For this purpose, the Chebyschev's inequality states with the notations used in this paper

$$P\left(\left|E - \overline{E}_u\right| \geq \lambda\sigma_E\right) \leq \frac{1}{\lambda^2} \tag{91}$$

Fig. 14 Feasible and non-feasible regions

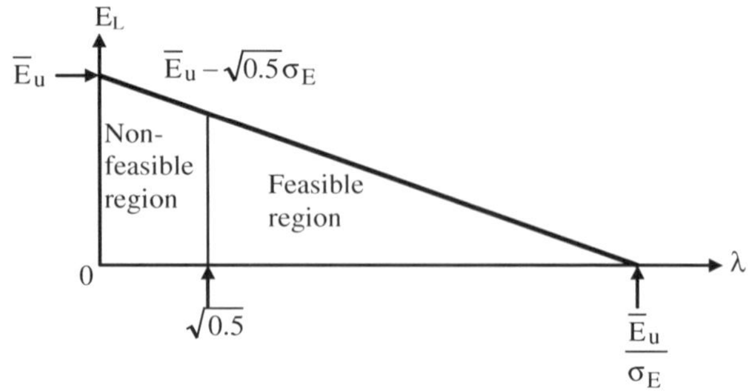

Note that corresponding to the one-, two-, and three-standard deviation bounds λ equals 1, 2 and 3, respectively. In the wind energy calculation values that are greater than or equal to an upper limit, i.e., in the upper tail of the PDF are considered and by assuming almost a symmetrical PDF. By consideration from Fig. 13 Eq. (91) can be rewritten for wind energy generation purpose as

$$P\left[(E - \overline{E}_u) \leq \lambda\sigma_E\right] \leq \frac{1}{2} - \frac{1}{2\lambda^2} \tag{92}$$

In fact, this expression shows the basic equation for wind energy risk. Figure 13 indicates schematically the notations used in the previous expressions.

Since at maximum $1/2\lambda^2$ can assume the value of one which implies that the minimum value of λ is equal to $\sqrt{0.5}$. As stated above for reliable wind energy generations in practice, the lower limit defined as E_L must always be exceeded and hence it can be written as

$$E_L = \overline{E}_u - \lambda\sigma_E \tag{93}$$

The graphical representation of this equation is shown in Fig. 14. It is obvious that the minimum and maximum values of λ emerge, respectively, as $\sqrt{0.5}$ and $\overline{E}_u/\sigma_E$ which can be calculated from the data at hand by using Eqs. (82) and (87).

In the same figure the feasible and non-feasible regions of the E_L and λ parameters are shown graphically. In cases when $\overline{E}_u/\sigma_E$ is less than $\sqrt{0.5}$ reliable then generation of wind energy is not possible Eq. (93) can be rearranged such that

$$\lambda = \frac{E_u - \overline{E}_L}{\sigma_E} = U_E \tag{94}$$

which is the definition of the standard energy value? Hence, the standard reliability statement of the wind energy can be stated as

$$P(E \geq E_L) = 1 - \frac{\sigma_E^2}{2(E_L - \overline{E}_u)^2} = 1 - \frac{1}{2U_E^2} \tag{95}$$

Since, $0 < P(E > E_L) < 1$ then this last expression defines the variability domain of U_E as $\sqrt{0.5} < U_E < +\infty$. Knowing that the risk is the complementary

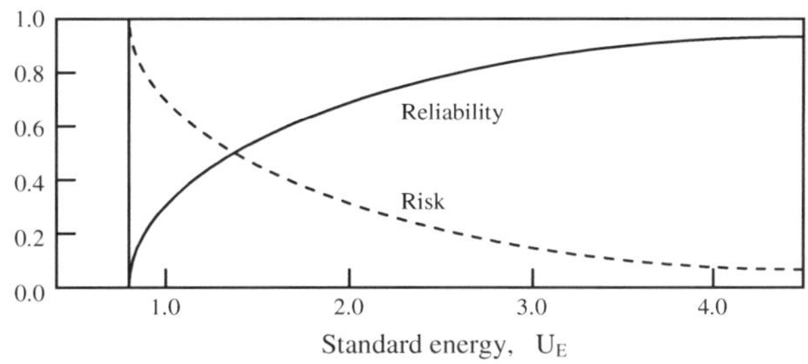

Fig. 15 Reliability and risk curves

statement, the standard risk and reliability curves are drawn with the help of Eq. (95) as in Fig. 15. The following sequences of steps are necessary in finding the lower energy limit for a given reliability (or risk) level

a. Calculation of $\overline{E}_u$ and σ_E from Eqs. (82) and (87) provided that the meteorological data are available.
b. Prepare a special $\lambda - E_L$ graph as in Fig. 2 for the station concerned. Find the feasible and non-feasible parameter space separation value as $E^* = \overline{E}_u - \sqrt{0.5}\sigma_E$.
c. A level of reliability in general as α must be adopted or in particular 90 or 95 % are suitable in practical applications.
d. Enter the vertical axis in Fig. 15 with this reliability value and read from the horizontal axis the standard energy value, U_E.
e. Calculate from Eq. (95) the lower limit as $E_L = \overline{E}_u - \sigma_E U_E$.
f. Control whether $0 < E_L < E^*$ or not. If not, then go to step (c) and adopt a lower reliability value and repeat the subsequent steps until the control is acceptable. Consequently, the reliability (risk) level of wind energy generator at a given site can be obtained.

There is also another procedure, this time to find the reliability (or risk) corresponding to a given lower limit according to the following steps:

(a) After finding the parameters as in step (a) in the previous procedure adopt a lower wind energy limit as E_L.
(b) Calculate the standard energy value from Eq. (95).
(c) Check whether $U_E > \sqrt{0.5}$, if not go to step (a) and increase E_L value and find the corresponding reliability (risk) values from Fig. 15.

10.4 Application

The methodology developed in the previous sections is implemented for the daily meteorological data available at Gökçeada island station in the northwestern part of Turkey (Fig. 16).

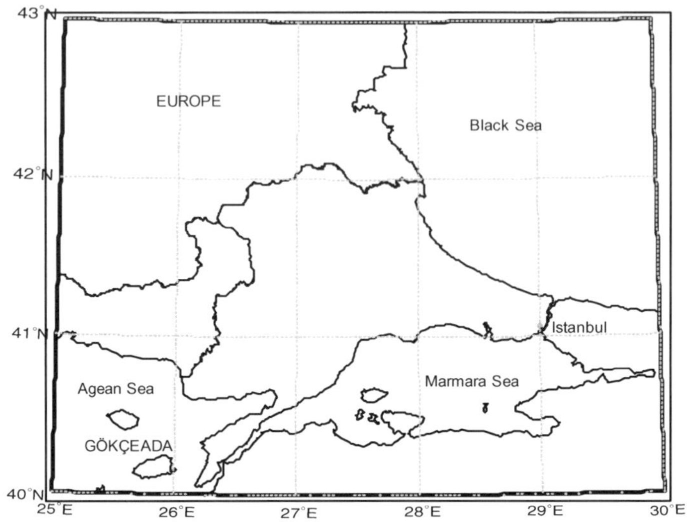

Fig. 16 Gökçeada location map

Table 8 Wind energy reliability (risk) calculations at Gökçeada station

Month	$\overline{E}$ (W/m^{-2})	σ_E (W/m^{-2})	$\overline{E}_u/\sigma_E$	E^*	E_L	
					90 %	95 %
January	1085.4	100.2	10.83	1014.5	857.7	754.4
February	151.6	19.5	7.75	137.8	107.3	87.2
March	151.0	21.4	7.04	135.8	102.4	80.4
April	184.7	26.4	6.99	166.0	124.7	97.6
May	88.8	8.3	10.70	82.9	70.0	61.4
June	107.0	13.8	7.75	97.2	75.7	61.5
July	132.9	8.3	15.90	127.0	113.9	105.2
August	173.0	11.8	14.55	164.6	146.0	133.7
September	153.4	13.1	11.65	143.8	122.5	109.8
October	142.0	10.2	13.90	134.8	118.8	108.3
November	290.8	54.0	5.38	252.6	168.2	112.6
December	137.2	47.5	2.88	103.6	29.4	−19.6

This is the windiest area within Turkey and all the wind energy studies within the country first concentrates on the data available at this station. Various authors have studies the wind potential in this area but none of them were concerned with the reliability (risk) calculations but applied directly the formula given in Eq. (10) with standard atmosphere conditions where ρ was considered as constant. Also the concern was either in the wind speed frequency distribution parameters according to traditional Weibull function [3, 16, 25, 32] or in the wind energy statistical parameters such as arithmetic average and variance.

The arithmetic average and standard deviation of the wind energy for various months at Gökçeada station are presented in the second and third columns of Table 8. The execution of the steps in the previous section in finding the reliability (risk) values are performed in detail but the conclusive values for each month are presented in the same table.

The unit energy values corresponding to 90 and 95 % reliability levels are read off from Fig. 15 as 2.25 and 3.3, respectively. Due to increase in the reliability level there is decrease in the lower reliable limit value, E_L. In fact, in December it is not possible to generate reliable wind energy at 95 % reliability level. The biggest E_L value for a given reliability levels the more will be the wind energy generation. From this point of view, the priorities according to the order of significance go to months January, November, August, April, March, etc.

11 Areal Assessment of Wind Speed and Topography

Dynamic meteorological structures in the troposphere and the topographic features cause temporal and spatial variability in wind speed. In large-scale wind energy utilization establishment feasibility depends on the areal wind variability in addition to sitting, sizing, operation and maintenance policies. Corotis et al. [13] have used the auto-correlation and cross-correlation analysis in order to confirm the existence of significant correlation in the wind records at a single site for a period of 8–12 h within a day; 10–17 days within one month and between sites for similar time lags and separations up to 100 km or more. They have shown that the daily averages are highly correlated above the correlation coefficient 0.8 for distances up to about 100 km in winter and summer seasons for seven northern Illinois sites.

On an areal basis, in order to predict the winds at one site from records at others, it is necessary, to have detailed information on terrain and weather patterns. Although the cross-correlation function definition can give a direct indication of the dependence of variations from the mean at any two sites, it suffers from the following drawbacks.

1. Auto-correlation and cross-correlation formulations require symmetrically (normal, Gaussian) frequency distribution of wind speed data for reliable calculations. It is well established in the literature that wind speeds accord with a Weibull, Gamma or logarithmic normal frequency distributions which are all skewed. However, the Point Cumulative Semi-Variogram (PCSV) technique proposed in this paper does not have symmetricity requirement for wind speed data.
2. Correlation function measures the variation around the arithmetic average values of the wind speed data at individual sites. However, in the regional calculations, a measure of relative variability between two sites is necessary. For this purpose, the Semi-Variogram (SV) or Cumulative Semi-Variogram (CSV) concepts are developed and their modification as the PCSV is presented and used for the regional assessment of the wind variability in this section.

Barros and Estevan [6] presented a method for evaluating wind power potential from a 3 months long wind record at a site and data from an areal network of wind systems. Their key assumption was that "wind speed has some degree of areal correlation" which is logical, but they failed to present an effective method for the objective calculation of the areal variability, except by employing cross- and auto-correlation techniques. Their statement does not provide an objective measure of spatial correlation. Skibin [40] raised the following questions:

1. What is "a reasonable areal correlation?"; are the correlation coefficients between the weekly averages of wind speed a good measure of it? Answers to these questions are necessary by any objective method. In this section, PCSV technique is proposed to answer these questions.
2. Do the calculated averages represent the actual ones?
3. How applicable to the siting of wind generators the results obtained by the use of spatial correlation coefficients?

In deciding about the effectiveness of the wind speed measurement around a site, the topographic and climatic conditions must be taken into consideration. Especially, any method of wind speed prediction should consider the topographic and climatologic features. The smaller the area of influence, the more homogeneous orographic, weather and climatologic features, and consequently, the simplest is the model. However, large areas more than 1,000 km in radius around any site may contain different climates with different troughs and ridges, high and low pressure areas with varying intensities. Furthermore, in heterogeneous regions with varying surface properties (such as land–sea–lake–river interfaces) and variable roughness parameters, the local wind profile and wind potential will be affected significantly. The wind energy potential is highly sensitive to height variations of hills, valleys and plains. The reasons for wind speed variations are not only of orographical origin, but also different flow regimes (i.e. anabatic–katabatic influences compared with hill top conditions, upwind compared with leeside sites, flow separation effects). All these effects will lose their influence further away from the siting point. It can be expected that smaller distance from the site corresponds to a larger correlation. It has been noticed that a small region had a higher correlation coefficients between the sites [6]. Additionally, the wind speed variance is actually needed for more accurate predictions of wind energy potential.

Barchet and Davis [5] have stated that better estimates were obtained when the radius of influence was about 200 km from the site. However, this information is region dependent and there is a need to develop an objective technique whereby the radius of influence can be estimated from a given set of sites.

The wind speed measured at a site is determined mainly by two factors; the overall weather systems (which usually have an extent of several hundred kilometers) and the nearby topography within few kilometers of the station. The application of measured wind speed statistics for calculating the wind energy potential in an area requires therefore, extrapolation of wind speed statistics to points at which measurements were not taken.

The collective effect of the terrain surface and obstacles, leading to an overall retarding of the wind near the ground, is referred to as the roughness of the terrain. Orographic elements such as hills, cliffs, ridges and escarpments exert an additional influence on the wind. Roughness and orography are amongst the main factors that affect the wind speed.

The main purpose of this section is to present Point Cumulative Semi-Variogram (PCSV) technique to depict the regional variation in the wind speed, in the topography of nearby sites, and hence, to assess the roughness, spatial dependence, wind climatologic features and wind energy potential. Contrary to the correlation functions, the PCSV's are rather robust and valid for any distribution function. In fact, the central limit theorem of classical statistics states that whatever the underlying probability distribution function of a random variable, its successive summations or averages approach a normal distribution [20].

11.1 Point Cumulative Semivariogram

Most often quantification of degree of regional variability is measured by statistical variance, covariance or correlation function. They cannot account completely for the regional dependence due to either non-normal (skewed) distribution functions as mentioned above and/or irregularity of wind generator sites. The variability of any phenomenon within an area can best be measured by comparing the relative changes, i.e., deviation between any pairs of site values. For instance, if any two sites, d apart have wind speeds S_i and S_{i+d} then the relative variability (deviation) can simply be written as $(S_i - S_{i+d})$. The closer the two speed values the smaller the deviation. In the case of m wind speed measurement sites in an area, there will be $N = m(m - 1)/2$ possible different deviation values. Each deviation will have positive or negative sign, and if a regional variability measure is proposed as the summation of these deviations, it will not reflect the real situation. This is because in the summation positive and negative values might cancel each other and the resulting total deviation becomes rather small. In order to define an objective measure of regional variability on the basis of deviations, it is better to consider deviation-squares which are all positive. Hence, the deviation square, $D^2(d)$ dependent on the distance, d is defined as

$$D^2(d) = (S_i - S_{i+d})^2 \tag{96}$$

This is also referred to as the structure function of the regionally varying wind speeds. Equation (96) assumes that the smaller the distance, d, the smaller will be the structure function. This regional variability in the wind speed may be a product of an active weather phenomenon and surface roughness, i.e., orography in the case of wind energy potential assessments.

The classical semivariogram (SV) technique is defined from Eq. (96) as the arithmetic average of all available structure functions within the study area [35].

$$\gamma_{d} = \frac{1}{2N(d)}\sum_{i=1}^{N(d)}(S_{i} - S_{i+d})^{2} \tag{97}$$

where γ_d is the SV value at distance d, and N(d) is the total number of equally spaced observations. It is noticed that in Eq. (97) the arithmetic average of structure functions is divided by 2 which again by definition comes from theoretical requirements.

The SV provides a measure of spatial dependence among a multitude of sites as an alternative to the auto-covariance of a time series. Its change with distance on a Cartesian coordinate system is referred to as the SV function. The SV technique is approximately suitable for irregularly-spaced data, but it has practical difficulties [44]. On the other hand, an alternative cumulative semivariogram concept (CSV) proposed by the same author alleviates most of the SV drawbacks in identifying the spatial dependence structure of regionalized variables (ReVs). However, the CSV definition is similar to the SV and the only difference is the successive summations procedure. The CSV has all of the advantages of SV and additionally it provides an objective way to derive theoretical models of the regional dependence behavior of the regionalized variables (ReVs).

The point Cumulative Semi-Variogram (PCSV) function is proposed as the CSV calculated for a single point (site). The PCSV identifies the areal variability of wind speed around a single site rather than the whole region. It presents the areal effect of all the sites within the study area on a particular site. Consequently, the number of PCSV's is equal to the number of available sites. Each PCSV provides a basis for nearby station variability interpretations and their mutual comparisons at different sites lead to invaluable information for describing the heterogeneity of the ReV in an area. The treatment of available wind speed data at m sites according to the following steps leads to sample PCSV for a particular site.

1. Calculate the arithmetic average, $\overline{S}$ and standard deviation, S_s for the area from wind speed data. Standardize all wind speed data by subtracting from each wind speed record the average and then divide this difference by the standard deviation. Hence, the standardized wind speed value at site i become

$$s_{i} = \frac{S_{i} - \overline{S}}{S_{S}} \tag{98}$$

2. Calculate distances between the desired site and the remaining sites. If there are m sites the number of distances is m − 1, d_i (i = 1, 2,..., m − 1).

3. For each pair calculate the squared differences between wind speeds similar to Eq. (96); hence each distance will have corresponding squared difference, $(s_D - s_i)^2$ where s_D and s_i are the regionalized variables at the desired and i-th sites, respectively. Consequently, there are (m − 1) squared differences of wind speed. Take the half values of these squared differences in order to comply with Eq. (97).

4. Rank the distances in ascending order and plot distances d_i versus corresponding successive cumulative sums of half-squared differences. Hence, a

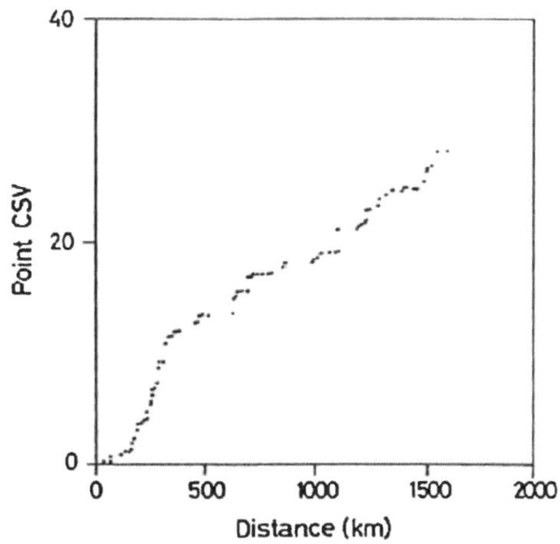

Fig. 17 A representative PCSV

non-decreasing function will be obtained which is named as the sample PCSV for the desired site. A representative PCSV is shown in Fig. 17. All these steps imply that the PCSV, $\gamma(d_i)$ can be expressed as

$$\gamma(\mathbf{d_i}) = \frac{1}{2}\sum_{i=1}^{m-1}(s_D - s_i)^2 \tag{99}$$

5. Application of these steps in turn for each site leads to m sample PCSVs of wind speeds.

The sample PCSVs are potential information sources in describing the areal wind speed variation characteristics around each site. Among these characteristics are the radius of influence, areal dependence and structural behavior of the regionalized variable near the site such as the nugget (sudden changes) and sill effects, heterogeneity as will be explained in the following section. All what is applied for the wind speed data can be applied similarly for the site elevations leading to elevation sample PCSVs. Hence, the comparison of these two sets of PCSVs for the same station provides five different categories (A, B, C, D and E) as shown in Fig. 18. As mentioned above since wind speed and elevation data are standardized prior to PCSV calculations the sample PCSVs do not have dimensions on the vertical axis. This gives opportunity to show wind speed and elevation sample PCSVs on the same Cartesian coordinate system for the same site.

Each category implies different weather and orographic conditions around the station considered. The following general descriptive interpretations are valid for each category [47].

Category A: Wind speed and station elevation PCSVs fall on almost each other at all distances with small deviations. In practice, such deviations may be tolerated up to within ±10 % error limits. In this category, areal wind speed pattern is entirely dependent on the orographic features and weather conditions of wind climatology are governed by the topography in the lower planetary boundary layer.

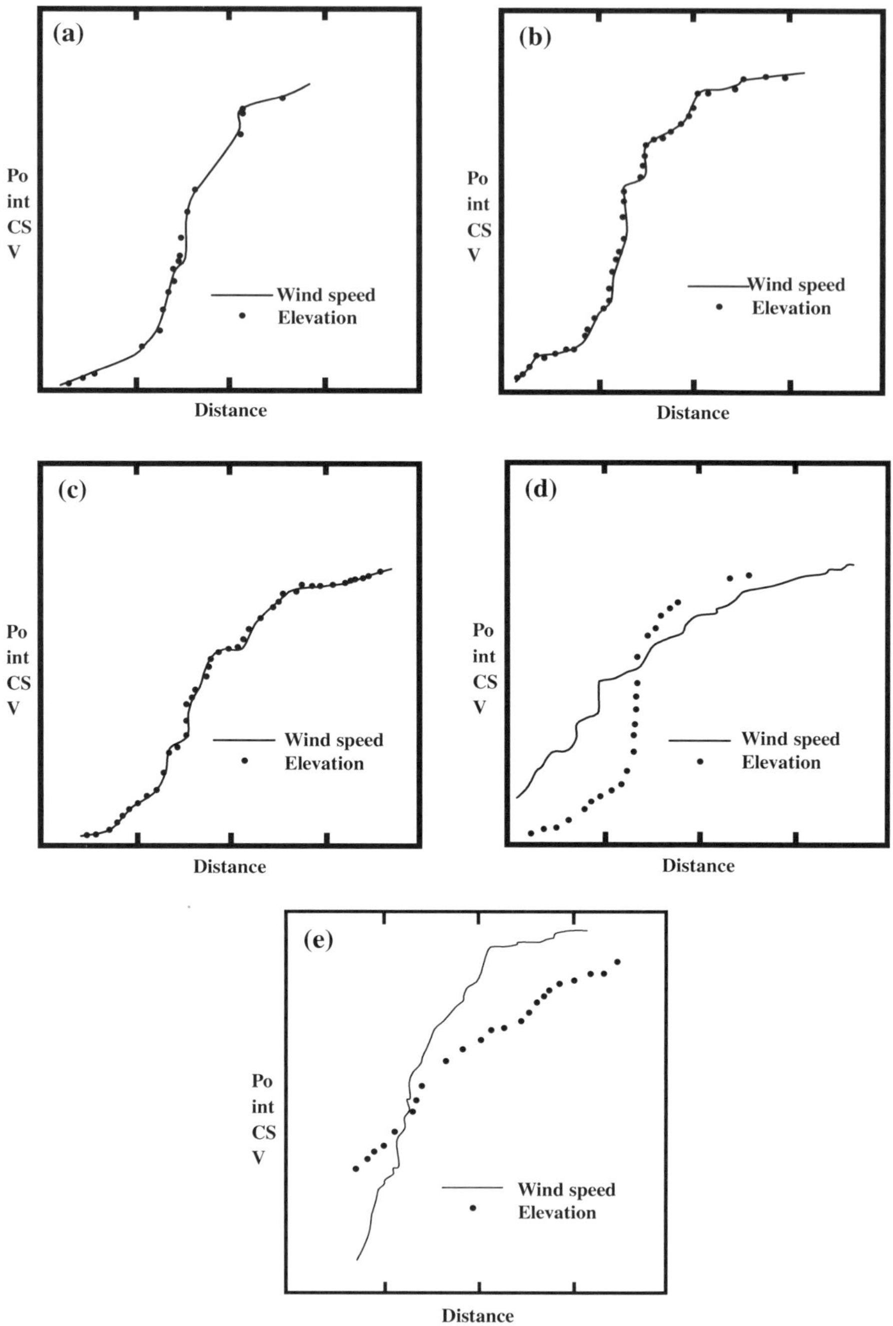

Fig. 18 Five possible categorizations

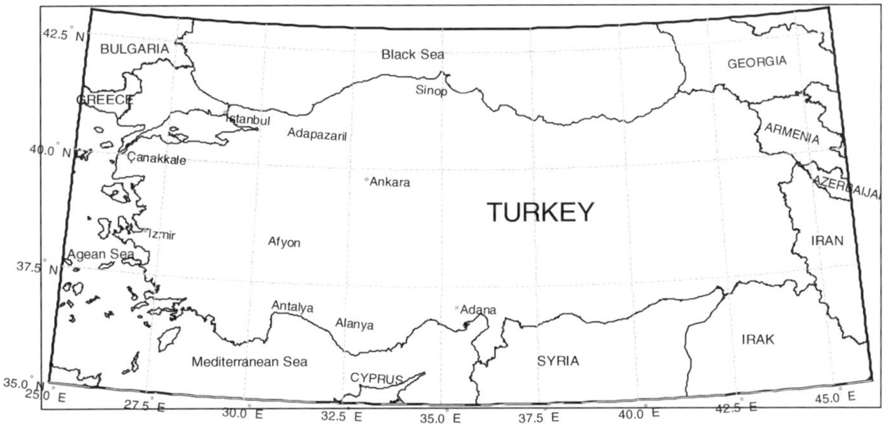

Fig. 19 Station locations

Category B: Elevation PCSV is completely below the wind speed PCSV which implies more dominant air shed born effects than the topographic influence on the wind speed. Consequently, wind speed regional variability is more effective than the topographic roughness. Compared with category A, this case gives rise to more wind energy generation possibilities.

Category C: This is opposite to category B and the wind speed PCSV is smaller than the elevation PCSV at all distances. Hence, less regional variability exists in the wind speed than the surrounding topography. Since the topographical effects are more dominant wind energy generation is expected to be comparatively smaller than categories A and B.

Category D: Herein, to a certain distance around the station wind speed, hence air shed born weather effects are dominant over topographic variation. However, opposite situation occurs at large distances.

Category E: This is contrary to category D and the station site may be surrounded by high hills, and consequently, the wind energy generation is comparatively smaller than any other category.

11.2 Application

The sample PCSV methodology is implemented to 35 wind speed measurement sites scattered all over Turkey as shown in Fig. 19 [30, 31].

The location and wind speed characteristics of each site are presented in Table 9 for 10 major centers in Turkey.

The sample PCSVs for 35 station sites considered in this study are presented in Fig. 20. It is to be noticed that all the horizontal axis have the same scale however it has not been possible to do the same for the vertical axis.

Table 9 Wind characteristics in Turkey

Station name	Latitude	Longitude	Elevation (m)	Wind speed (m/s)	Category
1. Adana	36.98	35.30	20	1.4	E
2. Adapazarı	40.68	30.43	31	1.7	C
3. Afyon	38.75	30.53	1034	2.7	B
4. Alanya	36.55	32.00	7	1.9	D
5. Ankara	39.95	32.88	891	1.8	B
6. Antalya	36.86	30.73	51	2.7	B
7. Çanakkale	40.13	26.40	3	3.9	C
8. Istanbul	40.97	29.08	33	3.2	C
9. Izmir	38.40	27.17	25	3.6	E
10. Sinop	42.03	35.17	32	3.6	A

Although there are many interpretation facilities throughout the sample PCSVs several of them are presented in the following. By considering the basic definition and becoming more familiar with PCSV after some interpretive application studies, the reader may add many other significant interpretations for his/her purpose concerning wind energy generation possibilities at any site.

1. Wind speed and elevation PCSVs at sites have rather different appearances from each other which indicate that the wind speed variability over the area around the site is heterogeneous. For instance, PCSVs for Adana and Alanya appear significantly different from each other. A first glance through the whole sample PCSVs gives the impression that, in general, all five distinctive categories explained in the previous section are available at different sites in Turkey. Some of the sample PCSVs has a convex curvature initial portion (at small distances) then either single or multiple broken straight lines follow at large distances. This means that at small distances the spatial dependence is strong but with the increase of distance such a dependence decreases. Especially, wind speed PCSVs at Adana, Alanya, Ankara, Antalya, have such behaviors. Such patterns are not common in the elevation PCSVs except at Ankara where both wind speed and elevation PCSVs have initial curvatures with PCSVs starting right from the zero distance. Some other group of individual PCSVs does not have any curvature but many broken straight lines and have an intercept on the horizontal distance axis. This is a common appearance in the elevation PCSVs and abundance of broken straight lines indicates the heterogeneity involved around the reference station at a variety of distances.
2. Mutual considerations of wind speed and elevation PCSVs at each station together with the aforementioned five categories (**A**, **B**, **C**, **D** and **E**) give anyone the ability to categorize each site as shown in the last column of Table 9. The stations with the same category can be regarded as homogeneous collectively and hence whole Turkey can be divided into five distinct wind speed homogeneity regions.

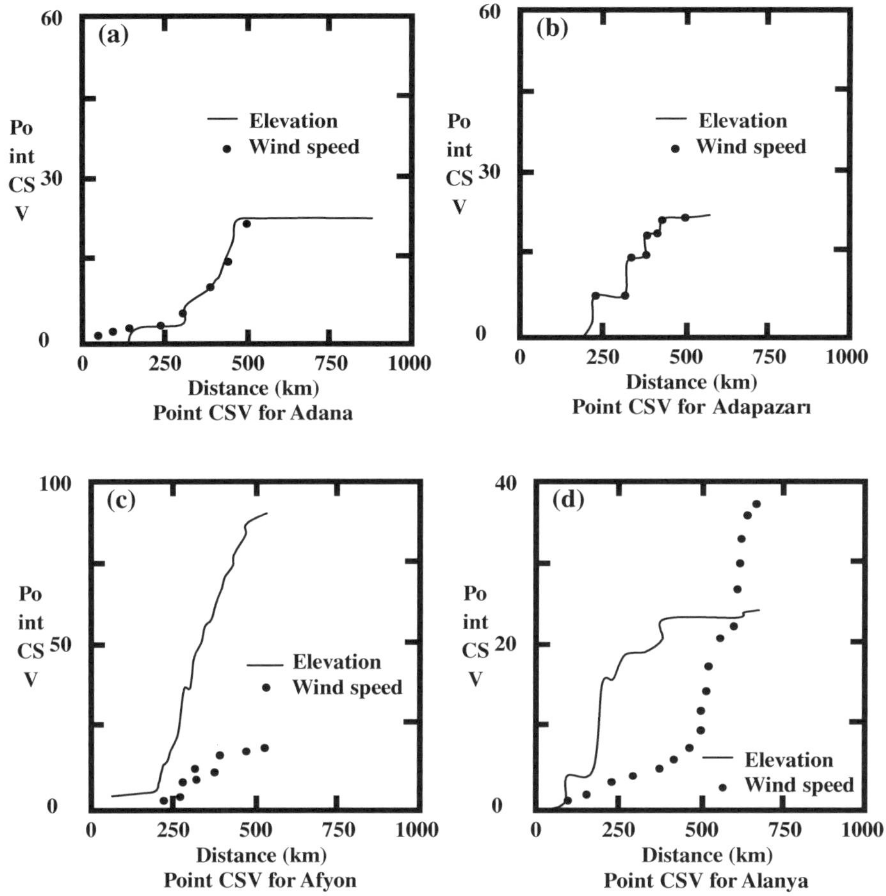

Fig. 20 10-Site PCSVs. **a** Point CSV for Adana. **b** Point CSV for Adapazan. **c** Point CSV for Afyon. **d** Point CSV for Alanya. **e** Point CSV for Antalya. **f** Point CSV for Ankara. **g** Point CSV for Çanakkale. **h** Point CSV for Sinop. **i** Point CSV for Göztepe. **j** Point CSV for Izmir

3. Some of the sample PCSVs does not pass through the origin. This indicates that wind speed or elevation variability is not a regionally smooth process. For instance, in the case of wind speed PCSVs, they are under the control of some topographic and/or meteorological factors. This further implies that in spatial wind speed characteristics, homogeneous conditions do not prevail and meteorological factors (wind direction, pressure, temperature and humidity) influence the regional wind speed variability concurrently and sequentially in addition to the topographic factors.
4. Some of the sample PCSVs has intercept, on the horizontal (distance) axis very distinctively for the elevation PCSV as in Fig. 20. Hence, from Eq. (99) $s_D \cong s_i$ which implies structural control in the regional topographic behavior?

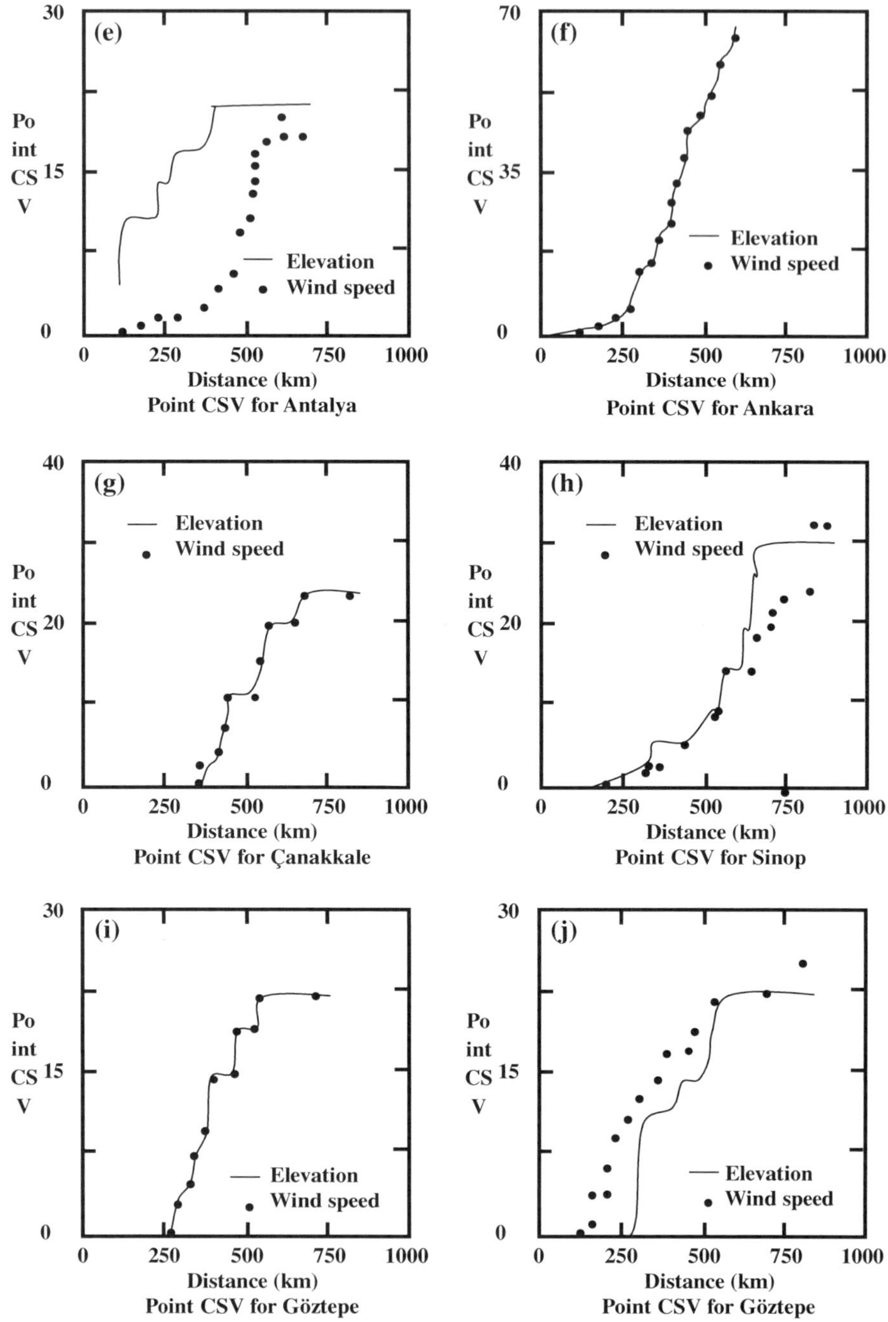

Fig. 20 (continued)

5. A group of sample PCSVs passes through the origin (see Ankara PCSVs). Such a property on PCSV diagram implies the continuity in the regional variability (herein, wind speed and elevation). Continuity means that there are no nugget effects or distinctive discontinuities within the regional variable concerned.
6. Some of the sample PCSVs has "curvature portions" at moderate distances. In fact, such a range corresponds to the distance scale, i.e., radius of influence, as defined in turbulent flow [51]. The radius of influence is defined as the radius at which the PCSV reaches a particular magnitude, say, c and it can be simply stated with the notation from Eq. (4) as

$$\lim_{d_i \to \infty} \gamma(d_i) = c \tag{100}$$

The initial curvature implies that the ReVs at these sites have regional dependencies which weaken towards the end of curvature distance range [48]. Since curvatures are convex the implication is that there is a positive regional structural dependence. Furthermore, the curvature implies that the ReV has areal serial dependence, i.e., not only external factors but also the ReV generating phenomena in the troposphere contribute to the regional dependence structure.

7. The sample CSVs help to identify the underlying generating mechanism of areal phenomenon. Likewise, the sample PCSVs provide clues about the wind speed generating mechanism around the site concerned. For instance, if the sample PCSV passes through the origin with straight line portion only, then the regional variable complies with an independent (white noise) process with no regional dependence. However, when the sample PCSV is in the form of a straight line but does not pass through the origin then a moving average process describes the underlying generating mechanism of regional variability [48].

Curvature in the CSV implies, in general, auto-regressive processes among which the Markov and ARIMA processes theoretical CSVs are presented in a work for a set of parameters [48]. Generally, the existence of a straight line following an initial curvature portion indicates that the underlying generating process of the regionalized variable agrees with a regional Markov process, whereas, if the curvature continues at reduced rates at large distances then an ARIMA process is the convenient model. It is necessary to state herein that future studies are necessary for a complete picture of theoretical PCSVs in cases of other stochastic processes which will help to identify the most suitable one for the regionalized wind speed data.

8. In the case of a single straight line following an initial curved section, the slope of the large-distance straight line portion is related to the standard deviation of the wind speed or elevation variability.

12 Wind Energy Evaluation from Wind Speed Time Series

Although the amount of wind energy is economically insignificant for the time being in many parts of the world, mankind took the advantage of its utilization since early times whenever he/she found the chance to provide power for various tasks. Among these early utilizations are hauling of water from a lower elevation to a higher level, grinding grains in mills by water power and other mechanical power applications for many centuries. It is still possible to see at some parts of the world these type of marginal benefits from the wind speed. Perhaps the wind-mills in Holland exemplify the most publicized application of wind power. All of the previous activities have been technological and the scientific investigation of wind power formulations and accordingly development of modern technology appeared after the turn of twentieth century. In recent decades the significance of wind energy originates from its friendly behavior to the environment so far as the air pollution is concerned although there are, to some extent, noise and appearance pollution in the modern wind-farms. Due to its cleanness, wind power is sought wherever possible for conversion to the electricity with the hope that the air pollution as a result of fossil fuel burning will be reduced [11]. In some parts of USA, up to 20 % of electrical power is generated from wind energy [41]. In fact, after the economic crises in 1973 its importance increased by forcing the economic limitations and today there are set up wind-farms in many western European countries [1, 19, 52].

Among the renewable energy sources wind provides a potential alternative to fossil fuel sources and recently due to the economical feasibilities at low costs the wind energy generation became attractive for power generation. As the issue of fossil fuel pollution of the lower atmospheric layers leading to carbon dioxide increase as consequently the global warming (greenhouse) effect there is an increased interest in alternative, renewable and less expensive energy resources. Hence, in recent years studies concerning the solar and especially wind energy have appeared more frequently in the scientific literature. Almost all the wind power investigations relied mostly upon arithmetic average of the wind speed which represents the simplest central tendency measure of the wind power density, i.e., probability distribution function [34, 49]. On the other hand, many authors based the wind power estimates on elaborated wind speed statistics including the standard deviation, variance, skewness coefficient and kurtosis. In these studies several probability density functions (PDF) such as the log-normal, Gamma (Pearson-type III), Weibull and Rayleigh distributions are tried as possible useful fit to empirical wind speed frequency. Especially, Justus et al. [26, 28, 29] advocated for the use of the two-parameter Weibull distribution in wind velocity applications. Consequently, their suggestion has been taken in granted in many parts of the world during the wind power calculations. Unfortunately, in some of the studies without testing the wind speed data for the suitability to the two-parameter Weibull distributions the calculations are carried out automatically. In some researches not the genuine wind velocity data but their cubes are assumed to have two-parameter Weibull distribution as noticed by Hennessey [23]. Auwera et al. [4] showed that a three-parameter Weibull distribution fits the wind speed data in a more refined manner than the two-parameter Weibull PDF.

12.1 Classical Wind Energy Calculation Formulations

Since wind speed is a kinematics variable accordingly the wind energy is closely related to the kinetic energy. In fact, by considering various physical definitions and the specific density, ρ, of the air the very basic wind energy, E_U, is obtained as in Eq. (10). In the application of this formula the following assumptions must be kept in mind.

1. The standard air density is assumed temporally and spatially constant as $\rho = 1.23$ g/cm^3. This further implies that the air is dry with no water vapor and therefore the temperature and pressure changes do not effect the wind energy calculation.
2. This wind energy is valid per unit time per area. So far area is concerned with the turbine blade sweeping area and therefore related to the turbine design but the time is closely related either to the time intervals in a time series or the wind energy calculation time interval such as the question of what is the wind potential of the wind at a given site for one day, week, month or year.
3. The wind speed in this formulation is in fact the instantaneous velocity. However, for some finite time period it is necessary to consider the average speed because in this case the wind velocity can be considered as constant. It is very important to notice at this stage that rather than the cube of the average speed the average speed cube value must be substituted in Eq. (10).
4. Although in nature the wind speed changes rather randomly except from the arithmetic average value none of the other significant statistical parameters such as the variance, skewness, kurtosis, etc. are employed in Eq. (10).

In order to alleviate some of the drawback in the application of Eq. (10) some other alternative are proposed in the literature. For instance rather than the use of the arithmetic wind speed value in order to include the variance of the wind speed over the considered time interval the wind speed is expressed first in terms of the average speed $\overline{\mathbf{V}}$ and fluctuation term, $\boldsymbol{\varepsilon}$, around it. The fluctuation term is in fact the deviations from the average wind speed in the interval and therefore it can assume positive as well as negative values. In fact the average of this fluctuation term is equal to zero with a constant variance, $\boldsymbol{\sigma}_{\varepsilon}^2$. It is assumed at this stage that the fluctuation term has a normal probability distribution function which implies that the skewness coefficient is equal to zero. Hence, the substitution of $V = \overline{V} + \varepsilon$ into Eq. (10) after some algebraic manipulation leads to

$$E = \frac{1}{2}\rho\left(\overline{V}^3 + 3\overline{V}\sigma_\varepsilon^2\right) \tag{101}$$

This expression considers uniform change in the wind speed between two successive wind speed measurement instances and in the applications there arise the question of whether adopting the wind speed at the initial time instant or at the final instant as the constant wind speed. However, adaptation of the wind speed as the arithmetic average of the initial and final wind speed measurements provide another constant speed. Hence, the change in the wind speed during the time period is not considered but rather the variance is taken into consideration.

A further improvement in the wind energy calculations can be obtained by considering the dependecce of the air density on the pressure and temperature changes and consequently on the basis of the gas state equation one can write that

$$E_U = \alpha\left(\frac{P}{T}\right)V^3 \tag{102}$$

where α is a constant and equal to 0.5/R or 0.174 where R is the universal gas constant equal to 2.87 for dry air when P is in millibars, T in degrees Kelvin and ρ in kg/cm^3. The meteorological variables on the right hand side are conventionally measured at any meteorology station, but Eu cannot be measured directly. Hence, the questions are to find ways of estimating wind energy from these measurements whenever the basic meteorological data are available. A common point of critics for all these formulations is that the wind speed is not constant over the interval considered but either increases or decreases.

12.2 Suggested Wind Calculation Formulation

None of the aforementioned wind energy formulations are directly applicable to a given time series of wind speed. For instance, one of the common mistakes in use of Eq. (10) for calculating the total wind energy during a time duration τ is the direct substitution of average wind velocity $\overline{V}$. Hence, Eq. (10) takes the following form

$$E_T = \frac{1}{2}\rho\overline{V}^3\tau \tag{103}$$

which includes a great mistake in that the average wind speed is assumed as the instantaneous wind speed and uniformly effective over the time duration considered? Another alternative for the use of the same Eq. (10) in total wind energy calculation is the substitution of average of the velocity cube which leads to

$$E_T = \frac{1}{2}\rho\overline{V^3}\tau \tag{104}$$

which again requires the averages of the velocity cubes during the time duration considered? This is logically better than the use of Eq. (103) but extreme wind speeds are included very extremely by taking the cube of the speed and it will effect unexpectedly the overall wind energy calculation. In fact this is the special case of Eq. (101) when the wind speed variance is assumed as zero which is tantamount to saying that the wind velocity throughout considered time duration is uniform with no fluctuations. This is not a representative actual case for wind energy calculations.

Another alternative in the total wind energy calculation is the use of Eq. (101) by taking into consideration the variance in the wind speed variations around the mean wind speed value. In this equation after determining the basic time period τ the mean wind velocity, variance of the wind speeds over this period and the average of the wind speed cubes over the same period must be calculated. Hence, the total energy calculation formula becomes as

$$E_T = \frac{1}{2}\rho\left(\overline{V}^3 + 3\overline{V}\sigma_\varepsilon^2\right)\tau \tag{105}$$

This expression takes into account the average deviations from the wind speed time series. Its total wind energy yield is more than Eq. (104). Representative wind speed time series and the standard deviation are shown in Fig. 1. However, still better approximations to the total wind energy calculation can be obtained by considering the basic definition of Eq. (1) that it is valid for unit time. Hence, during an infinitesimally small time period, dt, there will be infinitesimally small energy, dE, as

$$dE = \frac{1}{2}\rho V^3 dt \tag{106}$$

However, it is obvious from a given time series that during two successive time intervals namely t_1 and t_2 there are wind speeds V_1 and V_2, respectively and the change of velocity between these two time instances is assumed to be linear. The slope, say α, of such a segment is assumed constant. Hence, infinitesimally small time dt corresponds to infinitesimally small speed dV and by definition $dV = \alpha dt$, or $dt = dV/\alpha$ the substitution of which into Eq. (106) leads to

$$dE = \frac{1}{2}\rho V^3 \frac{dV}{\alpha} \tag{107}$$

Consequently, with the condition of remaining within $t_1 - t_2$ time interval this equation can be integrated and the result is the total wind energy E_T after some algebra one can obtain

$$E_T = \frac{1}{8}\frac{\rho}{\alpha}\left(V_2^4 - V_1^4\right) \tag{108}$$

On the other hand by definition $\alpha = (V_2 - V_1)/(t_2 - t_1)$ and if the basic regular time interval in the time series is assumed as a unit bid and equal to 1 then

$\alpha = (V_2 - V_1)$ and the substitution of this into the last expression leads after the necessary manipulations to

$$E_T = \frac{1}{8}\rho(V_2 + V_1)(V_2^2 + V_1^2) \tag{109}$$

This is the total amount of wind energy generated during one unit time duration of the basic time series. If there are many basic time intervals within a given time duration then the total wind energy can be obtains as the summation of similar expression in Eq. (109). This is the fundamental total wind energy calculation equation suggested and used in this paper. In fact, when the wind speed is assumed as uniform, i.e., $V_2 = V_1$ then Eq. (109) reduces to Eq. (10). By definition this formulation takes into account every possible deviation in the form of linear variation in the wind speed it is expected to yield the greatest wind energy amounts among other formulations.

References

1. Anderson M (1992) Current status of wind forms in the UK. Renew Energy Syst
2. Ang AH, Jang WH (1975) Probability concepts in engineering planning and design. Wiley, New York
3. Aslan Z, Menteş S, Tolun S (1993) Gökçeada rüzgar enerji potansiyelinin belirlenmesi (Wind power determination at Gökçeada station). First national clean energy symposium. Meteorology Department, İstanbul Technical University, pp 104–112 (in Turkish)
4. Auwera LD, De Meyer F, Malet LM (1980) The use of the Weibull three-parameter model for estimating mean wind power densities. J Appl Meteorol 19:819–825
5. Barchet WR, Davis WE (1983) Estimating long-term mean winds from short-term wind data, report 4785, Battelle, PNL, p 21
6. Barros VR, Estevan EA (1983) On the evaluation of wind power from short wind records. J Appl Meteorol 22:1116–1123
7. Bennett M, Hamilton PM, Moore DJ (1983) Estimation of low-level winds from upper air data. IEE Proc 130(9):517–520, Pt A
8. Benjamin JR, Cornell CA (1970) Probability statistics and decision making for civil engineers. McGraw-Hill, New York
9. Bergey KH (1980) The Lanchester-Betz limit. J Energy 3(6):382–384
10. Betz A (1926) Schraubenpropeller mit geringstem Energieverlust. Göttinger Nachrichten, mathematisch-physikalische Klasse, pp 193–213
11. Clarke A (1988) Wind farm location and environmental impact. Network for Alternative Technology and Technology Assessments C/O EEDU, The Open University, UK
12. Conradsen K, Nielsen LB (1984) Review of Weibull statistics for estimation of wind speed distribution. J Climatol Appl Meteorol 23:1173–1183
13. Corotis RB, Sigl AB, Cohen MP (1984) Variance analysis of wind characteristics for energy conversion. J Appl Meteorol 23:1477–1479
14. Cox DR, Miller HD (1965) The theory of stochastic processes. Methuen, London
15. Davenport AG (1963) The relationship of wind structure to wind loading. In: Proceedings of conference wind effects on structures, National Physics Laboratory, London, England, pp 19–82

16. Doğan V (1991) Gökçeada için rüzgar enerji potansiyelinin istatistiksel metotlarla incelenmesi statistical investigation of Gökçeada wind potential. Istanbul Tech Univ J:31 (in Turkish)
17. Dunn PD (1986) Renewable energies: sources, conversion and application. Peter Peregrinus Ltd, p 373
18. Erasmus DA (1984) A boundary layer wind flow model for areas of complex terrain. Unpublished Ph.D. Thesis, University of Hawaii, p 975
19. EWEA (1991) Time for action: wind energy in Europe. European Wind Energy Association
20. Feller W (1968) An introduction to probability theory and its application. Wiley, New York, p 509
21. Gustavson MR (1979) Limit to wind power utilization. Science 204(4388):13–17
22. Haltiner GJ (1971) Numerical weather prediction. Wiley, New York, p 317
23. Hennessey JP Jr (1977) Some aspects of wind power statistics. J Appl Meteorol 16:119–128
24. Holton JR (1972) An introduction to dynamic meteorology. Academic Press, New York, p 319
25. İncecik S, Erdoğmuş F (1994) An investigation of wind power potential in western coast of Anatolia. Renew Energy
26. Justus CG, Hargraves WR, Yalçin A (1976) Nationwide assessment of potential output from wind powered generators. J Appl Meteorol 15:673–678
27. Justus CG, Mikhail A (1967) Height variation of wind speed and wind distribution statistics. Geophys Res Lett 3:261–264
28. Justus CG, Hargraves WR, Mikhail A (1976b) Reference wind speed distributions and height profiles for wind turbine design and performance evaluation applications, ERDA ORO/5107-76/4
29. Justus CG (1978) Wind and wind system performance. Franklin Institute Press, Philadelphia
30. Kadıoğlu M (1997) Trends in surface air temperature data over Turkey. Int J Climatol 17:511–520
31. Kadıoğlu M, Şen Z (1998) Power-law relationship in describing temporal and spatial precipitation pattern in Turkey. Theor Appl Climatol 59:93–106
32. Kahraman G (1993) Gökçeada'da Uğurlu ve Doruk tepe istasyonlari rüzgar enerji potansiyelinin incelenmesi. Istanbul Tech Univ J:64
33. Kaminsky FC, Kirchhoff RH (1988) Bivariate probability models for the description of average wind speed at two heights. Sol Energy 40(1):49–56
34. Malet LM (1978) Element d'appreciation de l'energie eolienne en Belgique. IRM Publishing Services B, no. 59
35. Matheron G (1963) Principles of geostatistics. Econ Geol 58:1246–1266
36. Papoulis A (1965) Probability, random variables, and stochastic processes. McGraw Hill, New York, p 583
37. Petersen EL, Troen I, Frandsen S, Hedegard K (1981) Wind atlas for Denmark. Riso National Laboratory, DK-4000 Roskilde, Denmark, p 229
38. Reed JW (1979) Wind power climatology of United States—Supplement, SAND78-1620
39. Schlatter TW (1988) Past and present trends in the objective analysis of meteorological data for nowcasting and numerical forecasting. 8th conference on numerical weather prediction. American Meteorological Society, pp 9–25
40. Skibin D (1984) Comment "On the evolution of wind power from short wind records". J Appl Meteorol 23:1477–1479
41. Sorensen HA (1983) Energy conversion systems. Wiley, New Jersey
42. Sutton OG (1953) Micrometeorology: a study of physical process in the lowest layers of the Earth's atmosphere. McGraw-Hill, New York
43. Şahin AD (2004) Progress and recent trends in wind energy. Prog Energy Combust Sci Int Rev J 30:501–543
44. Şen Z (1989) Cumulative semivariogram models of regionalized variables. Int J Math Geol 21:891–903

45. Şen Z, Şahin AD (1997) Regional assessment of wind power in western Turkey by the cumulative semivariogram method. Renew Energy 12(2)
46. Şen Z (1997) Statistical investigation of wind energy reliability and its application. Renew Energy 10:71–79
47. Şen Z (1999) Terrain topography classification for wind energy generation. Renew Energy 16:904–907
48. Şen Z (1992) Standard cumulative semivariograms of stationary stochastic processes and regional correlation. Int J Math Geol 24(4):417–435
49. Şen Z (1996) Statistical investigation of wind energy reliability and its application. Renew Energy 10:71–79
50. Şen Z (2008) Solar energy fundamentals and modelling techniques: atmosphere, environment, climate change and renewable energy. Springer, London, p 376
51. Taylor GI (1925) Eddy motion in the atmosphere. Phil Trans R Soc Lond A 215(1):1–26
52. Troen I, Petersen EL (1989) European wind atlas, commission of the European communities. Riso National Laboratory, Roskilde, p 655
53. Wark K, Warner CF (1981) Air pollution. Its origin and control. Harper Collins Publishers, New York, p 526

Part II
Grid Integration of Wind Power Systems

Integration of Wind Power Production in a Conventional Power Production System: Stochastic Models and Performance Measures

Maria Teresa Vespucci, Marida Bertocchi, Asgeir Tomasgard and Mario Innorta

Abstract A stochastic programming model for the daily coordination of hydro power plants and wind power plants with pumped storage is introduced, with hourly wind power production uncertainty represented by means of a scenario tree. Historical data of wind power production forecast error are assumed to be available, which are used for obtaining wind power production forecast error scenarios. These scenarios are then combined with information from the weather forecast, resulting in wind power production scenarios. Ex-ante and ex-post measures are considered for assessing the value of the stochastic model: the ex-ante performance evaluation is based on the Modified Value of Stochastic Solution for multistage stochastic programming, introduced independently in Escudero (TOP 15(1):48–66, 2007) and Vespucci (Ann Oper Res 193:91–105, 2012); the ex-post performance evaluation is defined in terms of the Value of Stochastic Planning, introduced in Schütz (Int J Prod Econ, 2009), that makes use of the realized values of the stochastic parameter. Both measures indicate the advantage of using the stochastic approach.

M. T. Vespucci (✉) · M. Bertocchi
Department of Management, Economics and Quantitative Methods,
via dei Caniana 2 24127 Bergamo, Italy
e-mail: maria-teresa.vespucci@unibg.it

M. Bertocchi
Department of Mathematics, Statistics, Computing and Applications,
University of Bergamo, Via Dei Caniana 2 24127 Bergamo, Italy
e-mail: marida.bertocchi@unibg.it

A. Tomasgard
Department of Industrial Economics and Technology Management,
Norwegian University of Science and Technology, 7491 Trondheim, Norway
e-mail: asgeir.tomasgard@iot.ntnu.no

M. Innorta
Department of Information Technology and Mathematical Methods,
University of Bergamo, Via Marconi 5a 24044 Dalmine, BG, Italy
e-mail: mario.innorta@fastwebnet.it

P. M. Pardalos et al. (eds.), *Handbook of Wind Power Systems*, Energy Systems,
DOI: 10.1007/978-3-642-41080-2_5,

1 Introduction

Due to the significant changes toward deregulation and competition in the electricity industry and market, electricity firms are exposed to significantly higher risks and opportunities than in the previous monopolistic context. At the same time electricity production technologies have been developed based on renewable resources (photovoltaic solar power, wind power, geothermal power, biomass and waste power) as an alternative to the use of conventional sources (coal, natural gas, oil), with the aim of reducing fuel consumption and pollution. With the introduction of power production from renewable energy sources the need of suitable decision-support models for coordinating conventional and new technologies in energy production systems has greatly increased and the operation scheduling models have been generalized so as to take into account the interactions between conventional and new technologies. In particular the integration of large amounts of wind power into the electricity system can cause transmission grid instabilities, due to the unbalance between local power demand and wind power production, characterized by intermittency, see [1] and the references therein. The most traditional method to the integration of intermittent power is the use of other power plants to provide operational and capacity reserve. Hydropower is the technology that plays a key role in providing grid balancing services, see [2–6]. In order to investigate the technical and economic issues related to the integration of large amounts of wind power into the electricity system, stochastic programming models have been introduced, see [7–10].

In [11] the problem of a power producer who aims at maximizing his own profit by efficiently coordinating wind power production with the operation scheduling of hydro plants has been analyzed on the basis of historical data. In the operation scheduling models presented in this chapter, we consider a power producer who owns both hydropower plants and wind power plants. In our operation scheduling model, the hourly wind power production is a parameter depending on wind speed forecasts and the time horizon considered is the day, as weather forecasts can be valid up to 24–36 h. We develop two stochastic scheduling models where the uncertainty of the hourly wind power production values in the next 24 h is represented by means of a scenario tree, with a probability value associated to each scenario, generated by quantile regression and by ARIMA techniques. In order to assess the convenience of using the stochastic programming approach, we use both the ex-ante measure, introduced independently in [11] and in [12], and the ex-post measure, suggested by [13] to evaluate the value of stochastic planning.

This chapter is organized as follows. In Sect. 2 the stochastic model of the daily operation scheduling model for a generation system consisting of pumped storage hydro plants and wind power plants is developed, based on a node representation of the scenario tree for the uncertainty of the hourly wind power production. In Sect. 3 the two scenario generation techniques for errors forecasting, namely quantile regression and ARIMA, are discussed. In Sect. 4 we discuss ex-ante and ex-post performance measures of the proposed model. This comparison shows the losses that the producer may suffer if he implements the deterministic solution instead of the stochastic one.

2 The Stochastic Model

In this section we introduce the linear stochastic programming model for the daily coordination of wind power plants and hydro power plants with pumped storage, taking into account wind power production uncertainty. The scheduling problem is considered within a daily time scope, as (24–36 h), and the time unit is one hour.

The power producer is assumed to have hourly commitments deriving from bilateral contracts with customers. If his own production is not sufficient to satisfy the hourly commitments, the producer may purchase electricity on the spot market; analogously, if his hourly production exceeds the amount required to satisfy bilateral contracts, the producer may either sells the excess production on the spot market, or use it for pumping water toward up-stream reservoirs, therefore storing potential energy for later use. The power producer is assumed to aim at profit maximization, which implies that he will tend to sell electricity on the spot market in hours with high market price; he will instead reduce hydro production and pump water in up-stream reservoirs in hours with low market price.

Hourly spot prices are exogenous model parameters, as the power producer is assumed to be a price taker. In order to study the effect of wind power production uncertainty on power producer's decisions the hourly electricity sell price λ_t $[Euro/MWh]$ and purchase price $\mu_t \geq \lambda_t$ are assumed to be known with certainty, i.e. the forecast of the hourly wind power production for the next 24 h is the only stochastic parameter. The stochasticity of energy prices will be included in a planned future development of the model.

On the basis of forecasts on power production output for each hour of the following day, the producer has to find the profit maximizing schedule of the hydro production system, taking into account technical constraints and the restriction imposed by the medium term scheduling on the amount of water available for production in the considered day: at the end of the day a minimum water volume, determined by the medium term scheduling, must be guaranteed in every reservoir.

A large number of references in the literature deals with hydropower production planning via stochastic programming, see among others [14–18]. In this chapter we are mainly interested in assessing the impact of wind power production stochasticity on the daily operation of the hydropower system. Therefore, we do not consider other sources of stochasticity.

In the following the representation of wind power production uncertainty and the models we have used for the hydro production system and the interactions with the spot market are modeled.

2.1 The Stochastic Parameter Representing the Hourly Wind Power Production Forecast

The uncertainty on the hourly wind power production is represented by means of a scenario tree, see reference papers [19] and [20], whose structure is described by a set of nodes $\{1,\ldots,n,\ldots,N\}$ and a pointer *pred(n)*, that associates every node n, $2 \le n \le N$, to its ancestor. A probability value p_n is associated to every node n, $1 \le n \le N$, as the stochastic model is written in terms of the scenario tree nodes.

The day is divided in K stages, with T_k representing the set of hours belonging to stage k, $1 \le k \le K$. At stage 1 a vector of dimension $|T_1|$ is associated to the root node of the scenario tree and represents the known wind power production in hours 1 to $|T_1|$. At stage k, $2 \le k \le K$, wind power production uncertainty is represented by the set of nodes N_k: each node is associated to a vector of dimension $|T_k|$ expressing the wind power production forecast in hours $|T_{k-1}| + 1$ to $|T_{k-1}| + |T_k|$. The stochastic parameter $WP_{t,n}[MWh]$ represents the hourly wind power production forecast in node n and hour t. The number of scenarios represented by the scenario tree equals the number of its leaf nodes and the 24-dimensional vectors containing the hourly wind power production forecast in one scenario can be traced from the leaf node back to the root node by means of the pointer *pred(n)*. For instance, the scenario tree in Fig. 1 represents 27 scenarios and contains 40 nodes, the first 13 of which are branching nodes: $N_1 = \{1\}$ in stage 1, $N_2 = \{2,3,4\}$ in stage 2, N_3 contains nodes 5–13 in stage 3 and N_4 contains nodes 14–40 in stage 4.

2.2 The Model of the Hydroelectric System

The hydroelectric system consists of a number of cascades, i.e. sets of hydraulically interconnected hydro plants, pumped storage hydro plants and reservoirs. It is mathematically represented by a directed multi-graph, see Fig. 2, where nodes represent water storages (reservoirs) and arcs represent water flows (either power generation, or pumping, or spillage).

Let the set of nodes and the set of arcs be denoted by J and I respectively and let the interconnections among water storages and water flows be represented by the arc-node incidence matrix A ($A_{i,j} = -1$, if arc i leaves node j; $A_{i,j} = 1$, if arc i enters node j; $A_{i,j} = 0$, otherwise). The energy coefficient k_i [MWh/10^3 m^3] is associated to arc $i \in I$, with $k_i > 0$, if arc i represents generation, $k_i < 0$, if arc i represents pumping, and $k_i = 0$, if arc i represents spillage. For $i \in I, j \in J, t \in T_k, n \in N_k$ and $1 \le k \le K$ the decision variables of the stochastic model for the hydro system are

- $q_{i,t,n}$ [10^3 m^3/h], the water flow on arc i in hour t in node n: this variable represents either a turbined volume, if arc i represents generation, or a pumped volume, if arc i represents pumping, or a spilled volume, if arc i represents spillage) in node n;

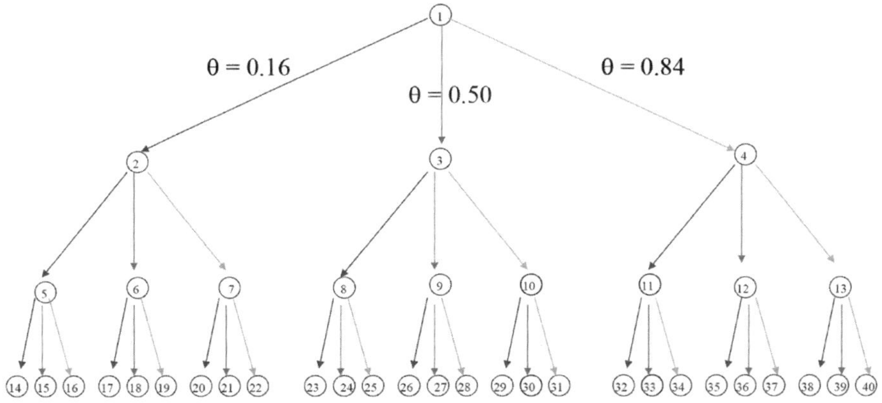

Fig. 1 Structure of the four-stage scenario tree obtained by the quantile regression method

Fig. 2 The hydro system

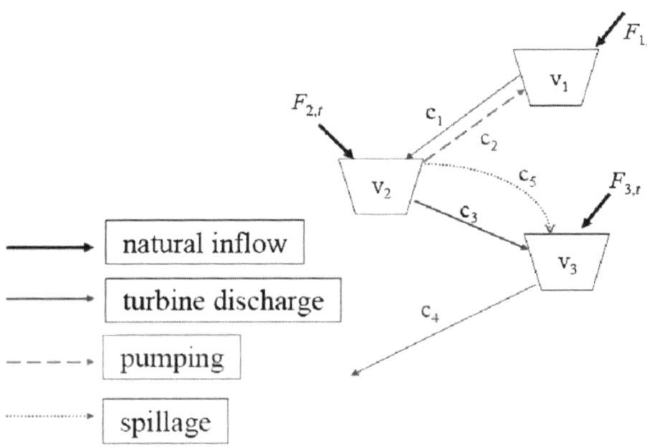

- $v_{j,t,n}$ $[10^3 \text{ m}^3]$, the storage volume in reservoir j at the end of hour t in node n.

The feasible values of $q_{i,t,n}$ and of $v_{j,t,n}$ are defined by the following technical constraints that describe the hydroelectric system

- for $1 \leq k \leq K, t \in T_k$ and $n \in N_k$

$$0 \leq q_{i,t,n} \leq \overline{q}_i \quad i \in I \tag{1}$$

$$0 \leq v_{j,t,n} \leq \overline{v}_j \quad j \in J \tag{2}$$

$$v_{j,t,n} = v_{j,t-1,\rho} + F_{j,t} + \sum_{i \in I} A_{i,j} \cdot q_{i,t,n} \quad j \in J \tag{3}$$

- for $n \in N_K$

$$v_{j,T,n} \geq \underline{v}_{j,T} \qquad j \in J \tag{4}$$

Constraints (1) require that in every stage k the water flow $q_{i,t,n}$ on arc $i \in I$ in hour $t \in T_k$ and in node $n \in N_k$ is nonnegative and bounded above by the maximum flow $\bar{q}_i$ $[10^3\ \mathrm{m}^3/\mathrm{h}]$ that can be either turbined, or pumped, or spilled in arc i.

Constraints (2) ensure that in every stage k the storage volume $v_{j,t,n}$ of reservoir $j \in J$ at the end of hour $t \in T_k$ in node $n \in N_k$ is nonnegative and bounded above by the maximum storage volume $\overline{v}_j$ $[10^3\ \mathrm{m}^3]$ in reservoir j.

The mass balance equations (3) relate the storage volumes in two subsequent hours, $t-1$ and t. If hours $t-1$ and t belong to stage k, then $\rho = n$; if hour $t-1$ belongs to stage $k-1$ and hour t belongs to stage k, then $\rho = pred(n)$. The initial storage volume in reservoir j is represented by the parameter $v_{j,0,1}$ $[10^3\ \mathrm{m}^3]$. The parameter $F_{j,t}[10^3\ \mathrm{m}^3/\mathrm{h}]$ represents the natural inflow in reservoir $j \in J$ in hour t and is assumed to be known with certainty in the daily planning horizon. In this chapter we do not consider extreme weather scenarios, like sudden floods or period of storms, that may happen in some geographical areas. Constraints (3) require the storage volume in reservoir j at the end of hour t in node n to be equal to the storage volume at the end of the previous hour plus the sum of inflows in hour t (i.e. turbine discharge from upstream hydro plants, pumped volumes from downstream hydro plants, spilled volumes from upstream reservoirs) minus the sum of outflows in hour t (turbine discharge to downstream hydro plants, pumped volumes to upstream hydro plants and spilled volumes to downstream reservoirs).

Constraints (4) require that in every scenario the storage volume of reservoir j at the end of hour T is bounded below by $\underline{v}_{j,T}$ $[10^3\ \mathrm{m}^3]$, the minimum storage volume required at the end of the current planning period, so as to provide the required initial storage volume at the beginning of the following planning period.

2.3 The Interactions with the Spot Market and the Objective Function

In every hour t of the planning period the power producer must satisfy a load, represented by the parameter $L_t[\mathrm{MWh}], 1 \leq t \leq T$, that derives from bilateral contracts with customers. If the total hourly production is less than L_t, an amount of energy must be bought on the spot market in order to satisfy the load demand. In the daily horizon this demand is known with certainty, while in papers dealing with longer time horizon as [21] stochasticity may become a relevant issue. The amount of energy to be bought is represented by the nonnegative decision variable $buy_{t,n}$. On the other hand, if the total hourly production exceeds the load from bilateral contracts, the excess quantity can be sold on the spot market and/or used for pumping. The amount of energy to be sold is represented by the nonnegative decision variable $sell_{t,n}$, while the amount of energy used for pumping is given by $\sum_{i \in I,\, k_i < 0} |k_i| \cdot q_{i,t,n}$. Therefore, the constraints

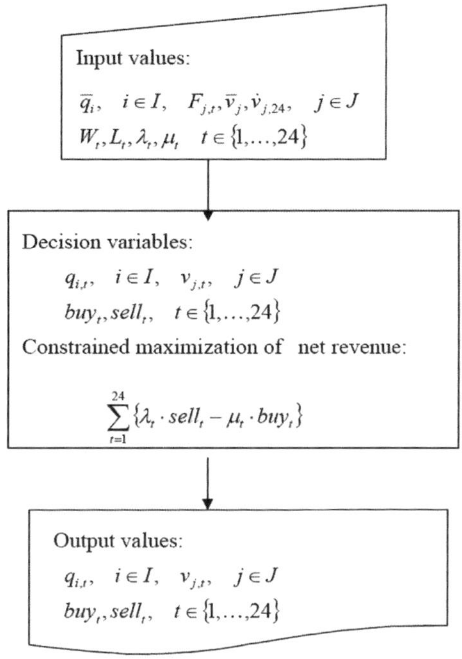

Fig. 3 Flow chart describing inputs, outputs and procedure of the deterministic approach

$$\sum_{i \in I} k_i \cdot q_{i,t,n} + WP_{t,n} + buy_{t,n} - sell_{t,n} = L_t \tag{5}$$

$$buy_{t,n} \geq 0, \quad sell_{t,n} \geq 0 \tag{6}$$

for $1 \leq k \leq K, \quad t \in T_k$ and $n \in N_k$, describe the interactions with the spot market.

The power producer aims at maximizing the expected daily net revenues, expressed as

$$\sum_{k=1}^{K} \sum_{n \in N_k} \left[p_n \cdot \sum_{t \in T_k} (\lambda_t \cdot sell_{t,n} - \mu_t \cdot buy_{t,n}) \right]. \tag{7}$$

Summarizing, in the stochastic linear programming model for the optimal coordination of wind power plants and hydro power plants with pumped storage the objective function (7) is to be maximized under constraints (1–6). For the sake of clarity, we report the input variables, the procedure and the output variables of the deterministic and stochastic models respectively in Figs. 3 and 4.

The scheduling of the hydro-wind production system determined by the stochastic model may substantially differ from the one determined by the deterministic model, as it is shown in the following example. Let the hydro production system consist of three hydro plants, one of which with pumped storage, see Fig. 2. The wind power production in the first six hours is known to be zero, i.e.

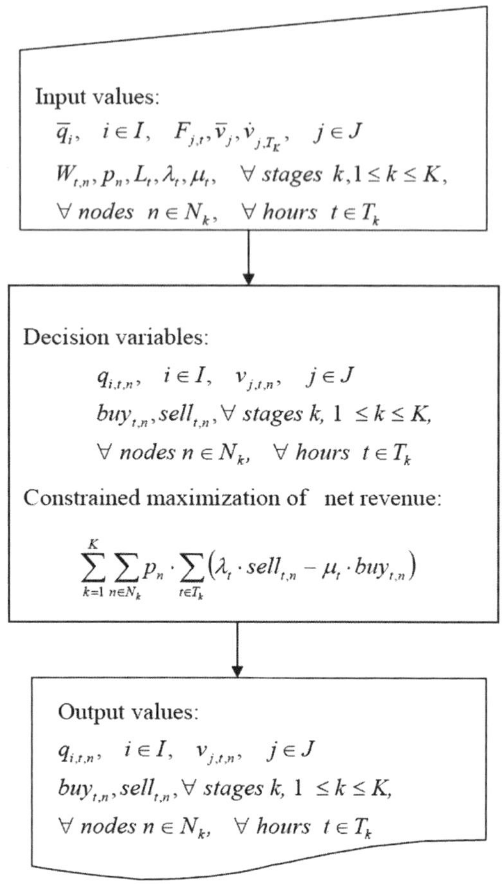

Fig. 4 Flow chart describing inputs, outputs and procedure of the stochastic approach

$WP_{t,1} = 0$ for $1 \leq t \leq 6$, while for $7 \leq t \leq 12$ it is predicted to be either $WP_{t,2} = 10$, with probability $p_2 = 0.3$, or $WP_{t,3} = 400$, with probability $p_3 = 0.7$. The deterministic model solves the problem on the basis of the average prediction $WP_t = 283$, for $7 \leq t \leq 12$, and yields the solution reported in Fig. 5a, where $q_{i,t}$ represents a turbined volume, for $i = 1, 3, 4$, and a pumped volume, for $i = 2$. In Fig. 5b the turbined and pumped volumes of the stochastic solution are shown. It can be noticed that the stochastic solution suggests pumping in 5 out of the 6 first-stage hours, while no pumping is suggested by the deterministic solution. Moreover, in the stochastic solution only the hydro plant located at the end of the cascade generates electricity, i.e. more potential energy is saved in the upstream reservoirs. Figure 6a and b show, for a low wind power production scenario, the impact of using in first-stage hours the deterministic solution and the stochastic solution respectively. From the graphs it can be noticed that, in order to satisfy commitments deriving from bilateral contracts, a large quantity of energy must be bought in second-stage hours, when in first-stage hours the deterministic solution

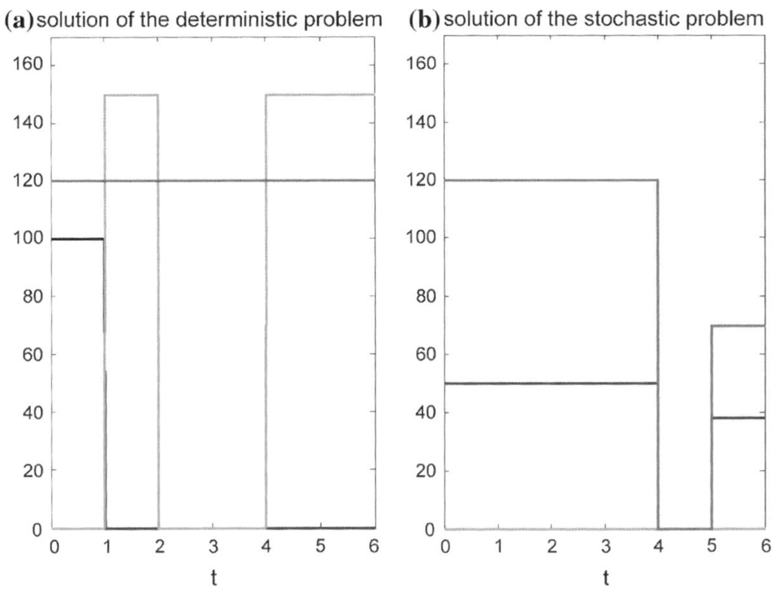

Fig. 5 Optimal values of water flows $q_{i,t}$, $1 \le i \le n$ and $1 \le t \le 6$, computed by the deterministic model. **a** (*black line*) $q_{1,t} = 100, t \in (0, 1]$ and 0 otherwise; (*blue line*) $q_{2,t} = 0, t \in (0, 6]$; (*green line*) $q_{3,t} = 100, t \in (0, 1] \cup [3, 4]$ and 150 otherwise; (*red line*) $q_{4,t} = 120, t \in (0, 6]$ and by the stochastic model. **b** (*black line*) $q_{1,t} = 100, t \in (0, 6]$; (*blue line*) $q_{2,t} = 50, t \in (0, 4], q_{2,t} = 0, t \in (4, 5], q_{2,t} = 38, t \in (5, 6]$; (*green line*) $q_{3,t} = 0$, $t \in (0, 6]$; (*red line*) $q_{4,t} = 120$, $t \in (0, 4]$, $q_{4,t} = 0, t \in (4, 5]$, $q_{4,t} = 70$, $t \in (5, 6]$

is used; on the other hand, no energy must be bought in second-stage hours, when the stochastic solution is used. An analogous comparison is made in Fig. 6c and d for a high wind power production scenario: more energy can be sold in second-stage hours, when in first-stage hours the stochastic solution is used.

3 Scenario Generation

Depending on the available data, different approaches have been introduced for generating scenarios representing the uncertainty of wind power production. In [11] wind power production scenarios are obtained by first generating scenarios of wind speed, on the basis of historical wind speed data, and then using the wind farm power curve to compute the power production scenarios corresponding to the wind speed scenarios.

In this chapter we develop an alternative procedure for generating wind power production scenarios that combines information from weather forecast and from time series of wind power production forecast errors. This methodology is suitable to be used when data series on predicted power production and on measured power

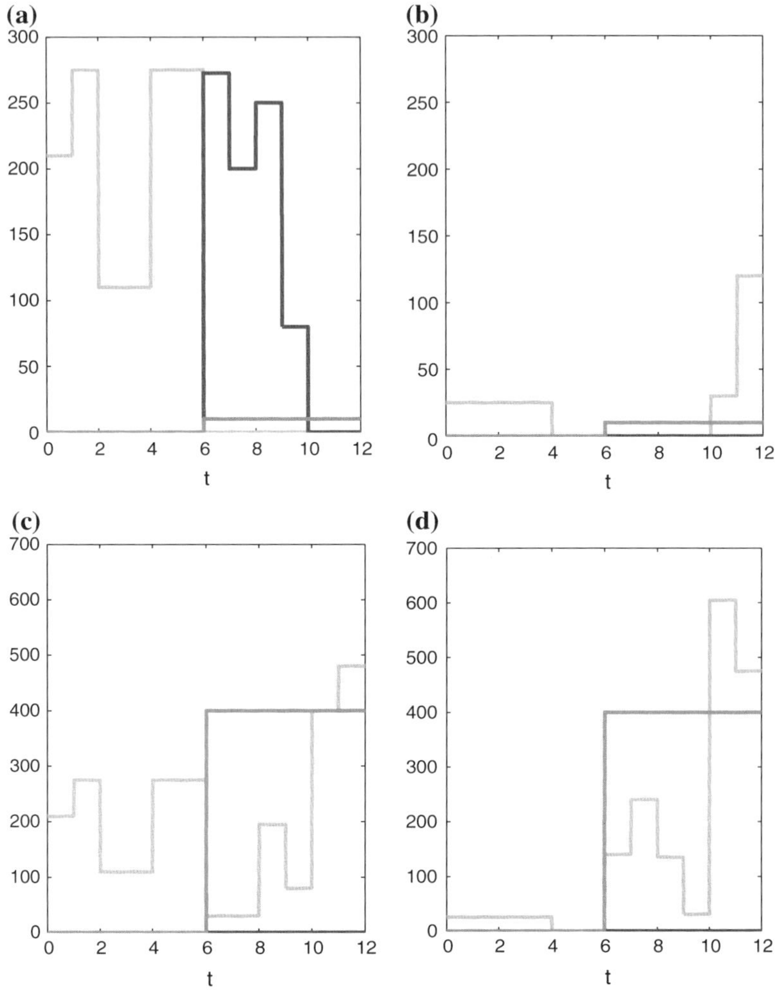

Fig. 6 Wind power production in hours 1–12 (*red line*) and optimal values of sold energy (*green line*) and bought energy (*blue line*). **a** deterministic model, low wind power production. **b** stochastic model, low wind power production. **c** deterministic model, high wind power production. **d** stochastic model, high wind power production

production are available, as well as a wind power forecast system based on a tool for simulating and forecasting meteorological phenomena. In [22] a wind power forecast system has been developed based on the high-resolution Regional Atmospheric Modeling System (RAMS), see [23], for simulating and forecasting meteorological phenomena, and on a set of post processors that reduce the systematic errors and compute the predicted power of a wind farm. In the forecast system described in [22] boundary conditions from a weather forecast

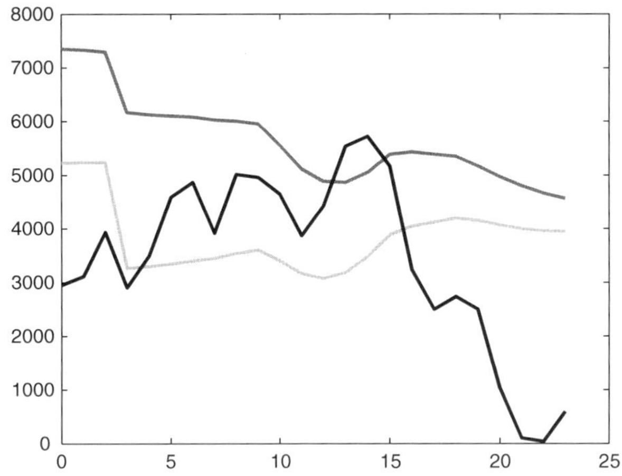

Fig. 7 WP_t^M (*black line*), WP_t^R (*blue line*) and WP_t^C (*red line*) on December 16, 2002

deterministic model are fed into RAMS, in order to generate the wind forecast hourly time series for the day ahead period; a correction of the wind speed, aiming at reducing systematic errors, is then performed by using a Model Output Statistic, see [24]; finally, the resulting wind speed is used to compute the predicted power through the wind farm power curve. In order to test the performances in predicting the power production, this wind power forecast system has been tested on data related to the KLIM wind farm, located in Denmark, which have been supplied in the frame of the ViLab exercise, promoted by the European POW'WOW project [25]. A data set produced in the ViLab exercise, referred to the period from 16 December 2002 to 30 April 2003, has been used for generating the wind power production scenarios used in this work. For every hour t the following information are contained in this data set:

1. WP_t^M: measured wind power production in hour t;
2. WP_t^R: forecast of wind power production in hour t obtained by first predicting the wind speed in hour t by RAMS and then applying the wind farm power curve;
3. WP_t^C: forecast of wind power production in hour t obtained by first predicting the wind speed in hour t by RAMS, then correcting the forecast computed by RAMS, in order to reduce systematic errors, and finally applying the wind farm power curve.

In the first hours of the daily forecast period, the values of WP_t^C are closer to WP_t^M than the values of WP_t^R, as it can be noticed in Figs. 7 and 8 where the values of WP_t^M, WP_t^R and, WP_t^C are shown for 16 December 2002 and 17 December 2002 respectively. It can be noticed, however, that a relevant forecast error is still present in some hours. Based on this evidence and on the papers [8] and [26], we introduce a procedure for generating scenarios of the forecast error e_t, defined as

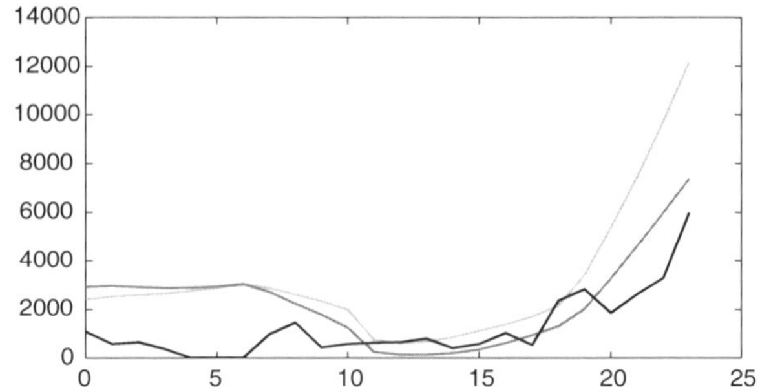

Fig. 8 WP_t^M (*black line*), WP_t^R (*blue line*) and WP_t^C (*red line*) on December 17, 2002

$$e_t = WP_t^M - WP_t^C. \tag{8}$$

We consider two alternative methods for estimating the forecast errors, namely the Quantile Regression method [27] and the Autoregressive Integrated Moving Average (ARIMA) method, see [28] and [29].

3.1 The Quantile Regression Method

Let the power production forecast error e_t be described by the model

$$e_t = \alpha + \sum_{d=1}^{D} \beta_d \cdot e_{t-d}, \tag{9}$$

where D is the number of time lags. By the quantile regression approach, the values of the model parameters α and $\beta_d, 1 \leq d \leq D$, can be computed so as to identify the hyperplane $H(\theta)$, $0 \leq \theta \leq 1$, such that θ % of sample data lies above the hyperplane and $(1-\theta)$% of sample data lies below the hyperplane. This is done by determining the values of α and $\beta_d, 1 \leq d \leq D$, that solve the problem

$$\begin{aligned} \min \theta \cdot & \sum_{t: e_t \geq \alpha + \sum_{d=1}^{D} \beta_d \cdot e_{t-d}} \left(e_t - \alpha - \sum_{d=1}^{D} \beta_d \cdot e_{t-d} \right) \\ & + (1-\theta) \cdot \sum_{t: e_t < \alpha + \sum_{d=1}^{D} \beta_d \cdot e_{t-d}} \left(\alpha + \sum_{d=1}^{D} \beta_d \cdot e_{t-d} - e_t \right) \end{aligned} \tag{10}$$

i.e. by minimizing the sum of magnitudes of the differences between the observed value e_t and the value predicted by the model, with positive differences weighted by θ and negative differences weighted by $1-\theta$.

Table 1 Optimal quantile regression coefficients for $D = 1$

$D = 1$	$\theta = 0.16$	$\theta = 0.50$	$\theta = 0.84$
α_θ	−1220.16	−11.19	1,111.17
β_θ	0.81	0.91	0.93

In order to construct the balanced scenario tree shown in Fig. 1, where at each stage the branching nodes are equally probable, the interval [0, 1] has been divided in the subintervals [0, 1/3], [1/3, 2/3] and [2/3, 1], which are represented by their mean values, 0.16, 0.50 and 0.84 respectively. The quantile regression hyperplanes $H(\theta)$, for $\theta = 0.16$, 0.50 and, 0.84 have been computed: by construction any point of the three hyperplanes may occur with probability 1/3. The model parameters α and β_d, for the three considered values of θ, have been computed using the time lag values $D = 1, 2, 3$. The best fit of data was obtained with the time lag $D = 1$ and the forecast error scenario tree has been generated with the values of α_θ and β_θ reported in Table 1.

The forecast errors in the first-stage hours are assumed to be known and are computed as

$$e_{t,1} = WP_t^M - WP_t^C, \quad t \in T_1. \tag{11}$$

Uncertainty in the second-stage hours is represented by nodes $n = 2, 3, 4$, that are associated to quantile regression models with $\theta = 0.16$, 0.50 and, 0.84 respectively. In the three nodes the forecast error values of the first hour are computed as

$$e_{t,n} = \alpha_\theta + \beta_\theta \cdot e_{t-1,1} \tag{12}$$

and the forecast error values of the remaining second-stage hours are

$$e_{t,n} = \alpha_\theta + \beta_\theta \cdot e_{t-1,n} \tag{13}$$

This procedure is repeated for all branching nodes in the scenario tree reported in Fig. 1. The wind power production forecast $WP_{t,n}$ is finally obtained by computing

$$WP_{t,n} = WP_t^C + e_{t,n} \tag{14}$$

In Fig. 9 the wind power production scenario together with the measured one are reported for comparison: we notice that the wind power production scenarios are within a reasonable distance from the measured wind power production.

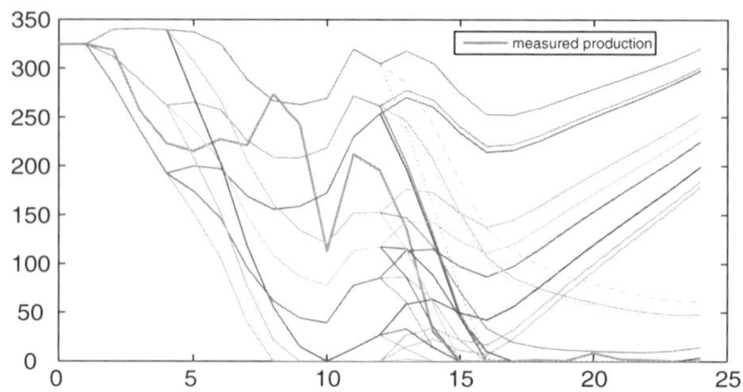

Fig. 9 Measured wind power production (*purple line*) and wind power production scenario tree obtained with quantile regression

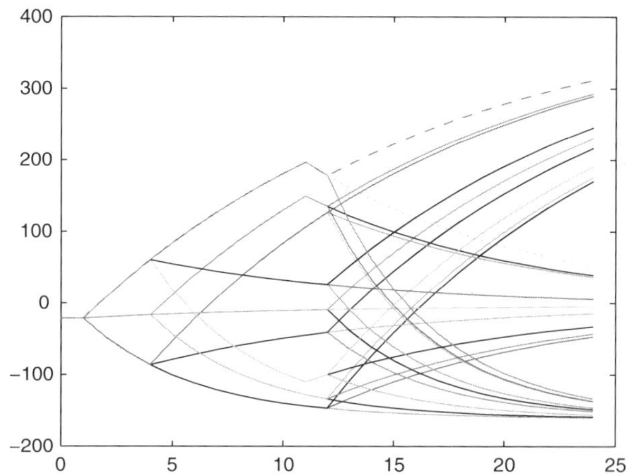

Fig. 10 Errors of wind power obtained with quantile regression

3.2 The Autoregressive Integrated Moving Average Method

The historical power production forecast errors have alternatively been assumed to be described by the autoregressive integrated moving average model (Fig. 10).

$$e_t = \gamma + \sum_{d=1}^{D} \delta_d \cdot e_{t-d} + \sum_{c=1}^{C} \eta_c \cdot \varepsilon_{t-c} \qquad (15)$$

In fitting the model the lowest standard deviation was obtained with ARIMA(1,0,1), i.e. no integration (no differencing of data), one lag ($C = 1$) for the autoregressive part and one lag ($D = 1$) for the moving average part, with coefficients $\gamma = 148.678$, $\delta_1 = 0.822093$ and $\eta_1 = 0.142345$. By Eq. (15) forecast error scenarios are generated, which are then used for obtaining power production forecast scenarios by the formula

$$WP_t = WP_t^C + e_t \qquad (16)$$

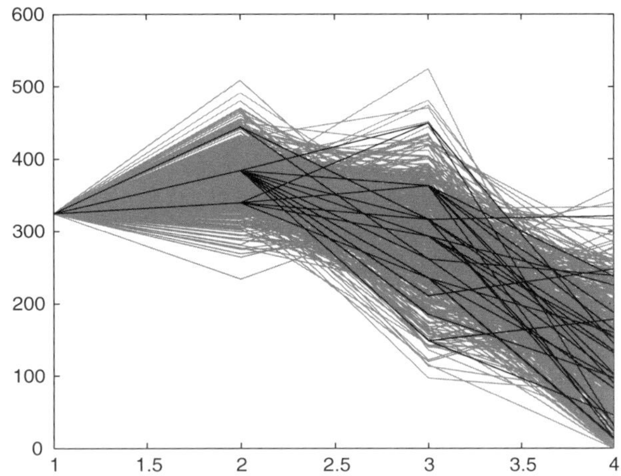

Fig. 11 500 scenarios of hourly wind power production obtained by the ARIMA method (*red lines*); reduced scenario tree obtained by backward scenario reduction technique (*black lines*)

Finally, by applying the backward scenario reduction technique introduced by Pflug and Hochreiter [30], 27 scenarios have been obtained, in order to compare the quantile regression approach and the ARIMA approach on the basis of the information contained in an equal number of scenarios. In Fig. 11 the scenarios of hourly wind power production obtained by the ARIMA method are shown, as well as the reduced scenario tree obtained by applying the backward scenario reduction technique.

4 Performance Measures

Two approaches have been introduced in the literature in order to assess the value of modeling uncertainty. The first approach gives rise to the *ex-ante performance measures*, like the Value of the Stochastic Solution (VSS) for two-stage problems, introduced in [31]. As scenario tree is assumed to represent the future realizations of the stochastic parameter, ex-ante performance measures compare the result obtained by the stochastic model, in which the scenario tree represents the stochastic parameter uncertainty, and the result obtained by the corresponding deterministic model, where the uncertain parameter is assigned the best ex-ante knowledge, i.e. the average scenario computed on the scenario tree. This measure for assessing the value of using the stochastic model is available before uncertainty is revealed.

The second approach, see [13], gives rise to the *ex-post performance measures*: these can be computed only after uncertainty has been revealed, as they make use of the realized values of the stochastic parameter. Here the actual result of having solved the problem by the stochastic model is computed.

4.1 Ex-ante Performance Evaluation

In order to assess the value of modeling uncertainty, the following procedure has been introduced in [31] for evaluating the Value of the Stochastic Solution of a two-stage model:

1. solve the stochastic model, obtaining the so called here-and-now solution optimal value (HNS);
2. solve the Expected Value (EV) problem, or Mean Value problem, i.e. the deterministic problem where at each stage the stochastic parameter takes the expected value computed on the scenario tree;
3. solve the Expected Mean Value (EEV) problem, i.e. the problem obtained by substituting into the stochastic problem the optimal values of the first-stage decision variables of the EV problem;
4. compute the Value of Stochastic Solution (VSS): for a maximization problem it is defined as

$$VSS = HNS - EEV . \tag{17}$$

In order to deal with stochastic problems with more than two stages, we have extended the definition of VSS by introducing the Modified Value of Stochastic Solution (MVSS), defined as

$$MVSS = HNS - MEEV , \tag{18}$$

i.e. MVSS is the difference between the (optimal) objective value of the here-and-now solution and the (optimal) objective value of the Modified Expected Mean Value (MEEV) problem, which, for a four-stage stochastic problem, is computed by the following procedure. Let $\mathcal{T}$ denote the scenario tree used for computing the here-and-now solution, let $\mathcal{T}_n$, $n \in N_k$, $k = 1, 2, 3$, denote the subtree of $\mathcal{T}$ with root in node n and let $\mathcal{T}^{(r)}$, $r = 1, 2, 3$, denote the scenario tree obtained from $\mathcal{T}$ by substituting the subtrees $\mathcal{T}_n$, $n \in N_r$, with the expected values of the stochastic parameter computed on $\mathcal{T}_n$. Let $SP^{(r)}$ denote the problem where the uncertainty of the stochastic parameter is represented by the scenario tree $\mathcal{T}^{(r)}$. In order to compute the (objective) value of MEEV, problem $SP^{(1)}$ is solved first; then problem $SP^{(2)}$ is solved, where the first-stage variables of $SP^{(2)}$ are assigned the optimal values determined by $SP^{(1)}$; finally problem $SP^{(3)}$ is solved, where the first-stage and the second-stage variables of $SP^{(3)}$ are assigned the optimal values determined by $SP^{(1)}$ and by $SP^{(2)}$ respectively. This procedure reproduces the sequence of decisions taken in the real situation. Indeed, suppose the planning problem is going to be solved with the uncertainty represented by only one scenario and suppose the future behavior of the stochastic parameter is described by the scenario tree $\mathcal{T}$. When only first-stage values of the stochastic parameter are revealed, the best prediction of the stochastic parameter values in stages 2–4 is

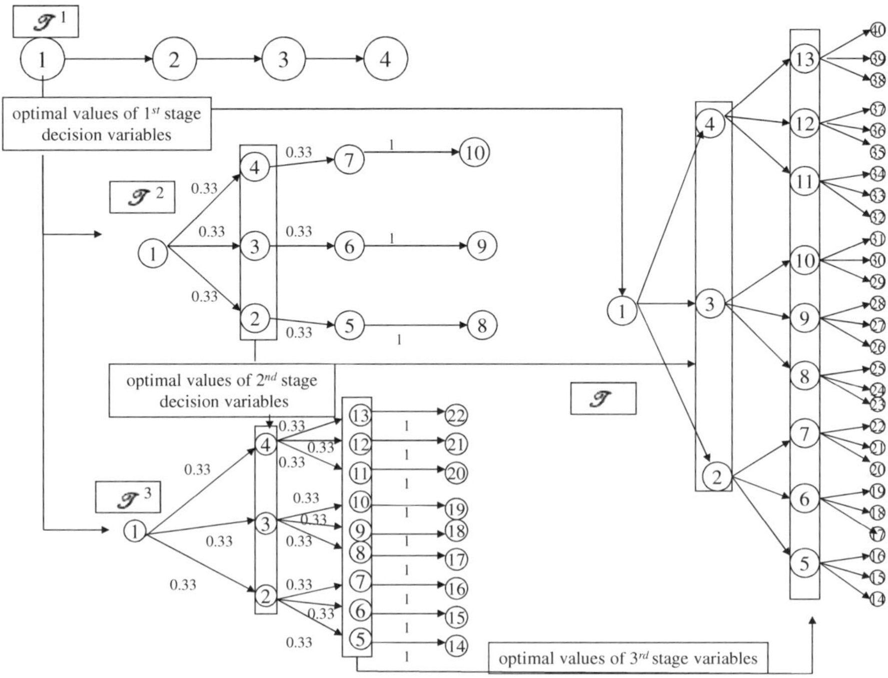

Fig. 12 Computation of the Modified Expected Mean Value (MEEV) objective function

represented by $\mathcal{T}^{(1)}$ and problem $SP^{(1)}$ is solved. When the second-stage values of the stochastic parameter are revealed, these values are, with probability $p(n)$, those represented in node n, $n \in N_2$: therefore, with probability $p(n)$, the problem will have to be solved taking into account that

- the known values of the stochastic parameter are given in node n, $n \in N_2$,
- the stochastic parameter in stages 3 and 4 takes the expected value of subtree $\mathcal{T}_n$,
- first-stage variables must take the optimal value computed at stage 1.

This is done by solving problem $SP^{(2)}$, with the first-stage variables constrained to be equal to the optimal values determined by $SP^{(1)}$. At stage 3 an analogous procedure is performed (Fig. 12).

4.2 Ex-post Performance Evaluation

In [13] an evaluation procedure for a two-stage stochastic model is introduced, which is based on the use of the realized values of the stochastic parameter and yields the so-called Value of Stochastic Planning (VSP). The problem in [13] is to be solved every week, with a time horizon of four weeks, the first of which is in stage 1, while the following three are in stage 2. A yearly planning is obtained by a rolling horizon procedure: a four-week problem is solved for weeks 1–4, then a second four-week problem is solved for weeks 2–5, and so on. Therefore, 52 four-week problems are solved, spanning a time horizon of 55 weeks. The procedure for computing the VSP requires considering different types of problems:

1. solve the 52 four-week stochastic problems, where in each problem the first-stage values of the stochastic parameter are the realized ones; compute the sum, over the 52 problems, of the profit pertaining to the first week (P_{STOCH});
2. solve the 52 four-week mean value problems; compute the sum, over the 52 problems, of the profit pertaining to the first week (P_{EV});
3. solve the 52 four-week deterministic problems where the stochastic parameters are assigned their realized values; compute the sum, over the 52 problems, of the profit pertaining to the first week (P_{PIR});
4. solve the 55 week deterministic problem where the stochastic parameters are assigned their realized values; compute the sum of the profit pertaining to the first 52 weeks (P_{PI}).

Problem types STOCH, EV and PIR are solved R = 52 times and the VSP is assessed on the basis of the ratios

$$\frac{P_{STOCH}}{P_{PI}}, \qquad \frac{P_{EV}}{P_{PI}}, \qquad \frac{P_{PIR}}{P_{PI}}. \tag{19}$$

In the hydro-wind coordination problem the subsequent problems to be solved, as uncertainty is revealed, must all consider a time horizon up to the end of the day, as constraint (4) requires that in every scenario the storage volume of reservoir j at the end of the last hour of the planning period is bounded below by $\underline{v}_{j,T}$. Table 5 shows how the time horizon has been divided in stages (sets T_k^r) in each of the problems considered. The profit values P_{STOCH}, P_{EV} and, P_{PI}, are defined as

$$P_{STOCH} = \sum_{r=1}^{R} \sum_{t \in T_1^r} (\lambda_t \cdot sell_{t,1} - \mu_t \cdot buy_{t,1}) \tag{20}$$

$$P_{EV} = \sum_{r=1}^{R} \sum_{t \in T_1^r} (\lambda_t \cdot sell_{t,1} - \mu_t \cdot buy_{t,1}) \tag{21}$$

$$P_{PI} = \sum_{t=1}^{24} (\lambda_t \cdot sell_{t,1} - \mu_t \cdot buy_{t,1}). \tag{22}$$

As all problems must consider a time horizon up to the end of the day, the profit value computed by the problem type PIR coincides with that computed by the problem type PI. Note that in problem type PI the 24 h can be thought as all belonging to node 1. The values P_{STOCH} and P_{EV} express the total daily profit that

Table 2 Hydro reservoir data: capacity, initial and minimum final storage volumes, natural inflow

Reservoir	$\bar{v}_j$	$v_{j,0}$	$\underline{v}_{j,T}$	$F_{j,t}$
V_1	1,000	100	0	1
V_2	2,000	1,000	500	1
V_3	2,000	1,000	500	1

Table 3 Hydro arc data: energy coefficient and capacity

Arc	k_i	$\bar{q}_i$
c_1	1.0	100
c_2	−1.7	50
c_3	1.1	150
c_4	0.9	120
c_5	0.0	50

the power producer obtains if he reschedules his own production decisions, as new information becomes available, by the stochastic model and by the mean value model respectively. The value of P_{PI} represents the profit that the power producer would obtain if he had perfect knowledge of the future.

5 Numerical Results

In this section we discuss the numerical results obtained by solving the stochastic model on the previous discussed cases studies. The simulation framework is based on ACCESS 2000, for data input, on EXCEL 2000 and MATLAB release 12, for data output, and on GAMS release 21.5, for modeling and solving the optimization problem by using the CPLEX solver. In both the cases considered, the hydro system is composed by one cascade with three reservoirs and three hydro plants, one of these is a pumped storage hydro plant as shown in Fig. 2 (see Tables 2 and 3 for input data of the hydro system).

5.1 Case Study 1

In case study 1 the uncertainty on the hourly wind power production is represented by the scenario tree in Fig. 1, characterized by an initial stage comprising 1 h, followed by three proper stages (i.e. stages with branching) of three, eight and twelve hours respectively. In each node the following information are contained: the node number n, $1 \leq n \leq 40$, the forecast on wind power production in hours $t \in T_k$ (i.e. in the hours associated to the stage which node n belongs to) and the

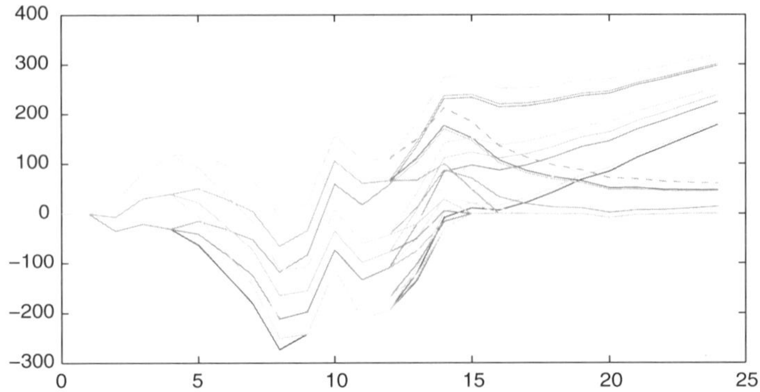

Fig. 13 The difference between the wind power obtained by quantile regression along the scenarios and the measured wind power

Table 4 Table containing the MVSS for the two types of scenarios

Profit value (in Euro)	Quantile regression scenarios	ARIMA scenarios
EV model	3,701.27	3,672.73
HNS model	3,626.65	3,669.87
MEEV model	1,714.97	3,612.03
MVSS	1,911.68	57.84

node probability p_n. The hourly wind power production values at each node n have been generated on the basis of the forecasting procedure based on quantile regression method on prediction errors described in the previous section. As already mentioned we have constructed a balanced tree where all nodes at the same stage have the same probability. In Fig. 13 we report the difference along the various scenarios of the wind power obtained by quantile regression method and the measured wind power production.

We use the procedure suitable for multi-stage problems described in the previous section, based on the idea of reproducing the decision process as the uncertainty reveals.

Table 4 illustrates the ex-ante comparisons of deterministic and stochastic models using the two types (ARIMA and quantile regression) of scenarios.

A large difference appears in the value of MVSS for the two types of scenarios. This may suggest that the quantile regression scenarios describe better the volatility in wind power production along time. Indeed, from Figs. 9 and 11 it can be noticed that the scenarios produced by the quantile regression approach are less concentrated around the average scenarios, that is more extreme situations may be represented.

Another measure of the role of randomness of the model parameters is given by the Expected Value of Perfect Information (EVPI) (see [31]). EVPI is defined as the difference between the expected value of the *wait-and-see solution* (WSS), computed by finding the optimal solution for each possible realization of wind power production, i.e. for each scenario, and the optimal objective value of the stochastic model, HNS, as follows:

$$EVPI = WSS - HNS \tag{23}$$

For the quantile regression scenarios, $EVPI = 3657.39 - 3626.65 = 30.74$, this means that one should be willing to pay 30.74 in return for complete information in advance about the wind power production, which represents about 1 % of the optimal profit.

This case study shows that the quantile regression technique is appropriate to describe the uncertainty of wind power production.

5.2 Case Study 2

The advantage of using the stochastic model with respect to the mean value model may also be assessed on the basis of realized wind power production values and this is the aim of this section. We perform further testing of the stochastic model described in Sect. 2 in order to produce ex-post performance evaluation as described in Sect. 4.2. Based on the results of case study 1, we focus our analysis on the scenario tree obtained by the quantile regression method.

In this case study the wind power production of the first 3 h of the planning period is assumed to be known. Every three hours the scheduling model is solved on a time horizon up to midnight. The three problem types (STOCH, EV, PI) are solved 8 times on a time horizon with a decreasing number of hours. Each instance consists of a first-stage of 3 h and successive stages as described in Table 5.

The scenarios are generated by the regression quantile technique described in Sect. 3.1, where each stage follows the scheme in Table 5. For the configuration corresponding to $r = 1$, we get the following ex-ante measures $MVSS = 35.34$ and the EVPI is in percentage 46.80. This result has been obtained considering a higher load demand. We note that they go in opposite direction with respect to case study 1 where a better value of MVSS was obtained together with a worst value of EVPI obtained when the load was lower. We compute the MVSS and EVPI for various load values and we report for an intermediate load value the following results: MVSS is 28.83 and EVPI in percentage equals to 48.98.

Table 6 reports the values obtained by the three problem types.

If we assume as a basis for the high load the optimal profit $P_{PI} = 1332.83$ for the 24 h coming from realized wind production, we can compute the following percentages

$$\frac{P_{STOCH}}{P_{PI}} = 90.75\ \%, \quad \frac{P_{EV}}{P_{PI}} = 59.56\ \%. \tag{24}$$

Table 5 Elements of sets T_k^r used in the computation of VSP

r	T_1^r	T_2^r	T_3^r	T_4^r	Number of scenarios
1	{1, 2, 3}	{4, 5, 6}	{7, ..., 12}	{13, ..., 24}	27
2	{4, 5, 6}	{7, 8, 9}	{10, ..., 15}	{16, ..., 24}	27
3	{7, 8, 9}	{10, 11, 12}	{13, ..., 18}	{19, ..., 24}	27
4	{10, 11, 12}	{13, 14, 15}	{16, ..., 18}	{19, ..., 24}	27
5	{13, 14, 15}	{16, 17, 18}	{19, 20, 21}	{22, 23, 24}	27
6	{16, 17, 18}	{19, 20, 21}	{22, 23, 24}	–	9
7	{19, 20, 21}	{22, 23, 24}	–	–	3
8	{22, 23, 24}	–	–	–	1

Table 6 Optimal profit for stochastic model, mean value model, perfect information model

Load	P_{STOCH}	P_{EV}	P_{PI}
High	1,209.48	793.89	1,332.83
Intermediate	1,404.08	1,017.98	1,452.83

Analogously, if we assume as a basis for the intermediate load the optimal profit $P_{PI} = 1452.83$ for the 24 h coming from realized wind production, we can compute the following percentages

$$\frac{P_{STOCH}}{P_{PI}} = 96.64\,\%, \quad \frac{P_{EV}}{P_{PI}} = 70.07\,\%. \tag{25}$$

The ex-post performance measures confirm a strong superiority of the stochastic approach with respect to the mean value model.

6 Conclusions

In this chapter we have introduced a stochastic multi-stage linear model for the daily hydro-wind power system scheduling problem with scenarios on hourly wind power production. In order to study the influence of scenario generation on the optimal solution, we studied two approaches for scenario generation, the quantile regression and the autoregressive integrated moving average techniques. We got, from the value of stochastic solution, that the quantile regression scenarios describe the uncertainty better than the ARIMA scenarios. We also give two types of performance measures of the multi-stage stochastic model, the ex-ante and the ex-post measures. We show the difference among the two measures and discuss a case where the ex-post measures gives further information.

Acknowledgments This research was partly supported by the research grants Fondi di Ateneo 2009–2010 of the University of Bergamo (coordinated by M. Bertocchi and L. Brandolini) and by research grant of Accordo Regione Lombardia Metodi di integrazione delle fonti energetiche rinnovabili e monitoraggio satellitare dell'impatto ambientale, CUP: F11J10000200002 (coordinated by A. Fassó). We also acknowledge RSE in Milan, for providing data.

References

1. Moura PS, de Almeida AT (2010) Large scale integration of wind power generation. In: Rebennack S et al (eds) Handbook of power systems I, energy systems. Springer, Berlin, pp 95–119
2. Castronuovo ED, Lopes JAP (2004) On the optimization of the daily operation of a wind-hydro power plant. IEEE Trans Power Syst 19:1599–1606
3. Castronuovo ED, Lopes JAP (2004) Optimal operation and hydro storage sizing of a wind-hydro power plant. Int J Elec Power 26:771–778
4. Castronuovo ED, Lopes JAP (2004) Bounding active power generation of a wind-hydro power plant. In: International conference on probabilistic methods applied to power systems, IEEE, New York, 705–710
5. Denault M et al (2009) Complementarity of hydro and wind power: improving the risk profile of energy inflows. Energy Policy 37(12):5376–5384
6. Matevosyan J, Sonder L (2007) Short-term hydropower planning coordinated with wind power in areas with congestion problems. Wind Energy 10(3):195–208
7. Nørgård P, Giebel G., Holttinen H, Söder L., Petterteig A (2004) Fluctuations and predictability of wind and hydropower, WILMAR deliverable D2.1, Risø-R-1443
8. Barth R, Söder L, Weber C, Brand H, Swider D (2006) Documentation methodology of the scenario tree tool, WILMAR Deliverable D6.2 (b), Institute of Energy Economics and the Rational Use of Energy (IER), University of Stuttgart, Stuttgart
9. Garcia-Gonzalez J et al (2008) Stochastic joint optimization of wind generation and pumped-storage units in an electricity market. IEEE Trans Power Syst 23:460–468
10. Meibom P, Barth R, Brand H, Weber C (2007) Wind power integration studies using a multistage stochastic electricity system model. In: IEEE power engineering society, pp 1–4
11. Vespucci MT, Maggioni F, Bertocchi M, Innorta M (2010) A stochastic model for the daily coordination of pumped storage hydro plants and wind power plants. Ann Oper Res. doi:10.1007/s10479-010-0756-4
12. Escudero LF, Garin A, Merino M, Perez G (2007) The value of the stochastic solution in multistage problems. TOP 15(1):48–66. doi:10.1007/S11750-007-00005-4
13. Schütz P, Tomasgard A (2009) The impact of flexibility on operational supply chain planning. Int J Prod Econ. doi:10.1016/j.ijpe.2009.11.004
14. Dentcheva D, Römisch, W (1998) Optimal power generation under uncertainty via stochastic programming. In: Stochastic programming methods and technical applications. Lecture notes in economics and mathematical systems, vol 458 Springer, New York, pp 22–56
15. Fleten SE, Kristoffersen T (2008) Short-term hydropower production planning by stochastic programming. Comput Oper Res 35(8):2656–2671
16. Latorre J, Cerisola S, Ramos A (2007) Clustering algorithms for scenario tree generation: application to natural hydro inflows. Eur J Oper Res 181(3):1339–1353
17. Nowak M, Römisch W (2000) Stochastic Lagrangian relaxation applied to power scheduling in a hydro-thermal system under uncertainty. Ann Oper Res 100(1–4):251–272
18. Wallace SW, Fleten SE (2003) Stochastic programming models in energy. In: Ruszczynski A, Shapiro A (eds): stochastic programming. Handbooks in operations research and management science vol 10. Elsevier, Amsterdam, pp 637–677

19. Dupačová J, Consigli G, Wallace SW (2000) Scenarios for multistage stochastic programs. Ann Oper Res 100(1–4):25–53
20. Kaut M, Wallace SW (2007) Evaluation of scenario-generation methods for stochastic programming. Pac J Optim 3(2):257–271
21. Philpott A, Craddock M, Waterer H (2000) Hydro-electric unit commitment subject to uncertain demand. Eur J Oper Res 125:410–424
22. Alessandrini S, Decimi G, Palmieri L, Ferrero E (2006) A wind power forecast system in complex topographic conditions. In: Proceedings of the European wind energy conference and exhibition ewec 2009
23. Pielke R, Cotton W, Walko R, Tremback C, Lyons W, Grasso L, Nicholls M, Moran M, Wesley D, Lee T, Copeland J (1992) A comprehensive meteorological modeling system: RAMS. Meteorol Atmos Phys 49:69–91
24. von Bremen, L (2007) Combination of deterministic and probabilistic meteorological models to enhance wind farm forecast. J Phys Conf Ser 75(1): 012050
25. Kariniotakis G, Pinson P, Marti I, Lozano S, Giebel G (2007) POW'WOW virtual laboratory for wind power forecasting: ViLab, In: EWEC'07 Conference, Milan, Italy 7–10 May 2007
26. Söder L(2004) Simulation of wind speed forecast errors for operation planning of multi-area power systems. In: Proceedings of international conference on probabilistic methods applied to power systems, IEEE: doi:10.1109/pmaps.2004.243051, pp. 723–728
27. Koenker R, Bassett G Jr (1978) Regression quantiles. Econometrica 46(1):33–50
28. Engle RF, Granger CWJ (1987) Co-integration and error-correction: Representation, estimation and testing. Econometrica 55:251–276
29. Davidson J (2000) Econometric Theory. Blackwell Publishing
30. Pflug G, Hochreiter R (2007) Financial scenario generation for stochastic multi-stage decision processes as facility location problem. Ann Oper Res 152(1):257–272
31. Birge J, Louveaux F (2000) Introduction to stochastic programming. Springer, New York

Grid Integration of Wind Power Systems: Modeling of Wind Power Plants

Mithun Vyas, Mohit Singh and Surya Santoso

Abstract In the United States, wind power is expected to make up a significant portion of future generation portfolios. A scenario in which wind power will supply 20 % of U.S. peak demand by 2030 has been examined and found feasible [1]. A challenge facing power system planners and operators, in the near future, is the grid integration of large amounts of wind power. To determine the impacts of large wind power plants on system stability, reliable computer models are necessary. However, wind turbine models are not readily available in most dynamic simulation software. The diversity and manufacturer-specific nature of technologies used in commercial wind turbines exacerbates the modeling problem. A solution to this problem is to develop a generic, manufacturer-independent modeling framework that can be implemented in any software capable of simulating power system dynamics.

1 Introduction

1.1 Motivation

In the United States, wind power is expected to make up a significant portion of future generation portfolios. A scenario in which wind power will supply 20 % of U.S. peak demand by 2030 has been examined and found feasible [1]. A challenge facing power system planners and operators, in the near future, is the grid integration of large amounts of wind power. To determine the impacts of large wind power plants on system stability, reliable computer models are necessary. However, wind turbine models are not readily available in most dynamic simulation software. The diversity and manufacturer-specific nature of technologies used in commercial wind turbines exacerbates the modeling problem. A solution to this

M. Vyas · M. Singh · S. Santoso (✉)
The University of Texas at Austin, Austin, USA
e-mail: ssantoso@mail.utexas.edu

P. M. Pardalos et al. (eds.), *Handbook of Wind Power Systems*, Energy Systems,
DOI: 10.1007/978-3-642-41080-2_6,

problem is to develop a generic, manufacturer-independent modeling framework that can be implemented in any software capable of simulating power system dynamics. This chapter describes the development of generic models for:

1. A fixed-speed wind turbine,
2. A wind turbine employing the principle of rotor resistance control, and
3. A turbine with a doubly-fed induction machine employing flux-vector control (DFIG).

The focus of this chapter is on wind turbines which use induction generators, since they comprise the largest installed base of wind turbines and also have more significant effects on the bulk power system than other machines. A detailed description of the wind turbine models is provided along with details of their implementation on two different software platforms, widely used in industry and academia namely PSCAD/EMTDC and MATLAB/SIMULINK. While the central purpose of these models is to study the interaction between the wind turbine and the power system, they may also be used to examine the interaction of aerodynamic, mechanical, and electrical functions within the wind turbine.

1.2 Wind Turbine Technologies

Fixed-speed wind turbines are so named because they operate with less than 2 % variation in turbine rotor speed. They employ squirrel-cage induction machines directly connected to the power grid. The rotor blades are attached to the hub at a fixed pitch, and are designed in such a manner that the air flow over the blades changes from streamline flow to turbulent flow at high wind speeds. This limits the kinetic power extracted from the wind at high wind speeds in order to protect the induction machine and drive train between overheating and overspeeding. Turbines using this design are known as stall-regulated. A side-effect of stall regulation is that energy capture from the wind is sub-optimal. In contrast, variable-speed wind turbines are designed to operate at a wide range of rotor speeds. The rotor speed may vary with the wind speed, or with other system variables, depending on the design employed. Typically in variable-speed turbines, the blades are not rigidly fixed to the hub, and can be rotated a few degrees to turn them out of or into the wind. Additional speed and power controls allow these turbines to extract more energy from a wind regime than would be possible with fixed-speed turbines. For DFIG turbines, power converters are needed to interface the wind turbine and the grid. One important advantage of converter-based systems is that they allow independent real and reactive power control.

Fixed-speed wind turbines are low-cost, robust, reliable, simple to maintain, and proven in the field [2]. A large number of fixed-speed wind turbines have been installed over the past decade-and-a-half, and more continue to be installed [3]. While variable-speed wind turbines form the bulk of new installed capacity, a niche for fixed-speed wind turbines still exists. Therefore, it can be expected that fixed-speed wind turbines will continue to play a role in the power systems of the future. While there are many wind turbine dynamic models available in the literature, the focus is largely on modeling variable-speed wind turbines [4–9]. These models often oversimplify the mechanical drive train and aerodynamics, since the aim is to evaluate power and rotor speed control mechanisms. In the model developed here, a modular approach is used to represent each of the turbine's functions. One block represents the aerodynamics, another the mechanical drive train, and a third represents the electrical generator. A control block may also be included. The blocks are integrated to form the complete wind turbine model, which is implemented in PSCAD/EMTDC and MATLAB/SIMULINK. This model is a platform on which more advanced variable-speed wind turbine models can be developed.

Variable-speed turbines use wound-rotor induction machines as generators, and control over output power is achieved through control of the rotor resistance, or through the use of power electronic converters in the rotor circuit in DFIG turbines. This chapter explains the principle behind rotor resistance control in detail, and discusses different control strategies for achieving optimal power extraction. The DFIG technology is also discussed and a model is developed and validated using available wind power plant field data.

1.3 Background on Wind Turbine Modeling

Modeling of wind turbine generator systems (WTGS) can be broadly classified into:

1. Static modeling
2. Dynamic modeling

Static models of WTGS can be used for steady state analysis or quasi-steady state analysis such as load flow studies, power quality assessment, short circuit calculations whereas a dynamic model of WTGS is needed for various types of system dynamic analysis e.g. stability study, control system analysis, optimization techniques to mention just a few. Considering the static models of a WTGS, they are characterized by a simple voltage source (V), a voltage and real power source (V, P) or a real and reactive power source (P, Q). The choice of model used depends on specific application and the type of WTGS [10]. The tree diagram of Fig. 1 shows the model types and their applications. In this chapter the focus is on functional models designed for studying transient stability.

In general, a WTGS can be equipped with either a synchronous or asynchronous generator, it can be directly connected to the grid or connected through a power electronic converter. It may use aerodynamic torque control (blade pitching, stall control) and/or generator torque control (varying the rotor resistance, flux-vector control) for output power optimization. The possibilities stated give rise to a very general model framework, whose block diagram in shown in Fig. 2. This general framework is used to represent each wind turbine technology that is

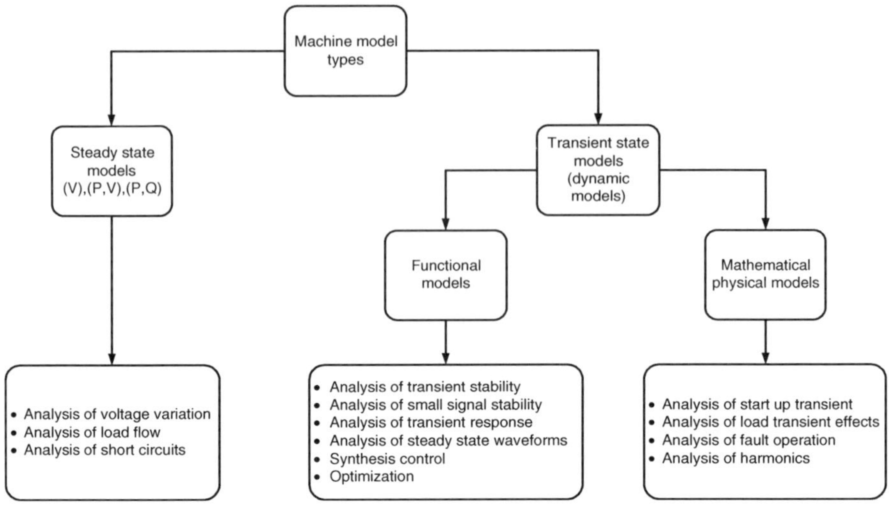

Fig. 1 Model types and their applications

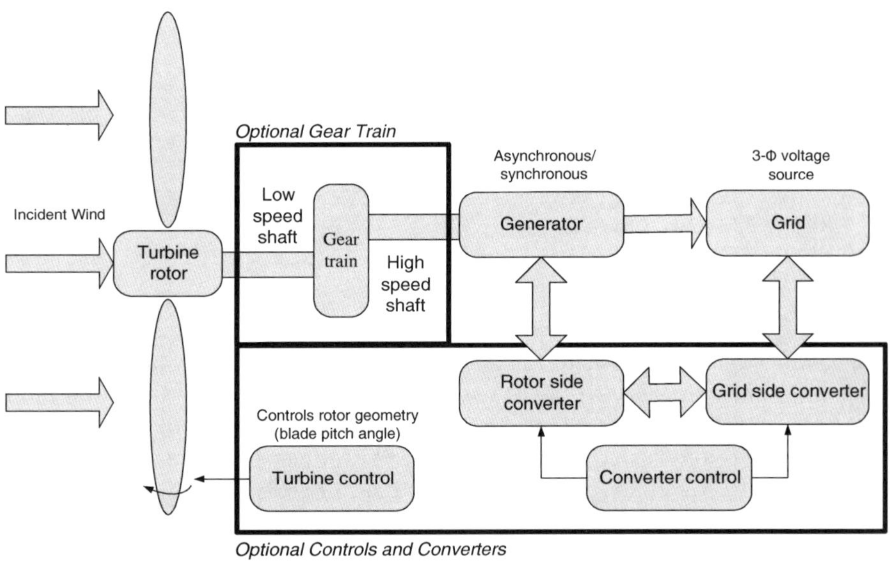

Fig. 2 Functional block diagram for a generic wind turbine generator system

modeled in this chapter, with suitable modifications for each technology. This general framework is software-independent. In this chapter, each block of the framework is discussed. The physical theory behind each block is presented, and implementation of each block is also described. The development of the complete models is achieved by combining these blocks.

1.4 Chapter Organization

In the next section, the theory behind the blocks comprising the general framework is discussed. The following section describes the implementation of the wind turbine models for each of the three technologies mentioned previously, using the general framework. The implementation is carried out using PSCAD/EMTDC and MATLAB/SIMULINK platforms. The practical modeling issues, such as tuning of controllers for the various subsystems, are also discussed in this section. This section also includes the modeling of a DFIG WTGS with validation.

2 Modeling Concepts and Related Theory

Wind turbines are designed to capture the kinetic energy present in wind and convert it to electrical energy. An analogy can be drawn between wind turbines and conventional generating units which harness the kinetic energy of steam. From a modeling standpoint, a fixed-speed wind turbine consists of the following components:

1. Turbine rotor and blade assembly (prime mover)
2. Shaft and gearbox unit (drive-train and speed changer)
3. Induction generator
4. Control system

The interaction between each of the components listed above determines how much kinetic energy is extracted from the wind. Figure 3 illustrates the interaction between the wind turbine components in a basic fixed speed wind turbine. Modeling of the electrical subsystems is fairly straightforward, as power system modeling software usually includes a built-in induction machine model. However, modeling of the aerodynamics and mechanical drive-train is more challenging. These components are modeled based on the differential and algebraic equations that describe their operation. The following subsections describe the modeling of the four components listed above.

2.1 Aerodynamic Model

The aerodynamic block consists of three subsystems: tip-speed ratio calculation, rotor power coefficient (C_p) calculation, and aerodynamic torque calculation. Wind speed (V_{wind}) and pitch angle (β) are user-defined inputs. Since the model is intended to study the dynamic response of wind turbines to grid events, the assumption is usually made that the wind speed stays constant during the grid

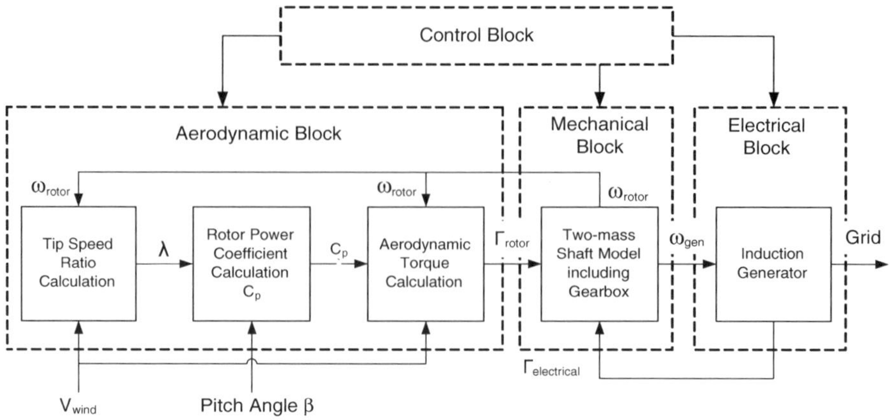

Fig. 3 Block diagram for a fixed-speed stall-regulated wind turbine

event. However, the models allow the wind speed input signal to be set to any value at the start of the simulation run-time and also to be modified during the run. It is also possible to use a time-series of actual wind speed data.

2.1.1 Power Available in Wind Stream and its Extraction

The kinetic energy (KE) in any object of mass m moving with a velocity v is given by

$$KE = \frac{1}{2} m v^2 \tag{1}$$

A wind turbine is an electromechanical energy conversion device, that captures kinetic energy available from wind. This kinetic energy is turned into mechanical energy of the rotor and eventually into electrical energy from the generator. Power available in moving air is given as follows

$$P_{wind} = \frac{d(KE)}{dt} = \frac{1}{2} m' v^2 \tag{2}$$

where m' is the mass flow rate. For a wind turbine rotor sweeping an area A of radius R, power available in that area is given by Eq. (3)

$$P_{wind} = \frac{1}{2} \rho A v^3 \tag{3}$$

where ρ is the air density, $A = \pi R^2$ and v is the velocity of the moving air particles or in general wind. To determine the power extracted by a wind turbine rotor, Betz model (1926) is widely used. Betz model is not only used to find the power from an ideal turbine rotor, but also to find the thrust of the wind on the ideal rotor and

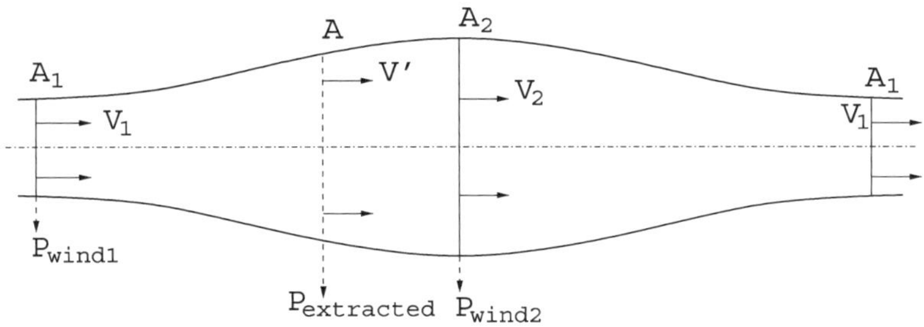

Fig. 4 Wind flow conditions before and after the converter

the effect of the rotor operation on the local wind field. This simple model is based on linear momentum theory. The analysis assumes a control volume whose boundaries are the surface of a stream tube and its two cross sections. The turbine in the analysis is represented by a uniform *actuator disk* or *converter* which creates a discontinuity of pressure in the stream tube of air flowing through it. Betz analysis further assumes that [12].

- Air is homogeneous, incompressible and has achieved steady state fluid flow,
- There is no frictional drag,
- Number of blades on the rotor are infinite,
- Uniform thrust occurs over the disk or rotor area,
- A non-rotating wake, and
- The static pressure far upstream and far downstream of the rotor are equal to the undisturbed ambient static pressure;

Figure 4 shows the wind flow conditions for an energy converter. The power extracted from wind using such a converter, is given by the difference in moving air particle power before and after the converter. The power extracted by the energy converter is given by Eq. (4)

$$P_{extracted} = P_{wind1} - P_{wind2} = \frac{1}{2}\rho(A_1 v_1^3 - A_2 v_2^3) \tag{4}$$

Figure 4 describes the change in wind velocity before and after the converter. To achieve ideal efficiency in energy conversion it is required that the air velocity after the converter (v_2) becomes zero. This is physically impossible, because that would render a need for the wind velocity before the converter to be zero and the air to be still. A more practical energy converter, will have air pressure increase just before the converter, which would simultaneously result in air velocity decrease, thus exerting a force given by Eq. (5)

$$F = m'(v_1 - v_2) \tag{5}$$

Thus, the power extracted from wind is given by Eq. (6)

$$P_{extracted} = Fv' = m'(v_1 - v_2)v' \tag{6}$$

By comparing the two equations obtained for $P_{extracted}$ [Eqs. (4)–(6)], and assuming that the mass flow rate through the converter is constant, the air velocity through the converter is the average wind velocity $v' = \frac{1}{2}(v_1 + v_2)$. Then the power extracted from the converter can be computed as

$$P_{extracted} = \frac{1}{4}\rho A(v_1^2 - v_2^2)(v_1 + v_2) \tag{7}$$

The term rotor power coefficient can now be defined (since, $P_{extracted} < P_{wind}$) as follows

$$C_p = \frac{P_{extracted}}{P_{wind}} \tag{8}$$

It is the ratio of power extracted from the rotor to the power available from wind, also known as rotor performance coefficient and sometimes referred as Betz factor. As described earlier, Betz created a (1D) model based on linear momentum theory, along with some assumptions for the analysis. The power coefficient can achieve a maximum value of 0.593, when $\frac{v_2}{v_1} = \frac{1}{3}$. This is the maximum theoretically possible value of C_p. Due to aerodynamic losses, actual value of power coefficient never achieves 0.593. In practice three effects are accounted for decrease in maximum achievable value of C_p.

1. Rotation of wake behind the rotor
2. Finite number of possible rotor blades and their associated tip losses, and
3. Non-zero aerodynamic drag [12]

In the next section, the relation between C_p and tip speed ratio (λ_r) at a particular value of blade pitch angle (β) will be presented. This relation can be used to develop $C_p - \lambda_r$ curves. These are used to determine the rotor power for any combination of wind speed and rotor speed. These curves provide immediate information on the maximum value of C_p and optimum tip speed ratio. The data for such a relationship is found from turbine tests and modeling [12].

2.1.2 Relation of Power Coefficient with Pitch Angle and Tip Speed Ratio

An empirical relation between C_p (rotor power coefficient), tip speed ratio (λ_r) and blade pitch angle (β) is used for developing a look-up table that provides a value of C_p for a given value of wind speed and tip speed ratio. Blade pitch angle can be defined as the angle between the plane of rotation and blade chord line. Tip speed ratio is defined as the ratio of the blade-tip linear speed to the wind speed [12]

$$\lambda_r = \frac{\omega_{rot}R}{v_1} \tag{9}$$

where R is the rotor radius and ω_{rot} is the angular velocity of the rotor.

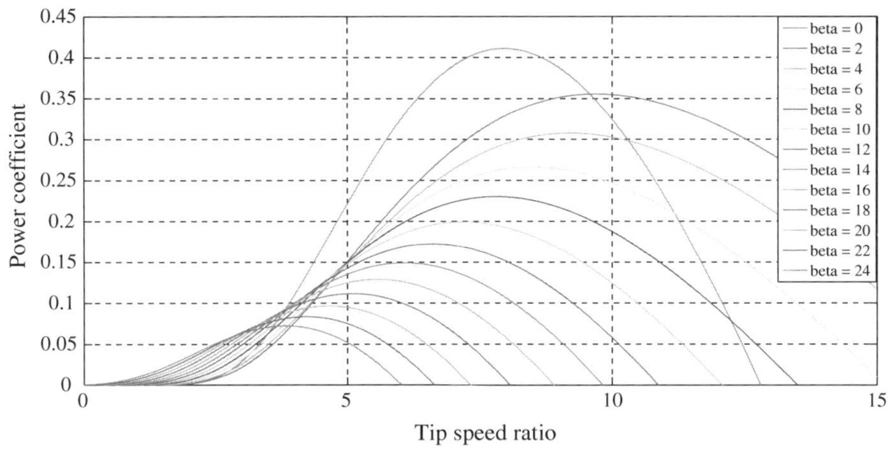

Fig. 5 Power coefficient C_p as a function of tip speed ratio λ_r

Shown below is one such empirical relation between C_p, λ_r, and β. Equation (10) is used to generate a look-up table for C_p. When provided with the values of λ_r and β, the corresponding value of C_p can be found. The $C_p(\lambda_r)$ curve obtained from the equation works only for positive values of pitch angle β.

$$C_p(\lambda, \beta) = c_1\left(c_2\frac{1}{\Lambda} - c_3\beta - c_4\beta^x - c_5\right)e^{-c_6\frac{1}{\Lambda}} \tag{10}$$

$$\frac{1}{\Lambda} = \frac{1}{\lambda + 0.08\beta} - \frac{0.035}{1+\beta^3} \tag{11}$$

while the coefficients $c_1 - c_6$ are proposed as equal to : $c_1 = 0.5, c_2 = 116, c_3 = 0.4, c_4 = 0, c_5 = 5, c_6 = 21$ [10]. Once, C_p is determined, aerodynamic torque of the rotor can be computed using Eqs. (3), (8) and (12). A mechanical model for the drive train developed in next Sect. 2.1.3 is used to determine the angular speed of the generator, ω_{gen} and angular speed of the turbine rotor, ω_{rot}. For all the models developed ω_{gen} is provided as an input to the induction machine.

$$P_{extracted} = \tau_{rot}\omega_{rot} \tag{12}$$

Figure 5 shows the C_p versus λ_r characteristics obtained from Eq. (10), note that only positive values of blade pitch angle can be used with these curves.

2.1.3 Blade Pitching

Blade pitch angle control is used to directly vary the power coefficient of a wind turbine. As it determines the operating power coefficient, it can be effectively used to control the mechanical output power of the rotor. A reduction in mechanical power of the rotor can be achieved by reducing or minimizing the angle of attack above its critical value. By limiting the power coefficient, power extracted from the wind is limited. This kind of power control is also known as *pitch control*. Pitch control can be used to serve different purposes such as

- Optimizing the power output of the wind turbine, by maximizing the mechanical power output for a given wind speed, this is generally applied for low and moderate wind speeds below rated wind speed.
- Preventing excess mechanical power output in strong winds above rated wind speeds. This keeps a check on the mechanical power and keeps it below the rated value in strong winds.
- To prevent disconnected wind turbines from turning [11].

There are two common ways in which pitch angle control can be used for regulating the power output of wind turbines

- Active pitch control: For variable-speed pitch-regulated wind turbines, wind turbine operation and power output can be affected either by speed changes or blade pitch angle changes. Below rated power, such machines operate at variable speed to optimize the tip speed ratio at fixed pitch. After rated power output is achieved generator torque control is used to maintain output power, while pitch control is used to maintain rotor speed. At high wind speeds, power output of the generator can be maintained constant, with an increase in rotor speed. This increased energy available from the wind is stored as kinetic energy in the rotor. This results in reduced aerodynamic torque and thus deceleration of the rotor. If the wind speed continues to remain high, aerodynamic efficiency of the rotor can be reduced by changing the pitch, resulting in reduced rotor speed.
- Passive pitch control : In case of passive control wind speed is used to provide the actuator power, which adjusts the blade pitch angle to shape the power curve of the wind turbine. In such wind turbine designs, the effects of change in rotor speed or wind speed are related to change in blade pitch angle.

2.2 Mechanical Drive Train Model

The drive train of a wind turbine generally consists of turbine rotor, low speed rotor shaft, gearbox with transmission ratio a, high speed shaft of the generator and the generator itself (either synchronous or induction). In case of wind turbines using synchronous generators, usually the design calls for a generator with a high pole count, thus reducing the mechanical speed of the generator shaft. The gearbox can then be omitted from the drive train. More than 90 % of the drive train moment of inertia is accounted for by the rotor (blades and hub) [10]. The generator accounts for 6–8 %, while the remaining parts account for 2–4 % of the total moment of inertia. Since the generator's torsional stiffness is very high, approximately two orders of magnitude higher than that of the rotor shaft, and

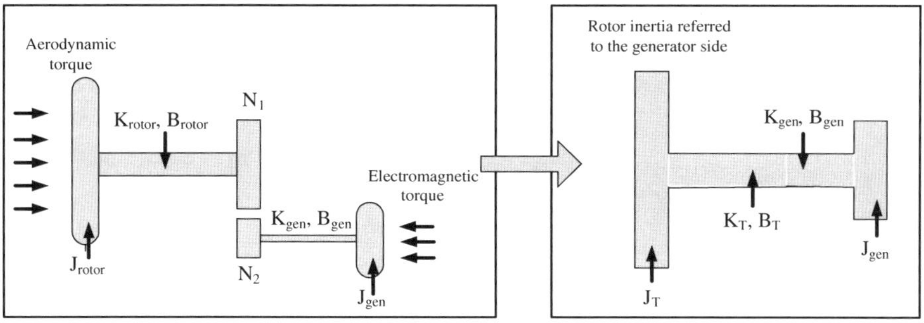

Fig. 6 Two-mass model for the drive train

Fig. 7 Two-mass model for the drive train with opposing torque action

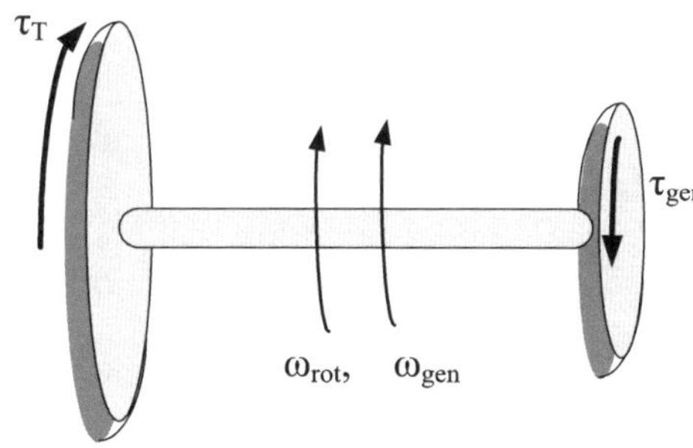

about fifty times higher than the hub with blades, the torsional vibration on the drive train elements cannot be ignored. Their characteristics (frequency and amplitude) can affect wind turbine performance. Hence, it is impossible to model the drive train as a lumped single mass. Typically the masses of the rotor and the generator are much larger than the mass of the gearbox. If we neglect the mass of the gearbox, the properties (stiffness constant and torsional constant) of the two shafts can be combined into one equivalent shaft resulting in a two-mass model as shown in Fig. 6. Moreover, the equivalent shaft of the two-mass model is not infinitely stiff, and thus the model cannot be generally reduced to a one-mass model. Hence, a *two-mass* model is preferred.

Note from Fig. 7 that the aerodynamic torque from the rotor is counteracted by the electromagnetic torque from the generator. Also note from Fig. 6 that rotor speed ω_{rot}, torque τ_{rot} and moment of inertia J_{rot} are all referred to the generator side using the gear transmission ratio a.

By balancing the torque for each mass, differential equations formed can be solved to determine the rotor, generator speeds ω_{rot} and ω_{gen} respectively. For each rotating mass, the product of moment of inertia J and angular acceleration θ'' must equal the sum of the torques acting on the mass.

For the turbine rotor torque, equation can be written as

$$J_T \theta_T'' = \tau_T - B_{eqv}(\omega_T - \omega_G) - K_e qv(\theta_T - \theta_G) \tag{13}$$

For the generator torque, equation can be written as

$$J_G\theta_G'' = -\tau_G + B_{eqv}(\omega_T - \omega_G) + K_eqv(\theta_T - \theta_G) \tag{14}$$

subscript T used in Eqs. (13) and (14) denotes the rotor parameters referred to the generator side of the gearbox and subscript G denotes generator parameters.

2.3 *Modeling of Induction Machines*

The induction machine has typically been favored for use in wind turbines due to the fact that induction generators do not need to be synchronized with the grid. Since wind turbines operate under varying wind speed conditions, resulting in varying shaft speeds, conventional synchronous generators cannot be easily used for this application. In a conventional synchronous machine connected to a steam turbine, it is possible to control real and reactive power output independently of each other by varying the steam flow rate and the excitation respectively. This decoupling effect cannot be achieved in fixed-speed and rotor-resistance control based technologies. In a DFIG turbine, the decoupling of real and reactive power is achieved through the use of power electronics and flux-vector control. In this subsection the considerations for modeling an induction machine and the concept of flux-vector control are introduced.

2.3.1 Introduction

The winding arrangement of a conventional 2-pole, 3-phase, wye-connected symmetrical induction machine is shown in Fig. 8. The stator windings are identical with equivalent turns N_s and resistance r_s. The rotor windings can be approximated as identical windings with equivalent turns N_r and resistance r_r. The model assumes the air-gap is uniform and the windings are sinusoidally distributed.

In Fig. 8, the winding of each phase is represented by an elementary coil. One side of the coil is represented by a $\otimes$ indicating that the assumed positive direction of current is down the length of the stator (into the plane of the paper). The other side of the same coil is represented by a $\odot$ which indicates that the assumed positive direction of current is out of the plane of the paper. The axes *as*, *bs* and *cs* represent the positive directions of the magnetic fields produced due to the currents flowing in the stator windings of phase *a*, *b* and *c* respectively. These directions are obtained using the right hand rule on the phase windings. Similarly axes *ar*, *br* and *cr* with respect to the rotor windings are shown. These rotor axes are fixed to the rotor and rotate with it at an angular velocity of ω_r. The angular displacement of the rotor with respect to the positive *as* axis is θ_r. In the stationary *abc* reference frame, the relationships between voltages, currents and flux linkages of each phase for this machine can be written from Fig. 9.

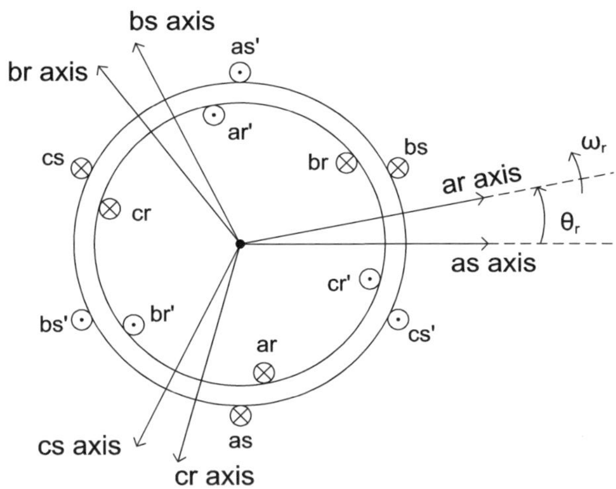

Fig. 8 Schematic winding diagram

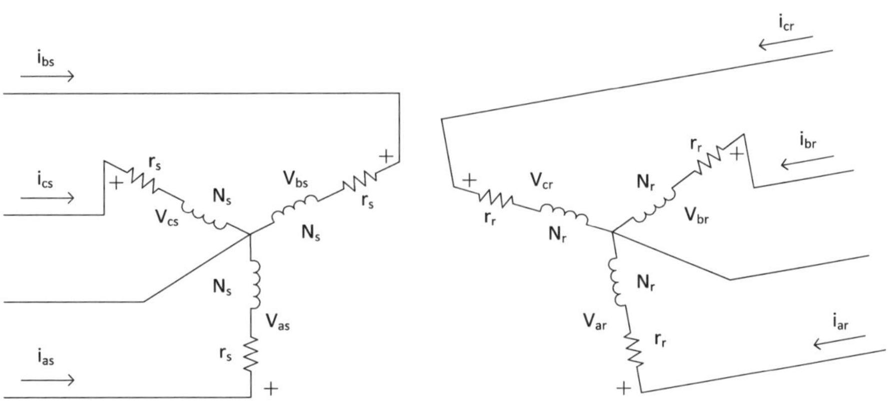

Fig. 9 Equivalent circuit (2-pole, 3-phase, wye-connected IM)

2.3.2 Reference Frame Theory and the Clarke and Park Transforms

It is known that for rotating machine inductances are functions of the rotor speed, due to which the coefficients of the differential equations (voltage equations) describing machine operation vary with time, except when the rotor is stationary. It is difficult to develop machine models that can be used for dynamic studies, using these complex equations. These time-varying equations can be written in a time-invariant form by choosing a frame of reference that is rotating at the appropriate speed. Referring machine variables to a rotating frame can, not only reduce the complexity of modeling the machine but also serve as a tool for better understanding of machine operation. Two such transformations to be used for

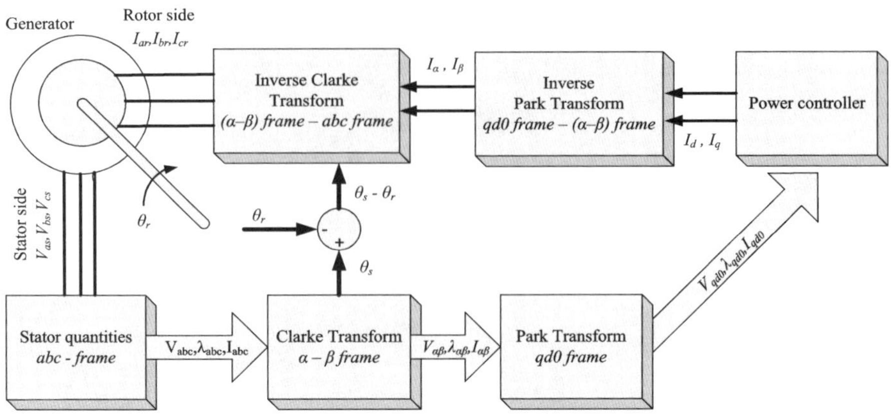

Fig. 10 Block diagram for abc-$\alpha\beta$-qd0 transform used for DFIG

developing a doubly-fed induction generator based wind turbine model are the Clarke and Park transforms. These are two different transformations used to achieve independent active and reactive power control for induction generators. When used in conjunction, these transforms convert stator *abc* quantities to $\alpha - \beta$ quantities (stationary two axis frame also known as $(\alpha - \beta)$ frame—Clarke transform) and eventually to the rotating *qd0* frame (Park Transform) as shown in Fig. 10.

Transformation from *abc* Frame to *qd0* Frame

In the stationary *abc* reference frame, the relationships between the voltages, currents and flux linkages of each phase for an induction machine can be written as follows:

$$\vec{V}_{abcs} = \vec{r}_s \vec{i}_{abcs} + \frac{d(\vec{\lambda}_{abcs})}{dt} \tag{15}$$

$$\vec{V}'_{abcs} = \vec{r}_s \vec{i}'_{abcs} + \frac{d(\vec{\lambda}'_{abcs})}{dt} \tag{16}$$

where, λ is the flux linkage, subscripts s and r stand for variables and parameters associated with the stator and rotor side respectively, Eq.(16) represents machine parameters when referred to the rotor side. The flux linkages in the Eqs. (15)–(16) can be written as

$$\vec{\lambda}_{abcs} = \vec{L}_{sr} \vec{i}_{abcs} + \vec{L}'_{sr} \vec{i}'_{abcr} \tag{17}$$

$$\vec{\lambda}'_{abcr} = \vec{L}'^{T}_{sr} \vec{i}'_{abcs} + \vec{L}'_{r} \vec{i}'_{abcr} \tag{18}$$

The resultant voltage equations from Eqs. (15)–(18) are as follows

$$\vec{V}_{abcs} = \left(\vec{r}_s + \frac{d\vec{L}_{sr}}{dt}\right)\vec{i}_{abcs} + \frac{d\vec{L}'_{sr}}{dt}\vec{i}'_{abcr} \tag{19}$$

$$\vec{V}'_{abcr} = \frac{d\vec{L}'^{\mathrm{T}}_{sr}}{dt}\vec{i}'_{abcs} + \left(\vec{r}'_r + \frac{d\vec{L}'_r}{dt}\right)\vec{i}'_{abcr} \tag{20}$$

As can be seen in Eqs. (19) and (20) voltages, inductances and currents are in the stationary *abc* reference frame. They are thus time-variant. Analysis and modeling of time-variant equations is cumbersome. Using the Clarke and Park transforms these time-variant quantities can be converted into time-invariant quantities. Applying Park transform, the *abc* frame quantities are converted in *qd*0 frame quantities. *qd*0 frame is rotating at the synchronous frequency.

$$\vec{V}_{qd0s} = \vec{r}_s\vec{i}_{qd0s} + \omega_{qds}\vec{\lambda}_{dqs} + \frac{d\vec{\lambda}_{qd0s}}{dt} \tag{21}$$

$$\vec{V}'_{qd0r} = \vec{r}'_r\vec{i}'_{qd0r} + (\omega_s - \omega_r)\vec{\lambda}'_{dqr} + \frac{d\vec{\lambda}'_{qd0r}}{dt} \tag{22}$$

where ω_s and ω_r are the rotational speed of the synchronously rotating *qd*0 frame and rotor frame respectively.

A wound rotor induction machine can be represented in a synchronously rotating *qd*0 reference frame as described above. Assuming that the stator currents are balanced, a resultant stator magnetic field (H_{total}) with a constant magnitude and rotating at synchronous speed (ω_s) is produced [13]. Using Clarke's transform, θ_s can be obtained and *qd*0 frame rotated at synchronous speed ω_s. Now, since the angular speeds of the stator magnetic field and the *qd*0 rotating frame are identical, stator magnetic field vector $\vec{\lambda}_{total}$ is fixed with respect to the *q*- and *d*-axes of the *qd*0 rotating frame. If the *q*-axis of the rotating *qd*0 frame is oriented in such a manner, so that it aligns perfectly with the $\vec{\lambda}_{total}$, field along the *q*-axis would be of zero magnitude. Figure 11 shows MATLAB plots for the stator magnetic field in stationary *abc*, $\alpha\beta$ and rotating *qd*0 frames. Figure 12 shows the alignment of equivalent stator flux along the *q*-axis.

Since, λ_{total} is aligned along the *q*-axis,

$$\lambda_{qs} = \lambda_{total} \tag{23}$$

and

$$\lambda_{ds} = 0 \tag{24}$$

substituting Eqs. (23)–(24) in Eqs. (21)–(22), V_{ds} and V_{qs} are obtained

$$V_{ds} = -\omega_s\lambda_{qs} = \omega_s\lambda_{total} = constant \tag{25}$$

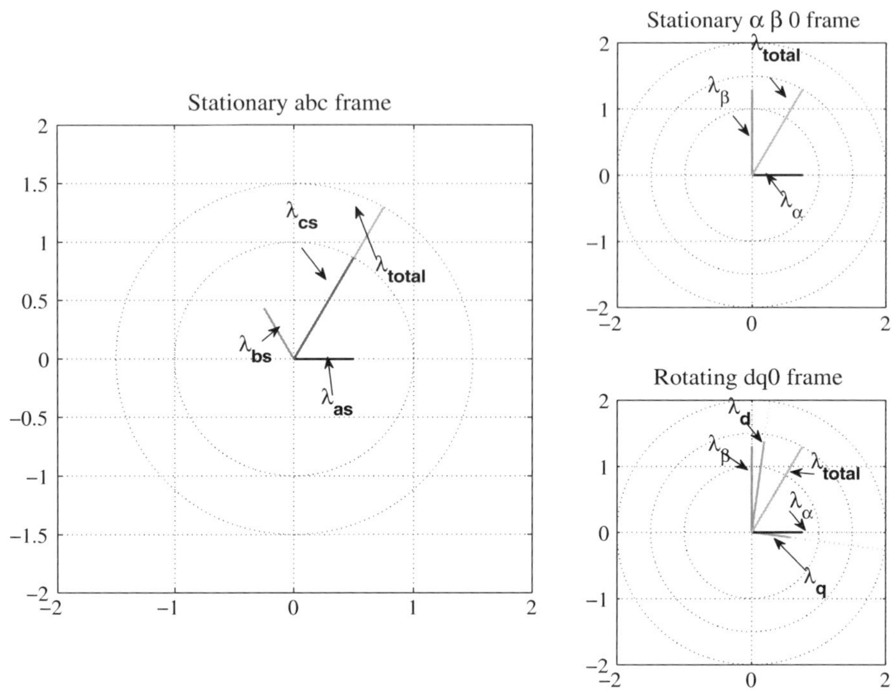

Fig. 11 Transformation from *abc* to *rotating* *qd*0 frame

and

$$V_{qs} = 0 \tag{26}$$

From Eq. (25) speed of the stator field ω_s is constant, hence V_{ds} is time invariant and V_{qs} is almost negligible, with $\lambda_{ds} = 0$, the stator q-axis current can be obtained as

$$i_{qs} = \frac{\lambda_{qs} - L_m i'_{qr}}{L_{ls} + L_M} \tag{27}$$

Similarly, the stator d-axis current can be obtained as

$$i_{ds} = \frac{-L_m i'_{dr}}{L_{ls} + L_M} \tag{28}$$

From these results it can be seen that the stator currents are linearly dependent on the rotor currents. Inductance and flux quantities in Eqs. (27) and (28) are time-invariant, thus the stator $qd0$ axis currents can be controlled by adjusting the rotor q-axis and d-axis currents appropriately. The next step is to show that the real and reactive power output of the machine can be decoupled, and control over real and reactive power can be achieved through controlling rotor q- and d-axis currents

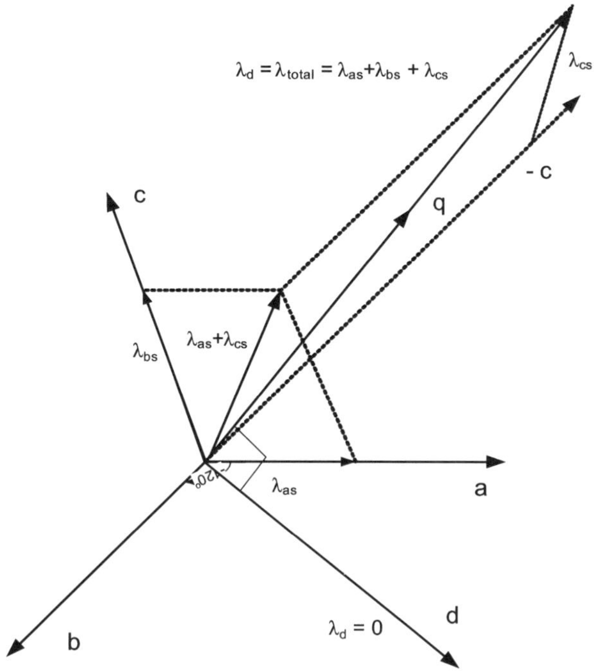

Fig. 12 Aligning equivalent stator flux λ_{total} along q-axis

respectively. The real and reactive power in the stator windings can be derived as follows:

$$S = V_s I_s^* \tag{29}$$

$$V_s = V_{qs} + jV_{ds} \tag{30}$$

$$I_s = I_{ds} + jI_{qs} \tag{31}$$

Thus, apparent power S is given by

$$S = P_s + jQ_s = (V_{qs}I_{ds} + V_{ds}I_{qs}) + j(V_{ds}I_{ds} - V_{qs}I_{qs}) \tag{32}$$

$$P_s = \frac{3}{2}(V_{qs}i_{ds} + V_{ds}i_{qs}) \tag{33}$$

$$Q_s = \frac{3}{2}(V_{ds}i_{ds} - V_{qs}i_{qs}) \tag{34}$$

Since, $V_{qs} = 0$, Eqs. (33) and (34) can be written as

$$P_s = \frac{3}{2}V_{ds}i_{qs} \tag{35}$$

$$Q_s = \frac{3}{2} V_{ds} i_{ds} \tag{36}$$

From Eqs. (35), (36), (27) and (28) active and reactive power equations can be simplified as follows

$$P_s = \frac{-3}{2} \omega_s \lambda_{qs} \left(\frac{\lambda_{qs} - L_m i'_{qr}}{L_{ls} + L_M} \right) \tag{37}$$

$$Q_s = \frac{3}{2} \omega_s \lambda_{qs} \left(\frac{L_m i'_{dr}}{L_{ls} + L_M} \right) \tag{38}$$

From Eqs. (37)–(38) it can be noted that, quantities like λ_{qs}, ω_s, L_{ls}, L_M, L_m are all time invariant quantities, thus Eqs. (37)–(38) can be further simplified as

$$P_s = (k_{ps1} - k_{ps2}) i'_{qr} \tag{39}$$

$$Q_s = k_{qs} i'_{dr} \tag{40}$$

where k_{ps1}, k_{ps2}, k_{qs} are the respective constants of active and reactive power equations. It can be clearly seen from Eqs. (39)–(40) that stator active power P_s can be independently controlled by q-axis rotor current, while stator reactive power Q_s can be independently controlled by d-axis rotor current in an induction machine.

3 Wind Turbine Model Development and Implementation

This section describes wind turbine models developed in PSCAD/EMTDC and MATLAB/SIMULINK platforms. The first model developed was a fixed speed wind turbine model, it produces rated active power at rated wind speed (one wind speed only). As the wind turbine operates at a constant angular speed (rpm), maximum power is obtained at one wind speed only. It should be noted that blade pitch angle is kept constant for the model. Hence, the efficiency of such a wind turbine at varying wind speeds is less. The blade pitch angle for such wind turbines is a preset value, determined by wind speed in the area of installation. The blade pitch angle at which maximum power is obtained varies for different C_p versus λ_r characteristics. For simulation purposes, the rated wind speed was set at 14 m/s, with cut-in speed of 6 m/s and cut-out speed of 20 m/s. The blade pitch angle was set to −6.1667°. With the basic model at hand, fixed speed model is further developed into a variable speed wind turbine model [14]. The advantage of a variable speed wind turbine is that, the torque speed characteristics of the machine can be manipulated, to obtain maximum/rated power at varying wind speed. To put it very precisely, a variable speed wind turbine has larger generator speed variations than the fixed speed wind turbine. It is capable of producing maximum torque, thus maximum power at different generator speeds.

To achieve rated power output above rated wind speed, different control strategies are implemented. Now, rated power above rated wind speed can be produced in two ways

- Pitch control
- Rotor resistance based control

In case of the first method, the operating, blade pitch angle is varied to obtain rated power at any wind speed above rated, this method does not manipulate the torque speed characteristics of the machine. It can be visualized as fixed speed wind turbine, which operates at variable pitch angles achieved by calculating the optimum pitch angle for a given wind speed and output power, and then physically changing the rotor blade pitch angle, while the turbine is in operation. The second method entails, more control over the torque speed characteristics of the machine, as the rotor resistance is varied, generator speed changes, and the machine operates with new torque speed curve depending on the output torque and thus the power requirement.

As discussed earlier, in case of utilizing an induction machine for wind turbines, it can be directly connected to the grid, or through a power electronic converter. When the rotor and stator side of the machine are switched using converters (rectifier and inverter), such a system is called doubly-fed induction generator system (DFIG). Using a DFIG provides independent active (P) and reactive (Q) power control of the machine. When using a variable-speed rotor resistance control or variable-speed pitch control strategy, desired active power can be obtained, but there is no control over the reactive power absorbed or generated by the machine. In further sections, a detailed model development procedure for fixed speed, variable speed and DFIG based wind turbine system has been discussed. Given below is a list of wind turbine models. All models employ induction machine, which is rated at $V_{ll} = 690$ V, S $= 1.8$ MVA. A detailed machine specification including stator and rotor resistances and inductances can be found in the appendix section. The rated power output of the turbine was set to 1.5 MW.

The different configurations of wind turbine models, that were implemented are listed below

- Fixed speed wind turbine model
- Variable speed wind turbine model
  - Rotor resistance control

    Constant power strategy
    Constant current strategy
- Doubly Fed induction generator model

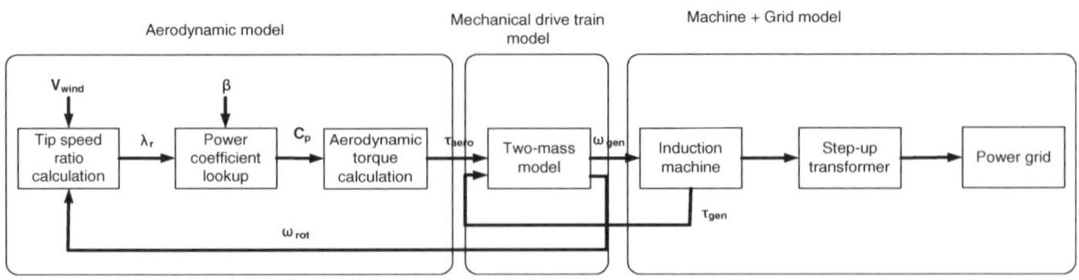

Fig. 13 Generic model for fixed speed wind turbine

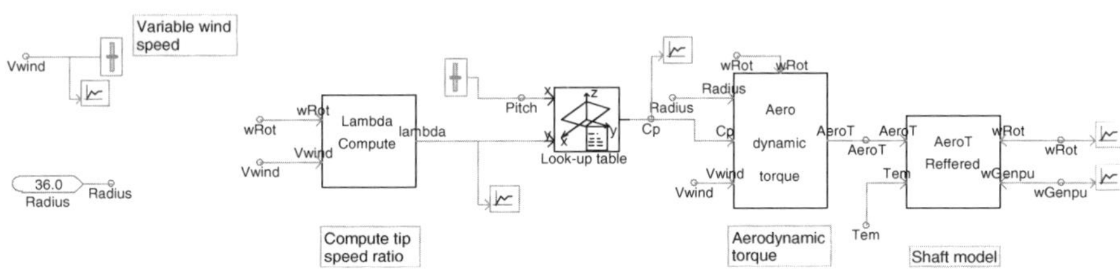

Fig. 14 PSCAD simulation model for a WT

3.1 Fixed Speed Wind Turbine

For fixed speed wind turbine, over the entire wind sweep from cut-in speed of 6 m/s to cut-out speed of 20 m/s, generator speed does not vary much, hence the name *fixed speed wind turbine*. The aerodynamic model developed for a fixed speed wind turbine, is used for all the other wind turbine models. The only function of the aerodynamic model is to provide the speed input to the generator. As the generator speed input varies with wind speed, the power output of the generator varies accordingly. In case of the fixed speed wind turbine, the output power profile builds up with increase in wind speed from cut-in wind speed of 6 m/s, peaks at 14 m/s (rated wind speed) and then drops later due to passive stalling of the rotor blades. Figure 13 shows the specific block diagram representation of the fixed speed wind turbine which is modeled, based on the more general diagram given in Fig. 3. Figure 14 shows the simulation model components in PSCAD.

3.1.1 Fixed-Speed Wind Turbine Model in PSCAD/EMTDC

As can be seen from the block diagram, wind speed variable V_{wind} and rotor speed ω_{rot} are used to compute the tip speed ratio (λ_r) given by the relation in Eq. (9), where $R = 36$ m. Using λ_r and the a blade pitch angle as inputs to a lookup table, corresponding value of power coefficient (C_p) is computed. The relation between C_p, λ_r and blade pitch angle (β) used for all models are given in Eqs. (10) and (11). The C_p versus λ_r characteristics obtained are shown in Fig. 5. Once, C_p is

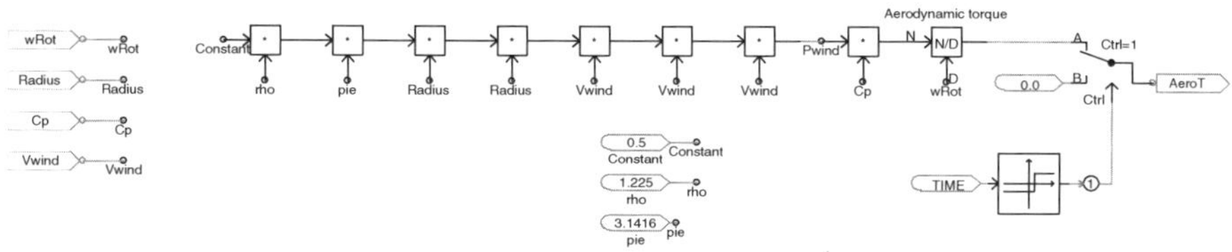

Fig. 15 Aerodynamic torque computation

obtained, it is then used by the aerodynamic torque calculation block to calculate the instantaneous aerodynamic torque of the rotor. Shown in Fig. 15 is the internal block diagram of the aerodynamic torque calculation block.

Since, the rotor shaft is a low speed shaft rotating between 15 and 20 rpm, a gear train has to be included, which is then connected to the high speed shaft of the induction generator rotating at a base frequency of 125.667 rad/sec, for a 6 pole machine converts to 1,200 rpm. To model the gear train, incorporating the rotor slow speed shaft, gear train and generator high speed shaft as a two-mass model. The two-mass model is governed by three differential equations, Eqs. (41), (42) and (48) and a gear ratio of $a = 70$ was assumed. A detailed list of constants used for modeling the gear train and the rotor, generator shafts is provided in the Appendix section. Given below is the set of differential equations used to model the gear train and rotor generator shafts as a two-mass.

$$\frac{dX_1}{dt} = \omega_{rotr} - \omega_{gen} \tag{41}$$

$$\frac{d\omega_{rotr}}{dt} = \frac{\tau_{aeror} - B_{eqv}(\omega_{rotr} - \omega_{gen}) - K_{eqv}X_1}{J_{rotr}} \tag{42}$$

$$\frac{d\omega_{gen}}{dt} = \frac{-\tau_{gen} - B_{eqv}(\omega_{rotr} - \omega_{gen}) + K_{eqv}X_1}{J_{gen}} \tag{43}$$

$$J_{rotr} = \frac{J_{rot}}{a^2} \tag{44}$$

$$B_{eqv} = \frac{B_{rot}}{a^2} + B_{gen} \tag{45}$$

$$K_{eqv} = \frac{\frac{K_{rot}}{a^2} K_{gen}}{\frac{K_{rot}}{a^2} + K_{gen}} \tag{46}$$

Equations (41), (42) and (48) can be solved with initial conditions of the integrator set for $\omega_{rot} = \omega_{gen} = 125.66\,\text{rad/sec}$ from Eq. (47), (48). τ_{aeror} is the aerodynamic torque referred to the generator shaft, obtained by dividing it by the gear ratio. J_{rotr} is the moment of inertia of the rotor referred to the generator shaft. Electromagnetic torque output of the induction machine τ_{gen} is converted from its per unit equivalent by multiplying it by the rated generator torque = 15,914.67 Nm

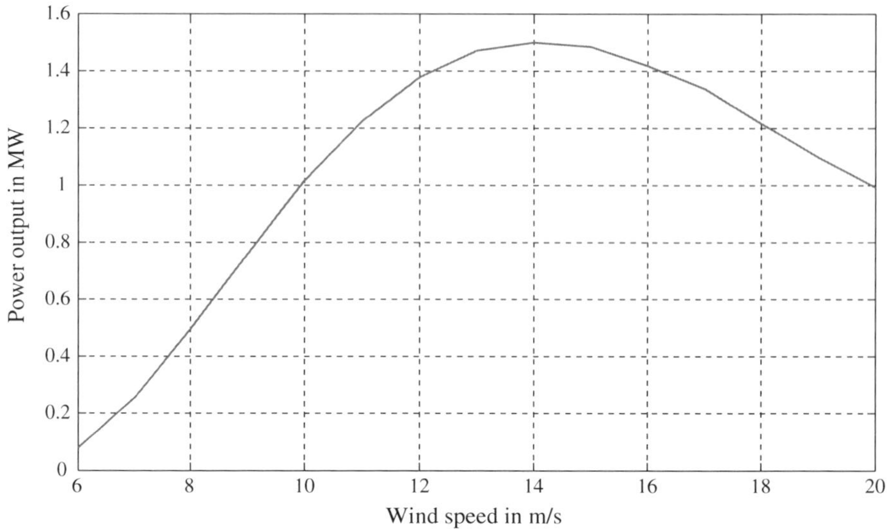

Fig. 16 Power profile for fixed speed wind turbine model

[refer Eq. (12)]. A negative value of τ_{gen} is applied to the two-mass model, as it operates against the rotor torque. Before feeding ω_{gen} to the machine, it is converted to its per unit equivalent by dividing it by the rated speed of 125.667 rad/sec.

$$N = \frac{120f}{P} \tag{47}$$

$$\omega_{gen} = \frac{2\pi N}{60} \tag{48}$$

where N is the generator speed in rpm, P is the number of poles and f is the synchronous frequency.

The PSCAD machine model is directly connected to the grid. A step up transformer connected in *delta–wye* is used to connect the stator terminals to a three-ϕ voltage source (representing the grid). Once the model is ready, the blade pitch angle has to be set to obtain rated power of 1.5 MW. It was observed that, a maximum power of 1.5 MW was obtained at $\beta = -6.166°$. β is then kept fixed and rated wind speed is set at which machine outputs 1.5 MW. The model is then run at wind speed ranging from 6 to 20 m/s. The power profile for the model is then obtained as shown in Fig. 16.

A look at the power profile of the wind turbine shows that rated power of 1.5 MW is obtained at a fixed wind speed of 14 m/s and, fixed pitch of −6.166°. As the wind speed varies power produced varies roughly as the cube of the wind speed. At rated wind speed the electrical power generated becomes equal to the rating of the turbine, and then stalling takes place above the rated wind speed. This is achieved by making use of post-stall reduction in lift coefficient and associated

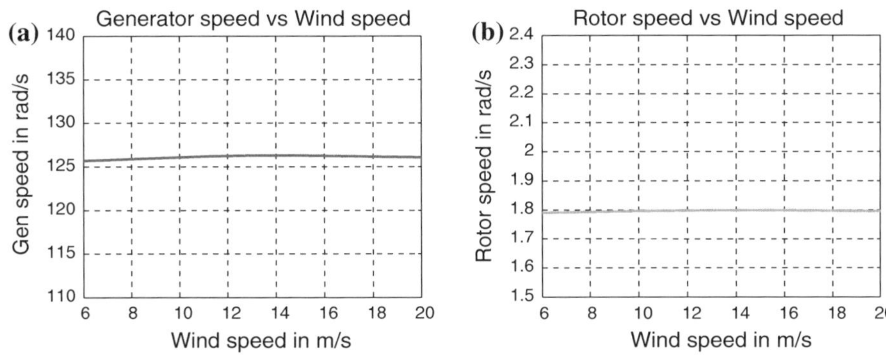

Fig. 17 Rotor and generator speed variation with wind speed. **a** Generator speed variation with wind speed. **b** Rotor speed variation with wind speed

increase in drag coefficient, which places a ceiling on the output power as wind speed increases. As can be seen from Fig. 16 power output of the generator falls below 1.5 MW at any wind speed above 14 m/s. It can be also noted that, the output of the generator drops significantly almost 0.079 MW at a wind speed of 6 m/s. This stalling of the wind turbine is attributed to the increase in angle of attack as wind speed increases, and increasingly large part of the blade enters the stall region. The stalling effect reduces the rotor efficiency and puts a cap on the output power. Stall regulated machines generally suffer from the disadvantage of uncertainties in aerodynamic behavior post-stall, which can result in inaccurate power levels and blade loading at rated wind speed and above.

For a fixed speed wind turbine, rotor speed and thus the generator speed variation as wind speed varies are very less. As can be seen from Fig. 17a, b, the generator speed reaches a maximum of 126.281 rad/sec at 14 m/s and then decreases due to passive stalling. The overall slip variation is a maximum of $-0.49\ \%$. Results obtained from the PSCAD model are later compared with those from a similarly developed MATLAB/SIMULINK model to demonstrate that the model can be implemented in different platforms.

A torque *Vs* slip plot for the model shows that, the torque rise is very steep. As the wind speed increases, the generator speed does not increase, as shown in Fig. 17a it attains a maximum value at 14 m/s. The generator torque achieves maximum value at 14 m/s and then drops above rated speed. Figure 18 shows the torque slip characteristics of the machine during the entire wind speed sweep from 6 to 20 m/s.

3.1.2 Fixed-Speed Wind Turbine Model in MATLAB/SIMULINK

For the purpose of demonstrating the generality of the model shown in Fig. 13, the results obtained from the PSCAD/EMTDC model of fixed speed wind turbine are compared with those from a parallel MATLAB/SIMULINK model. A similar approach for modeling the wind turbine was used. Initially an aerodynamic model

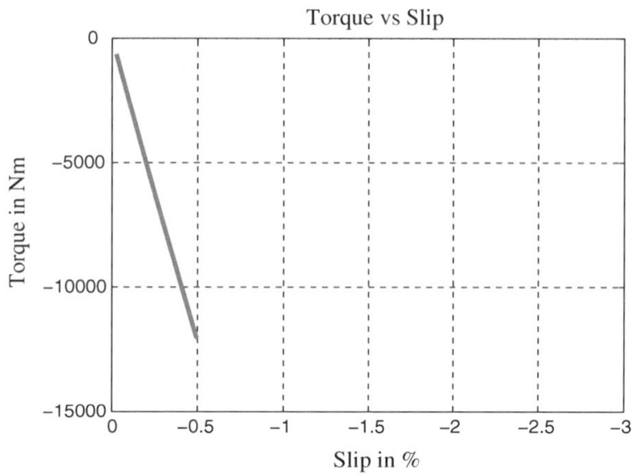

Fig. 18 Torque slip characteristics

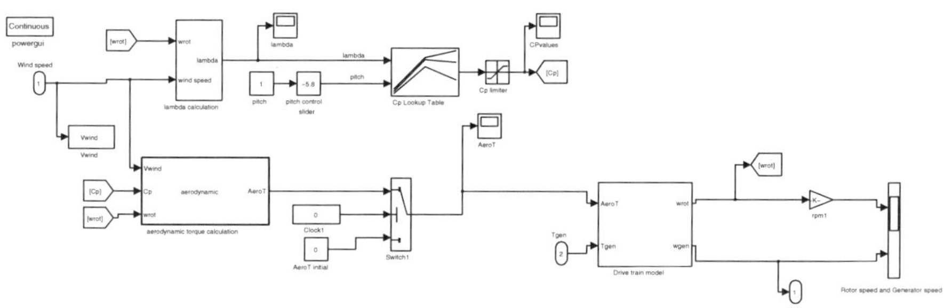

Fig. 19 Aerodynamic model in SIMULINK

simulating the rotor blades, rotor shaft, gear train and generator shaft was modeled in SIMULINK. Generator speed output ω_{gen} was then fed to a induction machine model, working in the squirrel cage mode (rotor circuit shorted). The built-in machine model in SIMULINK provides a number of options for machine specifications. The machine can be customized very well according to the model demand. It provides a greater depth in terms of setting the reference frame with options of stationary, synchronous and rotor frames. The machine stator is connected to a three phase RLC voltage source through a *delta-wye* connected 0.69/34.5 kV step up transformer. X/R ratio for the voltage source is set at 10. Figure 19 shows the internal block diagram for the aerodynamic model developed in SIMULINK. The results from both the models were seen to match closely.

The aerodynamic model in Fig. 20 is provided a ramp input for wind speed, the simulation time is set for 100 s with a ramp rate of 0.14 m/s^2, with an initial wind speed of 6 m/s, which means that at 100 s, the wind speed reaches a peak of 20 m/s. The rotor blade pitch angle is set to $-5.8°$ and an empirical lookup table is used to determine C_p using tip speed ratio (λ) and pitch angle (β) as inputs. Once, C_p is

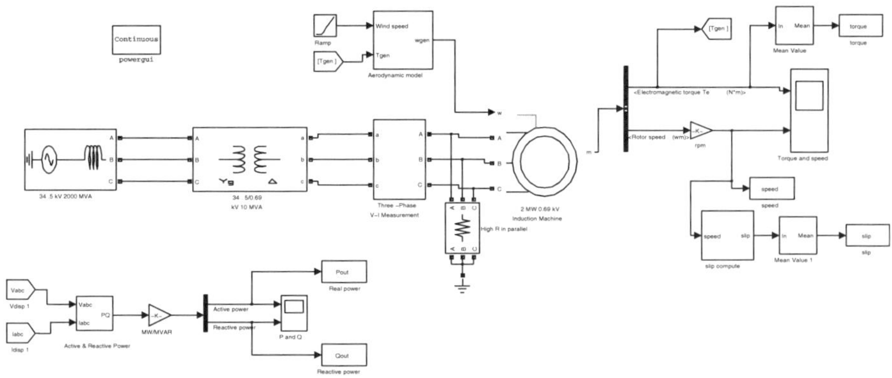

Fig. 20 Fixed speed wind turbine model in SIMULINK

determined a subsystem block calculating the power available from the wind uses C_p to determine the power extracted from the wind and thus, the aerodynamic torque of the rotor blades.

This input is then used in the two-mass model of the rotor shaft, gear train and generator shaft to solve the differential equations for generator speed (ω_{gen}) and rotor speed (ω_{rot}). The generator speed (ω_{gen}) thus obtained is then fed to the built in asynchronous machine model in SIMULINK, with rotor circuit shorted and neutral grounded (squirrel cage mode). The electromagnetic torque (T_{em}) obtained from the machine model is then fed back to the two-mass model. Figure 20 shows the entire SIMULINK model of a fixed speed wind turbine, a multimeter measuring the stator voltage and currents is used, whose outputs are V_{abc} and I_{abc}. The voltage and currents measured are used to determine the active and reactive power flowing out of the stator circuit of the machine.

The active and reactive power profile obtained from the SIMULINK model of the turbine are shown in Fig. 21. It can be seen that the active power peaks at 14 m/s, while there is no control over the reactive power generated. It remains negative, indicating that the machine constantly absorbs some reactive power from the grid. A typical fixed speed wind turbine power profile is obtained from the model, which is comparable to the PSCAD model described earlier in the previous section. After the wind speed crosses its rated value, the active power output of the machine drops to almost 1.2 MW.

A drop similar to the active power output of the machine can be seen in the torque profile too (see Fig. 22), at 14 m/s T_{em} reaches a peak value of approx −12 kNm, as the wind speed increases beyond 14 m/s T_{em} drops. A look at the speed profile of the machine, shows that it does not vary much during the entire wind speed sweep from 6 to 20 m/s it holds almost a constant value above 1,200 rpm and reaches a high of approx 1,205 rpm, when the wind speed reaches its rated value of 14 m/s. Overall, ω_{gen} variation with wind speed is very less. It can be observed from the torque slip characteristics of Fig. 23, that slip reaches a maximum value of −0.49 % at a maximum torque of approx −12,000 Nm. Thus, a peak torque (producing peak active

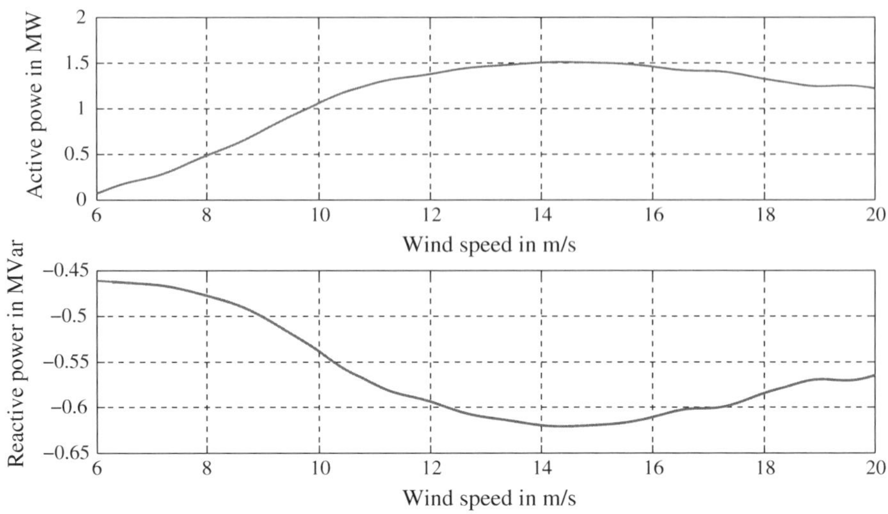

Fig. 21 Active and reactive power profile for fixed speed wind turbine model in SIMULINK

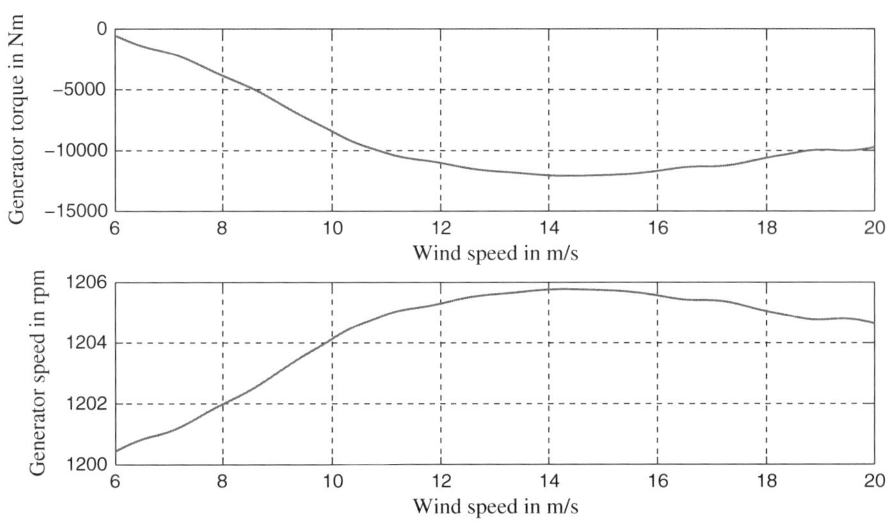

Fig. 22 Generator torque and speed profile in SIMULINK

power) occurs at only one slip or speed value of the induction machine. The working and functional characteristics of a fixed speed wind turbine have been shown through models built on two different platforms. The results obtained are similar in many respects. The power profile, torque, speed and torque-slip characteristics obtained clearly show the stalling effect after rated wind speed. With the two models at hand, working of a fixed speed wind turbine and further grid integration of such a turbine can be studied in detail.

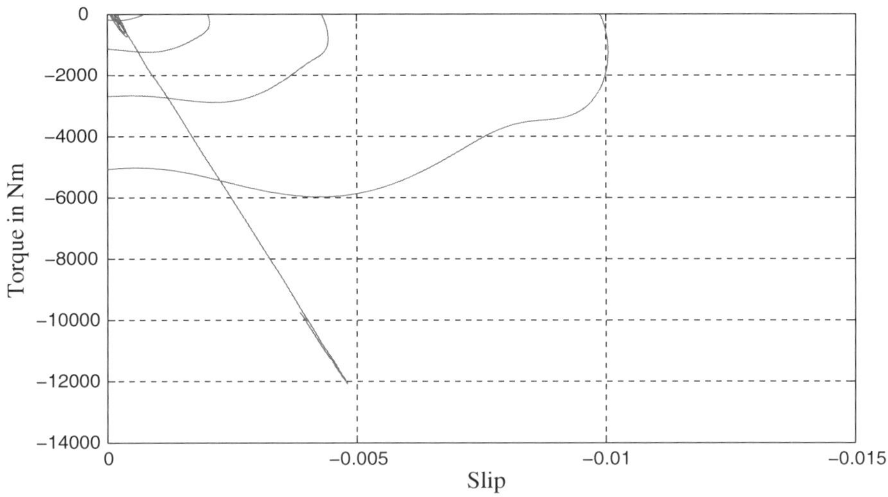

Fig. 23 Torque slip characteristics of induction machine in SIMULINK

3.2 Variable Speed Wind Turbine

As described in the previous section, for a fixed-speed wind turbine, there is no active control over the power output of the machine, once the rotor blade pitch angle is set. In variable-speed machines however, it is possible to control the output power using torque control. Various possible torque control methods exist to achieve constant power output above rated wind speed in variable-speed wind turbines. Two of these torque control methods are implemented in this chapter. These are:

- Aerodynamic torque control
- Generator torque control

Aerodynamic torque output from the rotor is determined by tip speed ratio and C_p, rotor geometry (blade pitch and aileron settings), wind speed, yaw error and rotor drag. Since, there is no control over the wind speed, other parameters have been used to control the aerodynamic torque. Any change in the tip speed ratio changes the rotor efficiency thus changing the rotor torque. A change in rotor geometry i.e. varying the rotor pitch angle results in a change in lift and drag forces thereby changing the torque output. Pitching the blade can regulate the torque output either by reducing the angle of attack or increasing it, as in case of stalling. Rotor blades for pitch-regulated wind turbines are designed to operate at maximum efficiency (maximum power production) for relatively high angles of attack (Fig. 24). At these high angles of attack, change in rotor blade position (typically moving the turbine into the stall region) is accomplished with more exact control, is faster and results in a quieter overall operation. The downside is that inducing

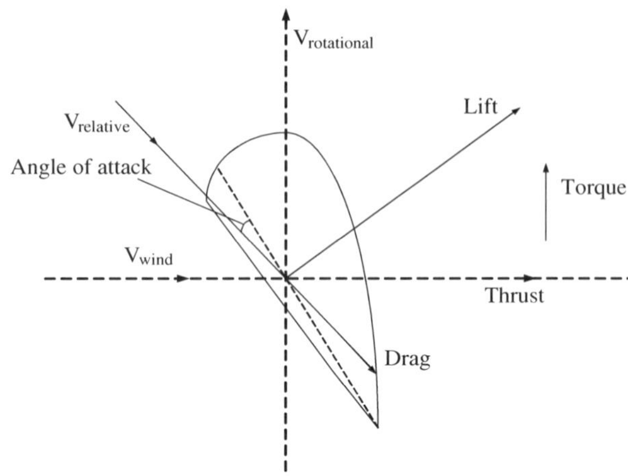

Fig. 24 Blade geometry of horizontal axis wind turbine [12]

stalling from the very start results in unsteady loads, less accuracy in control, and greater thrust on the turbine due to unsteady nature of the stalled flow.

In case of generator torque control, torque of the generator can be either changed through the design characteristics or by the use of power converters. As demonstrated in the fixed speed wind turbine model, grid connected generators operate over a very small or no speed range and provide the required torque at or near synchronous speed, which depends on the type of machine (induction or synchronous). For a grid connected induction generator change in ω_{gen} is a small percentage of the synchronous speed, this results in low torque spikes and softer response. In contrast, for a synchronous generator any forced torque change results in an instantaneous compensating torque, which can result in higher torque and power oscillations.

An induction generator can very rapidly achieve any desired value of target torque by the use of a power converter. The converter determines the frequency, phase angle and value of the currents to be injected into the machine windings, this allows the machine to be set to any desired value of torque, thus controlling the power output of the generator.

3.2.1 Pitch Control

As explained above, aerodynamic torque control can be achieved by changing the rotor blade geometry (blade pitch angle β) for varying wind speeds. Pitch control is somewhat analogous to steam governor action in a synchronous machine, as both mechanisms control the mechanical input power to the generator. It can be visualized as fixed speed operation with an optimum pitch angle to produce maximum power at any wind speed above rated.

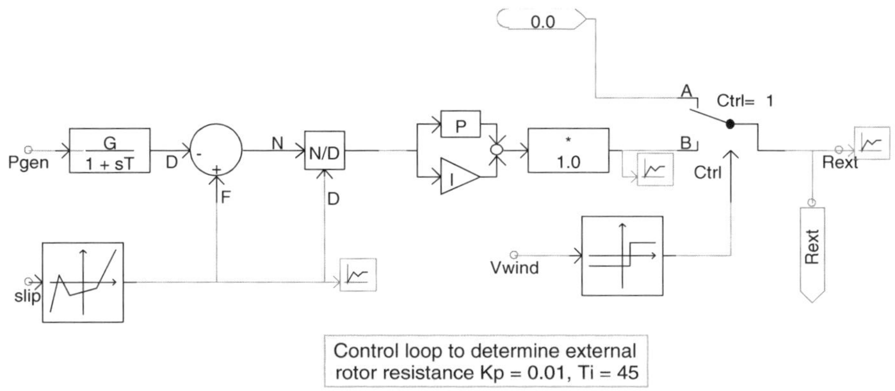

Fig. 25 R_{ext} estimation module in PSCAD using PI controller

3.2.2 Rotor Resistance Control

This section describes the simulation results of a variable speed wind turbine using PSCAD/EMTDC. PSCAD has been used to model and simulate the turbine. A built in machine model of a wound rotor induction machine is used to implement constant power and constant current strategy. The rotor pitch angle is set to −6.483° (rated pitch) to obtain a maximum power output of 1.5 MW at 14.2 m/s (rated wind speed). The wind turbine uses a 6-pole, 690 V, 1.8 MVA wound rotor induction machine as a generator.

Constant power strategy to maintain constant power above rated wind speed

Constant power strategy aims to maintain a constant power output of the WTGS above rated wind speed in the stall region. It was observed for the fixed speed WTGS, that output power falls as wind speed exceeds the rated value. With the use of PI controllers, rotor resistance of induction machine can be varied in such a manner, that active power output remains constant. To maintain constant active power output, reference value of active power is compared with actual power generated. The error signal is then fed to a PI controller. The output of the PI controller is the new value of single phase rotor resistance. Rotor resistance value thus calculated is equal for all three-phases. To obtain a rated slip of 2.25 % internal rotor resistance of 0.048 Ω is included in the rotor circuit. Figure 25 shows the PSCAD block diagram for the PI control implemented [14].

A rated slip of 2.25 % is obtained at 14.2 m/s and 1.5 MW output power. As the wind speed increases above the rated wind speed output power of the generator tends to fall. To maintain the output power constant calculated value of R_{ext} is included in the rotor circuit, to increase the torque and thus the output power. To calculate the exact value of R_{ext}, actual generated power is compared with rated power (1.5 MW) and the corresponding error is converted into per unit (base as

Table 1 Look-up table for PI controller tuning

Parameter	Rise time	Settling time	Overshoot	Steady-state error
K_p	Decrease	Small change	Increase	Decrease
K_i	Increase	Increase	Increase	Eliminate
K_d	Indefinite	Decrease	Decrease	None

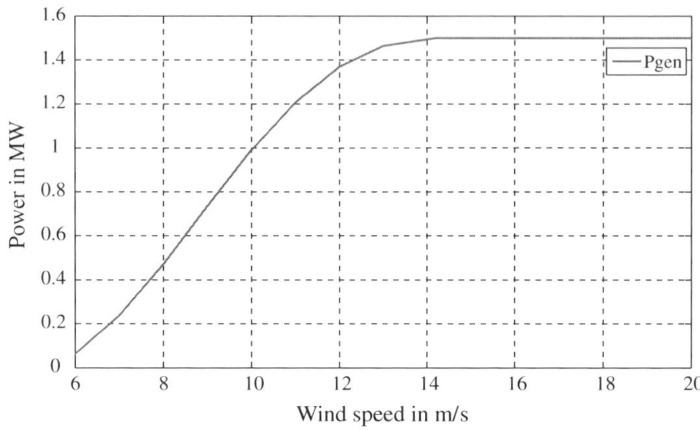

Fig. 26 Wind power profile for a variable speed wind turbine using rotor resistance control

rated power) and error signal is fed to the PI controller. Once the output of the PI controller converges, R_{ext} is obtained.

Ziegler Nichols tuning algorithm [15] was used to tune the PI controller. Tuning of the PI controller was done as follows:

1. Critical gain K_c was found by setting a very high value of integral time constant $T_i = 10^6$ s. At critical gain $K_c = 0.026$ the output of the PI loop starts to oscillate sustainably, below K_c the output just manages to converge and achieve a constant value. Further integral gain K_i can be calculated using the formula shown. $K_i = \frac{1.2 \times K_c}{P_c} = 0.0226$ where $K_p = 0.45 \times K_c = 0.011$ and $P_c = 0.6$ s is the oscillation period of the PI controller output. $T_i = \frac{1}{K_i} = 45$ s.
2. Ziegler Nichols method is an iterative process. Using the values obtained above as initial values, further fine tuning of the controller was achieved following the Table 1. The table shows the effect of increasing or decreasing the proportional and integral gain K_p and K_i and was effectively used as guide in fine tuning the PI controller.

It is clear from the above Fig. 26 that power remains constant above rated speed of 14.2 m/s. The real power excursions with step changes in wind speed from 14.2 m/s (rated) to 15.2 m/s to 16.2 m/s are obtained using PSCAD model. Graph in Fig. 27 shows the real power excursions while the wind speed is changed in

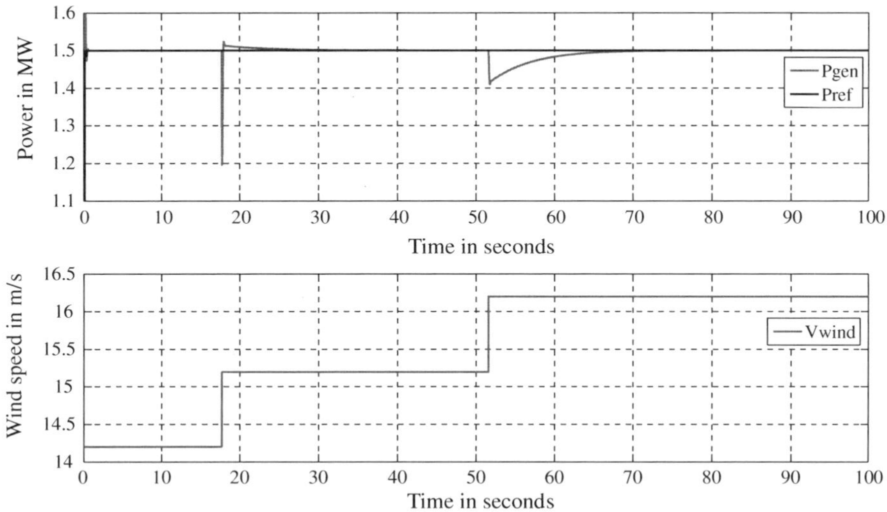

Fig. 27 Power excursion: constant power strategy using PI control from 14.2 to 15.2 to 16.2 m/s

steps of 1 m/s, starting from 14.2 to 16.2 m/s. It can be seen that an undershoot for a wind speed change from 14.2 m/s(rated) to 15.2 m/s is quite large (1.5–1.2 MW). It can be attributed to the proportional gain and integral gain of the integrator. The PI controller when tuned for low undershoot and overshoot increases the settling time. It is also evident from the graph that the undershoot is very less (1.5–1.41 MW) for a step change from 15.2 to 16.2 m/s.

Power excursion for change in wind speed from 18 to 19 m/s (see Fig. 28) is more than that for 15.2–16.2 m/s, with a variation from 1.5 to 1.38 MW. It can also be seen that the undershoot increases over further step changes from 18 to 19 m/s (1.5–1.33 MW) and 19–20 m/s (1.5–1.26 MW) simultaneously the settling time goes down. Better results with reduced undershoots are demonstrated by the PID controller, at the cost of increased settling time, which seems like a reasonable trade-off.

It is observed that the power output of the generator is maintained constant at 1.5 MW above rated wind speed. As the slip increases, the frequency of rotor current I_{rrms} increases.

Torque slip characteristics in operating region

The torque slip characteristics plotted in Fig. 29 shows a large variation of slip from 6 to 14.2 m/s. This increased slip variation is attributed to the internal rotor resistance $R_{int} = 0.048\ \Omega$ and can be changed to vary the rated slip from 2 to 2.5 %. As the slip variation is large over a range of increasing electromagnetic torque (negatively), the wind turbine can be controlled using PI controller to achieve a constant power output of 1.5 MW.

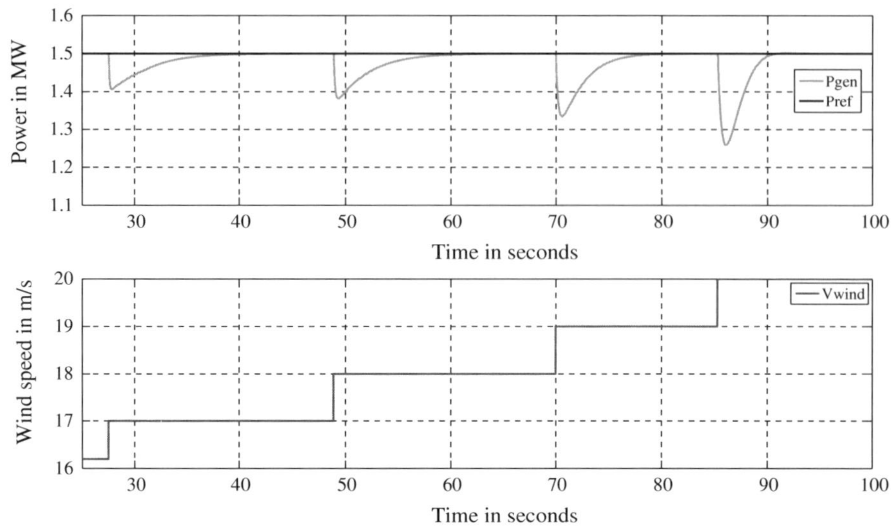

Fig. 28 Power excursion: constant power strategy using PI control from 17 to 18 to 19 to 20 m/s

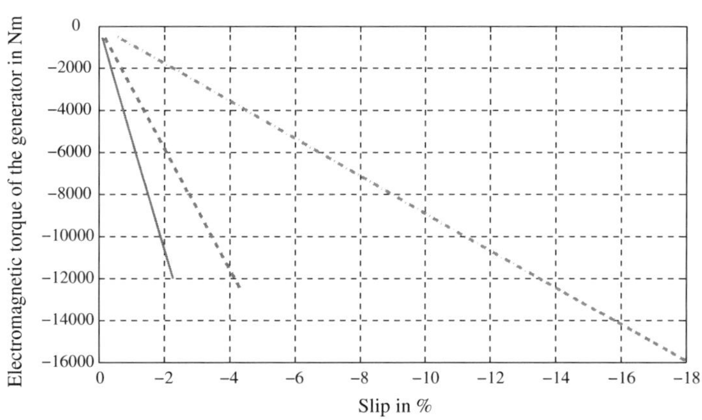

Fig. 29 Torque slip characteristics with varying R_{ext} at 8, 16 and 20 m/s wind speed

Another way of plotting the torque speed characteristics is by the use of electromagnetic torque formula as follows (Fig. 30),

$$\tau_e = \frac{3}{\omega_s} \cdot \frac{V_s^2}{\left(R_s + \frac{R_{rs}}{s}\right)^2 + (X_{ls} + X_{lrs})} \cdot \frac{R_{rs}}{s} \tag{49}$$

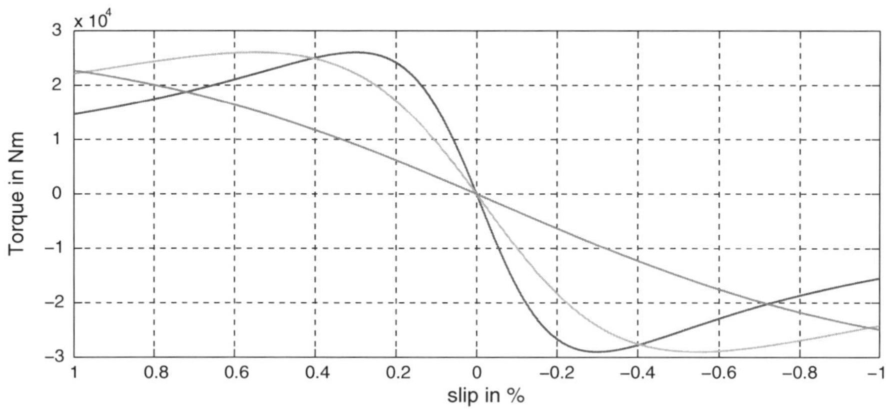

Fig. 30 Complete torque slip characteristics with $R_{ext} = 0, 0.039, 0.22\ \Omega$ at 8, 16, 20 m/s wind speed respectively

where

τ_e	Electromagnetic torque of the machine
V_s	Generator terminal voltage
ω_s	synchronous speed
R_s	stator resistance
R_{rs}	Rotor resistance referred to the stator side
X_{ls}	Stator leakage reactance
X_{lrs}	Stator leakage reactance referred to the stator side

To improve the transient response of the machine, for step changes in wind speed, a PID controller can also be used. PID controllers are tuned in a similar fashion as demonstrated for the PI controller.

A differentiator is included in the PI control loop to reduce the overshoot and settling time, with a flip side of increased system instability. Results obtained for the PID controller show improvements not only on reduced undershoot and no overshoot, but also highly accurate steady state value. As can be seen from the power excursion shown in Fig. 27 the undershoot goes as low as 1.2 MW and an overshoot of 1.52 MW. Opposed to this, the power excursion of the PID controller has improved undershoot of 1.38 MW and overshoot totally eliminated. This improvement can be attributed to the inclusion of the differentiator and reduced values of integrator time constant.

As visible from the Fig. 31, overshoot is almost eliminated and undershoot reduced with the use of PID controller. For a step change in wind speed from 14.2 to 15.2 m/s, the output falls till 1.38 MW and then tracks the rated value. Similarly, the undershoot for a change in wind speed from 15.2 to 16.2 m/s has been reduced to 1.41 MW.

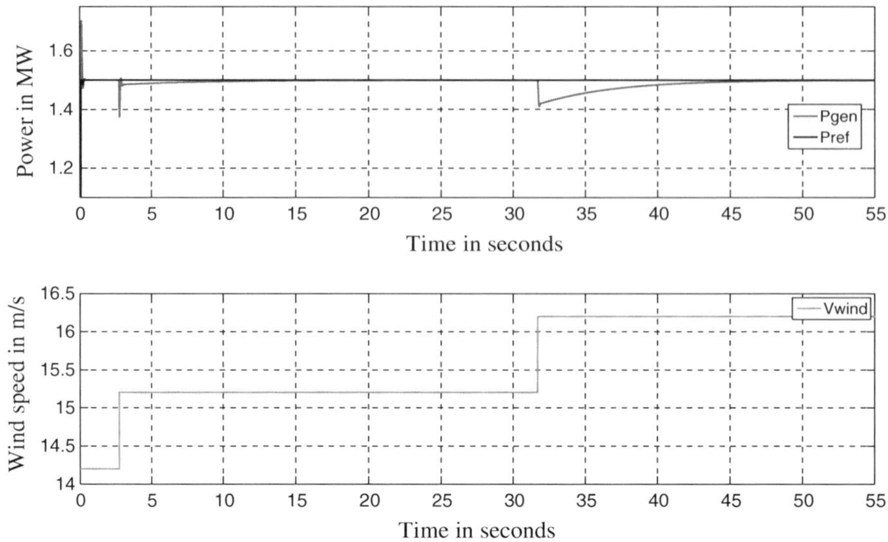

Fig. 31 Power excursion: constant power strategy using PID control from 14.2 to 15.2 to 16.2 m/s

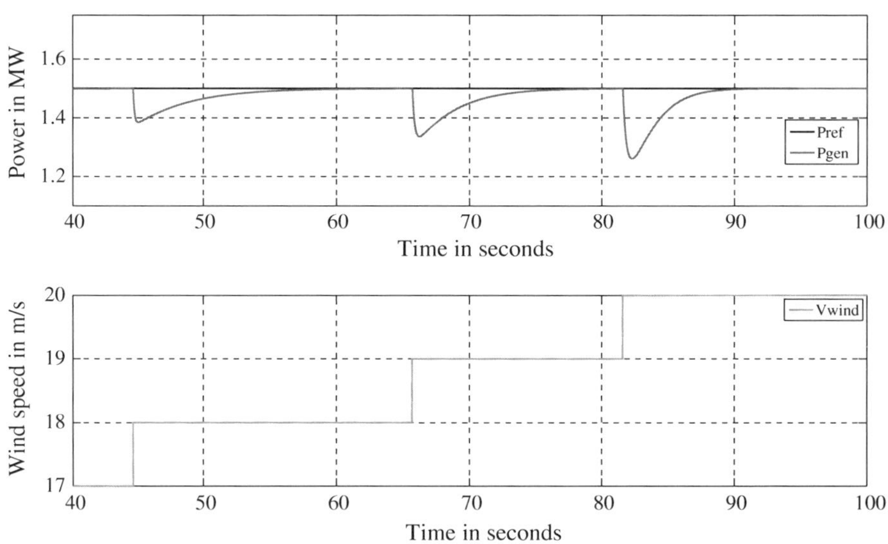

Fig. 32 Power excursion: constant power strategy using PID control from 17 to 18 to 19 to 20 m/s

A similar power excursion profile can be seen in Fig. 32 for step change from 17 m/s to 18 m/s to 19 m/s to 20 m/s. The profile obtained is similar to that obtained with PI controller. Thus, a PID controller proves to be better in reducing the overshoots than the PI controller.

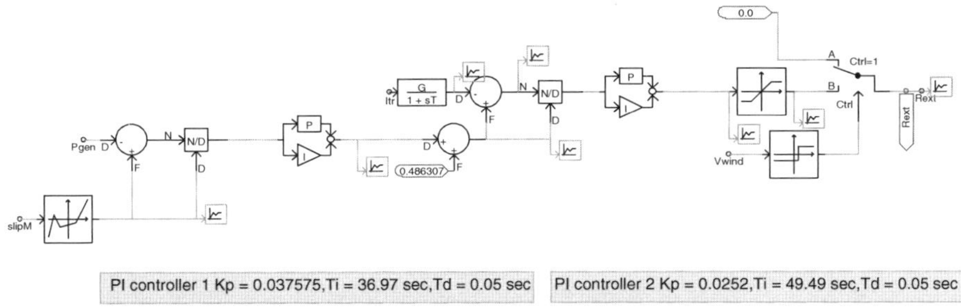

Fig. 33 R_{ext} estimation module in PSCAD using built-in PI controller

Constant current strategy to maintain constant power above rated wind speed

Another method to implement variable speed wind turbine model is constant current technique. In this method rotor current is not allowed to fluctuate beyond a bandwidth. Initially, constant current was implemented using an error signal obtained from the $I_{rref} - I_{ract}$. Where, I_{rref} is the rms value of rotor current at rated wind speed (with rated pitch) and I_{ract} is the rms value of rotor current at any wind speed above rated. As we try to maintain the current in the rotor same as the rated value of rotor current, it was observed that the power output falls, above rated wind speed. The total variation in output power was 1.5–1.45 MW, for a range of wind speed from 14.2 to 20 m/s. Though the rotor current was maintained constant at rated value, the error current method did not provide a constant output power as desired for a variable speed wind turbine. This is due to the phase angle of the rotor current, which was not accounted, while maintaining constant current (only magnitude was considered).

To maintain output power constant, another error loop of $P_{ref} - P_{act}$ was added to the overall control loop calculating R_{ext}, where $P_{ref} = 1.5$ MW and P_{act} is the actual power generated at any wind speed above rated. Error signal is then given to the PI controller. Output of the PI controller when added to the $I_{rated} = I_{ref}$. This I_{ref} is then compared with I_{act} and output fed to the PI controller. Output of the PI controller forms the R_{ext} estimated. This method works, and is found to converge to constant power output eventually. Wind power profile remains the same as shown in Fig. 26), since the output power converges to 1.5 MW for all wind speeds above rated.

Initially a PI controller is used to implement constant current strategy. Tuning of the cascaded PI controllers in done using the same tuning algorithm as demonstrated for constant power strategy.

Value of the integral time constant set for PI loop 1 (Fig. 33) is not used as the final value. This is chiefly because, the final value of T_i (integral time constant) also depends on proper tuning of PI loop 2 (Fig. 33).

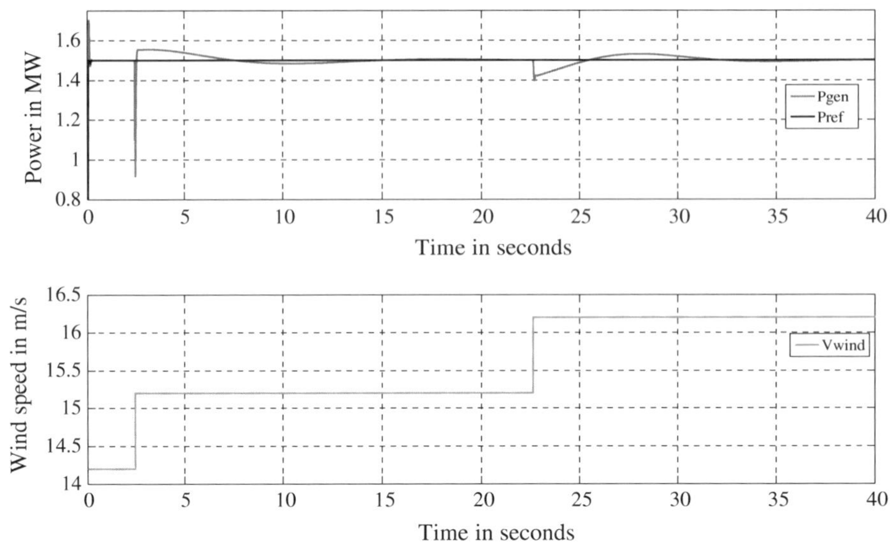

Fig. 34 Power excursion: constant current strategy using PI control from 14.2 to 15.2 to 16.2 m/s

Power excursions for step change in wind speed from 14.2 to 15.2 to 16.2 m/s has been shown in Fig. 34. The settling time increases, with increased oscillations before the output settles.

The undershoot in this case is quite high 0.5 MW, while you can also see an overshoot till 1.9 MW. This is mainly because, there are two PI controllers, minimizing error in power output and then rotor current to obtain R_{ext} so as to achieve constant power output of 1.5 MW. The power excursion for step change in wind speed from 15.2 to 6.2 m/s is far better, as the undershoot and overshoot is very less 1.4 and 1.55 MW respectively.

Though the power output converges after oscillating for some time, still the power excursions obtained using constant current strategy are worse as compared to constant power strategy. While obtaining a reference current for each wind speed, the rotor current oscillates a lot more as compared to the rotor current oscillations in constant power strategy.

Power excursion shown Fig. 35 is for step change in wind speed from 17 m/s to 18 m/s to 19 m/s to 20 m/s. With a step change in wind speed from 17 to 18 m/s, the output power oscillates from 1.38 to 1.57 MW, similarly larger oscillations can be seen for step wind speed change from 18 to 19 m/s and from 19 to 20 m/s. Even with increased oscillations, the power output converges and settles down to a constant value of 1.5 MW after some time.

Comparing the two methods used to achieve constant power, constant power strategy is clearly better than constant current. It is mainly due to reduced oscillations during power excursion for step changes in wind speed. Even the rotor current oscillations are less for constant power strategy. The phase angle of the

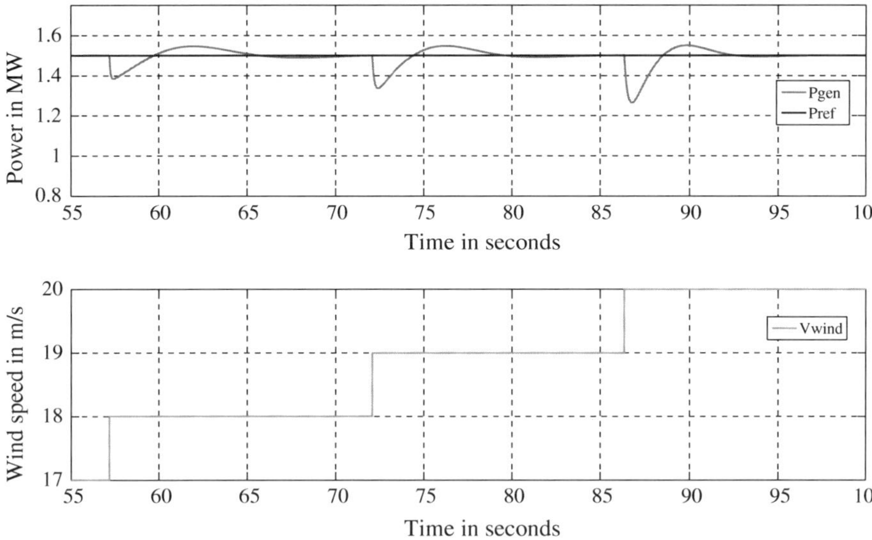

Fig. 35 Power excursion: constant current strategy using PI control from 17 to 18 to 19 to 20 m/s

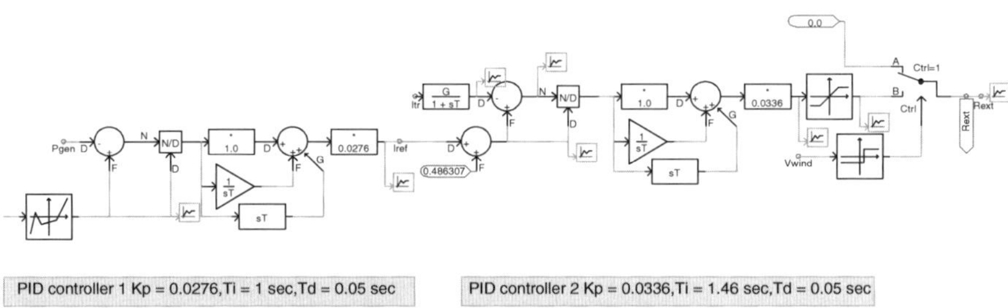

Fig. 36 R_{ext} estimation module in PSCAD using PID controller

rotor current has not been accounted for, while keeping the current magnitude constant. As we are considering only the current magnitude, results obtained using constant current strategy are not as accurate, as obtained using constant power strategy. To improve the accuracy of rotor resistance estimation, PID controllers are employed, it can be seen from the results obtained, that overshoots and undershoots have been reduced. Tuning of the PID controller is difficult as compared to the PI controller. Mostly because, inaccurate gain K_d (differential gain) can make the system unstable (Fig. 36).

Value of the integral time constant set for PID loop 1 was not used as the final value. This is chiefly because, the final value of T_i also depends on proper tuning of PID loop 2.

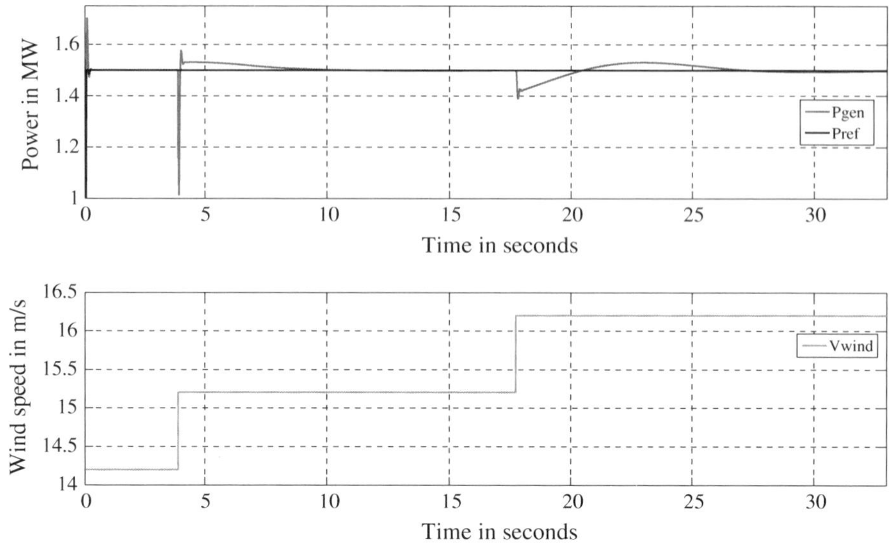

Fig. 37 Power excursion: constant current strategy using PID control 14.2 to 15.2 to 16.2 m/s

Power excursions obtained for the PID controller are better than the PI controller, the undershoot has reduced and the output falls till 1.23 MW for step change in wind speed from 14.2 to 15.2 m/s. Similarly, the oscillation period and the settling time of the power output has also been reduced for the change in wind speeds shown in Fig. 37.

It is fair to conclude that the performance of self built PID controller is better than built in PI controller, even during wind speed change from 17 to 18 to 19 to 20 m/s there are less oscillations. Figure 34 can be compared with Fig. 37, while Fig. 35 can be compared with Fig. 38 to observe the difference, still the output fluctuates far more as compared to the constant power strategy implemented previously.

Constant power strategy and constant current strategy were successfully implemented with PI as well as PID controller for a variable speed wind turbine. The output power was maintained at 1.5 MW for any wind speed above rated wind speed of 14.2 m/s. If we compare the performance of PID controllers with PI controllers, it was observed that PID controllers, helped in reducing the undershoots, overshoots and oscillations in output power response. It was also observed that constant power strategy is more favorable for faster response with almost no oscillations as compared to constant current strategy.

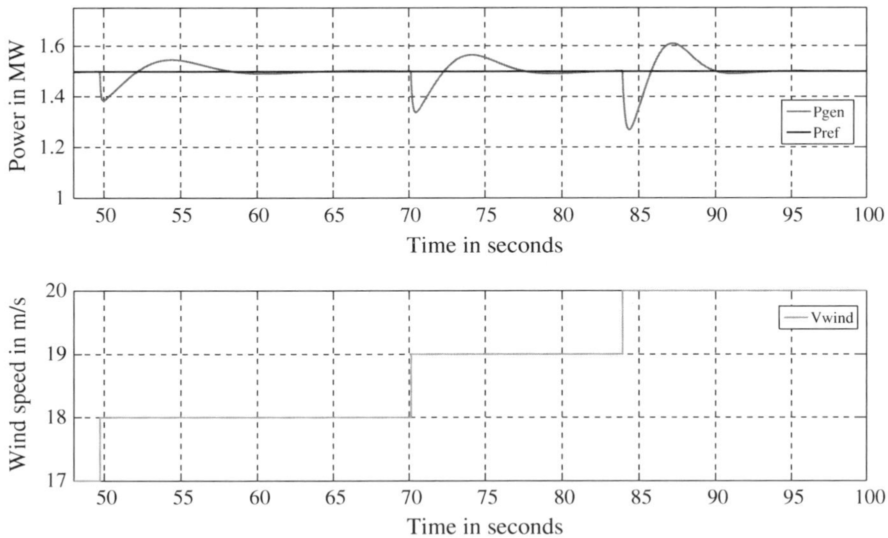

Fig. 38 Power excursion: constant current strategy using PID control from 17 to 18 to 19 to 20 m/s

3.2.3 Hybrid Control

In the case of fixed speed wind turbine, the pitch angle (β) is set such that power output reduces with increase in wind speed beyond the rated speed. This is due to passive stalling above rated wind speed. The power extracted from the wind can be obtained from Eqs. (3)–(8). In case of rotor resistance control, external resistances are added to the rotor circuit to vary the slip or generator speed at which maximum generator torque is obtained. Since external resistance is implemented electronically, it responds fast to rises in wind speed [14]. However, due to the inclusion of the extra resistance, rotor thermal losses can be several hundreds of kW. A solution to this problem is adjusting the value of the blade pitch angle for the purpose of power control since C_p is dependent on pitch angle as well. Since pitching rate is slow due to the high inertia of the rotor blades, rotor resistance can be included in the circuit only until the time the pitch is re-adjusted.

Figure 5 shows the C_p versus λ curves for different pitch angles. In low to medium wind speeds the pitch angle is controlled to allow the wind turbine to operate at its optimum condition (maximum C_p condition). In high wind speed region, the pitch angle is increased to shed some of the available wind power. Figure 24 shows the blade geometry of a horizontal axis wind turbine. With increase in the pitch angle, the angle of attack decreases, decreasing the lift force resulting in reduced power output. Similarly, a reduction in the pitch angle increases the power output. Therefore, at low wind speeds the pitch angle is set low whereas at high speeds the angle is increased to relatively higher values.

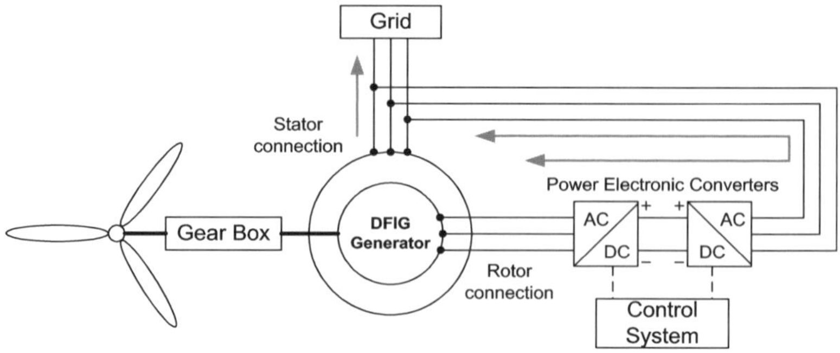

Fig. 39 DFIG wind turbine schematic

3.3 Doubly Fed Induction Generator

It has been already shown in the case of rotor resistance control that, variable wind speed turbines provide better optimization of output power produced by the rotor. In the case of rotor resistance control, changing the rotor resistance changes the slip and thus required torque is produced at varying wind speeds above rated wind speed. When induction machine operation is controlled by the use of power converters to achieve variable speed operation, it is observed that independent active and reactive power control can be achieved. The DFIG is a wound rotor induction generator in which the rotor windings are connected to the grid through power converters. Two VSIs are used for such a connection, linked using a DC link capacitor. With the use of a DFIG, it is possible to transfer power in both directions across the inverter-converter pair. This enables the generator to operate above and below the synchronous speed. Operating the machine over synchronous speed initiates a power flow from the rotor circuit to the grid, while operating the machine below synchronous speed (sub synchronous speed) initiates a power flow from the grid connected stator circuit to the rotor circuit. The amount of active and reactive power transferred to the grid and the machine slip are controlled by rotor current injection into the rotor circuit. For this purpose, reference frame theory is used to obtain $qd0$-axis rotor currents, which can independently control the active and reactive power output of the machine.

On comparison with traditional induction generator, DFIG configuration has many advantages:

- Provides the ability to achieve independent active and reactive power control.
- Supports grid voltage by controlling the reactive power produced or absorbed, which helps maintain a stable grid voltage.

A schematic representation of a DFIG wind turbine system is shown in Fig. 39. In DFIG turbines, the induction generator is a wound-rotor induction machine. Because only part of the real power output flows through the rotor circuit, the power rating of the converter need only be about 20–30 % of the rated turbine

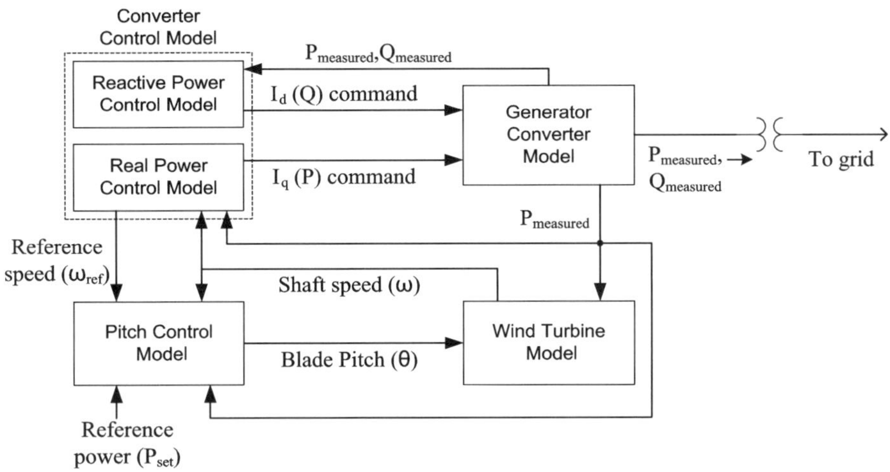

Fig. 40 DFIG model structure

output. A control system is employed to regulate currents in the rotor to extract the maximum possible power from the wind.

The reference wind power plant has a nameplate rating of 204 MW and consists of 136 DFIG wind turbines, each rated at 1.5 MW. Any other WPP can be modeled by changing model parameters. It is connected to the transmission system at 138 kV. The collector system employs 34.5 kV feeders and is adjacent to the transmission station. The wind power plant is equipped with a voltage regulator that controls voltage at the transmission station, relying on the reactive power capability of the WTGs only. There is no additional reactive compensation within the wind power plant.

The subsystems of the developed general wind power plant model are modeled after the subsystems of a typical real-world DFIG turbine. These functionalities include the independent control of real and reactive power and the control of generator speed and blade pitch angles. They are summarized as follows:

- Generator/converter model
- Converter control model
- Wind turbine mechanical model
- Pitch control model
- Collector system

Figure 40 shows the connection of these subsystems and the signals they exchange.

The model is meant to represent the aggregate terminal behavior of the typical WTGS in the wind power plant. In an actual wind power plant, a local grid collects the output from the wind turbines into a single point of connection on the grid. As a wind power plant is usually made up of several identical machines, it is a reasonable approximation to parallel all the turbines into a single equivalent large turbine behind a single equivalent impedance. The power rating of the single equivalent wind turbine is equal to the combined rated power ratings of all wind turbines on the farm. The single machine equivalent also assumes that all the machines generate the same power output at the same time. This assumes that the geographic area occupied by the wind farm is small enough that the wind profile over it is uniform.

A point to be noted is that the electrical dynamic performance of DFIGs at fundamental frequency is dominated by the converter. The combined electrical behavior of the generator and converter in the DFIG is largely like that of a current-regulated voltage source inverter, which may be simplified for modeling purposes as being equivalent to a regulated current source. Therefore, the generator and converter can be combined and modeled as a single current source. This current source is controlled using flux-vector control to obtain the desired real and reactive power flows.

Flux-vector control allows the decoupled control of real and reactive power. For decoupled control over real and reactive power output a controller based on flux-vector control was modeled. As mentioned before, wound rotor induction machines are used in DFIG wind turbines. In the stationary *abc* reference frame, the relationships between the voltages, currents and flux linkages of each phase for a machine of this type are time variant. Analysis in this reference frame is cumbersome, so time variant quantities are made time invariant by transforming them into an appropriate rotating reference frame, i.e. the rotating $qd0$ reference frame. The currents flowing in the stator are assumed to be balanced. These currents produce a resultant stator magnetic field which has a constant magnitude and is rotating at synchronous speed. Since the angular speeds of the stator magnetic field and the $qd0$ rotating frame are identical, the vector of the stator magnetic field is fixed with respect to the q- and d-axes of the $qd0$ rotating frame. The d-axis of the reference frame is oriented in such a way that it aligns with the vector of the stator magnetic field. The real and reactive power can be controlled by adjusting the stator q- and d-axis current. The stator q- and d-axis currents can be controlled by adjusting the rotor q- and d-axis currents. The stator real and reactive power can thus be written as:

$$P_s = k_{ps} \cdot i'_{qr} \tag{50}$$

$$Q_s = -k_{qs1} + k_{qs2} \cdot i'_{qr} \tag{51}$$

where k_{ps}, k_{qs1}, and k_{qs2} are the respective constants for the stator real and reactive power. Equations 50 and 51 clearly show that the stator real and reactive power can be controlled by the rotor q- and d-axis currents independently. In both the positive-sequence model and the three-phase model, the reference q- and d-axis currents are generated by the converter control block, as shown in Fig. 40. In the three-phase model, these q- and d-axis currents are converted back to three-phase currents using the inverse Park transform [13] prior to injection into the collector

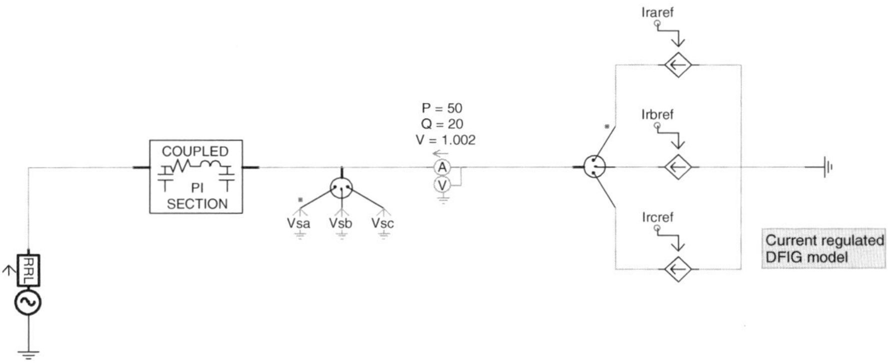

Fig. 41 DFIG model in PSCAD

system. The other subsystems, namely, the converter control model, wind turbine mechanical model, and pitch control model have been modeled in the same manner as in [22].

3.3.1 PSCAD Model

This section describes the simulation results for a current regulated representation of doubly fed induction generator (DFIG) using PSCAD/EMTDC (Fig. 41). The real and reactive power of a wind turbine generator can be independently controlled using a doubly-fed induction generator. A current regulated representation of DFIG is modeled and principle of flux vector control applied to show independent P and Q control. Steps involved in developing the model and implementing vector control, along with results obtained have been shown.

1. Perform Clarke Transform (abc-$\alpha\beta$): Stator voltage V_{sa}, V_{sb}, V_{sb} are converted from three-axis (abc) quantities to two-axis quantities ($\alpha\beta$) V_α and V_β by performing Clarke transform. Obtained two-axis voltages are integrated to obtain corresponding flux values (λ_α, λ_β). Instantaneous value of stator flux λ_s, its magnitude and angular position are determined. A sample run is performed for $P_{genref} = 50$ MW, $Q_{genref} = 20$ MVAR to obtain a sample value of instantaneous stator flux magnitude and angular position. $\lambda_{total} = |\lambda_s| =$ constant, whereas the angular position being an instantaneous value it keeps on varying during the simulation run time. Thus, we obtain a constant magnitude rotating vector λ_s (Fig. 42).
2. Perform Park Transform ($\alpha\beta$ – dq0): A synchronously rotating frame $qd0$ at synchronous speed ω_s is constructed and stator flux aligned along d axis to obtain $|\lambda_d| = |\lambda_s| =$ constant, and $|\lambda_q| = 0$.

Stator d and q voltages are also computed and found to be $V_d = 0$ kV, $V_q = 28.2107$ kV (Fig. 43).

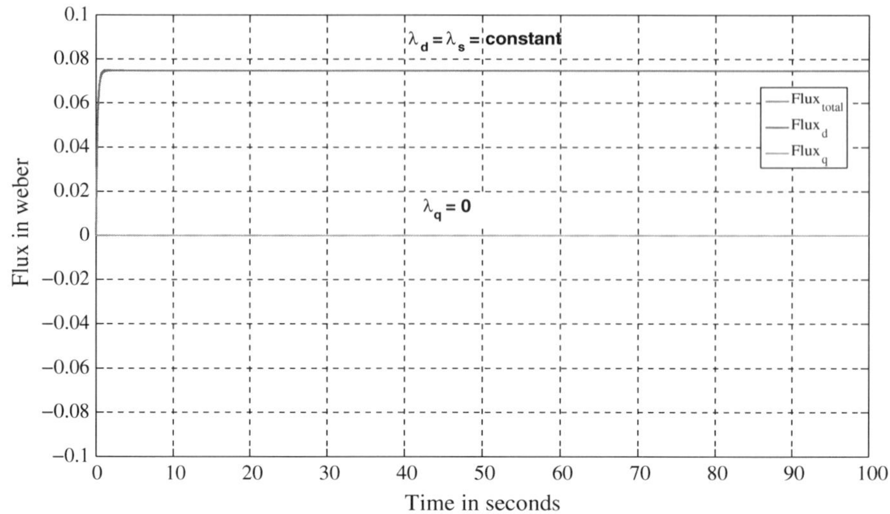

Fig. 42 $|\lambda_d| = |\lambda_s|$ = constant and $|\lambda_q| = 0$

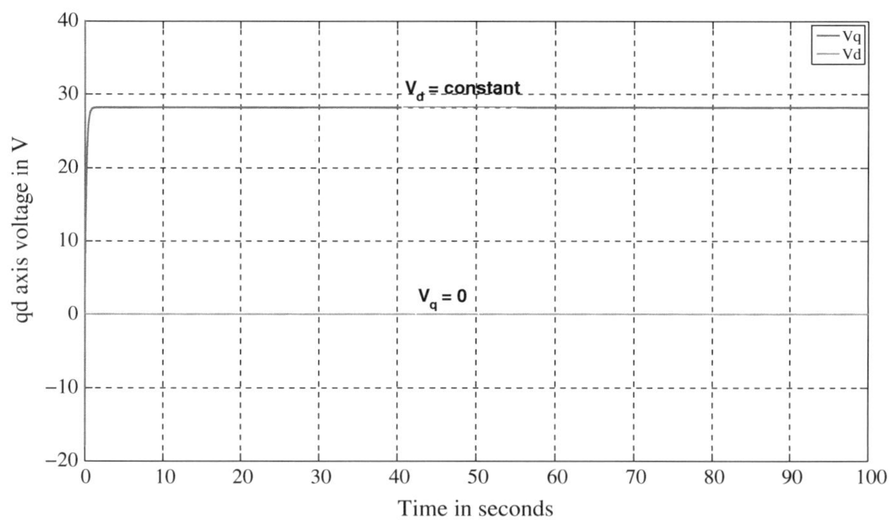

Fig. 43 Voltage along d-q axis

3. Assuming that available wind power is sufficient to generate the desired level of apparent power, reference values for real and reactive power are set to sample values $P_{genref} = 50$ MW, $Q_{genref} = 20$ MVAR. Value of reference currents to be given to the regulated current source is found out, by comparing the generated power with reference power values.

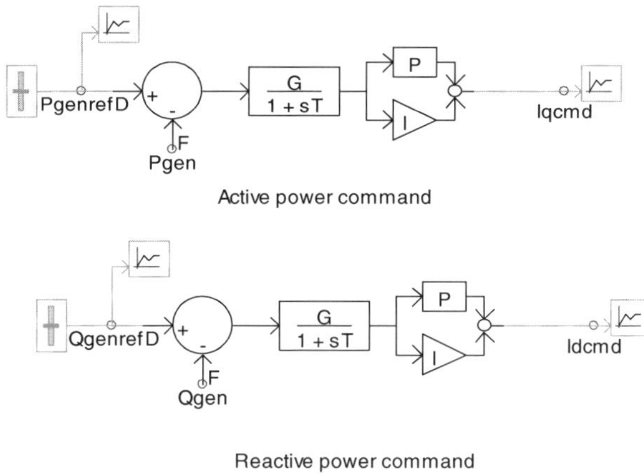

Fig. 44 Computing reference currents I_d and I_q

4. The reference or command currents thus obtained along *dq-axis* (I_d, I_q) can be converted into currents along *αβ-axis* (I_α, I_β) through inverse park transform and eventually into currents along *abc-axis* (I_{raref}, I_{rbref}, I_{rcref}) using inverse Clarke transform. It can be shown through real and reactive power equations developed in vector control theory, that the active power from the generator is controlled by I_q, while reactive power by I_d (Fig. 44).
5. The output is measured for different values of desired power and power generated is found to track accurately the desired value. Further, to establish decoupled or independent control of real and reactive power, $Q_{genref} = 20$ MVAR is kept constant and $P_{genref} = 50$–400 MW with steps of 50 MW and corresponding P_{gen} and Q_{gen} values are registered. It is clearly seen from Fig. 46 that with step change of 50 MW in real power, reactive power remains constant as we increase the desired real power from 50 to 400 MW. There are small overshoots during transient period, but the output settles really fast. Thus, it is demonstrated that, reactive power change is independent of real power demand. It can be satisfactorily established from Fig. 46 that step changes of 20 MVAR in reactive power have no change in the real power output of the DFIG. Thus, Fig. 46 confirm independent P and Q control (Fig. 45).

3.3.2 MATLAB/SIMULINK Model

Modeling in SIMULINK involved developing the model in parts:

1. Developing the abc-$\alpha\beta$-qd0 and inverse transformation blocks.
2. Using controlled current sources to inject reference currents at stator or grid frequency.

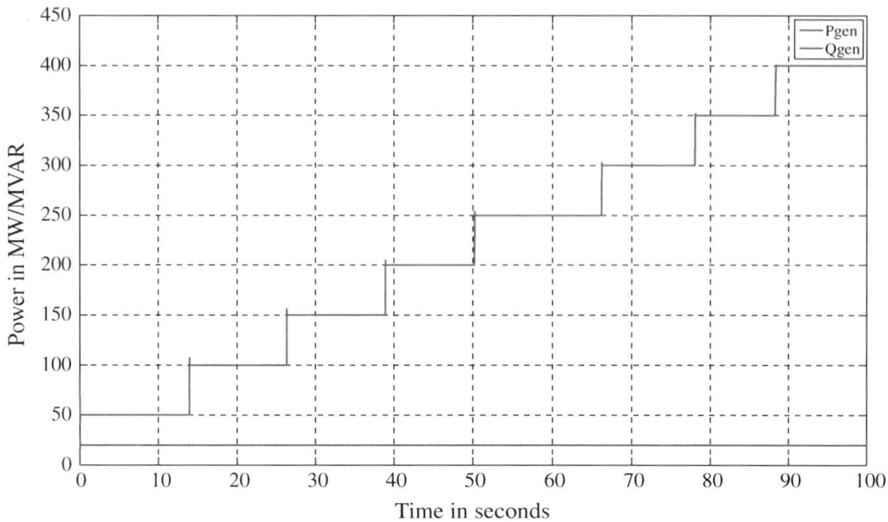

Fig. 45 $P_{genref} = P_{gen} =$ 50–400 MW in steps of 50 MW, $Q_{genref} = Q_{gen} =$ 20 MVAR

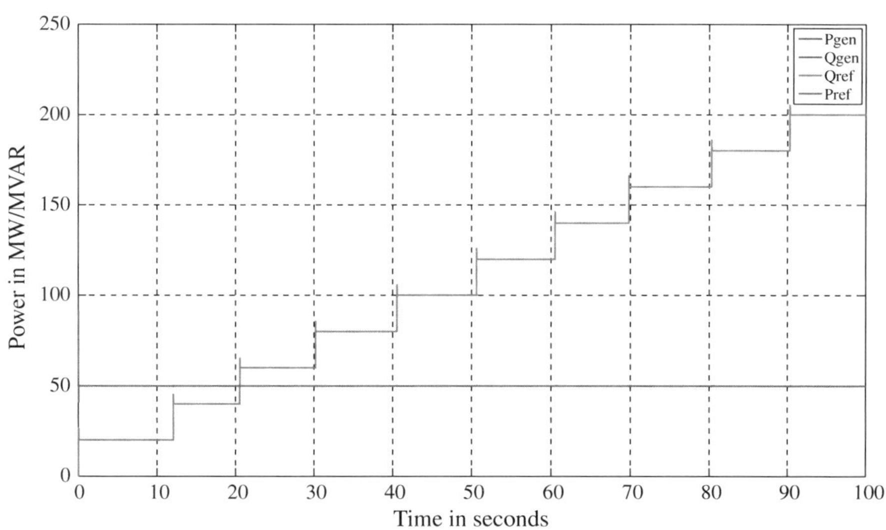

Fig. 46 $Q_{genref} = Q_{gen} =$ 50–400 MW in steps of 50 MW, $P_{genref} = P_{gen} =$ 50 MW

3. Tuning the PI controllers to compute I_d and I_q command currents by comparing P_{ref} and Q_{ref} with P_{gen} and Q_{gen} respectively.
4. Demonstrate successful power tracking and decoupled control of P and Q.

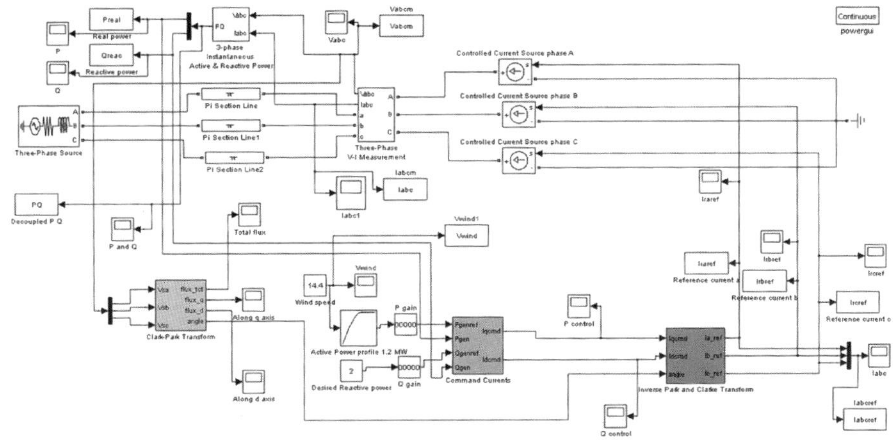

Fig. 47 Block diagram of DFIG model in SIMULINK

5. Develop and use a look up table approach to compute reference P and reference Q for a 1.8 MVA machine, 1.5 MW rated active power at rated wind speed of 14.2 m/s.

Stator voltage in the *abc* frame and at the grid frequency of 60 Hz and are measured and transformed to $\alpha\beta$ stationary frame to obtain two axis voltages V_α and V_β from V_{abc}. Further, the voltage signals are integrated to obtain the flux along $\alpha\beta$ axis. A rotating qd0 frame is then constructed and q-axis aligned with stator field. The newly created qd0 frame rotates at grid frequency or stator frequency. With *qd*0 frame constructed and sample input values for voltage showing proper alignment of q-axis with stator field, the next part of modeling the DFIG includes, using current controlled sources (representing induction machine) to inject currents into the grid.

As such controlled sources are available in SIMULINK. Using the current source blocks available in SIMULINK and an entire block modeling a three phase voltage source as grid and *pi* sections representing transmission lines are used. Given below is the specification for grid parameters and transmission line parameters used initially (Figs. 47 and 48).

Grid voltage = Vs = 34.5 kV
Grid frequency = fs = 60 Hz
MVA base = 100
X/R ratio = 10

Transmission line parameters for a three phase line

R = 0.2568 Ω/km
L = 2×10^{-3} H/km
C = 8.6×10^{-9} F/km
Length of the line = 100 km

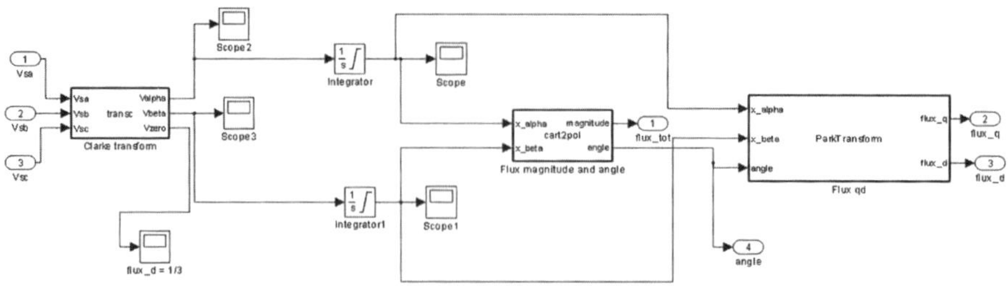

Fig. 48 Block diagram for $abc - \alpha\beta - qd0$ tranformation

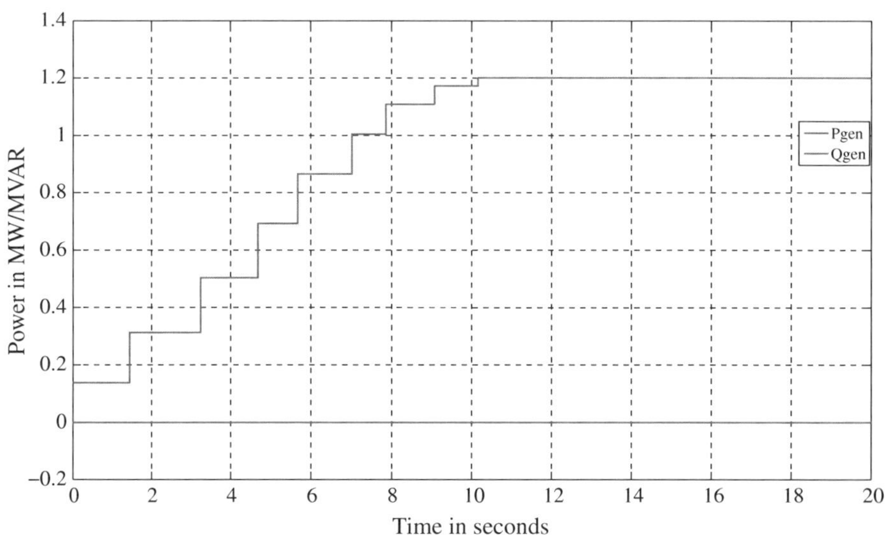

Fig. 49 Active power excursion

3-ϕ multi-meter available in SIMULINK was used to measure I_{abc} and V_{abc}. An in built complex power measurement unit is used to measure active and reactive power. PI controller is I_{qcmd} and I_{dcmd} command currents, by comparing the actual values of active power and reactive power with the reference values respectively. Tuning of the PI controllers is performed using Ziegler Nichols method as explained in Sect. 3.2.2.

Once the PI controllers are tuned, I_{qcmd} and I_{dcmd} command currents are transformed using Inverse Park and Inverse Clarke transform to obtain I_{aref}, I_{bref}, I_{cref} reference currents for the controlled current sources. As shown in Fig. 49 P and Q can be controlled independently to achieve "Power factor correction" or "voltage regulation", i.e. the wind turbine can produce reactive power to maintain the voltage constant or keep the reactive power output to 0 MVAR. To demonstrate

Table 2 Power tracking

Wind speed (m/s)	Active power P (MW)
6	0.138232
7	0.312485
8	0.504339
9	0.692401
10	0.864456
11	1.00413
12	1.10796
13	1.17216
14.4 (rated)	1.2
15	1.2
16	1.2
17	1.2
18	1.2
19	1.2
20	1.2

active power tracking for a given wind turbine Table 2 is used to plot the power profile. Table 2 shows the amount of active power to be generated for a particular wind speed. Wind turbine used to obtain the power profile shown in the table below is rated at 14.4 m/s at +0.165° pitch angle and rated active power for the generator is 1.2 MW (machine rated for 1.8 MVA). The maximum rated power obtained for the given C_p versus λ_r a characteristic is 1.2 MW (Fig. 50).

3.4 DFIG Model Validation

The aim of the validation process is to demonstrate and prove that the DFIG does indeed behave like a real-world DFIG wind power plant, especially during fault conditions. A fault case is used to test the time-domain model. Actual data for three-phase voltages and currents (at the bus where the collector system is connected to the grid) has been provided in the data from the real-world wind power plant. Also, the reactive power demand was set to zero, but the real power (dependent on the wind speed) was set to a differing constant value. The validation procedure for the three-phase model is detailed in Fig. 51.

3.4.1 Calculation of Real and Reactive Power in MATLAB Script

The following calculations are carried out in the MATLAB script mentioned in Fig. 51. The voltage $[v_{abcs}]$ and current $[i_{abcs}]$ extracted from the time-domain model are converted from values on the stationary *abc* frame to equivalent values on the rotating *qd*0 reference frame. This is done using the Park Transform shown below:

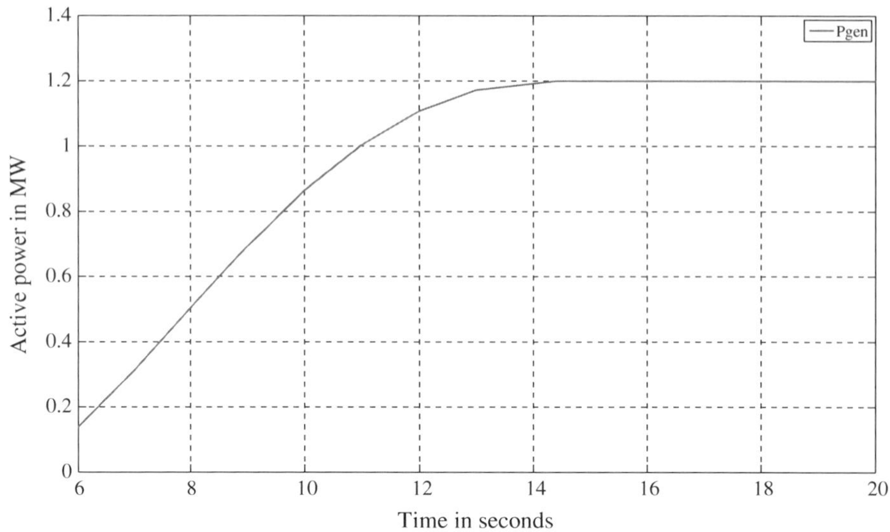

Fig. 50 Wind active power profile for the DFIG

$$[T_{qd0}] = \frac{2}{3}\begin{bmatrix} cos\theta_q & cos(\theta_q - \frac{2\pi}{3}) & cos(\theta_q + \frac{2\pi}{3}) \\ sin\theta_q & sin(\theta_q - \frac{2\pi}{3}) & sin(\theta_q + \frac{2\pi}{3}) \\ \frac{1}{2} & \frac{1}{2} & \frac{1}{2} \end{bmatrix} \tag{52}$$

The following transformation equations can then be used [13] to convert phase voltage and current quantities into the $qd0$ domain.

$$[v_{qd0s}] = [T_{qd0}] \cdot [v_{abcs}] \tag{53}$$

$$[i_{qd0s}] = [T_{qd0}] \cdot [i_{abcs}] \tag{54}$$

It is possible to align the d- and q-axis quantities in such a manner that the d-axis voltage is zero. The calculation of real and reactive power can then be performed using the following equations (where subscript s signifies stator quantities):

$$P_s = \frac{3}{2}(v_{qs} \cdot i_{qs}) \tag{55}$$

$$Q_s = -\frac{3}{2}(v_{qs} \cdot i_{ds}) \tag{56}$$

3.4.2 Validation Based on Pre-fault Data

The first stage of the validation was to calculate the pre-fault real and reactive power for each phase in the phasor-domain using one cycle of pre-fault voltage

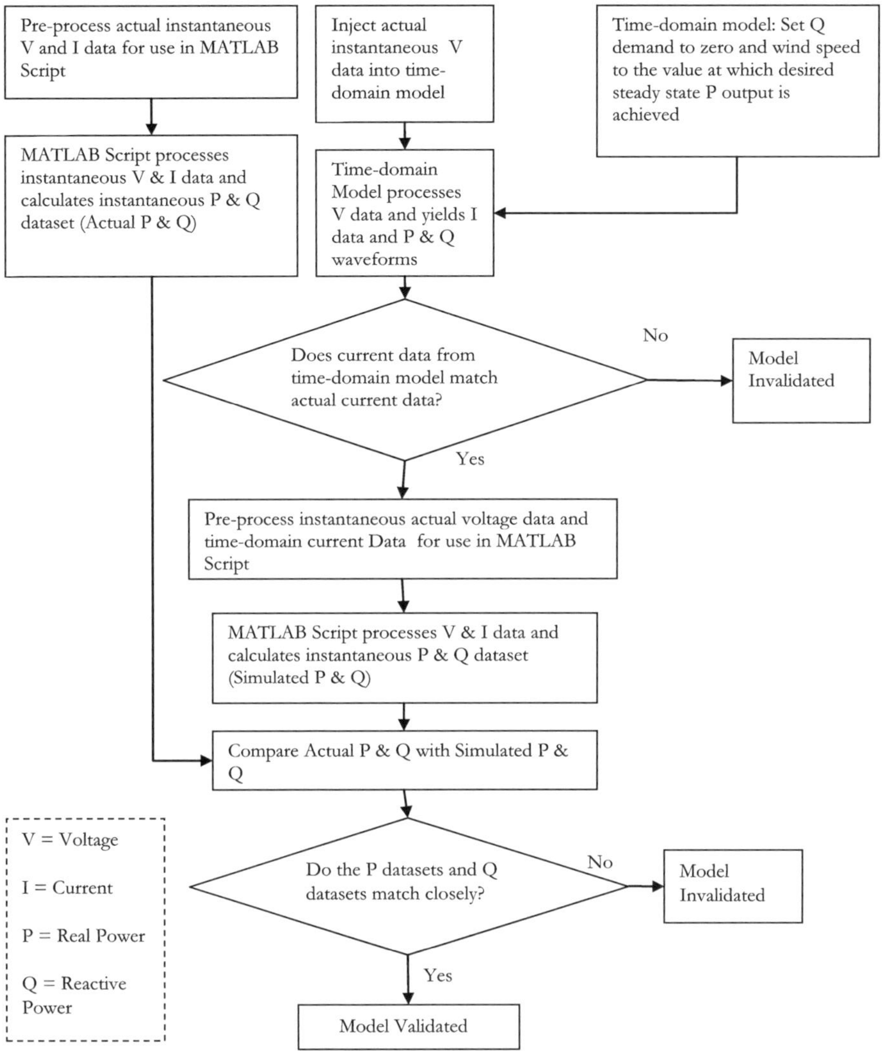

Fig. 51 Validation procedure flowchart

and current from the actual fault data. Since the system is in steady state and the voltages are balanced, one cycle of the steady state voltage and current for phase A is adequate for calculations. The sign convention used is to consider real and reactive power flowing out of the wind plant model to be positive and into the wind plant model to be negative. Once the real and reactive powers in the phasor-domain were obtained, the next stage of the validation was to use the script to compare the actual data and time-domain model output data. The real power and reactive power values generated from the actual data and the data extracted from

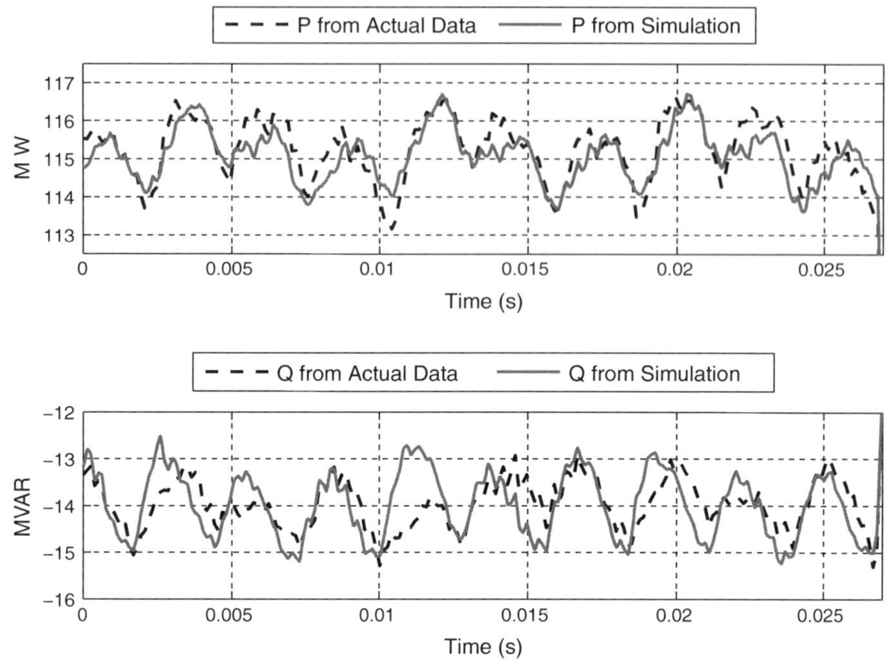

Fig. 52 Steady-state: reactive power comparison: actual versus time domain model

the time-domain model match closely. They also match with the values obtained from phasor-domain calculations (see Fig. 52).

The calculations in phasor-domain for phase A were as follows:

$V_{rms} = 80.74$ kV
$I_{rms} = 559.38$ kA
$(V_{ph} - I_{ph}) = -8.96°$
$P_{1\phi} = V_{rms} \bullet I_{rms} \bullet \cos(V_{ph} - I_{ph}) = 44.61$ MW
$Q_{1\phi} = V_{rms} \bullet I_{rms} \bullet \sin(V_{ph} - I_{ph}) = -7.03$ MVAR
$P_{3\phi} = 3 \bullet (44.61)$ MW $= 133.84$ MW
$Q_{3\phi} = 3.(-7.03)$ MVAR $= -21.10$ MVAR

3.4.3 Fault-Time Validation

To evaluate the performance of the time-domain model during fault conditions, the validation process described earlier is used. The real and reactive power datasets 1 and 2 were generated and plotted together in order to compare the closeness of the match (See Fig. 53). The results show a close match, both in magnitude and phase. The wind power plant model provides a good approximation of the behavior of an actual wind plant under steady state and fault conditions. There are

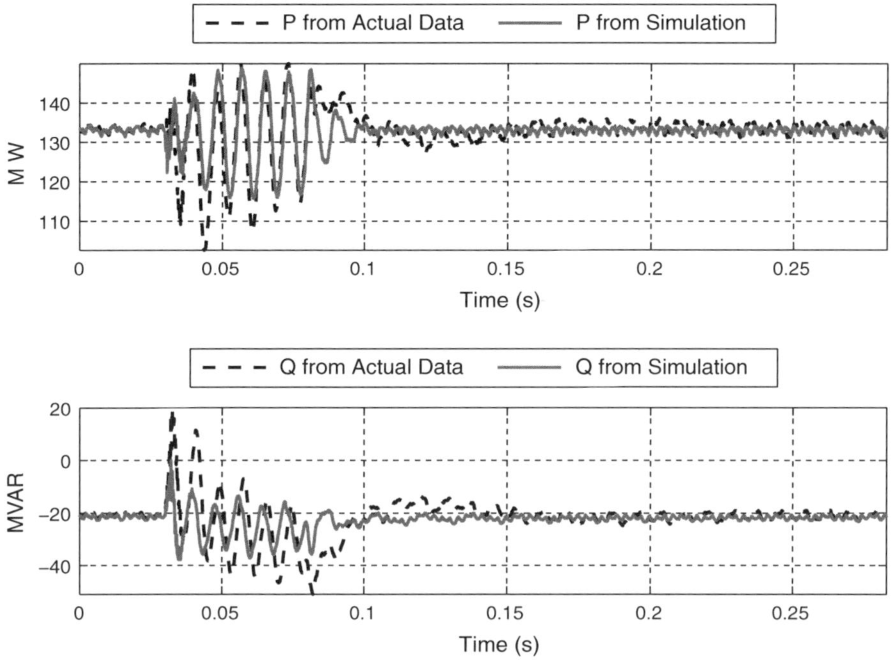

Fig. 53 Comparison between actual and simulation-based real power and reactive power during fault condition

some small discrepancies between the real and reactive power plots obtained from the actual real and reactive power dataset (actual P and Q) and simulated real and reactive power dataset (simulated P and Q). These discrepancies may be due to the simplifications made to the time-domain model in order to preserve its general nature, since the induction generator and the power electronic converter are not explicitly modeled.

The three-phase WPP model has been comprehensively validated using instantaneous voltage, current, real power and reactive power fault data. A method to calculate WPP real and reactive power output from available voltage and current data at the POI has been presented. The results show that, as expected, the three-phase model is better able to reproduce the fault dynamics observed in the actual case.

4 Conclusion

This chapter presents a direct-connect fixed-speed wind turbine model, a variable-speed rotor-resistance control wind turbine model, and a DFIG wind turbine model with validation. The modeling includes the wind turbine aerodynamic rotor and drive train representations which are often oversimplified. The wind turbine and wind farm models presented can be used for educational activities in wind power integration studies. Furthermore, they can be used to perform many real-world studies, such as power curve generation, wind power integration, short-circuit, dynamic interactions between wind turbines, power system dynamic stability and so on.

Appendix

Machine Specifications

Poles	6
Rated voltage (l–l)	690 V
Rated power	1.8 MVA
Base angular frequency	376.99 rad/s
Stator/rotor turns ratio	0.379
Angular moment of inertia	0.578 s
Stator rotor resistance	0.0054 p.u.
Wound rotor resistance	10^{-6} p.u.
Magnetizing inductance	6.83309 p.u.
Stator leakage inductance	0.08 p.u.
Rotor leakage inductance	0.04782 p.u.

Mechanical Data for Shaft Model

J_{rot} Rotor moment of inertia (kg mm)	J_{rot} = 4,950,000 kg mm
J_{gen} Generator moment of inertia (kg mm)	J_{gen} = 80
J_{q2} Gearbox moment of inertia (kg mm)	J_{q2} = 15 kg mm
K_{rq1} Spring constant rotor shaft (Nm/rad)	K_{rq1} = 9,800,000 Nm/rad
K_{q2g} Spring constant generator shaft (Nm/rad)	K_{q2g} = 2,950,000 Nm/rad
D_{rot} Damping rotor (Nms/rad)	D_{rot} = 0 Nms/rad
D_{rot} Damping gearbox (Nms/rad)	D_{q2} = 2.4 Nms/rad
D_{rot} Damping generator (Nms/rad)	D_{gen} = 0 Nms/rad
D_{rot} Damping rotor shaft (Nms/rad)	D_{rq1} = 13,500 Nms/rad
D_{rot} Damping generator shaft (Nms/rad)	D_{q2g} = 30 Nms/rad
f_n Nominal frequency (Hz)	f_n = 60 Hz
P_{gn} Nominal mechanical power (MW)	P_{gn} = 1.5 MW
a Gear ratio	$a = 70$
p Generator pole pairs	$p = 3$

Nomenclature

λ_r	Tip speed ratio
ρ	Air density
λ	Flux linkage
f	Frequency
P	Real power
Q	Reactive power
V	Voltage
I	Current
L	Inductance
R	Resistance
β	Pitch angle of blades
β_0	Initial Pitch angle of blades
β_q	Angle measured from the positive stationary a-phase axis to the rotating q-axis
ω	Angular velocity
τ	Torque
J	Moment of Inertia
B	Damping constant
K	Shaft stiffness
N	Gear ratio
θ	Twist in shaft

Superscripts and Subscripts

$'$	Parameter referred to stator
s	Stator quantity
r	Rotor quantity
d	d-axis quantity
q	q-axis quantity
abc	Parameter in abc reference frame
$qd0$	Parameter in $qd0$ reference frame
l	Leakage quantity (used with inductance)
m	Mutual quantity (used with inductance)
rms	Root mean square quantity
ph	Phase quantity
1ϕ	Single-phase quantity
3ϕ	Three-phase quantity
G, gen	Generator
T, rot	Rotor
eqv	Equivalent value (generator and rotor combined)

References

1. 20 % Wind Energy by 2030: Increasing Wind Energy's Contribution to U.S. Electricity Supply, U.S. D.O.E., July 2008
2. Ackermann T (ed) (2005) Wind power in power systems. Wiley, New York
3. Hansen AD, Hansen LH (2007) Wind turbine concept market penetration over 10 years (1995–2004). Wind Energy 10(1):81–97 (Wiley Online Library)
4. Slootweg JG, Polinder H, Kling WL (2001) *D*ynamic modeling of a wind turbine with doubly fed induction generator. In: Proceedings of 2001 power engineering society summer meeting, 2001. IEEE, vol 1. pp. 644–649
5. Uctug MY, Eskandarzadeh I, Ince H (1994). Modeling and output power optimization of a wind turbine driven double output induction generator. Electric power applications, IEE proceedings, vol 141. pp 33–38
6. Kim S-K, Kim E-S, Yoon J-Y, Kim H-Y (2004) PSCAD/EMTDC based dynamic modeling and analysis of a variable speed wind turbine. In: Proceedings of 2004 power engineering society general meeting, 2004. IEEE, vol 2. pp 1735–1741
7. Slootweg JG, Kling WL (2003) Aggregated modeling of wind parks in power system dynamics simulations. In: Proceedings of 2003 power tech conference proceedings, 2003 IEEE Bologna, vol 3. p 6
8. Delaleau E, Stankovic AM, Dynamic phasor modeling of the doubly-fed induction machine in generator operation. www.ece.northeastern.edu/faculty/stankovic/Conf_papers/lsiwp03.pdf
9. Gagnon R, Sybille G, Bernard S, Pare D, Casoria S, Larose C (2005) Modeling and real-time simulation of a doubly-fed induction generator driven by a wind turbine. www.ipst.org/TechPapers/2005/IPST05_Paper162.pdf
10. Lubosny Z (2003) Wind turbine operation in electric power systems: advanced modeling. Springer, Berlin
11. Akhmatov V (2005) Induction generators for wind power. Multiscience Publishing Company, Essex, UK
12. Manwell JF, McGowan JG, Rogers AL (2003) Wind energy explained: theory design and applications. Wiley, New York England Reprinted with corrections August
13. Krause PC (1986) Analysis of electric machinery. McGraw Hill Co, New York
14. Burnham DJ, Santoso S, Muljadi E (2009) Variable rotor resistance control of wind turbine generators. In: Power and energy society general meeting, 2009. PES' 09. IEEE, 26–30 July 2009
15. Hang CC, Astrom KJ, Ho WK (1991) Refinements of the Ziegler-Nichols tuning formula. Control Theory and Applications, IEE Proceedings D, vol 138, No. 2
16. E. Muljadi and C.P. Butterfield, Pitch-controlled variable-speed wind turbine generation. In: Industry applications conference, 1999. Thirty-fourth IAS annual meeting. conference record of the 1999 IEEE, vol 1. pp 323–330. 3–7 October 1999
17. Singh M, Santoso S (2007) Electromechanical and time-domain modeling of wind generators. Power engineering society general meeting, 2007. IEEE, 24–28 June 2007
18. Singh M, Faria K, Santoso S, Muljadi E (2009) Validation and analysis of wind power plant models using short-circuit field measurement data. Power & energy society general meeting, 2009. PES '09. IEEE, 26–30 July 2009
19. Santoso S, Hur K, Zhou Z (2006) Induction machine modeling for distribution system analysis—A time domain solution. Transmission and distribution conference and exhibition, 2005/2006 IEEE PES, pp 583–587. 21–24 May 2006
20. Muljadi E, Butterfield CP, Ellis A, Mechenbier J, Hochheimer J, Young R, Miller N, Delmerico R, Zavadil R, Smith JC (2006) Equivalencing the collector system of a large wind power plant. IEEE power engineering society, annual conference, Montreal, Quebec, 2006 IEEE PES

21. Muljadi E, Pasupulati S, Ellis A, Kosterov D (2008) Method of equivalencing for a large wind power plant with multiple turbine representation. In: 2008 IEEE power and energy society general meeting-conversion and delivery of electrical energy in the 21st century
22. Muljadi E, Ellis A (2008) Validation of wind power plant dynamic models. IEEE power engineering society general meeting, Pittsburgh, 20–24 July 2008
23. Muljadi E, Butterfield CP, Parsons B, Ellis A (2007) Effect of variable speed wind turbine generator on stability of a weak grid. IEEE Trans Energy Conversion 22(1):29–36
24. Muljadi E, Nguyen TB, Pai MA (2008) Impact of wind power plants on voltage and transient stability of power systems. In: Energy 2030 conference, 2008. ENERGY 2008. IEEE, 17–18 Nov 2008

Deterministic Approaches for the Steady-State Analysis of Distribution Systems with Wind Farms

P. Caramia, G. Carpinelli, D. Proto and P. Varilone

Abstract Wind farms have several impacts on the steady-state behavior of an electrical distribution system, and these impacts must be taken into consideration. This chapter dealt with deterministic approaches for the steady-state analysis of distribution systems with wind farms, considering both balanced and unbalanced systems. The steady-state analysis was performed by using appropriate algorithms to solve the load-flow non-linear equation system. Several models of wind farms are illustrated and included in the load-flow analysis, i.e., fixed-speed, semi-variable-speed and variable-speed wind generation systems were considered. Numerical applications on a 17-bus balanced system and an IEEE 34-bus unbalanced test distribution system are presented and discussed considering various wind farm models.

P. Caramia
Department of Engineering, Università degli Studi di Napoli Parthenope,
Centro Direzionale, Is. 80143 Naples, Italy
e-mail: pierluigi.caramia@uniparthenope.it

G. Carpinelli (✉) · D. Proto
Department of Electrical Engineering and Information Technology,
Università degli Studi di Napoli Federico II, Via Claudio 21, Naples, Italy
e-mail: guido.carpinelli@unina.it

D. Proto
e-mail: danproto@unina.it

P. Varilone
Department of Electrical and Information Engineering,
Università degli Studi di Cassino e del Lazio Meridionale,
Via G. Di Biasio 43, Cassino, Italy
e-mail: varilone@unicas.it

P. M. Pardalos et al. (eds.), *Handbook of Wind Power Systems*, Energy Systems,
DOI: 10.1007/978-3-642-41080-2_7,

1 Introduction

In recent years, the number of dispersed generation (DG) units connected to distribution networks has continued to grow, and projections are that this trend will continue into the future trend. In fact, the needs for saving energy and reducing environmental impacts, together with technology evolution and increased customer demand for highly reliable electricity, provide incentives for the proliferation of generation units connected to distribution systems that are close to the customers. Other important drivers that are leading to this proliferation are strictly linked to the new liberalized electricity markets. Such generators, with their comparatively small size, short lead times, and different technologies, allow various components of the electricity market (e.g., utilities, independent producers, and customers) to respond in a flexible way to changing market conditions. In addition, these generators make it possible to sell ancillary services, such as reactive power and back-up power.

Even though DG units can be based on different types of primary energy sources, the presence of wind generators in distribution systems is increasing, and wind power is now producing a significant fraction of the energy produced by renewable energy sources all over the world.

The foreseeable extensive use of wind turbine generator units (WTGUs) in the future requires that distribution system engineers properly account for their impact in the system. In fact, their interconnection with the system significantly alters the characteristics of the distribution systems, traditionally designed with the assumption of a passive network. The consequence of the presence of WTGUs (and, more generally, of all dispersed generators) is that the assumption of a passive network is no longer valid; instead, the network becomes active, which generates a number of new technical considerations that must be addressed, such as distribution network planning and operation, especially protection coordination, steady-state analysis, and power quality issues.

Then, in this context, there is a growing imperative to study the problems of active distribution networks including WTGUs in order to understand and quantify the technical impact that a high penetration of these generators may have on the operation and performance of these systems. In particular, publications in the relevant literature have paid extensive attention to the steady-state analysis at the power frequency of distribution systems with wind generators. The analysis is achieved by using the well-known load-flow method, i.e., a non-linear equation system is solved with a proper algorithm, allowing us to determine the electrical state of the distribution system and, after that, e.g., the voltage profile and the system losses. Accordingly, WTGU models to be included in the load-flow assessment have been deeply investigated, and several models have been proposed in the relevant literature with different details [1–15].

Both deterministic load flows, which assume that all the input data are known with certainty, and probabilistic load flows, which assume that some input data are affected by uncertainties, have been proposed. This chapter deals with deterministic load flows, while the companion chapter [16] dealt with the probabilistic approaches.

The chapter is organized such that a brief summary of the mathematical formulation of the load-flow equations for distribution systems without wind farms is provided first (Sect. 2). Both balanced and unbalanced distribution systems are considered; in fact, distribution systems can operate with unbalanced loading conditions and can be characterized by the presence of feeders with missing phases, and the steady-state analysis should account for also any unbalances in the system. Then, the mathematical formulation of the load flow for distribution systems with wind farms is presented (Sect. 3); the main models of WTGUs that appeared in the relevant literature are presented first and, then, the problem of their inclusion in the load-flow equations is discussed, once again considering both balanced and unbalanced distribution systems. Finally, numerical applications are conducted on a 17-bus balanced test distribution system and on an IEEE 34-bus unbalanced test distribution system. For each, the various WTGU models were tested and compared.

Since this chapter dealt only with the load flow in distribution systems with wind farms, the cases that involve transmission systems and offshore wind farms were not analyzed.

2 Load-Flow Equations for Distribution Systems Without Wind Farms

In the next sub-sections, the load-flow equations for balanced and unbalanced distribution systems are briefly reviewed. We only refer to mathematical formulations that have been further developed to include wind farm models.

2.1 Load-Flow Equations for Balanced Systems

Let us refer to a balanced power system in which the busbars from 1 to n_{load} are load busbars and those from $n_{load}+1$ to n_{bus} are generator busbars (the last busbar is the slack busbar). In the load busbars, the active and reactive powers are assigned (PQ busbars), while in the generator busbars without the slack, the active power and the voltage amplitude are assigned (PV busbars). A single-phase representation of the system is adequate, and the steady state of the system is described by the following non-linear equation system (load-flow equations) [17]:

$$\begin{aligned} P_{\rm i}^{sp} &= V_i \sum_{j=1}^{n_{bus}} V_j \left[G_{ij} \cos\theta_{ij} + B_{\rm ij} \sin\theta_{ij} \right], \quad {\rm i} = 1, \ldots, n_{bus} - 1 \\ Q_{\rm i}^{sp} &= V_i \sum_{j=1}^{n_{bus}} V_j \left[G_{ij} \sin\theta_{ij} - B_{\rm ij} \cos\theta_{ij} \right], \quad {\rm i} = 1, \ldots, n_{load}, \end{aligned} \tag{1}$$

where:

P_i^{sp}, Q_i^{sp} are the active powers specified at each load and generator busbar without the slack and the reactive powers specified at each load busbar, respectively

V_i, δ_i are the voltage magnitudes and arguments

G_{ij}, B_{ij} are the conductance and susceptance of the i–j term of the admittance matrix

n_{bus}, n_{load} are the system bus number and the load bus number, respectively

θ_{ij} $= \delta_i - \delta_j$.

The non-linear Eq. (1) represent the active balances at load busbars and generator busbars without the slack and the reactive balance at the load busbars.

Some active and reactive powers in (1) cannot be assigned but must be expressed as a function of the bus voltages. This would be the case, for example, when loads have active and reactive powers that are voltage-dependent; this dependence can be easily incorporated in (1). If bus k includes only loads with such a dependence, the following expression holds:

$$\begin{aligned} P_{0k}\left(\frac{V_k}{V_{0k}}\right)^{\alpha} &= V_k \sum_{j=1}^{n_{bus}} V_j\left[G_{kj}\cos\theta_{kj} + B_{kj}\sin\theta_{kj}\right] \\ Q_{0k}\left(\frac{V_k}{V_{0k}}\right)^{\beta} &= V_k \sum_{j=1}^{n_{bus}} V_j\left[G_{kj}\sin\theta_{kj} - B_{kj}\cos\theta_{kj}\right] \end{aligned}, \qquad (2)$$

where P_{0k} and Q_{0k} are active and reactive load powers at reference voltage V_{0k} (typically the rated voltage), and α and β are load-dependent constants.

The conventional Newton–Raphson algorithm (NRA) or the Gauss–Seidel algorithm (GSA) can be applied to solve equation system (1) or (2). However, it has been shown in the relevant literature that, in some cases, the NRA and GSA may become inefficient for the steady-state analysis of distribution systems, because these systems have particular characteristics, such as a radial structure and high R/X ratios. In these cases, robust and efficient methods have been applied to deal with the special characteristics of these systems. In the following, some of these methods will be briefly discussed with reference to a radial distribution system.

The proposed methods can be classified in two main categories:

1. Methods based on proper modifications of the conventional NRA or the conventional GSA [18, 19]
2. Methods based on forward/backward sweep processes [20–26].

The sweep-based algorithms seem to be especially useful in analyzing distribution networks, because they are characterized by low computational effort, high robustness, and low memory requirements. Due to the radial structure of the networks, these algorithms calculate the network currents, powers, and voltages through iterative sweeps in backward and forward directions along the line

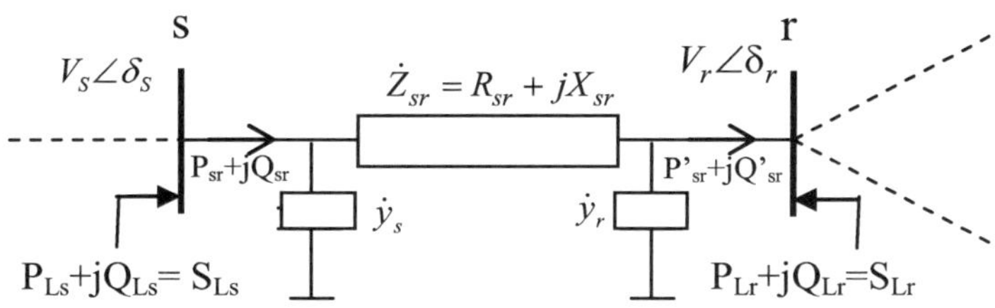

Fig. 1 Generic line section of a radial distribution system

sections, each section characterized by a parent node on the source side (sending bus) and a child node on the load side (receiving bus) (Fig. 1). In the following, the subscripts s and r denote sending and receiving ends, respectively.

The forward sweep mainly consists of node voltage calculations by using either Kirkhoff's law [20] or by bi-quadratic equations [21–26] from the sending end to the receiving end of a feeder or lateral; the backward sweep consists mainly of a power and/or current summation from the receiving end to the sending end of the feeder or lateral. Usually, the procedures assign tentative initial voltage values at all busbars; during the iterative procedure, the substation distribution bus voltage is kept constant, while the voltages in the other busbars are updated. For the sake of simplicity, we consider the case of only one PV bus.

In particular, in [20], the Kirchoff's formulation is applied, and, at first, the so-called nodal *current injections* $\overline{I}_r^{(k)}$ for each receiving node r are calculated:

$$\overline{I}_r^{(k)} = \left(\dot{S}_{Lr}\Big/\overline{V}_r^{(k-1)}\right)^* - \dot{Y}_r\overline{V}_r^{(k-1)}, \tag{3}$$

where $\overline{V}_r^{(k-1)}$ is the voltage at node r at the $(k-1)$th iteration, $\dot{S}_{Lr}$ is the load power injection at node r, $\dot{Y}_r$ is the sum of all the shunt admittances at node r and where symbol $*$ refers to the complex conjugate operator.

Then, the backward sweep is performed, calculating the branch currents starting from the receiving to the sending bus using the following equation:

$$\overline{J}_{sr}^{\,k} = -\overline{I}_r^{(k)} + \sum_{m\in M}\left(\overline{J}_{rm}\right)^k, \tag{4}$$

where $\overline{J}_{sr}^{\,k}$ is the current in the branch between bus s and bus r, $\overline{I}_r^{(k)}$ is given by (3), $\sum_{m\in M}\left(\overline{J}_{rm}\right)^k$ is the sum of the currents leaving node r, and M is the set of all the lines connected to the receiving bus.

Finally, the receiving voltage is calculated starting from the sending bus using the following equation (forward sweep):

$$\overline{V}_r^k = \overline{V}_s^k - \dot{Z}_{sr}\overline{J}_{sr}^{\,k}, \tag{5}$$

where $\overline{V}_r^k$ is the voltage at node r, $\overline{V}_s^k$ is the voltage at node s, and $\dot{Z}_{sr}$ is the series impedance of the branch between bus s and bus r.

In [21], the sweep method calculates the active and reactive branch power starting from the receiving busbars and going to the sending busbars and includes the losses (backward sweep), whereas the node voltages are calculated with a bi-quadratic equation starting from the sending busbars and going to the receiving busbars (forward sweep).

The bi-quadratic equation is given by:

$$V_r^4 + \left[2(P_r R_{sr} + Q_r X_{sr}) - V_s^2\right] V_r^2 + \left(P_r^2 + Q_r^2\right)\left(R_{sr}^2 + X_{sr}^2\right) = 0, \tag{6}$$

where V_r, P_r, and Q_r are the voltage, active power, and reactive power of the receiving busbar, respectively, and V_s is the voltage of the sending busbar. P_r and Q_r include the powers and the losses of all the nodes fed by the receiving bus.

In [22], an iterative procedure, similar to the one applied in [21], is used with various accurate load models (voltage-dependent load models). The equation between the receiving and sending bus voltages to be applied in the backward sweep is:

$$V_r = \left[V_s^2 - 2(P_r R_{sr} + Q_r X_{sr}) + \left(P_r^2 + Q_r^2\right)\left(R_{sr}^2 + X_{sr}^2\right)/V_s^2\right]^{1/2}. \tag{7}$$

In [23], a sweep-based algorithm is proposed that uses a polynomial equation on the forward step and ladder equations on the backward step. The polynomial equation is derived in terms of the hyperbolic parameters of the loss-line model and using exponential load models; it is given by:

$$\begin{aligned} &A_{sr}^2 V_r^4 + 2A_{sr} V_r^2 Z_{sr}\left(P_{L0} V_r^{\alpha}\cos(\theta_Z - \delta_A) + Q_{L0} V_r^{\beta}\sin(\theta_Z - \delta_A)\right) \\ &\quad - V_s^2 V_r^2 + \left(P_{L0}^2 V_r^{2\alpha} + Q_{L0}^2 V_r^{2\beta}\right) Z_{sr}^2 = 0, \end{aligned} \tag{8}$$

where A_{sr} is the magnitude of $\dot{A}_{sr} = \cosh(\gamma_{sr}), \gamma_{sr} = \sqrt{\dot{Z}_{sr}\dot{y}_{sr}}$, $\dot{Z}_{sr,}\dot{y}_{sr}$ are the line series impedance and shunt admittance, respectively, ($\dot{y}_{sr} = 2\dot{y}_s = 2\dot{y}_r$ in Fig. 1), Z_{sr} is the magnitude of $\dot{Z}_{sr}$, θ_Z and δ_A are the phase angles of $\dot{Z}_{sr}$ and $\dot{A}_{sr}$, respectively, P_{L0} and Q_{L0} are the active and reactive load powers at the nominal voltage, α and β are the load exponents [See Eq. (2)].

Once the node voltages are calculated in forward direction using polynomial Eq. (8), new active and reactive powers of loads and, then, load currents are obtained; finally, the backward process is started and is performed by calculating new values of voltages using known line impedances and the aforementioned load currents.

Reference [24] also proposes a method for calculating the load flow solution of distribution systems. In the backward sweep, the sending-end power is calculated at each branch, considering load power and losses. In the forward sweep, the receiving-end voltage is calculated from the sending voltage using the powers obtained in the backward sweep and using the following equation:

$$V_r = \sqrt{(V_s - \Delta V')^2 + \Delta V''^2}, \tag{9}$$

where $\Delta V'$ and $\Delta V''$ are given by $\Delta V' = \frac{R_{sr}P_s + X_{sr}Q_s}{V_s}$ and $\Delta V'' = \frac{X_{sr}P_s - R_{sr}Q_s}{V_s}$, respectively. Once again, P_r and Q_r include the power of all the nodes fed by the receiving bus and the losses.

In [25], an algorithm similar to the one proposed in [24] was used with a slightly different equation for the voltage calculation.

Finally, the algorithm proposed in [26] is used to calculate the receiving-end powers, P_r and Q_r, including the load powers, which depend on load voltage (exponential load model) and the active and reactive losses. The magnitude of the voltage at the receiving end is calculated by starting from the sending end in the forward sweep by using the following equation:

$$V_r = \left\{ \frac{K_{sr} \pm \left[K_{sr}^2 - 4\left({R_{sr}}^2 + {X_{sr}}^2\right)P_r^2 \sec^2\phi_r\right]^{1/2}}{2} \right\}^{1/2}, \tag{10}$$

with $K_{sr} = V_s^2 - 2P_r(R_{sr} + X_{sr}\tan\phi_r)$, and $\phi_r = \tan^{-1}(P_r/Q_r)$.

2.2 Load-Flow Equations for Unbalanced Systems

In the case of unbalanced systems, phase coordinates are used so that voltages, powers, and currents are represented by a three-element vector (one for each phase); also, a three-phase representation of all system components is used so that generators, transformers (with all the possible winding connections) and lines are represented by admittance sub-matrices.

Referring to an unbalanced power system in which the busbars from 1 to n_{load} are load busbars, those from $n_{load} + 1$ to $n_{load} + n_g$ are generator terminal busbars, and those from $n_{load} + n_g + 1$ to $n_{load} + 2n_g = n_{bus}$ are generator internal busbars (with the last terminal and internal busbars being slack busbars), the most general three-phase load flow equations can be expressed as [17]:

$$\begin{aligned}
(P_i^p)^{sp} &= V_i^p \sum_{k=1}^{n_{bus}} \sum_{m=1}^{3} V_k^m \left[G_{ik}^{pm}\cos\theta_{ik}^{pm} + B_{ik}^{pm}\sin\theta_{ik}^{pm}\right] \\
(Q_i^p)^{sp} &= V_i^p \sum_{k=1}^{n_{bus}} \sum_{m=1}^{3} V_k^m \left[G_{ik}^{pm}\sin\theta_{ik}^{pm} - B_{ik}^{pm}\cos\theta_{ik}^{pm}\right] \quad p = 1,2,3 \quad \text{and} \quad i = 1,\ldots,n_{load}
\end{aligned} \tag{11a}$$

$$
\begin{aligned}
(P_i^p)^{sp} &= V_i^p \sum_{k=1}^{n_{bus}} \sum_{m=1}^{3} V_k^m \left[G_{ik}^{pm} \cos\theta_{ik}^{pm} + B_{ik}^{pm} \sin\theta_{ik}^{pm} \right] \\
(Q_i^p)^{sp} &= V_i^p \sum_{k=1}^{n_{bus}} \sum_{m=1}^{3} V_k^m \left[G_{ik}^{pm} \sin\theta_{ik}^{pm} - B_{ik}^{pm} \cos\theta_{ik}^{pm} \right] \\
\left(P_j^{gen}\right)^{sp} &= \sum_{p=1}^{3} V_j^P \sum_{k=1}^{n_{bus}} \sum_{m=1}^{3} V_k^m \left[G_{jk}^{pm} \cos\theta_{jk}^{pm} + B_{jk}^{pm} \sin\theta_{jk}^{pm} \right] \\
(V_i)^{sp} &= f\left(\overline{V}_i^1, \overline{V}_i^2, \overline{V}_i^3\right) \quad p = 1,2,3, \quad i = n_{load} + 1, \ldots, n_{load} + n_g - 1 \\
&\text{and} \quad j = n_{load} + n_g + 1, \ldots, n_{bus} - 1
\end{aligned}
\tag{11b}
$$

$$
\begin{aligned}
(P_i^p)^{sp} &= V_i^p \sum_{k=1}^{n_{bus}} \sum_{m=1}^{3} V_k^m \left[G_{ik}^{pm} \cos\theta_{ik}^{pm} + B_{ik}^{pm} \sin\theta_{ik}^{pm} \right] \\
(Q_i^p)^{sp} &= V_i^p \sum_{k=1}^{n_{bus}} \sum_{m=1}^{3} V_k^m \left[G_{ik}^{pm} \sin\theta_{ik}^{pm} - B_{ik}^{pm} \cos\theta_{ik}^{pm} \right] \\
(V_i)^{sp} &= f\left(\overline{V}_i^1, \overline{V}_i^2, \overline{V}_i^3\right) \quad p = 1,2,3 \quad \text{and} \quad i = n_{load} + n_g
\end{aligned}
\tag{11c}
$$

being

G_{ik}^{pm}, B_{ik}^{pm}	terms of conductance matrix [**G**] and susceptance matrix [**B**], respectively, relating busbar i with phase p and busbar k with phase m
$(P_i^p)^{sp}$, $(Q_i^p)^{sp}$	specified active and reactive powers, respectively, at busbar i with phase p
V_i^p, δ_i^p	voltage magnitude and argument, respectively, at busbar i with phase p
θ_{ik}^{pm}	$= \delta_i^p - \delta_k^m$.

Equation (11a) represent the phase active and reactive power balances at load busbars, Eq. (11b) represent the phase active and reactive power balances at generator terminal busbars and the active power and voltage regulation balances at generator busbars without the slack, and Eq. (11c) represent the phase active and reactive power balances at the slack generator terminal busbar and the voltage regulation balance at the slack generator busbar. For more details about the equations, see Ref. [17].

The conventional Newton–Raphson algorithm (NRA) or the Gauss–Seidel algorithm (GSA) can also be used to solve equation system (11a, b, c). However, for unbalanced systems, it has also been shown in the relevant literature that the NRA or the GSA may, in some cases, become inefficient, and robust and efficient methods have been applied based on the same 1. and 2. categories introduced for balanced system in Sect. 2.1.

As an example of methods based on proper modifications of the conventional GSA, in [10], two matrices, the bus-injection to branch-current matrix [**BIBC**] and the branch-current to bus-voltage matrix [**BCBV**], were introduced, and simple multiplications were used to obtain the power flow solution.

In particular, the [**BIBC**] matrix links the vectors of branch currents $\left[\overline{\mathbf{B}}\right]$ and bus current injections $\left[\overline{\mathbf{I}}\right]$ as:

$$\left[\overline{\mathbf{B}}\right] = [\mathbf{BIBC}]\left[\overline{\mathbf{I}}\right]. \tag{12}$$

In (12), the constant matrix [**BIBC**] is a matrix that has non-zero entries of +1 only; it was developed on the basis of the topological structure of distribution feeders. In particular, if a three-phase line section is between bus i and j, it is represented in the [**BIBC**] matrix by a 3×3 identity matrix.

The [**BCBV**] matrix links the vectors of branch currents $\left[\overline{\mathbf{B}}\right]$ and the difference between bus voltages and no-load bus voltages $\left[\Delta\overline{\mathbf{V}}\right]$:

$$\left[\Delta\overline{\mathbf{V}}\right] = [\mathbf{BCBV}]\left[\overline{\mathbf{B}}\right]. \tag{13}$$

In (13), the constant matrix [**BCBV**] is also developed on the basis of the topological structure of distribution feeders. In particular, if a three-phase line section is between bus i and j, it is represented in the [**BCBV**] matrix by a 3×3 complex $\left[\dot{\mathbf{Z}}_{sr}\right]$, including the neutral or ground effects.

By combining (12) and (13), the result is:

$$\left[\Delta\overline{\mathbf{V}}\right] = [\mathbf{BCBV}][\mathbf{BIBC}]\left[\overline{\mathbf{I}}\right] = [\mathbf{DLF}]\left[\overline{\mathbf{I}}\right]. \tag{14}$$

Then, the solution of the unbalanced three-phase distribution load flow is obtained by iterative calculations:

$$\begin{gathered} \overline{I}_i^k = \left(\frac{P_i + jQ_i}{\overline{V}_i^k}\right)^* \\ \left[\Delta\overline{\mathbf{V}}^{k+1}\right] = [\mathbf{DLF}]\left[\overline{\mathbf{I}}^k\right], \end{gathered} \tag{15}$$

with k iteration number and symbol * refers to the complex conjugate operator.

Examples of methods based on forward/backward sweep algorithms were developed in [27, 28], which are the natural extension of the algorithm proposed in [20] for balanced systems. In particular, in [27], the steps of the algorithm are the same; the extension to the unbalanced systems consists of substituting the Equations from (3) to (5) with the following three-phase equations:

$$\begin{bmatrix} \overline{I}_r^1 \\ \overline{I}_r^2 \\ \overline{I}_r^3 \end{bmatrix}^k = \begin{bmatrix} \left(\dot{S}_r^1/\overline{V}_r^1\right)^* \\ \left(\dot{S}_r^2/\overline{V}_r^2\right)^* \\ \left(\dot{S}_r^3/\overline{V}_r^3\right)^* \end{bmatrix}^{k-1} - \begin{bmatrix} \dot{Y}_r^1 & 0 & 0 \\ 0 & \dot{Y}_r^2 & 0 \\ 0 & 0 & \dot{Y}_r^3 \end{bmatrix} \begin{bmatrix} \overline{V}_r^1 \\ \overline{V}_r^2 \\ \overline{V}_r^3 \end{bmatrix}^{k-1}, \tag{16}$$

$$\begin{bmatrix} \overline{J}_{ls}^{1} \\ \overline{J}_{ls}^{2} \\ \overline{J}_{ls}^{3} \end{bmatrix}^{k} = -\begin{bmatrix} \overline{I}_{r}^{1} \\ \overline{I}_{r}^{2} \\ \overline{I}_{r}^{3} \end{bmatrix}^{k} + \sum_{m \in M} \begin{bmatrix} \overline{J}_{rm}^{1} \\ \overline{J}_{rm}^{2} \\ \overline{J}_{rm}^{3} \end{bmatrix}^{k}, \tag{17}$$

$$\begin{bmatrix} \overline{V}_{r}^{1} \\ \overline{V}_{r}^{2} \\ \overline{V}_{r}^{3} \end{bmatrix}^{k} = \begin{bmatrix} \overline{V}_{s}^{1} \\ \overline{V}_{s}^{2} \\ \overline{V}_{s}^{3} \end{bmatrix}^{k} - \left[\dot{\mathbf{Z}}_{\mathrm{sr}}\right] \begin{bmatrix} \overline{J}_{ls}^{1} \\ \overline{J}_{ls}^{2} \\ \overline{J}_{ls}^{3} \end{bmatrix}^{k}. \tag{18}$$

In particular, in (16), for the receiving bus and the phase p, S_r^p, $\overline{I}_r^p$, $\dot{Y}_r^p$, and $\overline{V}_r^p$ are the scheduled (known) apparent power, the current injection, the admittance of all shunt elements and the phase-voltage, respectively; k is the iteration number. In (17), which is used in the backward sweep, once again for the receiving bus and the phase p, $\overline{J}_{ls}^p$ and $\overline{J}_{rm}^p$ are the current flow on line section l and the mth line connected to the receiving bus; M is the set of all the lines connected to the receiving bus. Finally, in (18), which is used in the forward sweep, $\left[\dot{\mathbf{Z}}_{\mathrm{sr}}\right]$ is the three-phase series impedance matrix of the line between sending and receiving busbars. The above algorithm was extended in [29] for use in analyzing convergence properties. A similar approach has been also used in [28].

3 Load-Flow Equations for Distribution Systems with Wind Farms

Load-flow equations for both balanced and unbalanced distribution systems can easily be integrated with the equations that describe the models of the Wind Turbine Generator Units (WTGUs), as will be shown in the next sub-sections. These models depend on the whole wind energy conversion system and usually are classified by the following two criteria. The first criterion refers to the presence or absence of power electronic devices; with such a classification we have systems [3]:

1. without power electronics (i.e., induction generators directly connected to the distribution system);
2. with partially-rated power converters (i.e., induction generators with a rotor-resistance converter or doubly-fed induction generator);
3. with full-scale power electronic devices (i.e., induction generators or synchronous generators with static converters with active and reactive powers control).

The second classification, which is used in this chapter, refers to the speed characteristics and includes [1]:

1. fixed speed WTGUs (induction generators directly connected to the distribution system that are driven by wind turbines with either a fixed turbine blade angle or a pitch controller to regulate the blade angle);
2. semi-variable-speed WTGUs (induction generators with a rotor-resistance converter);
3. variable-speed WTGUs (doubly-fed induction generators, synchronous/induction generators with full-scale static converters).

3.1 Load-Flow Equations for Balanced Systems

In this sub-section, the main models of WTGUs that have appeared in the relevant literature are presented first considering the classification based on speed characteristics, and, then, their inclusion in the load-flow equations is discussed.

3.1.1 Fixed-Speed WTGUs

This category includes the models of induction generators that are directly connected to the distribution system. Strictly speaking, the rotor speed of an asynchronous generator varies, but the variations are very small, so such generators can be considered as fixed-speed WTGUs. Even though the favorite type of wind generator today is the doubly-fed induction generator (Sect. 3.1.3), the installed capacity of WTGUs is overwhelmingly based on asynchronous generators. For this reason perhaps, the fixed-speed WTGU is the most analyzed in the relevant literature in terms of models to be included in the load-flow. Some of the main proposed models will be analyzed in this sub-section. They can be classified essentially in two different categories that depend on the way the WTGU is represented in the load-flow equations; these models are identified as the PQ-bus model and the RX-bus model. In the PQ-bus model, the WTGU is described by its active and reactive powers; in the RX-bus model, the WTGU is described by an impedance once the generator parameters and the slip are known. In the following, some of the most popular PQ-bus and RX-bus models are shown.[1]

In [4], starting from the well-known equivalent circuit of an asynchronous generator in steady-state condition (Fig. 2a), both a PQ-bus model and an RX-bus model were proposed. In Fig. 2a, $\overline{V}$ is the bus voltage, X is the sum of the rotor X_r and stator X_s leakage reactances, X_m is the magnetizing reactance, X_c is the reactance of the capacitor bank used for power factor improvement, R is the sum of the stator R_s and rotor R_r resistances, and s is the slip.

[1] It should also be noted that a so-called PX-bus model was proposed in [14], with P being the active power and X being the generator's non-linear magnetizing reactance. However, this model was not particularly popular.

Fig. 2 Equivalent circuits of an asynchronous generator **a** Γ circuit **b** T circuit

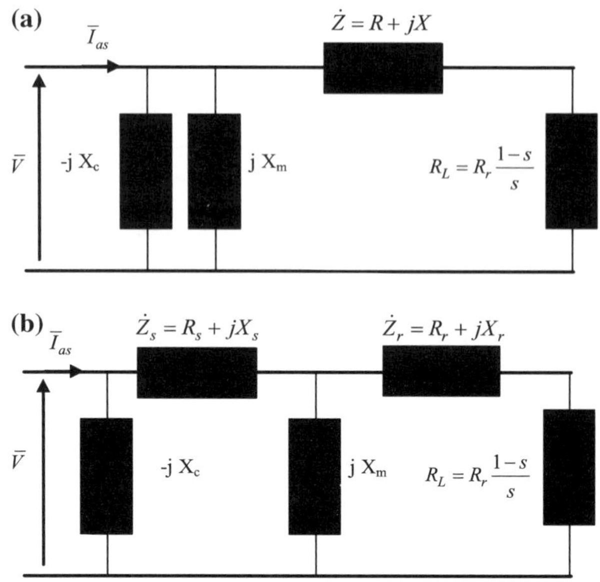

Neglecting the active power losses in the PQ-bus model, the generated active power P is assumed to be equal to the mechanical power P_m (wind turbine mechanical power output). The mechanical power is assumed constant and calculated as a function of the wind speed by means of the power curve for the turbine.

The reactive power Q is calculated using a quadratic equation that depends on the active power. The following expressions are proposed (Fig. 2a)[2]:

$$Q = -\left(V^2 \frac{X_c - X_m}{X_c X_m} + X \frac{V^2 - 2RP_m}{2(R^2 + X^2)} - X \frac{\sqrt{(V^2 - 2RP_m)^2 - 4P_m^2(R^2 + X^2)}}{2(R^2 + X^2)}\right) \tag{19a}$$

$$Q = -\left(V^2 \frac{X_c - X_m}{X_c X_m} + \frac{X}{V^2} P_m^2\right) \tag{19b}$$

The expression (19a) is obtained by applying Boucherot's theorem to the equivalent circuit of an asynchronous generator in Fig. 2a, while the approximated expression (19b) is obtained by neglecting the resistance R and the voltage drop on the leakage reactance X. Both expressions clearly show the dependence of reactive power on active power and voltage.

[2] It is important to highlight that the signs of the expressions of the active and reactive powers of WTGUs reported in this section are coherent with the load flow equations recalled in the Sects. 2.1 and 2.2.

In [4], the above mentioned model was included in a Newton-type algorithm; in particular, the expression (19b) was used. All busbars with WTGUs are considered to be classical PQ-bus. The active power (positive) is specified off-line [Eq. (1)] and, then, considered constant in the iterative load-flow calculations, while the reactive power [Eq. (19b)] is dependent on the bus voltage so that it is not specified off-line and considered constant, but it is updated in a sequential way; in practice, in contrast to a conventional PQ-bus in which reactive power is assigned off-line and remains constant during iterations, the reactive power is considered to vary across iterations, and its value is updated in each iteration on the basis of the value of the current bus voltage. It is, however, specified that we can also consider the voltage in the reactive power to be a constant value. Obviously, in this case, both active and reactive powers can be specified off-line and, then, remain unchanged during the iterations; this assumption, in normal operating conditions, does not seem to introduce significant errors. This model was also applied in [13].

In the RX model, as previously mentioned, the WTGU is represented by an impedance that is based on the steady-state model of the asynchronous generator in Fig. 2b. The result is:

$$\dot{Z}_g = (R_s + jX_s) + \frac{jX_m\left(\frac{R_r}{s} + jX_r\right)}{\frac{R_r}{s} + j(X_r + X_m)}, \tag{20}$$

with an added shunt capacitor of reactance X_c.

For the sake of simplicity, we refer to only one WTGU, and the following procedure is proposed to solve the load-flow equations, including the aforementioned WTGU model:

1. Let $s = s_r$ in each asynchronous machine, with s_r being the rated slip. Calculate the first value of $\dot{Z}_g$ with (20).
2. Calculate the admittance $\dot{Y}_g = 1/\dot{Z}_g$ and include it in the system admittance matrix.
3. Run the load-flow. Using the voltages at the bus where the WTGU is located and the current slip value, calculate the rotor current $\bar{I}_R$ and the mechanical power of the machine ($P_m = -R_r \frac{1-s}{s} I_R^2$, Fig. 2b).
4. With the current slip value and the assigned wind speed, calculate the tip speed ratio, the power coefficient, and, then, the wind turbine power.
5. Compare the mechanical power values obtained in steps 3. and 4. If the mismatch is lower than a pre-specified tolerance, stop the algorithm; otherwise, calculate an updated value of the slip and go to step 2.

The updated value of the slip is calculated using a Newthon-type iterative procedure that, for assigned bus voltage (the value obtained in step 3. of the above procedure), solves an equation that equates the wind turbine power and the mechanical power of the machine expressed in the unknown slip. In practice, the state of the system is obtained by using two sequential iterative processes, i.e., A. the classical load-flow analysis in which the WTGU bus is considered a PQ-bus with active and reactive powers equal to zero and B. the calculation of the slip of the machine using the balance between the mechanical power of the wind turbine and the electrical machine. In the discussion of the paper [4] reported in [5], it is suggested to use the unified solution method instead of the sequential method proposed by the authors. The unified solution method simply solves all equations together (load-flow equations and the equation that equates the mechanical power of the machine and the power taken from the wind).

The unified solution method is used in [6]. The WTGU is represented as a PQ-bus with active and reactive power that varies during the iterations; starting from the asynchronous equivalent circuit in Fig. 2a, in which stator resistance and capacitor susceptance are neglected, the following expressions for the two powers are used:

$$\begin{aligned} P &= -\frac{V^2 R_r s}{R_r^2 + s^2 X^2} \\ Q &= -\left(\frac{V^2}{X_m} + \frac{sPX}{R_r}\right) \end{aligned} \quad . \tag{21}$$

It should be noted that Eq. (21) are obtained considering the unknown machine slip both in active and reactive power outputs, followed by the further use of the equation obtained by equating the mechanical power of the machine and the wind turbine power.

The algorithm used to solve the load-flow equations is the well-known Newton–Raphson algorithm.

In [2], a new WTGU model was proposed for incorporation in a sweep-based algorithm. Starting from the equivalent circuit of the asynchronous generator of Fig. 2a, in which a resistance R_m is included in the magnetizing branch and the presence of the capacitor is neglected, the following expressions for the active and reactive power outputs are obtained:

$$\begin{aligned} P &= -\left(\frac{V^2}{R_m} - \frac{R^2 P_m}{Z^2} + \frac{RV^2}{2Z^2} - \frac{R\sqrt{(2RP_m - V^2)^2 - 4P_m^2 Z^2}}{2Z^2} + P_m\right) \\ Q &= -\left(\frac{V^2}{X_m} - \frac{XRP_m}{Z^2} + \frac{XV^2}{2Z^2} - \frac{X\sqrt{(2RP_m - V^2)^2 - 4P_m^2 Z^2}}{2Z^2}\right). \end{aligned} \tag{22}$$

In [2] it seems that, for a specified wind speed, the mechanical power P_m in (22) is assigned off-line, independently from the slip value, and its sign is negative since it is the mechanical power that is transferred from the rotor to the stator.

It should be noted that the reactive power expressions are the same as those in Eq. (19a or 19b), while the active power output P is different from the one used in the PQ-bus model proposed in [4], because it includes the active power losses and the mechanical power P_m.

The following procedure is used for the steady-state analysis of a distribution system with WTGU:

1. initialize the bus voltages of the distribution system and the mechanical power of the WTGUs for an assigned wind speed. The iteration count i = 1.
2. Calculate active and reactive power of the WTGUs using Eqs. (22).
3. Calculate loads and currents of the WTGUs using the equation:

$$\overline{I} = \left(\frac{P + jQ}{\overline{V}}\right)^*, \tag{23}$$

   where $\overline{V}$ is the bus voltage at the ith iteration, and P and Q are the active and reactive powers of the loads and the WTGU, respectively.
4. Calculate active and reactive power transferred from each branch, including losses, and each line current using the WTGU's loads and powers, respectively.
5. Calculate each line receiving end bus voltages using forward voltage formulation of the particular sweep-based load-flow algorithm.
6. If the convergence tolerance is satisfied, go to step 7; otherwise, set i = i + 1, and go to step 2.
7. Print results.

In [2], the effects of WTGUs on power losses and the voltage profile of two distribution systems were evaluated by comparing the results obtained from a number of sweep-based algorithms from the references [20–27].

Two WTGU models that are substantially similar to the one represented by Eq. (22) were proposed in [12]. These models simply use different expressions for the receiving voltage to be included in the equations that define the active and reactive power outputs. The models are included in a sweep-based, load-flow algorithm and used also to simulate the distribution system in the MATLAB/ Simulink SimPowerSystems Blockset.

Finally, as a last example of fixed-speed WTGUs, Ref. [1] proposed a PQ-bus model in which the active power output P is assumed to be equal to the value furnished by the manufacturer (in the form of the power curve) for an assigned wind speed value.

Since the active power output can be expressed as a function of the bus voltage and slip as (Fig. 2b without the capacitor); it results:

$$P = \frac{\left[R_s\left(R_r^2 + s^2(X_m + X_r)^2\right) + sR_rX_m^2\right]V^2}{\left[R_rR_s + s\left(X_m^2 - (X_m + X_r)(X_m + X_s)\right)\right]^2 + \left[R_r(X_m + X_s) + sR_s(X_m + X_r)\right]^2}, \tag{24}$$

and, since P and all the machine parameters are known, Eq. (24) can be rewritten as a quadratic equation to determine the unknown slip s and solved as:

$$s = \min\left|\frac{-b \pm \sqrt{b^2 - 4ac}}{2a}\right|$$

where

$$\begin{aligned} a &= PR_s^2(X_m + X_r)^2 + P[X_m X_r + X_s(X_r + X_m)]^2 - V^2 R_s (X_r + X_m)^2 \\ b &= 2PR_s R_r X_m^2 - V^2 R_r X_m^2 \\ c &= PR_r^2(X_m + X_s)^2 + P(R_r R_s)^2 - V^2 R_r^2 R_s. \end{aligned} \tag{25}$$

Finally, once again on the basis of the equivalent circuit in Fig. 2b without a capacitor, the output reactive power can be expressed as a function of the bus voltage and the calculated slip as:

$$Q = -\frac{\left[X_m X_r s^2 (X_m + X_r) + X_s s^2 (X_m + X_r)^2 + R_r^2 (X_m + X_s)\right] V^2}{\left[R_r R_s + s(X_m^2 - (X_m + X_r)(X_m + X_s))\right]^2 + \left[R_r(X_m + X_s) + sR_s(X_m + X_r)\right]^2} \tag{26}$$

It should be noted that the active and reactive power expressions (24) and (26) are different from the previously shown expressions (22) (obtained under various hypotheses from the equivalent circuit in Fig. 2a), because they are obtained now from the equivalent circuit in Fig. 2b in which the magnetizing branch is located between the parameters of the stator and the rotor.

Reference [1] included the above-mentioned PQ-bus model in a sweep-based algorithm, converting the powers in complex current injections. It should be noted that, in [2], it was shown that Eq. (25) has multiple solutions and has a very complicated structure with multiple formulas that cause a computational burden in the load-flow analysis.

3.1.2 Semi-Variable Speed WTGUs

This category includes WTGUs with pitch-controlled wind turbines and induction generators with wound rotors that are connected to an external resistance that is varied by means of a power electronic converter. Thanks to the presence of two controllers, i.e., a pitch controller and a resistance controller, the active power output of the WTGU is equal to maximum power at wind speed below rated and equal to rated power above rated wind speed. The power values can be derived from the power curve furnished by the manufacturer. The computation of reactive power output requires attention, because the rotor resistance is variable and unknown, as is the machine slip. Fortunately, Divya and Rao in [1] suggested a very simple procedure to overcome the problem. They reduced the two unknowns (slip and rotor resistance) to only one unknown by considering that the expressions of active and reactive power outputs can be written as a function of only unknown $R_{eq} = R_r/s$. In fact, by manipulating Eq. (24) so as to express active power output as a quadratic function of only the new unknown R_{eq} and solving for it, the following result is obtained:

$$R_{eq} = \min\left|\frac{-b_1 \pm \sqrt{b_1^2 - 4a_1c_1}}{2a_1}\right|$$

where

$$\begin{aligned}
a_1 &= P\left[R_s^2 + (X_m + X_s)^2\right] - V^2R_s^2 \\
b_1 &= 2PR_sX_m^2 - V^2X_m^2 \\
c_1 &= PR_s^2(X_m + X_s)^2 + P\left(X_m^2 - (X_m + X_r)(X_m + X_s)\right)^2 - V^2R_s(X_m + X_r)^2.
\end{aligned} \tag{27}$$

Once R_{eq} is known, the reactive power output can easily be calculated with the following equation:

$$Q = -\frac{\left\{R_{eq}^2(X_m + X_s) - \left[X_m^2 - (X_m + X_r)(X_m + X_s)\right](X_m + X_r)\right\}V^2}{\left\{R_{eq}R_s + \left[X_m^2 - (X_m + X_r)(X_m + X_s)\right]\right\}^2 + \left[R_{eq}(X_m + X_s) + R_s(X_m + X_r)\right]^2}. \tag{28}$$

Reference [1] also included the above-mentioned PQ-bus model in a sweep-based algorithm, converting the powers in complex current injections as well.

3.1.3 Variable-Speed WTGUs

This category includes WTGUs with pitch-controlled wind turbines and with either doubly-fed induction generators (DFIG) or asynchronous and synchronous generators with back-to-back converters (GBBC).

Recall that doubly-fed induction generators are asynchronous machines that are directly connected to the grid with a wound rotor and slip rings; a converter is connected to the rotor circuit through the slip rings. A speed variation of ±30 % around the synchronous speed can be obtained by controlling this converter. In addition, both active and reactive output powers can be controlled.

Generators with back-to-back converters are connected to the grid through a full-power, voltage-source converter. Both active and reactive output powers can be controlled.

The WTGU controllers usually operate so that the active power output is equal to maximum power at wind speed below rated and equal to rated power above rated wind speed. The power values can be derived from the power curve furnished by the manufacturer. The reactive power can be controlled by converters and, then, assumed as assigned on the basis of the set point; in some cases, the controllers operate so that the power factor is maintained at a fixed value.

In [7] and in [1], then, both WTGUs were modeled as an equivalent PQ-bus.

The active power output, as previously shown, is assumed to be equal to maximum power at wind speeds below rated and equal to rated power above rated wind speed. For the reactive power output, the following expression applies:

$$Q = -Q^{sp}$$

or

$$Q = -P\frac{\sqrt{1-\cos^2(\varphi)^{sp}}}{\cos(\varphi)^{sp}}, \tag{29}$$

where Q^{sp} and $\cos(\varphi)^{sp}$ are the specified values of reactive power and power factor, respectively. The same model is applied in [13] in which the specified reactive power is assumed to be zero.

Reference [7] used the well-known Newton–Raphson algorithm to solve the load-flow equations, including the WTGU models.

In [1], the above-mentioned PQ-bus model was included in a sweep-based algorithm, converting the powers in complex current injections as well.

In [7] as in [13], it was noted that the above WTGU might also be controlled such that the output voltage is kept constant. In this case, the WTGU can be modeled as a PV-bus (i.e., constant active power output and constant bus voltage). This model can easily be included in the load-flow equations and solved with a classical Newton–Raphson algorithm.

3.1.4 Variable-Speed WTGUs in Abnormal Voltage Conditions

In both WTGUs considered in Sect. 3.1.3, the active and reactive power outputs can be limited, since the controllers have limiters that restrict the possible values of the active component, the reactive component, and the total current magnitudes. The limiters have no influence at normal operating conditions (bus voltages around ±10 % of the rated value), because the machine is designed so that none of the above limits is violated. However, in some cases, these limits can be violated (mainly in distribution systems in which poor voltage regulation exists), and the use of limiters should be considered.[3]

In [1], an algorithm was proposed to take into account limiters with different control objectives. They considered that, in the most general case, each component of the current (active component, reactive component, and total current magnitudes) should not exceed their admissible value. Then, the model to be included in the load-flow should take into account that each current component must be calculated and checked to determine if it violates the limit. In the absence of a limit violation, the procedure of Sect. 3.1.3 is applied. When there is a limit violation,

[3] The fixed-speed and semi-variable-speed WTGUs are usually tripped in case of abnormal voltage conditions.

the active and reactive power outputs must be recalculated in order to satisfy the limits. The procedure to be implemented in case of violation is different in the case when the violation refers to the total current and maintaining the reactive power output constant has a high priority (in this case, the active power output is reduced in order to maintain the specified reactive power). The procedure is also different for the case in which the violation refers to the active or reactive current component. For more details about the procedure for accounting for these violations, see Ref. [1].

Once again, in [1], the above-mentioned procedure was included in a sweep-based algorithm, converting the powers in complex current injections as well.

3.2 Load-Flow Equations for Unbalanced Systems

As previously shown, several researchers have dealt with the models of WTGUs for the case of balanced systems, practically considering all the possible WTGU configurations. In the relevant literature, only a few papers have been published that deal with unbalanced systems. Some of these publications will be reported in this sub-section.

In [8], a three-phase model was proposed for fixed-speed WTGUs (induction generators connected directly to the distribution system).

Each asynchronous machine is represented in the three-phase load-flow equations by equations that represent the real and imaginary currents at each of the three phases of the bus where the WTGU is located. The balance equations are:

$$
\begin{aligned}
I^{p}_{asR,i} &= \sum_{k=1}^{n_{bus}} \sum_{m=1}^{3} V^{m}_{k} \left[G^{pm}_{ik} \cos\theta^{pm}_{ik} - B^{pm}_{ik} \sin\theta^{pm}_{ik} \right] \\
I^{p}_{asI,i} &= \sum_{k=1}^{n_{bus}} \sum_{m=1}^{3} V^{m}_{k} \left[G^{pm}_{ik} \sin\theta^{pm}_{ik} + B^{pm}_{ik} \cos\theta^{pm}_{ik} \right],
\end{aligned}
\tag{30}
$$

where $I^{p}_{asR,i}$ and $I^{p}_{asI,i}$ are the real and imaginary components, respectively, of the phase currents injected in the network phases by the asynchronous machine. For an explanation of the other symbols in (30), see Sect. 2.2.

The real and imaginary components of the phase currents injected by the asynchronous machine in (30) can be obtained by starting from the well-known equivalent circuit of the three-phase induction machine without capacitor shown in Fig. 2b. The circuit of Fig. 2b applies to both positive and negative sequence networks; the only difference between the two is the value of the "equivalent resistance" R_L as defined by:

$$
\begin{aligned}
R_{Ld} &= \frac{1 - s_d}{s_d} R_r \\
R_{Li} &= \frac{1 - s_i}{s_i} R_r,
\end{aligned}
\tag{31}
$$

where the positive sequence slip s_d is given by:

$$s_d = \frac{n_s - n}{n_s}, \tag{32}$$

with n_s synchronous speed and n rotor speed; as is well known, the negative sequence slip s_i can be expressed as a function of the positive sequence slip s_d as:

$$s_i = 2 - s_d. \tag{33}$$

Then, the sequence impedances depend only on the positive sequence slip s_d of the asynchronous machine.

The phase currents of the asynchronous machine can be expressed as:

$$[\mathbf{I}_{as}] = [\mathbf{A}][\mathbf{Y_1}][\mathbf{A}]^{-1}[\mathbf{V_\Delta}] \tag{34}$$

where:

$$[\mathbf{A}] = \begin{bmatrix} 1 & 1 & 1 \\ \exp(j\frac{4}{3}\pi) & \exp(j\frac{2}{3}\pi) & 1 \\ \exp(j\frac{2}{3}\pi) & \exp(j\frac{4}{3}\pi) & 1 \end{bmatrix} \tag{35}$$

$$[\mathbf{Y_1}] = \begin{bmatrix} \frac{1}{\sqrt{3}}\exp(-j\frac{1}{6}\pi)\dot{Y}_{gd} & 0 & 0 \\ 0 & \frac{1}{\sqrt{3}}\exp(j\frac{1}{6}\pi)\dot{Y}_{gi} & 0 \\ 0 & 0 & 1 \end{bmatrix} \tag{36}$$

In (34), $[\mathbf{V_\Delta}]$ is the vector of the line-to-line voltages (expressed as the difference between two phase voltages) at the bus feeding the asynchronous machine, and, in (36), the sequence admittances $\dot{Y}_{gd}, \dot{Y}_{gi}$ are given by:

$$\begin{aligned} \dot{Y}_{gd} &= \frac{\dfrac{1}{R_s + jX_s}\left[\dfrac{R_r + R_{Ld} + j(X_r + X_m)}{(R_r + R_{Ld} + jX_r)jX_m}\right]}{\dfrac{1}{R_s + jX_s} + \left[\dfrac{R_r + R_{Ld} + j(X_r + X_m)}{(R_r + R_{Ld} + jX_r)jX_m}\right]} \\ \dot{Y}_{gi} &= \frac{\dfrac{1}{R_s + jX_s}\left[\dfrac{R_r + R_{Li} + j(X_r + X_m)}{(R_r + R_{Li} + jX_r)jX_m}\right]}{\dfrac{1}{R_s + jX_s} + \left[\dfrac{R_r + R_{Li} + j(X_r + X_m)}{(R_r + R_{Li} + jX_r)jX_m}\right]}, \end{aligned} \tag{37}$$

where the "equivalent resistances" are given by (31). Analogous relationships can be obtained by considering the equivalent circuit of the simplified induction machine, which is characterized by the magnetizing reactance applied to the input terminals before the stator parameters.

As the analysis of Eq. (34) indicates, a further unknown exists, i.e., the machine positive sequence slip. Therefore, an additional equation must be added, i.e., (neglecting the active power losses) the balance equation that links the mechanical power, given by the power curves of the wind turbines, to the three-phase active powers P_{el} generated by the asynchronous machine, given by:

$$P_{el} = \text{real}\left([\mathbf{V}_f]^T[\mathbf{I}^*_{as}]\right), \tag{38}$$

where:

$$[\mathbf{V}_f] = [\mathbf{T}_r][\mathbf{V}_\Delta] = [\mathbf{A}][\mathbf{T}][\mathbf{A}]^{-1}[\mathbf{V}_\Delta] \tag{39}$$

$$[\mathbf{T}] = \begin{bmatrix} \frac{1}{\sqrt{3}}\exp(-j\frac{1}{6}\pi) & 0 & 0 \\ 0 & \frac{1}{\sqrt{3}}\exp(j\frac{1}{6}\pi) & 0 \\ 0 & 0 & 1 \end{bmatrix} \tag{40}$$

and $\left[\mathbf{I}^*_{as}\right]$ is the matrix of the conjugated of the phase currents present in the matrix $[\mathbf{I}_{as}]$ given by (34).

In [8], the model was included in a three-phase load flow solved by a classical Newton-Raphson algorithm. No problem should arise in also including the model in a sweep-based algorithm.

The same authors in [9] represented the WTGU as an equivalent, three-phase PQ-bus in the frame of a load flow solved by a Newthon-Raphson algorithm, using PQ balance equations obtained starting from the current expressions presented above.

In [10], a very simplified model was proposed for induction generators that are directly connected to an unbalanced distribution system. This model assumes that the reactive power for each phase is a function of its active power, that is:

$$Q^1 = -a_0 - a_1 P - a_2 P^2 \tag{41}$$

where a_0, a_1, and a_2 are experimentally obtained constants [4].

In [10], the model was included in the three-phase load flow shown in Sect. 2.2.

In [29], Zhu and Tomsovic classified the dispersed generators as PQ or PV nodes. For PQ nodes, the models are identical to those used for constant power load models except that, obviously, the current is injected in the bus. In the case of PV nodes, if the computed reactive power is out of the reactive generation limit, then the reactive power generation is set to this limit, and the generation unit is considered to be a PQ bus.

4 Numerical Applications

Numerical applications are organized in three different sections:

- in Sect. 4.1, the performance of different WTGUs is analyzed for assigned bus voltages and wind speed;
- in Sect. 4.2, the impact of wind generating units on a 17-bus balanced system is assessed by means of load-flow analysis;
- in Sect. 4.3, the analysis conducted in Sect. 4.2 is repeated for the IEEE 34-bus unbalanced test distribution system.

Table 1 Electrical parameters of the induction generators

	Type of wind generating units		
	Stall-regulated, fixed-speed [1]	Pitch-regulated, fixed-speed [30]	Semi-variable speed [1]
Nominal power (MW)	1.0	0.6	1.0
Nominal voltage (V)	690	690	690
R_s (p.u.)	0.007141	0.009199	0.005671
X_s (p.u.)	0.21552	0.09452	0.15250
R_r (p.u.)	0.00630	0.008191	0.00462
X_r (p.u.)	0.088216	0.113421	0.096618
X_m (p.u.)	3.3606	4.0957	2.8985
X_c (p.u.)	3.3606	4.0957	2.8985

Since no convergence problems were experienced for any of the numerical examples, a classical Newthon-Raphson algorithm was used.

4.1 Performance of Different Wind Turbine Generating Units

The performance of stall-regulated (fixed-speed), pitch-regulated (fixed-speed) and semi-variable-speed WTGUs was analyzed in terms of active (reactive) power generated (absorbed), varying both the wind speed and the terminal supply voltage. The circuit parameters of the induction generator are reported in Table 1 [1, 30]; for all the considered WTGUs, the reactance X_c of the capacitor bank is assumed to be equal to the magnetizing reactance X_m of the corresponding induction generator.

The models of the WTGUs illustrated in Sect. 3.1 are considered and applied. In particular, for the stall-regulated fixed-speed WTGU the RX model is applied.

For the pitch-regulated fixed-speed WTGU the Eq. (26) is applied and for the semi-variable speed WTGU the Eq. (28) is applied to calculate the absorbed reactive power; the injected active power, instead, is assumed to be equal to the value furnished by the manufacturer (in the form of the power curve).

The performance of the variable-speed WTGU is not considered here since this type of WTGU is characterized by a generated active power equal to the maximum power at wind speeds below the rated speed and equal to the rated power above the rated speed; then, for an assigned wind speed value, the active power output for this type of WTGU can be assumed to be equal to the value furnished by the manufacturer in the form of the power curve; instead, the absorbed reactive power can be controlled by a converter and then assumed as assigned on the basis of the set point.

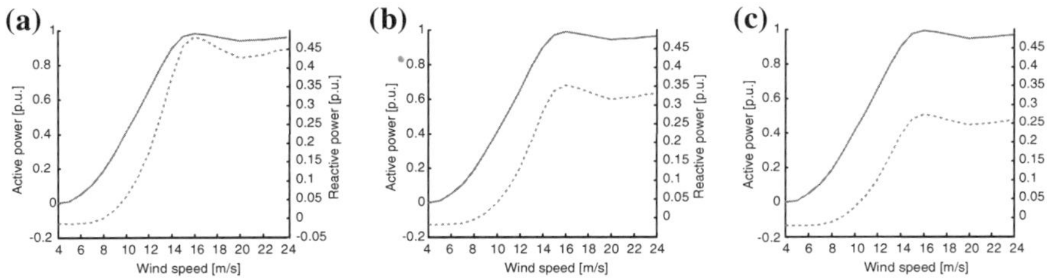

Fig. 3 Active (continuous line) and reactive (dashed line) power of a stall-regulated, fixed-speed WTGU (P = 1 MW) versus wind speed: **a** V = 0.9 p.u., **b** V = 1.0 p.u., **c** V = 1.1 p.u

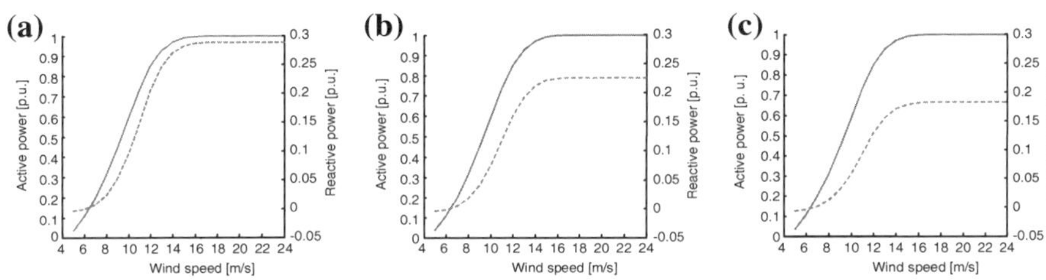

Fig. 4 Active (continuous line) and reactive (dashed line) power of a pitch-regulated, fixed-speed WTGU (P = 0.6 MW) versus wind speed: **a** V = 0.9 p.u., **b** V = 1.0 p.u., **c** V = 1.1 p.u

The variation of active power (blue line) and reactive power (red line) with wind speed for three different terminal voltages values (0.9, 1.0, and 1.1 p.u) are shown in Figs. 3, 4, and 5.

In the case of the stall-regulated, fixed-speed WTGU, it is evident that:

- the active power increases with increases in wind speed from cut-in to nominal wind speed, and, beyond nominal speed, it decreases marginally; in addition, for a given wind speed the active power marginally increases as voltage increases.
- the reactive power demand increases as wind speed increases up to the nominal speed, and, beyond the nominal wind speed, it decreases slightly. The variation of the terminal voltages, different from active power, has a greater impact on reactive power. For a given wind speed, as terminal voltage increases, the reactive power demand is reduced significantly.

In the case of the pitch-regulated, fixed-speed WTGU:

- the active power is not influenced by variations in the terminal voltage, but it varies with wind speed in accordance with the power curve of the machine.
- the reactive power demand increases as the voltage amplitude decreases, and it increases as wind speed increases up to the nominal wind speed, after which the reactive power remains constant.

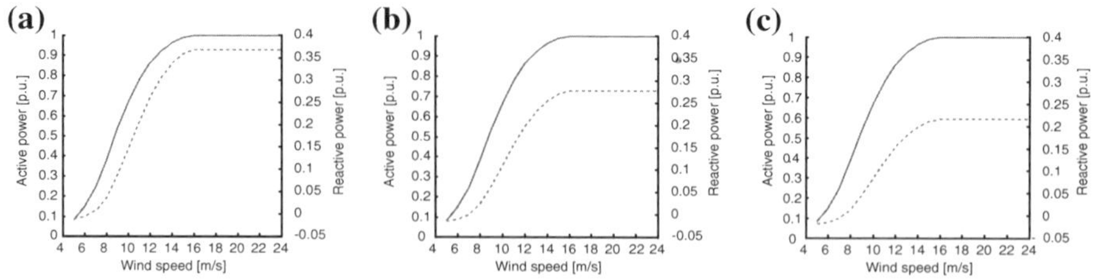

Fig. 5 Active (continuous line) and reactive (dashed line) power of a semi-variable speed WTGU (P = 1.0 MW) versus wind speed: **a** V = 0.9 p.u., **b** V = 1.0 p.u., **c** V = 1.1 p.u

In the case of the semi-variable speed WTGU:

- the active power is not influenced by variations in the terminal voltage; however, it increases with wind speed up to the nominal value following the power curve of the machine.
- the reactive power demand has the same behavior as the pitch-regulated, fixed-speed WTGU with respect to voltage and wind speed variations.

4.2 Load Flow of a Balanced Distribution System

The load-flow analysis for wind-driven generating units is applied to the 17-bus, balanced, three-phase network illustrated in Fig. 6 [31]. This system contains sixteen busbars at 12.5 kV and one busbar (#1) at 138 kV. The main data of the considered test system are reported in Tables 2 and 3; the system base is 10 MVA. Four case studies were considered with increasing wind-generation penetration:

Case I	No wind-driven generating unit
Case II	1.0-MW, semi-variable-speed WTGU at bus #9 where the wind speed = 10 m/s and 1.0-MW GBBC, variable-speed WTGU at bus #13 where the wind speed = 15 m/s
Case III	1.0-MW, semi-variable-speed WTGU at bus #9 where the wind speed = 10 m/s, 1.0-MW, GBBC, variable-speed WTGU at bus #13 where the wind speed = 15 m/s, 1.0-MW-DFIG, variable-speed WTGU at bus #15 where the wind speed = 13 m/s
Case IV	1.0-MW, semi-variable-speed WTGU at bus #9 where the wind speed = 10 m/s, 1.0-MW, GBBC, variable-speed WTGU at bus #13 where the wind speed = 15 m/s, 1.0-MW, DFIG, variable-speed WTGU at bus #15 where the wind speed = 13 m/s and 1.0-MW, stall-regulated, fixed-speed WTGU at bus #17 where the wind speed = 11 m/s.

Each WTGU is connected to the busbar of the distribution system through a 1.2-MVA transformer characterized by a reactance of 0.12 p.u. (referred to its base). The parameters of the induction generator of the stall-regulated, fixed-speed WTGU and the semi-variable-speed WTGU are reported in Table 1.

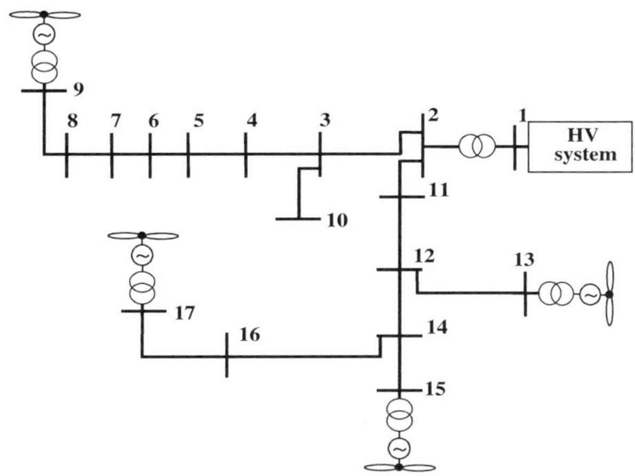

Fig. 6 17–busbar-balanced, 3-phase, test-distribution system [31]

Table 2 Bus data

Bus code	Linear loads		Capacitors
	P (MW)	**Q** (MVAr)	$\mathbf{Q}_{\mathrm{C}}$ (MVAr)
1	0	0	−1.2
2	0	0	–
3	0.2	0.12	−1.05
4	0.4	0.25	−0.6
5	1.5	0.93	−0.6
6	3.0	2.26	−1.8
7	0.8	0.5	–
8	0.2	0.12	−0.6
9	1.0	0.62	–
10	0.5	0.31	–
11	1.0	0.62	−0.6
12	0.3	0.19	−1.2
13	0.2	0.12	–
14	0.8	0.5	–
15	0.5	0.31	–
16	1.0	0.62	−0.9
17	0.2	0.12	–

Once again the models of the WTGUs illustrated in Sect. 3.1 are considered and applied.

For the stall-regulated, fixed-speed WTGU, both the RX and PQ models were considered.

For the semi-variable-speed WTGU Eq. (28) is applied to calculate the absorbed reactive power and the injected active power is assumed to be equal to the value furnished by the manufacturer (in the form of the power curve).

For the variable-speed WTGU (both the GBBC and DFIG configurations), the models proposed in [1] were applied; in particular the absorbed reactive power is

Table 3 Line data

Starting bus	Ending bus	R (%)	X_l (%)	X_c (%)
HV system		0.050	0.354	0
1	2	0.031	6.753	0
2	3	0.431	1.204	0.0035
3	4	0.601	1.677	0.0049
4	5	0.316	0.882	0.0026
5	6	0.896	2.502	0.0073
6	7	0.295	0.824	0.0024
7	8	1.720	2.120	0.0046
8	9	4.070	3.053	0.0051
3	10	1.706	2.209	0.0043
2	11	2.910	3.768	0.0074
11	12	2.222	2.877	0.0056
12	13	4.803	6.218	0.0122
12	14	3.985	5.160	0.0101
14	15	2.910	3.768	0.0074
14	16	3.727	4.593	0.0100
16	17	2.208	2.720	0.0059

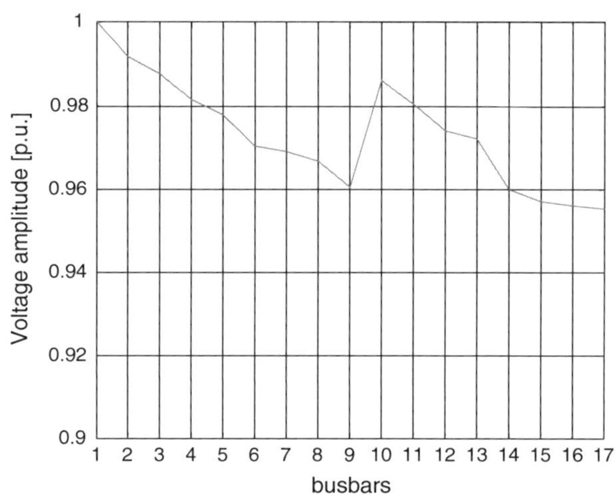

Fig. 7 Voltage profile in the absence of WTGUs: Case I

fixed at a value of zero (Eq. (29), with $\cos \varphi = 1$) and the injected active power is assumed to be equal to the value furnished by the manufacturer.

Figures 7, 8, 9, and 10 show the voltage profile obtained in all the considered cases; in particular, Fig. 10a, b refer to case IV, in which the stall-regulated, fixed-speed WTGU is modeled by the RX and PQ models, respectively.

Table 4 reports the summary of the load-flow results at busbars characterized by the presence of WTGUs. Finally, Table 5 reports the reduction of the real

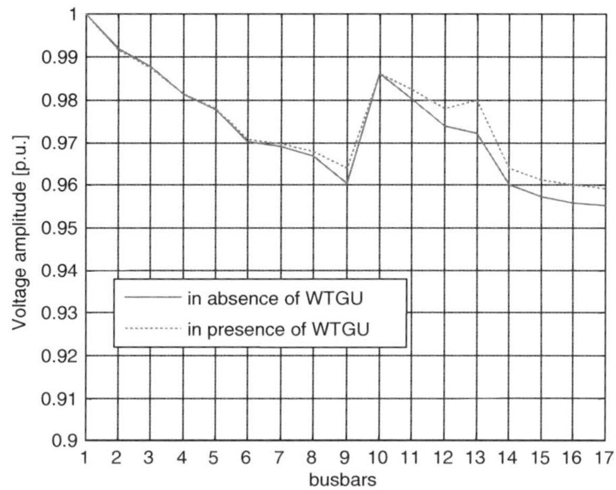

Fig. 8 Comparison between voltage profile in the absence of WTGUs (Case I) and in the presence of WTGUs at basbars #9 and #13 (Case II)

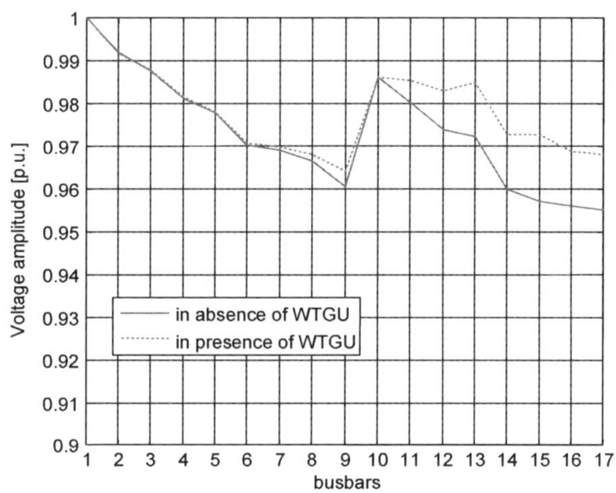

Fig. 9 Comparison between voltage profile in the absence of WTGUs (Case I) and in the presence of WTGUs at basbars #9, #13 and #15 (Case III)

power losses due to the presence of WTGUs (cases II, III, and IV) as compared to the power losses in absence of WTGUs (case I).

As was foreseeable, with the increase of the active power generated by WTGUs, the voltage magnitude increases, and the system power loss decreases. In particular, the increases in the voltage amplitude are more significant in the branches where the power generated by WTGUs is relevant with respect to the power-load demands.

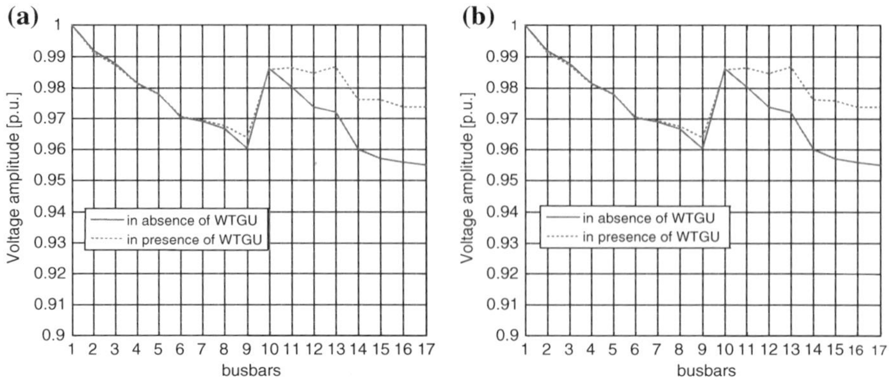

Fig. 10 Comparison between voltage profile in the absence of WTGUs (Case I) and in the presence of WTGUs at basbars #9, #13, #15, and #17 (Case IV). **a** RX model and **b** PQ model for the stall-regulated, fixed-speed WTGU at busbar #17

Table 4 Summary of load-flow results at WTGU busbars

Case	Bus	WTGU typology	Wind speed (m/s)	P (MW)	Q (MVAr)	Voltage (p.u.)
II	9	Semi-variable	10	0.67	0.12	0.96
	13	Variable GBBC	15	0.88	0	0.98
III	9	Semi-variable	10	0.67	0.12	0.96
	13	Variable GBBC	15	0.88	0	0.98
	15	Variable DFIG	13	1.0	0	0.97
IV	9	Semi-variable	10	0.67	0.12	0.96
	13	Variable GBBC	15	0.88	0	0.98
	15	Variable DFIG	13	1.0	0	0.97
	17	Stall regulated fixed speed (RX model)	11	0.52	0.08	0.97
IV	9	Semi-variable	10	0.67	0.12	0.96
	13	Variable GBBC	15	0.88	0	0.98
	15	Variable DFIG	13	1.0	0	0.97
	17	Stall regulated fixed speed (PQ model)	11	0.53	0.09	0.97

Table 5 Power loss reduction

Case	Percentage
II	24
III	43
IV[a]	51
IV[b]	51

[a] RX model, [b] PQ model for stall-regulated, fixed-speed WTGU

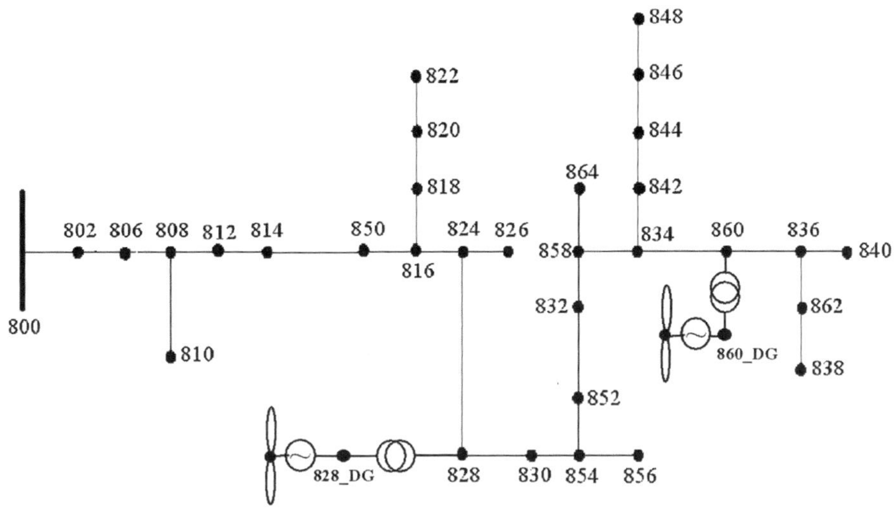

Fig. 11 IEEE 34-bus distribution test system [32]

4.3 Load Flow of an Unbalanced Distribution System

The load-flow analysis in presence of wind-driven generating units is also applied to a slightly modified version of the unbalanced, IEEE 34-bus test system shown in Fig. 11 [32]. The IEEE 34-bus test system is an actual distribution system with 83 system nodes and a voltage level of 24.9 kV. The only substation on the network is located above node 800 with the transformer from 69 to 24.9 kV; all the other nodes are just joints of branches.

This system contains a mixture of single-phase and three-phase lines and loads. The system lines 808–810, 816–818, 818–820, 820–822, 824–826, 854–856, 858–864, and 862–838 are single-phase lines, and the remaining lines are three-phase. The active and reactive load phase powers are reported in Table 6; the complete network data and parameters can be found in [32]. The analyzed network is characterized by the absence of voltage regulators. In this distribution system, one stall-regulated, fixed-speed WTGU of 330 kW was connected at bus #828, and one stall-regulated, fixed-speed WTGU of 330 kW was connected at bus #860; the complete generator data can be found in [4].

Figure 12 shows the phase voltage profile obtained in the absence and in the presence of WTGUs; Table 7 reports the summary of the load-flow results at busbars characterized by the presence of WTGUs.

The presence of wind-driven generators increases the magnitude of the voltages at all the buses; in addition, in the considered case, the real power losses of the system were reduced by about 52 %.

Table 6 Active and reactive load-phase powers

Node	P_a (kW)	Q_a (kVAr)	P_b (kW)	Q_b (kVAr)	P_c (kW)	Q_c (kVAr)
806	–	–	30	15	25	14
810	–	–	16	8	–	–
820	34	17	–	–	–	–
822	135	70	–	–	–	–
824	–	–	5	2	–	–
826	–	–	40	20	–	–
828	–	–	–	–	4	2
830	16.9	8.0	9.9	4.9	24.6	9.8
832	139.1	69.6	137.6	68.8	137.0	68.5
858	7	3	2	1	6	3
834	4	2	15	8	13	7
844	152.4	116.5	143.0	111.2	143.5	111.6
846	–	–	25	12	20	11
848	20	16	43	27	20	16
860	36	24	40	26	130	71
836	30	15	10	6	42	22
840	27.3	16.2	31.3	18.2	9.3	7.2
838	–	–	28	14	–	–
864	2	1	–	–	–	–
856	–	–	4	2	–	–

Fig. 12 Comparison between the voltage profile in the absence of WTGUs and in presence of WTGUs at busbars #828 and #860: **a** phase a, **b** phase b, and **c** phase c

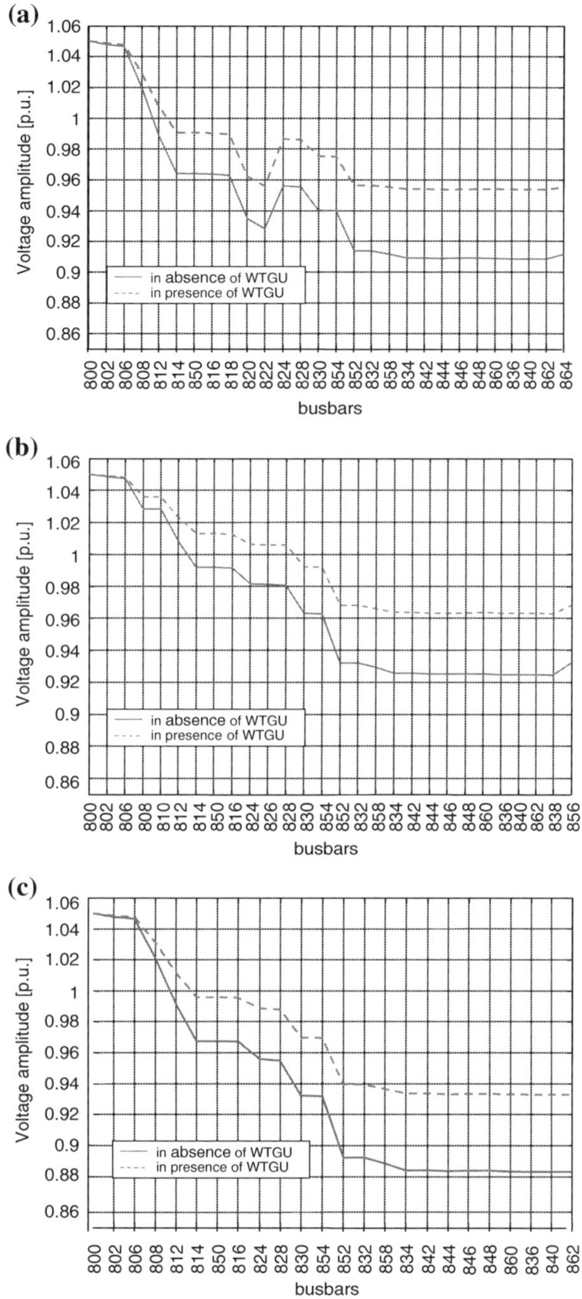

Table 7 Summary of load-flow results at WTGU busbars

Bus	WTGU typology	Wind speed (m/s)	Phase	P (MW)	Q (MVAr)	Voltage (p.u.)
			a	0.115	0.072	0.986
828	Stall-regulated, fixed-speed	15	b	0.117	0.075	1.00
			c	0.118	0.073	0.988
			a	0.114	0.073	0.954
860	Stall-regulated, fixed-speed	15	b	0.119	0.077	0.964
			c	0.117	0.070	0.933

5 Conclusions

In this chapter, some deterministic approaches for the load-flow analysis of balanced and unbalanced distribution systems with wind farms have been analyzed, considering fixed-speed, semi-variable-speed, and variable-speed wind generation systems.

Numerical applications of test distribution systems were discussed, considering various wind farm models.

The main conclusion of the chapter is that the steady-state analysis of a distribution system is mandatory in order to determine the voltage profile and the losses, which are strongly influenced by the presence of the wind farm. More accurate studies can be conducted using probabilistic approaches that are capable of taking into account the unavoidable time-varying nature of the wind speed and of the loads on the distribution system. These approaches are the subject of the companion chapter [16], in which a comparison of the features of deterministic and probabilistic approaches is presented.

References

1. Divya KC, Nagendra Rao PS (2006) Models for wind turbine generating systems and their application in load flow studies. Electr Power Syst Res 76(9–10):844–856
2. Eminoglu U, Dursun B, Hocoaglu MH (2009) Incorporating of a new wind turbine generating system model into distribution systems load flow analysis. Wind Energy 12(4):375–390
3. Blaabjerg F, Chen Z, Kjaer SB (2004) Power electronic as efficient interface in dispersed power generation systems. IEEE Trans Power Deliv 19(5):1184–1194
4. Feijòo AE, Cidràs J (2000) Modeling of wind farms in the load flow analysis. IEEE Trans Power Deliv 15(1):110–1115
5. Feijòo AE, Cidràs J (2001) Modeling of wind farms in the load flow analysis. IEEE Trans Power Deliv 16(4):951 (Discussion of the paper)
6. Liu B, Zhang Y (2008) Power flow algorithm and practical contingency analysis for distribution systems with distributed generation. Eur Trans Electr Power (Published online)
7. Zhao M, Chen Z, Blaabjerg F (2009) Load flow analysis for variable speed offshore wind farms. IET Renew Power Gener 3(2):120–132

8. Carpinelli G, Pagano M, Caramia P, Varilone P (2007) A probabilistic three-phase load flow for unbalanced electrical distribution systems with wind farms. IET Renew Power Gener 1(2):115–122
9. Caramia P, Carpinelli G, Pagano M, Varilone P (2005) A probabilistic load flow for unbalanced electrical distribution systems with wind farms. In: cigre symposium on power systems with dispersed generation, Athens, 13–16 April 2005
10. Teng JH (2008) Modelling distributed generations in three-phase distribution load flow. IET Gener Transm Distrib 2(2):330–340
11. Ellis A, Muljadi E (2008) Wind power plant representation in large-scale power flow simulations in WECC. In: Proceedings of IEEE power engineering society, general meeting, Pittsburgh, 20–24 July 2008
12. Eminoglu U (2009) Modeling and application of wind turbine generating system (WTGS) to distribution systems. Renew Energy 34(11):2474–2483
13. Coath G, Al-Dabbagh M (2005) Effect of steady-state wind turbine generator models on power flow convergence and voltage stability limit. In: Proceedings of Australasian universities power engineering conference (AUPEC 2005), vol 1. www.itee.uq.edu.au/~aupec/aupec05/AUPEC2005/Volume1/S113.pdf
14. Pecas Lopes JA, Maciel Barbosa FP, Cidras J (1991) Simulation of MV distribution networks with asynchronous local generation systems. In: Proceedings of 6th Mediterranean electrotechnical IEEE conference, vol 2. LJubljana, pp 1453–1456
15. Diduch CP, Rost A, Venkatesh B (2006) Distribution system with distributed generation load flow. In: Proceedings of large engineering systems conference on power engineering, Halifax, pp 55–60
16. Bracale A, Carpinelli G, Di Fazio A, Russo A (2014) Probabilistic approaches for steady state analysis in distribution systems with wind farms. In: Pardalos PM, Pereira MVF, Rebennack S, Boyko N Handbook of wind power systems. Energy Systems. Springer, Berlin. doi: 10.1007/978-3-642-41080-2_7
17. Arrillaga J, Arnold CP, Harker BJ (1990) Computer analysis of power systems. Wiley, New York
18. Zhang F, Cheng S (1997) A modified newthon method for radial distribution system power flow analysis. IEEE Trans Power Syst 12(1):389–397
19. Teng JH (2002) A modified gauss-seidel algorithm for three phase power flow analysis in distribution networks. Electr Power Energy Syst 24(2):97–102
20. Shirmohammadi D, Hong HW, Semlyen A, Luo GX (1988) A compensation-based power flow method for weekly meshed distribution and transmission networks. IEEE Trans Power Syst 3(2):753–762
21. Cespedes RG (1990) New method for the analysis of distribution networks. IEEE Trans Power Deliv 5(1):391–396
22. Haque MH (1996) Load flow solution of distribution systems with voltage dependent load models. Electr Power Energy Syst 36(3):151–156
23. Eminoglu U, Hocoaglu MH (2005) A new power flow method for radial distribution systems including voltage dependent load models. Electr Power Syst Res 76(1–3):106–114
24. Semlyen A, Luo GX (1990) Efficient load flow for large weekly meshed networks. IEEE Trans Power Syst 5(4):1309–1316
25. Rajicic D, Ackovski R, Taleski R (1994) Voltage correction power flow. IEEE Trans Power Syst 9(2):1056–1062
26. Satyanarayana S, Ramana T, Sivanagaraju S, Rao GK (2007) An efficient load flow solution for radial distribution network including voltage dependent load models. Electr Power Compon Syst 35(5):539–551
27. Cheng CS, Shirmohammadi D (1995) A three-phase power flow method for real-time distribution system analysis. Trans Power Syst 10(2):671–679
28. Thukaram D, Banda HMW, Jerome J (1999) A robust three phase power flow algorithm for radial distribution systems. Electr Power Energy Syst 50:227–236

29. Zhu Y, Tomsovic K (2002) Adaptive power flow method for distribution system with dispersed generation. Trans Power Deliv 17(3):822–827
30. Khadraoui MR, Elleuch M (2008) Comparison between optislip and fixed speed wind energy conversion systems. In: Proceedings of 5th international multi-conference on systems, signals and devices, Philadelphia University, Amman, Jordan
31. Chang WK, Grady WM, Samotji MJ (1993) Meeting IEEE-519 harmonic voltage and voltage distortion constraints with an active power line conditioner. In: IEEE-PES winter meeting, New York
32. Kersting WH (2001) Radial distribution test feeders. In: IEEE power engineering society winter meeting, 28 Jan–1 Feb, vol 2. pp 908–912 (website http://ewh.ieee.org/soc/pes/dsacom/testfeeders.html)

Probabilistic Approaches for the Steady-State Analysis of Distribution Systems with Wind Farms

A. Bracale, G. Carpinelli, A. R. Di Fazio and A. Russo

Abstract This chapter deals with probabilistic approaches for the steady-state analysis (probabilistic load flow) of distribution systems with wind farms. The probabilistic analysis is performed taking into account the randomness of both the distribution system loads and the wind energy production. Several approaches are presented to obtain the probability functions of state and dependent variables (e.g., voltage amplitudes and line flows). These approaches are mainly concentrated on wind farm probabilistic models, using one of the classical probabilistic techniques (e.g., Monte Carlo simulation, convolution process, and special distribution functions) to perform the probabilistic load flow. Numerical applications on a 17-bus balanced test distribution system and on an IEEE 34-bus unbalanced test distribution system are presented and discussed, considering the various wind farm models.

A. Bracale
Department of Engineering, Università degli Studi di Napoli Parthenope,
Centro Direzionale, Is. C4, Napoli, Italy
e-mail: antonio.bracale@uniparthenope.it

G. Carpinelli (✉)
Department of Electrical Engineering and of Information Technologies,
Università degli Studi di Napoli Federico II, Via Claudio 21, Napoli, Italy
e-mail: guido.carpinelli@unina.it

A. R. Di Fazio
Department of Electrical and Information Engineering,
Università degli Studi di Cassino e del Lazio Meridionale,
Via G. Di Biasio 43, Cassino, Italy
e-mail: a.difazio@unicas.it

A. Russo
Energy Department, Politecnico di Torino, Corso Duca degli Abruzzi 24, Turin, Italy
e-mail: angela.russo@polito.it

P. M. Pardalos et al. (eds.), *Handbook of Wind Power Systems*, Energy Systems,
DOI: 10.1007/978-3-642-41080-2_8,

1 Introduction

There is a growing imperative to study the problems of active distribution networks, including wind farms, and, in the relevant literature, significant attention has been paid to the steady-state analyses of distribution systems with wind generators. For the most part, these analyses have been conducted using the well-known load flow method. Both deterministic load flows, which assume that all the input data are known with certainty, and probabilistic load flows, which assume some input data are affected by uncertainties, have been proposed. This chapter deals with probabilistic load flow, while the companion chapter [1] deals with the deterministic approaches.

The importance of the probabilistic approaches is based on the fact that long- and medium-term system planning and the short-term operation (i.e., one day ahead operation) of electrical distribution systems are affected significantly by the uncertainties in wind energy production. Thus, as the market penetration of wind plants increases, the uncertainty of power system operation and management will also increase. This indicates a need for adequate probabilistic tools for long-term and daily operation of power systems, that may provide important advantages as a less expensive network in power planning and a better management of congestions in power operating.

In addition, problems in distribution systems with wind farms, such as undervoltages and overvoltages, are well known. It is especially important to avoid overvoltages, which can occur in the case of low loads and high wind energy production, due to the risk of incurring damage to system components. Using deterministic load flow analyses to ascertain such conditions can be inadequate, because these analyses should be conducted on some specific load and wind energy production conditions, which are not always easy to identify [2].

Probabilistic load flow is certainly a powerful tool to study a distribution system in the presence of uncertainties. This approach provides the probability that a system variable, such as the node voltage, will have a certain value. Usually, the contributions to the power input at each bus, i.e., the load value and the wind generating capacity, are taken into account as random variables so that a proper probabilistic characterization of these variables can be determined. Once the input random variables have been characterized from a probabilistic point of view, a proper probabilistic technique must be applied to obtain the characterization of the output random variables, e.g., voltages, losses, and line flows.

In this chapter, a brief review of the mathematical formulation of probabilistic load flow for both balanced and unbalanced distribution systems without wind farms is presented (Sect. 2). Then, some of the main probabilistic models for the steady-state analysis of wind farms [2–17] are presented and included in the mathematical formulation of the probabilistic load flow for distribution systems with wind farms (Sect. 3). Finally, numerical applications on a 17-bus balanced distribution system and an IEEE 34-bus unbalanced test distribution system are presented and discussed.

2 Probabilistic Load-Flow for Distribution Systems without Wind Farms

In the companion chapter [1], it was shown that, for both balanced and unbalanced distribution systems, the load flow equations comprise a non-linear equation system that can be synthetically represented, for a given network configuration, as:

$$\boldsymbol{Y_b} = g(\boldsymbol{X}), \tag{1}$$

where $\boldsymbol{X}$ is the output vector of state variables (amplitudes and arguments of voltages); $\boldsymbol{Y_b}$ is the input variable vector (i.e., active and reactive power injections); and g is the non-linear function. In the case of probabilistic load flow, $\boldsymbol{X}$ and $\boldsymbol{Y_b}$ are output and input random variable vectors, respectively.

Starting from the equation system (1), a probabilistic analysis of a distribution system can be performed in several ways. In the following subsections we are dealing with:

- the non-linear Monte Carlo simulation (NLMC);
- the linear (L) or direct current (DC) Monte Carlo simulation (LMC, DCMC);
- the convolution process (CP); and
- the special distribution approach (SD).

The first technique applies the Monte Carlo simulation to the non-linear load flow equations. The second technique applies the Monte Carlo simulation either to the non-linear equation system linearized around the expected value region or to the DC load flow equations. The third technique is based on a convolution process applied after the previous linearization. The last technique is based on the use of special distribution functions that are used to approximate the probability functions of the output random variables of interest.

In this chapter, we refer to only the main approaches that, to the best of our knowledge, have been further developed with the inclusion of wind farm models.

2.1 Non-linear Monte Carlo Simulation

The NLMC procedure consists of solving the load flow equations several times, each time assuming as $\boldsymbol{Y_b}$ vector elements in (1) one set of the input random variables generated according to their assigned probability density functions (pdfs). The process is repeated a sufficient number of times to obtain adequate accuracy in the estimate of the output random variable pdfs. A useful stopping criterion can be based on the use of a coefficient of variation tolerance [18].

2.2 Linear or DC Monte Carlo Simulation

The LMC simulation procedure consists of solving, in each step of a classical Monte Carlo simulation process, a linear equation system instead of the non-linear equation system (1), with obvious saving in terms of computational efforts. This technique requires two pre-simulation steps in order to find the best linear system to be included in the MC simulation.

The first step consists of solving a deterministic load flow using the mean values $\mu(\boldsymbol{Y_b})$ of the input random variables as input data, that is:

$$g(\boldsymbol{X_0}) = \mu(\boldsymbol{Y_b}). \tag{2}$$

Once the state vector solution $\boldsymbol{X_0}$ is known, the load flow equations (1) are linearized around this point in the second pre-simulation step, as follows:

$$\boldsymbol{X} \cong \boldsymbol{X}_0 + \boldsymbol{A}[\boldsymbol{Y_b} - \mu(\boldsymbol{Y_b})], \tag{3}$$

where the matrix $\boldsymbol{A}$ is the inverse of the Jacobian matrix evaluated in $\boldsymbol{X_0}$.

The linear equations (3) are included in the Monte Carlo simulation to calculate approximate pdfs of the output random variables starting from the pdfs of the input random variables.

It should be noted that since the load flow equations are linearized around an expected value region (the point $\boldsymbol{X_0}$), any movement away from this region produces an error. These errors could increase with the variance of the input random variables pdfs, with an entity linked to the non-linear behavior of the equation system.

For the LMC simulation, the closed form relationships that link the state vector $\boldsymbol{X}$ to any one dependent variable of interest can also be included. For example, the link:

$$\boldsymbol{D} = g_D(\boldsymbol{X}) \tag{4}$$

between the state vector $\boldsymbol{X}$ and the dependent variable vector $\boldsymbol{D}$ can also be considered to obtain the statistical characterization of the dependent variables (i.e., the line flows).

The DCMC technique consists of solving, in each step of a classical Monte Carlo simulation process, the DC power flow equations instead of the non-linear equation system (1), once again with saving in terms of computational efforts.

Assuming the voltage magnitudes equal to the nominal value and the losses equal to zero, the DC power flow can be formulated as [19]:

$$\boldsymbol{P} = \boldsymbol{B}\delta, \tag{5}$$

where $\boldsymbol{P}$ is the vector of the nodal power injections, $\boldsymbol{B}$ is the susceptance matrix, and $\boldsymbol{\delta}$ is the vector of arguments of voltages.

2.3 Convolution Process

The CP requires a three-steps procedure.

The first two steps are identical to the pre-simulation steps performed in the LMC simulation of Sect. 2.2; they are finalized to obtain a linear form of the load flow equations so that each output random variable is expressed as a linear combination of the input random variables.

The third step only consists of the convolution process applied to the linear equations (3) in the form:

$$f(X_i) = X_{0i} + f(z_{i1}) * f(z_{i2}) * \cdots * f(z_{in}), \tag{6}$$

where:

- f represents the pdf;
- X_{0i} represents the ith term of the vector $\boldsymbol{X_0}$;
- $*$ represents the convolution;
- z_{ij} represents the (i,j) term $A_{ij}\,[Y_{bj} - \mu(Y_{bj})]$;
- A_{ij} represents the (i,j) term of the matrix $\boldsymbol{A}$; and
- Y_{bj} represents the jth term of the vector $\boldsymbol{Y_b}$.

Equation (6) can be evaluated using numerical methods based on Laplace transforms or, more efficiently, in terms of execution time and precision, transforming the equations into the frequency-domain using Fast Fourier Transform [20].

As known, the computational efforts of the convolution process depend on the number of normally distributed input random variables; in fact, all normally distributed functions can be easily grouped in one unique normal equivalent, since only the expected value and the covariance matrix are required to define this function. Therefore, Eq. (6) contains discrete or other pdfs and this normal equivalent, with obvious computational advantages.

Using the convolution process to evaluate the pdfs of the dependent variables, such as the active and reactive line flows, requires the linearization of the non-linear link (4) between state variable vector $\boldsymbol{X}$ and dependent variable vector $\boldsymbol{D}$. To do this, we can evaluate the vector $\boldsymbol{D_0}$ such that:

$$\boldsymbol{D}_0 = g_D(\boldsymbol{X}_0). \tag{7}$$

Linearizing equations (4) around the point $\boldsymbol{X_0}$ produces:

$$\boldsymbol{D} \cong \boldsymbol{D}_0 + \boldsymbol{A_D}(\boldsymbol{X} - \boldsymbol{X_0}), \tag{8}$$

where: $\boldsymbol{A_D} = \left[\left.\frac{\partial g_D}{\partial \boldsymbol{X}}\right|_{X=X_0}\right]\boldsymbol{A}$.

Similar to Eqs. (3), (8) expresses each random element of the vector $\boldsymbol{D}$ as a linear combination of the random elements of the state vector $\boldsymbol{X}$. A convolution process similar to the one given by (6) can be applied to Eq. (8) in order to approximate the pdfs of the dependent variables.

Equation (6) can be applied also to Eq. (5) when the DC load flow equations are considered.

2.4 Special Distributions Approach

The SD approach consists of approximating the pdfs of the output random variables of interest with pdfs for which analytical expressions are univocally determined once only some moments or cumulants of the pdfs to be approximated are known[1]. In this way, the problem of obtaining the output random variable pdfs is confined to the problem of obtaining only their first moments or cumulants.

This technique is based on the following three steps.

In the first step, the non-linear equation system (1) is linearized around an expected value region, so each random element of the state vector is expressed as a linear combination of the elements of the input random vector; alternatively, the DC power flow equations can be used.

In the second step, the first moments or cumulants of the pdfs of the output random variables are evaluated with the well-known, closed-form relationships that properly use the linear links obtained in the first step.

In the third step, the pdfs of the random variables of interest, i.e., the amplitudes of the voltages, are approximated by applying an approximate distribution.

The first two steps are trivial. With reference to the third step, once the moments or the cumulants of the true marginal pdf to be approximated are known, several approaches can be applied to analytically describe the form of the pdfs.

In [21], the Pearson distributions were applied for a probabilistic power flow analysis. The Pearson distributions are a family of pdfs that has been demonstrated to represent a very high number of pdfs observed in measurements and as output of analytical approaches. They are univocally defined on the basis of the first four moments of the pdf to be approximated.

In [22], the Gram–Charlier expansion series, which assumes that a pdf can be approximated as a series in the derivatives of the normal pdf, was applied to probabilistic power flow. It has interesting properties and is computationally inexpensive; however, for a non-Gaussian pdf, it seems to have convergence problems, so alternative tools are needed for probabilistic power flow. Also, the approach based on the Cornish–Fisher expansion has been successfully applied in [7]. The Cornish–Fisher expansion is related to the Gram–Charlier series. This

---

[1] As a reminder, given a random variable X with the pdf $f(x)$, the moment of order n is defined as $m_n = \int_{-\infty}^{\infty} x^n f(x)\, dx$, and the cumulant of order n is defined as $\kappa_n = \left.\frac{d^n \Psi(t)}{dt^n}\right|_{t=0}$, where $\Psi(t)$ is the cumulant-generating function, i.e., the logarithm of the moment-generating function, if it exists.

approach provides an approximation of a quantile α of a cumulative distribution function $F(x)$ in terms of the quantile of a normal $N(0, 1)$ distribution φ and of the cumulants of $F(x)$. The theoretical deduction of this expansion, which is quite complex, can be found in [23] or [24]. Using the first five cumulants, the expansion is given by the following equation:

$$\begin{aligned} x(\alpha) \approx \xi(\alpha) &+ \frac{1}{6}\left[\xi^2(\alpha) - 1\right]\,\kappa_3 + \frac{1}{24}\left[\xi^3(\alpha) - 3\,\xi(\alpha)\right]\,\kappa_4 \\ &- \frac{1}{36}\left[2\xi^3(\alpha) - 5\,\xi(\alpha)\right]\,\kappa_3^2 + \frac{1}{120}\left[\xi^4(\alpha) - 6\,\xi^2(\alpha) + 3\right]\,\kappa_5 \\ &- \frac{1}{24}\left[\xi^4(\alpha) - 5\,\xi^2(\alpha)\right]\kappa_2\kappa_3 + \frac{1}{324}\left[12\xi^4(\alpha) - 53\,\xi^2(\alpha)\right]\,\kappa_3^2, \end{aligned} \tag{9}$$

where $x(\alpha) = F^{-1}(\alpha)$, $\xi(\alpha) = \varphi^{-1}(\alpha)$, and κ_r is the cumulant of order r of the cumulative distribution function $F(x)$.

Although the convergence properties of the Cornish–Fisher series are difficult to demonstrate [25] and are somehow related to Gram–Charlier series, their behavior for non-Gaussian pdfs is better than the latter, as was demonstrated in [7].

Also Von Mises functions were applied in [26–28] to handle discrete distributions. Recently, in [29] the Point Estimate Method with Gram–Charlier distributions was applied as an alternative to the Monte Carlo simulation.

3 Probabilistic Load-Flow for Distribution Systems with Wind Farms

The probabilistic load-flow shown in Sect. 2 can easily be extended to include the presence and uncertainties of Wind Turbine Generation Units (WTGUs). Several approaches that have been presented in the relevant literature were focused on wind farm probabilistic models, and they used one of the probabilistic techniques shown in Sect. 2 to determine the probabilistic load flow of distribution systems with wind farms included. In this section, some of the main approaches are shown: they are denominated on the basis either of the wind farm probabilistic models or of the probabilistic technique applied to perform the load flow. They are:

- *the Convolution-based approach;*
- *the Markov-based approach;*
- *the Universal Generating Function-based approach*;
- *the Bayesian-based approach;*
- *the Special Distributions-based approach*; and
- *the hybrid approaches.*

In addition, we note that, from a theoretical point of view, all the load flow equations of a distribution system with wind farms presented in the companion chapter [1] can easily be included in the *Non-linear Monte Carlo simulation* procedure shown in Sect. 2.1 in order to obtain the pdfs of the output variables of interest (see the numerical applications in Sect. 4). Fixed-speed, semi-variable-speed, and variable-speed wind generation systems can be considered. Obviously, the pdfs of the loads and wind speeds (or wind power productions) must be assigned.

With reference to loads, usually the Gaussian pdf is considered adequate.

For wind speed, the Weibull or Rayleigh distributions are usually applied. However, even if the wind speed is considered constant for the same wind farm, its layout can provoke different wind speeds to be considered for different WTGUs, especially if the direction of the wind is taken into account, and this direction coincides with the alignment of the WTGUs. When this occurs, a WTGU located behind another WTGU, seen from the direction from which the wind is coming, will be affected by the wake created by the upwind WTGU. In this case, the effective wind speed for the downwind WTGU can be calculated by the model proposed in [30] and shown in [31]. In particular, when the wind is parallel to the generators, the wind speed given is valid for the first WTGU, but for the other turbines positioned behind the first WTGU, the wind speed is calculated as the product of the wind speed at the WTGU located in front of it and a constant K. The value of K depends on different variables, such as the turbine thrust coefficient, the diameter of the rotor, the axial distance between the WTGUs, and the wake decay constant, which is a function of the hub height and the roughness length. For more details about the procedure, see [31].

Moreover, when a correlation exists among loads at different busbars or among wind speeds in different locations, the procedures shown in the Appendix A.1 can be applied in the frame of the Monte Carlo procedure to generate the correlated random input variables.

3.1 Convolution-Based Approach

In [3], which is one of the first papers on probabilistic load flow in distribution systems with wind farms, a probabilistic model for a WTGU was developed.

As input, the WTGU model assumes the wind speed to be a random variable represented by a normal pdf[2], and it is based on the active and reactive power curves reported in Fig. 1. With reference to the active power, the following equations can be written:

[2] In the authors' opinion, the use of a normal pdf instead of the more popular Weibull pdf is justified by the considerations that they are interested to short-term predictions. In any case, they consider that the proposed approach can be extended to medium- and long-term predictions if the uncertainties of wind and loads are properly modelled.

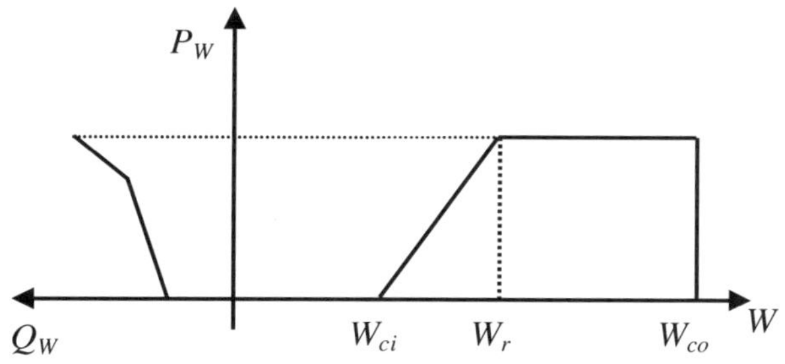

Fig. 1 Wind turbine active and reactive power versus wind speed

$$\begin{aligned} P_W &= 0 && \text{for } 0 \le W \le W_{ci} \text{ and } W \ge W_{co} \\ P_W &= f(W) && \text{for } W_{ci} \le W \le W_r \\ P_W &= P_{\max} && \text{for } W_r \le W \le W_{co}, \end{aligned} \tag{10}$$

where $f(W)$ is approximated as a linear function (or parabolic or cubic function), and W_{ci}, W_r, and W_{co} are the cut-in, rated, and cut-off characteristic values of the WTGU, respectively.

The corresponding active power probabilities are obtained, starting from the wind speed probabilities deduced from the normal distribution, by applying the following equations:

$$\begin{aligned} p(P_W = 0) &= 1 + \varphi(W_1) - \varphi(W_3) \\ p(P_W = f(W)) &= \frac{1}{\sqrt{2\pi}\sigma_W K_1} e^{-\frac{\left[\frac{(P_W - K_2)}{K_2} - \mu_W\right]^2}{2\sigma_W^2}} \\ p(P_W = P_{\max}) &= \varphi(W_3) - \varphi(W_2), \end{aligned} \tag{11}$$

where φ is the cumulative distribution function of the standard normal distribution. In (11):

$$\begin{aligned} &W_1 = \frac{W_{ci} - \mu_W}{\sigma_W}, \quad W_2 = \frac{W_r - \mu_W}{\sigma_W}, \quad W_3 = \frac{W_{co} - \mu_W}{\sigma_W}, \quad K_1 = \frac{P_{\max}}{W_r - W_{ci}}, \\ &K_2 = -K_1 W_{ci} \end{aligned} \tag{12}$$

and the linear approximation $f(W) = K_1 W + K_2$ is assumed; in (12), μ_W and σ_W are the mean value and the standard deviation of the wind normal distribution, respectively.

Reactive power values and probabilities are obtained by considering the link between active and reactive powers shown in Fig. 1.

Once the probabilistic model of the wind power production has been developed and assuming a normal distribution for the load powers, the convolution process shown in Sect. 2.3 can be applied. In fact, from a theoretical point of view, the power flow equations can be linearized in order to obtain the unknown quantities (i.e., the amplitudes of the voltages) as linear functions of the nodal active and reactive power injections; then, the convolution technique can be performed to obtain the unknown pdfs. Actually, due to the radial structure of the distribution system, the authors in [3] applied a simpler linearization. Assuming that the voltage at the medium voltage (MV) bus is fixed by the voltage regulator, the voltage at each bus is expressed analytically as the difference between this value and the voltage drop from the MV bus; this analytical link is linearized so that the voltage magnitudes at each bus are expressed as a linear function of the power injections.

3.2 Markov-Based Approach

The Markov-based approach is based on the assumption that the wind speed can be considered as a stochastic process[3] with a continuous state space (linked to wind speed values) and a continuous parameter space (linked to time). This stochastic process is modelled approximately by using the discrete Markov process (Markov's chains)[4] [4, 14–16].

In particular, in [4], a simple Markov chain-based model was applied for the evaluation of the pdfs of distribution system busbar voltages. In particular, the authors proposed a Markov chain model that is able to generate a wind speed time series[5]; this series is used as input to the power curve to convert wind speed to generated power for use in the probabilistic power flow analysis.

The starting point is represented by the measurements of wind speed. Since averaging periods of 1 h are generally used for wind speed measurements because this allows the variations due to turbulence and weather fronts to be separated, a time resolution of 1 h is often used. In any case, the hourly wind speed measurement data can also be divided into several time periods to determine the seasonality variations.

In order to apply the Markov approach, some specific characteristics concerning the number of states, the transition probabilities, and the possible transitions

---

[3] As is well known, a stochastic process is defined as a model of a system that develops randomly in time according to probabilistic laws.

[4] The use of the discrete Markov process to model wind speed (and then the wind farm) was proposed in the relevant literature both for the probabilistic power flow analysis and for the reliability analysis of distribution networks that have wind farms.

[5] We note that not all authors consider the use of time series to be a strictly probabilistic approach. In fact, strictly speaking, unlike the Monte Carlo simulation, the input data are not derived from pdfs, but the time series of load and wind generation are directly applied. In the frame of this method, the steady-sate of the distribution system is simulated during a suitable time period (i.e., 1 week or 1 year). Using the active and reactive power curves of the WTGU (Fig. 1), the output power series can be obtained. From the load and wind generation time series, the corresponding state and dependent variables can be obtained by performing subsequent load flow calculations for each point; this approach was applied also in [9, 10].

among states must be assessed. In particular, the first-order Markov model[6] proposed in [4] relies on the following assumptions.

First, the states of the wind speed must be identifiable. Then, on the basis of field measurement data, the wind speed sample record is represented by a discrete number of N wind speed states, each corresponding to an opportune range of values. The states can be fixed using wind speed intervals of constant or variable amplitude; states with variable intervals can be used in order to take into account the non-linear control characteristic of WTGUs.

Second, the wind speed model must be stationary (i.e., the probability of making a transition from one state to another is time-invariant) and the process must lack memory (i.e., the probability of being in a given state is independent of all the past states, except for the immediately preceding one). These two requirements imply that the transition probabilities between the states are constant.

Finally, the probability of a transition from a given wind speed state to another state is directly proportional to the steady state probability of the new state.

Once the number of states and the wind speed intervals for each state are defined for Markov model, the wind speed data are used to obtain an approximation of the transitional probability matrix that, as is well known, characterizes the discrete Markov process; in particular, when N states are considered, the transitional probability matrix $\boldsymbol{TM}$ is defined as:

$$\boldsymbol{TM} = \begin{bmatrix} p_{11} & p_{12} & \cdots & p_{1N} \\ p_{21} & p_{22} & \cdots & p_{2N} \\ \vdots & \vdots & \ddots & \vdots \\ p_{N1} & p_{N2} & \cdots & p_{NN} \end{bmatrix}, \tag{13}$$

where p_{ij} is the probability of a transition between states i and j.

From the wind speed data, it is possible to obtain an approximation of the transitional probability matrix defined in (13); the maximum likelihood estimate for the matrix entries is:

$$p_{ij} = \frac{n_{ij}}{\sum_{j=1}^{N} n_{ij}}, \tag{14}$$

where n_{ij} is the number of transitions from state i to state j encountered in the measured wind speed data.

The approximation of $\boldsymbol{TM}$ that is obtained is used to generate a wind speed time series by using the following procedure.

First, the initial state i is selected. Then, a random number x between zero and unity is generated and compared to the ith row of the matrix to find the next state; in particular, if x is lower than (or equal to) the first element in the row ($x \leq p_{i1}$), the next state is state 1. If x is greater than the first element in the row and lower

[6] In addition to the first-order Markov chain model, a second order model has been proposed to generate wind speed time series. For example, see [14].

than (or equal to) the sum of the first two elements in the row ($p_{i1} < x \leq p_{i1} + p_{i2}$), the next state is state 2. Thus, in general, if x is greater than the sum of the first $k-1$ elements in the row and lower than (or equal to) the sum of the first k elements in the row ($\sum_{j=1}^{k-1} p_{ij} < x \leq \sum_{j=1}^{k} p_{ij}$), the next state is state k. The process is then repeated to find further states, using the latest state as the initial state.

Eventually, the generated wind speed time series are used as input to the wind turbine power curve to convert wind speeds to wind power outputs that are used in the probabilistic power flow analysis performed applying the NLMC procedure as probabilistic technique.

3.3 Universal Generating Functions-Based approach

In [5], an approach to model wind farms (WFs) for reliability assessment was proposed; this approach can be easily applied also in the frame of a probabilistic load flow. In particular, WFs are modeled as multi-state systems[7] (MSSs), the probability distribution of which can be determined by the Universal Generating Functions (UGFs) technique. In general, each element of an MSS is modeled by a discrete random variable, for which the probability mass function, named the performance distribution (PD), is represented by an UGF (or u-function). Thus, the PD of the whole MSS is obtained by combining u-functions and proper composition operators, as is shown later.

The UGF technique does not focus on the wind state space (as in the Markov techniques) or on a specific wind trial (as in the Monte Carlo simulation); rather, it focuses on some kind of 'projection' of the available power output space of the wind farm onto the wind space. As a result of this different point of view, the approach based on UGFs is flexible and effective, and it can easily be used for systems that have a large number of states with low computational burden. Then, in this subsection, we refer to the WF model instead of a single WTGU to better highlight the potentiality of the technique illustrated.

Let us consider a WF composed of N types of WTGUs. It is modeled as an MSS composed of the wind element, which is the energy primary source, and the machine subsystem, which is composed of N types of WTGUs. The inputs to the WF model are the record of k_W measured data of wind speed and the N power curves of the WTGUs. In general, referring to the jth type of WTGU, the generated power P_j is a non-linear function of the continuous variable wind speed W (Fig. 1), which includes the contribution of all the WTGs of the jth type. The output of the WF model is the PD of the power injected into the grid by the WF, used as input to the probabilistic load flow, as explained below.

[7] A multi-state system (MSS) is a system that performs its mission with various levels of efficiency, referred as performance rates.

Let the random variable G_W be the performance rate of the wind speed. The PD(G_W) can be represented in a polynomial form by a u-function $u_W(z)$ as:

$$u_W(z) = \sum_{h=1}^{k_W} p_{W_h} z^{g_{W_h}}, \tag{15}$$

where the discrete variable z represents all the k_W possible states of the wind speed, p_{W_h} is the probability of occurrence, and the exponent is the performance rate g_{W_h} of the wind element, related to the hth state. The former is evaluated by using the frequency approach; the latter coincides with the value of measured data of wind speed.

Let the random variable G_{P_j} be the performance rate of the jth type of WTGU. The PD $\left(G_{P_j}\right)$ can be represented in terms of a u-function, as:

$$u_{P_j}(z) = \sum_{m=1}^{k_{F_j}} p_{P_{m_j}} z^{g_{P_{m_j}}}. \tag{16}$$

The random variables G_{P_j} are statistically dependent on G_W. Consequently, the state of the wind element affects both the performance rate $g_{P_{m_j}}$ and the probability of occurrence $p_{P_{m_j}}$ of the mth state of G_{P_j}. In particular, variables G_{P_j} are functions of the random variable G_W according to Eq. (10) or similar. Furthermore, the random variables G_{P_j} are mutually independent for a given state of G_W.

The link between the random variables G_{P_j} can be expressed by the following deterministic function, named the structure function:

$$\Psi\left(G_{P_1}, \ldots, G_{P_j}, \ldots, G_{P_N}\right) = \sum_{j=1}^{N} G_{P_j}. \tag{17}$$

To evaluate the output of the WF model using the UGF approach, the following steps must be performed.

The *first step* is to partition the performance set $\mathbf{g}_W$ of the wind element on the basis of all active power curves. Referring to the jth type of WTGU, the partition operator $\boldsymbol{\pi}_{P_j}(\mathbf{g}_W)$ is defined as:

$$\boldsymbol{\pi}_{P_j}(\mathbf{g}_W) = \left\{\mathbf{g}_W^{m_j}, \quad m_j \in \left[1, \ldots, k_{P_j}\right]\right\}, \tag{18}$$

with $\mathbf{g}_W^{m_j} = \left\{g_{W_h} : G_{P_j} = g_{P_{m_j}}\right\}$.

Then, the overall partition $\boldsymbol{\pi}(\mathbf{g}_W)$ of the WF, defined as:

$$\boldsymbol{\pi}(\mathbf{g}_W) = \left\{\mathbf{g}_W^i, \quad i \in [1, \ldots, M]\right\}, \tag{19}$$

is obtained by recursively applying the partition operator for each WTGU type:

$$\boldsymbol{\pi}(\mathbf{g}_W) = \boldsymbol{\pi}_{P_1}(\boldsymbol{\pi}_{P_2}(\ldots(\boldsymbol{\pi}_{P_M}(\mathbf{g}_W))\ldots)). \tag{20}$$

The *second step* is rewriting the PD $\left(G_{P_j}\right)$, accounting for $\boldsymbol{\pi}(\mathbf{g}_W)$ in terms of the conditional u-function $\bar{u}_{P_j}(z)$ as:

$$\bar{u}_{P_j}(z) = \sum_{m=1}^{k_{F_j}} \bar{\mathbf{p}}_{P_{m_j}} z^{g_{Fm_j}}, \tag{21}$$

where $\bar{\mathbf{p}}_{P_{m_j}} = \left\{ p_{P_{m_{j|1}}}, \ldots, p_{P_{m_{j|i}}}, \ldots, p_{P_{m_{j|M}}} \right\}$.

The generic term $p_{P_{m_{j|i}}}$ is either unity or null, according to whether the ith subset $\mathbf{g}_W^i$ in the partition $\boldsymbol{\pi}(\mathbf{g}_w)$ does or does not correspond to the image $g_{P_{m_j}}$ in the performance space of the jth type of WTGU.

The *third step* consists of introducing the random variable G_P, which represents the machine subsystem and whose PD can be expressed in terms of conditional u-function $\bar{u}_P(z)$ as:

$$\bar{u}_P(z) = \sum_{m=1}^{k_F} \bar{\mathbf{p}}_{P_m} z^{g_{Fm}}, \tag{22}$$

where $\bar{\mathbf{p}}_{P_m} = \left\{ p_{P_{m|1}}, \ldots, p_{P_{m|i}}, \ldots, p_{P_{m|M}} \right\}$.

In practice, Eq. (22) can be evaluated by applying the composition vectorial $\overset{\mathrm{O}}{\underset{\psi}{\otimes}}$ operator over the N u-functions $\bar{u}_{P_j}(z)$:

$$\begin{aligned}\bar{u}_P(z) &= \overset{\mathrm{O}}{\underset{\psi}{\otimes}}\left(\bar{u}_{P_1}(z), \ldots, \bar{u}_{P_j}(z), \ldots, \bar{u}_{P_N}(z)\right)\\ &= \sum_{m=1}^{k_{F_1}} \cdots \sum_{m=1}^{k_{F_j}} \cdots \sum_{m=1}^{k_{F_N}} \circ\left(\bar{\mathbf{p}}_{P_{m_1}}, \ldots, \bar{\mathbf{p}}_{P_{m_j}}, \ldots, \bar{\mathbf{p}}_{P_{m_N}}\right) z^{\psi\left(g_{P_{m_1}}, \ldots, g_{P_{m_j}}, \ldots, g_{P_{m_N}}\right)},\end{aligned} \tag{23}$$

where:

$$\begin{aligned}\circ\left(\bar{\mathbf{p}}_{P_{m_1}}, \ldots, \bar{\mathbf{p}}_{P_{m_j}}, \ldots, \bar{\mathbf{p}}_{P_{m_N}}\right) = \Big\{ & p_{P_{m_{1|1}}} \cdot \ldots \cdot p_{P_{m_{j|1}}} \cdot \ldots \cdot p_{P_{m_{N|1}}}, \ldots, p_{P_{m_{1|i}}} \cdot \ldots \cdot p_{P_{m_{j|i}}} \cdot \ldots \cdot\\ & p_{P_{m_{N|i}}}, \ldots, p_{P_{m_{1|M}}} \cdot \ldots \cdot p_{P_{m_{j|M}}} \cdot \ldots \cdot p_{P_{m_{N|M}}} \Big\}.\end{aligned} \tag{24}$$

The *fourth step* consists of evaluating the PD of the random variable G, representing the output of the WF model. The related u-function $u(z)$ can be expressed as:

$$u(z) = \sum_{t=1}^{k} p_t z^{g_t}, \tag{25}$$

and can be obtained as:

$$u(z) = \sum_{m=1}^{k_F} \left(\sum_{h=1}^{k_W} p_{W_h} p_{P_{m|\mu\left(g_{W_h}\right)}} \right) z^{g_{P_m}}. \tag{26}$$

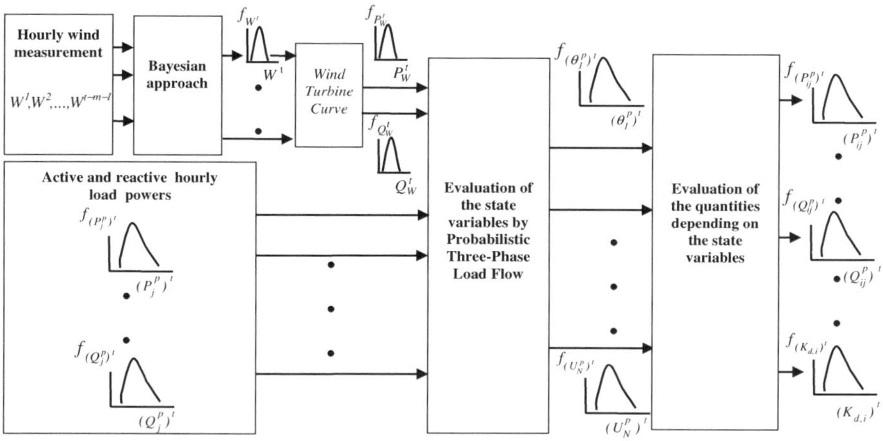

Fig. 2 Bayesian method to evaluate hourly steady state operating conditions of electrical distribution systems that include WTGUs [6]

The auxiliary function $\mu\left(g_{W_h}\right)$ in Eq. (26) is introduced to reduce the computational burden associated with the evaluation of $p_{P_{m|i}}$; it assigns the integer i enumerating the subset $\mathbf{g}_W^i$ in $\pi(\mathbf{g}_W)$ to the generic term g_{W_h}, to which g_{W_h} belongs.

Reactive power and related probabilities values can be obtained by considering the link between active and reactive powers shown in Fig. 1. Eventually, the generated power outputs can be used in the probabilistic power flow analysis.

3.4 Bayesian-Based Approach

In [6], a probabilistic method (Fig. 2) was proposed to forecast the steady-state operating conditions of an unbalanced electrical distribution system "at hour t," starting from the knowledge of the hourly wind speed measurements at the sites where the wind farms are installed, which are collected before the $t - m$ hour (i.e. $m = 1$ or 2 or 4 if the forecast is effected 1 or 2 or 4 h before t hour) and the pdfs of the active and reactive phase load powers. In this subsection we assume $m = 1$ for the sake of clarity.

The hourly wind speeds measured at the sites where the wind farms are installed are used to obtain the pdfs f_{W^t} of the wind speeds "at hour t" by a Bayesian approach. Then, the wind speed pdfs are used as inputs to the power curves to convert wind speeds to generate power outputs that are used with the load pdfs in a probabilistic, three-phase, load flow analysis of an unbalanced distribution system.

With particular reference to the Bayesian approach for the prediction of the pdfs f_{W^t} of wind speeds related to hour t, the random variable representing wind speed w was modelled using the following Weibull pdf:

$$f_w(w|\eta\ \beta) = \frac{\beta}{\eta}\left(\frac{w}{\eta}\right)^{\beta-1} e^{-\left(\frac{w}{\eta}\right)^{\beta}}, \tag{27}$$

where η is the scale parameter, and β is the shape parameter.

Then, the prior random variables have to be fixed [32–34]. The first prior variable is the shape parameter β. Even though the scale parameter η could be considered the second prior random variable, it is most useful to introduce new random variables that permit a more adequate predictive model. To do this, the scale parameter η is expressed in terms of the mean value of wind speed μ by the following relationship:

$$\eta = \frac{\mu}{\Gamma\left(1+\frac{1}{\beta}\right)}, \tag{28}$$

where $\Gamma(z)$ is the Gamma function. Then, the auto regressive (AR) model of the first order for calculating the wind speed mean value μ^t from the knowledge of the wind speed value at adjacent hour w^{t-1} is applied. This AR model of the first order uses the following equation:

$$\log(\mu^t) = \alpha_1 + \alpha_0 \log(w^{t-1}), \tag{29}$$

where α_0 and α_1 are the coefficients of the AR model (29).

If we look at the equations in (27), (28) and (29), it appears that the new prior random variables needed to apply the Bayesian approach (other than the shape parameter β) are α_0 and α_1. The prior distributions of the parameters are assumed to be normal and equal to $\alpha_0 \sim$ N(0, 10^4), $\alpha_1 \sim$ N(0, 10^4), and $\beta \sim$ N(0, 10^4), respectively. The choice of a large variance for the prior distributions guarantees that the measured data **w**, containing n samples of w collected before $t-1$ hour, influence the parameters of the posterior distributions more than the prior distributions would.

In order to obtain an approximation of the posterior distributions of parameters α_0, α_1, and β, an analytical expression of the unnormalized posterior pdf is needed. In particular, once the prior distributions for α_0, α_1 and β are fixed and once the distribution of the observed quantities **w** is known, they can be used to write the following equation for the unnormalized posterior pdf:

$$q(\alpha_0\alpha_1\beta|\mathbf{w}) = \left\{\prod_{t=3}^{n+1}\left[\frac{\beta}{\eta^{t-1}(\alpha_0,\alpha_1)}\left(\frac{w^{t-1}}{\eta^{t-1}(\alpha_0,\alpha_1)}\right)^{\beta-1} e^{-\left(\frac{w^{t-1}}{\eta^{t-1}(\alpha_0,\alpha_1)}\right)^{\beta}}\right]\right\} e^{-\left(\frac{0.5\alpha_0^2}{10^4}\right)} e^{-\left(\frac{0.5\alpha_1^2}{10^4}\right)} e^{-\left(\frac{0.5\beta^2}{10^4}\right)}, \tag{30}$$

where $\eta^{t-1}(\alpha_0, \alpha_1)$ can be evaluated by applying Eq. (28), where the wind speed mean value μ^{t-1} depends on the parameters α_0 and α_1 through Eq. (29).

The unnormalized posterior distribution analytical expression (30) is used in a Markov Chain Monte Carlo (MCMC) approach that uses the Metropolis-Hasting (MH) algorithm to obtain an approximation of posterior distributions for the parameters α_0, α_1 and β [34–36]. In practice, in the MCMC simulation, for each (trial) sample of α_0, α_1 and β obtained from prior distributions, a four steps procedure is applied.

In the first step, the mean value μ^{t-1} is calculated by applying relationship (29) using α_0, α_1 samples and w^{t-2}.

In the second step, η^{t-1} is calculated by applying relationship (28) using μ^{t-1} obtained in the first step and the β sample.

In the third step, an unnormalized posterior distribution value is calculated, applying relationship (30) using η^{t-1} obtained in the second step, the β sample, and the n samples of observed quantities $\mathbf{w}$.

In the fourth step, the unnormalized posterior distribution value obtained in the third step is used in the Metropolis-Hasting (MH) algorithm to obtain the samples of the posterior distributions of α_0, α_1, and β.

Once the samples from the posterior distributions of α_0 and α_1 are known, these samples and the wind value w^{t-1} are used to calculate the pdf of the mean value of wind speed at "hour t" μ^t, once again applying Eq. (29). The pdf of the scale parameter distribution η^t is calculated from the knowledge of the μ^t pdf and the posterior distribution of β, once again applying Eq. (28). Finally, using the posterior distribution of β and the η^t pdf, the predictive distribution of the wind speed w^t can be obtained. In fact, the predictive distribution of the wind speed w^t is calculated using an algorithm that generates a sample from the Weibull distribution for each sample of the posterior distributions of parameters β and η^t.

3.5 Special Distributions-Based Approach

The approach shown in Sect. 2.4 can easily be extended to include the presence of WTGUs. To do this, the random input, i.e. the active and reactive power injections, at busbars where a WTGU is installed, must include the wind production. In [7], the probabilistic load flow in the presence of wind production was performed by applying the Cornish–Fisher expansion. In particular, the procedure was aimed at determining the statistical characterization of line power flows. Four steps were performed.

In the first step, a DC load flow with the expected value of the wind power injected to the system is solved. This gives the expected value of system variables and of line power flows.

In the second step the moments and cumulants of the cumulative distribution function of the wind power injections are calculated.

In the third step the cumulants of the random variables of the system variables and of power flows through the lines of interest based on the linearity of the DC load flow are calculated.

In the fourth step, the Cornish–Fisher expansion (9) is used to find the value of the cumulative distribution function of the system power in the lines of interest.

In [37] the Point Estimate method proposed in [29] was extended to include wind farms uncertainties.

3.6 Hybrid Approaches

In [2], the probabilities of undervoltages and overvoltages are calculated using an hybrid procedure. This procedure is performed as shown later synthetically.

Initially, a load flow of the distribution system with wind farms is performed with expected values of loads and wind power productions, all assumed to be independent, random, Gaussian variables.

Then, a test is conducted to detect the risk of undervoltages and overvoltages by performing two deterministic load flows with the following values for loads and generated power productions:

$$\begin{aligned} P_W &= \mu_{P_W} - C_{sd}\,\sigma_{P_W}, \\ P_{load} &= \mu_{P_{load}} + C_{sd}\,\sigma_{P_{load}} \end{aligned} \tag{31}$$

for undervoltages. The following values for loads and generated power productions:

$$\begin{aligned} P_W &= \mu_{P_W} + C_{sd}\,\sigma_{P_W}, \\ P_{load} &= \mu_{P_{load}} - C_{sd}\,\sigma_{P_{load}} \end{aligned} \tag{32}$$

are for overvoltages. In (31) and (32), $\mu_{P_{load}}$, $\sigma_{P_{load}}$, μ_{P_W}, σ_{P_W} are mean values and standard deviations of the load and wind powers, respectively; in the numerical application, a C_{sd} value of 3.5 is suggested as a compromise between computational efforts and accuracy of the results.

Finally, if risks of undervoltages or overvoltages are detected, a Monte Carlo simulation procedure is applied for this case, assuming that, as previously specified, both load and generated power are random Gaussian variables.

A further hybrid method based on cumulants, special functions, and convolution approaches also was proposed in [8]. The probabilistic modelling of the wind farm is based on a three-parameter Weibull pdf, considered to be more appropriate than the commonly used two-parameter one, since it seems to better represent the higher wind speed values; starting from the Weibull pdf, the power curve is used to obtain the probability distribution of the wind power analytically. Different probabilistic load flow techniques are, then, performed that also are able to take into account generator and branch outages. The techniques separate discrete and continuous random variables; the continuous variables are treated by the cumulant method, while discrete distributions are treated separately with the Von Mises method [26–28]. After the separate steps, the convolution is applied to the continuous and discrete output distributions to obtain the resulting output random variable probability functions.

4 Numerical Applications

In this Section, for the sake of conciseness, only the main methods described in the previous Sections are applied to analyze the probabilistic impact of WTGUs; first, a 17-bus balanced distribution system is considered, and, then, an IEEE 34-bus unbalanced distribution system is considered. In particular, with reference to the balanced system, we apply:

1. in Sect. 4.1, the *Non-linear Monte Carlo simulation approach;*
2. in Sect. 4.2, the *Markov-based approach;*
3. in Sect. 4.3, the *Generating functions-based approach;* and
4. in Sect. 4.4, the *Special Distributions-based approach* (in particular the *cumulant-based approach*).

With reference to the unbalanced distribution system, we apply:

5. in Sect. 4.5, the *Non-linear Monte Carlo simulation;* and
6. in Sect. 4.6, the *Bayesian-based approach.*

4.1 Non-linear Monte Carlo Simulation for Balanced Distribution Systems

The *Non-linear Monte Carlo Simulation* is applied to the probabilistic steady-state analysis of the 17-bus, balanced, three-phase network illustrated in Fig. 6 of the companion chapter [1] and reported in Appendix A.2. This system contains 16 busbars at 12.5 kV and one busbar (#1) at 138 kV. The main data of the considered test system are reported in Tables 2 and 3 of the companion chapter [1]. The active power and reactive power have been modeled as random Gaussian variables. The values reported in Table 2 have been assumed as the mean values of active power and reactive power, whereas the standard deviation has been assumed to be equal to 10 % of the mean value.

The following cases have been considered:

Case (I)	No wind generating units
Case (II)	1.0 MW semi-variable speed WTGU at bus #9
	1.0 MW GBBC variable speed WTGU at bus #13
Case (III)	1.0 MW semi-variable speed WTGU at bus #9
	1.0 MW GBBC variable speed WTGU at bus #13
	1.0 MW DFIG variable speed WTGU at bus #15
	1.0 MW stall-regulated fixed speed WTGU at bus #17

Details about the WTGU models and further data can be found in Sect. 4 of Ref. [1].

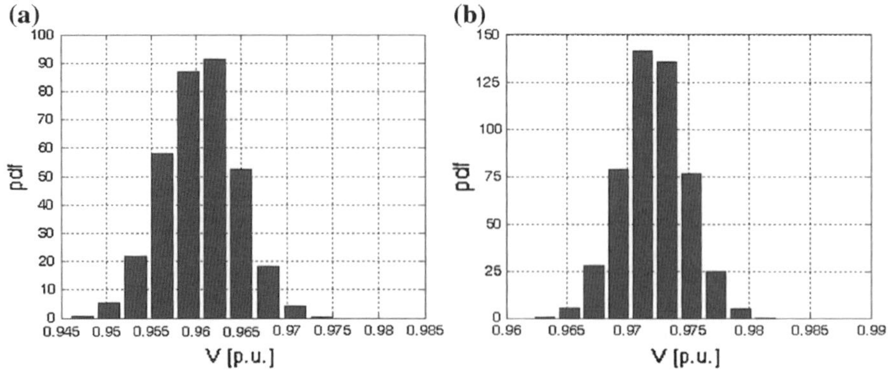

Fig. 3 Case I: pdf of voltage amplitude at busbar #9 (**a**) and busbar #13 (**b**)

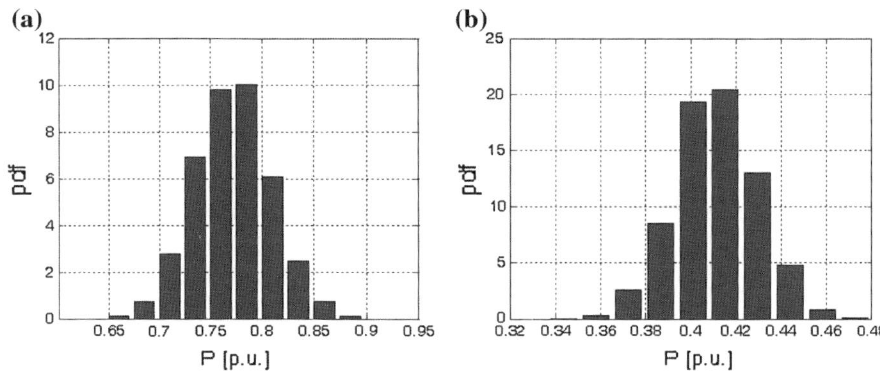

Fig. 4 Case I: pdf of active power flow in line 2–3 (**a**) and in line 2–11 (**b**)

A Rayleigh pdf for the wind speed is assumed. In particular, in all the considered cases, the mean value of the wind speed has been assumed equal to 8.1 m/s.

In this sub-section, we show:

Case (I) The pdfs of voltage amplitude at the selected busbars #9 and #13 and the pdfs of active power flow in the line between busbars #2 and #13 and in the line between busbars #2 and #11 (Figs. 3, 4).

Case (II) The pdfs of the powers injected by the WTGUs, the pdfs of voltage amplitude at the selected busbars #9 and #13, the profile of mean value of voltage amplitude at all busbars compared with the profile in the absence of WTGU, and the pdfs of active power flow in the line between busbars #2 and #13 and in the line between busbars #2 and #11 (Figs. 5, 6, 7, 8).

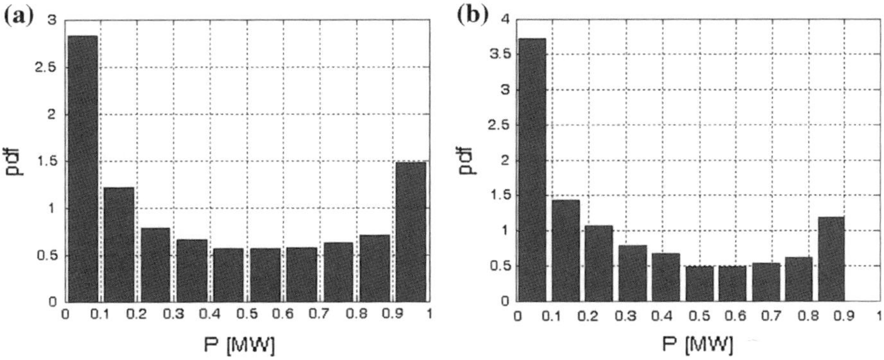

Fig. 5 Case II: pdf of the power injected by the WTGU connected at busbar #9 (**a**) and busbar #13 (**b**)

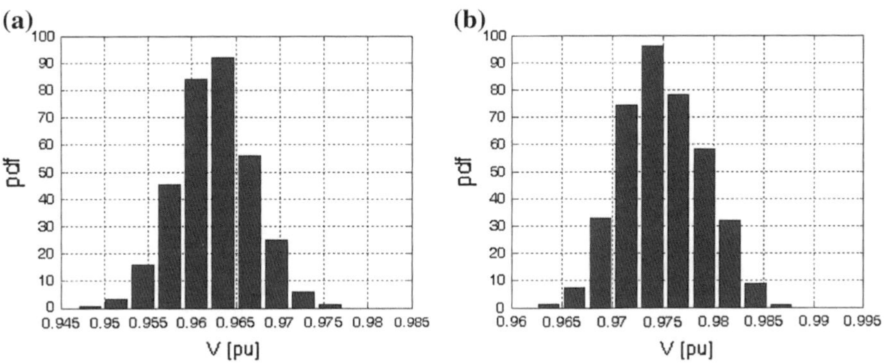

Fig. 6 Case II: pdf of voltage amplitude at busbar #9 (**a**) and busbar #13 (**b**)

Fig. 7 Case II: Comparison between voltage profile in the absence of WTGUs (Case I) and the voltage profile in the presence of WTGUs at busbar #9 and busbar #13

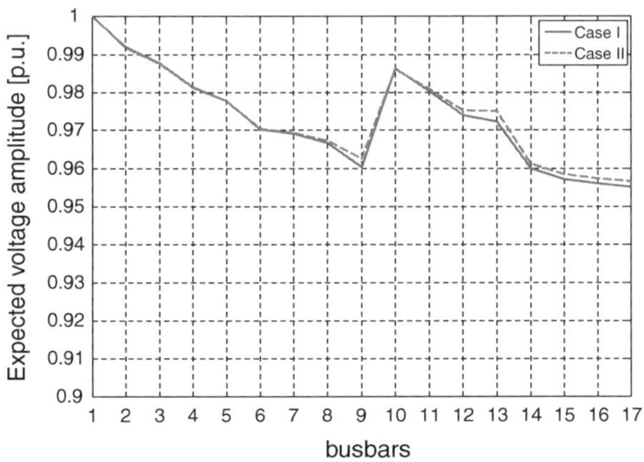

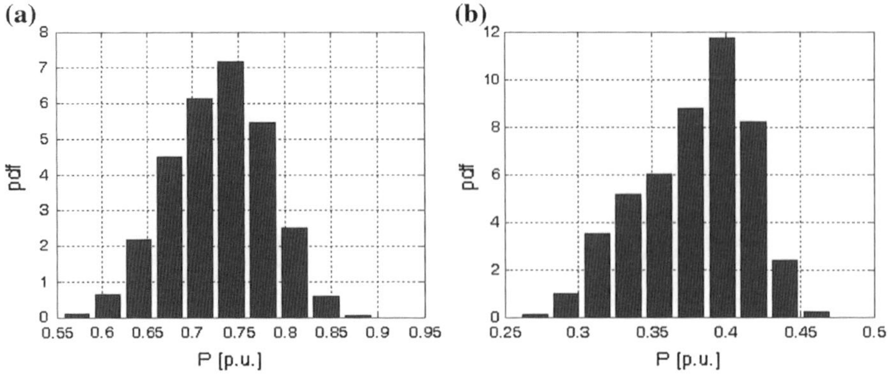

Fig. 8 Case II: pdf of active power flow in line 2–3 (**a**) and in line 2–11 (**b**)

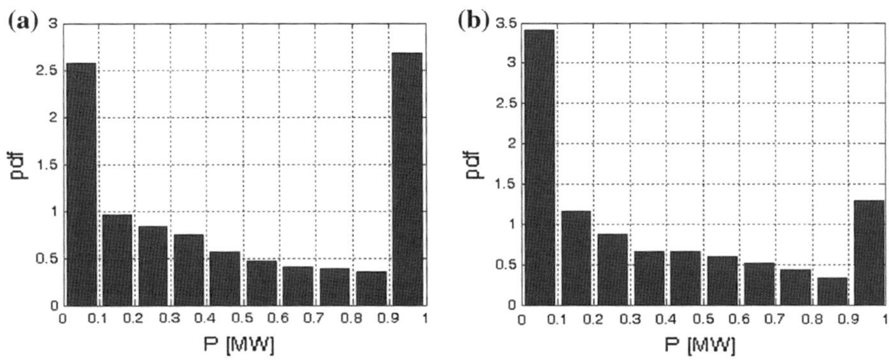

Fig. 9 Case III: pdf of the power injected by the WTGU connected at busbar #15 (**a**) and busbar #17 (**b**)

Case (III) The pdfs of the powers injected by the WTGUs, the pdfs of voltage amplitude at the selected busbars #9 and #13, the profile of mean value of voltage amplitude at all busbars compared with the profile in the absence of WTGU, and the pdfs of active power flow in the line between busbars #2 and #13 and in the line between busbars #2 and #11 (Figs. 9, 10, 11, 12).

Comparing Figs. 7 and 11, it is apparent that, as the penetration of wind energy production increases, the expected voltage amplitudes at all busbars are higher with respect to Case (I). The same consideration can be derived from the analysis of Figs. 3, 6, and 10.

From the analysis of the pdfs of Figs. 4, 8 and 12, the impact of wind energy production on the active power flow in the lines of the distribution system can be

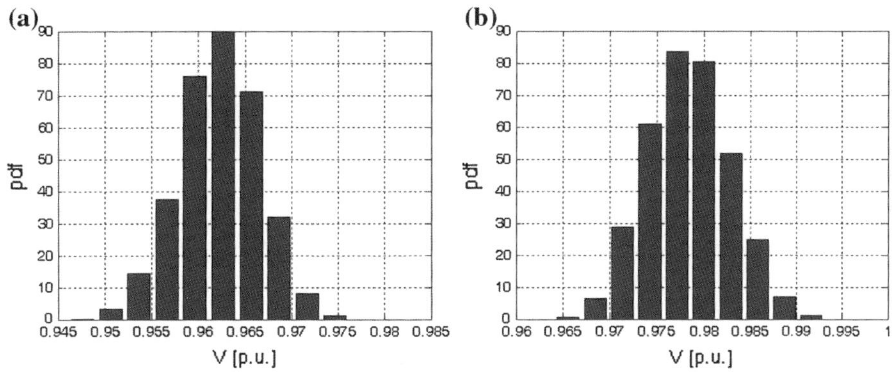

Fig. 10 Case III: pdf of voltage amplitude at busbar #9 (**a**) and busbar #13 (**b**)

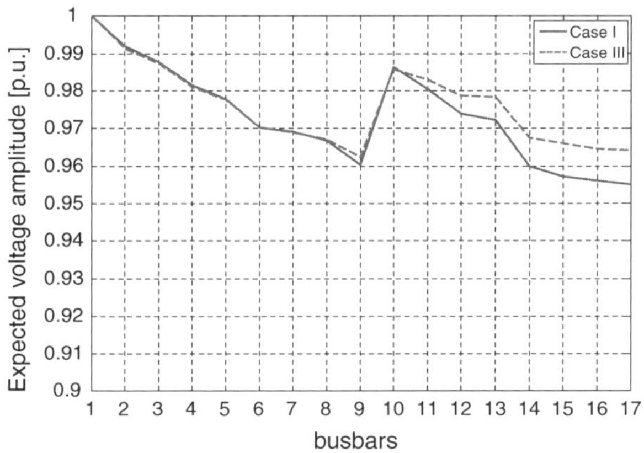

Fig. 11 Case III: Comparison between the voltage profile in the absence of WTGUs (Case I) and the voltage profile in the presence of WTGUs at busbar #9, busbar #13, busbar #15, and busbar #17

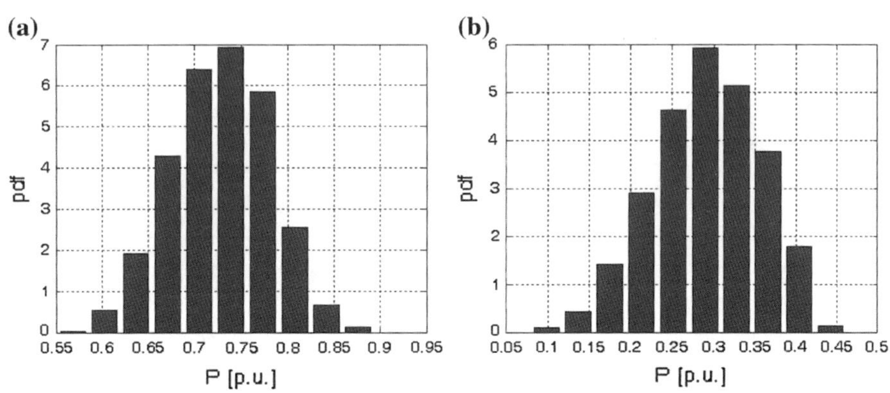

Fig. 12 Case III: pdf of active power flow in line 2–3 (**a**) and in line 2–11 (**b**)

Table 1 Power loss reduction

Case	Reduction of power losses with respect to Case I (%)
II	11.7
III*	30.1

* PQ model for stall-regulated fixed speed WTGU [1]

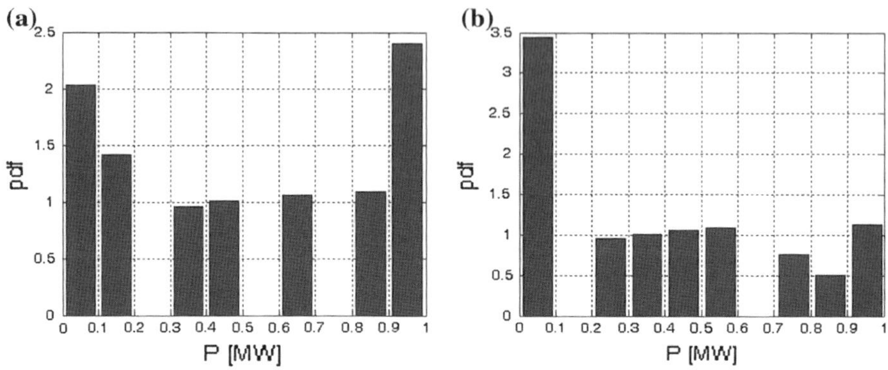

Fig. 13 Case III: pdf of the power injected by the WTGU connected at busbar #15 (**a**) and busbar #17 (**b**)

investigated. The larger the penetration of WTGU is, the lower the active power flowing in the lines becomes, as foreseeable.

Finally, Table 1 reports the reduction of the real power losses due to the presence of WTGUs (Cases II and III) with respect to the power losses in the absence of WTGUs (Case I). A significant reduction of the losses is clearly apparent; this reduction is growing with the wind energy production.

4.2 Markov-Based Approach for Balanced Distribution System

We have also considered the use of the Markov-based approach to evaluate the balanced distribution system.

The case considered here is Case III already described in Sect. 4.1, but, in this subsection, the wind speed is a time series obtained by applying the Markov-based approach of Sect. 3.2. The wind speed measurements used in the Markov-based approach were obtained from the Royal Netherlands Meteorological Institute (http://www.knmi.nl/samenw/hydra/index.html) and refer to the data measured at wind stations in Texelhors, The Netherlands, from January 2001 to September 2009.

We show, for the sake of conciseness, only the pdfs of the powers injected by the WTGUs at two busbars #15 and #17, the pdfs of the voltage amplitudes at two

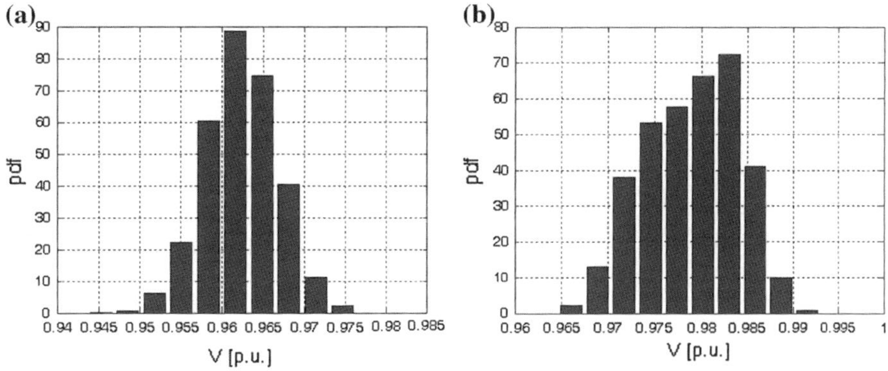

Fig. 14 Case III: pdf of the voltage amplitudes at busbar #9 (a) and busbar #13 (b)

Fig. 15 Case III: Comparison between the voltage profile in the absence of WTGUs (Case I) and the voltage profile in the presence of WTGUs

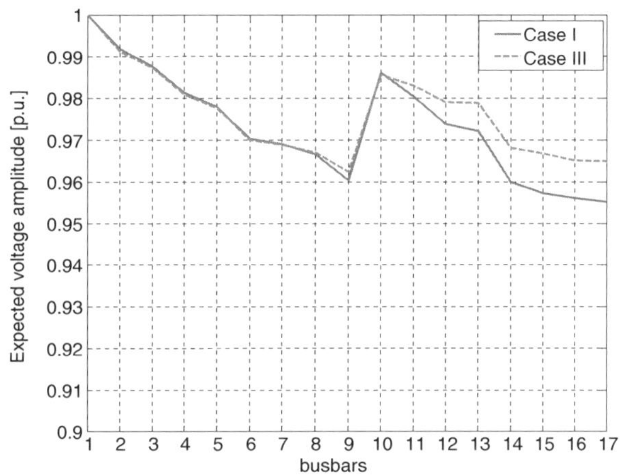

busbars #9 and #13, and the profile of the mean values of voltage amplitude at all busbars compared with the profiles in the absence of WTGU (Figs. 13, 14, 15).

In the considered case, the real power losses of the system were reduced with respect to Case (I) by about 30.5 %.

4.3 Universal Generating Functions-Based Approach for Balanced Distribution System

The balanced distribution system has been also considered to apply the Generating functions-based approach shown in Sect. 3.3.

The case considered here is the Case III already described in Sect. 4.1. Concerning the wind energy source, once again the long-term wind speed data from

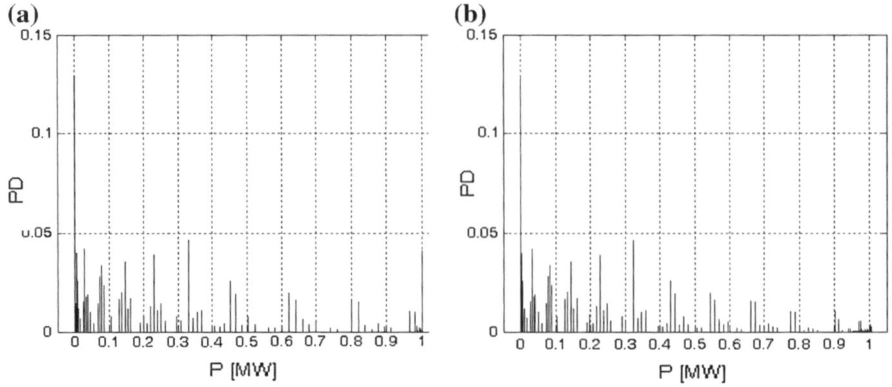

Fig. 16 Performance distributions of the generated power of WTGUs at busbar #15 (**a**) and busbar #17 (**b**)

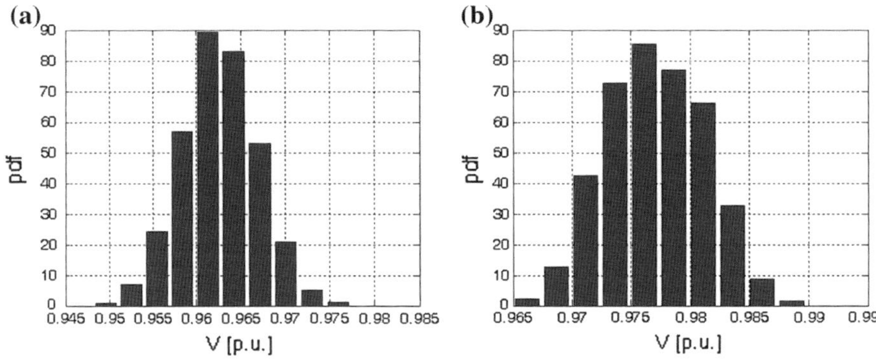

Fig. 17 Case III: pdf of voltage amplitude at busbar #9 (**a**) and busbar #13 (**b**)

Fig. 18 Case III: Comparison between voltage profile in the absence of WTGUs and in presence of WTGUs

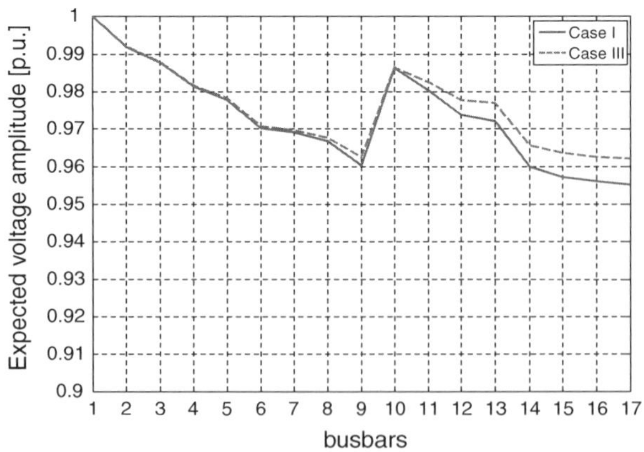

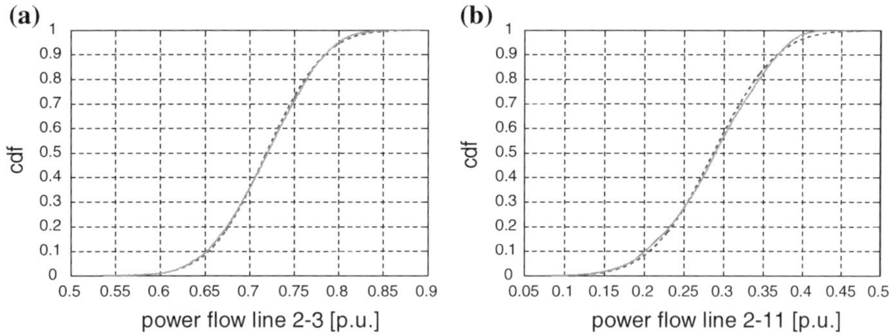

Fig. 19 Case III: cdf of power flow in lines 2–3 (**a**) and 2–11 (**b**) obtained with Monte Carlo simulation and with Cornish–Fisher expansion method (*dotted line*)

the Royal Netherlands Meteorological Institute (http://www.knmi.nl/samenw/hydra/index.html) have been used, as in the previous subsection.

We show, for the sake of conciseness, only the PDs of the powers injected by the WTGUs at two busbars #15 and #17, the pdfs of the voltage amplitudes at two busbars #9 and #13, and the profile of mean value of voltage amplitude at all busbars compared with the profile in the absence of WTGU (Figs. 16, 17, 18).

The reduction of power losses is 20.64 % with respect to the Case (I); this reduction is lower than the one obtained in the Markov-based approach (30.5 %) because the probability functions of the generated powers with the Generating functions-based approach have a mean value lower than the one obtained with the Markov-based approach.

4.4 Special Distributions Approach

The balanced distribution system also was studied by means of a DC load flow, and the results of a Monte Carlo simulation applied to the DC load flow were compared with the results obtained with the Cornish–Fisher expansion-based procedure shown in Sect. 3.5.

In the DC load flow analysis, the voltage amplitude was assumed to be equal to 1 p.u. at all busbars, and line resistances were assumed to be negligible, therefore no voltage and power losses are reported below.

Once again, the case considered here is the Case (III) already described in Sect. 4.1, and the wind speed was modeled as a Rayleigh random variable as in Sect. 4.1.

In the following, the cumulative distribution functions (cdfs) and the probability distribution functions (pdfs) of active power flow in two selected lines are shown.

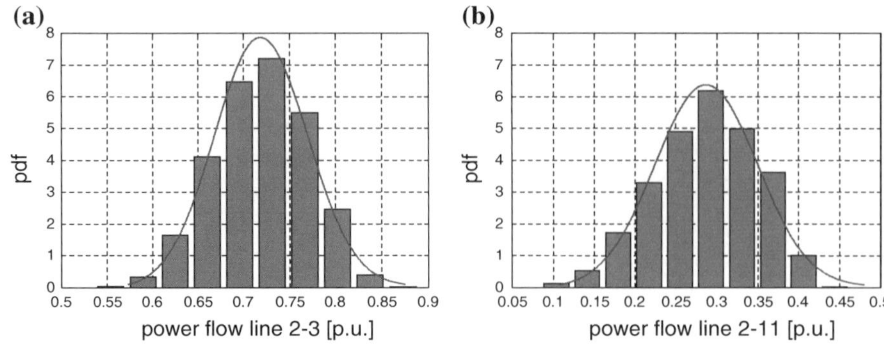

Fig. 20 Case III: pdf of power flow in lines 2–3 (**a**) and 2–11 (**b**) obtained with Monte Carlo simulation (histogram) and with Cornish–Fisher expansion method

Figures 19 and 20 show that there is a good fit.

Comparing the pdfs of the power flows shown in Fig. 20 with the analogous pdfs obtained with the non-linear Monte Carlo Simulation method (i.e., the pdf of Fig. 12), we can observe that, as expected, the values of power flows range in similar intervals in spite of the approximations.

4.5 Non-linear Monte Carlo Simulation for Unbalanced Distribution System

The *Non-linear Monte Carlo Simulation* was applied to the probabilistic steady-state analysis of the slightly modified version of the unbalanced, IEEE 34-bus test system illustrated in Fig. 11 of the companion chapter [1] and reported in Appendix A.2. This test system is an actual distribution system with 83 system nodes and a voltage level of 24.9 kV. The active and reactive load phase powers are reported in Table 4 of the companion chapter, while the complete network data and parameters can be found in [38]. The values reported in Table 4 were assumed to be the mean values of the active and reactive load phase powers, whereas the standard deviation was assumed equal to be 10 % of the mean values.

The following WTGUs have been considered: one 330 kW, stall-regulated fixed speed WTGU at bus #828 and one 330 kW, stall-regulated, fixed speed WTGU at bus #860 [30].

The wind speeds at bus #828 and #860 were characterized by uncorrelated Rayleigh pdfs with mean values equal to 7.20 and 8.18 m/s, respectively.

Figure 21 shows the mean values of the phase voltage profiles obtained in the absence and in the presence of WTGUs at busbars #828 and #860. Figure 22 shows the pdfs of the phase voltage at busbar #836 in the presence of WTGUs.

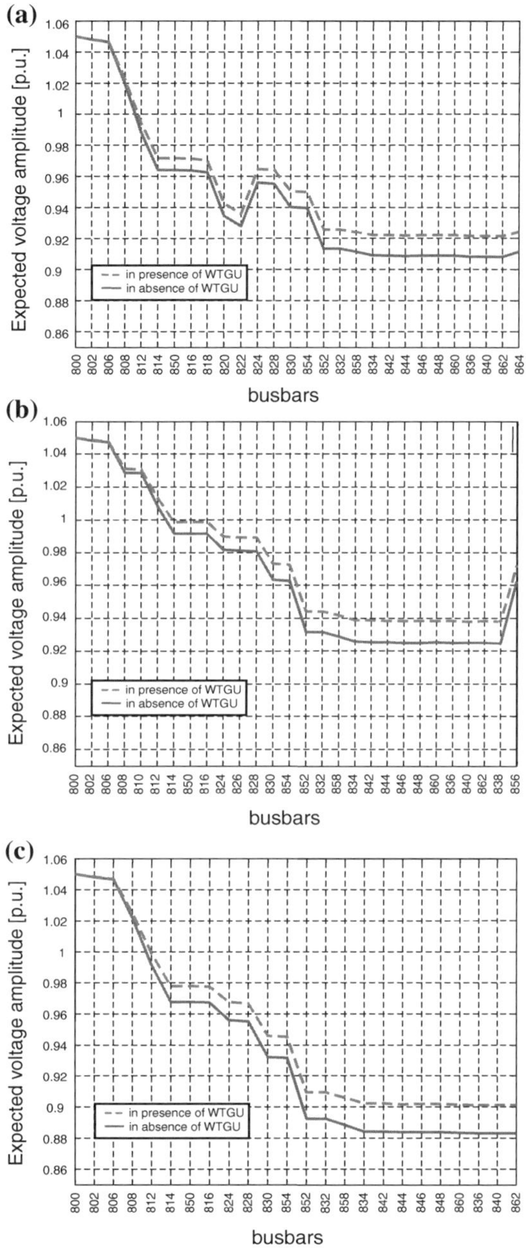

Fig. 21 Comparison between the voltage profiles in the absence of WTGUs and in the presence of WTGUs at busbar #828 and busbar #860: **a** phase a, **b** phase b, and **c** phase c

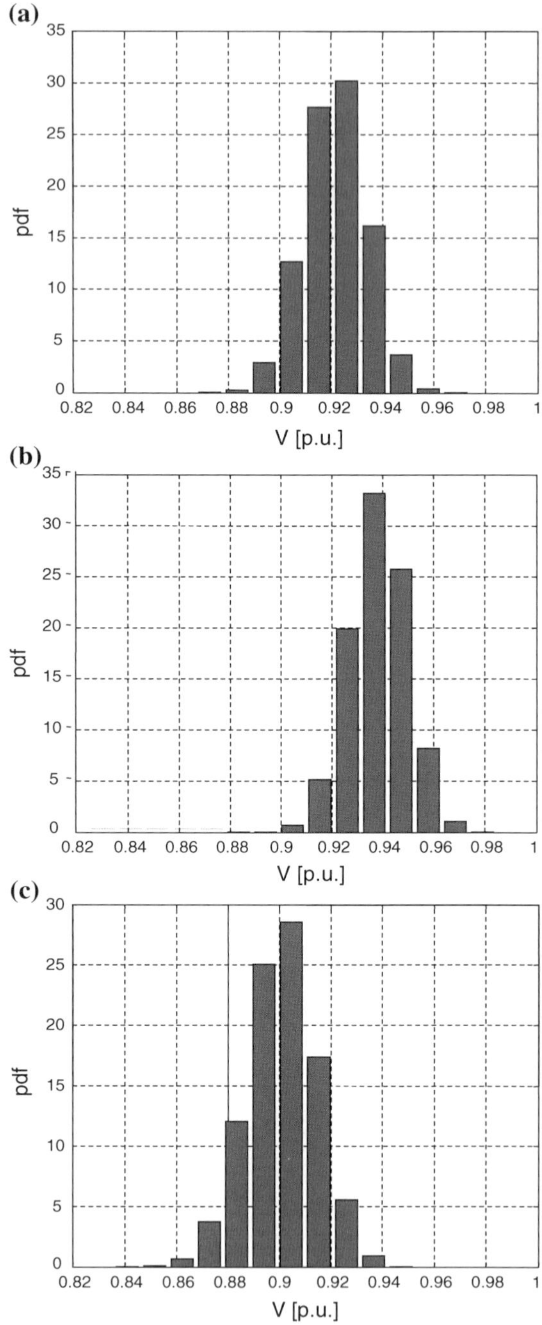

Fig. 22 Pdf of voltage amplitude at busbar #836 in the presence of WTGUs at busbar #828 and busbar #860: **a** phase a, **b** phase b, and **c** phase c

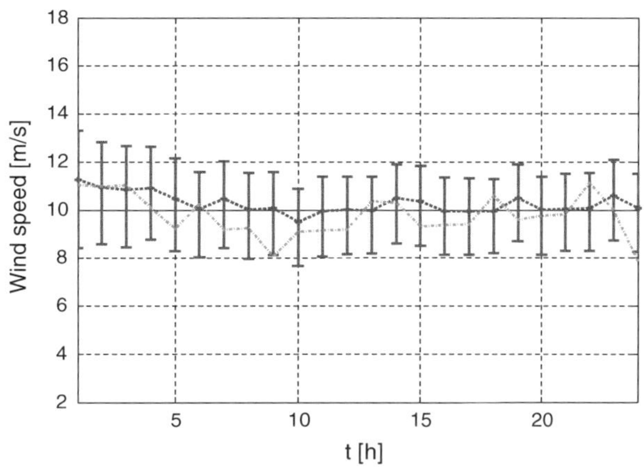

Fig. 23 Predicted (*dot* and *dashed line*) and measured (*dashed line*) hourly wind speed at busbar #828 (wind station at De Kooy)

It is apparent that the presence of wind-driven generators also increases the magnitude of the voltages at all the buses in this case; in addition, in the considered case, the real power losses of the system were reduced by about 19 %.

4.6 Bayesian-Based Approach for Unbalanced Distribution System

The Bayesian probabilistic approach was applied to evaluate the performance of the same unbalanced distribution system in the previous subsection.

The distribution system, the loads, and the generated power are the same as reported in the previous subsection. The results reported in this subsection were a synthesis of the ones published in [6].

The hourly steady-state analysis of the distribution system was conducted with reference to an autumn day.

The pdfs of the hourly wind speed were obtained by applying the Bayesian approach illustrated in Sect. 3.4. In particular, the prediction wind speed pdf of the first hour was based on the observed wind speed samples for 6 days before ($n = 144$ h; $m = 1$). These data, which were obtained from the Royal Netherlands Meteorological Institute (http://www.knmi.nl/samenw/hydra/index.html), refer to the data measured at the wind stations at Texelhors and De Kooy from September 19–25, 2004.

Figure 23 compares, as an example, the hourly forecasted wind mean values (and the spread between minimum and maximum values of the pdf) with the actual wind values obtained from the wind station at De Kooy; good predictive behavior clearly appears, as was the case for the wind station at Texelhors.

Figures 24, 25, 26 and 27 report the mean values and the standard deviations (expressed in percentage of the mean values) of the phase voltages during 24 h at

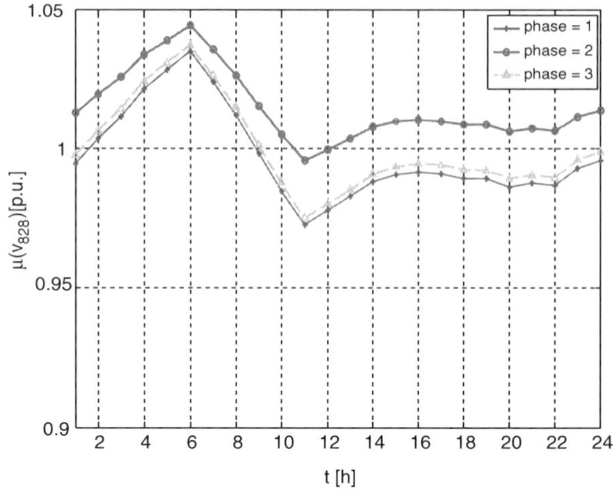

Fig. 24 Mean values of the phase voltage at busbar #828 over 24 h

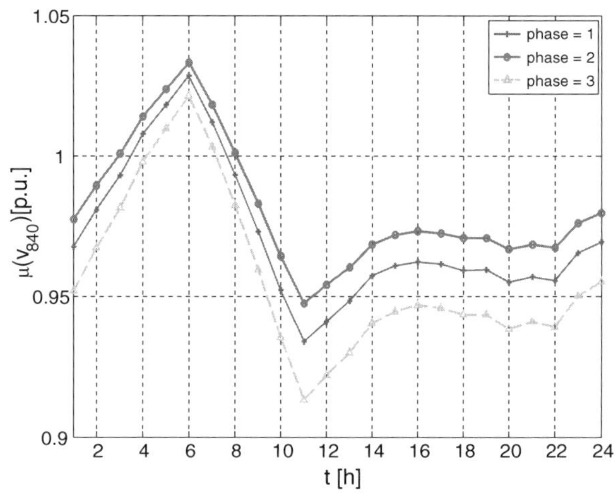

Fig. 25 Mean values of the phase voltage at busbar #840 over 24 h

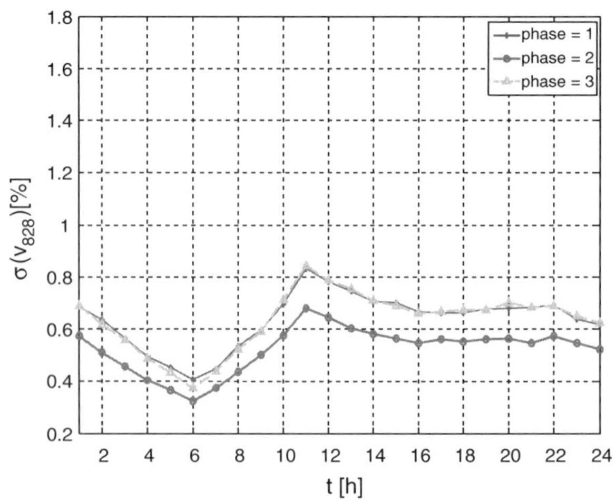

Fig. 26 Standard deviation of the phase voltage at busbar #828 over 24 h

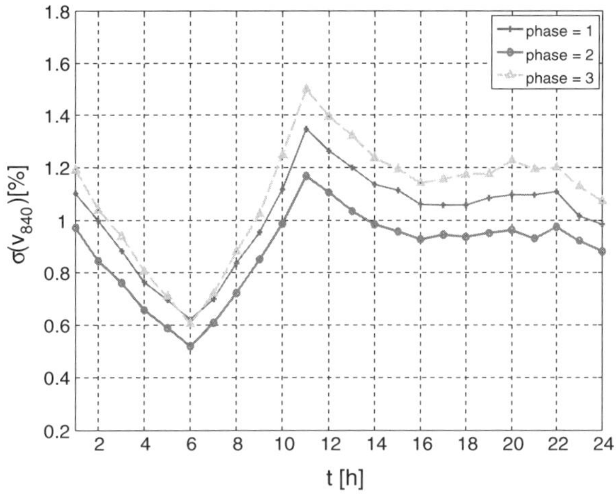

Fig. 27 Standard deviation of the phase voltage at busbar #840 over 24 h

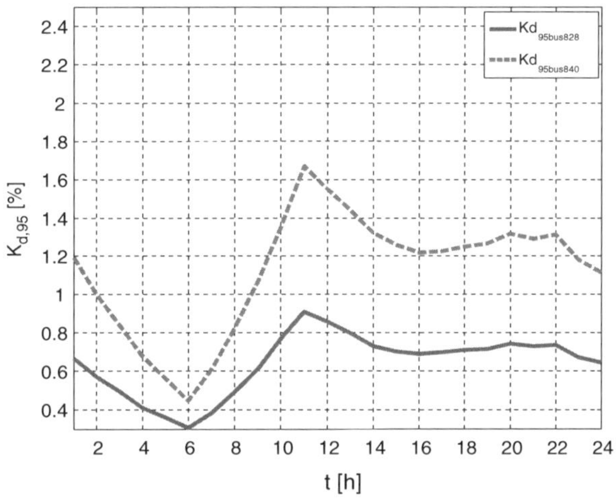

Fig. 28 95th percentile of the unbalance factor at busbar #828 and busbar #840 over 24 h

busbars #828 and #840. The mean value of the phase voltage is influenced by the load profile, assuming the minimum value in correspondence with the maximum load ($h = 11$) and the maximum value in correspondence with the minimum load ($h = 6$). The standard deviations are expressed as percentages of the mean value and assume the maximum and minimum values are in correspondence with the maximum and minimum loads, respectively.

To provide an idea of the unbalances characterizing the operating of the electrical distribution system, Fig. 28 shows the 95th percentile of the unbalance factor, Kd_{95}, at busbars #828 and #840 over the 24-h study period. The unbalance factor follows the load profile. Busbar #828 shows a mean value of the unbalance factor less than 1 % during the 24 h, while busbar #840 shows many values greater than 1 % and a maximum value or about 1.7 % in hour 11.

5 Conclusions

In this study, some probabilistic approaches for the load-flow analysis of balanced and unbalanced distribution systems with wind farms were analysed. Various probabilistic models of the wind farm were presented and used in several techniques, e.g., Monte Carlo simulation, convolution process, and special distribution functions, to perform probabilistic analyses.

Numerical applications of balanced and unbalanced test distribution systems were discussed, considering various wind farm models.

The main conclusion of the chapter is that the probabilistic steady-state analysis of a distribution system is mandatory in order to take into account the unavoidable uncertainties of wind speed and of the loads on the distribution system. It ensures secure operation of the distribution system, mainly by assessing the effects of significant wind energy penetration in the market.

Appendix A.1

A.1.1 Simulation of Gaussian Correlated Random Variables

In order to generate an approximate n-vector $\boldsymbol{R}$ of Gaussian random variables characterized by the mean value $\mu(\boldsymbol{R})$ and by the covariance matrix $cov(\boldsymbol{R})$, the following procedure was applied [39]:

1. Evaluate the eigenvalues $l_1, \ldots, l_n$ and the corresponding eigenvectors $\boldsymbol{\Psi}_1, \ldots, \boldsymbol{\Psi}_n$ of the matrix $cov(\boldsymbol{R})$;
2. Generate an n-vector $\boldsymbol{\Gamma}$ of uncorrelated Gaussian variables characterized by zero mean values and variances equal to $l_1, \ldots, l_n$;
3. Generate the vector $\boldsymbol{R}$ as:

$$\boldsymbol{R} = \mu(\boldsymbol{R}) + [\boldsymbol{\Psi}_1, \ldots, \boldsymbol{\Psi}_n]\boldsymbol{\Gamma}. \tag{33}$$

A.1.2 Simulation of Rayleigh Correlated Random Variables

In order to generate random numbers of N_V random variables $\omega_1, \ldots, \omega_{N_V}$, which are Rayleigh distributed and correlated, the following approximate procedure can be applied [13]:

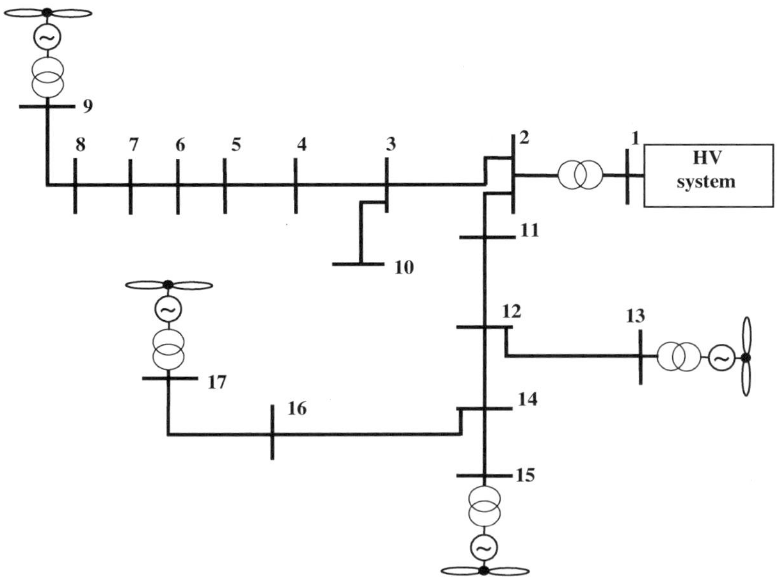

Fig. 29 17-bus balanced, 3-phase, test-distribution system

1. With $\mu_1, \ldots, \mu_{N_V}$ as the mean values, $\sigma_1, \ldots, \sigma_{N_V}$ as the standard deviations, and ρ_{ij} $(i, j = 1, \ldots, N_V,\ i \neq j)$ as the correlation factors, the covariance matrix can be built as:

$$\boldsymbol{\Omega}_\omega = \begin{bmatrix} \sigma_1^2 & \cdots & \rho_{1N_V} \\ & \ddots & \\ \rho_{N_V 1} & \cdots & \sigma_{N_V}^2 \end{bmatrix}, \tag{34}$$

and a lower triangular matrix $\mathbf{L}$ can be determined so that :

$$\boldsymbol{\Omega}_\omega = \mathbf{L}\,\mathbf{L}^T \tag{35}$$

2. For each random variable ω_i, a vector of random numbers $\boldsymbol{\omega}_i^*$, distributed according to a Rayleigh pdf with a mean value of μ_i, is generated;
3. Starting from ω_i^*, a new variable is determined according to the following relationship:

$$\mathbf{z}_i^* = \frac{\boldsymbol{\omega}_i^* - E\left(\boldsymbol{\omega}_i^*\right)}{\sigma\left(\boldsymbol{\omega}_i^*\right)} \tag{36}$$

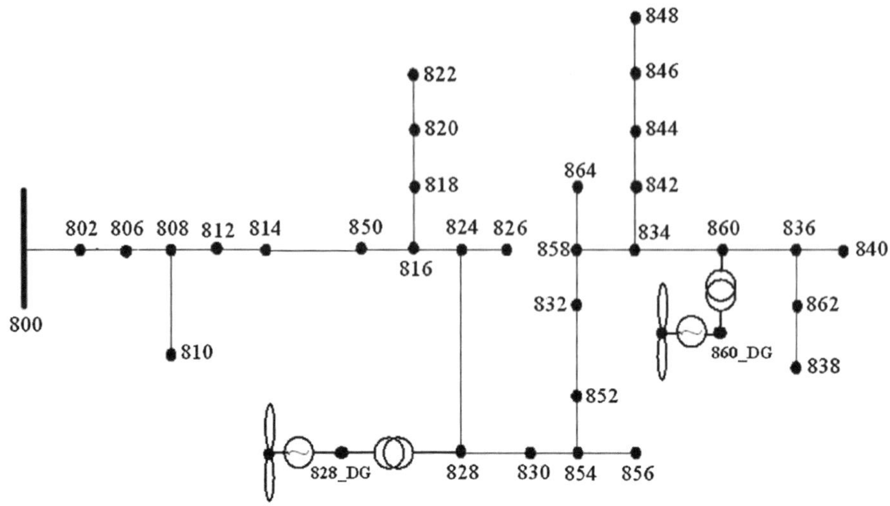

Fig. 30 IEEE 34-bus test-distribution system

4. The random numbers $\mathbf{r}_i^*$ of correlated Rayleigh variables can be determined as:

$$\mathbf{r}_i^* = \left| \mathbf{L}\left(\mathbf{z}_i^*\right)^T + \mu_i \right|. \tag{37}$$

Appendix A.2

The 17-bus balanced test distribution system and the IEEE 34-bus unbalanced test distribution system are shown in Figs. 29 and 30.

References

1. Caramia P, Carpinelli G, Proto D, Varilone P Deterministic approaches for steady state analysis in distribution systems with wind farms. In: Pardalos PM, Pereira MVF, Rebennack S, Boyko N (eds) Handbook of wind power systems: operation, modeling, simulation and economic aspects. Springer, Berlin
2. Jorgensen P, Tande JO (1998) Probabilistic load flow calculation using Monte Carlo Techniques for distribution network with wind turbines. In: 8th international conference on harmonics and quality of power ICHQP'98, Athens (Greece), 14–16 Oct 1998

3. Hatziargyriou ND, Karakatsanis TS, Papadopoulos M (1993) Probabilistic load flow in distribution systems containing dispersed wind power generation. IEEE Trans Power Syst 8(1):159–165
4. Masters CL, Mutale J, Strbac G, Curcic S, Jenkins N (2000) Statistical evaluation of voltages in distribution systems with embedded wind generation. IEE Proc Generat Transm Distrib 147(4):207–212
5. Di Fazio AR, Russo M (2008) Wind farm modelling for reliability assessment. IET Renew Power Gener 2(4):239–248
6. Bracale A, Caramia P, Carpinelli G, Varilone P (2008) A probability method for very short-term steady state analysis of a distribution system with wind farms. Int J Emerg Electr Power Syst 9(5), Article 2
7. Usaola J (2009) Probabilistic load flow with wind production uncertainty using cumulants and Cornish–Fisher expansion. Int J Electr Power Energy Syst 31(9):474–481
8. Zhaohong BIE, Gan L, Liu L, Xifan W, Xiuli W (2008) Studies on voltage fluctuations in the integration of wind power plants using probabilistic load flow. In: IEEE power and energy society general meeting—conversion and delivery of electrical energy in the 21st century, pp. 1–7, 20–24 July 2008
9. Boulaxis NG, Papathanassiou SA, Papadopoulos MP (2002) Wind turbine effect on the voltage profile of distribution networks. Renew Energy 25:401–415
10. Boehme T, Wallace AR, Harrison GH (2007) Applying time series to power flow analysis in networks with high wind penetration. IEEE Trans Power Syst 22(3):951–957
11. Carpinelli G, Pagano M, Caramia P, Varilone P (2007) A probabilistic three-phase load flow for unbalanced electrical distribution systems with wind farms. IET Renew Power Gener 1(2):115–122
12. Chen P, Chen Z, Bak-Jensen B, Villafafila R (2007) Study of power fluctuation from dispersed generations and loads and ist impact on a distribution network through a probabilistic approach. In: 9th international conference on electrical power quality and utilization, Barcelona (Spain), 9–11 Oct 2007
13. Feijòo AE, Cidràs J, Dornelas JLG (1999) Wind speed simulation in wind farms for steady-state security assessment of electrical power systems. IEEE Trans Energy Convers 14(4):1582–1588
14. Papaefthymiou G, Klockl B (2008) MCMC for wind power simulation. IEEE Trans Energy Convers 23(1):234–240
15. Amada JM, Bayod-Rújula ÁA (2007) Wind power variability model: Part I—foundations. In: 11th international conference on electrical power quality and utilization, Barcelona (Spain), 9–11 Oct 2007
16. Conlon MF, Mumtaz A, Farrell M, Spooner E (2008) Probabilistic techniques for network assessment with significant wind generation. In: 11th international conference on optimization of electrical and electronic equipment (OPTIM 2008), Braşov, Romania, 22–24 May 2008, pp 351–356
17. Bracale A, Carpinelli G, Proto D, Russo A, Varilone P (2010) New approaches for very short-term steady-state analysis of an electrical distribution system with wind farms. Energies 3(4):650–670. doi:10.3390/en3040650
18. Pereira VF, Balu NJ (1992) Composite generation/transmission reliability evaluation. Proc IEEE 80(4):470–491
19. Anders GJ (1990) Probability concepts in electric power systems. Wiley, New York
20. Allan RN, Leite da Silva AM, Burchett RC (1981) Evaluation methods and accuracy in probabilistic load flow solutions. IEEE Trans Power Apparatus Syst (PAS) 100(5):2539–2546
21. Caramia P, Carpinelli G, Di Vito G, Varilone P (2003) Probabilistic techniques for three-phase load flow analysis. In: IEEE/PES power tech conference, Bologna (Italy), June 2003
22. Zhang P, Lee T (2004) Probabilistic load flow computation using the method of combined cumulants and Gram–Charlier expansion. IEEE Trans Power Syst 19(1):676–682
23. Kendall MG, Stuart A (1958) The advanced theory of statistics, vol 1. Charles Griffin & Co., Ltd., London, UK

24. Cornish EA, Fisher RA (1937) Moments and cumulants in the specification of distributions. Revue de l'Institut International de Statistics 4:307–320
25. Jaschke SR (2001) The Cornish–Fisher-expansion in the context of delta–gamma normal approximations. Discussion Paper 54, Sonderforschungsbereich 373. Humboldt-Universität, Berlin. http://www.jaschke-net.de/papers/CoFi.pdf
26. Santabria L, Dillon TS (1986) Stochastic power flow using cumulants and Von Mises functions. Int J Electr Power Energy Syst 1(8):47–60
27. Hu Z, Wang X (2006) A probabilistic load flow method considering branch outages. IEEE Trans Power Syst 21(2):507–514
28. Sanabria LA, Dillon TS (1998) Power system reliability assessment suitable for a deregulated system via the method of cumulants. Int J Electr Power Energy Syst 20(3):203–211
29. Caramia P, Carpinelli G, Varilone P (2010) Point estimate schemes for probabilistic three-phase load flow. Electr Power Syst Res 80(2):168–175
30. Mortensen NG, Landberg L, Troen I, Petersen EL (1993) Wind atlas analysis and application program (WAsP)
31. Feijòo AE, Cidràs J (2000) Modeling of wind farms in the load flow analysis. IEEE Trans Power Syst 15(1):110–111
32. Gelman A, Carlin JB, Stern HS, Rubin DB (1995) Bayesian data analysis. Chaoman & Hall, London (UK)
33. Black TC, Thompson JW (2001) Bayesian data analysis. Computing Sci Eng 03(4):86–91
34. Lenk P (2001) Bayesian inference and Markov Chain Monte Carlo. University of Michigan, USA
35. Congdon P (1993) Applied Bayesian modelling. Queen Mary University of London (UK), Wiley, London
36. Chib S, Greenberg E (1995) Understanding the metropolis-hastings algorithm. Am Stat 49(4):327–335
37. Bracale A, Carpinelli G, Caramia P, Russo A, Varilone P (2010) Point estimate schemes for probabilistic load flow analysis of unbalanced electrical distribution systems with wind farms. IEEE/PES 14th international conference on harmonics and quality of power (ICHQP), 26–29 Sept 2010, Bergamo (Italy)
38. Kersting WH (2001) Radial distribution test feeders. In: IEEE power engineering society winter meeting, vol 2, 28 Jan–1 Feb 2001, pp 908–912. http://ewh.ieee.org/soc/pes/dsacom/testfeeders.html)
39. Fukunaga L (1972) Introduction to statistical pattern recognition. Academic Press, London

Advanced Control Functionalities for Grid Integration of Large Scale Wind Generation

Fernanda Resende, Rogério Almeida, Ângelo Mendonça and João Peças Lopes

Abstract Many system operators have issued grid code requirements to ensure a reliable and secure operation of power systems with high integration levels of wind generation. Then, wind farms are required to behave as much as possible like conventional power plants equipped with synchronous generators, providing thus active and reactive power regulation and fault ride through capability. In addition, it is necessary to assure that problems of small signal stability do not arise. This chapter deals with advanced control functionalities to improve the performance of double fed induction generators, regarding frequency control and fault ride through capability, and with robust tuning of classical power system stabilizers installed in double fed induction generators to provide damping as an ancillary service. Since fault ride through is a matter of major concern for many system operators, the control functionalities of a static compensator based solution to provide fault ride through capability of wind farms equipped with fixed speed induction generators are also addressed. The effectiveness of the above mentioned advanced control functionalities is demonstrated through numerical simulations.

F. Resende (✉) · R. Almeida · Â. Mendonça · J. P. Lopes
INESC-Porto, Porto, Portugal
e-mail: fresende@inescporto.pt

J. P. Lopes
Faculty of Engineering of Porto University, Porto, Portugal

R. Almeida
Federal University of Amapá, Amapá, Brazil

F. Resende
Lusófona University of Porto, Porto, Portugal

P. M. Pardalos et al. (eds.), *Handbook of Wind Power Systems*, Energy Systems,
DOI: 10.1007/978-3-642-41080-2_9,

1 Introduction

The widespread growth of wind power capacity installed worldwide required the development of larger and more robust Wind Turbine Generation Systems (WTGS). Requirements in terms of power control ability during normal operation and in case of grid abnormalities have increased significantly, leading with the variable speed concept with Double Fed Induction Generator (DFIG) systems and with WTGS equipped with full-scale power converters. In the DFIG configuration, only the rotor of the induction generator is connected to the ac grid through a partial-scale frequency converter, which is responsible to control the rotor speed in order to allow a wide speed range of operation. Typically, the variable speed range is ±30 % around the synchronous speed and the frequency converter rating is only 25–30 % of the configuration rated power, which makes this concept attractive and popular for an economic view point [29]. As a result the DFIG has been the most commonly used generator type [18].

Following the increasing wind power integration levels, many system operators have updated their grid codes, through the establishment of specific and stricter technical requirements for wind farms operation [15]. In many systems, wind generation has priority over conventional power plants usually equipped with synchronous generators, which could not be dispatched in order to accommodate the surplus of wind generation. As a result, several negative impacts can arise regarding the system technical and operational characteristics, focusing particularly on the reduction of system dynamic stability and security of supply and on system frequency and voltage regulation capabilities. Therefore, the most common grid code requirements focus mainly the control ability of wind farms regarding both active and reactive power control to provide frequency and voltage regulation, respectively, and Fault Ride-Through (FRT) capability [54].

Synchronous generators may be equipped with Power System Stabilizers (PSS) in order to provide damping to low frequency power oscillations occurring among their rotors. As the large surplus of wind generation in some areas may increase the power flows between weakly interconnected grids, the damping levels will be reduced since the existing PSS may no longer be able to provide the additional damping necessary in all wind generation conditions [33]. Therefore, wind farms should be able to supply damping as an ancillary service [23, 24].

Grid code requirements have been the major drivers for the fast developments of WTGS Wind turbine generation systems—WTGS technologies over the past few years [20]. Their compliance has challenged the manufacturers to develop conventional control capabilities for wind turbines and therefore for wind farms, exploiting the variable speed concept. For this purpose, advanced control functionalities have been developed and embedded into the control systems of the individual WTGS frequency converters, providing thus enlarged control capabilities, which play a key role to support the most stringent grid code requirements issued by many system operators.

Requirements in terms of active and reactive power control, under normal operating conditions, have been addressed through an hierarchical control structure comprising both a central control level and a local control level supported by an advanced SCADA system [4, 20]. Depending on the network status, the system operator may require adjustments in outputs of active and reactive power of the wind farm. For this purpose a two level hierarchical control and management system has been exploited. The central control level is responsible to control the wind farm output power by sending out active and reactive power references to the local control level. The main parameters from the individual WTGS and from the Point of Common Coupling (PCC) are recorded by the data acquisition system and used to determine the optimum active and reactive power set points for each WTGS in order to meet the system operator requests. The local control level addresses the control system of individual WTGS and ensures that the references sent from the wind farm central control level are achieved.

FRT capability has been the major concern of network operators since the effects of grid faults may propagate over very large geographical areas leading with the disconnection of wind farms. Therefore a serious threat will be created regarding the network security of supply because a large amount of wind generation could be lost simultaneously. In order to prevent this problem WTGS must remain connected during and after severe grid disturbances, ensuring fast restoration of active power to pre-fault levels as soon as the fault is cleared and injecting reactive current in order to support the grid voltage during disturbances, providing thus fast voltage recovery after fault clearing.

In spite of the current use of variable speed WTGS, many wind farms equipped with Fixed Speed Induction Generators (FSIG), the standard technology installed in the early 1990s, are still in operation in many countries and can fulfil FRT requirements to a different degree. The dynamic behaviour of FSIG is dominated by the grid connected induction generator which always draws reactive power from the grid. Although shunt capacitor batteries are typically used to compensate the reactive power requirements during steady state operation, these devices exhibit rather poor performance during fault conditions, since their reactive power injection capability decreases significantly during voltage drops. Thus, in the event of voltage dip the generator torque and active power output can be reduced considerably resulting in rotor acceleration and further rotor instability. Also, the machine operation at increased slip values results in increased reactive power absorption, particularly after fault clearance and partial restoration of the system voltage. This effectively prevents voltage recovery and may affect other neighbouring generators whose terminal voltage remains depressed. Since the dynamic behaviour of the induction generator itself can not be improved, FRT capability can be assisted by external static compensation devices connected at the wind farm terminals [46]. For this purpose, STATic synchronous COMpensator (STATCOM) based solutions are considered the most effective approaches for these situations [11]. However, a STATCOM rating approximately 100 % of the total wind farm installed capacity is required to guarantee a successful voltage recovery [8, 19, 36, 46].

This chapter deals with the advanced control functionalities, which have been developed to enhance the performance of DFIG control capabilities regarding the compliance with grid code requirements, particularly FRT capability and primary frequency control. The problem of PSS robustly tuning in systems with large share of wind generation is also addressed in this chapter, taking into account that PSS installed in DFIG can be used to provide additional damping to the electromechanical modes of oscillation. Moreover, the control functions of an external STATCOM based solution to provide FRT capability of wind farms equipped with FSIG are also addressed. The performance of these advanced control functionalities as well as the performance of the robustly tuned PSS is evaluated through numerical simulations performed in small test systems. The results obtained are presented and discussed.

In Sect. 2 the mathematical model of the DFIG configuration is presented together with the main issues regarding DFIG operation and control; Sect. 3 describes control functions based on fuzzy control that can be used to improve FRT capability of DFIG; Sect. 4 presents a robust primary frequency control approach for DFIG; A suitable approach to perform robust tuning of PSS installed in DFIG is addressed in Sect. 5; Sect. 6 deals with the commonly used control approaches of STATCOM when used to provide FRT capability of wind farms equipped with FSIG. Final remarks are presented in Sect. 7.

2 DFIG Dynamic Modelling and Operation

The DFIG concept corresponds to a variable speed WTGS equipped with a Wound Rotor Induction Generator (WRIG) and a partial scale frequency converter connected to the rotor windings through slip rings. The stator is directly connected to the ac grid whereas the rotor is connected through the frequency converter, being the DFIG fed from both the rotor and stator sides, as it can be observed from the typical electrical scheme depicted in Fig. 1. Thus, the DFIG configuration allows supplying electrical power to the grid from the stator and exchanging electrical power with the grid through the rotor circuit. Also, controlling the active power flow on the rotor circuit, both in magnitude and direction, allows the variable speed operation feature over a wide speed range, from sub synchronous to super synchronous speed. Therefore, the frequency converter comprises two Insulate Gate Bipolar Transistors (IGBT) based Pulse Width Modulation (PWM) converters, the Rotor Side Converter (RSC) and the Grid Side Converter (GSC), connected back-to-back through a dc-link capacitor and controlled independently of each other [40, 44], as it can be observed from Fig. 1.

During normal operating conditions, the RSC control system is responsible to control the active and reactive power fed into the grid though the stator circuit by controlling the rotor current components while the GSC control system is responsible to control the dc-link voltage based on the active power balance between the RSC and the ac grid and, additionally, to ensure the DFIG operation

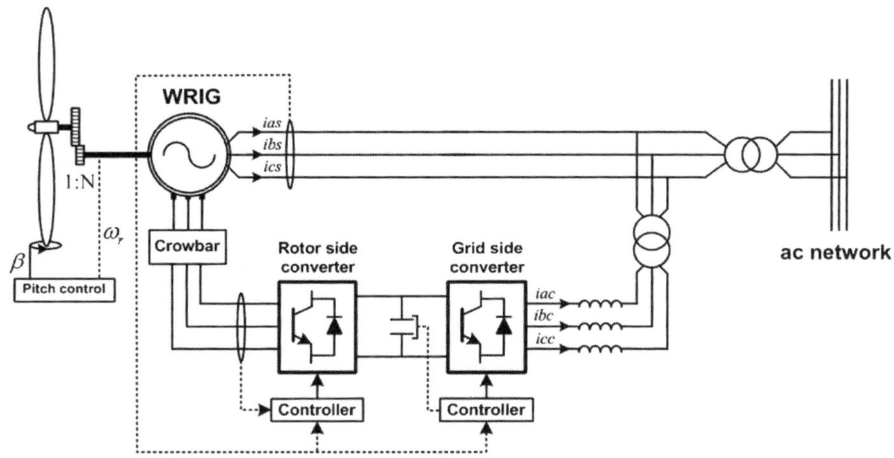

Fig. 1 Typical configuration of the DFIG electrical part

with a given power factor. However, since the RSC can compensate a bigger reactive power demand than the GSC, it is usually assumed that the reactive power demand is mainly controlled by means of the RSC [1, 7, 34].

In case of a grid fault, high transient currents arise in the grid connected WRIG stator, which are transferred to the rotor leading to over currents and over voltages that can damage the frequency converter. In order to avoid these over currents an ac crowbar protection system has been commonly used, being triggered by the dc-link voltage, which rises rapidly due to the rotor current peaks. Then, for dip voltage sags, the crowbar short-circuits the WRIG rotor and the DFIG configuration goes temporarily to an asynchronous machine operation mode, since RSC is blocked and the WRIG controllability is lost. In contrast to the RSC, the GSC is kept in operation. So, in order to provide voltage support under these circumstances, a co-ordinated voltage control of RSC and GSC is required. Thus, in the adopted control strategy, as long as the crowbar is triggered, the GSC will provide its maximum reactive power in order to improve the voltage level. When the crowbar is removed, the RSC is taking over the voltage control in order to re-establish immediately the voltage level. In this way the DFIG configuration will be able to comply with the requirements of FRT and grid support capability [1, 34].

The mathematical models commonly used to represent the DFIG dynamic behaviour with impact on the ac network during transient stability studies share a high degree of complexity. A generic model is based on the typical DFIG configuration illustrated in Fig. 1, comprising basically the following main components:

- The aerodynamic system;
- The mechanical system;
- The generator (WRIG);

- The frequency converter and the corresponding control systems;
- The pitch control system.

Suitable dynamic models for these main components and their interconnections are presented in the following sections. These models are intended for power stability studies. Therefore, the control functionalities of the frequency converter are addressed under the framework of the control systems of both RSC and GSC [13, 17, 31, 41].

2.1 The Aerodynamic System

The WTGS aerodynamic system, also called the wind turbine rotor, comprises the turbine blades, which are responsible for reducing the incoming wind speed and for transforming the absorbed kinetic energy into mechanical power. The basic physics related with this energy conversion as well as with the mathematical representation of the turbine rotor are reported in [1, 2].

The amount of the kinetic energy extracted from the incoming wind in the turbine rotor swept area will depend on the wind turbine efficiency usually known as power coefficient, C_p. In turn, the C_p will depend on the angle of attack between the plane of the moving rotor blades and the relative wind speed as seen from the moving blades. Thus, the angle of attack is typically expressed in term of the tip speed ratio, λ, commonly defined as the relationship between the blade tip linear speed and the incoming wind speed as

$$\lambda = \frac{\omega_{turb} R}{V_w}, \tag{1}$$

where ω_{turb}(rad-elec/s) is the wind turbine rotor speed, V_w(m/s) is the incoming wind speed and R(m) is the blade radius.

In pitch controlled wind turbines, the angle of attack can be modified according to the pitch angle of the blades, β, which is adjusted through the pitch control system. Therefore, the power coefficient is commonly expressed as a function of λ and β, $C_p(\lambda, \beta)$.

Then, from a physical point of view, the wind turbine rotor is typically represented through the relationship between the available wind power, P_{wind}, and the mechanical power output, P_m, as

$$P_m = C_p(\lambda, \beta)\, P_{wind} = \frac{1}{2} \rho C_p(\lambda, \beta) A V_w^3, \tag{2}$$

where P_m(W) is the mechanical power on the turbine shaft, A(m^2) is the swept area by the turbine blades and $\rho = 1{,}225$ (kg/m^3) is the air density.

Numerical approximations have been developed to calculate C_p as a function of both λ and β. The following approximation was adopted as suggested by [51].

$$C_p(\lambda,\beta) = 0.73\left(\frac{151}{\lambda_i} - 0.58\beta - 0.002\beta^{2.14} - 13.2\right)e^{-18.4/\lambda_i} \tag{3}$$

where

$$\lambda_i = \frac{1}{\frac{1}{\lambda - 0.02\beta} - \frac{0.003}{\beta^3+1}} \tag{4}$$

Equations (3) and (4) lead to the $C_p(\lambda,\beta)$ versus λ characteristic curves for several values of the pitch angle of the blades, β (in degrees), and can be used for different wind turbines since the difference between the corresponding power curves can be neglected in dynamic simulation studies as demonstrated in [51]. Thus, the mathematical representation of the wind turbine rotor is commonly approached by a C_p–λ–β characteristic implemented through a look up table. The link between the aerodynamic and mechanical systems will be either mechanical power or mechanical torque, T_m. They are related to each other through the wind turbine rotational speed, ω_{turb}(mec-rad/s), as $P_m = \omega_{turb}T_m$.

2.2 The Mechanical System

The wind turbine mechanical system consists of the drive train comprising the rotating masses, the gear box and the connecting shafts. Thus, the drive train is commonly represented through the corresponding inertia of the system.

The major sources of inertia lie in both the turbine and WRIG rotors. Since the tooth wheels of the gearbox contribute only with a small fraction, the gear inertia is often neglected and only the transformation ratio is included. Thus, a typical model for the mechanical system is a two-mass model with a connecting shaft [50]. However, due to the decoupling effect of the power electronic converters of variable speed WTGS, the shaft properties are hardly reflected at the ac network, as discussed in [51]. Thus, it was assumed that the wind turbine rotor is modelled as a lumped mass only, leading to a more damped behaviour than when a two-mass model is used [50]. Therefore, the effects of torsional oscillations and stress on the turbine shaft were not considered. Such an assumption, however, does not compromise the quality of the comparative analysis regarding the advanced control functionalities to be addressed, as stated in [5–7].

2.3 The Wound Rotor Induction Generator

Representing the WRIG by a simple voltage behind a transient reactance equivalent circuit is a conventional modelling technique in wind turbine based DFIG applications for system dynamic behaviour analysis. The WRIG can be regarded as a conventional induction generator with a nonzero rotor voltage and represented by a reduced order model, neglecting the dc component and the fast transients in the stator current. In addition, the WRIG is operated as a generator and it is considered that the stator currents are positive when flowing towards the network, as it can be observed from Fig. 1, and the active and reactive power are positive when fed into the grid.

The equations describing the WRIG dynamics are derived with the rotor variables referred to the stator and transformed into the synchronous d–q axis reference frame with the q-axis 90° ahead the d-axis in the direction of the rotation flux. According to [7, 13, 26], the electrical equations of the WRIG rotor can be written in per unit as follows:

$$\begin{cases} \frac{de_d}{dt} = -\frac{1}{T_0}\left[e_d - (X - X')i_{qs}\right] + s\omega_s e_q - \omega_s \frac{L_m}{L_{rr}} \upsilon_{qr} \\ \frac{de_q}{dt} = -\frac{1}{T_0}\left[e_q - (X - X')i_{qs}\right] - s\omega_s e_d + \omega_s \frac{L_m}{L_{rr}} \upsilon_{dr} \end{cases}, \tag{5}$$

where

- e_d and e_q are, respectively, the per unit direct and quadrature components of the voltage behind the transient reactance;
- i_{ds} and i_{qs} are, respectively, the per unit direct and quadrature components of the stator current;
- υ_{dr} and υ_{qr} are, respectively, the per unit direct and quadrature components of the rotor voltage;
- T_0 is the rotor open circuit time constant in seconds;
- X is the per unit open circuit reactance;
- X' is the per unit short circuit reactance;
- L_m is the per unit mutual magnetizing inductance between the stator and the rotor windings;
- L_{rr} is the per unit self-inductance of the rotor windings;
- ω_s is the rotational speed of the synchronous reference frame in rad/s;
- $s\omega_s = (\omega_s - \omega_r)$ is the slip frequency;
- s is the slip;
- ω_r is the generator rotor speed in rad/s;

To complete the WRIG model the differential equations given by (5) have to be combined with the rotor swing equation that provides the rotor speed state variable as

$$\frac{d\omega_r}{dt} = \frac{1}{J}(T_m - T_e - D\omega_r), \tag{6}$$

where T_m is the mechanical torque, T_e is the electromechanical torque, J is the total moment of inertia (kg m^2) accounting both the turbine and the generator rotor inertia and D is the damping factor.

According to [7], the electromagnetic torque can be calculated as

$$T_e = e_d i_{ds} + e_q i_{qs} \tag{7}$$

where i_{ds} and i_{qs} can be derived algebraically from the stator voltage equations given by (8).

$$\begin{cases} v_{ds} = -R_s i_{ds} + X' i_{qs} + e_d \\ v_{qs} = -R_s i_{qs} - X' i_{ds} + e_q \end{cases}, \tag{8}$$

The active and reactive power in both stator and the rotor can be calculated as

$$\begin{cases} P_s = T_e \omega_r = v_{ds} i_{ds} + v_{qs} i_{qs} \\ Q_s = v_{qs} i_{ds} - v_{ds} i_{qs} \end{cases} \tag{9}$$

$$\begin{cases} P_r = -sP_s = v_{dr} i_{dr} + v_{qr} i_{qr} \\ Q_r = v_{qr} i_{dr} - v_{dr} i_{qr} \end{cases} \tag{10}$$

The active power delivered to the grid is then

$$P_g = P_s + P_r \tag{11}$$

In turn, the mechanical power can be obtained as

$$P_m = \omega_r T_m = (1 - s) P_s \tag{12}$$

It is assumed that T_m, T_e and P_m are positive values for generation. Thus, during sub-synchronous operation $s > 0$, $P_r < 0$, $P_m < P_s$ and so the rotor absorbs power from the stator. On the other hand, during the super-synchronous operation, $s < 0$, $P_r > 0$, $P_m > P_s$ and the rotor produces power, being the power delivered to the grid through both the stator and rotor circuits.

2.4 The DFIG Control Scheme

The variable speed feature makes it possible to adjust the turbine rotational speed to its optimum value, thus maximizing the power coefficient and, therefore, the generated power for several wind speeds below the rated wind. The power speed wind turbine characteristic obtained from Eq. (2) has been commonly used for tracking the WRIG rotor speed to the optimum tip speed ratio, λ_{opt}. According to [4], the speed control aims to keep the DFIG operation according to the pre-defined maximum power extraction curve defined as:

$$P_{ref} = k_{opt} \omega_r^3, \tag{13}$$

where

$$k_{opt} = \frac{1}{2} \rho \frac{C_{p\,opt}}{\lambda_{\mathrm{opt}}^3} \pi R^5 \tag{14}$$

and

$$\omega_{r\,ref} = \frac{p}{2} G\omega_{turb\,opt} \tag{15}$$

is the rotor speed of the generator, being p the number of poles of the WRIG and G the gearbox transmission ratio. Thus, the optimal speed of the turbine shaft can be derived from the Eq. (1) and be rewritten as:

$$\omega_{turb\,opt} = \frac{\lambda_{opt} V_w}{R}, \tag{16}$$

where the value of λ_{opt} can be obtained from the roots of the derivative of Eq. (3) regarding the tip speed ratio.

The DFIG whole control scheme comprises two main control systems: The mechanical control of the wind turbine pitch angle and the electrical control on the frequency converter. Thus, in order to control the power balance and thereby also the rotational speed, some coordination between these two control systems is required. During partial load conditions the electrical control assures variable speed operation while the pitch angle of the blades is kept constant. Above the rated wind speed the pitch angle is increased in order to decrease the power coefficient and, therefore, to shed some of the aerodynamic power, being the wind turbine shaft speed controlled to its rated value and therefore the DFIG output is kept near its rated power.

The frequency converter control scheme has been commonly designed to regulate the rotor speed of the WRIG in order to track the optimal wind speed reference, $\omega_{turb\,opt}$, for maximum power extraction given the incoming wind speed. For this purpose, the electrical control scheme is based on the WRIG rotor current regulation on the stator flux oriented reference frame, in which the d-axis is aligned with the stator flux linkage axis, being the quadrature component of the stator voltage, v_{qs}, equal to the terminal voltage, V_s, and the direct component of the stator voltage v_{ds} equal to zero [28, 38]. This yields the following equations regarding the stator currents:

$$\begin{cases} i_{ds} = -\frac{1}{L_{ss}} \frac{V_s}{\omega_s} + \frac{L_m}{L_{ss}} i_{dr} \\ i_{qs} = \frac{L_m}{L_{ss}} i_{qr} \end{cases}, \tag{17}$$

where L_{ss} is the per unit self-inductance of the stator windings.

Therefore, the stator active and reactive powers can also be expressed in terms of the rotor currents as

$$\begin{cases} P_s = V_s \frac{L_m}{L_{ss}} i_{qr} \\ Q_s = V_s \frac{L_m}{L_{ss}} i_{dr} - \frac{V_s^2}{\omega_s L_{ss}} \end{cases} \tag{18}$$

The rotor currents can be derived using the rotor voltages expressed in terms of direct and quadrature components of the rotor currents as in (19).

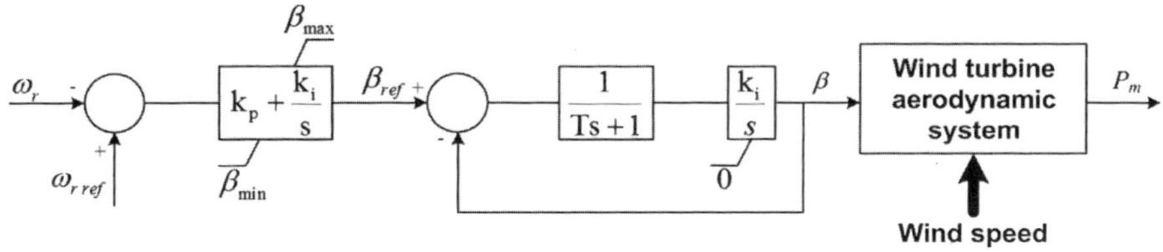

Fig. 2 The pitch control system

$$\begin{cases} v_{dr} = R_r i_{dr} - s\omega_s\left(L_{rr} - \frac{L_m^2}{L_{ss}}\right)i_{dr} + \left(L_{rr} - \frac{L_m^2}{L_{ss}}\right)\frac{1}{\omega_s}\frac{di_{dr}}{dt} \\ v_{qr} = R_r i_{qr} + s\omega_s\left(L_{rr} - \frac{L_m^2}{L_{ss}}\right)i_{qr} + \left(L_{rr} - \frac{L_m^2}{L_{ss}}\right)\frac{1}{\omega_s}\frac{di_{qr}}{dt} + \frac{sL_m V_s}{L_{ss}} \end{cases} \tag{19}$$

The Eq. (18) illustrate the decoupling between the WRIG stator active and reactive power control. Thus, the active power can be controlled through i_{qr} while the reactive power can be controlled through i_{dr}. Controlling the frequency converter includes the control systems of both the RSC and GSC, as it can be observed from Fig. 1. Thus, the DFIG control provided by the frequency converter is based on the following control strategy:

- The RSC controls independently the active and the reactive power on the WRIG stator side allowing the machine operation either for a given power factor according to grid requirements imposed by system operators or for voltage control in order to regulate the voltage at the point of common coupling of the WTGS by adjusting the reactive power supply.
- The GSC is controlled to keep the dc-link voltage constant regardless of the magnitude and direction of the active power flow through the rotor circuit. It can also represent a shunt reactive power compensator providing additional voltage control capabilities.

Thus, it was assumed that the RSC behaves like a voltage source inverter whereas the GSC behaves like a current source inverter [7].

2.5 The Pitch Control System

As stated before, the blade pitching control system is primarily used to limit the mechanical power for wind speeds above the rated speed and comprises both the speed controller and the actuator. The speed controller regulates the wind turbine shaft speed to its rated speed, $\omega_{r\,ref}$, and provides the pitch reference angle, β_{ref}, to the actuator, which is responsible to turn the turbine blades to this angle reference, according to the control scheme presented in Fig. 2.

The pitch controller is designed to operate within the limits of the pitch actuator and hence it cannot change the pitch angle too fast or beyond the limits, β_{min} and β_{max} [28]. This means that the blades can only be turned within certain physical limits, ranging from 0 to 90°, or even a few degrees to the negative side, and the pitch speed is usually less than 5° per second, although it may exceed 10° per second during emergencies [2].

The wind turbine pre-defined optimum power extraction curve can be established for a given k_{opt} associated to an optimum fixed blade angle, β_{opt}, that can be determined for an average short term wind speed forecast. Although, for wind speeds below the rated speed, the optimum pitch angle is in the range of few degrees above zero [2], the minimum limit of the blade angle is usually kept to its optimum value corresponding to $\beta_{\min} = \beta_{opt1}$ in Fig. 2. Also, the speed reference and, thereby, the blade angle reference has to be defined according to the mechanical power extracted from the incoming wind as follows:

$$\begin{cases} \omega_{r\,ref} = \frac{P_{ref}}{T_m} \Rightarrow \beta_{ref} = \beta_{opt1}, & \text{if } P_m \leq P_{\max} \\ \omega_{rref} = \frac{P_{\max}}{T_m} \Rightarrow \beta_{ref} = \beta_{\max}, & \text{if } P_m > P_{\max} \end{cases} \tag{20}$$

where $P_{\max}$ is the mechanical power rating.

2.6 The Rotor Side Converter Control System

The control functions performed by the rotor side converter are derived from Eqs. (18) and (19), aiming the fast dynamic control of both direct and quadrature components of the voltage applied to the rotor circuit in order to assure variable speed operation and independent control of the stator active and reactive power output. Thus, controlling the voltage magnitude allows the electromagnetic torque control, which must follow the speed reference provided by the optimum power-speed extraction curve. In turn, controlling the voltage phase allows the control of the reactive power exchanged with the grid, which in normal operation is set to a given reactive power reference in order to operate the DFIG with a given power factor according to the system operation requirements. In case of grid disturbances leading with voltage sags, if the current in the rotor circuit is not high enough to trigger the crowbar protection system, the RSC will provide voltage control on the point of the DFIG common coupling by adjusting the reactive power supply. Therefore two control approaches have been adopted to represent the RSC dynamics. Under normal operating conditions the RSC is controlled to perform active and reactive power control whereas following grid disturbances the RSC control is set to perform voltage control. These two control approaches are addressed in the following two subsections.

2.6.1 Rotor Speed and Terminal Voltage Control

Due to their simple structure and robustness, Proportional-Integral (PI) controllers have been commonly exploited to generate the $d - q$ components of the rotor

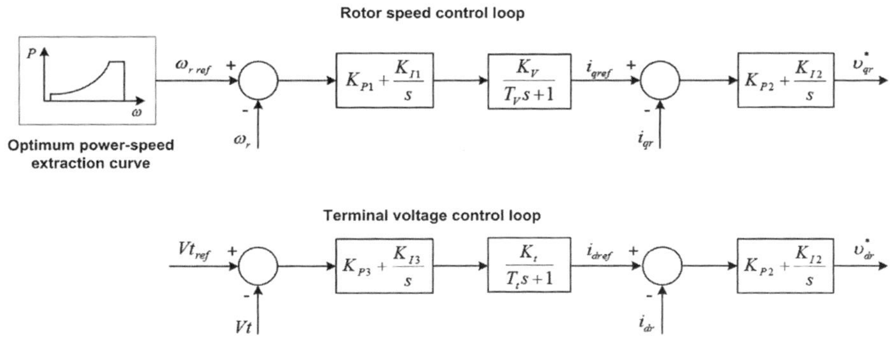

Fig. 3 Control scheme for rotor speed and terminal voltage control loops

voltage references. Thus, taking into account the Eq. (19), two control loops can be defined: The rotor speed control loop and the terminal voltage control loop, according to the control schemes that are presented in Fig. 3.

The speed error determines the quadrature component of the reference current through a PI controller. The required reference of the quadrature component of the rotor voltage can then be obtained through another PI controller. Regarding the terminal voltage control loop, the actual terminal voltage is compared with its reference value and the error is passed through a PI controller to obtain the reference value of the direct current component. This signal is then compared to the *d*-axis actual value and the error is sent to a second PI controller, which outputs the reference for the direct component of the rotor voltage.

Besides the above mentioned PI controllers, additional control blocks were included in order to represent the nonlinear and coupling terms that characterize the dynamic behaviour of the generator rotor according to Eq. (19). As discussed in [7], these control blocks improve the performance of the controllers during severe grid disturbances. The involved parameters, K_v, K_t, T_v and T_t should be derived from the machine parameters.

2.6.2 Active and Reactive Power Control

As already mentioned previously, during normal operation the RSC is set to control the active and reactive power independently. Thus, two control loops have been commonly adopted to derive the reference components of the rotor voltage [12]: The active power control loop and the reactive power control loop, as it can be observed from Fig. 4. These control loops are also defined based on Eqs. (18) and (19) exploiting PI controllers to derive the *d–q* components of the rotor voltage. However, in this control design, the additional control blocks representing the coupling and the nonlinear terms were not considered, since the involved time constants derived from the machine parameters are typically very small and therefore, under these circumstances, their effect on the RSC performance can be neglected.

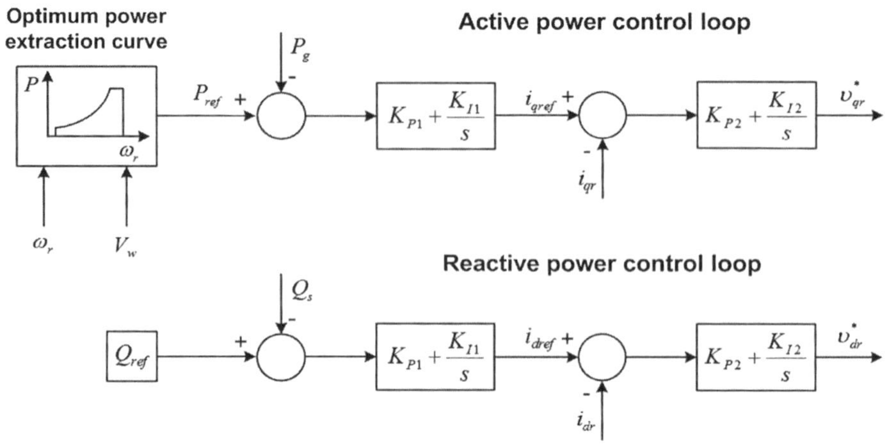

Fig. 4 Control scheme for active and reactive power control loops

The reference active power is provided by the maximum power extraction curve, given by Eq. (13), according to the actual generator speed in order to assure the DFIG operation at the maximum aerodynamic efficiency while the reactive power reference is set to a given set-point which can be controlled according to grid requirements imposed by system operators. The rotor voltage components are then obtained from two separate sets of PI controllers involving a cascade structure, in which the outer PI controllers are used to regulate the references of the rotor current components i_{dref} and i_{qref}, and the inner side PI controllers are used to regulate the υ_{qr}^* and υ_{dr}^* components, respectively.

The rotor reference voltage components provided by the RSC control scheme are transformed to *abc* coordinates through the Park's inverse transformation [26] and sent to the PWM signal generator of the converter in order to control the IGBT switching, as represented in Fig. 5.

It should be stressed that in power stability studies the power electronic interfaces are commonly represented by a fundamental frequency approach based on their control functions only, so that switching transients, harmonics and inverter losses are neglected since fast transient phenomena are not relevant for the purpose under analysis. However, converter losses and filters can be added by specifying the series impedance to the frequency converter [13, 17, 31, 41].

2.7 The Grid Side Converter Control System

Under normal operating conditions, the GSC is controlled to keep the dc-link voltage constant and to assure the converter operation with a unity power factor. When the DFIG is set to provide voltage control following grid disturbances, the

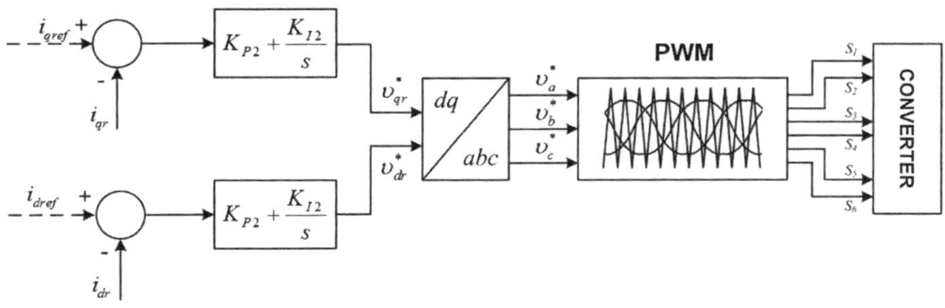

Fig. 5 Control scheme of the IGBT switching

reactive power demand is assigned to the GSC when the RSC is blocked, due to the co-ordinated voltage control, as already mentioned previously. Therefore, the reactive power reference for the GSC is set to its limit, so that the converter will contribute with its maximum reactive power to perform voltage support. Thus, the GSC control strategy comprises a fast inner current control loop to control the current injected to the grid following a predefined reactive power reference, and an outer slower control loop that controls the dc-link voltage, allowing that the energy delivered to the dc-link capacitor is transferred either to the utility system or to the RSC depending upon the DFIG operation in sub synchronous or super synchronous mode, respectively.

The current control loop is implemented using the instantaneous power theory proposed in [3] for active power filters control. Thus, according to Fig. 6, the instantaneous voltages and currents in three-phase circuits, v_{abc} and i_{abc}, respectively, expressed as instantaneous space vectors, are easily transformed to α–β–0 coordinates as follows:

$$\begin{bmatrix} v_0(t) \\ v_\alpha(t) \\ v_\beta(t) \end{bmatrix} = C \times \begin{bmatrix} v_a(t) \\ v_b(t) \\ v_c(t) \end{bmatrix} \quad \text{and} \quad \begin{bmatrix} i_0(t) \\ i_\alpha(t) \\ i_\beta(t) \end{bmatrix} = C \times \begin{bmatrix} i_a(t) \\ i_b(t) \\ i_c(t) \end{bmatrix} \tag{21}$$

where C is the Clark transformation given by

$$C = \sqrt{\frac{2}{3}} \times \begin{bmatrix} 1/\sqrt{2} & 1/\sqrt{2} & 1/\sqrt{2} \\ 1 & -1/2 & -1/2 \\ 0 & \sqrt{3}/2 & -\sqrt{3}/2 \end{bmatrix} \tag{22}$$

So, as described in [9], the instantaneous active and reactive powers are defined as:

$$\begin{bmatrix} p(t) \\ q(t) \end{bmatrix} = \begin{bmatrix} v_\alpha(t) & v_\beta(t) \\ -v_\beta(t) & v_\alpha(t) \end{bmatrix} \times \begin{bmatrix} i_\alpha(t) \\ i_\beta(t) \end{bmatrix} \tag{23}$$

where $p(t)$ is the instantaneous active power (W) and $q(t)$ is the instantaneous reactive power (VAr).

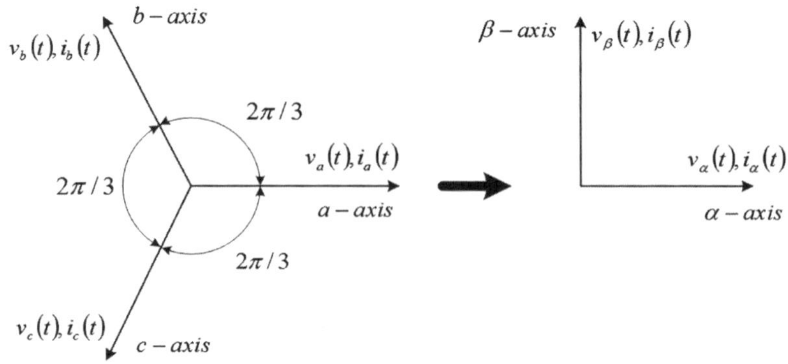

Fig. 6 *abc* to α–β coordinates transformation

The current reference signals i_a^*, i_b^* and i_c^* employed to turn on and to turn off the switches of the PWM based current source inverter are obtained instantaneously from Eq. (23) as follows:

$$\begin{bmatrix} i_{ca}^* \\ i_{cb}^* \\ i_{cc}^* \end{bmatrix} = \sqrt{\frac{2}{3}} \times \begin{bmatrix} 1 & 0 \\ -1/2 & \sqrt{3}/2 \\ -1/2 & -\sqrt{3}/2 \end{bmatrix} \times \begin{bmatrix} v_\alpha(t) & v_\beta(t) \\ -v_\beta(t) & v_\alpha(t) \end{bmatrix}^{-1} \times \begin{bmatrix} p_c(t) \\ q_c(t) \end{bmatrix} \tag{24}$$

where $p_c(t)$ and $q_c(t)$ are, respectively, the instantaneous active and reactive power on the GSC terminals, $v_\alpha(t)$ and $v_\beta(t)$ are the stator voltage in α–β coordinates, respectively.

In balanced three-phase systems the instantaneous active and reactive powers are constant and equal to the three-phase active and reactive powers, respectively [9]. Thus, $q_c(t)$ in Eq. (23) corresponds to the reactive power reference that assures the required power factor and $p_c(t)$ is the active power flow through the rotor circuit. Neglecting losses and the switching harmonic frequencies on the inverter output currents, the energy stored in the dc link capacitor can be obtained as [9]:

$$E_c = \int_{-\infty}^{t} (p_r(t) - p_c(t))dt = \frac{1}{2} C_{dc} v_{dc}^2(t) \tag{25}$$

where $p_r(t)$ is the RSC instantaneous power output, C_{dc} is the dc-link capacity (F) and $v_{dc}(t)$ is the dc capacitor instantaneous voltage (V).

The above equation shows that the dc-link voltage deviations depend on the power balance. If this voltage is kept constant $p_c(t) = p_r(t)$ and the active power exchanged with the grid corresponds to the RSC active power output. Thus, according to [5], the full control strategy of the rotor side converter is represented through the control scheme depicted in Fig. 7.

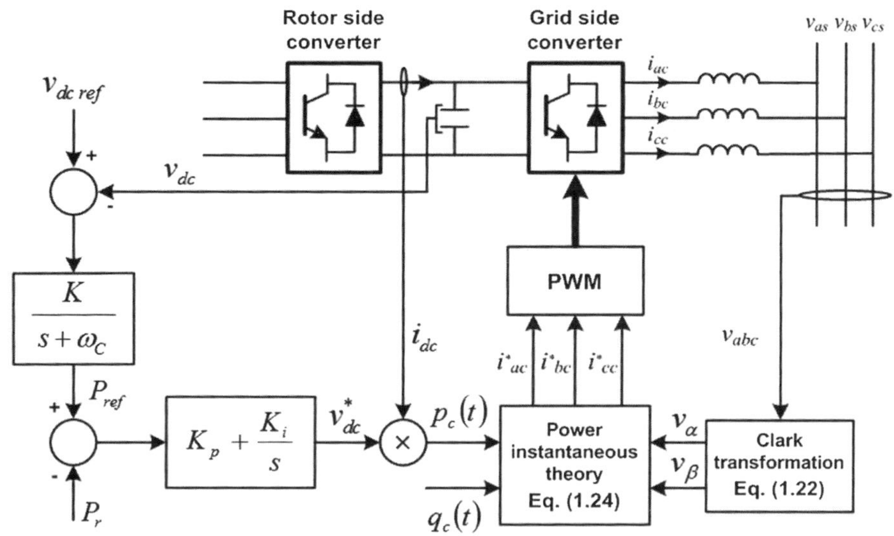

Fig. 7 The rotor side converter control scheme

As it can be observed from Fig. 7, the reference signal of the RSC output power is derived from the error between the reference value of the dc-link voltage and its current value through a controller defined as a proportional gain together with a low pass filter to eliminate the switching harmonics in the dc-link voltage.

3 Using Fuzzy Control to Improve DFIG Fault Ride Through Capability

To keep the RSC in operation plays an important role to improve FRT capability of DFIG and, therefore, the network dynamic behaviour following disturbances successfully eliminated, as reported in [39, 47, 53]. Therefore, the limitation of the generator currents under these circumstances has been a matter of concern regarding the RSC control design, since a robust control of the WRIG can prevent the crowbar trigger [7, 13]. However, the DFIG performance depends upon the suitable choice of the PI gains used in the RSC control system that is based on physical insights. Thus, tuning PI gains for performance optimization is a very hard task since the nonlinearities involved in the DFIG dynamics require a carefully tuning procedure for different operating conditions. Furthermore, the complexity level of this task will increase when the DFIG is connected to large power systems, due to the system dimension, or to an isolated system due to the stronger coupling between active and reactive powers [7].

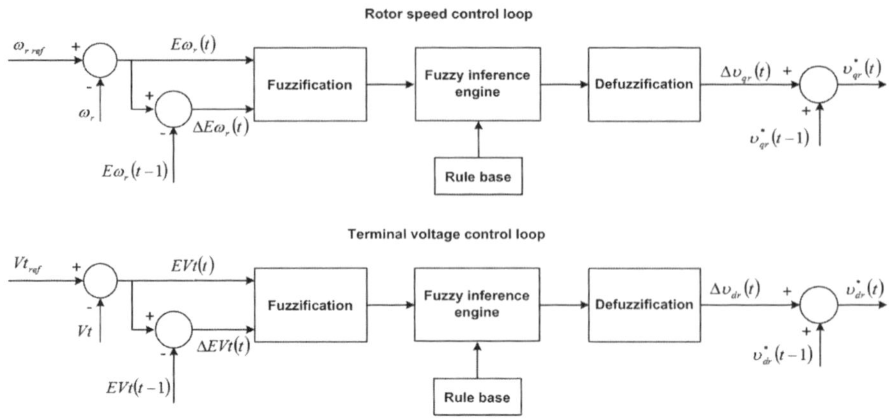

Fig. 8 Block diagram of the fuzzy controller

Fuzzy control provides a systematic way to control nonlinear processes based on the human experience, being considered as a heuristic approach that can improve the performance of closed loop systems. When properly designed, fuzzy controllers can perform better than conventional PI controllers when used for control purposes of nonlinear dynamic systems [27, 40].

Due to their robustness, the design of two fuzzy controllers is proposed in this section as an alternative control strategy of the RSC based on PI controllers. The performance of this approach is then evaluated by comparison with the performance of the conventional PI based approach. For this purpose, the whole DFIG dynamic model presented in Sect. 2 together with the designed fuzzy control approach were embedded in the full dynamic simulation tool and dynamic simulations were carried out considering the occurrence of disturbances.

3.1 Fuzzy Controllers Design

As the RSC control system is set to perform voltage control, the fuzzy control strategy comprises two controllers: The rotor speed controller and the terminal voltage controller. Their structures are similar, comprising the fuzzy inference engine and the defuzzification block [52] as depicted in Fig. 8. All the inputs and outputs are expressed in per unit referred to the system bases.

The number and the form of the membership functions defining the fuzzy inputs and outputs of both controllers were defined off-line and the universe of discourse for each variable was normalized to distinct values according with the behaviour of the corresponding variables observed during dynamic simulations. Standard triangular membership functions were adopted for both the input and output of the fuzzy sets of both fuzzy controllers. The designed fuzzy sets for rotor speed control and terminal voltage control are presented in Fig. 9.

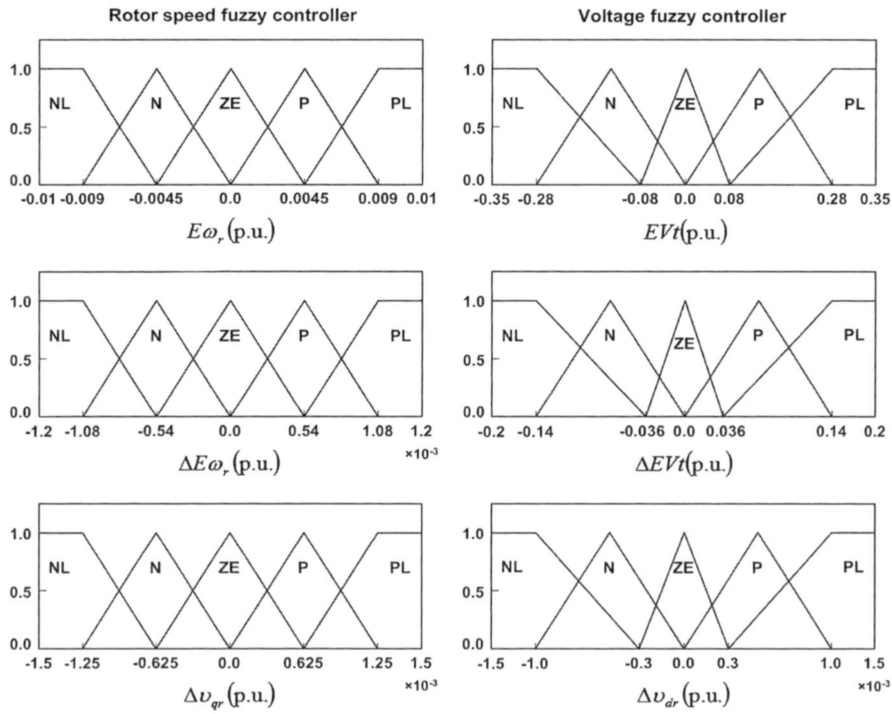

Fig. 9 Triangular membership functions for inputs and outputs fuzzy sets of fuzzy controllers

Table 1 Rule base for rotor speed fuzzy controller and reactive power fuzzy controller

		$\Delta E\omega_r$							ΔEVt				
		NL	N	ZE	P	PL			NL	N	ZE	P	PL
$E\omega_r$	NL	PL	PL	PL	P	ZE	EVt	NL	NL	NL	NL	N	ZE
	N	PL	PL	P	ZE	N		N	NL	NL	N	ZE	P
	ZE	P	P	ZE	N	N		ZE	N	N	ZE	P	P
	P	P	ZE	N	NL	NL		P	N	ZE	P	PL	PL
	PL	ZE	N	NL	NL	NL		PL	ZE	P	PL	PL	PL

As it can be observed from Fig. 9, the fuzzy sets were defined as: *NL*—Negative Large, *N*—Negative, *P*—Positive, *PL*—Positive Large and *ZE*—Zero. The control rules of the fuzzy controllers are commonly represented through a set of heuristically chosen rules. In this case, the designed fuzzy rules for both controllers are presented in Table 1.

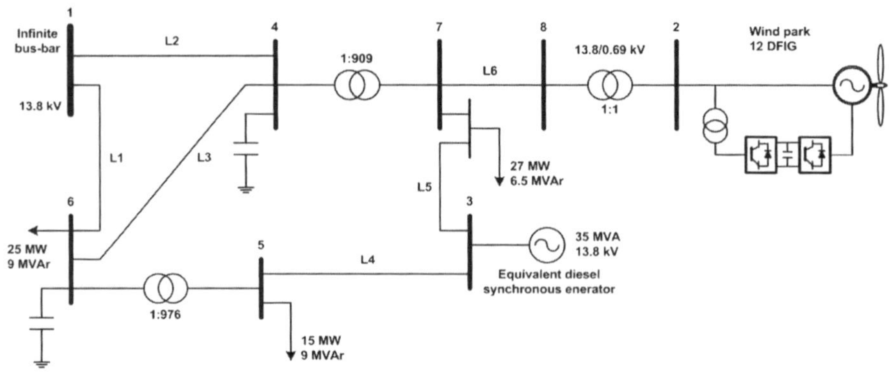

Fig. 10 Test system single line diagram

3.2 The Test System

In order to evaluate the impact of the RSC control strategy based on fuzzy controllers on the power system dynamic behaviour, comparative tests with the PI controllers have been performed through dynamic simulations considering the occurrence of a grid disturbance in the test system depicted in Fig. 10.

For simulation purposes, a dedicated dynamic power system simulation software package was developed under *MATLAB*™ environment. The full power system dynamic model involves the mathematical representation of the utility grid, the wind park equipped with 12 DFIG operating under similar conditions and the equivalent diesel synchronous generator connected to bus 3.

The general multimachine power system model developed in [52] is used to define, in a compact form, the transmission network and the stator equations of the machines. The synchronous generator of the equivalent diesel group is represented through a fourth order transient model [26] together with both the automatic voltage regulator (IEEE type 1) and the speed governor. Additionally, a simple first order model is used to represent the diesel engine dynamics, as reported in [52]. The wind park is represented through a dynamic equivalent model corresponding to the twelve DFIG, whose dynamics are represented through the mathematical model described in Sect. 2.

It must be stressed that RSC control aims to perform rotor speed and terminal voltage reference tracking and thereby it is based on both the speed and terminal voltage control loops presented in Sect. 2.6.1. Then, the PI gains were carefully adjusted using trial and error approaches in order to obtain minimum variations of the design variables, considering the simulation of the above mentioned disturbances taking place on test system depicted in Fig. 10. After the long process of tuning, the derived PI gains are presented in Table 2 together with the other parameters, which were obtained based on the machine characteristics.

Table 2 Parameters of the RSC controller using PI controllers

K_{P1}	K_{I1}	K_{P2}	K_{I2}	K_{P3}	K_{I3}	K_v	K_t	T_v	T_t
20.27	15.76	0.0443	0.0032	15.06	10.50	0.0321	0.0321	0.4434	0.4434

The DFIG operating conditions correspond to a constant wind speed of 15 m/s, such that the rotor speed equals 1.01 p.u. The terminal bus voltage is set to 1.02 p.u. Regarding the GSC, the reactive power set point is set to its maximum value in order to provide additional reactive power injection through the rotor circuit. The results obtained are presented in the following section.

3.3 Simulation Results and Discussion

In order to evaluate the performance of both PI and fuzzy controllers with respect to the rotor current behaviour, which could trigger the crowbar protection, a three-phase short circuit taking place on bus 6 of the test system was simulated at $t = 1$ s with a typical clearing time of 100 ms. It was assumed that the crowbar will be triggered if the rotor current increases above 1.25 kA (0.18 p.u. on the WRIG base) and that the DFIG will be disconnected from the grid when the terminal voltage is outside the range $0.7 < Vt < 1.15$ p.u. The definition of voltage and current limits was based on the rating of both the generator and the frequency converter. The results obtained following the short circuit occurrence are presented in Fig. 11.

As it can be observed from Fig. 11a, the fuzzy controllers would be able to keep the machine connected to the grid while the PI controllers will lead to the voltage protection actuation since the upper voltage limit was exceeded, as it can be observed from Fig. 12c. On the other hand, the rotor current will trigger the crowbar protection and therefore the DFIG operating conditions will be changed. If the protection regulation settings were stricter and the controllers were not able to damp the current oscillations, the crowbar could be triggered several times due to the current overshoot following the RSC re-connection attempt and, therefore, the DFIG performance will become worsened. The DFIG electromagnetic torque can be observed from Fig. 11b. Large oscillations can be experienced by the machine when controlled by PI controllers as a result of the crowbar operation. During the short circuit the fuzzy controllers perform better, in spite of the fact that the membership functions were not carefully adjusted as in the PI controller design.

For large control inputs, leading to large variations with respect to the reference values, fuzzy controllers usually perform better than PI controllers while similar performances have been achieved for small input variations [30]. Therefore, although no optimization methods were used for both kinds of controllers, fuzzy controllers present a better whole performance regarding systems subjected to severe disturbances as in power systems applications.

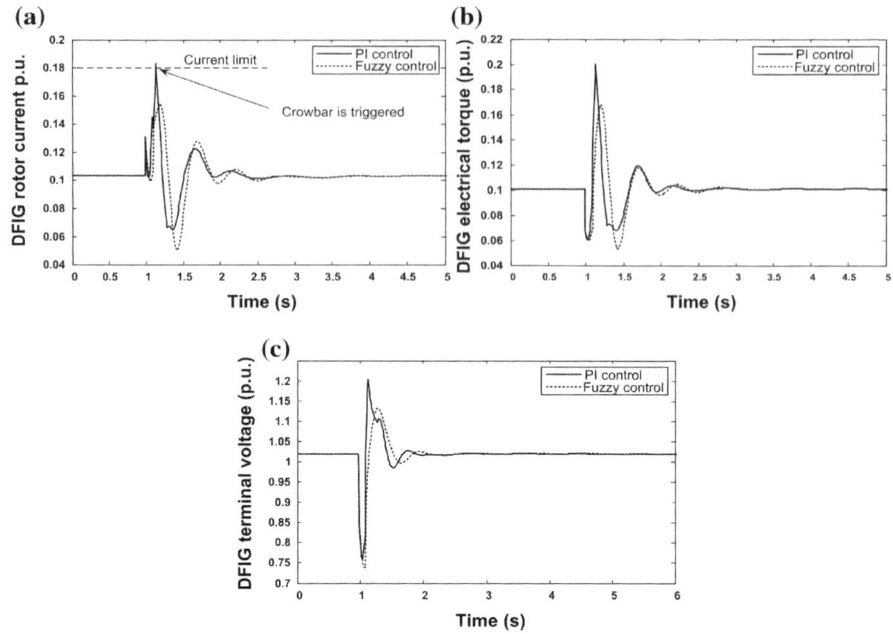

Fig. 11 DFIG behaviour following a three-phase short-circuit: (**a**) rotor current; (**b**) electrical torque; (**c**) terminal bus voltage

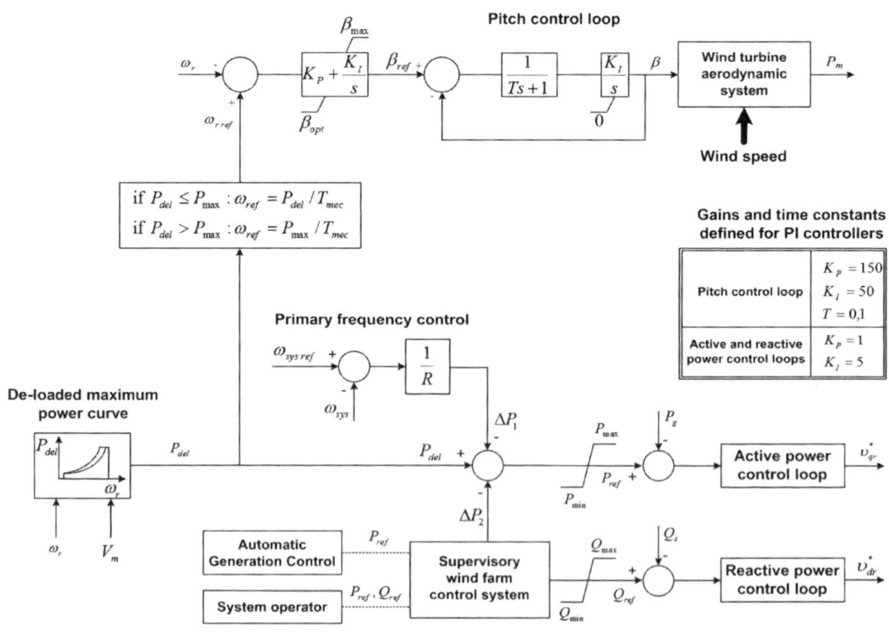

Fig. 12 Full frequency control strategy of the DFIG rotor side converter

Although PI controllers could play an important role regarding the DFIG behaviour following ac grid disturbances, they must be carefully designed. This is a very difficult task due to the presence of nonlinear and coupling terms on the mathematical equations representing the dynamics of DFIG. Also, a large amount of dynamic simulations is required to meet the acceptable range of PI gains. In contrast, fuzzy controllers were adjusted in just a few simulations.

4 Participation of DFIG in System Frequency Regulation

Regarding the potential of wind parks to participate effectively in primary frequency regulation, several research works have been developed aiming DFIG to contribute for inertia control and for primary frequency control [14, 20, 22, 25, 37].

Inertia control can be provided through a supplementary inertial control loop [14] exploiting the kinetic energy stored in the rotating mass of wind turbines, such that the additional amount of power supplied by the DFIG to the grid is proportional to the derivative of the system frequency. Nevertheless, the inertia control is similar to the one commonly used in synchronous generators [9, 25], being the droop loop used to change the active power injected by the DFIG proportionally to the frequency deviation. In [37] both frequency control schemes are exploited, but the droop loop is only activated when the system frequency exceeds certain limits. Moreover, in order to provide inertia control, the DFIG has to be de-loaded, meaning that there is a margin to increase the active power output during large low frequency periods. Thus, in [22], the pitch angle controller was exploited to develop a primary frequency control strategy that allows the DFIG to provide a proportional frequency response. For this purpose the active power injected by DFIG is adjusted through the regulation of the minimum pitch angle according to the frequency deviation.

In this section, a robust primary frequency control approach for DFIG is presented. Following a system frequency deviation an active power injection takes place through the initial leading action of the RSC. The pitch control system is further used in order to adjust the mechanical power accordingly. The amount of the injected active power is defined from the proportional frequency regulation loop together with a power reference adjustment obtained from a de-loaded power curve such that a new equilibrium point of operation can be obtained following frequency changes. The effectiveness of this control approach was tested in a small isolated power system when the wind generation plays an important role.

4.1 The DFIG Frequency Control Approach

In order to allow DFIG participation in frequency regulation the adopted control approach focuses on the RSC control strategy, which is designed to perform

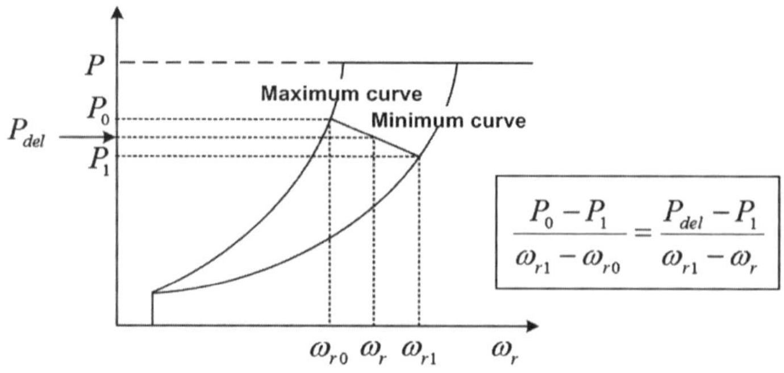

Fig. 13 Schematic diagram of de-loaded optimal active power curve

directly active and reactive power control, as presented in Sect. 2.6.2. The GSC operates at a unitary power factor and controls the dc-link voltage, according to the control scheme presented in Sect. 2.7.

Besides the active and reactive power control loops, the full control approach of the RSC comprises also the primary frequency control loop that defines the ΔP_1 set point, as a result of system frequency changes, a pitch control strategy and a control block associated to an external supervisory wind farm control system, as it can be observed from Fig. 12.

DFIG are required to respond to a request from the system operator, due to an Automatic Generation Control (AGC) demand or because the operational reasons related with the need to control power flows in the network area, through the supervisory wind farm control block. This request leads to an optimized adjustment of the active power output of the individual DFIG inside the wind park [4] defining thus the ΔP_2 set-point.

The active power control loop allows the DFIG tracking the power reference, P_{del}, obtained from the de-loaded maximum power curve [6], defined in terms of the DFIG reserve capacity to be available for inertia control. This reference power is also used to adjust the rotor speed through the pitch control loop, as illustrated in Fig. 13.

The adoption of a power reference derived from the de-loaded power extraction curve allows the increase of the DFIG active power generation when the system frequency decreases as a result of either a sudden load increase or a large generation facility loss. Thus, from Fig. 13, the power reference can be defined as:

$$P_{del} = P_1 + \frac{P_0 - P_1}{\omega_{r1} - \omega_{r0}}(\omega_{r1} - \omega_r), \tag{26}$$

where P_0 and P_1 are the maximum and the de-loaded active powers for a given wind speed, respectively, being ω_{r0} and ω_{r1} the minimum and the maximum rotor speeds referred to the generator side, respectively.

The relationship between P_0 and P_1 is defined as

$$P_1 = k_{del}P_0, \tag{27}$$

where $k_{del} = 1 - \frac{\%\ \text{de-loading}}{100}$, $P_1 = k_{opt1}\omega_{r1}^3$, $P_0 = k_{opt0}\omega_{r0}^3$ with k_{opt1} and k_{opt0} being the optimum constants given by Eq. (14) for both maximum and minimum power curves, respectively.

From the above equations, the reference power can be written as

$$P_{del} = \frac{(\omega_r - \omega_{r0})k_{del} + \left[\left(\frac{k_{del}k_{opt0}}{k_{opt1}}\right)^{1/3}\omega_{r0} - \omega_r\right]}{\left[\left(\frac{k_{del}k_{opt0}}{k_{opt1}}\right)^{1/3} - 1\right]\omega_{r0}} P_0 \tag{28}$$

Assuming that P_0 and ω_0 are previously known, the reference power depends basically of the rotor speed ω_r referred to the generator side according to Eq. (20).

Like in a conventional synchronous generator, the primary frequency control integrated into the rotor side active power control loop comprises a droop loop, characterized by a regulation R expressed in per unit in the system base, which is responsible to change the DFIG injected active power, P_g, following a system frequency deviation [6] as

$$P_g = P_{del} - \frac{1}{R}\left(\omega_{sys} - \omega_{sys\_ref}\right), \tag{29}$$

where ω_{sys} and ω_{sys_ref} are the system frequency and its rated value, respectively.

The injected power increment requires a new operating point defined through a new rotational speed and the corresponding mechanical power according to the turbine power curve for the incoming wind speed. So this control strategy is complemented with the pitch control system (used together with the frequency converter control) by adjusting the rotor speed referred to the generator side, ω_{r_ref}, according to the de-loaded maximum power curve taking into account the operational conditions, as presented in Fig. 12.

4.2 The Test System

The effectiveness of the proposed primary frequency approach was evaluated using a test system comprising 5 DFIG (660 kW each) installed in a wind park connected to an isolated grid represented by a single bus with a synchronous thermal unit rated at 10 MW, as it can be observed from Fig. 14.

The WRIG mathematical model described in Sect. 2.3 together with both the RSC full frequency control strategy depicted in Fig. 12 and the GSC control presented is Sect. 2.7, was embedded in the dedicated dynamic simulation package already used in Sect. 3.2. The thermal unit was also represented by the dynamic model adopted in Sect. 3.2, considering that the frequency droop of the generator is 4 % in the machine base and that the integral control loop allows the system to

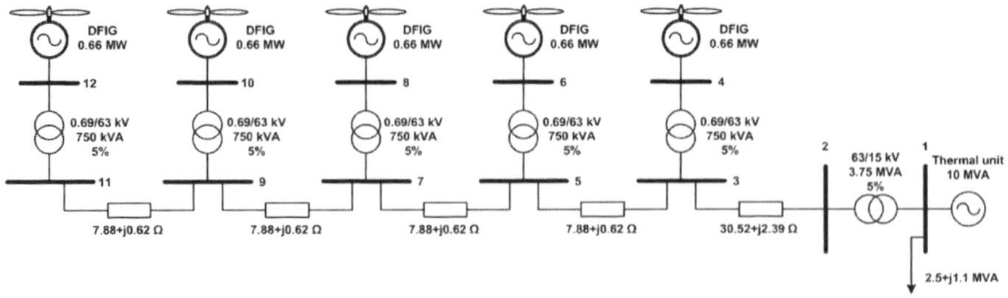

Fig. 14 Single line diagram of the test system

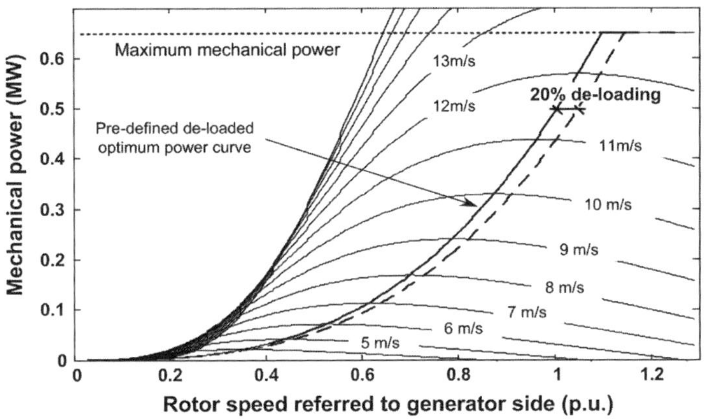

Fig. 15 De-loaded optimal power curve adopted for each DFIG

recover frequency to its nominal value following unbalances between load and generation, being the integral control gain of 4.5 p.u. in the system base power.

4.3 Simulation Results and Discussion

A de-loaded margin of 20 % was adopted for each DFIG, providing in this way a 20 % generation reserve available for the entire wind speed range, as it can be observed from Fig. 15.

For simulation purposes, it was assumed that the wind speed was 12 m/s in the entire wind park. Because of an operator request, all wind generators are operating in the minimum optimum curve (dotted line in Fig. 15), corresponding to 458 kW. In addition, the droop, *R*, of the DFIG primary frequency control loop was considered as 5 % with respect of the machine base. It was also assumed that the GSC reactive power reference is set to zero in order to assure the GSC operation with a unitary power factor.

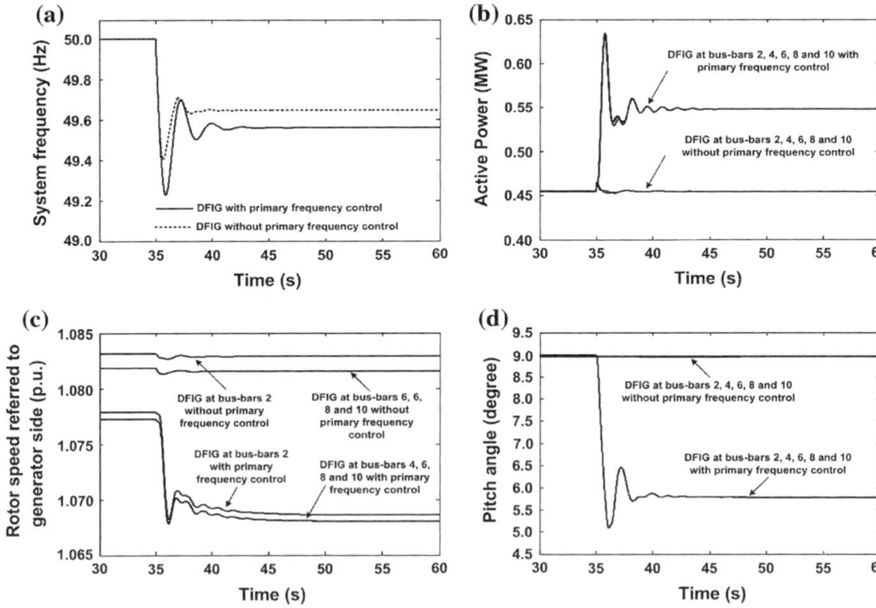

Fig. 16 DFIG dynamic behaviour considering the thermal unit without frequency control: (**a**) frequency behaviour; (**b**) active power; (**c**) rotor speed; (**d**) pitch angle

A change in load value was simulated at $t = 35$ s by increasing the load connected at bus 1 in 80 % leading with a system frequency deviation. In order to show the DFIG capability of keeping up part of the load increase it was assumed that the integral frequency control loop of the thermal unit was not in operation, being this situation similar to the initial periods following a system disturbance when only primary frequency control operates while expecting for AGC to correct the frequency drift.

The system frequency dynamic behaviour is presented in Fig. 16a, considering that the DFIG provides frequency control (dotted line) and considering that it is not participating in this control (full line). It can be clearly observed that the frequency excursion is smaller when the wind generators are participating in grid frequency control. The DFIG dynamic behaviour regarding active powers, rotor speeds and pitch angles with and without primary frequency control is presented in Fig. 16a, b and c, respectively.

When DFIG do not participate in frequency control, only minor changes in active power outputs and rotor speed behaviours can be observed, as a result of induced changes in the electrical torques of these units resulting from changes in network voltages following load changes. When frequency control is activated, the DFIG increases fast the active power output in order to compensate the system load increase. Therefore, the DFIG rotor speed decreases following the de-loaded extraction power curve. As a result, the pitch angle is decreased, allowing the

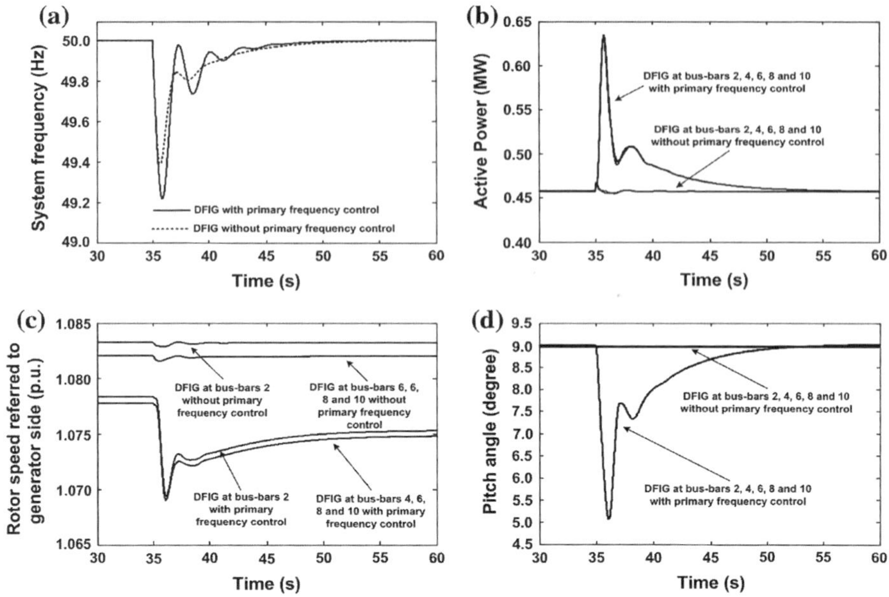

Fig. 17 System frequency behaviour considering the thermal unit with frequency control

increase of the mechanical power extracted from the incoming wind speed. Therefore, after the initial transients, these variables stabilize in a different value than the initial one. The results allow concluding that the system behaviour is considerably improved in terms of the frequency deviations when DFIG participates in frequency control.

In order to highlight the DFIG dynamic behaviour, similar results considering the presence of the integral control loop in the speed governor of the thermal unit are presented in Fig. 17. Grid frequency behaviour is presented in Fig. 17a, showing that the integral control action of the thermal unit governor brings the system frequency to its nominal value. Figure 17b, c and d show the behaviour of active powers, rotor speeds and pitch angles of the DFIG.

From these results one can conclude that when DFIG participates in primary frequency control the frequency deviations are smaller and better damped due to the fast operation of the RSC. Thus, the presence of DFIG operating with primary frequency control can contribute to increase the system robustness with particular benefits for isolated power systems. On the other hand, DFIG can provide frequency control as an ancillary service to help supporting frequency when large disturbances occur. Such a service can be required during operation scenarios where conventional power plants have a smaller presence, like during valley hours with a large amount of wind generation.

5 Robust Tuning of PSS to Installed in DFIG

Power system stabilisers installed in DFIG can be used to provide additional damping to the electromechanical modes of oscillation, which are related to low frequency (0.1–2 Hz) power oscillations occurring among the rotors of the synchronous machines as a result of either some controller settings or large power flows through weak transmission lines. However, DFIG will contribute effectively for this purpose only if the oscillation mode is controllable through power injection on the DFIG terminals and observable through the PSS input signals. Moreover, the performance of PSS installed in DFIG is significantly affected by the system operating conditions resulting from different wind speeds [33]. Therefore, it is desirable to find a trade-off solution that will provide a set of parameters for each PSS able to assure an acceptable damping level for all operating scenarios. This solution is commonly known as a robust solution.

Due to the large number of operating scenarios and taking into account that PSS installed in DFIG add damping indirectly, a robust solution may be quite difficult to achieve. Additionally, some care should be taken in order to assure that the PSS does not change the modes of oscillation of the DFIG affecting its performance. Thus, in this section, a heuristic approach is presented in order to perform a simultaneous robust tuning of the several PSS installed in DFIG involving a very large set of scenarios that represent different operating conditions. For this purpose an optimization problem will be solved pursuing the minimization of PSS action so that the minimum acceptable damping is achieved for a representative set of scenarios. The mathematical model of DFIG equipped with PSS as well as the algorithm to perform PSS robust tuning is presented in this section. The effectiveness of the robust solution is evaluated using a suitable test system. The results obtained are also presented and discussed.

5.1 Model of DFIG with PSS

The mathematical model of DFIG with PSS is based on the DFIG model presented in Sect. 2, representing the fundamental dynamics of a WRIG, a wind turbine with pitch angle control and a RSC control system based on both active and reactive power control loops, as depicted in Fig. 18. The dynamics of the GSC were not considered.

As it can be observed from Fig. 18, the PSS output is a stabilizing signal to be summed to the quadrature component of the rotor voltage in the active power control loop. This signal is responsible to create a generator damping torque in phase with the rotor speed through the active power modulation [3], where the DFIG acts as an amplifier of the controller similarly to synchronous machines when used for the same purpose [48].

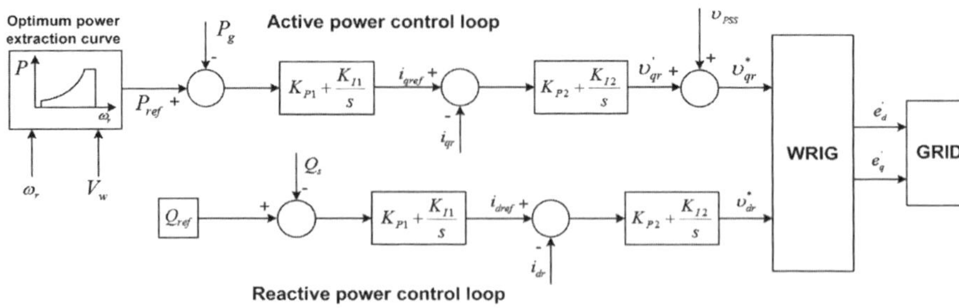

Fig. 18 Control scheme of DFIG with PSS

Assuming that the pitch angle is constant during the studied period, the DFIG state space model suitable for modal analysis is derived from its detailed model through the linearization of the differential equations describing both the WRIG fundamental dynamics and the ones related with the RSC based PI controllers, being given by [33]:

$$\Delta \dot{x} = A_1 \Delta x + B_1 \Delta \upsilon_s + C_1 \Delta P + D_1 \Delta \upsilon_r, \tag{30}$$

This equation is then defined as a function of state variables x, terminal voltage υ_s, active power P and rotor voltage υ_r. The terms related with active power and rotor voltage must be eliminated after deriving the following algebraic equations:

$$\begin{cases} \Delta P = A_2 \Delta x + B_2 \Delta \upsilon_s + D_2 \Delta \upsilon_r \\ \Delta \upsilon_r = A_3 \Delta x + B_3 \Delta \upsilon_s + C_3 \Delta P \end{cases} \tag{31}$$

After some mathematical transformations the DFIG state space model can be written in the canonical form as:

$$\begin{aligned} \Delta \dot{x} &= A \Delta x + B \Delta \upsilon_s \\ \Delta i &= C \Delta x + D \Delta \upsilon_s \end{aligned} \tag{32}$$

This kind of models regarding synchronous machines and DFIG are combined with the network equations in a multi-machine system model and the state matrix is then used to perform standard modal analysis [33]. The damping conditions associated to each mode of oscillation are determined from the state matrix eigenvalues exploiting QR method routines from MATLAB™.

The configuration of PSS to install in DFIG is presented in Fig. 19.Here K is the PSS gain, T_w is the time constant of the washout filter used to prevent the PSS from responding to steady-state changes of the input signal and the time constants T_1–T_4 are used in the lead-lag blocks in order to provide the necessary phase compensation. Since the PSS output is added to the quadrature component of the rotor voltage, a limitation block is also included in the PSS scheme, being the maximum and minimum limits established according to the machine operation.

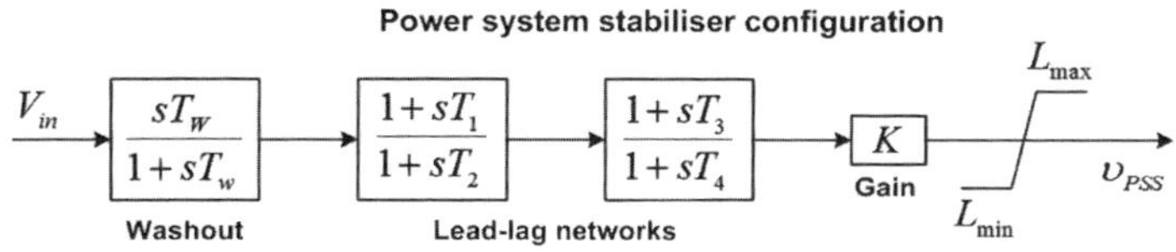

Fig. 19 Power system stabiliser configuration to be installed in DFIG

Regarding the PSS inputs, residue analysis should be used in order to select the most adequate signals among rotor speed, voltage, electrical power and electrical frequency. However, from the physical insights acquired when dealing with this problem, the input signals that perform best are voltage and frequency. In fact, due to the decoupling between grid and DFIG provided by the RSC controller, rotational speed and electrical power are less sensitive to oscillations that occur in the grid and usually do not contain sufficient information about the mode of oscillation to damp by the PSS.

5.2 Methodology for PSS Robust Tuning

The importance of PSS tuning and how this task may be affected by the increase of wind power integration has been stressed and several methods have been used for this purpose. However, they are not suitable to find a robust solution for a very large number of scenarios. Meta-heuristics based optimization algorithms can find robust solutions but require too much computational effort [32]. In contrast, sensitivities based methods require less computational effort but do not assure that a robust solution will be found [42]. Therefore, the methodology adopted for PSS robust tuning is based on a meta-heuristic approach.

The PSS parameters are determined by solving an optimization problem with capability to deal with several operational requirements which are modelled in the objective function or considered as constrains. The mathematical formulation is derived from the method presented in [32] that pursues the minimization of the PSS gain in order to provide the minimum acceptable damping. As a result a low stabilizer gain is obtained reducing thus the effect of the limitation block in the stabilizer output [42] in contrast with the alternative formulations, which aim the damping maximization [46]. For a given scenario k, the problem can then be formulated through the following objective function aiming the minimization of a quality index, Q_k as [33]:

$$\min Q_k(X) = \sum_{i=1}^{nm} \sum_{j=1}^{ng} \left(w_{ji} K_j + WN_j \alpha_j \right), \tag{33}$$

subjected to

$$\varsigma_i \geq \varsigma_{i\,\min} \tag{34}$$

$$|\Delta f_i| \leq \Delta f_{i\min} \tag{35}$$

$$X_{j\max} \geq X_j \geq X_{j\min}, \tag{36}$$

where X represents the solution of the problem corresponding to the values of the parameters for the several PSS to be installed, nm is the number of modes, ng is the total number of generators, w is a weighting factor, K is the PSS gain, W is a penalty factor to weight phase compensation, N is the number of lead-lag blocks, α is the filtering ratio, ς is the damping and Δf is the eigenvalue frequency change. The subscript i is the order of the mode of oscillation while the subscript j represents the order of the DFIG.

Thus, the objective function given by (33) comprises the sum of the PSS gains with a penalty because of phase compensation, subjected to the set of constrains given by Eqs. (34)–(36), which require, respectively, that damping should be higher than a given threshold, frequency should be within a given range and control variables should be within allowable physical limits.

The transfer function of the PSS expressed in terms of the control variables is given by [33]:

$$V_{PSS}(s) = K\frac{sT_W}{1+sT_W}\left(\frac{1+sT_n}{1+sT_n/\alpha}\right)^N V_{in}(s), \tag{37}$$

where K is the PSS gain and T_W is the washout time constant.

The control variables that constitute the solution of the problem are, for each PSS, the gain, the number of lead-lag blocks N and the time constant values, $T_n = T_1$, and $\alpha = T_1/T_2$, of each of these blocks. Limits are also imposed to the gains of the stabilisers and to the time constants of the lead-lag blocks. The washout time constant is specified previously to the PSS design and it is not adjusted. The weighting factors w_{ji} are used to provide an indication of the best locations to install PSS. As the location is not a concern it was assumed that $w_{ji} = 1$.

Considering a number of operating scenarios nc, the PSS tuning problem is envisaged as multi-criteria minimization problem having nc attributes corresponding to the quality of the PSS performance in each scenario. A trade-off solution will be found in the non-dominated or Pareto-optimal border of the set of solutions in the attribute space. However, the recognition of this border is a difficult problem and several strategies may be adopted for this purpose. A simple one is the minimization of the weighted sum of the attributes, as

$$\min Q = \sum_{k=1}^{nc} z_k Q_k, \tag{38}$$

where z_k denotes the relative importance of scenario k.

This is equivalent of finding, over the Pareto border, the point closest to the ideal solution using the space metric L_1, although other metrics could be used.

Considering that all the scenarios have equal importance, all $z_k = 1$, the objective function given by Eq. (33) can be re-written as

$$\min Q = \min\{\max[Q_k]\} \tag{39}$$

The PSS control parameters derived from Eq. (39) corresponds to project the PSS with its worst behaviour as best as possible.

In order to obtain the PSS robust solution, an Evolutionary Particle Swarm Optimization (EPSO) tool was exploited. EPSO was built over the concepts of both evolutionary strategies and particle swarm optimization in an attempt to combine the best of both techniques [35] to handle a set of particles in order to find a suitable solution for PSS to install in DFIG. Each particle involves a set of parameters comprising both the strategic and the object parameters. The set of object parameters represents a potential robust solution for PSS. Thus, given a set of particles, for a given iteration, the general algorithm of EPSO can be described as follows [33]:

1. Replication of each particle r times;
2. Mutation of the strategic parameters of each particle;
3. Reproduction of the mutated particles by generating an offspring according to the particle movement rule;
4. Evaluation of the fitness of each particle by calculating the corresponding Q value;
5. Selection of the best particles, according to the criterion given by the objective function, to form the new generation.

It should be noted that each particle evaluation requires the calculation of the state matrix eigen-values and the further assessment of the modes of oscillation. On the other hand, the number of performed evaluations, N_1, grows linearly with the number of scenarios, as

$$N_1(nc) = r \times it \times p \times nc, \tag{40}$$

where r is the number of replications, it is the number of iterations, p is the number of particles and nc is the number of scenarios.

Since the number of operating scenarios is very high, a set of representative scenarios has to be identified before running this proposed PSS tuning procedure. Thus, the following section proposes a suitable procedure for identification of representative operating scenarios.

5.3 Identification of Representative Operating Scenarios

The integration of wind power increases largely the diversity of system operating conditions and, thereby, the resulting number of operating scenarios. Moreover, the effectiveness of the PSS to install in the DFIG depends on the amount of wind generation as a result of variations of the number of wind turbines in operation and of the wind speed variability. These factors make the identification of a reduced set of relevant scenarios a very hard task.

The operating scenarios can be generated automatically using a structured Monte Carlo procedure. However, the large number of operating scenarios to be generated will lead to a computational burden that may render unfeasible the proposed PSS robust tuning procedure. Therefore, in a second phase, the set of all possible operating scenarios has to be reduced to a much smaller number of scenarios representing the most unfavourable operating conditions, the so-called set of worst case scenarios. Thus, a suitable solution derived from this set of scenarios will be acceptable for all the operating scenarios.

In order to select the set of worst case scenarios, some engineering judgment is required taking advantage of the well known fact that transmission of bulk power through weak transmission lines tends to originate low damped modes of oscillation. This is, however, a trial and error approach that can be tedious when the number of scenarios is very high.

To speed up the identification of the worst case scenarios without ignoring operating conditions that will be less obvious but relevant for PSS tuning, a heuristic approach should be followed, as proposed in [33]. Two different sets are created. The set A will include all the scenarios while the set B will be empty and eventually will contain some of the worst scenarios. Then, the next steps are followed:

1. Calculate the fitness for each scenario according with the Eq. (33);
2. Select the scenario with worst performance corresponding to the highest values of the fitness;
3. Include the selected scenario in set B if it is not a member of this set. Otherwise the solution was found and the algorithm stops;
4. Identify the PSS robust solution by tuning simultaneously all PSS using the operating scenarios grouped in set B;
5. Evaluate the robust solution considering all the operating scenarios;
6. If the worst fitness is better than a given limit, previously specified, the solution was found and the algorithm stops. Otherwise check the number of iterations;
7. Stop the algorithm if the number of iterations exceeds its maximum value. Otherwise return to step 1.

The result of this algorithm will be a robust solution for the PSS tuning problem, that is, a suitable solution for all operating scenarios grouped in set B. However, if the algorithm stops in step 6, because the maximum number of iterations was achieved, it is not assured that a robust solution was found. Nevertheless, that solution should lead to a reasonable damping improvement. In both cases the number of performed evaluations by the proposed heuristic is given by:

$$N_2(a) = \sum_{i=1}^{a} N_1(i) + nc(a+1), \tag{41}$$

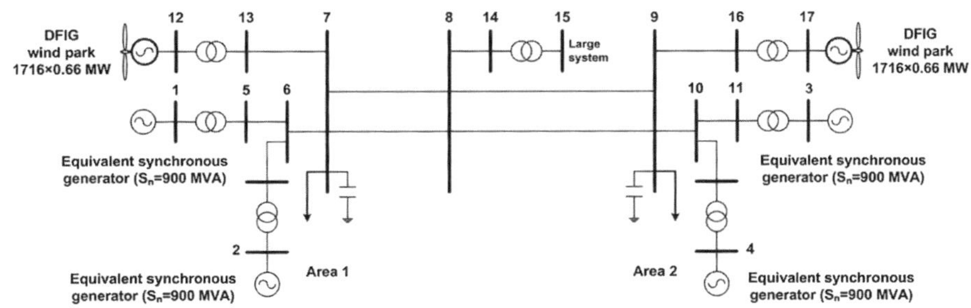

Fig. 20 Single line diagram of the test system

where N_2 is the total number of evaluations and a is the number of iterations performed by the heuristic and corresponds to the number of scenarios in set B.

Equation (41) comprises two components: the number of evaluations required to tune the PSS parameters using the operating scenarios in the set B and the number of evaluations performed to assess the adequacy of the robust solution regarding the scenarios of the set A. It should be noted that the first component has a much larger weight in the total number of the performed evaluations.

5.4 The Test System

The performance of the algorithm presented in Sect. 5.2 was evaluated using the test system depicted in Fig. 20, comprising two areas connected to a large system. Each area contains two synchronous equivalent generators, one wind park equipped with DFIG and represented through an equivalent machine, an equivalent load and a capacitor bank. The generator connected to bus 15 corresponds to the large system represented as an infinite bus.

The synchronous generators represent a group of strongly coupled generators. The equivalent model consists of a six-order model, neglecting the magnetic saturation, together with voltage regulators represented by the IEEE type 1 model [26]. The DFIG with PSS are represented by the model presented in Sect. 5.1. Loads and capacitor batteries are modelled as constant impedances.

5.5 Generation of the Operating Scenarios

The scenarios to be considered are generated by the automatic procedure presented in Sect. 5.2, taking into account several issues with a significant influence on the damping levels. Then, variation levels with respect to the reference values defined for the base case [32] were considered as follows:

1. Unit commitment of synchronous generators and wind farms. Generators connected to buses 1 and 2 are out of service to accommodate wind generation while generators connected to buses 3 and 4 are kept in operation. Regarding wind generation, it was also considered that both wind farms can be out of operation, one half of the DFIG are in operation and all the DFIG are in operation.
2. Load level. A variation of 25 % around the reference value was considered.
3. Generation of synchronous machines. A variation of 25 % around the reference value was also considered.
4. Wind speed corresponding to different values of active power. Variations in the range between 7 and 10 m/s for the wind park connected to bus 12 and variations in the range between 8 and 14 m/s for the wind farm connected to bus 17 were considered.
5. Grid configuration. It was considered that one of the lines between buses 7 and 8 is switched off and one of the lines between buses 8 and 9 is switched off.

In order to generate the operating scenarios the following procedure was then adopted:

1. The operating conditions of the base scenario were changed according to the defined ranges forming a combined set of operating scenarios. Thus, starting from the base scenario, the unit commitment was changed and the resulting scenarios were stored. Then, the loads of all the stored scenarios were changed generating, thus, new scenarios that were also stored. In the same way, the active power of synchronous generators as well as the wind speed and the grid configuration were modified.
2. A power flow was performed for all the operating scenarios in order to eliminate unfeasible scenarios.
3. The modes of oscillation for all the feasible scenarios were calculated and analysed.

This procedure generated a set of 1,300 operating scenarios requiring, thus, further selection. The heuristic approach presented before was used for selecting the worst case scenarios. In these scenarios the global level of wind generation reaches 50 % in some cases, being the wind power generation in area 1 around 70 %, whereas in area 2 it is approximately 45 %.

5.6 Results and Discussion

The electromechanical modes of oscillation for these scenarios are presented in Fig. 21a, showing that the minimum damping of each scenario varies between 5 and 18 %. The low damping is mainly due to a mode associated with the synchronous generator connected to bus 2.

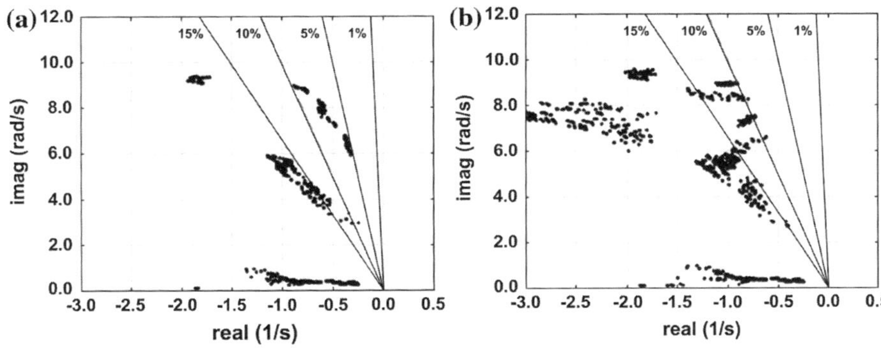

Fig. 21 Pole map: (**a**) without PSS installed in DFIG; (**b**) with PSS installed in DFIG

Table 3 Characterisation of the operating conditions for the worst case scenarios

nc	P_L (MW)	P_{G12} (MW)	P_{G2} (MW)	P_{G3} (MW)	$F_{7–8}$ (MW)	$F_{8–9}$ (MW)
1	2334.0	358.6	525.0	593.3	113.7	−594.5
2	2917.5	240.2	700.0	593.3	−23.4	395.1
3	2334.0	240.2	350.0	593.3	−178.1	−594.5
4	2917.5	240.2	525.0	719.1	−196.4	−388.3
5	2917.5	700.3	700.0	719.1	1109.7	581.1

PSS installed in synchronous generators connected to buses 1 and 3 are considered to be in operation. However, as these PSS were tuned for scenarios without wind generation, the damping level is not adequate since many modes of oscillation exhibit damping levels lower than 10 and 15 %. Another reason for the low damping levels is related with the fact that one synchronous generator equipped with PSS was out of service in order to accommodate the surplus of wind generation in the system.

PSS to install in DFIG connected to buses 12 and 17 were tuned using the algorithm presented in Sect. 5.2 in order to find a robust solution able to increase damping in all modes of oscillation to levels above 15 % with frequency variations less than 5 % for all the operating scenarios. The selected worst case scenarios are characterised in Table 3, through the system total load, production of generators connected to buses 2 and 12 and the active power flow through the interconnection lines. The damping and frequency of the less damped modes of oscillation with and without PSS installed in DFIG are also presented in Table 4.

With the PSS tuning algorithm, a robust solution was found with less than 160,000 evaluations considering 2 replications of each particle ($r = 2$), a constant number of particles ($p = 20$) and the number of iterations was defined to be $it = 250$. The obtained PSS parameters are presented in Table 5, where de generator number corresponds to the number of the connecting bus.

The modes of oscillation for all the scenarios can be observed from Fig. 21b.

Table 4 Characterisation of the modes of oscillation for the worst case scenarios

Without PSS in DFIG				With PSS in DFIG		
c	ς (%)	f(Hz)	Gen.	ς (%)	f(Hz)	Gen.
1	7.23	1.19	2	10.20	1.19	2
2	5.59	1.03	2	13.24	0.99	2
3	8.39	0.47	3	10.24	1.43	2
4	5.59	1.08	2	9.31	1.05	2
5	10.10	0.54	1	14.53	0.46	1

Table 5 Value of the PSS parameters

Gen.	K	N	T	α
1	6.06	2	0.5155	27.28
3	7.72	2	0.4716	21.83
12	0.12	0	–	–
17	0.95	2	0.9968	0.27

The obtained solution is able to increase the damping levels for more adequate values. However, the target value of 15 % was not achieved for all the modes of oscillation, since it is not possible to obtain a better solution for scenario 4 without reducing the damping levels of the other scenarios as a result of the fact that DFIG do not participate in electromechanical oscillations thereby contributing only indirectly to increase the damping levels. On the other hand, the selected operating scenarios represent extremely unfavourable operating conditions for PSS.

Finally, the PSS performance was also assessed through non-linear time domain simulations using the PSS/E dynamic simulation tool, considering a step change with amplitude of 0.01 p.u. applied to the reference voltage of the AVR associated to generator 2 (Fig. 22) and, in a second simulation example (Fig. 23), the occurrence of a three-phase short circuit, taking place at bus 8, with duration of 150 ms. Because the number of scenarios is high, results for one case are presented next in order to illustrate the system response. It should be noted that the generator number corresponds to the bus number. Also, the active power output is referred to the system base, $S_b = 100$ MVA.

As it can be observed from Figs. 22 and 23, the installation of PSS in DFIG tuned with the proposed approach lead to quite adequate results. The damping improvement is clearly observed in synchronous generator connected to bus 2 (Generator 2) and also in the power flow through the line between buses 7 and 8.

The results obtained allow concluding that although not participating in the electromechanical oscillations, DFIG can contribute effectively to provide additional damping. However, the integration of wind generation increases the system operation complexity, thereby increasing the difficulty to perform PSS tuning. Under these circumstances, the heuristic based approach used to select the representative set of operating scenarios plays a key role regarding the reduction of the computational burden of PSS robust tuning. Using the representative set of

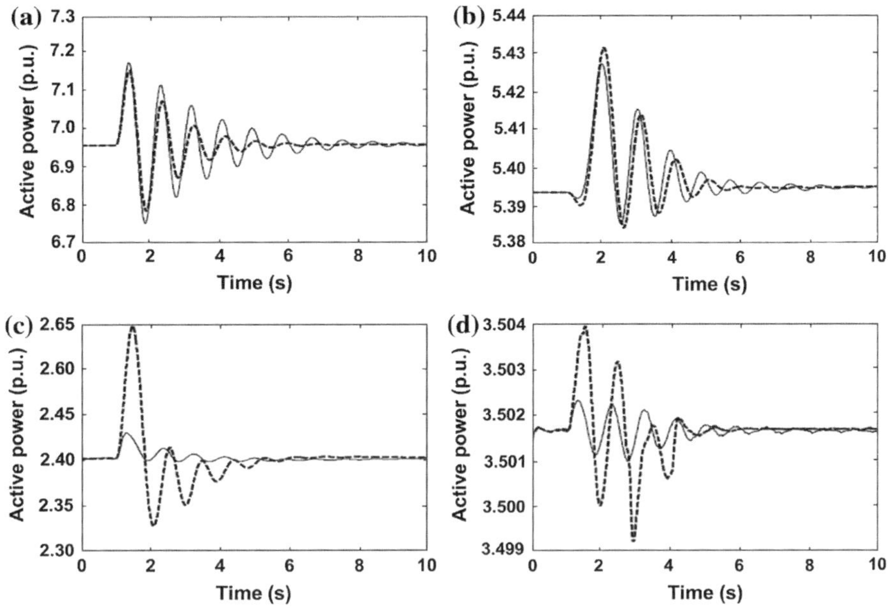

Fig. 22 Response to a step change in reference voltage of generator 2 without PSS installed in DFIG (*full line*) and with PSS installed in DFIG (*dashed line*): (**a**) active power output of generator 2; (**b**) active power output of generator 3; (**c**) active power output of generator 12; (**d**) active power output of generator 17

scenarios corresponding to the worst operating conditions, the best solution found by the EPSO based approach correspond to the robust solution for PSS.

6 External FRT Solutions Used in Wind Parks Equipped with FSIG

The STATCOM based solution connected to wind farm terminals is adopted to provide FRT capability of wind farms equipped with FSIG. For this purpose this static compensator device is controlled to regulate the wind farm terminal bus voltage through the reactive power injection into the grid. Thus, when system voltage drops following a grid disturbance, the STATCOM injects immediately reactive power in an attempt to limit the voltage dip. In this section, the commonly used control functions are presented and the STATCOM performance is evaluated through numerical simulations considering the occurrence of both balanced and unbalanced faults taking place in the utility ac grid.

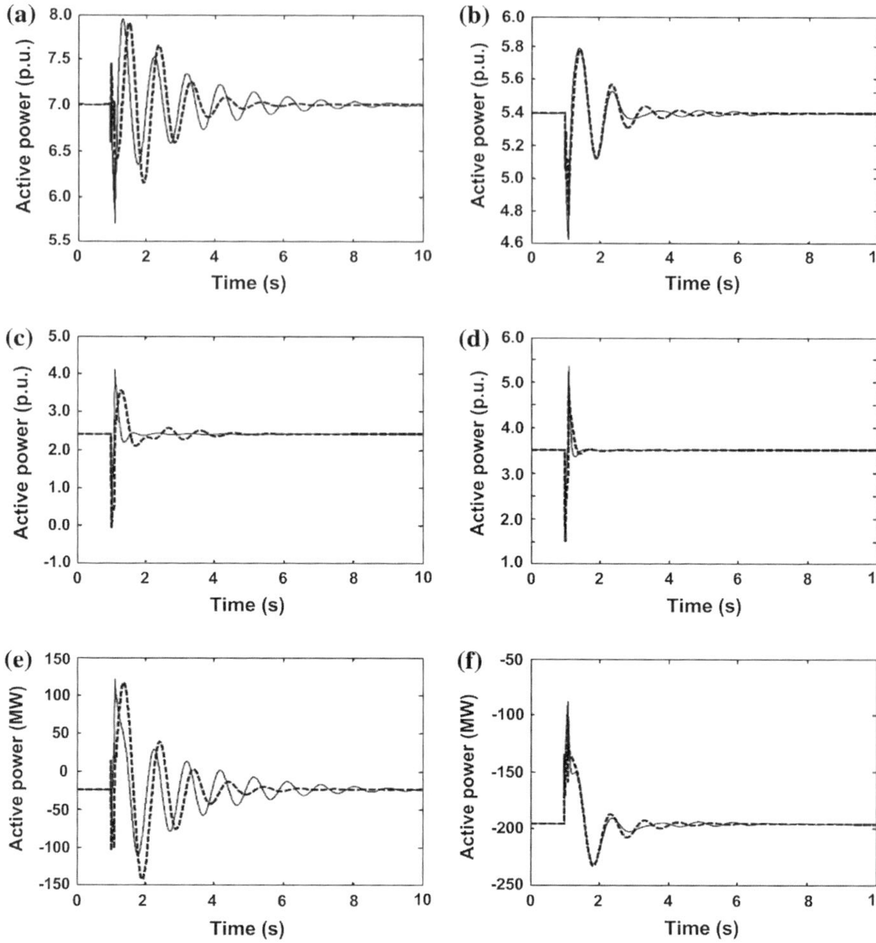

Fig. 23 Response to a 150 ms short circuit in bus 8, without PSS installed in DFIG (*full line*) and with PSS installed in DFIG (*dashed line*): (**a**) active power output of generator 2; (**b**) active power output of generator 3; (**c**) active power output of generator 12; (**d**) active power output of generator 17; (**e**) active power flow in line 7–6; (**f**) active power flow in line 9–8

6.1 The STATCOM Control System

The STATCOM consists of a three-phase Voltage Source Converter (VSC) shunt connected to the wind farm terminal bus through a coupling transformer [21], as it can be observed from Fig. 24.

The VSC using IGBT based PWM inverter uses a PWM technique to synthesize a sinusoidal waveform from the dc voltage source corresponding to voltage at the dc link. Power electronic components are switched on and off at high frequency (several kHz) and, therefore, high frequency harmonics and ripples will be limited

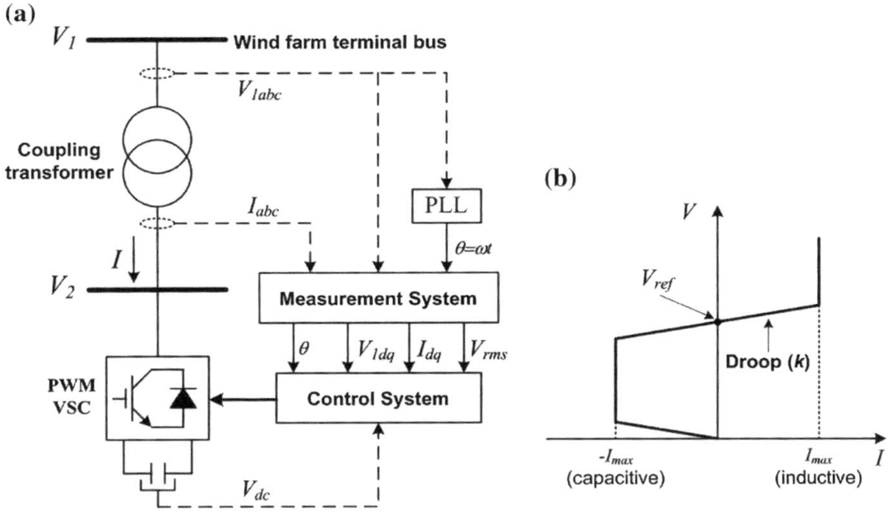

Fig. 24 The STATCOM: (**a**) structure; (**b**) *V–I* characteristic

by connecting filters at the ac side of the VSC [10]. As a consequence of the high switching frequency, a small time step would be required for time domain simulations. Since the VSC behaviour is not a matter of concern in this research, a simplified modelling approach was adopted, such that the VSC reproduces an ideal the reference voltage provided from the control system.

According to Fig. 24a, the active and reactive power transfer, P and Q, respectively, between the VSC and the grid can be expressed as follows:

$$\begin{cases} P = \frac{V_1 V_2 \sin \delta}{X} \\ Q = \frac{V_1 (V_1 - V_2 \cos \delta)}{X} \end{cases}, \tag{42}$$

where X is the reactance of both the interconnection transformer and the VSC filter and δ is the angle of V_1 with respect to V_2.

In steady state operation $\delta = 0$ and only reactive power is flowing. The VSC absorbs reactive power if V_2 is lower than V_1. Otherwise it generates reactive power.

As the VSC is controlled to regulate V_1, it implements the *V–I* characteristic presented on Fig. 24b. Thus, the STATCOM can operate with its rated current even at reduced voltages and, in addition, as long as the reactive current stays within its minimum and maximum values imposed by the converter rating, $-I_{\max}$ and $I_{\max}$, respectively, the voltage is regulated according to its reference value, V_{ref}. However, a voltage droop (usually between 1 and 5 % at maximum reactive power output) is normally used, being the *V–I* characteristic described by

$$V = V_{ref} + k \times I, \tag{43}$$

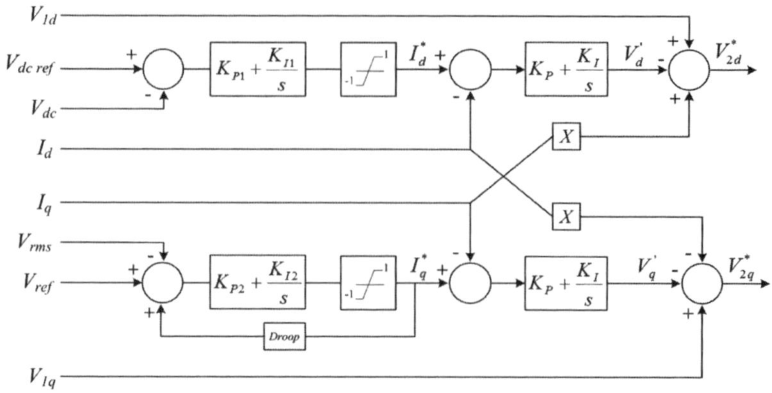

Fig. 25 The VSC control scheme

where V is the positive sequence voltage (p.u.), I is the reactive current (p.u./P_{nom}) and P_{nom} is the three-phase nominal power of the STATCOM.

The VSC control system is designed based on the vector control technique, which has been commonly used to deal with fast dynamic requirements and to provide decoupling control ability [10, 16, 19, 36, 43, 45, 49]. Thus, using the d–q reference frame rotating at the grid frequency, ω, with the q-axis 90° ahead the d-axis, being the d-axis aligned with the position of the wind farm terminal bus voltage, the active and reactive power can be controlled independently using I_d and I_q, respectively, so that the reactive power flow can be controlled via I_q whereas the dc-link voltage can be controlled via I_d [10]. In order to synchronize on the positive sequence component of the three-phase terminal voltage V_1, and to provide the angle $\theta = \omega t$ to the Park transformation [26] as well as to its inverse, a Phase Looked Loop (PLL) is employed, as it can be observed from Fig. 24.

The VSC control scheme comprises two control loops based on PI controllers, as illustrated in Fig. 25. The first control loop is responsible for keeping constant the dc-link voltage through a small active power exchange with the ac network in order to compensate the active power losses regarding both the transformer and the inverter. The second control loop controls the terminal voltage through the reactive power exchange with the ac network.

For these purposes, the PI controllers involve a cascade structure, as it can be observed from Fig. 25. The outer side PI controllers are used to regulate the dc-link voltage V_{dc}, and the magnitude of the direct component of the terminal voltage V_{rms}, providing thus active and reactive current references, respectively, I_d^* and I_q^*. These current references are limited between 1 p.u. capacitive and -1 p.u. inductive. The PI2 controller has a droop characteristic to allow some variations around the terminal voltage, according to Eq. (43). The inner side PI controllers are used to regulate the active and reactive currents (I_d and I_q) by providing

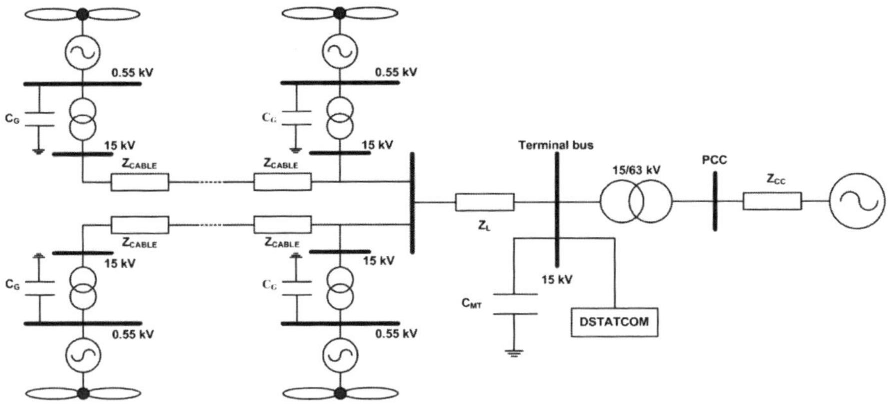

Fig. 26 The test system single line diagram

voltage references V_{2d}^* and V_{2q}^*, respectively, which, after a *dq*-to-*abc* transformation, are sent to the PWM signal generator of the VSC in order to control the magnitude and the phase of the terminal voltage V_2.

6.2 The Test System

The test system presented in Fig. 26 was used to evaluate the STATCOM performance regarding FRT capability wind farms equipped with FSIG when facing external faults using an *EMTP–RV*® dynamic simulation tool.

This test system is a 10 MW wind farm connected to the utility system. The wind farm comprises 20 SCIG with a rating of 500 kW. Each wind turbine is connected to the wind farm internal network of 15 kV through a 630 kW, 0.55/15 kV transformer. Low Voltage (LV) capacitor batteries for individual SCIG reactive power compensation are connected to its terminal bus in order to assure the reactive power requirements under unload conditions. Another capacitor bank connected on the Medium Voltage (MV) side of the wind farm substation is used to provide the additional reactive power required by the SCIG when operated under load conditions and reactive power must be injected into the ac network under normal operating conditions in order to comply with a given power factor requested by system operators. In order to provide FRT capability, this wind farm is also equipped with a STATCOM with 10 MVA of rating power. The wind farm is connected to the High Voltage (HV) network by means of a 10 MVA, 15/63 kV transformer. The short-circuit power at PCC is 200 MVA.

The STATCOM model was built under *EMTP-RV*® environment (version 1.0.2), since it is not available in this version library. The induction machine model is based on a fourth-order state space model [52] built in the library in *EMTP-RV*®.

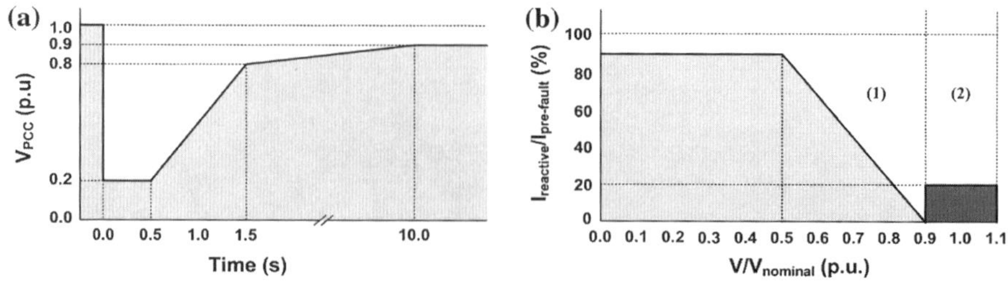

Fig. 27 Characteristic curves of Portuguese grid codes: (**a**) time–voltage characteristic; (**b**) characteristic curve of reactive current delivered by wind farms during/after voltage dips *1* fault and recovery region; *2* normal operation region

The wind speed is reproduced through the mechanical torque which is applied to the induction generator model. The utility system is represented by a constant voltage source connected in series with the corresponding equivalent impedance of Thevenin. The remaining components are represented through the models available on *EMTP-RV*® library.

6.3 Simulation Results and Discussion

The FRT capability of wind farms equipped with FSIG was evaluated considering the Portuguese proposal of grid codes for wind farms connection [16]. Regarding FRT requirements, wind power plants must fulfil the following minimum technical requirements:

- To remain in operation during voltage dips, if the wind farm terminal bus voltage remains above the curve depicted in Fig. 27a during the fault and after its clearance for the time limits also defined according to Fig. 27a;
- To deliver reactive power during voltage dips in order to provide voltage support for the network. The requested reactive power is indexed to the reactive current flowing through PCC before the fault occurrence, being the wind farm obliged to stay in the white region of the curve depicted in Fig. 27b.

In order to demonstrate the STATCOM contribution to provide FRT capability of wind farms equipped with FSIG, the most severe fault conditions regarding voltage recovery were considered. Thus it was assumed that the test system wind farm is operated with a $tg(\phi) = 0.2$, near its rated power, corresponding to the maximum amount of reactive power drawn to SCIG. Then, a three-phase short-circuit is simulated on the ac network, far from the wind farm, at $t = 1$ s and with a clearance time of 500 ms. The voltages at the PCC and at the wind farm terminal bus are depicted in Fig. 28 when the wind farm is operated with and without the STATCOM.

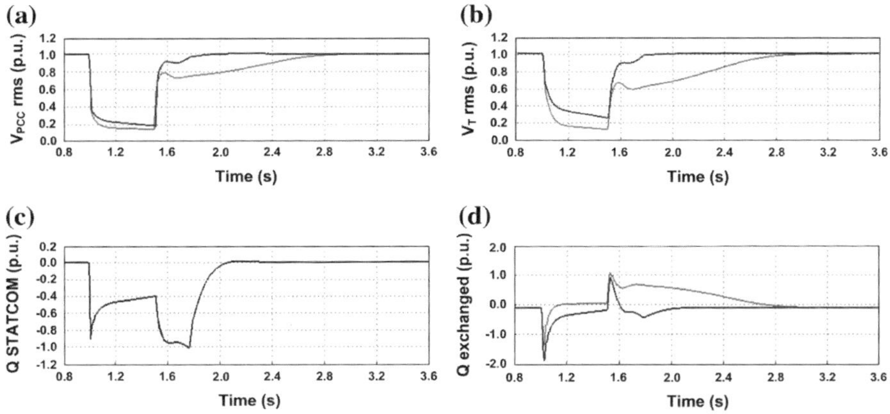

Fig. 28 System dynamic behaviour following a balanced fault: (**a**) voltage at the PCC with STATCOM (*blue*) and without STATCOM (*red*); (**b**) terminal bus voltage with STATCOM (*blue*) and without STATCOM (*red*); (**c**) reactive power injected from the STATCOM; (**d**) reactive power exchange to the ac grid with STATCOM (*blue*) and without STATCOM (*red*)

Without the STATCOM, the voltage at the PCC drops below 0.2 p.u. during the fault, leading with the wind farm disconnection according to the considered FRT requirements. The STATCOM based solution avoids this situation, since the voltage at the PCC remains above 0.2 p.u., as it can be observed from Fig. 28a, contributing thus to FRT enhancement. In addition, after the fault is cleared, the voltage recovers faster and within the time limits required by the FRT requirements. Moreover, a better improvement is achieved regarding the terminal bus voltage by means of reactive power injection into the grid from the STATCOM during the fault and after its clearance, as it can be observed from Fig. 28b.

The STATCOM control system reacts immediately to a sudden voltage dip, saturating the controller and injecting the maximum reactive current into the grid, as illustrated in Fig. 28c. However, it should be noted that the reactive power injection is limited by the voltage drop and therefore the STATCOM provides a little contribution for voltage support during the fault. In contrast, after the fault is cleared, the reactive power injected from the STATCOM corresponds to its rated power contributing to a faster voltage recovery and a clear increase of the system stability margin. In addition, the magnetization process of FSIG is dynamically supported making it possible to decrease the rotor speed and to reduce the reactive power consumption from the grid, as it can be observed from Fig. 28d.

When the STATCOM is used to perform voltage control, since the VSC control system is based on the ac-positive sequence voltage measured at the STATCOM connection point, the same amount of reactive power is injected on the three-phases. The results obtained from numerical simulation demonstrate the effectiveness of this commonly used control technique under balanced conditions, such as following three-phase short-circuits, in which the voltages on the three phases experiment similar amplitudes of voltage drop.

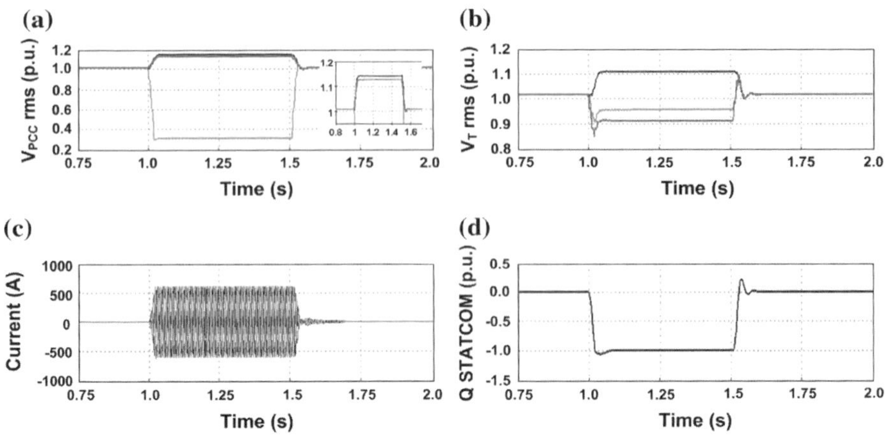

Fig. 29 System dynamic behaviour following an unbalanced fault: (**a**) voltage at the PCC—phase a (*red*), phase b (*blue*) and phase c (*green*); (**b**) terminal bus voltage—phase a (*red*), phase b (*blue*) and phase c (*green*); (**c**) STATCOM reactive current—phase a (*red*), phase b (*blue*) and phase c (*green*); (**d**) reactive power injected from the STATCOM

Although the balanced faults are the most severe ones, its occurrence is extremely rare. In contrast, unbalanced faults, which occur when one or two phases are shorted to ground or to each other, happen most often leading with negative sequence components in the ac grid voltages. Thus, the STATCOM dynamic performance to provide FRT of wind farms equipped with FSIG should also be evaluated when facing unbalanced faults. For this purpose, a single phase short-circuit taking place on phase *a* was simulated on the HV network at $t = 1$ s and with a clearance time of 500 ms, considering that all the SCIG are operated near one half of their ratings and the MV capacitor battery assures the wind farm operation with $tg(\varphi) = 0.4$ in order to comply with power factor requirements. The results obtained are presented in Fig. 29.

The 15/63 kV transformer of type ΔY influences on how the grid fault appears on the wind farm terminal bus, as it can be observed from Fig. 29a, b. Although only the phase *a* was short to ground, both phases *a* and *c* experimented voltage dips, even with a much reduced magnitudes. However, the STATCOM injects a balanced three-phase reactive current into the ac grid, as it can be observed from Fig. 29c, leading with considerable over voltages experimented by the non-faulted phases, as it can be observed from Fig. 29a, b. In fact, the voltage at the PCC, regarding phase *b* and phase *c*, experiments over voltages above 1.1 p.u. and a similar situation is verified regarding the wind farm terminal bus voltage on phase *c*. After the fault elimination, transient over voltages take place on phase *a* and phase *b* of the wind farm connection point, as it can be observed from Fig. 29a. It should also be noted that the behaviour of the ac-positive sequence of the terminal bus voltage allows the STATCOM to inject its maximum reactive current, according to the *V–I* characteristic depicted in Fig. 24b, and therefore the reactive power output corresponds to its rated power. In contrast with the three-phase short-circuit simulated previously, the terminal bus voltage drop is much reduced allowing the maximum reactive power injection into the ac grid, as it can be observed from Fig. 29d.

The results obtained demonstrate that the STATCOM provides voltage support during voltage dips that arise following the occurrence of external short circuits, reducing thus the voltage drops, and contributes to fast voltage recovery after the fault clearing. Therefore, the STATCOM can be considered as an effective mean to improve FRT capability of existing wind farms equipped with FSIG. However, in case of unbalanced faults, such as a single-phase short circuit, over voltages can arise on the non-faulted phases, which can trigger the over voltage protections and subsequent wind farm disconnection, missing the FRT capability improvement. In order to avoid such situations, complementary control procedures to define the amount of reactive power injection should be derived. For this purpose, negative sequence of the voltage at the STATCOM connection point should also be taken into account. On the other hand, when possible, an adjustment of the settings of the over voltage protections should be performed.

7 Final Remarks

An effective integration of large scale wind generation leads to a new challenge clearly stated as control wind power in a similar way to conventional power plants equipped with synchronous generators. Therefore, the stability of fundamental electric parameters, frequency and voltage, has to be assured. Active power control is required for frequency regulation and reactive power control is required for voltage regulation. In addition, wind parks should comply with FRT requirements, according to the grid codes imposed by transmission system operators.

For this purpose, the variable speed feature of WTGS plays a key role. Regarding DFIG, advanced control functionalities have been developed aiming to improve the DFIG performance regarding the compliance with grid code requirements, such as FRT capability and primary frequency control. The main focus is put on the RSC, since it responsible for the generator operation and control in terms of speed and independent control of active and reactive power outputs. Also, classical PSS should be included on the RSC control system allowing DFIG equipped with PSS to provide damping contributions, enhancing thus the system small signal stability in case of conventional generation with PSS installed be replaced by wind based generation.

Wind farms equipped with FSIG require the adoption of external solutions to provide FRT capability. Although STATCOM based solutions are an effective mean for this purpose, the need of including complementary control procedures based on the negative sequence of the voltage at the STATCOM connection point should be evaluated when facing unbalanced faults in order to prevent the occurrence of over voltages in the non-faulted phases.

Discussion of advanced control functionalities to be implemented in control systems of power electronic converters of variable speed WTGS and STATCOM together with dynamic modelling issues will represent new insights for evaluating the power system robustness of operation following the increasing integration levels of wind based generation.

Acknowledgments The authors would like to acknowledge to FCT, Fundação para a Ciência e Tecnologia (Portuguese Foundation for Science and Technology), by its support within the Grants SFRH/BD/18469/2004 and SFRH/BPD/64022/2009.

References

1. Achmatov V (2002) Variable speed wind turbines with doubly fed induction generators. Part IV: Uninterrupted operation features at grid faults with converter control coordination. Wind Eng 27:519–529
2. Akhmatov V (2003) Analysis of dynamic behaviour of electric power systems with large amount of wind power. PhD Thesis, Technical university of Denmark, ISBN 87-91184-18-5
3. Akagi H, Kanazawa Y, Nabae A (1984) Instantaneous reactive power compensators comprising switching devices without energy storage components. IEEE Trans Ind Appl (IA-20):625–630
4. Almeida RG, Castronuovo E, Peças Lopes JA (2006) Optimum generation control in wind parks when carrying out system operator requests. IEEE Trans Power Syst 21:718–725
5. Almeida RG, Lopes JA (2005) Primary frequency control participation provided by doubly fed induction wind generators. In: Proceedings of 15th power systems computation conference, pp 1–7
6. Almeida RG, Peças Lopes JA (2007) Participation of doubly fed induction wind generators in system frequency regulation. IEEE Trans Power Syst 22:944–950
7. Almeida RG, Lopes JP, Barreiros JAL (2004) Improving power system dynamic behaviour through doubly fed induction machines controlled by static converter using fuzzy control. IEEE Trans Power Syst 19:1942–1950
8. Aten M, Martinez J, Cartwright PJ (2005) Fault recovery of a wind farm with fixed-speed induction generators using a STATCOM. Wind Eng 29:365–375
9. Barbosa P, Rolim L, Watanabe E, Hanitsch R (1998) Control strategy for grid connected DC–AC converters with load power factor correction. In: IEE proceedings of generation, transmission and distribution, vol 145, pp 487–491
10. Chen Z, Hu Y, Blaabjerg F (2007) Stability improvement of induction generator-based wind turbine systems. IET Renew Power Gener 1:81–93
11. Do Bomfim A, Taranto G, Falcão D (2000) Simultaneous tuning of power system damping controllers using genetic algorithms. IEEE Trans Power Syst 15:163–169
12. Ekanayake J, Holdsworth L, Jenkins N (2003) Control of DFIG wind turbines. Power Eng J 17:28–32
13. Ekanayake JB, Holdsworth L, XueGuang W et al (2003) Dynamic modeling of doubly fed induction generator wind turbines. IEEE Trans Power Syst 18:803–809
14. Ekanayake J, Jenkins N (2004) Comparison of the response of doubly fed and fixed speed induction generator wind turbines to changes in network frequency. IEEE Trans Energy Convers 19:800–802
15. Erlich I, Wrede H, Feltes C (2007) Dynamic behaviour of DFIG-based wind turbines during grid faults. In: Proceedings of power conversion conference 2007, vol 1, pp 1195-1200

16. Estanqueiro A, Castro R, Flores P, Ricardo J, Pinto M, Rogrigues R, Peças Lopes JA (2007) How to prepare a power system for 15% wind energy penetration: The Portuguese case study. Wind Energy 11:75–84
17. Feijoo A, Cidras J, Carrillo C (2000) A third order model for the doubly-fed induction machine. Elect Power Syst Res 56:121–127
18. Feltes C, Engelhardt S, Kretschmann J, Fortmann J, Koch F, Erlich I (2009) Comparison of the grid support capability of DFIG-based wind farms and conventional power plants with synchronous generators. In; Proceedings of IEEE power and energy society general meeting 2009, vol 1, pp 1–7
19. Gaztañaga H, Etxeberria-Otadui I, Ocnasu D, Bacha S (2007) Real-time analysis of the transient response improvement of fixed-speed wind farms by using a reduced-scale STATCOM prototype. IEEE Trans Power Syst 22:658–666
20. Hansen A, Sorensen P, Iov F, Blaabjerg F (2006) Centralized power control of wind farm with doubly fed induction generators. Renew Energy 31:935–951
21. Hingorani NG, Gyugyi L (2000) Understanding FACTS: concepts and technology of flexible AC transmission systems. Wiley-IEEE Press, Hardcover
22. Holdsworth L, Ekanayake J, Jenkins N (2004) Power system frequency response from fixed speed and doubly fed induction generators-based wind turbines. Wind Energy 7:21–35
23. Hughes F, Anaya-Lara O, Jenkins N, Strbac G (2005) Control of DFIG-based wind generation for power network support. IEEE Trans Power Syst 20:1958–1966
24. Hughes F, Anaya-Lara O, Jenkins N, Strbac G (2006) A power system stabilizer for DFIG-based wind generation. IEEE Trans Power Syst 21:763–772
25. Koch F, Erlich I, Shewarega F (2003) Dynamic simulation of large wind farms integrated in a multimachine network. In: Proceedings of IEEE power engineering society general meeting, vol 1, pp 2159–2164
26. Kundur P (1994) Power system stability and control. McGraw-Hill, New York
27. Lee C (1990) Fuzzy logic in control systems: fuzzy logic controllers. IEEE Trans Syst Man Cybern 20:404–418
28. Lei Y, Mullane A, Lightbody G, Yakamini R (2006) Modeling of the wind turbine with a doubly fed induction generator for grid integration studies. IEEE Trans Energy Convers 21:257–264
29. Li H, Chen Z (2008) Overview of different wind generation systems and their comparisons. IET Renew Power Gener 2:123–138
30. Li H, Philip C, Huang H (2001) Fuzzy neural intelligent system. CRC Press, New York
31. Machmoum M, Poitiers F, Darengosse C, Queric A (2002) Dynamic performance of a doubly-fed induction machine for a variable-speed wind energy generation. In: Proceedings of international conference on power systems technology 2002, vol 4, pp 2431–2436
32. Mendonça A, Peças Lopes J A (2003) Robust tuning of PSS in power systems with different operating conditions. In: Proceedings of IEEE bologna power tech conference, vol 1, pp 1–7
33. Mendonça A, Peças Lopes JA (2009) Robust tuning of power system stabilisers to install in wind energy conversion systems. IET Renew Power Gener 3:465–475
34. Michalke G, Hansen AD (2010) Modelling and control of variable speed wind turbines for power system studies. Wind Energy 13:307–322
35. Miranda V (2005) Evolutionary algorithms with particle swarm movements. In: Proceedings of 13th international conference on intelligent systems applications to power systems, vol 1, pp 6–21
36. Molinas M, Suul JA, Undeland T (2007) Improved grid interface of induction generators for renewable energy by use of STATCOM. In: Proceedings of ICCEP'07, vol 1, pp 215–222
37. Morren J, Haan S, Kling W, Ferreira J (2006) Wind turbines emulating inertia and supporting primary frequency control. IEEE Trans Power Syst 21:433–434
38. Novotny DW, Lipo TA (2000) Vector control and dynamics of AC drives. Oxford University Press, New York

39. Nunes M, Bezerra U, Peças Lopes JA, Zurn H, Almeida RG (2004) Influence of the variable speed wind generators in transient stability margin of the conventional generators integrated in electrical grids. IEEE Trans Energy Convers 9:692–701
40. Pena R, Clare JC, Asher GM (1996) Doubly fed induction generator using back-to-back PWM converters and its application to variable-speed wind energy generation. In: IEE proceedings of electric power applications, vol 143, pp 231–241
41. Poller MA (2003) Doubly-fed induction machine models for stability assessment of wind farms. In: Proceedings of 2003 IEEE bologna power tech conference, vol 3, pp 1–6
42. Pourbeik P, Gibbard M (1998) Simultaneous coordination of power systems stabilizers and FACTS devices stabilizers in a multimachine power system for enhancing dynamic performance. IEEE Trans Power Syst 13:473–479
43. Qi L, Langston J, Steurer M (2008) Applying a STATCOM for stability improvement to an existing wind farm with fixed-speed induction generators. IEEE power and energy society general meeting—conversion and delivery of electrical energy in the 21st century, vol 1, pp 1–6
44. Qiao W, Zhou W, Aller JM, Harley RG (2008) Wind speed estimation based sensorless output maximization control for a wind turbine driving a DFIG. IEEE Trans Power Electron 23:1156–1169
45. Rao P, Crow ML, Yang Z (2000) STATCOM control for power system voltage control applications. IEEE Trans Power Delivery 15:1311–1317
46. Resende FO, Peças Lopes JA (2009) Evaluating the performance of external fault ride through solutions used in wind farms with fixed speed induction generators when facing unbalanced faults. In: Proceedings of the IEEE bucharest power tech 2009, vol 1, pp 1–6
47. Rodriguez J et al (2002) Incidence of power system dynamics of high penetration of fixed speed and doubly fed wind energy systems: study of the Spanish case. IEEE Trans Power Syst 19:1089–1095
48. Rogers GJ (2000) Power system oscillations. Kluwer Academic Publishers, Boston
49. Rogers GJ, Shirmohammadi D (1987) Induction machine modelling for electromagnetic transient program. IEEE Trans Energy Convers (EC-2):622–628
50. Salman SK, Teo ALJ (2003) Windmill modelling considerations and factor influencing the stability of grid connected wind power based embedded generator. IEEE Trans Power Syst 18:793–802
51. Slootweg JG, Haan SWH, Polinder H, Kling L (2003) General models for representing variable speed wind turbines in power system dynamics simulations. IEEE Trans Power Syst 18:144–151
52. Stavrakakis G, Kariniotakis G (1995) A general simulation algorithm for the accurate assessment of isolated diesel-wind turbines systems interaction—part I: a general multimachine power system model. IEEE Trans Energy Convers 10:577–583
53. Stavrakakis G, Kariniotakis G (1995) A general simulation algorithm for the accurate assessment of isolated diesel-wind turbines systems interaction—part II: implementation of algorithm and case studies with induction generators. IEEE Trans Energy Convers 10:584–590
54. Tsili M, Papathanassiou S (2009) A review of grid code technical requirements for wind farms. IET Renew Power Gener 3:308–332

Network Stability Under High Wind Power Penetration

Emmanuel S. Karapidakis and Antonios G. Tsikalakis

Abstract This chapter addresses Network stability issues associated with high wind power penetration in a power system. Nowadays several countries and regions all over the world, either already operate under high wind power penetration or they have concrete plans to increase it. Thus, their grids and operation strategies should cope reliable with this wind penetration level. Some of the main identified solutions that will help the power systems to overcome the current network stability constraints are presented.

1 Introduction

A few decades ago electrical networks could afford to be overdesigned. However, in the last decade and especially nowadays power systems should operate under great complexity and stressed conditions. Increased complexity can be attributed to one and combination of the following reasons:

- Deregulation Market
- Environmental restrictions
- Load demand growth
- Increased power quality requirements
- Increased Renewable Energy Sources (RES) penetration
- Differentiation in reliability requirements of the customers. Now there is not load but customers with increased requirements
- Congestion or issues related to reinforcement of the grid.

E. S. Karapidakis · A. G. Tsikalakis (✉)
Department of Natural Resources and Environment, Branch of Chania/Technological Educational Institute of Crete, Crete, Greece
e-mail: atsikal@power.ece.ntua.gr; atsikalakis@isc.tuc.gr

P. M. Pardalos et al. (eds.), *Handbook of Wind Power Systems*, Energy Systems,
DOI: 10.1007/978-3-642-41080-2_10,

A significant concern under these conditions is that power systems might exhibit a different case of unstable behavior, characterized by both voltage and frequency instability [1, 2]. As a consequence, network stability should be a main concern for modern power system planning and operation, especially when intermittent power sources like wind are incorporated. Additionally, while the stability of a power system follows a non-linear behavior, as the stress conditions increase, this nonlinearity becomes more pronounced.

Currently, the vast majority of the electricity demand is produced in large centralized facilities, mainly conventional (*fossil fuel and/or nuclear powered*) and hydropower plants, transmitting electricity under High Voltage over long distances. These plants have emerged mainly due to improved economies of scale and fuel availability on the spot.

Weakly interconnected or Autonomous power systems, like the ones operating in isolated areas or mostly on islands, face increased complexity related to their operation and control [3, 4]. In most of these systems, the real cost of electricity production is much higher than in large interconnected systems due to the high installation, operating and maintenance costs (*economies of scale*). Security is also a major concern, since mismatches in generation and load and/or unstable system frequency control may more frequently lead to system failures than in interconnected systems [5, 6].

Wind power is one of the major sources of renewable energy with remarkable contribution to the installed capacity of electrical power systems. Since 2008, wind power has presented the highest annually installed capacity compared to any other type of power source, according to the European Wind Energy Association (EWEA). Only during 2009–2010 additional 20 GW were added in Europe. In this chapter, the impacts of large scale wind power generation on the dynamic performance especially of weak and/or islands power systems are presented.

Although under ongoing energy policies wind power exploitation appears particularly attractive, the integration of a substantial amount of wind power into electrical power systems needs careful consideration, so as to maintain high degree of reliability and security of the system operation. The main problems identified concern operational scheduling (mainly unit commitment) due to high wind power forecasting uncertainties, as well as steady state and dynamic operation disorder [7]. These problems may considerably limit the amount of wind generation that can be connected to the power systems, increasing the complexity of their operation. Thus, next to the more common angle and voltage stability concerns, frequency stability must be ensured, [8]. This depends on the ability of the system to restore balance between generation and load in case of a severe system upset with minimum loss of load.

Dynamic simulation studies are the first step in determining the level of wind power penetration in power systems. Analytical studies are required in order to derive security rules and guidelines for the optimal operation of each system [9]. Simulations of a power system dynamic performance mainly cover voltage and frequency calculations under several abnormal operating conditions, start-up or sudden disconnection of wind generation, wind fluctuations and short circuits on the transmission and distribution network. In order to ensure the maximum exploitation of the available renewable power sources and to operate systems with increased wind power penetration in the most economic and secure way, advanced energy management systems (EMS) are needed [10].

Moreover, in order to operate optimally within the new market conditions, the price of providing a given level of security has to be accounted for. This is directly linked to the provision of remedial actions, in case of insecure situations, [11]. Especially, for Dynamic Security, unlike Steady State security, remedial actions can only be preventive leading to load shedding or generation rescheduling. Consequently, the cost of load shedding has to be balanced with the cost of providing adequate spinning reserves in order to avoid it.

2 Definition and Classification of Network Stability

Power system's stability has been recognized as an important problem for secure system operation since the 1920s in Steinmetz [12] and AIEE [13]. Generally, transient stability is the main concern on the majority of the power systems. As power systems have evolved, new operation technologies and controls in highly stressed conditions have emerged. More precisely, voltage stability and frequency stability have become greater concerns than in the past. A clear understanding of different types of stability and how they are interrelated is essential for the satisfactory design and operation of power systems [1].

Power system stability is similar to the stability of any dynamic system having fundamental mathematical underpinnings. Precise definitions of stability can be found in the literature dealing with the rigorous mathematical theory of stability of dynamic systems. A circumstantial definition of Power System Dynamic Security and the corresponding types of Power System Stability is provided in Kundur et al. [6]. In Fig. 1 a concise classification of power system stability is provided.

Additionally, another significant issue is the relationship between the concepts of power system reliability, security, and stability of a power system:

- Power System Reliability: it refers to the probability of a required operation condition achievement, with few interruptions over an extended time period.
- Power System Security: it refers to the risk assessment of system ability to survive disturbances (taking into account the probability of these contingencies) without any power supply interruption.
- Power System Stability: it refers to the dynamic operation after a severe or moderate disturbance, consequently to an initial steady state operating condition.

As it is clearly mentioned in Kundur et al. [6], the analysis of security relates to the determination of the power system robustness to imminent disturbances. When a power system undergoes any operational change, it is important that as soon as

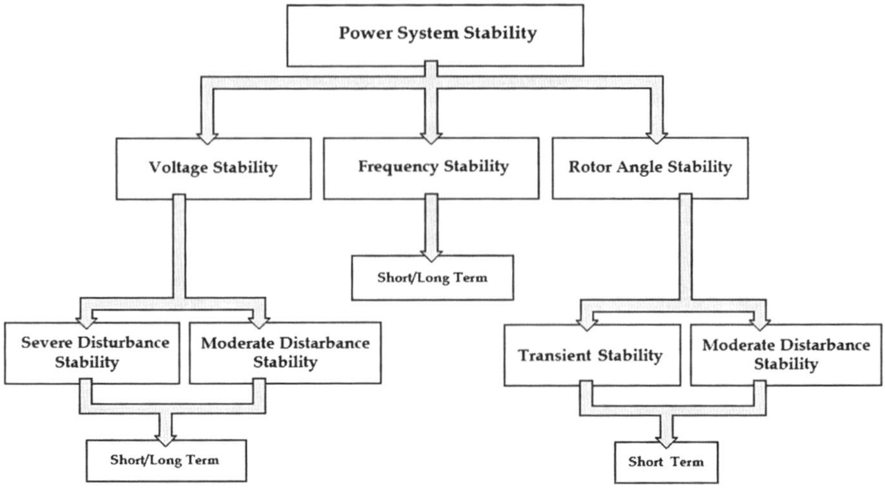

Fig. 1 Categories of power system stability

these changes are complete, the updated operating state of the system does not violate any technical constraint whatsoever. This implies that, in addition to the acceptance of the updated operating conditions, the system should survive the transition to these conditions. The above characterization of system security clearly highlights two aspects of its analysis:

- Static Security Analysis: It involves steady-state analysis of post-disturbance system conditions to verify that no voltage constraints are violated.
- Dynamic Security Analysis: It involves examining different categories of system stability described in Section III.

The common approach for dynamic security assessment has been a deterministic one. The power system is designed and operated to withstand a set of contingencies selected on the basis that they have significant possibility to occur in practice, they are usually defined as the loss of any single element in a power system either spontaneously or proceeded by a single, double, or three phase fault. This method is generally called as the N-1 criterion as it examines the behavior of an N-component network following the loss of any one of its components. In addition, emergency controls, such as generation tripping, load shedding, and controlled islanding, may be used to withstand such events and prevent widespread blackouts.

Concluding, dynamic behavior assessment requires a full representation of the examined network, while stability is directly connected with the electrical distance between generation and loads, thus depends on the network structure. Under deregulated energy markets with diversity of participants, the deterministic approach may not be appropriate. There is a need to account for the probabilistic nature of system conditions and events, and hence quantify and manage risk. The trend will be to expand the use of risk-based security assessment. In this approach, the probability of the system becoming unstable and its consequences are examined, and the degree of exposure to system failure is estimated. This approach may be computationally intensive but current computing and analysis tools make it possible.

3 Technical Issues Associated with High Wind Power Penetration

It is well known that the high wind capacity, either installed in the latest years or to be installed in near future, introduces a new set of technological challenges for power system operators and planners [14, 15]. The current power systems should be capable of coping with this large quantity of wind generation, without risking the occurrence of serious events or even "blackouts". New monitoring and control operation tools, increased performance of the wind turbines and/or a change in the power system conventional mode of operation and planning strategy should be identified [16]. This chapter presents some current methodologies for overcoming the usual existing barriers to high wind power penetration in power systems.

3.1 Technical Barriers to High Wind Penetration

Among the various renewable resources, wind power is assumed to have the most favorable technical and economical prospects. When deployed in small scale, as was done traditionally, the impact of wind generators on power system stability is minimal. In contrast, as the penetration level increases, the dynamic performance of the power system can be affected. The main identified technical barriers to high wind power penetration are presented below.

1. **Transmission System Capacity**

The first barrier to high wind generation in a power system is the limited grid capacity, referring mainly to the transmission system capacity. In many cases, especially within European region, wind farms developers invest themselves on building the interconnection lines to the existing network and even pay the cost for the possible distribution grid reinforcement. Future transmission system development plans should take into account wind farms installations or other RES investments [17].

2. **Unit Commitment and Security of Supply**

Solution of the unit commitment problem replies to the question which generating resources should be used to meet the electrical load at as lower cost as possible. The solution should also take into account both user defined uncertainties and the technical constraints of the generating units and the power system. Energy policy constraints, such as the obligation of buying the whole production coming from

RES, may be included as well. It is well known that the existence of a large amount of RES in a power system increases both the complexity of the unit commitment procedure and the instability of power supply. Some efforts in facing this complexity have been made in Tsikalakis et al. [18]. More precisely, regarding wind power, the following issues have been identified.

Wind power penetration threshold: Wind power provides rather unstable power generation due to wind speed variability. It is commonly accepted that there is a threshold above which, increasing the wind power penetration also increases the power reserve requirements of a system [14]. This fact has been addressed in detail for some power systems or control areas, e.g. Nordpool [19]. In these cases the associated costs are much lower than expected up to a certain level and only representative for very high penetrations. Furthermore, the upper limit of wind power penetration is highly dependent on the system generation mix; the slower the units response, the lower this limit is expected to be.

Wind power variability: Another issue strongly related to the wind generation is variability and the extremely difficult task to forecast wind power production within time intervals useful for power system operation [20, 21]. This creates uncertainty to the operators of the power system regarding secondary and tertiary control. In the very short term, wind gusts may cause significant frequency content of the power delivered to the system, mainly in the range of flicker emission (from 0.1 to 20 Hz). Those fluctuations could degrade the power quality in the surroundings of wind parks [22]. Therefore limits should successfully be defined mainly through international standards in order to guarantee an acceptable level of quality [23].

Wind power reliability: Another major problem with a case of high wind power generation is the possibility of its sudden disconnection from the grid as a response to a network disturbance. More precisely, voltages dips are usually originated by short circuits and may lead to the islanding of some parts of the network. Therefore, it would be useful for the network, the existing wind generators to withstand these disturbances and to remain connected to the grid. For these cases, a technical characteristic known as fault ride-through (FRT) capability of wind turbines is required.

3.2 Wind Power Increased Performance

Wind power systems should contribute as much as possible to the reliability of the power system they are connected to. This is feasible via enhancement of wind turbines with advanced characteristics.

1. **Wind Systems Advanced Characteristics**

Under the new operation conditions that have been previously described for the modern power systems, the wind system should follow with some innovative characteristics as they are presented hereupon.

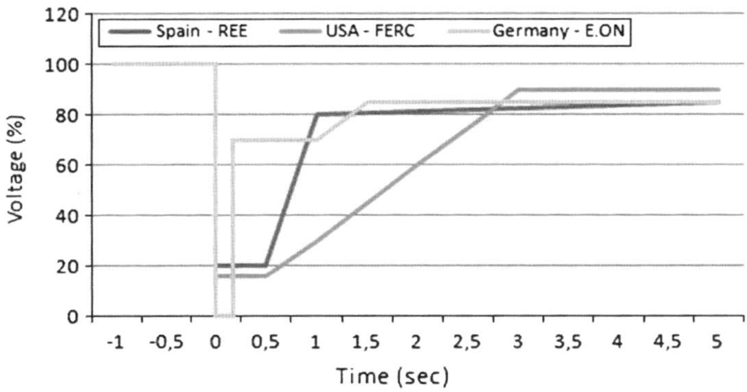

Fig. 2 FRT characteristic response curves required to the wind parks

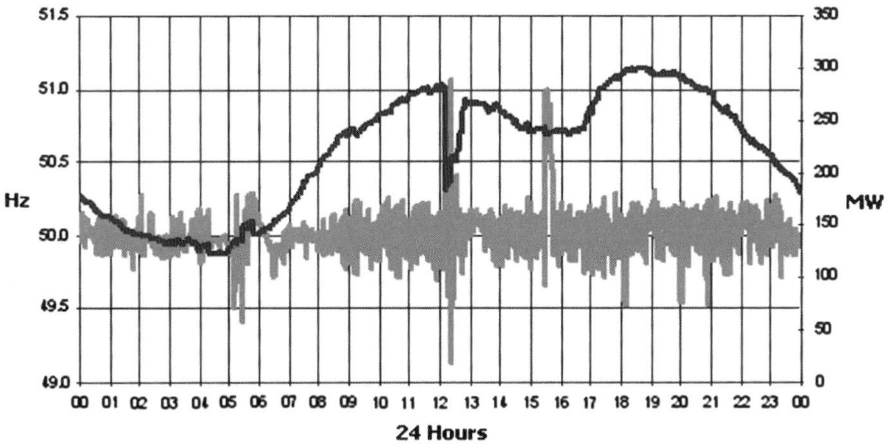

Fig. 3 Frequency dips in a case of wind parks disconnection in Crete's power system

Low Voltage Fault Ride-Through: Nowadays, recognizing the large potential of wind energy, but also revealing an extreme concern towards its growth and future development, most power systems operators have issued grid codes requiring the wind turbines and power plants to contribute with some basic operation functionalities. The more publicized one is the FRT capability, whose characteristics for the E-ON, FERC and Spanish grid codes are depicted in Fig. 2.

Participation in the primary frequency control: Another important issue is the wind turbine response to extreme frequency deviations as the one depicted in Fig. 3. Primary frequency regulation means adopting a specific frequency control and a deload operation strategy below the maximum extraction power curve. If wind generators with such capabilities are used, a considerable contribution can be obtained to reduce the impact of frequency dip [24]. Such control strategy can aid to the frequency regulation, especially in windy regions and power systems with reduced hydro power regulation capability. Isolated very windy power systems with traditional frequency control are a typical example for the application of this recent functionality of the wind turbines.

The use of power electronics will definitely contribute to the power system robustness, by avoiding grid electronic interfaced wind generators disconnection.

2. Wind Power Control and Curtailment

The replacement of large conventional power plants by hundreds of wind generation units spread over the transmission and distribution system requires the development of new concepts for monitoring, controlling and managing these generation resources. This new approach to the power system requires to bear in mind not only network operational restrictions, but also market procedures in compliance with the current liberalized regulations.

Innovative strategies and equipments are already being implemented in some European countries. The capacity of a wind park may be limited by the grid capacity to accept the total power the wind park. However, very seldom do wind turbines manage to reach their nominal capacity due to wind speed values other than the V_{rated} (i.e. rated wind speed) value. Therefore, in order to optimise the grid capacity use and not delay the wind power injection to the grid, some Transmission System Operators (TSOs) have allowed the installation of what is commonly named "overcapacity" of wind parks. Under these circumstances, the wind park interconnection line has only available 70–80 % capacity when compared to the rated power of the park. When the wind power tends to the nameplate capacity, the wind turbines and the whole park start to regulate the power production in order not to exceed the predefined maximum value for the congested area. Such operating experience with congested HV networks and increased wind power capacity has been gathered by the Hellenic Transmission System Operator as described in Kabouris and Vournas [25] for the region of Thrace in Northern Greece.

As anticipated, wind park overcapacity will be allowed provided that a control of production is performed to avoid an injection of power larger than the initially defined by grid technical constraints. Since the percentage of time that a wind park can exceed 80 % of its nameplate capacity is very low, the energy loss in this situation is negligible. Therefore this approach is economically beneficial both for the wind park developers and the power system operators, compared to grid reinforcement a time-consuming process. Such control can be achieved via Programmable Local Controllers (PLCs) as described in Kabouris and Hatziargyriou [26].

3. Wind Generation Aggregation

The spatial aggregation of the wind generation has several positive side effects as it enables to take advantage of one of the most basic characteristics of the wind resource which is the lack of spatial correlation in what concerns the fast wind fluctuations (typically above 0.1 Hz) and consequent smoothing effect of the injected power [27]. Many studies [28, 29] have been implemented for studying aggregated wind generation cases.

Nevertheless, what could be a negative characteristic may turn to be extremely beneficial for the power system operation. Of course, in order to profit from that effect, it is required that the wind parks share a common grid interconnection, otherwise large power fluctuation even though not felt by central dispatches, they may affect local or regional parts of the transmission network. The smoothing effect is also not present when a whole country (or power system) is immersed in high (or low) pressure atmospheric circulations or passed by large frontal surfaces.

The need to monitor remotely the state and level of generation of wind power plants was recognized both by the manufacturers of wind turbines and the International Electrotechnical Commission (IEC) whose Technical Committee 88—Wind Turbines started the development of a new international standard on communications (IEC 61400-25-XX).

4 Case of Autonomous Power Systems

Autonomous or isolated power systems are all the small and medium size power systems where no interconnection exists with conterminous and/or continental systems. These power systems face increased problems related to their operation and control [30, 31]. In most of these systems, dynamic performance is a major concern, since mismatches in generation and load and/or unstable system frequency control might lead to system failure much easier compared to interconnected systems.

Renewable sources and especially wind power exploitation appear particularly attractive in these cases [32]. However, the integration of a substantial amount of wind power in isolated systems needs careful consideration, so as to maintain high levels of reliability and security of the system operation [4]. The main problems identified concern operational scheduling (mainly unit commitment) due to high production forecasting uncertainties, as well as steady state and dynamic operating problems. These problems may considerably limit the amount of wind generation that can be connected to the island systems, increasing the complexity of their operation [18]. Thus, next to the more common angle and voltage stability concerns, frequency stability [33] must be ensured. This depends on the ability of the system to restore balance between generation and load following a severe system disturbance with minimum loss of load.

4.1 Autonomous Power System of Crete

Grete is the largest Greek island with approximately 8.500 Km^2 and the fifth in Mediterranean sea with more than 600,000 inhabitants, tripling in summer period. Additionally, considerable annual increase of electricity demand, up to 7 %, has been noted during the last decade. As a result, the annual energy consumption

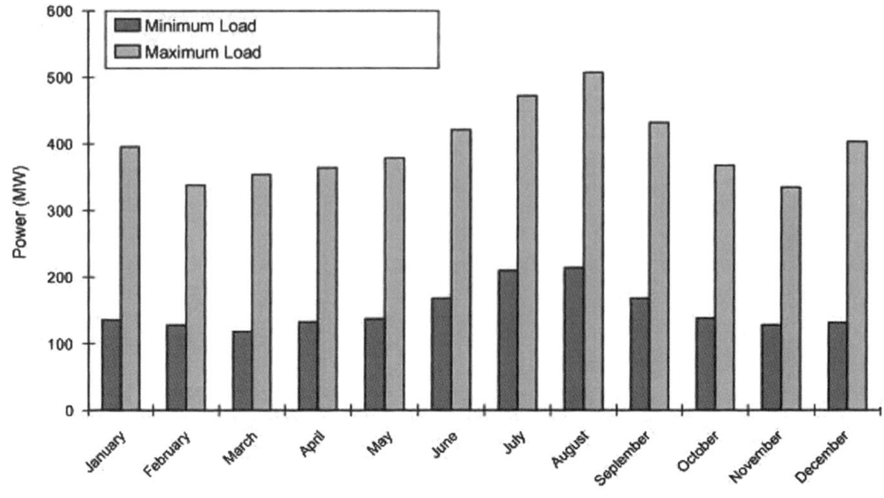

Fig. 4 Monthly variations of min and max load demand

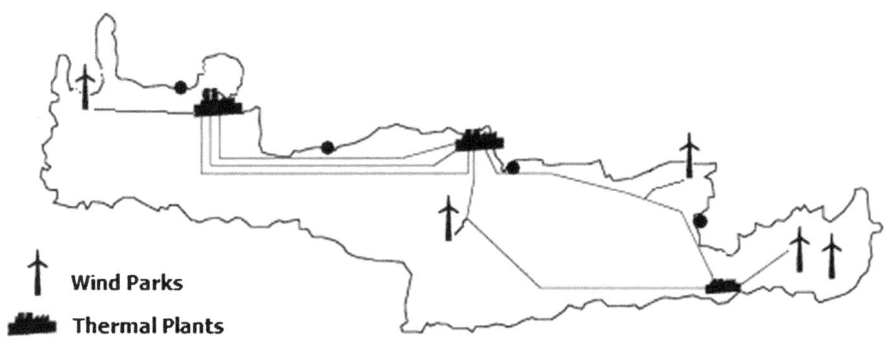

Fig. 5 Power plants and wind parks locations

during 2011 surpassed the 3 TWh compared to the modest 280 GWh of year 1975. Additionally, comparing the mean hourly load demand variation all year round, there is considerable demand fluctuation between months and seasons, as shown in Fig. 4. The annual peak load demand occurs on a warm and usually humid summer day while during the minimum spring valley hours demand is approximately equal to 25 % of the corresponding daily peak loads.

Island's electricity generation system is based mainly on three oil-fired thermal power units, located as it is shown in Fig. 5. The total capacity of the three power plants is 693 MW for winter and 652 MW for summer operation. Additionally, there are 24 wind parks installed with nominal power of 183.54 MW across the island. These WPs are connected to the grid through MV/HV substations of 20/150 kV. The steam and diesel units mainly supply the base-load. The Gas turbines normally

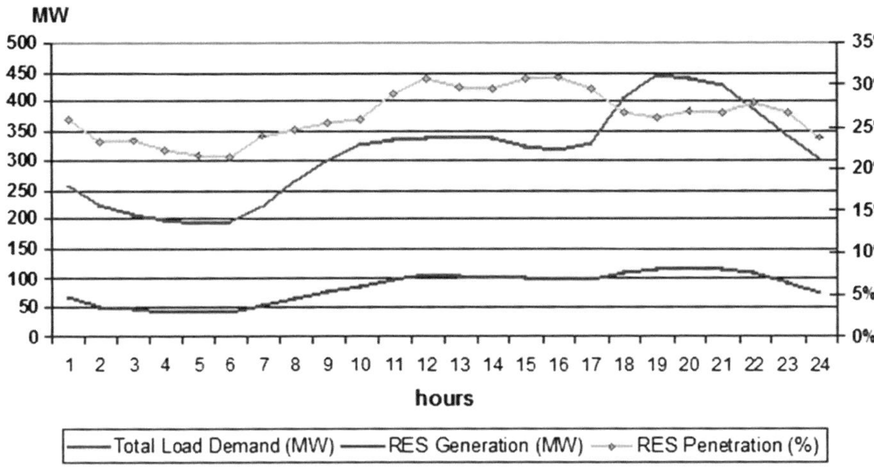

Fig. 6 Wind power penetrations in power system of Crete

supply the daily peak load or the load that cannot be supplied by the other units in outage conditions. These units have a high running cost that increases significantly the average cost of the electricity being supplied.

In Fig. 6 both the wind power production and the overall production during a specific day of 2008 is presented. In this case the portion of the corresponding wind generation varies between 22 % and 32 % of the total power supply that is considered as a significant high penetration for an autonomous system such as the Cretan one. Remarkable wind park installation activity started in 1992 and in 2000, the installed capacity had reached 67.35 MW, contributing 10 % of the annual demand, with both economic and environmental benefits for the operator of the power system [34]. Since 2000 the wind power penetration has been steadily above 10 % and by 2012, 183.54 MW had been installed providing 17 % of the annual demand; additional 55 MW have been authorized by the Regulatory Authority of Energy of Greece.

Even though minimum load demand is greater than current system technical minimum (approximately 100 MW) during the low consumption periods, there are periods, mainly in spring, that combination of wind power production and the technical minimum of the steam units may exceed the island demand. As a result, wind power is curtailed to maintain the secure operation of the steam units.

Taking into account the current wind parks and the prospect of many PV installations (102 MW authorization, 74 MW already installed) even now Cretan power system operators have to tackle with significant dispersed generation and high RES penetration. Thus autonomous power system of Crete has been an excellent representative for dynamic performance estimations, a conclusion drawn already since 1999 [29].

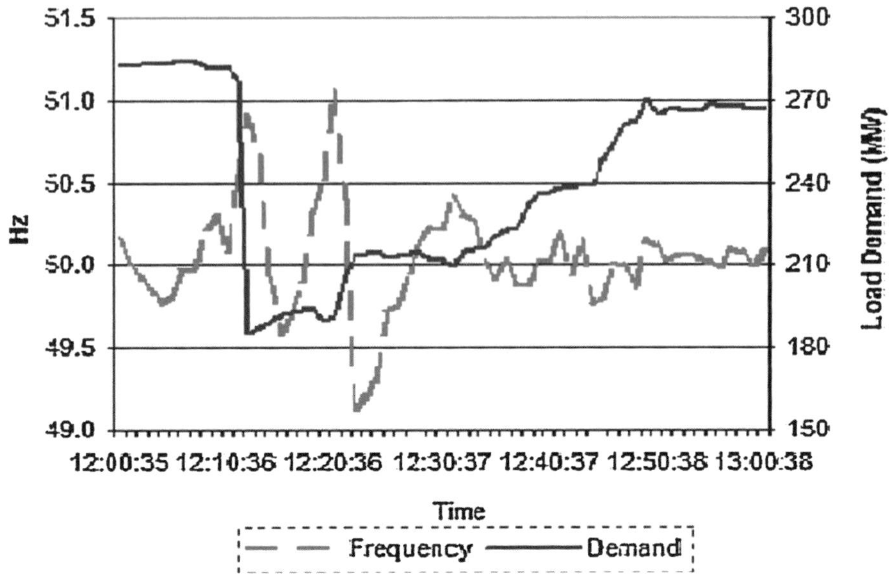

Fig. 7 Load shedding and Frequency fluctuations (Color figure online)

4.2 Power System Dynamic Performance

Many wind parks have been installed at the eastern part of the island, where the most favorable wind conditions exist. As a result, in case of faults on some particular lines, these wind parks are disconnected. Figure 7 depicts a real situation of a short-circuit occurrence (12:10:30) leading the frequency in spite of the fast load shedding (*blue line*) to vary between 49.1 and 51 Hz (*red line*). The frequency oscillations are the main characteristics of an autonomous power system like Crete's.

In order to investigate and assess the dynamic behavior of an examined power system, accurate simulations are needed (some information on Sect. 4.4). The models used for the presentation of the system components were chosen taking into account that the transient phenomena under study last between 0.1 and 10 s approximately.

4.3 Sudden Wind Power Variations

The aim of this analysis was to record any sudden variation of the wind speed and of the power production of the wind farm. In Figs. 8 and 9, two kinds of sudden variations, sudden loss of wind power and sudden increase of wind power respectively are provided as recorded by the SCADA system of the Transmission System Operator. This should be taken into account in selection of spinning reserves policy when considering increased penetration of wind power especially

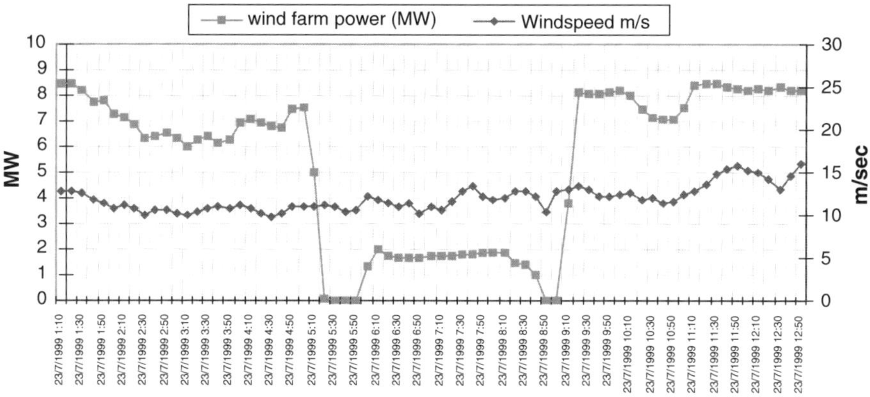

Fig. 8 Sudden power loss

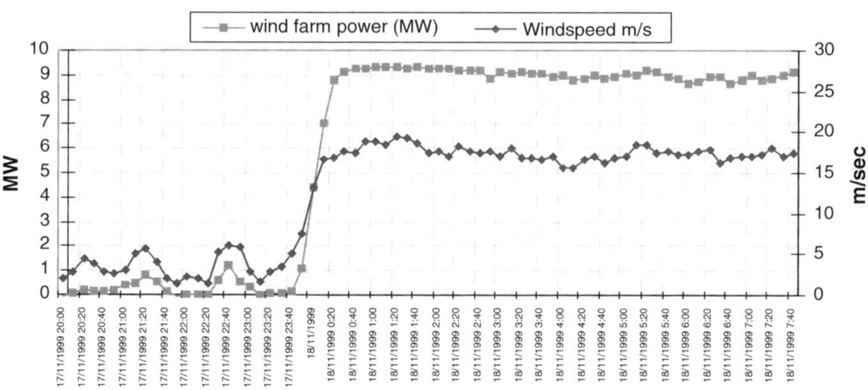

Fig. 9 Sudden increase of the produced wind power

in Autonomous power systems. Sudden loss of wind power may lead to frequency due to temporal lack of production. On the other hand, sudden wind power increase may lead to excess production which, under low loading may create problems in the slow response units to reduce their output.

The moving average of the wind farm power and the wind speed were calculated for short term (48 data points—every half an hour) and medium term (12 data points—every 2 h) and then compared. When a significant deviation between the short term and the medium term moving average of the power was recorded and caused by a deviation of the wind speed, a "sudden variation" has occurred.

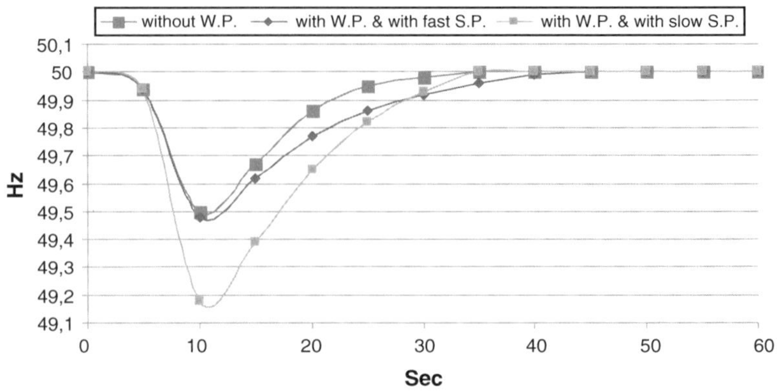

Fig. 10 Frequency deviation

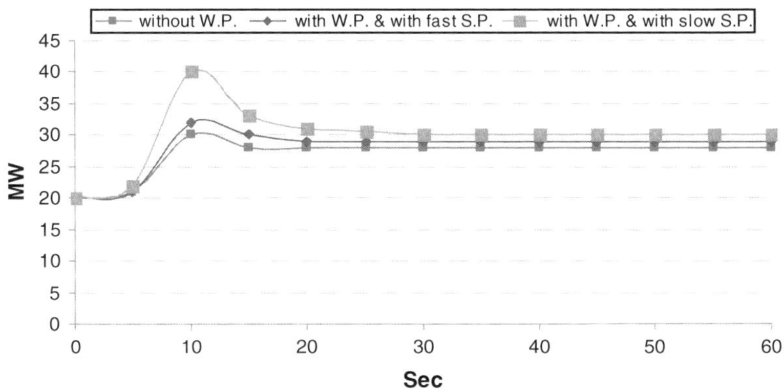

Fig. 11 Power generation change

4.4 Dynamic Security Assessment

EUROSTAG program [35], Power World Simulator [36] and Matlab [37] have been used for the simulation of the transient operation of the examined power system, under several operating conditions, using modeling like the ones described in [38, 39]. Conventional units trip, disconnection of wind generators as well as wind velocity fluctuations are the main disturbances under investigation. More precisely, the following cases are presented:

Conventional Generator Trip: The system was examined for a case of power unit disconnection (Gas Turbine), which was producing 20 MW. In Fig. 10 the changes in frequency for three different operating conditions are shown. Firstly, the system is considered to operate without wind turbines and it seems to be quite stable. Secondly, the system is considered to operate with 28 % of wind power, equal to 46 MW and with the fast conventional units such as diesel machines and gas

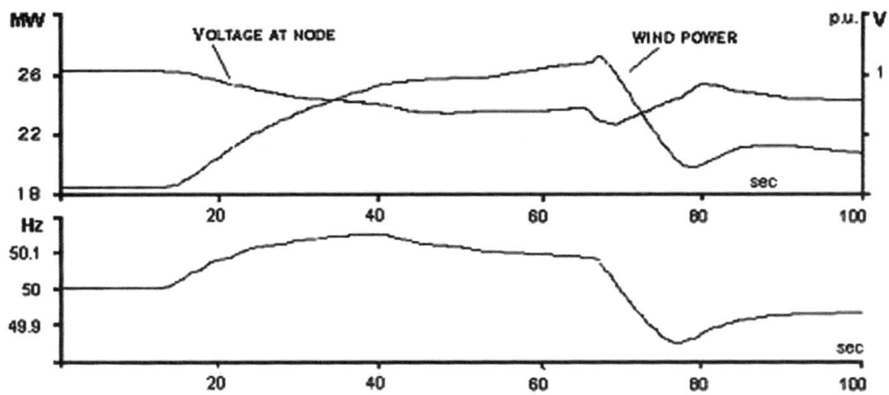

Fig. 12 Frequency and voltage variation

turbines to be in operation (fast spinning reserve). In this case, the system seems to be stable again. The lower value of the frequency is almost the same as in the previous case.

Correspondingly, in Fig. 11 the power output responses of a diesel unit in the previous three different operating conditions are presented.

Thirdly, the system is again considered to operate with the same high percent of wind power but with the slow machines, such as steam turbines, to cover the main spinning reserve (slow spinning reserve). In this case, the lower frequency value, which is equal to 49.14 Hz, surely causes the operation of wind parks protection devices, leading the system to collapse after the total wind power disconnection. Therefore, it is obvious that in case of large wind power penetration, the operation of the diesel machines and the gas turbines is necessary for the dynamic security of the system.

Wind Power Variability: In Fig. 12 the variation of the frequency and the voltage at the main wind park substation, are shown. The frequency follows the wind power changes, while the voltage profile follows an opposite trend. It can be seen that in case of normal wind power fluctuation, when the wind parks are not suddenly disconnected, and with sufficient spinning reserve, the power system remains satisfactorily stable.

Unit Commitment Generation Mix: A maximum wind power penetration of 30 % has been used by the system operators as the respective security margin. However, extensive transient analysis studies are conducted in order to assess the dynamic behavior of the system under various disturbances. Different combinations of the generating units have shown that a fixed security margin does not guarantee the system security and it distorts its economical operation. Thus, under the same contingency, the system is proven to collapse with wind power penetration lower than 30 %, while survives with higher penetrations.

Figure 13 depicts such an example providing the change of frequency caused by the outage of a Gas turbine, producing 23 MW for two different operating conditions. Case 1 corresponds to a total load of 207.2 MW supplied as described in the following Table 1.

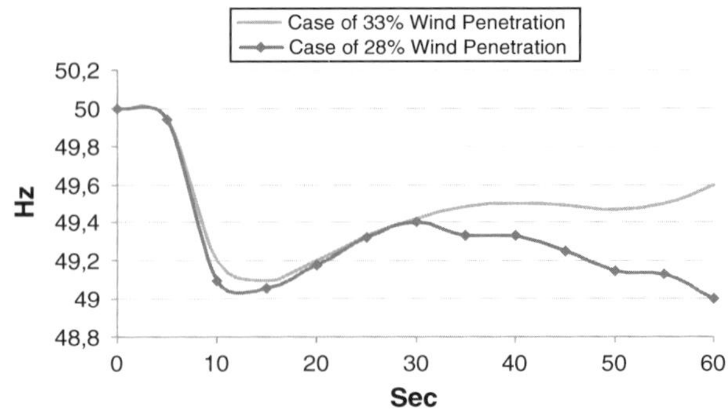

Fig. 13 Simulation results of Crete power system

Table 1 Case 1 units production and spinning reserve data

Production	Spinning reserve
27 MW by combined cycle	18 MW
56.8 MW by the new steam turbines	18.2 MW
21.3 MW by diesel	27.9 MW
10.1 MW by the remaining gas turbine	6.1 MW spinning reserve of max 16.2 MW

Table 2 Case 1 units production and spinning reserve data

Production	Spinning reserve (MW)
27.57 MW of combined cycle	17.43
69.3 MW of new steam turbines	5.7
23.4 MW of diesel	25.8

Wind power is 69 MW, corresponding to 33.3 % penetration. It can be seen that the frequency undergoes a severe transient reaching a lowest value of 49.1 Hz, however the system restores its balance in about 50 s. Case 2 corresponds to a lower load of 199 MW supplied as described in Table 2.

Wind power contributes 55.73 MW corresponding to 28 % penetration. Although the wind power penetration is lower than the security margin adopted the system does not manage to regain its stability and is finally led to frequency collapse. The difference is attributed to the fact that in the first case the spinning reserve is higher (70.2 MW) and provided by faster units (Gas Turbines), while in the second case by slower units (48.93 MW). The need for spinning reserve optimization can be clearly justified.

4.5 Preventive Dynamic Security

In this paragraph a method for on-line preventive dynamic security of isolated power systems is presented [39]. The method is based on decision trees (DTs) that provide the necessary computational speed for on-line performance and the flexibility of providing preventive control. Emphasis is placed on the on-line use of the method to test the dynamic security of each generation dispatch scenario and thus to provide corrective advice via generation re-dispatch. Moreover, the algorithm implemented provides the flexibility of displaying the cost of each re-dispatch. In this way, the method can help in objective decision-making. Results from the application of the system on actual load series from the island of Crete, where the proposed system is in trial operation, are presented.

A dispatch algorithm approximating actual operating practices followed in the Control system of Crete is applied next in order to complete the pre-disturbance Operating Points (OPs). For a given load demand P_L and wind power P_W, the total conventional generation P_C is equal to:

$$P_C = P_L + P_{Losses} - P_W \tag{1}$$

P_C is dispatched to the units in operation, depending on their type and their nominal power. The various thermal units are grouped according to their type. The attributes characterizing each Operating Point comprise the active power and spinning reserve of all conventional power units. Ten variables are selected as initial attributes. Five attributes correspond to the active production of the conventional unit groups and five attributes to the spinning reserves, respectively. For each of the Operating Points produced, two characteristic disturbances have been simulated using:

- Outage of a major gas turbine
- Three-phases short-circuit at a critical bus near the Wind Parks.

The first of these disturbances happens very frequently, while the second is particularly severe leading to the disconnection of most wind parks. For each Operating Point the maximum frequency deviation and the rate of change of frequency are recorded. Both of these parameters are checked against the values activating the under-frequency relays used for load shedding and the OPs are labeled accordingly. The security criteria were:

If fmin < 49 Hz **and** df/dt > 0.4 **then**
The system is insecure **else** is secure

Economic dispatch analysis determines the power setpoints of the online generating units (2), so as to meet the system load and losses at least cost.

$$P_C = P_1 + P_2 + \ldots + P_i + \ldots + P_n \tag{2}$$

where
P_C is the total conventional generation,
P_i is the generation of the ith unit
n is the number of units

Traditional dispatch algorithms tackle this problem as a constrained optimization problem and base its solution on the concept of equal incremental cost, also known as the Lambda Iteration algorithm: The total production cost of a set of generators is minimized, when all the units operate at the same incremental cost. In order to ensure that the operation set-points proposed by economic dispatch algorithm will provide a dynamically secure operation state of the system following pre-specified disturbances, the rules extracted by the relevant decision trees (if–then-else rules) can be used as additional constraints in the above optimization problem.

The presented approach provides the flexibility of displaying the cost of security, i.e. the cost associated with each re-dispatch. This is easily provided as the difference between the operating cost of the original dispatch and the operating cost of the secure re-dispatch. These costs can be calculated from the cost functions of the generating units, once the unit productions have been determined.

In addition, the security cost can be compared to the cost of load shedding. The unsupplied electric energy can be easily calculated from the operating settings of the under-frequency relays and the load forecasted at each bus affected. Alternatively, it can be estimated from the pre-disturbance load and the forecasted load as a whole however its cost is more difficult to determine. For the dispatcher the cost of load shedding can be the price the regulator imposes for energy not served. In the traditional monopoly operation this cost can be the revenue lost due to the unsupplied electric energy, although this by no means reflects the true cost of load shedding. In any case, the total cost can be calculated from:

$$S_L = C * \int_{t=0}^{T} P_L(t)dt \tag{3}$$

where:

P_L is the load shed

C is the cost in (€/kWh)

T is the duration of load disconnection

4.6 Cost of Security

In this paragraph results from the application of the secure economic dispatch algorithm on actual load series of Crete are presented. In Fig. 14, the total load, the corresponding security classification (1 for secure and 0 for insecure) for the machine outage contingency and the operating cost in Euros of a characteristic day are plotted. In the upper diagram, it is shown that, approximately between 9:00 and 10:30, the system is insecure, i.e. at least a significant load shedding will take place.

In the lower diagram, the effects of the secure economic dispatch algorithm on the security classification and the system operating costs are shown. The increase of costs during the previously insecure period, provided by the increased and probably faster (more expensive) spinning reserve, is notable (The total net cost of the security provided is 5,939.25 €, which amounts to 19.2 % of the total extra

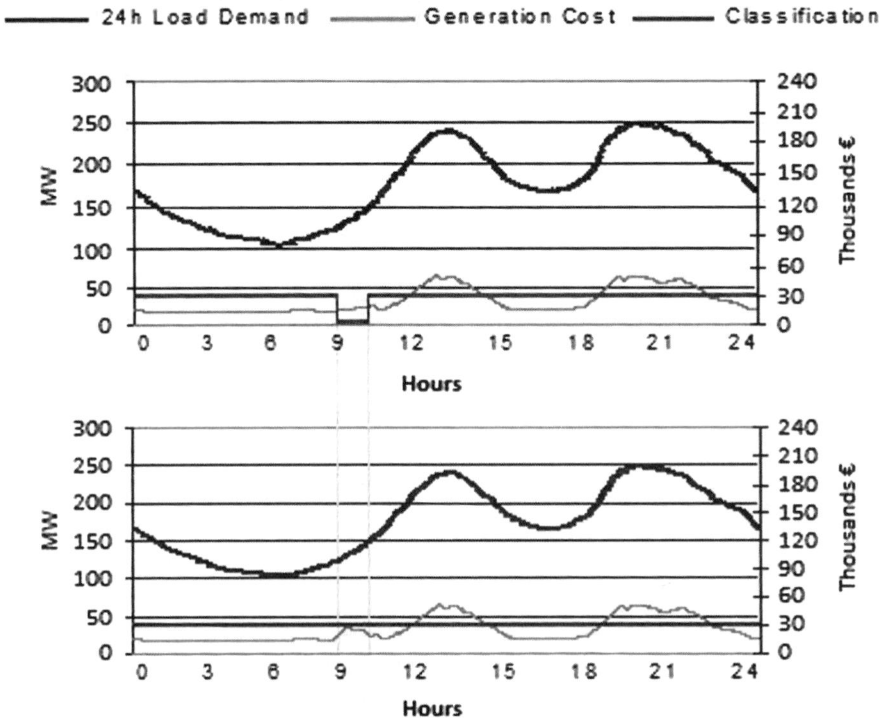

Fig. 14 Twenty four-hour diagrams illustrating load, security classification and operating cost

Fig. 15 System frequency deviation in case of first dispatch

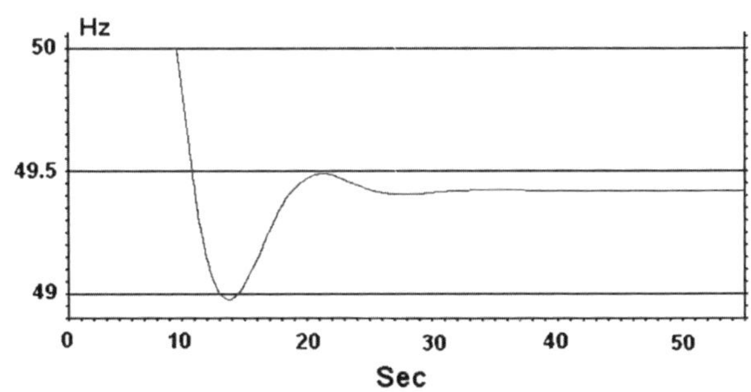

Fig. 16 System frequency deviation in case of second dispatch

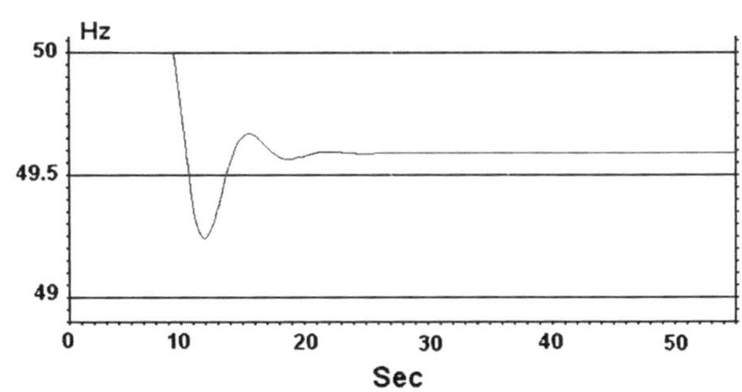

system operation cost of 30,935.41 €). The effect of the two dispatch scenarios on the system frequency deviation, in the case of the machine outage, as obtained by simulation programs, is shown in Figs. 15 and 16 [40]. It is clearly shown that the proposed re-dispatch will not cause load-shedding. The probability of the contingency occurrence however is not considered in this study.

5 Conclusion

Growing environmental concerns and attempts to reduce dependency on fossil fuel resources are bringing renewable energy resources to the mainstream of the electric power sector. Among the various renewable resources, wind power is considered to have the most favorable technical and economical prospects. When deployed in small scale, as was done traditionally, the impact of wind turbine generators on power system stability is minimal, if not positive. In contrast, as the penetration level increases, the dynamic performance of the power system can be affected.

Wind energy, being non-dispatchable, has different operational characteristics than conventional energy sources. Additionally, high levels of wind penetration create issues of power system control and interconnection issues. The objective of this chapter was to provide an approach for identifying the potential solutions for a power system with a high wind penetration levels. Another basic objective is to introduce the way that high wind penetration impacts on voltage dips, on transient frequency and on regulation/reserve requirements.

Review of several studies and their corresponding results indicate that the re-dispatch strategy has an important impact on the transfer margin of a system. The most severe contingencies in a system depend on the level of wind generation. Also, the maximum power transfer need not be at either minimum or maximum wind but could be at an intermediate wind level.

Furthermore, assessing the reactive power capability of a wind turbine is essential to analyze the effect of high penetration of wind parks on voltage response and stability. Generally wind generators have sensitive power electronic devices sensitive to high voltages and currents. Control enhancements should be developed to fully utilize the reactive power capability of the generator and the power electronic converters. These control enhancements include grid-side reactive power boost allowing the grid-side power converter to inject reactive power into the grid.

Finally, in this chapter the dynamic behavior of a power system with high percentage of wind power penetration (up to 40 %) was presented with emphasis given to the modeling of the system. More precisely, several simulations were performed to study the impact of the wind park on the dynamic behavior of a representative autonomous power system as Crete's power system. The most considerable disturbances that were investigated are the short circuit, the sudden disconnection of conventional power units as well as wind parks and the strong wind velocity fluctuations. Simulations have shown that the deviations of the power system voltage and frequency remain acceptable under most examined perturbations. However, the situation depends on the scheduling of the power units and the amount of allocated spinning reserve.

Significant replacement of conventional, based on synchronous generators, power generation, is expected with wind turbines that operate either asynchronous or variable-speed generators; hence, the dynamic performance of the power system will indeed be affected. Although wind turbines affect the transient stability of a power system, they should not be considered as a principal obstacle to an adequate secure and reliable operation. The stability of a power system can be maintained even if high penetration of wind power exist by additional system measures, control enhancement and preventive actions.

Finally, a method for on-line preventive dynamic security is proposed, in order to determine optimal reserves and to provide corrective advice considering dynamic security. Based on the Decision Trees classification new unit dispatch is calculated on-line, until a dynamically secure operating state is reached. This technique provides the flexibility of displaying the cost of each proposed solution weighted against the cost of load shedding; it forms therefore the basis for valuable decision-making aid. Results from the application of the method on actual load series from the island of Crete show the accuracy and versatility of the method. Moreover, the fast execution times required for on-line classification of the current operating state make the method suitable for large systems, as well.

Therefore, there is considerable impact of wind parks generation to the power system they are embedded to. This impact is generally proportional to the wind power penetration percentage (running active power injection and/or reactive power absorbedness). Furthermore most of the perturbations exclusively due to the operation of the wind generators do not affect significantly the operation of the power system. Finally, it should be noted that it is possible to operate a power system with a high level of wind penetration maintaining a high level of security. This is possible, if adequate spinning reserve of the conventional units is available. The issue of spinning reserve is particularly important; therefore it should be further investigated for the power system before applying specific numbers for penetration level.

References

1. Kundur P, Morison GK (1997) A review of definitions and classification of stability problems in today's power systems. Panel session on stability terms and definitions IEEE PES meeting, New York, 2–6 February
2. Cutsem T Van, Vournas C (1998) Voltage stability of electric power systems, Power electronics and power system series. Springer, New York, ISBN 978-0-7923-8139-6
3. Hatziargyriou N, Papadopoulos M (1997) Consequences of high wind power penetration in large autonomous power systems. CIGRE Symposium, Neptun, Romania, 18–19 September

4. Papathanasiou S, Boulaxis N (2006) Power limitations and energy yield calculation for wind farms operating in island systems. Renewable Energy 31(4):457–479 Elsevier
5. Hatziargyriou N., Contaxis G, Matos M, Pecas Lopes JA, Kariniotakis G, Mayer D, Halliday J, Dutton G, Dokopoulos P, Bakirtzis A, Stefanakis J, Gigantidou A, O'Donnel P, McCoy D, Fernandes MJ, Cotrim JMS, Figueira AP (2002) Energy Management and Control of Island Power Systems with Increased Penetration from Renewable Sources, Presented at the IEEE Power Engineering Society Winter Meeting
6. Kundur P, Paserba J, Ajjarapu V, Andersson G, Bose A, Canizares C, Hatziargyriou N, Hill D, Stankovic A, Taylor C, Van Cutsem T, Vittal V (2004) Definition and classification of power system stability. IEEE Trans Power Syst 19(2):1387–1401
7. Dialynas EN, Hatziargyriou ND, Koskolos NC, Karapidakis ES (1998) Effect of high wind power penetration on the reliability and security of isolated power systems. CIGRE Session, Paris, 30 August 1998
8. Hatziargyriou N, Karapidakis E, Hatzifotis D (1998) Frequency stability of power systems in large islands with high wind power penetration. Bulk Power Systems Dynamics and Control Symposium–IV Restructuring, Santorini, 24–28 August
9. Arrilaga J, Arnold CP (1993) Computer modeling of electrical power systems. Wiley, New York
10. Nogaret E, Stavrakakis G, Kariniotakis G (1997) An advanced control system for the optimal operation and management of medium size power systems with a large penetration form renewable power sources. Renewable Energy 12(2):137–149 Elsevier Science
11. La Scala M, Trovato M, Antonelli C (1998) On-line dynamic preventive control: An algorithm for transient security dispatch. IEEE Trans PWRS 13(2):601–610
12. Steinmetz CP (1920) Power control and stability of electric generating stations. AIEE Transactions, vol. XXXIX, Part II, pp 1215–1287
13. AIEE, Subcommittee on Interconnections and Stability Factors (1926) First report of power system stability. AIEE Transactions pp 51–80
14. Estanqueiro AI, de Jesus JMF, Ricardo J, dos Santos A, Lopes JAP (2007) Barriers (and Solutions...) to Very High Wind Penetration in Power Systems. In: Proceeding of the IEEE Power Engineering Society General Meeting, 24–28 June, pp 2103–2109
15. Ackermann T (2005) Wind power in power systems. Wiley, Stockholm (Royal Institute of Technology)
16. Georgilakis PS (2008) Technical challenges associated with the integration of wind power into power systems. Renew Sustain Energy Rev 12(3):852–863
17. Sucena Paiva JP,., Ferreira de Jesus JM, Castro R, Correia P, Ricardo J, Rodrigues AR, Moreira J, Nunes B (2005) Transient stability study of the Portuguese transmission network with a high share of wind power. XI ERIAC CIGRÉ, Paraguay, May 2005
18. Tsikalakis AG, Hatziargyriou ND, Katsigiannis YA, Georgilakis PS (2009) Impact of wind power forecasting error bias on the economic operation of autonomous power systems. J Wind Energy 12(4):315–331
19. Holttinen H (2004) The impact of large scale wind power production on the nordic electricity system. PhD Thesis dissertation, VTT Publications 554. Espoo, VTT Processes
20. Landberg L, Giebel G, Nielsen HA, Nielsen TS, Madsen H (2003) Short-term prediction–An overview. Wind Energy 6(3):273–280
21. Pinson P, Chevallier C, Kariniotakis G (2007) Trading wind generation with short-term probabilistic forecasts of wind power. IEEE Trans Power Syst 22(3):1148–1156
22. IEA Report (2005) Variability of wind power and other renewables: management options and strategies. IEA Publication, Paris
23. IEC 61400-21 (2001) Wind turbine generator systems–Part 21: Measurement and assessment of power quality characteristics of grid connected wind turbines. IEC Standard
24. De Almeida RG, Castronuovo ED, Lopes JAP (2006) Optimum Generation Control in Wind Parks When Carrying Out System Operator Requests. IEEE Trans Power Syst 21(2):718–725
25. Kabouris J, Vournas CD (2004) Application of interruptible contracts to increase wind-power penetration in congested areas. IEEE Trans Power Syst 19(3):1642–1649

26. Kabouris J, Hatziargryriou N (2006) Wind power in Greece–Current situation, future developments and prospects. IEEE Power Engineering Society General Meeting. Montreal, Canada
27. Lipman NH, Bossanyi EA, Dunn PD, Musgrove PJ, Whittle GE, Maclean C (1980) Fluctuations in the output from wind turbine clusters. Wind Eng 4(1):1–7
28. Soerensen P, Hansen AD, Rosas PAC (2002) Wind models for prediction of power fluctuations of wind farms. J Wind Eng Ind Aerodyn 90:1381–1402
29. Stefanakis J (1999) CRETE: An ideal case study for increased wind power penetration in medium sized autonomous power systems. Presented at the PES/IEEE Winter Meeting 1999
30. Smith P, O'Malley M, Mullane A, Bryans L, Nedic DP, Bell K, Meibom P, Barth R, Hasche B, Brand H, Swider DJ, Burges K, Nabe C (2006) Technical and economic impact of high penetration of renewables in an Island power system. CIGRE Session 2006, Paper C6–102
31. Kaldellis JK (2008) The wind potential impact on the maximum wind energy penetration in autonomous electrical grids. Renew Energy 33(7):1665–1677
32. Doherty R, O'Malley MJ (2006) Establishing the role that wind generation may have in future generation portfolios. IEEE Trans Power Syst 21(2006):1415–1422
33. Karapidakis ES, Thalassinakis M (2006) Analysis of wind energy effects in Crete's Island power system. In: 6th International world energy system conference. Turin, Italy, July 2006
34. Tsikalakis AG, Hatziargyriou ND, Papadogiannis K, Gigantidou A, Stefanakis J, Thalassinakis E (2003) Financial contribution of wind power on the island system of Crete. In: Proceeding of RES for Islands conference, Crete. pp 21–31
35. Meyer B, Stubbe M (1992). EUROSTAG: A single tool for power system simulation, transmission and distribution international, March 1992
36. Power World (2007) Power world user's guide. Power World Corporation, Simulator Version 13, 2001 South First Street Champaign, IL 61820
37. Power System Toolbox (2006) User's Guide, MATLAB 7 Package
38. Kazachkov YA, Feltes JW, Zavadil R (2003) Modeling wind farms for power system stability studies. IEEE Power engineering society general meeting, 13–17 July, Toronto, Canada
39. Slootweg JG, Kling WL (2004) Modelling wind turbines for power system dynamics simulations: An overview. Wind Eng 28(1)
40. Karapidakis ES, Hatziargyriou ND (2002) On-line preventive dynamic security of isolated power systems using decision trees. IEEE Trans Power Syst 17(2):297–304

Power System Operation with Large Penetrations of Wind Power

Eleanor Denny

Abstract The characteristics of wind power result in unique challenges for system operators when integrating large penetrations of wind generation into power systems. This chapter discusses some of the power system operational challenges associated with large penetrations of wind generation, such as increased reserve requirements and the costs associated with increases in the variable operation of conventional generators. A number of power system optimization techniques with wind generation are discussed, namely the fuelsaver approach, deterministic optimization, rolling commitment and stochastic optimization. Also, a discussion of certain flexibility solutions which may reduce the system costs of increased wind penetration levels is provided.

1 Introduction

Energy as a topic has recently been catapulted to the top of the global agenda. Environmental concerns, competitive pressures and security of supply make energy a multifaceted global issue and one of the greatest challenges facing mankind. The times of cheap abundant energy are rapidly coming to an end and the environmental impacts of decades of growth in energy consumption are becoming startlingly apparent. A secure, sustainable energy supply is of strategic importance to the well-being and health of a modern industrial society and requires significant innovation and research effort.

The electricity industry is heavily fossil fuel dependent, with 82 % of electricity produced by fossil fuels in 2008 [1]. This leaves the industry highly exposed to future price increases as global fossil fuel resources become more depleted. Also,

E. Denny (✉)
Department of Economics, Trinity College, Dublin 2, Dublin, Ireland
e-mail: dennye@tcd.ie

P. M. Pardalos et al. (eds.), *Handbook of Wind Power Systems*, Energy Systems,
DOI: 10.1007/978-3-642-41080-2_11,

the increasing concentration of the remaining resources in certain geographical areas may make fossil fuel prices more volatile.

Worldwide, the generation of electricity relies heavily on coal, the most carbon intensive of fuels, amplifying the sector's share in global carbon dioxide emissions. In 2007, 42 % of electricity was generated through the combustion of coal worldwide and countries such as Australia, China, India and South Africa produced between 68 and 95 % of their electricity and heat through coal [1]. As such the electricity industry is a major contributor of greenhouse gas emissions with the electricity and heat sector responsible for 41 % of worldwide carbon dioxide emissions in 2007 [2].

In a bid to reduce their reliance on fossil fuels, while simultaneously reducing electricity sector emissions, there has been a move by policy makers towards the promotion of clean renewable technologies for electricity generation. For example, EU leaders have committed to producing 20 % of final energy consumption from renewable sources by 2020, encompassing a 30 % renewable generation target for the electricity sector [3].

As one of the most advanced forms of renewable generation, much emphasis has been on wind, which has grown from 9,660 MW installed worldwide in 1998 to over 120,800 MW in 2008 [4]. As the market for wind energy has grown, the costs per kilowatt hour have reduced dramatically. Ambitious renewable energy targets together with reducing costs and successes to date, will ensure that wind energy continues to grow in electricity networks worldwide.

Wind generation, like solar, tidal and wave generation, exhibits 'variable' output. The output of these units depend upon weather conditions that cannot be controlled by the operator of the generator. This is known as being 'non-dispatchable'. For example, the amount of electricity generated by a wind turbine fluctuates as wind speed changes and that of a photovoltaic array with the intensity of sunlight. Thus, the control of their output is limited as operators can only reduce the potential output of such generators [5]. When significant penetrations of these forms of generation are connected to an electricity network, it can result in a requirement to alter the operation of the system to accommodate the variability of these generators.

As well as being variable, wind generation also faces a challenge of being relatively unpredictable. Since the underlying resource cannot be directly controlled, the renewable generation is high when conditions are favourable and low when unfavourable. Thus, forecasts of weather conditions are crucial when examining renewable generation sources. Unlike tidal power which can be predicted almost perfectly over long time horizons, wind generation, requires complex forecasting techniques which account for wind speed, wind direction, hub height, geographical surroundings, wind farm size, turbine dispersion, etc. Given the large number of factors which must be taken into account when forecasting wind generation, the margin for error can be significant and increases as the time horizon lengthens [6].

The function of power system operators is to supply electricity to customers in a reliable manner at a sustainable cost. Reliability of electricity supply is defined as "the ability to supply adequate electric service on a nearly continuous basis with few interruptions over an extended period of time" [7]. This involves ensuring that the generation meets the load at all times and that any short term gaps between load and generation are bridged quickly and precisely to maintain the integrity of the power system [8]. Generators are scheduled to meet the forecasted load and must alter their operating levels to follow the load as it fluctuates throughout the day. Since the output of some renewable generators, in particular wind generation, cannot be actively controlled and is difficult to predict accurately, this balancing of the generation and the load can become more challenging as the wind penetration levels increase [5].

The characteristics of wind power result in unique challenges for system operators when integrating wind generation into power systems, particularly when penetrations of wind generation become significantly large. It has thus been recognized that methodologies and tools must be developed in order to investigate the impact of large scale wind energy on power system operation. For example, Annex 25 of the International Energy Agency Research and Development Wind Agreement "Design and Operation of Power Systems with Large Amounts of Wind Power" was initiated in 2006 with the objective of providing information to facilitate the highest economically feasible wind energy penetration within electricity power systems worldwide [9]. Many other studies, such as [5, 10, 11] and [12], have highlighted the challenges associated with increased wind penetrations and [13] is a useful resource providing a comprehensive literature review of the salient studies conducted in North America.

Section 2 will discuss in further detail some of the power system operation challenges associated with increasing penetrations of wind power. Section 3 will discuss scheduling techniques of system operators for unit commitment with wind generation. Section 4 will discuss some power system characteristics which would facilitate increased penetrations of wind power.

Throughout this article, reference is made to the power system of Ireland as a case study. Located on the edge of the Atlantic Ocean, Ireland has a vast wind resource potential as evidenced in Fig. 1. In 2008, wind generation represented 11.1 % of electricity generation and in 2010 this is expected to exceed 15 % [14].[1] In fact, the island of Ireland as a single synchronous power system has arguably the largest penetrations of wind power in the world [15] and the Irish Government has recently set a target of 40 % of electricity from renewables by 2020, the majority of which is expected to come from wind generation [16].

The Irish power system is a small island system with no synchronous interconnection with other systems. As such it is a good case study system as it allows issues surrounding power system operation with large penetrations of wind generation to be evaluated in isolation and without being influenced by the operation of other systems. It also implies that any challenges associated with balancing

[1] On 22nd November 2009 Ireland achieved a world record of over 45 % of electricity demand produced from wind energy during the hours of 4 and 6 am while the installed wind capacity represented just 10 % of installed generation.

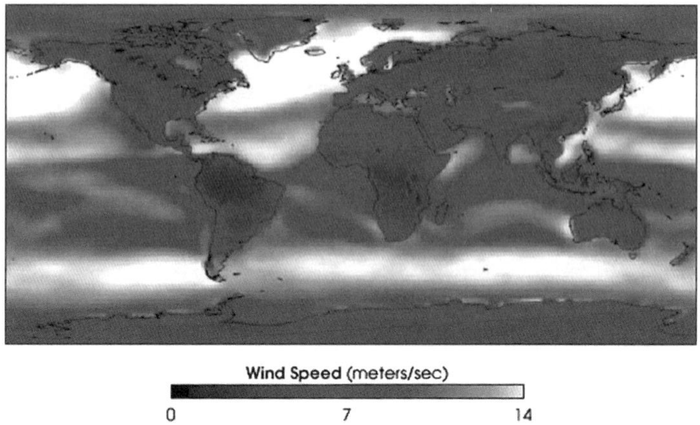

Fig. 1 Average global wind speeds [17]

load, wind output and conventional generation must be dealt with locally. As such the Irish system can be considered as an indicator of the challenges that other larger power systems may experience with large wind penetrations in the future.

2 Power System Operation Challenges with Large Penetrations of Wind

The interaction of wind energy and the electricity system is complex and there are significant challenges posed to system operators when large amounts of variable generation are introduced [11, 18]. This section discusses some of the main system operational challenges associated with increased penetrations of wind.

2.1 Reserve Provision

Under the European Union Directive 96/92/EC [19] the Transmission System Operator is responsible for "ensuring a secure, reliable and efficient electricity system and, in that context, for ensuring the availability of all necessary ancillary services". These ancillary services include the provision of sufficient reserve capacity to meet the load under unexpected system operating conditions [20]. As shown in [21] these unexpected system conditions can include the loss of a unit, a transmission line trip and unexpected load fluctuations. It was shown in [22] that an increase in the capacity of wind generation on an electricity system increases the uncertainty in the system, as wind generation is relatively unpredictable and non-dispatchable, which results in a requirement to carry additional reserve capacity in order to maintain system security.

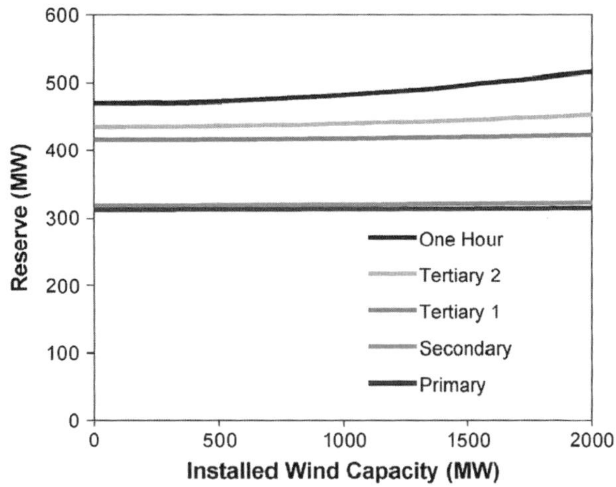

Fig. 2 Conventional reserve categories versus installed wind capacity for Ireland [27]

Depending on the electricity system, there are a number of reserve categories which a system operator must carry—from fast acting reserve, with a reaction time in the order of seconds, to reserve which responds more slowly [23, 24]. It has been shown in [25] and [26], that the standard deviation of wind power forecast errors over short time frames are small. As such [27], found that increasing wind power capacity has little effect on the reserve categories that operate over a short time frame (seconds to minutes), however, wind capacity causes a greater increase in the need for categories of reserve that act over longer periods of time. For example, Fig. 2 illustrates the different reserve categories required for the case study power system of Ireland with increasing wind generation [27]. In Fig. 2 the categories are defined as follows: 'One hour reserve' refers to reserve responding in the time frame of 20 min to 1 h; 'Tertiary 2' reserve responds in 5–20 min; 'Tertiary 1' in 90 s to 5 min; 'Secondary' in 15–90 s; and 'Primary' in 5–15 s.

In [28, 29] the reserve costs attributable to the stochastic nature of the wind energy output for a number of different US utilities were quantified. It was found that in many cases the costs of providing these services was in fact relatively small. The costs of additional reserve in the Pennsylvania, New Jersey and Maryland (PJM) market was of the order of \$0.05–0.30/MWh. Also [18] found that the additional reserve costs represented less than 4 % of the total additional system costs imposed by wind generation. A comprehensive summary of similar studies conducted for electricity utilities around the United States is provided in [12].

2.2 *Variable Operation of Conventional Power Stations*

In addition to the added reserve requirement with increases in wind generation, an increase in variable generation on an electricity system may require the system operator to alter how conventional generation is dispatched [30]. Conventional generation may be obliged to operate at lower operating levels in order to be available to ramp up to accommodate the inherent variability of the wind generation [5]. There may also be an increase in the number of start-ups and shut-downs of conventional units as system operators attempt to coordinate the following of the fluctuating load throughout the day and the variable output of the wind generation [31, 32].

Thermal units are designed to be at their most efficient when online and running at a stable load [33]. It has been shown in [34], that in general, units are optimized for continuous rather than cyclical operation and when operating in their normal range can operate for relatively long periods with relatively low risk of failure and loss of equipment life. When a generating unit is required to vary its output to meet the demand and balance the wind output, the components in the unit are subject to stresses and strains. This is known as cycling and includes ramping up and down and switching on and off.

When a unit is turned off and on, the boiler, steam lines, turbine and auxiliary components undergo large temperature and pressure stresses which result in damage. This damage accumulates over time and eventually leads to accelerated component failures, forced outages and a shortening of the life span of the unit [35]. It is estimated that one on–off cycle for a single unit can cost as much as $500,000 [36] when these additional wear and tear damages are included, thus an increase in cycling associated with increased wind generation could result in significant cost increases.

This wear and tear on the components of the generating units is exacerbated by a phenomenon known as creep-fatigue interaction. Creep is the change in the size or shape of a material due to constant stress on the material over time. This is likely in units which are operated at continuous output over long periods of time, such as baseloaded units. Creep stems from continuous exposure to elevated temperatures and high pressures [36]. Fatigue occurs when a material is exposed to fluctuating stresses resulting in fractures and failures. Fatigue is likely during cyclical operation when the materials experience large transients in both pressures and temperatures [35].

Older units which were designed and used for baseloaded operation over a number of years and are then forced to cycle on a regular basis are very susceptible to component failure through creep and fatigue damage interaction. This depletion of the life expectancy of conventional units is a serious issue when analysing wind generation penetration as it is likely that wind generation will alter the merit order and some units will switch from being baseloaded to being required to be more flexible. An analysis of the impact of wind generation on baseload unit cycling was conducted for a case study of the Irish power system by [32] and their results show that baseloaded combined cycle gas turbines (CCGT) are significantly adversely affected by increasing penetrations of wind generation as illustrated in Fig. 3.

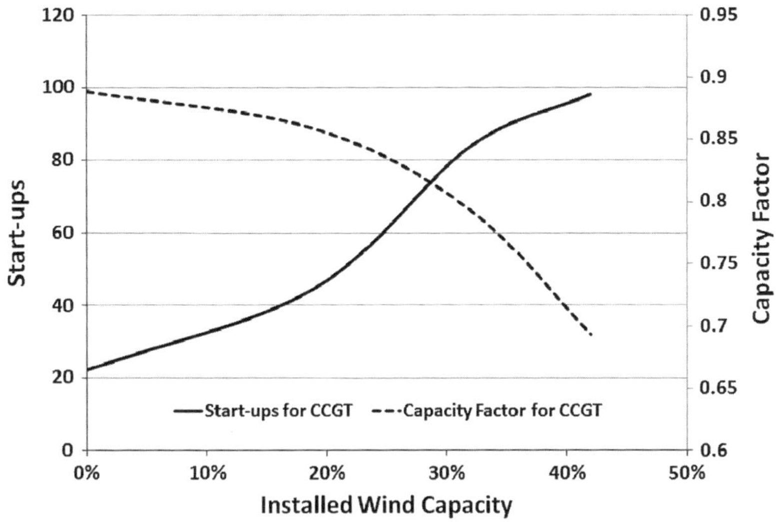

Fig. 3 Number of annual start-ups and capacity factor for an average 400 MW CCGT unit with increasing wind penetration [32]

2.2.1 Wind and Emissions

A further issue relating to the variable operation of conventional power stations is the impact of cycling on emissions. While wind generation can result in a reduction in the output of conventional units on the system, it is not necessarily the case that a reduction in output will result in a reduction in emissions [37].

Carbon dioxide (CO_2) is generated by the combustion of fuels containing carbon. The amount of carbon dioxide released is in direct relation with the amount of carbon in the fuel and the quantity of fuel burnt [38]. Thus a generation plant which burns a carbon intensive fuel will generate more carbon dioxide at increased levels of operation. Thus, if wind generation reduces the operating level of a carbon intensive unit, CO_2 emissions should fall. However, it is important to note that conventional power stations consume large levels of fuel when starting up. Thus, if wind generation results in an increase in start-ups, emissions of CO_2 could be adversely affected. In fact [31] showed that the carbon cost of a start-up can actually exceed the fuel cost of a start-up at a price of €30/ton CO_2.

Sulphur Dioxide (SO_2) is produced in a similar way to carbon dioxide and the emissions produced depend on the sulphur content of the fuel and the quantity of fuel burnt. It should be noted that natural gas contains a negligible amount of sulphur, thus emissions of SO_2 are not significant for gas turbines [39].

The production of Nitrogen Oxide (NO_X) differs from CO_2 and SO_2 production and does not depend solely on the nitrogen content of the fuel. It is also significantly affected by the flame temperature, the oxygen concentration and the residence time in the combustion chamber. The formation of NO_X can be attributed to four distinct chemical kinetic processes: thermal NO_X formation, prompt NO_X

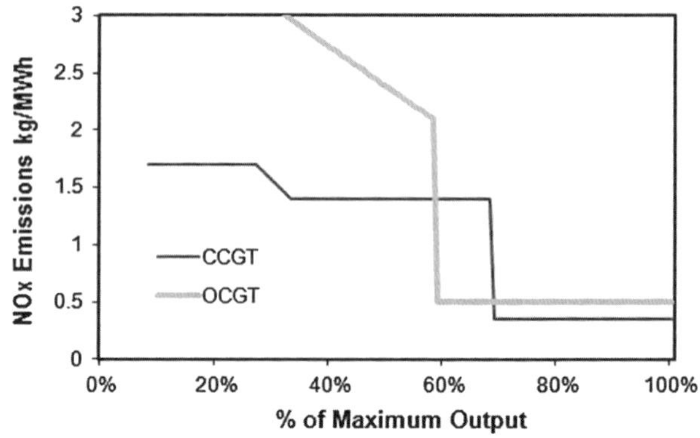

Fig. 4 Typical NO_X emissions from a CCGT and an OCGT [37]

formation, fuel NO_X formation and reburning. Thermal NO_X is formed by the oxidation of atmospheric nitrogen present in the combustion air. Prompt NO_X is produced by high-speed reactions at the flame front, and fuel NO_X is produced by oxidation of nitrogen contained in the fuel. The reburning mechanism reduces the total NO_X formation by accounting for the reaction of NO with hydrocarbons [40]. Fuel NO_X is the major source of NO_X emissions from the combustion of nitrogen bearing fuels such as heavy oils, coal and peat [41].

Thermal NO_X is the predominant source of NO_X emissions from a gas turbines. In order to achieve reduced emissions, gas turbine manufacturers have adopted lean premixed combustion as a standard technique. This premix (of fuel and air) achieves low levels of pollutant emissions without the need for additional hardware for steam injection or selective catalytic reduction [42]. Lean premixed combustion is limited by the presence of combustion instabilities which can produce turbine damage, flame instability and even flame extinction [43]. For this reason the fuel and air premix is not possible during startup and at reduced load levels (below about 65–70 % of maximum capacity). As a result, NO_X emissions in a CCGT increase significantly at lower loads. Figure 4 shows the NO_X characteristics of a CCGT and an open cycle gas turbine (OCGT).

From Fig. 4 it is clear that if a CCGT is forced to operate below approximately 70 % of its maximum rated capacity, its NO_X emissions will increase threefold. For an OCGT, operation below 60 %, will result in NO_X emissions increasing by up to six times. The NO_X characteristics of these units are important when examining the impact of wind generation on emissions. As discussed previously, wind generation may result in the system operator requiring more units to operate at lower efficiencies due to the increased reserve requirement. In addition, wind generation causes an increase in the cycling of marginal units. Thus, an increase in wind generation may result in more CCGT units operating at lower loads (as seen in Fig. 3), and as a result increasing their NO_X emissions. In other words, although wind generation itself does not produce any harmful emissions, it may in fact result in an increase in NO_X emissions through the reduced operation of gas fired units. This has been shown to be the case under certain system operation strategies, such as the fuelsaver strategy, discussed further in Sect. 3 [37].

2.3 Power System Stability and Other Issues

Related to the issue of reserve provision is the ability of wind generators to stay connected to the system after a fault i.e. fault ride-through capabilities. Currently many wind turbines have protection equipment installed which disconnects the turbine following a fault [44]. This sympathetic tripping can cause serious problems for system security, increasingly so with larger numbers of wind generators connecting to the system [45]. This is especially true for small, relatively isolated systems. Essentially, this sympathetic tripping aggravates a fault event resulting in increasing pressure on reserves and reliability standards.

In periods of low demand and high wind speeds (e.g. during windy nights) a significant contribution of wind power can be reached even when the overall share of wind power in the electricity supply is still modest. When the wind power penetration level is considerable it will no longer be feasible for wind turbines to disconnect during voltage or frequency disturbances, as this would lead to a large generation deficit and could lead to a reduction in the stability of the system [46, 47]. This issue has led to a review of many grid codes across Europe and now most include a necessity for wind turbines to have fault ride through capabilities [48, 49].

The most ideal sites for wind generation often coincide with remote locations which are frequently in weak areas of the electricity grid. This can result in connection challenges for wind generation with issues such as voltage control, harmonic emissions, short circuit levels and losses playing a significant role [50]. This has led to the possibility of distribution networks switching to 'active' rather than 'passive' systems through the installation of complex control systems and by equipping transformers with tap changers that can control voltage locally [51]. Rural locations for wind farms also cause problems for developers. For cost minimisation the location of plant is very important and there is often a trade-off between the windiest sites and easy access to the distribution network. There are large economies of scale associated with wind farm development which often cannot be realized in remote areas due to limitations on access.

Network congestion can also be an issue with high levels of wind power. An increase in wind generation, like any conventional generation, will increase the flows on the lines which may have a knock on effect on congestion [52]. Congestion limits the system operator's ability to make the most economic decisions in planning, sourcing and supplying electricity to customers and can result in increased costs. However, as stated by [53], 'no general rules apply to transmission congestion economics, which can only be studied on a case-by-case basis'. This is because congestion is highly system specific and depends on the location of the generators and loads on the network. Thus, without modelling the exact location of generators and the network characteristics, it is difficult to quantify congestion costs in a generic manner. Network reinforcement with increasing penetrations of wind generation can help ameliorates potential bottlenecks and congestion, and reduce costs.

This is not an exhaustive summary of the impacts of wind generation on power system operation but rather an illustration of some of the more direct impacts. Depending on the scope of an analysis, the external costs of wind generation could be much further reaching. It should also be noted that wind generation also provides significant benefits to the system such as a reduction in harmful emissions, fuel savings, capacity benefits, a hedge against imported fuel prices and supply concerns etc. However, for the purposes of this article we will concentrate on the system operational challenges and the following Sect. 3 will discuss some operational scheduling techniques used to minimize some of these costs and Sect. 4 will discuss system characteristics that can also reduce the system costs of wind.

3 Scheduling Tools

This section discusses some techniques for determining conventional generation schedules in the presence of large penetrations of wind generation. It begins with a simplistic approach which may be feasible when wind penetrations are low but is sub-optimal with larger wind penetrations. The second and third approaches are more sophisticated and result in lower system costs than the first approach.

3.1 Fuelsaver Approach to Unit Commitment

A simple approach to operating a power system with wind generation, known as the fuelsaver approach, has been presented in [54, 55]. Under this approach, wind generation is not considered in the scheduling of the plants and the unit commitment decisions are made ignoring any installed wind capacity. Once the commitment decision has been made, the wind generation is considered. If wind generation is available it is used and marginal conventional plants which were dispatched are deloaded to accommodate the wind generation. A conventional plant can be deloaded as far as its minimum but no plants are switched off. If wind production reaches a level such that no more conventional generation can be deloaded, then any further wind production is curtailed.

This operational strategy considers that the only benefit of wind generation is a fuel-saving one and it assumes that wind generation has a capacity value of zero. This is a simplistic approach and it allows issues of forecasting and reliability of wind production to be ignored.

The impact of operating the power system using a fuelsaver approach is discussed in [37] and it is shown that it is a highly inefficient way to operate the system with significant wind power penetration. The fuelsaver method of power system operation with wind generation results in an over commitment of conventional units, and these units will be running at low efficiencies. This has a knock on impact on the emissions and fuel savings of wind generation. In fact, under the fuelsaver approach, NO_X emissions can actually be greater than in the absence of any wind generation [37]. In addition, the fuelsaver approach results in large amounts of wind energy curtailment, of up to 30 % of the annual output at high levels of installed wind power [37]. Except in the case of very small penetrations of wind generation, operating the system in a fuelsaver manner is not recommended.

3.2 Deterministic Unit Commitment with Wind Generation

An improvement to the fuelsaver approach is to account for wind power forecasts in the unit commitment decision. One common software package which uses a deterministic approach to wind power is PLEXOS [56]. This is a unit commitment package which allows the user to model the plant on the system using various attributes such as the fuel use, startup time and costs, availability of reserve, etc. In the PLEXOS optimization model, the forecasted wind on the system is treated as a negative load, and is subtracted from the load. PLEXOS then uses either a mixed integer solver or rounded relaxation method to solve the unit commitment problem. However, the main limitation of this type of deterministic approach is that the wind forecast is considered as perfect and the stochastic nature of forecast errors are not taken into account and a surplus level of reserve capacity must be carried.

3.2.1 Rolling Unit Commitment

Models such as PLEXOS can be improved by conducting what is known as 'rolling planning', whereby the unit commitment is carried out more frequently, for example every 6 h instead of every 24 h [57]. For instance, 6-hour rolling commitment is carried out as follows: the first commitment is carried out, then the system is 'rolled' forward 6 h and the relevant details (wind forecasts, load forecasts, plant availability, and current state of system) are updated, before the system is once again committed. Figure 5 shows this in the case of the units being committed every 24 h and every 6 h over a 48 h period.

This 'rolling planning' approach means wind and load forecasts are being updated with greater frequency and more of the uncertainty of wind is captured in the model. This means more of the costs due to uncertainty will be minimized, leading to improved results and better performing schedules. However, while this 'rolling planning' approach is an improvement on the perfect forecast approach it still does not adequately account for the stochastic element in the errors of wind power forecasting. Thus, while a deterministic approach such as this may be utilized in systems with relatively low levels of installed wind generation, the omission of the stochastic element will cause more problems as wind penetrations grow.

Fig. 5 Example of rolling unit commitment [57]

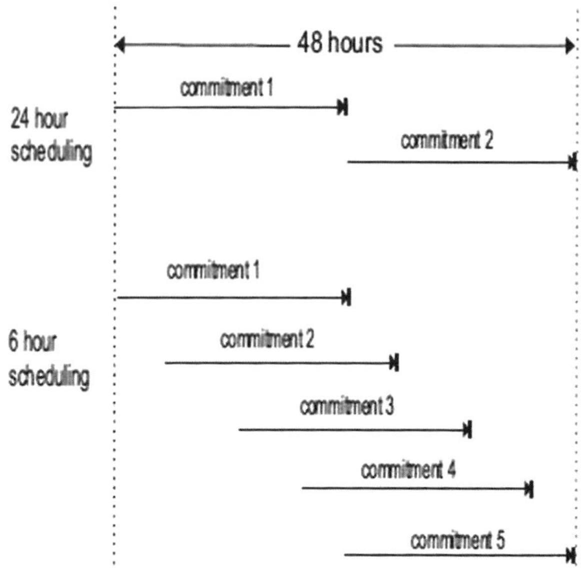

In fact [58] found that including the stochastic element of wind power forecasts in unit commitment modelling results in significantly better performing and less costly schedules than deterministic optimization.

3.3 Stochastic Unit Commitment with Wind Generation

A far superior approach than the fuelsaver and deterministic approaches to unit commitment with wind generation is to include the stochastic element of the wind power forecasts in the scheduling decisions. Large amounts of installed wind power add a significant stochastic element to the planning of the system, due to the uncertainty associated with wind power forecasts [25]. By explicitly taking into account the stochastic nature of wind in the unit commitment algorithm, more robust schedules will be produced.[2]

Stochastic optimization was used in unit commitment problems before wind generation became a significant concern for power system operation, as in [59] and [60]. In [59], a long term security-constrained stochastic unit commitment

[2] The issue of stochastic unit commitment with wind generation is discussed in detail in [58] and is summarized here.

(SCUC) model is described, which models unit and transmission line outages, as well as load forecasting inaccuracies. In [60], a method was developed to solve unit commitment problems when demand is not known with certainty. Both of these approaches show the benefits of using stochastic methods to solve the unit commitment problem. However, wind power as a stochastic input is not examined.

Stochastic security with wind generation is examined in [61], which formulates a market-clearing problem capable of accounting for wind power. The WILMAR project [62] developed a stochastic scheduling tool to examine the impact of the variability of wind in energy markets. The Wilmar Planning Tool was originally developed to model the Nordic electricity system and was later adapted to the Irish system as part of the All Island Grid Study [63]. It is currently employed in the European Wind Integration Study [64] and is considered to be the state of the art model for stochastic unit commitment with large penetrations of wind generation.

In the Wilmar model, the system is rescheduled as more precise wind and load forecasts are made available, giving a 'rolling planning' type of operation as discussed previously. Because more robust schedules are provided to cater for stochastic wind and load, the total expected costs of operating the system are lower than under the deterministic or fuelsaver approaches [58].

3.3.1 The WILMAR Stochastic Unit Commitment Model with Large Wind Penetrations

The main functionality of the WILMAR model is in two parts: a Scenario Tree Tool (STT) and a Scheduling Model. The STT is used to generate the scenarios that are used as inputs in the scheduling model. Possible future wind and load are represented by scenario trees, as shown in Fig. 6. The STT also produces time series for the forced unit outages. Each branch of the scenario tree corresponds to a different forecast of wind and load, as well as probability of occurrence. The wind and load scenarios are generated by Monte Carlo simulations of the wind and load forecast error, based on an auto-regressive moving average model describing the wind speed forecast error [58]. The high number of possible scenarios produced is then reduced using a scenario reduction approach, similar to [65].[3]

The Scheduling Model used is a mixed integer, stochastic optimization model [67] and the objective function being minimized, given in Eq. (1), is the expected cost of the system over the optimization period, covering all of the scenarios in Fig. 6.[4]

---

[3] More detailed information about the Scenario Tree Tool can be found in [66].

[4] Further details on the formulation of the unit commitment problem are given in [58].

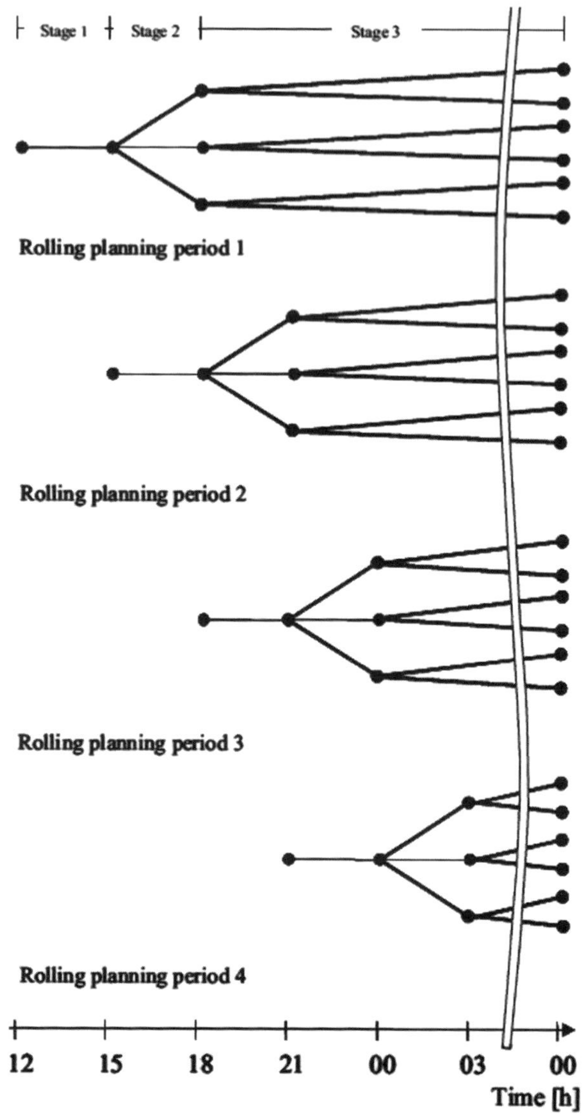

Fig. 6 Rolling planning with scenario trees [58]

$$
\begin{aligned}
V_{obj} = & \sum_{i \in I^{USEFUEL}} \sum_{s \in S} \sum_{t \in T} k_S F^{CONS}_{i,r,s,t} F^{PRICE}_{f,r,t} V^{ONLINE}_{i,t} \\
& + \sum_{i \in I^{START}} \sum_{s \in S} \sum_{t \in T} k_S F^{START}_{i,r,s,t} F^{PRICE}_{f,r,t} V^{ONLINE}_{i,t} \\
& - \sum_{i \in I^{START}} \sum_{s \in S} k_S F^{START}_{i,r,s,T_{END}} F^{PRICE}_{f,r,T_{END}} V^{ONLINE}_{i,T_{END}} \\
& + \sum_{i \in I^{USEFUEL}} \sum_{s \in S} \sum_{t \in T} k_S F^{CONS}_{i,r,s,t} F^{TAX}_{f,r,t} V^{EMISSION}_{i,t} \\
& + \sum_{s \in S} \sum_{t \in T} k_S L^{LOAD} \left(U^{QINTRA,+}_{r,s,t} + U^{QINTRA,-}_{r,s,t} \right) \\
& + \sum_{t \in T} k_S L^{LOAD} \left(U^{QDAY,+}_{r,t} + U^{QDAY,-}_{r,t} \right) \\
& + \sum_{s \in S} \sum_{t \in T} k_S L^{SPIN} U^{QSPIN,-}_{r,s,t} + \sum_{s \in S} \sum_{t \in T} k_S L^{REP} U^{QREP,-}_{r,s,t}
\end{aligned} \tag{1}
$$

A. Indices		*B. Parameters*		*C. Variables*	
F	Fuel	**EMISSION**	Rate of emission	**CONS**	Fuel consumed
i, I	Unit group	**END**	End time of optimization period	**U**	Relaxation variable
r, R	Region	**k**	Probability of scenario	**V**	Decision variable
s, S	Scenario	**L**	Infeasibility penalty	**ONLINE**	Integer on/off for unit
START	Units with start up fuel consumption	**LOAD**	Penalty for loss of load	**QDAY**	Day ahead demand not met
t, T	Time	**PRICE**	Fuel price	**QINTRA**	Intra day demand not met
USEFUEL	Units using fuel	**REP**	Penalty for not meeting replacement reserve	**QREP**	Replacement reserve not met
		SPIN	Penalty for not meeting primary reserve	**QSPIN**	Primary reserve not met
		TAX	Emission tax	**+, −**	Up, down regulation

The minimization covers fuel costs, carbon costs and startup costs. This is subject to constraints on units, such as startup time, minimum up and down times, ramping rates, and minimum and maximum generation, as well as interconnection constraints and losses, spinning and replacement reserve targets, and penalties for not being able to meet load or reserve targets.

It was found in [58] that utilizing a stochastic rather than a deterministic approach significantly reduces both the reserve requirement and the cycling of conventional units on the power system. In addition, [58] compared the use of currently available state of the art wind forecasts in a stochastic unit commitment model to perfect wind forecasts and found that using perfect wind forecasts reduces annual system costs by just 0.25–0.9 %. In other words, using a stochastic unit commitment technique is almost as good as having perfect wind power forecasts and perhaps efforts at improving wind power forecasting should not be a primary concern of system operators. In fact, if the cost of research, development and equipment to improve wind power forecasts were included, the cost of improved accuracy could be greater than the benefits it provides. Instead, system operators should focus efforts on employing stochastic unit commitment techniques with currently available forecasts.

4 System Planning: Flexibility

In general, the system costs of wind generation will be highly dependent on the underlying plant mix. Looking into the future, if it is envisaged that wind generation will play a major role in a plant portfolio, then the conventional plant mix should be optimized to accommodate this wind generation. In particular, a more *flexible* power system will be required [9].

The need for power system flexibility was recognized by the North American Electric Reliability Cooperation (NERC) in the formation of their Integration of Variable Generation Task Force (IVGTF) and by the International Energy Agency in their Grid Integration of Variable Renewable Energy (GIVAR) project. The GIVAR project is specifically aimed at assessing the power system flexibility requirements associated with increasing penetrations of renewable energy [68]. The project is broken into the following 5 strands: identification of the variability profiles of variable renewable energy technologies; development of a toolbox of measures to increase power system flexibility; definition of power system paradigms; development of a flexibility assessment method; and to review the costs and benefits of renewable energy integration.

A study conducted by [69] examines the optimal future conventional plant portfolios with high levels of installed wind generation. Their analysis shows that with increasing penetrations of wind generation, there is a reduction in the necessity for baseloaded generation and an increase in peaking capacity. In particular, the results point towards a reduction in coal fired generation and an increase in the need for open cycle gas turbines with increasing wind generation penetrations. Open cycle gas turbines have short start-up times and can ramp up and down with much greater ease than typical baseload units.

The system considered by [69] did not include any nuclear generation, however, for systems with large penetrations of inflexible baseload nuclear generation, flexibility will be a significant concern. Large penetrations of wind generation require system flexibility which cannot be provided by nuclear generation. As such, policy makers need to decide whether to pursue a policy of large scale renewable promotion or nuclear generation as the two are not necessarily complementary.

Increased interconnection to other systems has also been highlighted as facilitating the integration of wind generation by introducing more flexibility to help balance the variable wind output. The experiences of integrating wind generation in a number of different countries is discussed in [15], and in particular the challenges faced by isolated power systems are highlighted. The conclusions of this study illustrated the importance of interconnection in providing the range of services required to maintain a secure electricity system with large penetrations of wind generation, such as sufficient generation capacity, fast acting reserve provision, and control of excess wind generation.

Energy storage used in conjunction with renewable energy has also been suggested as a means to increase the use of renewable energy while maintaining a high quality of service reliability [70, 71]. The use of storage devices can help balance the wind generation output and can also be used to transfer energy from low-use periods to peak-use periods, allowing the system to operate at a more constant level and reducing energy supply costs. When increasing penetrations of electricity storage are being considered a key concern is the peak:off-peak price differential. This must be sufficient to cover the losses due to the round trip efficiency of the storage unit as well as the significant capital costs of storage. It should also be noted that on relatively small systems the storage unit can actually impact the electricity price by increasing the price during off-peak hours and decreasing it at peak times and thereby further decreasing its profit opportunities.

Active demand management can also provide a valuable source of flexibility to a power system. The traditional view of demand has been to place it in a different category to generation in the scheduling of power system operation [72]. However, if demand is considered as a flexible, responsive resource then these traditional divisions could be broken down and demand side resources utilized to improve overall system flexibility. Currently, it is mainly commercial and industrial customers that participate in demand side management schemes. As a result, the main types of demand which currently provide demand responses are on-site generation (diesel), refrigeration and deferral of production [73]. However, through the use of smart electricity meters, domestic customers can also lower their demand by reducing their discretionary loads, air conditioning or electric heating or deferring the usage of appliances until a later time or date. Any micro-generation connected within the home also offers itself as a means to reduce net consumption when required. The residential sector has been under utilized to date as a source of power system flexibility and into the future offers a potentially large resource to be exploited.

This article has provided a discussion of some of the main power system operation challenges associated with wind generation, as well as the optimal unit commitment techniques and system characteristics which can help minimize these costs. While each power system is different it is likely that if wind generation is to be pursued on a large scale the issues raised here will be relevant in many power systems worldwide. It is also clear that transmission system operators will be required to significantly alter traditional operating techniques to ensure wind generation can play a significant role in future electricity supply at minimal system cost.

References

1. International Energy Agency (2009) Key world energy statistics. Available http://www.iea.org
2. International Energy Agency Statistics (2009) CO_2 emissions from fuel consumption—highlights. Available http://www.iea.org/co2highlights
3. European Commission (2009) Directive 2009/28/EC of the European Parliament and of the Council of 23 April 2009: on the promotion of the use of energy from renewable sources and amending and subsequently repealing Directives 2001/77/EC and 2003/30/EC. Available http://www.europa.eu/
4. European Wind Energy Agency (2009) Global statistics. Available http://www.ewea.org/index.php?id=180
5. Gross R, Heptonstall P, Anderson D, Green T, Leach M, Skea J (2006) The costs and impacts of intermittency: an assessment of the evidence on the costs and impacts of intermittent generation on the British electricity network. A report of the technology and policy assessment function of the UK Energy Research Centre
6. Ummels B, Gibescu M, Pelgrum E, Kling K, Brand A (2007) Impacts of wind power on thermal generation unit commitment and dispatch. IEEE Trans Energy Conv 22:44–51
7. Kundur P, Paserba J, Ajjarapu V, Andersson G, Bose A, Canizares C, Vittal V (2004) Definition and classification of power system stability IEEE/CIGRE joint task force on stability terms and definitions. IEEE Trans Power Syst 19(3):1387–1401 (IEEE/CIGRE: The Institute of Electrical and Electronics Engineers (IEEE)/International Council on Large Electric Systems (CIGRE))
8. Kirschen D, Strbac G (2004) Fundamentals of power system economics. Wiley, West Sussex
9. International Energy Agency Annex 25 (2009) Power systems with large amounts of wind power. Available http://www.ieawind.org
10. Holttinen H et al (2006) Design and operation of power systems with large amounts of wind power, IEA collaboration. In: Nordic wind power conference. Espoo, Finland
11. ILEX, Strbac G (2002) Quantifying the system cost of additional renewable generation in 2020. Available http://www.dti.gov.uk
12. Smith J, Milligan M, DeMeo E, Parsons B (2007) Utility wind integration and operating impact state of the art. IEEE Trans Power Syst 22(3):900–908
13. Smith J (2006) North American literature and project review for all-island wind penetration study. Report prepared for SONI and ESB NG
14. Eirgrid (2009) Generation adequacy report 2010–2016. Available www.eirgrid.com
15. Söder L, Hofmann L, Orths A, Holttinen H, Wan Y, Tuohy A (2007) Experience from wind integration in some high penetration areas. IEEE Trans Power Syst 22:4–12
16. Commission for Energy Regulation Ireland (2008) Press release: CER announces unprecedented increase in renewable electricity. http://www.cer.ie
17. National Aeronautics and Space Administration (NASA) (2008) Visible earth: a catalogue of images and animations of our home planet. Available http://visibleearth.nasa.gov
18. Denny E, O'Malley M (2007) Quantifying the total net benefits of grid integrated wind. IEEE Trans Power Syst 22(2):605–615

19. European Commission (1996) Directive 96/92/EC of the European Parliament and of the Council of 19 December 1996 concerning common rules for the internal market in electricity. Available http://www.europa.eu
20. Wang P, Billington R (2004) Reliability assessment of a restructured power system considering reserve agreements. IEEE Trans Power Syst 19(2):972–978
21. Zhu J, Jordan G, Ihara S (2000) The market for spinning reserve and its impact on energy prices. In: IEEE PES winter meeting, Singapore, pp 1202–1207
22. Söder L (1993) Reserve margin planning in a wind-hydro-thermal power system. IEEE Trans Power Syst 8:564–571
23. Kundur P (2004) Power system stability and control. McGraw Hill Inc., New York
24. Rashidi-Nejad M, Song Y, Javidi-Dasht-Bayaz M (2002) Operating reserve provision in deregulated power markets. In: IEEE PES winter meeting, New York, pp 1305–1310
25. Marti I, Kariniotakis G, Pinson P, Sanchez I, Nielsen T, Madsen H, Giebel G, Usaola J, Palomares A, Brownsword R, Tambke J, Focken U, Lange M, Sideratos G, Descombes G (2006) Evaluation of advanced wind power forecasting models—results of the Anemos project. Available http://anemos.cma.fr
26. Kariniotakis G, Pinson P (2003) Evaluation of the MORE-CARE wind power prediction platform. Performance of the fuzzy logic based models. In: Proceedings of European Wind Energy Association Conference, Madrid, Spain
27. Doherty R, O'Malley M (2005) New approach to quantify reserve demand in systems with significant installed wind capacity. IEEE Trans Power Syst 20(2):587–595
28. Hirst E (2001) Interactions of wind farms with bulk-power operations and markets. Prepared for the project for Sustainable FERC Energy Policy
29. Hirst E (2002) Integrating wind energy with the BPA power system: preliminary study. Prepared for Power Business Line Bonneville Power Administration
30. Gardner P, Snodin H, Higgins A, Goldrick SM (2003) The impacts of increased levels of wind penetration on the electricity systems of the Republic of Ireland and Northern Ireland. Final Report to the Commission for Energy Regulation/OFREG by Garrad Hassan 3096/GR/04
31. Denny E, O'Malley M (2009) The impact of carbon prices on generation cycling costs. Energy Policy 37(4):1204–1212
32. Troy N, Denny E, O'Malley M (2010) Base-load cycling on a system with significant wind penetration. IEEE Trans Power Sys 25(2):1088–1097
33. Flynn D (2003) Thermal power plant simulation and control, vol 43, 1st edn., IEE power and energy seriesIEE, London
34. Subcommittee IEEE (1990) Current operating problems subcommittee: a report on the operational aspects of generation cycling. IEEE Trans Power Syst 5(4):1194–1203
35. Lefton S, Besuner P, Grimsrud G (1997) Understand what it really costs to cycle fossil-fired units. Power 141(2):41–42
36. Grimsrud P, Lefton S (1995) Economics of cycling 101: what do you need to know about cycling costs and why? APTECH engineering technical paper TP098, http://www.aptecheng.com
37. Denny E, O'Malley M (2006) Wind generation, power system operation, and emissions reduction. IEEE Trans Power Syst 21(1):341–347
38. Energy Information Administration (EIA) (2010) Energy glossary. Available http://www.eia.doe.gov/glossary
39. Energy Information Administration (EIA) (2000) Electric power annual 2000. Available www.eia.doe.gov
40. Kesgin U (2003) Study on prediction of the effects of design and operating parameters on NOx emissions from a leanburn natural gas engine. Energy Convers Manage 44:907–921
41. Li K, Thompson S, Peng J (2004) Modelling and prediction of NOx emission in a coal-fired power generation plant. Control Eng Pract 12:707–723
42. Mansour A, Benjamin M, Straub D, Richards G (2001) Application of macrolamination technology to lean, premix combustion. AMSE J Eng Gas Turbines Power 123(4):796–802

43. Cabot G, Vauchelles D, Taupin B, Boukhalfa A (2004) Experimental study of lean premixed turbulent combustion in a scale gas turbine chamber. Exp Thermal Fluid Sci 28:683–690
44. Mullane A, Lightbody G, Yacamini R (2005) Wind-turbine fault ride-through enhancement. IEEE Trans Power Syst 20(4):1929–1937
45. Xiang D, Ran L, Tavner PJ, Yang S (2006) Control of doubly fed induction generator in a wind turbine during grid fault ride-through. IEEE Trans Energy Conv 21(3):652–662
46. Lalor G, Mullane A, O'Malley M (2005) Frequency control and wind turbine technologies. IEEE Trans Power Syst 20(4):1905–1913
47. McArdle J (2004) Dynamic modelling of wind turbine generators and the impact on small lightly interconnected grids. Wind Eng 28(1)
48. Causebrook A, Fox B (2004) Decoding grid codes to accommodate diverse generation technologies
49. Johnson A, Urdal H (2004) Technical connection requirements for wind farms. Wind Eng 28(1)
50. Keane A, O'Malley M (2005) Optimal allocation of embedded generation on distribution networks. IEEE Trans Power Syst 20(3):1640–1646
51. IEE (2003) Embedded generation issues—a briefing note. An Energy and Environment Fact Sheet provided by the IEE
52. Ackermann T (2005) Wind power in power systems. Wiley, West Sussex
53. Milborrow D (2006) No limits to high wind penetration. Wind Power Mon 22(9)
54. Department of the Enterprise Trade and Investment (DETI) (2003) A study into the economic renewable energy resource in Northern Ireland and the ability of the electricity network to accommodate renewable generation up to 2010. Available www.energy.detini.gov.uk
55. Gardiner P, Snodin H, Higgins A, Goldrick SM (2003) The impacts of increased levels of wind penetration on the electricity systems of the Republic of Ireland and Northern Ireland (3096/GR/04)
56. PLEXOS for Power Systems (2006) Electricity market simulation. Available www.draytonanalytics.com
57. Tuohy, A, Denny E, O'Malley M (2007) Rolling unit commitment for systems with significant installed wind capacity. In: IEEE PES power technology Lausanne, Switzerland
58. Tuohy A, Meibom P, Denny E, O'Malley M (2009) Unit commitment for systems with significant wind penetration. IEEE Trans Power Syst 24(2):592–601
59. Wu L, Shahidehpour M, Li T (2007) Stochastic security-constrained unit commitment. IEEE Trans Power Syst 22(2):800–811
60. Takriti S, Birge J, Long E (1996) A stochastic model for the unit commitment problem. IEEE Trans Power Syst 11(3):1497–1508
61. Bouffard F, Galiana F (2006) Stochastic security for operations planning with significant wind power generation. IEEE Trans Power Syst 23(2):306–316
62. Wind Power Association (2010) Wind power integration in liberalised electricity markets (Wilmar) project. Available www.wilmar.risoe.dk
63. All Island Renewable Grid Study—Workstream 2B (2008) Wind variability management studies. Available http://www.dcmnr.gov.ie
64. European Wind Energy Association (2010) European wind integration study. Available http://www.windintegration.eu
65. Dupacova J, Grwe-Kuska N, Rmisch W (2003) Scenario reduction in stochastic programming: an approach using probability metrics. Math Program 96(3):493–511
66. Barth R, Söder L, Weber C, Brand H, Swider D (2006) Deliverable d6.2 (b) documentation methodology of the scenario tree tool. Institute of Energy Economics and the Rational Use of Energy (IER), University of Stuttgart, Stuttgart, Germany, 2006. Available www.wilmar.risoe.dk
67. Kall P, Wallace S (1994) Stochastic programming. Wiley, Chichester
68. Lannoye E, Milligan M, Adams J, Tuohy A, Chandler H, Flynn D, O'Malley M (2010) NERC integration of variable generation task force (IVGTF). Integration of variable generation: capacity value and evaluation of flexibility. Available http://www.nerc.com

69. Doherty R, Outhred H, O'Malley M (2006) Establishing the role that wind may have in future generation portfolios. IEEE Trans Power Syst 21(3):1415–1422
70. Barton J, Infield D (2004) Energy storage and its use with intermittent renewable energy. IEEE Trans Energy Convers 19(2):441–448
71. Brown P, Lopes J, Matos M (2008) Optimization of pumped storage capacity in an isolated power system with large renewable penetration. IEEE Trans Power Syst 23(2):523–531
72. Gellings C, Smith W (1989) Integration demand-side management into utility planning. Proc IEEE 77(6):1908–1918
73. Papagiannis G, Lettas N, Dokopoulos P (2008) Economic and environmental impacts from the implementation of an intelligent demand side management system at the European level. Energy Policy 36(1):163–180

Operational Reserve Assessment Considering Wind Power Fluctuations in Power Systems

Mauro Rosa, Manuel Matos, Ricardo Ferreira, Armando Martins Leite da Silva and Warlley Sales

Abstract Discussions on innovative criteria, operation strategies, and assessment tools are important insights when monitoring the security of supply considering renewable power sources for the years to come. In order to deal with the power fluctuations that come from wind uncertainties, this chapter explores a probabilistic methodology based on the chronological Monte Carlo simulation (MCS) to evaluate the long-term reserve requirements of generating systems considering wind energy sources. A new alternative to assess the power amount needs to adequately meet all assumed deviations is presented. Case studies with the IEEE-RTS 96 generating systems and some planning configurations of the Portuguese and Spanish generating systems are presented and discussed.

1 Introduction

The acceptance of wind energy around the world in the beginning of the twenty-first century has been extremely impressive. Only in this first decade, more than 50 countries started using wind generated electricity, highlighting its importance for the future [1]. In general, the countries' target is to make a transition to an emission free and sustainable energy system, through the use of renewable resources like wind. The most significant change towards renewable resources has

M. Rosa (✉) · M. Matos · R. Ferreira
INESC TEC (Formerly INESC Porto), Faculty of Engineering of the University of Porto, Porto, Portugal
e-mail: marosa@inescporto.pt

A. M. L. da Silva
Federal University of Itajubá, UNIFEI, Itajubá, Brazil

W. Sales
Federal University of São João Del Rei, UFSJ, São João del Rei, Brazil

P. M. Pardalos et al. (eds.), *Handbook of Wind Power Systems*, Energy Systems,
DOI: 10.1007/978-3-642-41080-2_12, © Springer-Verlag Berlin Heidelberg 2013

been an efficient use of existing fossil fuels, tending to rely more on fuels with lower carbon content (i.e. natural gas), and ultimately the use of oil for electricity production has been replaced, for instance, by wind power generation [2].

Although wind power has already achieved a suitable level of maturity, mainly due to its intensive use throughout the last decade, the developments over this period of time do not represent a complete range of control over this technology. Hence, just as with any new technology, a lot of challenges regarding the consistent integration of wind power into the system must be addressed. For instance, one of the most important issues around this type of technology is their forecasting properties, which gradually have been improved to manage wind uncertainty [3–5]. In general, power system planners and operators are already familiarized with a certain amount of variability and uncertainty, particularly those related to the system demand and, to a certain extent, to the conventional generation. The wind power unscheduled property causes, directly, an operational problem, since the output from wind generation cannot be dispatched as conventional sources and, indirectly, since it increases the costs of ancillary services required into the system operation, mainly due to the wind production variability [6].

Undoubtedly, the number of random variables and system complexities increases dramatically when wind power is added to the system operation, even though some of them have already been controlled, to a certain degree, over the last 10 years. However, it is mandatory to note that the issues related to wind power integration can vary from one system to another [7]. The decision to replace oil units to wind turbines can cause different effects on the different system configurations. The impact of wind power integration on a system where a large percentage of the generation comes from gas is somewhat different than that with coal generation [8]. Therefore, from the unit commitment point of view, the management units used to meet the load must consider the predominance of the available generation technology so that it is possible to meet the operational reserve requirements for systems with large wind integration. Different unit commitment strategies, which consider large amounts of wind production, can conduct to different risk-based levels, if a large percentage of the generation is based on gas or coal technologies.

In order to deal with the power fluctuation that comes from the wind uncertainties, this chapter explores a probabilistic methodology based on chronological Monte Carlo simulation (MCS) to evaluate the long-term reserve requirements of generating systems considering wind energy sources. The idea is to study the behavior of reliability indices (conventional and well-being), not only when a major portion of the wind energy sources is integrated into the system with predominant thermal power plants, but also considering other renewable technologies such as hydro and mini-hydro power sources, as well as solar technologies (solar thermal and photovoltaic) in much lower amounts. Case studies with the IEEE-RTS 96 generating systems and some planning configurations of the Portuguese and Spanish generating systems are presented and discussed.

2 Reserve Requirements Approaches

The evaluation of the risk-based level concerning capacity reserve requirements is an important aspect related to operation and expansion planning of generating systems. One of the first methodologies formally applied considering the idea of risk to calculate the capacity reserve was the (PJM) pennsylvania-jersey-maryland method [9]. During the last years, several variations have been proposed in relation to this method [10–12]. The main idea is that the risk index is the probability of the existing generation capacity not satisfying the expected load demand, during time period *T*, in which the operator may not replace any damaged unit nor initialize the operation of new ones [13]. Therefore, this index represents a probabilistic measurement associated with the scheduled generating reserve. Due to its operational characteristic, a considerable number of PJM method applications [10–12, 14] reveal that this system risk index is usually conditioned to a short period of time. Recently, some works have explored alternative methodologies considering the idea of this risk for long period of time [13, 15].

In general, there are three ways commonly used to assess the power amounts used as reserve needs: deterministic, probabilistic and/or hybrid approaches. The deterministic approaches are based on simple criteria, such as loss of the largest unit or a percentage of the load required in a given time *T*. Although they are simple to understand and easy to implement, they are problematic in a sense that they cannot provide a meaningful measurement of risk, but only a rule associated to a threshold whose definition is not always clear. Otherwise, probabilistic approaches can usually deal with the complex existing relations between different uncertain variables and essentially can better represent the stochastic behavior of power system [16]. The hybrid approach has been built combining deterministic perception with probability concepts named system well-being analysis [17–19]. This new framework reduces the gap between deterministic and probabilistic approaches by providing the ability to measure the degree of success of any operating system state. In a well-being analysis, success states are further split into healthy and marginal states, using the previously mentioned engineers' perception as criterion. In the last decade, well-being analysis has been applied not only in operational reserve assessment but also to areas such as generating adequacy systems and composite generation and transmission systems.

In fact, most utilities around the world have used deterministic rules in order to assess capacity reserve into interconnected power systems areas. In this case, the definition of secondary control reserve (spinning reserve) can vary from one country to another, and usually it depends on the security policy adopted by the system operator. For instance, the definition of the minimal spinning reserve capacity needs on several European countries follows the Union for the Coordination of Transmission of Electricity (UCTE) empirical formulation based on this equation [20, 21]:

$$R = \sqrt{a \times L_{\max} + b^2} - b \tag{1}$$

where,

- R is the spinning reserve recommended as a result of the equation;
- L_{max}, is the maximum forecasted load into a secondary control area;
- a and b are empirical constants that assume 10 and 150 MW, respectively.

As stated in the previous equation, the spinning reserve rule strongly depends on the maximum load forecast into power system control area, without directly considering other types of uncertainties such as unit forced outages, some emergency maintenance procedures, and mainly wind forecast errors. The latter aspect accounts for the developments in the wind power forecasting area [3, 8], where the main target is exactly to reduce the existing gap for balancing energy and reserve requirements in order to better integrate wind energy into the electricity supply system [8]. Other related techniques, such as the Denmark rule [22], unit commitment risk methodologies [23], and so on, could be cited as reference methodologies to define the power amounts used as secondary and/or tertiary reserve. However, this chapter intends to discuss the reserve requirements from the long-term perspective, and will concentrate efforts in the representation of a good quality uncertainty model, where all the chronological aspects are preserved [15]. In the next section, a chronological methodology based on sequential Monte Carlo simulation will be presented.

3 Probabilistic Evaluation of Power Systems

In power system literature, there are several techniques available that allow an accurate reliability assessment of the generating capacity. The basic consideration is to concentrate all generating units and loads in a single bus. The transmission lines constraints and failures are ignored, and the performance of the generating system is measured by comparison between the available generating capacities and the load at different snapshot times. The problem consists, basically, in measuring the ability of the generation system to meet the total load requirement, considering the load variations, the failure of units, as well as the unavailability of energetic resources, which can directly affect the generating capacity. If one knows the stochastic parameters λ and μ (i.e. failure and repair rates, respectively) of each generating unit, it is possible to calculate the probabilities that generating units are available or not (up or down) during a simulation process. The techniques used in generating adequacy assessment can be frequently divided into two basic categories: analytical and simulation. Generally, analytical approaches adopt the state space representation, but simulation can either adopt state space representation or chronological representation.

In the chronological or sequential representation, each subsequent system state is related to the previous set of sampled system states. As mentioned before, the chronological representation becomes a necessity when the operating system is history-dependent or time correlated, which is particularly fundamental to represent green power, such as wind and solar units and their respective time-dependent behaviors. At the same time, it can represent the impact on maintenance policies, ramping rates in thermal units, as well as complex correlated load models. Essentially in hydro electrical systems, when the reservoir has to be carefully controlled and, at any moment, the available power can depend on past water inflows, on past operation policies, and so on, the chronological representation becomes imperative [24]. Regarding the chronological representation, another attractive feature to bear in mind consists of making the development of the distributional aspects associated with system index mean values, as well as providing the most comprehensive range of reliability indices [25, 26].

The problem of calculating reliability indices is equivalent to the evaluation of the following expression [27]:

$$E(F) = \frac{1}{T}\int_{0}^{T} F(t)dt \tag{2}$$

where T is the period of simulation and $F(t)$ the test function to verify at any time t, if a related system state is adequate.

Two consecutive sampled system states differ from one state component only. This is the major difference between non-sequential and sequential models, which makes the Monte Carlo chronological simulation extremely expensive from the computational point of view [28]. Another critical constraint considered by non-sequential based models is the assumption of exponential distributions for all system state residence times.

3.1 Sequential Monte Carlo Simulation

The term sequential simulation means that the history of a system is simulated in fixed discrete time steps [29]. The sequential approach is based on sampling the probability distribution of the component state duration. It is used to simulate the stochastic process of the system operation through the use of its probabilities distributions, associated with mean-time-to-failure (MTTF) and mean-time-to-repair (MTTR) of each system component. Considering the two-state Markov Model, these are the operating and repair state duration distribution functions that are usually assumed to be exponential. Other distributions, such as *Weibull*, *Lognormal*, etc., can also be used to represent different behaviors. Therefore, in the discrete way, the problem of estimating reliability indices can be written as follows:

$$\tilde{E}[F] = \frac{1}{NY}\sum_{n=1}^{NY} F(y_n) \tag{3}$$

where NY is the number of simulated years, y_n is the sequence of system states x^k in the year n, and $F(y_n)$ is the function to calculate yearly reliability indices over the sequence y_n. The sequential approach can be summarized in the following steps:

1. Generate a yearly synthetic sequence of system states y_n by sequentially applying the failure/repair stochastic models of equipment and the chronological load model. Thus, the initial state of each component is sampled. Usually, in the first sample, it is assumed that all components are initially in the success or up state, even though other approaches may be used. The duration of each component residing in its present state is sampled from its probability distribution. Assuming an exponential probability distribution and using the inverse transform method [30], the duration of each component will follow:

$$T = -\frac{1}{\lambda}\ln(U) \tag{4}$$

where T is the time residence of each equipment, λ is the failure rate of the component if the present state is the up state or λ is the repair rate of the component if the present state is the down state, and U is a uniformly distributed random number sampled in the interval between [0,1];

2. Chronologically evaluate each system state x^k in the sequence y_n and accumulate the values;
3. In order to obtain yearly reliability indices, calculate the test function $F(y_n)$ over the accumulated values;
4. Estimate the expected mean values of the yearly indices as the average over the yearly results for each simulated sequence y_n;
5. The stop criterion is also based on the relative uncertainty of the estimates. Therefore, calculate β (coefficient of variation) using the following equation [20];

$$\beta = \frac{\sqrt{\tilde{V}\left(\tilde{E}(F)\right)}}{\tilde{E}(F)} \times 100\,\% \tag{5}$$

6. Verify if the degree of accuracy or confidence interval is acceptable. If the answer is yes, stop the simulation; otherwise, go back to step 1.

In the sequential approach, the system evaluation is conducted for each different system state in order to achieve the reliability index function. For instance, considering the *Loss of Load Expectation* (*LOLE*) index, $F(y_n)$ is the sum of the sampled duration of all failure states in y_n. In turn, if the $F(y_n)$ is the sum of energy not supplied associated with all failure states in y_n, $E[F]$ will represent the *Expected Energy Not Supplied* (*EENS*) index. Several other reliability indices can be easily achieved using the sequential approach.

3.2 Proposed Modeling of Components

The selection of an appropriate model to represent the stochastic behavior of power system units is an essential requirement in the adequacy evaluation of generation systems. Essentially, two aspects must be represented to each technology: firstly, the up and down cycle (failure/repair) of each unit or set of units; and secondly, the power availability of each unit or set of units, considering their natural resources, such as water inflows, wind speed, solar irradiations, and others. The two-state Markov model [16] and the multi-state Markov model [31] will be used to represent the conventional generation technology with large units, such as thermal and hydropower plants, as well as the distributed generation (DG) technology with small units concentrated in farms or aggregations, such as wind power, solar central receiver or photovoltaic, small hydro generators, and others. These representations are depicted in Fig. 1. More details about models can be found in [32].

Thermal power plants Based on the two-state Markov model, presented in Fig. 1a, the failure/repair cycles for all thermal power technologies, such as nuclear, coal, oil, gas, among others, are represented. The residence time to each state is assumed to be exponentially distributed, and can be calculated using Eq. (4). In thermal power plant representation, the available power is assumed to depend only on the unit unavailability. Clearly, any non-Markovian model could be used if the necessary parameters were available.

Hydropower plants Also based on two-state Markov model, presented in Fig. 1a, the failure/repair cycles for all hydro power units are represented. The residence time to each state is assumed to be exponentially distributed, and can be calculated using Eq. (4). In hydropower plant representation, the available power is assumed to depend on failure/repair cycle and on water storage of each reservoir. In other words, the level of each water storage reservoir defines the power available for each unit.

Nevertheless, the levels of water storage of each reservoir vary due to several aspects, such as water coordination policies, yearly or monthly hydrologic conditions, daily hydro-thermal coordination, and others. The full impact of water inflows on the hydropower available for each unit should consider a complex model involving stochastic dynamic programming tools linked to historical hydrologic series [33]. In order to simplify the proposed model, the power available for each unit can be represented through the use of two approaches, depending on the information level of each reservoir: first, to be proportional to the level of water storage in its respective reservoir and second, using a pre-established polynomial model linked to each reservoir. Due to the small variation of daily water storage levels, a monthly water storage level variation is assumed, which can better represent the water storage behavior with relative accuracy. Thus, the hydro power units are combined in several reservoirs containing monthly historical series. These series are defined for each hydraulic basin based on historical data. The aim is to capture the historical inflows, reservoir volumes, and type of operation.

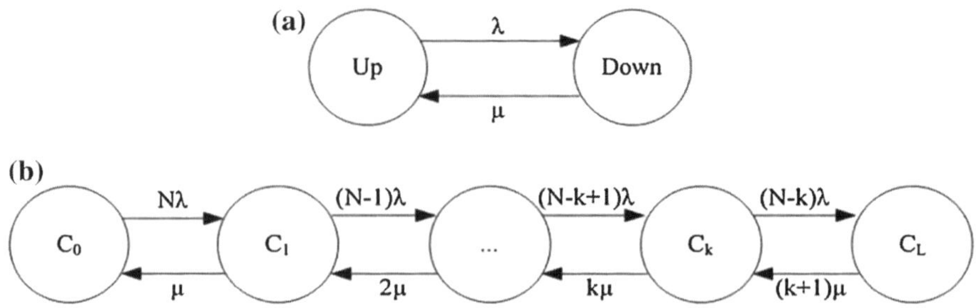

Fig. 1 (**a**) Two states Markov model, (b) multi-state Markov model

Therefore, the capacities of the hydro units will be defined for each month, according to the corresponding hydrological series.

Wind power farms Usually, in a wind power site, there are several generating units grouped into an equivalent multi-state Markov model, as shown in Fig. 1b. Only two stochastic parameters are necessary: unit failure and repair rates. Parameter N represents the number of generating units of the wind farm. If C is the unit capacity, the amount of power associated with the kth state is given by $C_k = (N - k) \times C$, $k = 0, \ldots, N$. The cumulative probability P_k (from 0 to k) associated with this state can be easily calculated. In order to reduce the number of these states during the chronological MCS, a simple truncation process sets the desired order of accuracy. Therefore, instead of $N + 1$ states, a much smaller number up to the capacity C_L will limit this model; e.g. $1 - P_L \leq$ tolerance.

The productions of the wind generating units will be defined for each hour, according to the hourly wind series for each geographic region. The wind series try to capture the wind speed and power conversion characteristics. Historical yearly series of per unit capacity fluctuations have to be provided per hour.

Small hydropower plants Small-hydro units are modeled similarly to the hydro generating units from the hydrological point of view, but they are grouped into multi-state units of Fig. 1b to simplify the modeling processing. Due to the lack of specific data in relation to the hydrological basin, where they are located as well as the technical characteristics of each small hydro unit, an equivalent reservoir is used to model the capacity variations with time, so some small unit clusters are promoted. In order to better represent the up and down cycles and their hydrologic fluctuations, they are grouped according to their main characteristics, such as capacity in MW, year unavailability (FOR), and mean time to repair (MTTR), as well as their geographic position. Therefore, each small hydro group can be seen as a small hydropower plant, which has similar units with the same characteristics, but different numbers of units. Obviously, if specific data are available, they can be properly considered. The benefits in terms of accuracy and the cost in terms of computing effort, as well as the levels of model detail, must always be balanced.

Solar power plants Solar energy can provide huge amounts of energy share, especially in regions with high solar irradiations [34]. There are different ways to access solar energy: either through power plants that use solar energy by concentrating direct irradiation, converting it into heat and finally into electricity (solar central receiver), or by using flat plate collectors or photovoltaic panels. In both cases, the up and down cycles are represented by a multi-state Markov model, as depicted in Fig. 1b. In order to better represent the up and down cycles and theirs capacity fluctuations, they are grouped according to their main characteristics, such as capacity in MW, year unavailability (FOR), and mean time to repair (MTTR), as well as their geographic position. Solar irradiations can be seen as the "fuel" for these devices. Hence, it is necessary to build yearly series that may be representative of the capacity fluctuations per hour, considering the geographic position [34].

Cogeneration power plants Cogeneration units are modeled similarly to the thermal units. However, like in the previous case, they are also clustered using multistate units of Fig. 1b. Moreover, an hourly utilization factor is specified, which models the actual cogeneration power used by the system. This factor varies during the year following the tariff attractiveness and/or the industry production cycle.

3.3 Proposed Load Model

Generally, based on the forecast supply and demand curves, a utility can examine its current and future state of generation, transmission, and distribution of energy. One of the difficulties in applying probabilistic methods to the power systems reliability assessment is the level of necessary details in some representations. For instance, specific load data is necessary to assess interruption costs involving several customer sectors [35]. Since load curves vary by season and by day of the week, reliability characteristics of a system may change throughout the days of the year. In some systems, summer electrical load is dominated by air conditioners, whereas winter electrical load is dominated by electric heating. In both cases, there are great differences in load shapes. In general, load characteristics are very important in reliability studies, so detailed load curves must be modeled [36]. On the other hand, it is known that any load model is an approximation of the actual future load. Its accuracy depends on the amount and quality of the available data [37]. The most detailed load curves span all 365 days of the year and consist of 8,760 hourly demand points.

Considering state space representation, load models based on Markovian assumptions have been used [37]. Usually, these models make it possible to keep track of the chronological aspects of the load shape of power systems reliability assessment, thus contributing to the reduction of the computational effort. However, in chronological representations, the standard chronological load model is used containing 8,760 levels, which correspond to each hour of a year. The chronological MCS will sequentially follow these load steps during the simulation process [30].

In load forecast studies, a fundamental topic that must be taken into consideration is the fact that the actual peak load generally differs from the forecast value within a certain error. In other words, the load forecast may assume a predication error. As the forecast is normally predicted on past experience, a level of uncertainty must be considered in order to insert a noise in the chronological representation. Two uncertainty levels, representing short and long-term load forecasting deviations, can be simulated through the MCS process. Gaussian or any other type of distribution may be used [16]. In the short-term representation, an hourly uncertainty is inserted during the simulation process, whereas, in the long-term representation, a yearly uncertainty is inserted during the same process.

4 Long-Term Assessment of Operational Reserve

Sequential Monte Carlo is chosen due to its suitable ability in preserving the relations between all important variables, dealing with time dependent characteristics. Therefore, as previously stated, a chronological load pattern, wind hourly power variation, hydro monthly conditions, and other time dependent events can be represented. Merging these variations with the sampling of unit failures and repair times leads to loss of load statistics, from which risk indices of generation adequacy are calculated. Loss of Load Probability (LOLP), LOLE or EENS are classic examples of those indices. Extension of the concept to check secondary reserve availability in each state is also included, in the process known in the literature as "well-being analysis" [17], where each simulated state is further classified as "healthy" (enough reserve margin), "marginal" (partial reserve margin) or "at risk" (loss of load). Probabilities of these three conditions and the expected number of hours per year in each one are useful indices to complement the classic ones. All previous risk indices are based on the following power balance equation:

$$R_{STA}(t) = G(t) - L(t) = 0 \tag{6}$$

where G represents the system available generation at hour "t", L is the total system load at hour "t", and R_{STA} is the static reserve at hour "t". Random variable G depends on the equipment availabilities and on the capacity fluctuations due to, for instance, hydrology and wind variations, etc. Random variable L depends on the short- and long-term uncertainties and also on the hourly variations.

4.1 Operational Reserve Model

In order to assess the performance of the operating reserve, new variables have to be defined as shown in Fig. 2. In this study, the primary (or regulation—P_{RC}) and secondary (or spinning) reserves are pre-defined values. Obviously, the spinning reserve amount can always be redefined, in case its associated performance is

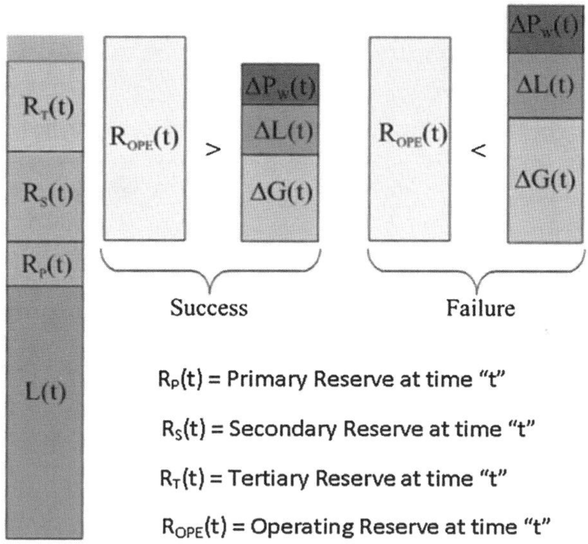

Fig. 2 Operating reserve assessment

below a pre-established acceptable value. The tertiary reserve (non-spinning or fast tertiary) is set up by those generators that can be synchronized within 1 h. This reserve is relevant in the present study and it constitutes the major concern for planners in this new environment characterized by a profusion of renewable sources. Therefore, the following power balance equation is set to assess the risk indices associated with the operating reserve:

$$R_{OPE}(t) = R_S(t) + R_T(t) < \Delta L(t) + \Delta P_W(t) + \Delta G(t) \tag{7}$$

where ΔL represents the short- and long-term load deviation at hour "t", ΔP_W represents the possible wind power capacity variation at hour "t", and ΔG represents the generating capacity variation due to forced outages at hour "t". From Fig. 2, it is possible to observe an extra amount of capacity at the top of the tertiary reserve. This is due to the discrete nature of unit generating capacities.

Equation (7) describes the risk of changes in the load, wind power capacity, as well as generating outages not duly covered by the amount of spinning reserve, and also by the generators that can be synchronized within 1 h. Therefore, the same traditional and well-being indices can be evaluated with this risk equation.

Although there are many reference values for LOLP or LOLE indices (i.e. static reserve), within the framework of generation planning, there are a few works with reference values for the operating reserve [13].

The implementation of the previous models is carried out through FORTRAN language (calculation mode). The convergence process is tracked by a coefficient of variation specified for the EENS index. Usually, when the convergence of EENS index is ensured, the others will converge as well. The probability distributions of all conventional and well-being, static and operating reliability indices are also evaluated.

4.2 Trade-off Between Spinning Reserve and Fast Tertiary Reserve

A new alternative to assess the power amount needs (secondary and fast tertiary reserve) to adequately meet all assumed power deviations is presented through the observation of the load, wind, and generation outage deviations as showed in Eq. (7). Based on some risk management concepts, the power amount that represents the secondary reserve criterion can be viewed as a value-at-risk [38], and the power amount that represents the fast tertiary reserve can be defined as a complement of the value-at-risk [39], since the fast tertiary reserve accounts for the power amounts exceeding the secondary reserve criterion. Therefore, let X denote a random variable defined as the summation of the load, wind, and generation outage deviations at hour "t" as shown in Fig. 2. Thus, the discrete variable X can be written as a probability function with possible values x_1, x_2, ..., x_n, as follows:

$$f(x_i) = P(X = x_i) \tag{8}$$

As $f(x_i)$ is defined as a probability, the following conditions must be verified:

$$f(x_i) \geq 0 \quad \forall x_i \in \Re \tag{9}$$

$$\sum_{i=1}^{n} f(x_i) = 1 \tag{10}$$

And the cumulative distribution function can be written as:

$$F(x) = P(X \leq x) = \sum_{x_i \leq x} f(x_i) \tag{11}$$

At this point, it is important to observe that x_i can assume positive and negative values, depending on the random process associated with the load forecasting error (short- and long-term), the wind forecasting error, as well as the random up and down cycles of generating units. Table 1 shows an illustrative example of this case, where the negative values can be interpreted as downward reserve and the positive values can be interpreted as upward reserve.

From the operational reserve point of view, firstly, only the positive values are assumed in order to define the power amount needs as secondary reserve criterion

Table 1 Reserve requirements illustrative example

	X_1	X_2	X_3	X_4	X_5
x_i (MW)	−20	−10	0	10	20
$f(x_i)$	1/8	2/8	2/8	2/8	1/8
$F(x)$	1/8	3/8	5/8	7/8	1
	Downward reserve			Upward reserve	

Table 2 Reserve requirements observation methodology

	X_1	X_2	X_3
x_i (MW)	(−P, 0]	(0, 10]	(10, 20]
$f(x_i)$	5/8	2/8	1/8
$F(x_i)$	5/8	7/8	1
	Upward reserve		

within a direct confidence interval. Therefore, all negative possible values to x_i are accumulated in 0 MW, as shown in Table 2. Another important issue is how to collect all observed values of the random variable X. During the simulation process it is possible to collect all values x_i since $x_i = x_t$, where "t" ranges between 1 and 8,760 h. After that, the values x_t can be organized into classes of power amount needs, and the higher value (usually at the right side of the classes) is used to represent X.

One of the first possible evaluations, which can be done through the use of this methodology, is to calculate the reserve needs expectation (*RNE*) value. Eq. (12) shows the expected value definition to the random variable X.

$$E(X) = RNE = \sum_{x_i} x_i f(x_i) \tag{12}$$

Before continuing, it is mandatory to remember some general aspects related to the load, wind, and generation outage deviations presented in the operational reserve proposed methodology. In fact, several factors can influence the reserve needs measured during this simulation process. However, some of them can directly influence the results, such as, the level of the load forecasting error defined for short- and long-term deviations, the wind forecasting error model, and the stochastic parameters of each generating unit. The latter has a significant participation in this methodology, since the definition of the primary and secondary reserve power amounts can directly influence the generation outage deviations.

Due to the very close correspondence between secondary and fast tertiary reserve, the estimated power amount for secondary reserve task must be robust and may automatically disregard outliers and rare power deviations. On the other hand, fast tertiary reserve can assume the task of covering outliers and rare power deviations, which can be interpreted as an alternative percentile measure or a complement of the value-at-risk assumed by secondary reserve. Thus, the secondary and fast tertiary reserve criteria, here named Secondary Reserve Criterion (S_{RC}) and Tertiary Reserve Criterion (T_{RC}), can be defined in the same way as value-at-risk and a complement of the value-at-risk [38], as follows:

$$S_{RC\alpha}(X) = \min\{x|\, F(x) \geq \alpha\}\ \text{MW} \tag{13}$$

$$T_{RC\alpha}(X) = \min\{x|\, \alpha < F(x) \leq \delta\}\ \text{MW} \tag{14}$$

where the secondary reserve of X (MW) associated with the confidence level $\alpha \in \,]0,1[$ is given by the minimum x such that the cumulative probability function evaluated in x is greater or equal to α.

As consequence, the fast tertiary reserve of X (MW) associated with the confidence level $\delta \in \,]0,1[$ can be written as the complement of the S_{RC}, and is given by the minimum x such that the cumulative probability function evaluated in x is smaller or equal to δ, such that δ is greater than α. Figure 3 shows a small example based on Table 2, where the *RNE* is 5 MW, the S_{RC} is 10 MW associated with a confidence level of 0.875 (or 87.5 %), and the T_{RC} is 20 MW. Note that, in this probability function there are only three values (impulses), where the complement of the S_{RC} (T_{RC}) is coincident to the last and the highest power deviation.

Obviously, the probability choices must be based on the secondary and fast tertiary reserve available in the generation plant. However, this methodology could be viewed as an alternative to define the trade-off between S_{RC} and T_{RC}, through the use of a well-known risk measure. In order to explore these concepts, the next section will discuss some aspects about these proposed methodologies in comparison with other approaches such as heurist or deterministic rules, as Eq. (1), and loss of the largest unit criterion.

5 Application Results

The proposed methodologies will be applied in two different ways. Firstly, it will follow a didactic purpose, where the IEEE-RTS 96 [40] is modified and used in the same way of Ref. [13] in order to show some specific details of the methodologies. For this purpose, the original configuration of the IEEE-RTS 96 is twice modified and renamed to cope with: (a) the power fluctuations of the hydro units (IEEE-RTS 96 H), and (b) wind hourly power fluctuations (IEEE-RTS 96 HW), where one of the coal units of 350 MW is replaced by 1,526 MW of wind power [13]. Secondly, a long term reserve requirements assessment is presented through the use of the Portuguese and Spanish system configurations (PGS and SGS, respectively) [41]. In these cases, some different problems linked to large systems are discussed, and not only wind power will be assessed, but also other renewable such as solar thermal and photovoltaic, cogeneration, and small hydro systems. The cases will show the results for the years 2008–2025.

5.1 IEEE-RTS 96 H and HW Configurations

The original configuration of the IEEE-RTS 96 consists of 96 generating units divided into five different technologies with a total installed capacity of 10,215 MW and the annual peak load of 8,850 MW. It is important to notice that

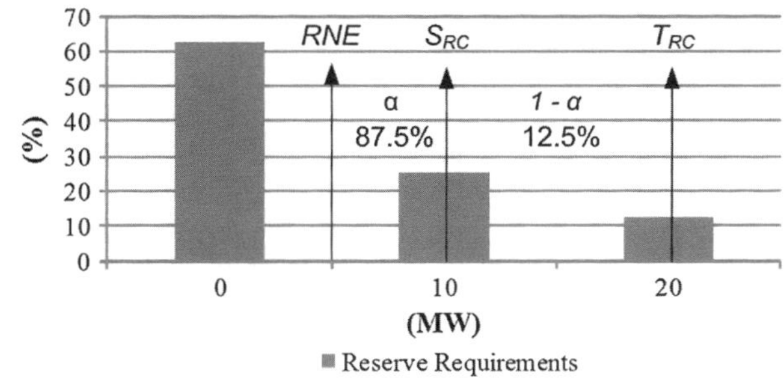

Fig. 3 Distribution function example

the static reserve corresponds to 16.3 % of the total installed capacity. From the renewable point of view, only 8.8 %, which consist of 900 MW on hydro power plants, are considered renewable and 9,315 MW are divided into thermal technologies. Figure 4a shows the capacity installed of each technology.

As previously stated, the first modification consists of the same original configuration considering hydro power fluctuation. However, the second modification consists of increasing the total installed capacity from 10,215 to 11,391 MW, replacing 350 MW of coal per 1,526 MW of wind power, as shown in Fig. 4b. The ratio between thermal coal and wind power technology (350/1526) is approximately 0.23 and consists of a capacity factor to cope with wind power fluctuations [13]. Therefore, the percentage of renewable also increases from 8.8 to 21.3 %.

The main modification of the IEEE-RTS 96 H and HW is the fluctuation capacity in the hydro and wind power production, which makes it possible to consider seasonal hydro monthly effects and hourly wind variations. In order to consider hydro fluctuation, five historical hydro series are considered for each basin, where the hydro system is divided into three different basins. Figure 5 shows the average monthly hydro fluctuation by each historical year.

Through a simple analysis, one can see that among all historical hydro series, year 2 presents an average volume of 74.05 %, which can be considered the worst hydrological condition. The best hydrological condition happens in year 3, where the basins reached an average volume of 80.38 % and, consequently, the best hydrological month is February, with an average volume of 89.16 %. For simulation purposes, the Base Case will be performed assuming that all historical hydro series have the same probability of occurring. However, in some variations, the worst and the best hydrological conditions can also be used. From the wind power fluctuation point of view, three important aspects must be highlighted:

- First, it is linked to its availability—for instance, some seasonal aspects identifying favorable and unfavorable production days, weeks, and months.
- Second, it is linked to its variability—for instance, to identify the possible range of the power availability.
- Third, it is linked to its forecasting properties—for instance, reducing the wind power output error or variability effects through the use of information, such as numerical weather prediction, regional weather knowledge and so on.

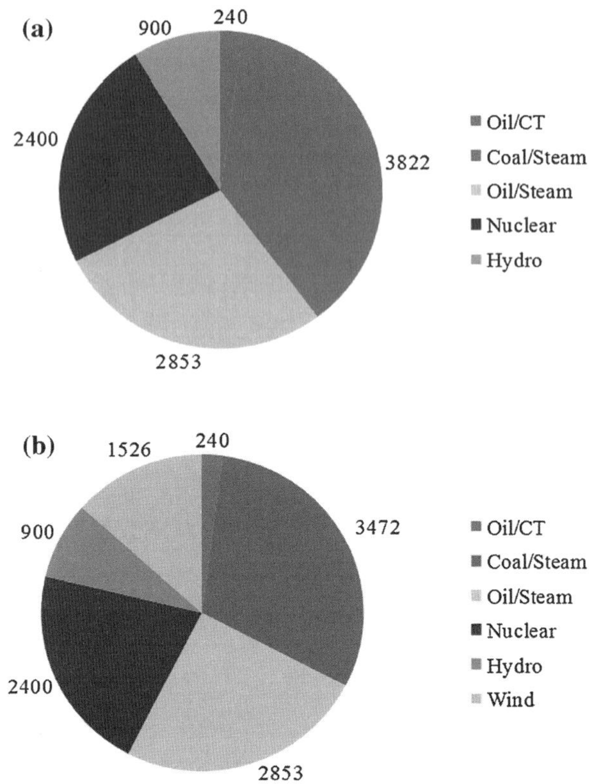

Fig. 4 IEEE-RTS H and HW based technologies

For the purpose of this study, the wind power subsystem consists of 763 units of 2 MW, distributed among three different areas (267-1, 229-2 and 267-3 units-area, respectively) with their own wind characteristics [13]. Each wind area is characterized by its own historical series, on an hourly basis, referring to the average power produced by a wind generator and mainly representing the power fluctuation. Figure 6 gives an idea of the monthly available power in p.u., where the worst condition happens in May, with an average availability of 0.1 p.u. ($\sim$153 MW), and the best condition happens in December, with an average availability circa 0.43 p.u. ($\sim$656 MW).

As previously stated, the range of the variability is an important aspect related to the power fluctuation throughout the year. IEEE-RTS 96 HW presents different ranges of variability per month. Although months like May and September have been classified as the lowest average wind power availability, they present a significant variability as shown in Fig. 7. On the other hand, December has the highest average wind power availability, although with one of the lowest variability rates. Several other related aspects about the wind and hydro power fluctuations are based on real systems, and were carefully integrated in the IEEE-RTS 96 HW in order to better represent real effects in the test system [13].

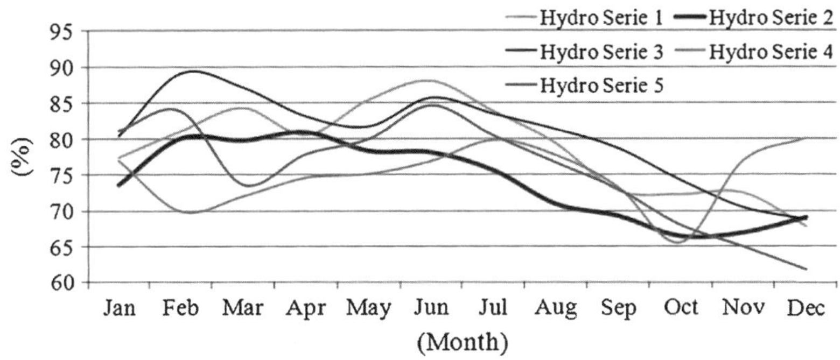

Fig. 5 Average hydro power fluctuations

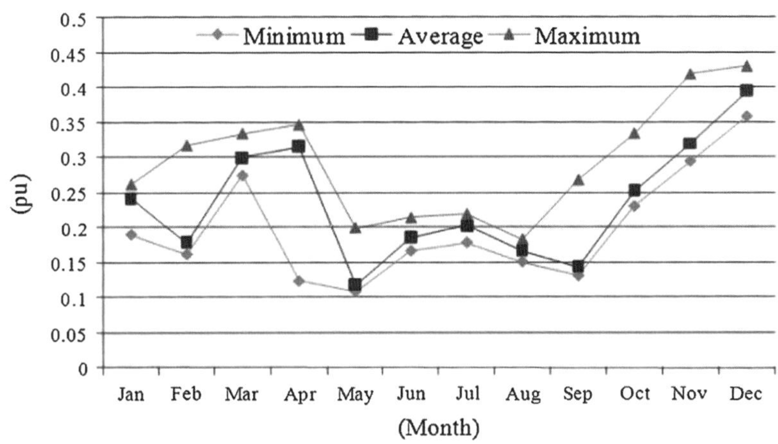

Fig. 6 Monthly wind fluctuation

Another major concern about the operational reserve assessment, presented in this study, is that the proposed methodology is strongly linked to some specific generating unit characteristics. For instance, the start-up times of each generator define the tertiary reserve possible to be used. As previously stated, the operational reserve depends on the defined amount of spinning reserve and also on the generating units that can be synchronized within 1 h. Table 3 shows some important characteristics of the generating units for the IEEE-RTS 96 H and HW.

As it is possible to see, the IEEE-RTS 96 H and HW consist of several generating units with start-up times higher than 1 h, which means that there are few options to deal with the proposed operational reserve assessment. However, it is possible to use this didactic system in order to highlight the needs for a modern generation plant with faster start-up unit technologies to cope with uncertainties that come from the new production paradigm such as renewable resources.

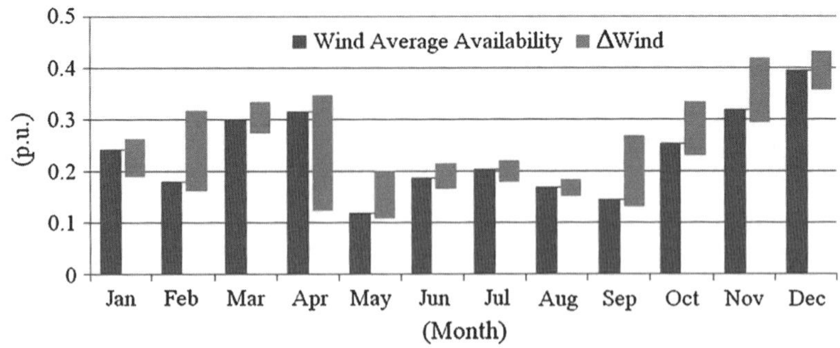

Fig. 7 Average wind power variability

Table 3 Generation of (a) deterministic and (b) stochastic data

(a)						(b)	
Group	Type	Cap. (MW)	No. of units	Cost ($/MWh)	S_{up} time (min)	λ (occ./year)	MTTR (h)
U2	Wind	2.0	763	0.00	5	4.56250	80.0
U12	Oil	12.0	15	30.52	120	2.97959	60.0
U20	Oil	20.0	12	44.26	60	19.46667	50.0
U50	Hydro	50.0	18	0.00	7	4.24424	20.0
U76	Coal	76.0	12	15.92	240	4.46939	40.0
U100	Oil	100.0	9	25.44	480	7.30000	50.0
U155	Coal	155.0	12	12.28	480	9.12500	40.0
U197	Oil	197.0	9	23.07	600	9.22105	50.0
U350	Coal	350.0	3	11.65	2880	7.61739	100.0
U400	Nuclear	400.0	6	6.58	60	7.96364	150.0

5.2 IEEE-RTS 96 H and HW—Application Results

The operational reserve evaluation presented in Eq. (7) is also extremely dependent on the wind power deviation, the load deviation, and the forced up and down cycles of the generating units. Therefore, to perform the simulations considering a good model of these deviations, the ΔP_w in Eq. (7) is assumed to be a forecasting by persistence, which means a deviation based on the difference between the hourly wind production $P_w(t)$ and $P_w(t+1)$. In the same way, to model the deviation ΔL in Eq. (7), two uncertainty levels with Gaussian deviations are used to represent short- and long-term load forecasting [16]. The results presented in this section, the 2 and 1 % load forecasting error, in short- and long-term respectively, are assumed. Finally, the forced up and down cycles of all generating units strongly depend on the stochastic characteristics of each generating units, which are presented in Table 3.

The first discussion will focus on the definition of the primary, secondary (or spinning), and tertiary reserve (or non spinning), which holds the operational reserve proposed methodology. Usually, the primary reserve is based on the largest power deviation to be handled in the system, which can be defined concerning some system characteristics as reliability, size of loads and generating units, interconnection with other systems and, mainly, it is strongly linked to the frequency regulation of each system. For instance, in Europe, it is proposed a share of the total of 3,000 MW along the interconnected European System. As previously stated, the primary (or regulation) reserve is a pre-defined value in the proposed methodology, and can always be redefined. For the purposes of this study, it will be set up as 85 MW to the IEEE-RTS 96 H and HW. Similarly, the secondary (or spinning) reserve is a pre-defined value. However, it is possible to define a confidence interval for the spinning power amount through some probabilistic inferences.

In general, some empirical rules as shown in Eq. (1) are used in order to define the power amount needs to cover power variations. The result achieved in Eq. (1) is a compromise between production and consumption based on half-hourly measurements of these variables. If applied on IEEE-RTS 96 H and HW, it achieves the following power amount:

$$R_{\text{sec}} = \sqrt{10 \times 8,850 + 150^2} - 150 = 183.16\ \text{MW} \tag{15}$$

Note that, in this case, the recommended secondary reserve value is defined regardless of the generating unit technology used in the system, and it depends only on the peak load into the secondary control area. Nevertheless, the most expressive secondary reserve recommendation is an amount equivalent to the largest unit in the system or secondary control area. In IEEE-RTS 96 H and HW, the largest unit is 400 MW.

5.2.1 Operational Reserve Discussions

In order to assess both criteria, Table 4 shows the conventional reliability indices: First, for static reserve, to establish some reference values to the IEEE-RTS 96 H and HW based on the previously described conditions (e.g. load uncertainty level). Second, for the operational reserve, considering a dispatch order based on the production costs presented in Table 3, where all wind and hydro power are primarily dispatched and a fast tertiary reserve is composed by those generating units that can be synchronized within 1 h. In this sense, Table 3 shows, through the use of the start-up time characteristics, that groups U20 and U400 could participate in the fast tertiary reserve. However, due to its attractive production cost characteristics and other security criteria, the group U400 is ordered after the hydro technology. Therefore, it is possible that this economical/security decision will probably affect the operational reserve performance, since it reduces the number of

Table 4 Static and operational reserve reliability indices

Indices	Conventional indices							
	$P_{RC} = 85$ MW; $S_{RC} = 183.16$ MW				$P_{RC} = 85$ MW; $S_{RC} = 400$ MW			
	Static		Operational		Static		Operational	
		(%)		(%)		(%)		(%)
IEEE-RTS 96 H								
LOLP	0.86E-04	3.95	0.19E-02	2.35	0.86E-04	3.95	0.20E-03	2.60
LOLE	0.7569	3.95	16.73	2.35	0.7569	3.95	1.795	2.60
EPNS	0.18E-01	5.71	0.1547	5.00	0.17E-01	5.71	0.32E-01	4.98
EENS	155.1	5.71	1355.0	5.00	155.1	5.71	280.6	4.98
LOLF	0.4580	1.00	27.01	1.77	0.4580	1.00	2.087	1.21
LOLD	1.653	3.73	0.6195	0.84	1.653	3.73	0.8601	2.57
IEEE-RTS 96 HW								
LOLP	0.63E-04	4.70	0.19E-02	3.08	0.63E-04	4.70	0.18E-03	2.44
LOLE	0.5476	4.70	16.33	3.08	0.5476	4.70	1.580	2.44
EPNS	0.12E-01	6.60	0.1416	4.23	0.12E-01	6.60	0.25E-01	4.98
EENS	106.9	6.60	1241.0	4.23	106.9	6.60	217.1	4.98
LOLF	0.3337	1.00	27.36	2.72	0.3337	1.00	2.105	1.07
LOLD	1.641	4.40	0.5968	100.00	1.641	4.40	0.7503	2.35

generating units that can participate in the fast tertiary reserve to 10 generating units of the group U20 (200 MW).

The stop criteria used in the simulation process is a coefficient of convergence $\beta = 5\ \%$ to the EENS index or 10.000 simulation years. This first simulation consists of the Base Case and will be performed assuming that all historical hydro series have the same probability of occurring in both cases, such as the wind series with the same probability of occurring as in the IEEE-RTS 96 HW.

From the static reserve point of view, both configurations seem to be very robust and can be considered adequate to meet the load requirements. In the IEEE-RTS 96 H, which considers the hydro fluctuation, the risk-based level can be measured through the presented reliability indices, where the LOLE index is 0.7569 h/year with the EENS of 155.1 MWh. The frequency of the failure events (LOLF) is 0.4580 occ./year with an average duration (LOLD) of 1.653 h. A vertical comparison (in Table 4) reveals that the system reliability increases when wind power is included in the IEEE-RTS 96 HW configuration. As it is possible to see, the LOLE index decreases to 0.5476 h/year, as the EENS index decreases to 106.9 MWh. The frequency of the failure events (LOLF) also decreases to 0.3337 occ./year. However, the average duration of each failure events remains almost the same, LOLD = 1.641 h, mainly due to the similar MTTR of the coal and wind generating units. The substitution of 350 MW of coal by 1,526 MW of wind accounts for these improvements, mainly due to the inclusion of several small generating units in place of a big generating unit. As expected, in the horizontal comparison (in Table 4), the static reserve is not affected by the secondary reserve criterion in both configurations, which explains the same results presented for static reserve in Table 4.

Another major concern regarding the performed simulation is the operational reserve behavior for both configurations. Comparing the results between the IEEE-RTS 96 H and HW systems in Table 4, the reliability performance improves when wind sources are added to the system. However, one of the most important comparisons must be performed horizontally in Table 4, where the two secondary reserve recommendations are compared. In the IEEE-RTS 96 H, the first criterion (S_{RC} = 183 MW) seems to be insufficient to cover the wind, load, and generation outage deviations. Although a LOLE index of 16.73 h/year could not represent a high risk-based level to the overall system (depending on the operator standards), it seems to be very far from the static reserve risk, which can mean that there is a lack of adequate generating units between both evaluations. On the other hand, the second criterion (S_{RC} = 400 MW) puts the operational reserve system performance very close to the static reserve risk, with an adequate level of LOLE of 1.795 h/year. However, there is no information about the reserve requirements needs and, essentially, if this amount of power adequately covers all power deviations without spillage resources.

In order to evaluate both secondary reserve criteria, Table 5 shows the performance of the system reliability considering that there is no fast tertiary reserve in both configurations. In these cases, only the secondary reserve criterion will try to meet the reserve requirements. Obviously, these results are only useful to identify if, on average, the criterion is adequate or not to the overall system.

Table 5 Secondary reserve reliability indices

Indices	Conventional indices			
	P_{RC} = 85 MW; S_{RC} = 183 MW		P_{RC} = 85 MW; S_{RC} = 400 MW	
	Secondary reserve	(%)	Secondary reserve	(%)
IEEE-RTS 96 H				
LOLP	0.2333E-01	1.01	0.1937E-02	1.97
LOLE	204.3	1.01	16.97	1.97
EPNS	1.871	1.46	0.1557	4.58
EENS	0.1639E+05	1.46	1364.0	4.58
LOLF	221.0	0.96	27.51	1.33
LOLD	0.9245	0.17	0.6169	100.0
IEEE-RTS 96 HW				
LOLP	0.259E-01	1.18	0.1941E-02	3.12
LOLE	227.1	1.18	17.00	3.12
EPNS	1.982	1.37	0.1403	3.67
EENS	0.1736E+05	1.37	1229.0	3.67
LOLF	249.3	1.18	28.78	2.55
LOLD	0.9111	1.02	0.5908	100.0

As expected, the first secondary reserve criterion of 183 MW proves to be insufficient to adequately meet the reserve requirements in the IEEE-RTS 96 H. The high LOLE index of 204.3 h/year and the high frequency (LOLF) of 221.0 occ./year, show that the system extremely depends on the fast tertiary reserve and, even so, the tertiary reserve available is also insufficient to achieve an operational reserve performance near the static reserve risk as seen before. On the contrary, the second secondary reserve criterion of 400 MW proves to be sufficient to meet the reserve requirements, and mainly it shows a lower dependence of the fast tertiary reserve. However, there is no information if this lower dependence of the fast tertiary reserve could be caused by an undue secondary reserve criterion and, consequently, leads to spillage resources and ancillary services.

Contrary to the static and operational reserve performances, which increase when wind production is added to the system, the secondary reserve performance degenerates when a large amount of wind production is included in the IEEE-RTS 96 HW. The LOLE index increases from 204.3 to 227.1 h/year, and this fact can be explained through the inclusion of the wind power deviations that come from the wind forecasting error, which raises the reserve requirements. Similarly to the previous configuration analyses, the first secondary reserve criterion of 183 MW proves to be insufficient to adequately meet the reserve requirements, making this configuration more dependent on the fast tertiary reserve than the IEEE-RTS 96 H. Once again, the secondary reserve criterion of 400 MW is sufficient to adequately meet the reserve requirements in IEEE-RTS 96 HW, with a lower dependence of the fast tertiary reserve. However, this lower dependence could be interpreted as an excessive power amount to the secondary reserve criterion and, consequently, spillage resources.

So far, both criteria seem to be inadequate to define the secondary reserve power amount for IEEE-RTS 96 H and HW. On one hand, the empirical rule seems to lead to an insufficient power amount and a high dependence of the fast tertiary reserve, causing a high risk-based level on the overall system. On the other hand, the largest unit used as a secondary reserve criterion seems to be sufficient to meet the load, wind, and generation outage deviations. However, there is no guarantee that it is an adequate power amount and does not lead to spillage resources.

5.2.2 Trade-off Between S_{RC} and T_{RC} Discussions

Thus, through the use of the *RNE* definition presented in Eq. (12), it is possible to calculate this index to the IEEE-RTS 96 H and HW using both secondary reserve criteria (183 and 400 MW). Table 6 shows the results of the *RNE*. Note that, as the other index presented before, for instance LOLE, it is possible to calculate the coefficient of convergence β for this expected value, highlighting a confidence interval on the result.

In the previous evaluations for IEEE-RTS 96 H, the two secondary reserve criteria (183 and 400 MW) have caused the system to have different generation

Table 6 RNE: reserve needs expectation to the IEEE-RTS H and HW

System	Secondary reserve criterion (MW)	*RNE* (MW)	β (%)
IEEE-RTS 96 H	183	45.66	0.23
	400	45.98	0.14
IEEE-RTS 96 HW	183	51.30	0.85
	400	51.46	0.85

outage deviations and, consequently, different reserve needs. Note that the *RNE* to the IEEE-RTS 96 H and HW using the same secondary reserve criteria is very close. However, as mentioned before, they are different due to the generating outage deviations presented in Eq. (12). Other interesting evaluation is that the *RNE* is lower than both defined criteria (183 and 400 MW), which could lead to the hazard interpretation, where the *RNE* would be used as a secondary reserve criterion. Obviously, it would be possible if the fast tertiary reserve available could be effectively enough to cover all deviation cases. However, it was previously discussed that a power amount of 183 MW, which is higher than *RNE,* has proved to be insufficient to cover all deviation cases with the fast tertiary reserve available in this system.

Therefore, the main idea of this section is to identify an adequate level of power amount as a secondary reserve criterion associated to a confidence interval and, consequently, to assess a risk-based level of fast tertiary reserve through the use of the individual and cumulative probability density function of this defined random variable *X*. Hence, based on the previous analyses, the assessment will be focused in the IEEE-RTS 96 H and HW with 85 MW as primary reserve and 400 MW as secondary reserve criteria. Table 7 shows the observations of *X* to the IEEE-RTS 96 H and HW.

Note that, for the IEEE-RTS 96 H, the $P(X = 0.0\ \text{MW}) = 58.49\ \%$, whereas for the IEEE-RTS 96 HW, the $P(X = 0.0\ \text{MW}) = 54.32\ \%$. It is important to remember that all negative values (downward reserve) are accumulated in zero, which can be interpreted as needless reserve. Therefore, it is possible to see that, when the wind production variability is included in the system, the reserve needs seem to increase, from $P(X > 0.0\ \text{MW}) = 1 - P(X = 0.0\ \text{MW}) = 41.51\ \%$ to $P(X > 0.0\ \text{MW}) = 1 - P(X = 0.0\ \text{MW}) = 45.68\ \%$. In fact, at the end of Table 7, it is possible to identify the most severe event of reserve needs to the IEEE-RTS 96 H when $X = 1{,}099.39$ MW with the probability of occurring of $P(X = 1{,}099.39\ \text{MW}) = 6 \times 10^{-5}\ \%$, whereas the most severe event of reserve needs registered to the IEEE-RTS 96 HW happens when $X = 965.95$ MW with the probability of occurring of $P(X = 965.95\ \text{MW}) = 23 \times 10^{-4}\ \%$. Obviously, the event $X = 1{,}099.39$ MW has a lower chance of happening than event $X = 965.95$ MW. However, it is highlighted that the loss of 350 MW has a significant impact in severity on the system, whereas the distributed effect caused by several wind power units (1,526 MW) reduces this severity, and can be viewed as a benefit to the system reliability as previously discussed, since an adequate

Table 7 IEEE-RTS 96 H and HW—individual probability $f(x_i)$ and cumulative probability $F(x)$

IEEE-RTS 96 H—$S_{Reserve}$ = 400 MW			IEEE-RTS 96 HW—$S_{Reserve}$ = 400 MW		
X (MW)	$f(x_i)$ (%)	$F(x) = \sum_{X_1 \leq X} f(x_i)$ (%)	X (MW)	$f(x_i)$ (%)	$F(x) = \sum_{X_1 \leq X} f(x_i)$ (%)
0.00	58.49574	58.49574	0.00	54.32408	54.32408
52.35	17.69306	76.18880	50.84	18.17577	72.49985
104.70	11.74604	87.93484	101.68	12.80593	85.30578
157.06	6.37706	94.31190	152.52	7.50431	92.81009
209.41	3.09909	97.41099	203.36	3.84985	96.65994
261.76	1.40803	98.81902	254.20	1.82236	98.48230
314.11	0.63585	99.45487	305.04	0.81596	99.29826
366.46	0.29665	99.75152	355.88	0.37481	99.67306
418.81	0.13520	99.88672	406.72	0.17903	99.85209
471.17	0.06413	99.95085	457.56	0.07885	99.93095
523.52	0.02904	99.97989	508.40	0.03840	99.96935
575.87	0.01158	99.99147	559.24	0.01870	99.98805
628.22	0.00534	99.99681	610.07	0.00802	99.99607
680.57	0.00209	99.99890	660.91	0.00273	99.99880
732.92	0.00063	99.99953	711.75	0.00055	99.99935
785.28	0.00022	99.99975	762.59	0.00021	99.99956
837.63	0.00009	99.99983	813.43	0.00021	99.99977
889.98	0.00005	99.99989	864.27	0.00000	99.99977
942.33	0.00004	99.99992	915.11	0.00000	99.99977
994.68	0.00002	99.99994	965.95	0.00023	100.00000
1047.03	0.00000	99.99994			
1099.39	0.00006	100.00000			

capacity factor of 0.23 (350 MW of coal replaced by 1,526 MW of Wind) can be correctly applied.

Once again, with Table 7, it is possible to identify an adequate secondary reserve criterion, dealing with the individual and cumulative probabilities of the reserve needs events. In order to present a simple example, two secondary reserve criteria will be evaluated. Firstly, to the IEEE-RTS 96 H, a S_{RC} based on the individual probability of reserve needs of $X = 314.11$ MW leads to $P(X = 314.11$ MW$) = 0.63$ %, and consequently to F(x) = 99.45 %, which means that this secondary reserve criterion can cover 99.45 % of the deviation cases ($S_{RC} = 314$ MW). Perhaps, it can be viewed as a conservative secondary criterion. However, it seems to be adequate mainly due to the reduced options for fast tertiary reserve available in this system. Note that, following the fast tertiary reserve criterion defined in Eq. (14), the T_{RC} could be defined with a confidence level $\delta = 0.9997$, considering that there are 200 MW available to the fast tertiary reserve. Secondly, to the IEEE-RTS 96 HW, a S_{RC} based on the individual probability of reserve needs of $X = 305.04$ MW leads to $P(X = 305.04$ MW$) = 0.81$ %, and consequently to F(x) = 99.29 %, which means that this S_{RC} can cover 99.29 % of the deviation cases, and the T_{RC} could be defined with a confidence level $\delta = 0.9996$, once again considering that there are 200 MW

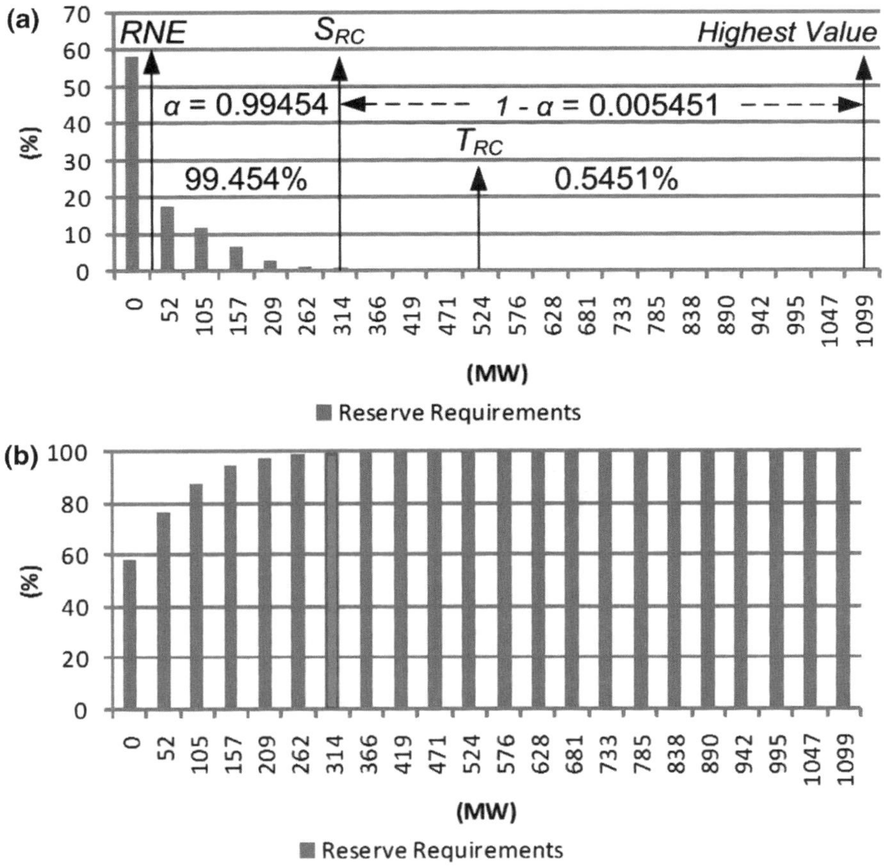

Fig. 8 IEEE-RTS 96 H—$P_{RC} = 85$ MW and $S_{RC} = 400$ MW **a** individual and **b** cumulative distribution density function of X

available to the fast tertiary reserve. Figures 8 and 9 show the individual and cumulative probability density function presented in Table 7.

Observe that, in both cases, the fast tertiary reserve has a small participation to cover the load, wind, and generation outage deviations since the secondary reserve criterion can cover more than 99 % of these deviations. The next analyses will focus on the use of these secondary reserve criteria.

As expected, neither empirical rules nor conservative recommendations based on power engineering practices can efficiently recommend a best trade-off between secondary and fast tertiary reserve criteria. Table 8 shows some illustrative results of the operational reserve based on S_{RC} and T_{RC} defined through the proposed methodology for both systems (IEEE-RTS 96 H and HW). As it can be observed, the proposed criteria is reduced approximately 90 MW of power in relation to the previous one (400 MW), promoting a good balance between secondary and fast

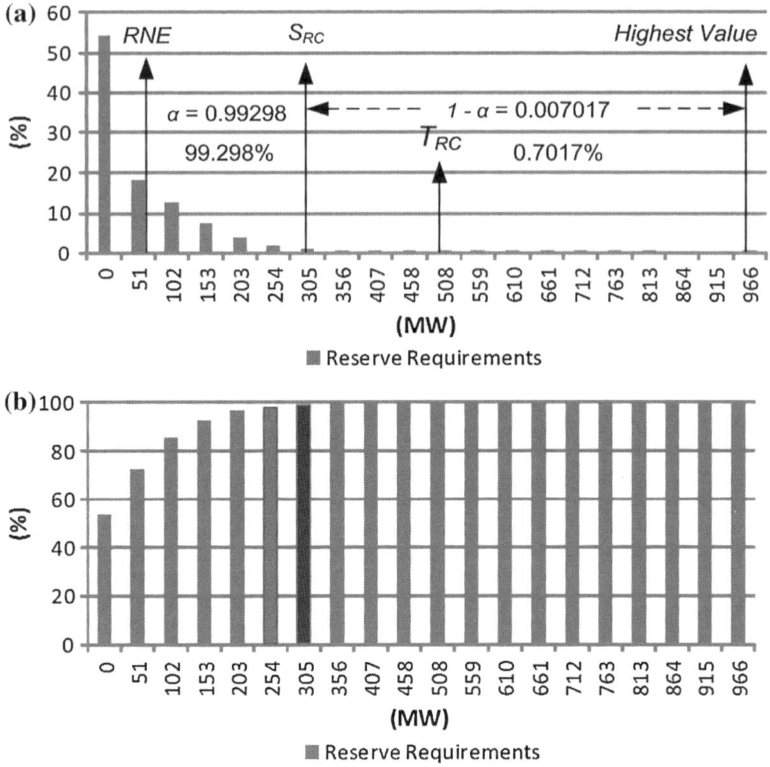

Fig. 9 IEEE-RTS 96 HW—$P_{RC} = 85$ MW and $S_{RC} = 400$ MW **a** individual and **b** cumulative distribution density function of X

Table 8 Operational and secondary reserve reliability indices

Indices	Conventional indices							
	IEEE-RTS 96 H				IEEE-RTS 96 HW			
	$P_{RC} = 85$ MW; $S_{RC} = 314$ MW				$P_{RC} = 85$ MW; $S_{RC} = 305$ MW			
	Operational		Secondary		Operational		Secondary	
		(%)		(%)		(%)		(%)
LOLP	0.41E-03	1.89	0.48E-02	2.16	0.47E-03	1.73	0.54E-02	1.79
LOLE	3.604	1.89	42.0	2.16	4.088	1.73	47.73	1.79
EPNS	0.45E-01	4.95	0.4052	3.60	0.45E-01	4.97	0.4502	1.98
EENS	391.0	4.95	3549.0	3.60	390.7	4.97	3.944	1.98
LOLF	5.660	0.88	59.54	1.88	6.968	0.81	66.78	1.64
LOLD	0.6368	1.63	0.70	100.0	0.5866	1.39	0.7147	100.0

tertiary reserve risk. In this case, it is possible to reduce the existing information gap between both reserve criteria and, consequently, to identify the system dependence of fast tertiary reserve, making sure that there are no spillage resources.

Bearing the previous operational reserve reliability indices in mind, to 183 MW and 400 MW as a secondary reserve criterion for both systems, it is possible to see that the LOLE index was reduced approximately from 16 to 4 h/year regarding to the first criterion (183 MW) and increased approximately from 1.7 to 4 h/year regarding to the second criterion (400 MW). In the same sense, the secondary reserve reliability indices, the LOLE index decreased approximately from 227 to 42 h/year regarding to the first criterion (183 MW) and increased approximately from 16 to 47 h/year regarding to the second criterion (400 MW). This sensitivity risk analysis reveals a balance between secondary and fast tertiary risk-based level concerning the fast tertiary reserve available. Another important point is that both criteria are adequately adjusted to its own variability, where the differences between IEEE-RTS 96 H and HW are compensated through the small power reduction (from 314 to 305 MW, respectively). Observe that, a more reduced secondary reserve criterion has been used for the IEEE-RTS 96 HW, promoting approximately the same results as those for the IEEE-RTS 96 H. Obviously, other criteria could be used in order to improve this risk analysis, mainly reducing the secondary reserve criterion and testing a more flexible fast tertiary reserve, for instance, with gas generating units. However, the target of this section is to present an alternative way to assess secondary and fast tertiary reserve.

5.3 Portuguese and Spanish Cases

The proposed algorithm has been tested under several conditions with different systems. Two cases using the Portuguese and Spanish Generating System (PGS and SGS, respectively) will be discussed, as follows. The first case will show the results using the PGS and SGS configurations for the year 2005. The idea is to evaluate these configurations to establish certain reliability parameters or standards for the years to come. The second case will illustrate the results obtained with the proposed algorithm considering *potential* configurations for the year 2020. This is part of the expansion plan studies of the PGS and SGS.

5.3.1 Portuguese and Spanish Configurations

For the year 2005, the PGS had 1,035 units with a total installed capacity of 12.59 GW, distributed as follows: 4.38 GW (Hydro); 5.43 GW (Thermal); 0.98 GW (Wind); 0.34 GW (Mini-hydro); and 1.46 GW (Co-generation). The annual peak load occurred in January and it was approximately 8.53 GW. Also, the amount of renewable power in the system is 45 % of the total capacity.

In 2020, this amount will correspond to 60 % of the total capacity, from this total 5.21 GW is of wind power.

For the year 2005, the SGS had 8,150 units with a total installed capacity of 70.2 GW, distributed as follows: 15.17 GW (Hydro); 37.1 GW (Thermal); 9.54 GW (Wind); 0.79 GW (Mini-hydro); and 7.6 GW (Co-generation). The annual peak load occurred in January and it was approximately 43.14 GW. Also, the amount of renewable power in the system is 36 % of the total capacity. In 2020, this amount will correspond to 47 % of the total capacity, from this total 28.16 GW is of wind power.

In 2005, there were 35 hydro power plants in Portugal and 174 in Spain. In this study, 6 hydrological basins in Portugal and also 6 in Spain were considered, and 16 years of monthly hydrological conditions were used (1990–2005). The driest year was 2005 for Portugal and 1992 for Spain, and the wettest year was 1996 for Portugal and 2003 for Spain. These monthly variations must be duly captured by the MCS-based reliability assessment proposed algorithm.

In 2005, there were 8 thermal power plants in Portugal and 74 in Spain (not including co-generation). About 655 wind-generating turbines (units) were in the PGS and 6,365 in the SGS. Bearing in mind the wind series, Portugal was divided into 7 regions and Spain into 18 regions. In order to model the deviation ΔP_W in (6), it was assumed a forecasting error by persistence. The error was then estimated by comparing the wind power availability in subsequent hours.

The actual 2005 hourly load curves were used for both PGS and SGS. Only uncertainty of the short-term forecasting was simulated using a Gaussian distribution. Sensitivity analyses were preliminarily carried out to validate the distribution parameters used in Portugal and Spain.

Several other data involving mini-hydro and co-generation units, maintenance, etc. have also to be processed. These data are not shown or discussed in this chapter due to the lack of space, but they represent relevant information and also sources of variations, which had to be carefully considered [41].

5.3.2 Discussions and Results

The proposed MCS algorithm is being tested with several possible configurations of the PGS and SGS. The idea is to determine the required amount of system generating reserve capacity to ensure an adequate power supply, bearing in mind not only the uncertainties from the equipment availabilities, but also the uncertainties due to the renewable power sources capacity variations. This study is being carried out considering a horizon of 20 years (2005–2025). Different scenarios involving not only hydro and wind unfavorable conditions, but also co-generation usage and maintenance strategies are being analyzed.

5.3.3 Analysis for the 2005 Configuration

In order to define reliability standards for the PGS and SGS, the *2005 Configuration* was used. Although several operating conditions or cases were tested, only four of them will be discussed. For the Base case, all historical hydrological and wind series were simulated with the *2005 Configuration* for both systems. In the H^+ case, the wettest hydrological year was considered, and in the H^- case, the driest hydrological was simulated. The HWM case considers that driest hydrological condition occurs simultaneously with all observed wind series having their capacities reduced by 50 %. Also, the usual amount of power on maintenance was increased by 20 %. Certainly, this is a very severe scenario.

Table 9 shows the traditional reliability indices for Portugal. As it can be seen from this table, the PGS configuration (2005) is adequate. As it could be expected, the worst condition occurs for "HWM", resulting in indices like: $LOLE_{STA} = 0.023$ h/year and the $LOLE_{OPE} = 0.032$ h/year. Under these "HWM" conditions, the performance can be considered perfectly acceptable.

Table 10 shows the well-being indices, for Spain. One can provide the following interpretation for this system, if everything goes wrong (i.e. HWM case) with the

Table 9 Conventional reliability indices—Portugal 2005

Case	Base	H+	H−	HWM
Indices	*Static reserve*			
LOLE (h/year)	0.006	0.001	0.012	0.023
EENS (MWh)	0.915	0.240	1.704	3.784
LOLF (occ./year)	0.006	0.002	0.011	0.021
LOLD (h)	1.036	0.870	1.057	1.076
	Operational reserve			
LOLE (h/year)	0.036	0.147	0.017	0.032
EENS (MWh)	3.760	12.66	2.542	5.304
LOLF (occ./year)	0.044	0.190	0.016	0.031
LOLD (h)	0.817	0.773	1.092	1.026

Table 10 Well-being indices—Spain 2005

Case	Base	H+	H−	HWM
Indices	*Static reserve*			
E_H (h)	8.760	8.760	8.758	8.754
F_H (occ./year)	0.147	0.000	1.068	3.568
E_M (h)	0.153	0.001	1.107	3.814
F_M (occ./year)	0.144	0.000	0.994	3.455
	Operational reserve			
E_H (h)	8.756	8.758	8.751	8.741
F_H (occ./year)	6.914	1.914	12.12	24.68
E_M (h)	4.188	1.169	7.508	16.20
F_M (occ./year)	6.909	1.917	12.19	24.60

2005 configuration. Bearing in mind the capacity analysis, the SGS will stay, in average per year, 8,754 h in healthy states, 3.814 h in marginal states, and 2.596 h in risk states. Bearing in mind the operating reserve, the SGS will stay, in average, 8,741 h in a healthy state, 16.20 h in a marginal state, and 2.832 h in risk states.

These are indeed very laow values for that particular stressing scenario. It has to be pointed out that the "HWM" scenario impacts more on the Spanish system than on the Portuguese system, since the SGS depended much more on wind sources in 2005: 13.6 % of the installed capacity in Spain, against 7.8 % in Portugal. In conclusion, both systems had very adequate composition for the year 2005. Moreover, the proposed simulation algorithm properly captured the performance of both capacity analysis and operating reserves.

The parameter β was set to 5 % in all tests for EENS index, and a maximum of 3,000 years were simulated. All simulations were carried out in a PC with 2.8 GHz. The CPU time was, in average, 0.4 h for the PGS and 7 h for the SGS. These huge CPU times indicated that convergence was difficult, due to lack of risk states in the simulation, which means that both systems were extremely reliable.

5.3.4 Analysis for Future Configurations

Initially, some sensitivity analyses were carried out with the 2005 configurations, by increasing the peak load of both systems in order to measure the capacity slackness in these configurations. Figure 10 shows the results for the index LOLE associated with the capacity analysis of the SGS ($LOLE_{STA}$). As it can be observed, the system will experience some difficulties in surviving more than 10 % of load growth from the 2005 configurations. Similar sensitivity tests were carried with the $LOLE_{OPE}$ indices, and a similar conclusion can be stated. In order to assess the risks associated with both capacity analysis and operating reserves in the period 2010–2025, the chronological MCS based algorithm is now being run for both the PGS and SGS.

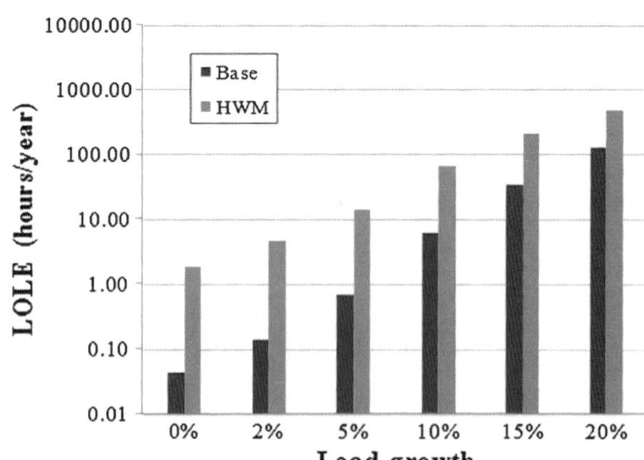

Fig. 10 Sensitivity Analyses—$LOLE_{STA}$ (h/year)

6 Final Remarks

Renewable energy technologies will take a greater share of the electricity generation mix in order to minimize the dependence on oil and the emission of CO_2. While contributions from renewable energy sources for electricity production is small, with the exception of hydro, their market penetration is growing at a much faster rate than any other conventional source. More renewable power sources cause, however, an increase in the number of random variables and operation complexities in the system, due to the *fluctuating* production levels of these sources. Therefore, the determination of the required amount of system capacity (both capacity analysis and operating reserves) to ensure an adequate supply becomes a very important aspect of generating capacity expansion analyses.

The dimensioning of operating reserve, spinning and non-spinning, plays an important role in systems with high penetration levels of renewable sources, mainly those from wind power, due to its natural volatility. Although there are many reference values for LOLE indices related with capacity analysis, there are no such standards for well-being indices or for operating reserves.

By testing recently operated generation arrangements, one can provide some preliminary values for the establishment of future standards. However, most generating systems, including the Portuguese and Spanish ones, have today a smaller amount of *fluctuating* capacity sources, like wind power, than they will have in the future; even considering that Spain has the second largest wind power capacity installed in the world.

Discussions on innovative criteria (e.g. the system has simultaneously to survive the worst hydrological and wind conditions), operation strategies, and assessment tools will be the new insights of generating capacity expansion planning considering renewable power sources for the years to come.

Acknowledgments The authors would like to thank to Luiz Manso and Leonidas Resende (UFSJ), and Diego Issicaba (INESC Porto) for helpful discussions and especially the authors would like to thank the Reserve Project Team of REN and REE who has provided helpful discussions and useful data systems.

References

1. Zervos A (2003) Developing wind energy to meet the Kyoto targets in the European Union. Wind Energy 6:309–319. doi:10.1002/we.93
2. Hl-Hallaj S (2004) More than enviro-friendly—renewable energy is also good for the bottom line. IEEE Power Energ Mag 2(3):16–22
3. Bessa R, Miranda V, Gama J (2008) Wind power forecasting with entropy-based criteria algorithms. In: Proceedings of PMAPS 2008—10th international conference on probabilistic methods applied to power systems, Rincón, Puerto Rico, May, 2008
4. Madsen H, Pinson P, Kariniotakis G, Nielsen HA, Nielsen TS (2005) Standardizing the performance evaluation of short-term wind prediction models. Wind Eng 29(6):475–489

5. Lange M (2005) On the uncertainty of wind power predictions—analysis of the forecast accuracy and statistical distribution of errors. J Sol Energy Eng Trans ASME 127(2):177–194
6. Ummels BC, Gibescu M, Pelgrum E, Kling WL, Brand AJ (2007) Impacts of wind power on thermal generation unit commitment and dispatch. IEEE Trans Power Syst 22(1):44–51
7. Ramakumar R, Slootweg JG, Wozniak L (2007) Guest editorial: introduction to the special issue on wind power. IEEE Trans Energy Convers 22(1):1–3
8. Ernest B, Oakleaf B, Ahlstrom ML, Lange M, Moehelen C, Lange B, Focken U, Rohrig K (2004) Predicting the wind—models and methods of wind forecasting for utility operations planning. Power Energy Mag 2(3):16–22
9. Anstine LT, Burke RE, Casey JE, Holgate R, John RS, Stewart HG (1963) Application probability methods to the determination of spinning reserve requirements for the Pennsylvania—New Jersey—Maryland interconnection. IEEE Trans Power App Syst 82(68):726–735
10. Billinton R, Chowdhury NA (1988) Operating reserve assessment in interconnected generating systems. IEEE Trans Power Syst 3(4):1487–1497
11. Khan ME, Billinton R (1995) Composite system spinning reserve assessment in interconnected systems. Proc Inst Elect Eng Gener Transm Distrib 142(3):305–309
12. Gooi HB, Mendes DP, Bell KRW, Kirschen DS (1999) Optimal scheduling of spinning reserve. IEEE Trans Power Syst 14(4):1485–1492
13. Leite da Silva AM, Sales WS, Manso LAF, Billinton R (2010) Long-term probabilistic evaluation of operating reserve requirements with renewable sources. IEEE Trans Power Syst 25(1):106–116
14. Gouveia EM, Matos M (2009) Evaluating operational risk in a power system with a large amount of wind power. Electr Power Syst Res 79:734–739
15. Matos M, Lopes JP, Rosa M, Ferreira R, Leite da Silva AM, Sales W, Resende L, Manso L, Cabral P, Ferreira M, Martins N, Artaiz C, Soto F, Lopez R (2009) probabilistic evaluation of static and operating reserve requirements of generating systems with renewable power sources: the Portuguese and Spanish cases. Int J Electr Power Energy Syst 31:562–569
16. Billinton R, Allan RN (1996) Reliability evaluation of power systems, 2nd edn. Plenum Press, New York
17. Billinton R, Fotuhi-Firuzabad M (1994) A basic framework for generating system operating health analysis. IEEE Trans Power Syst 9(3):1610–1617
18. Billinton R, Karki R (1999) Application of Monte Carlo simulation to generating system well-being analysis. IEEE Trans PWRS 14(3):1172–1177
19. Leite da Silva AM, Resende LC, Manso LAF, Billinton R (2004) Well-being analysis for composite generation and transmission systems. IEEE Trans Power Syst 19(4):1763–1770
20. UCTE operating handbook—Policies P1: load-frequency control and performance, March 2009. On line available: http://www.entsoe.eu
21. Rebours Y (2008) A comprehensive assessment of markets for frequency and voltage control ancillary services. PhD Thesis approved by the University of Manchester, England
22. Holttinen H (2005) Impact of hourly wind power variations on the system operation in the Nordic countries. Wiley, on line available: www.interscience.wiley.com
23. Shipley R, Patton A, Dennison J (1972) Power reliability cost vs. Worth. IEEE Trans Power Syst PAS-91:2204–2212
24. Jonnavithula A (1997) Composite system reliability evaluation using sequential Monte Carlo simulation. Ph.D. Thesis approved by the College of Graduate Studies and Research of the University of Saskatchewan
25. Leite da Silva AM, Schmitt WF, Cassula AM, Sacramento CE (2005) Analytical and Monte Carlo approaches to evaluate probability distributions of interruption duration. IEEE Trans Power Syst 20(3):1341–1348
26. Billinton R, Wangdee W (2006) Delivery point reliability indices of a bulk electric system using sequential Monte Carlo simulation. IEEE Trans Power Deliv 21(1):345–352
27. Rubinstein RY (1980) Simulation and Monte Carlo method. Wiley, New York

28. Rosa MA (2010) Multi-agent systems applied to assessment of power systems. Ph.D. Thesis approved by the Faculty of Engineering of University of Porto, Feb 2010
29. Rubinstein RY, Kroese DP (2008) Simulation and Monte Carlo Method, 2nd edn. Wiley, New York
30. Billinton R, Li W (1994) Reliability assessment of electric power system using Monte Carlo Methods. Plenum Press, New York
31. Leite da Silva AM, Manso LAF, Sales WS, Resende LC, Aguiar MJQ, Matos MA, Lopes JAP, Miranda V (2007) Application of Monte Carlo simulation to generating system well-being analysis considering renewable sources. Eur Trans Electr Power 17(4):387–400
32. Billinton R, Allan RN (1992) Reliability evaluation of engineering systems: concepts and techniques. Plenum Press, New York
33. Pereira MVF, Balu NJ (1992) Composite generation/transmission reliability evaluation. Proc EEE 80(4):470–491
34. Geuder N, Trieb F, Schillings C, Meyer R, Quaschning V (2003) Comparison of different methods for measuring solar irradiation data. In: Proceedings of international conference on experiences with automatic weather stations
35. Jonnavithula S (1997) Cost/benefit assessment of power system reliability. Ph.D. Thesis approved by the College of Graduate Studies and Research of the University of Saskatchewan
36. Schrock D (1997) Load shape development. Penn Well Publishing Company, Oklahoma
37. Manso LAF, Leite da Silva AM (2004) Non-sequential Monte Carlo simulation for composite reliability assessment with time varying loads. Mag Controle Automação 15(1):93–100, (in Portuguese)
38. Rockafellar RT, Uryasev SP (2000) Optimization of conditional value-at-risk. J Risk 2:21–42
39. Rockafellar RT, Uryasev SP (2002) Conditional value-at-risk for general loss distributions. J Bank Finance 26:1443–1471
40. The IEEE Reliability Test System 1996 (1999) A report prepared by the reliability task force of the application of probability methods subcommittee—IEEE reliability test system. IEEE Trans Power Syst 14(3):1–8
41. INESC Porto (2008) REN and REE: reserve project report, Oct 2008

Part III
Modeling, Control and Maintenance of Wind Facilities

A Review of Floating Wind Turbine Controllers

H. Namik and K. Stol

Abstract A review of the most important contributions made towards designing and testing controllers for floating wind turbines is presented in this chapter. These controllers range in complexity, description detail, simulation results and testing styles, and floating platform structure. Furthermore, they have not yet been compared against each other quantitatively using the same simulation conditions. However, the attributes of each control approach documented in the literature are listed in this chapter. Several approaches that deal with the reduction of platform pitch damping are discussed such as the use of individual blade pitching and tuned mass dampers.

1 Introduction

A recent trend in the wind turbine industry is to install the wind turbines offshore mainly due to better wind resource attributes when compared to onshore wind; these include [10, 31, 34]: stronger and steadier wind; less turbulence resulting in longer turbine life; less vertical shear (i.e. higher wind speeds at lower altitudes) and higher annual mean wind speed. On average offshore winds are 20 % faster. However, offshore winds also have greater extremes, interact with waves, and are harder to measure.

In addition to having better offshore wind resource attributes, placing wind turbines offshore has several design and location advantages; these include [13, 34]:

H. Namik (✉) · K. Stol
Department of Mechanical Engineering, The University of Auckland,
Auckland, New Zealand
e-mail: hhaz001@aucklanduni.ac.nz

K. Stol
e-mail: k.sol@auckland.ac.nz

P. M. Pardalos et al. (eds.), *Handbook of Wind Power Systems*, Energy Systems,
DOI: 10.1007/978-3-642-41080-2_13,

- Turbines operate at maximum efficiency since no restrictions are imposed due to noise and other regulations.
- Proximity to load centres (large cities) thus requiring less transmission distance and access less heavily used transmission lines.
- Reduced visual impact.
- Energy balance: An offshore turbine can recover the energy cost used during manufacture, transport, installation, operation, maintenance, and decommissioning in 3 months. Over its lifetime, the turbine can generate over 100 times that energy.
- Offset emissions: As the contribution of wind energy increases the emissions for power generation are reduced. For example, if 10 % of the UK's power comes from wind energy then UK's annual emissions are reduced by 15 %.
- Economics: Although going offshore has considerable increase in foundation cost, it was shown by the Danish wind industry that offshore wind turbines are feasible [34]. However, the foundation costs increase with increasing depth.

Placing the turbines offshore has the following challenges [14]: increased capital cost due to foundations, integration with the electrical network, and installation, operation and maintenance being subject to weather conditions.

The offshore wind energy potential is very large. For example, the UK offshore energy potential is estimated to be 986 TWh/year while demand is estimated to be 321 TWh/year (based on 2003 estimates); i.e. more than three times the demand [34]. The US energy potential for winds between 10 and 100 km offshore is estimated to be more than 900 GW which is more than the currently installed generation capacity; however, much of this potential is located in deep water that is deeper than 30 m [19].

Currently, the deepest installed fixed-bottom offshore wind turbine is the Beatrice wind farm off the coast of Scotland [1]; it is constructed in 44 m deep water. However, for water deeper than 60 m the most feasible option is to have floating wind turbines [24, 31]. A main challenge of floating wind turbines is that the motions induced by wind and wave conditions are almost impossible to eliminate. Therefore, the design of these turbines must take into account the added degrees of freedom due to platform motions. Hence, the control system can play an important role in reducing these motions.

There are three main floating wind turbine concepts. Each concept uses a different principle to achieve static stability. The three floating concepts (shown in Fig. 1) are: a buoyancy stabilised barge platform, a mooring line stabilised Tension Leg Platform (TLP), and a ballast stabilised spar-buoy. Of course, each concept has its advantages and limitations. In June of 2009, the world's first floating wind turbine, based on the spar-buoy concept with a 2.3 MW wind turbine, was installed off the coast of Norway in 220 m deep water [5]. Early comparisons of all three platforms used simple static or dynamic models that usually excluded the effect of the control system [11, 12, 23, 31, 39].

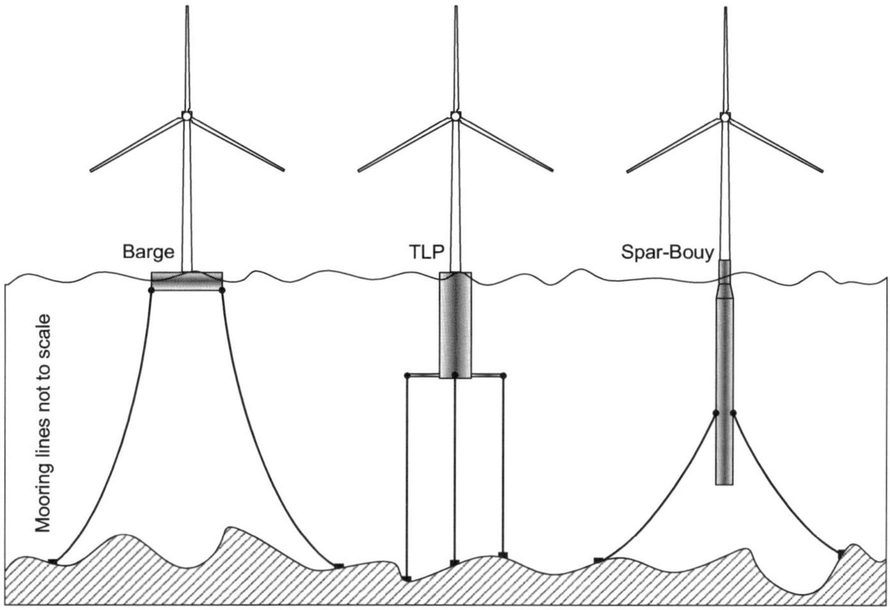

Fig. 1 The three main floating concepts

In this chapter, a brief review of the most important contributions made towards designing and testing controllers for floating wind turbines is presented. Section 2 describes some of the tools used to simulate and test the developed controllers on floating wind turbines. Properties of the main floating platforms and a nominal 5 MW wind turbines are also listed. Differences between collective and individual blade pitching mechanisms from a controls point of view are discussed in Sect. 3. Section 4 reviews several controllers implemented specifically on floating wind turbine platforms. Finally, main conclusions and an outlook into the future of floating controllers are presented in Sect. 5.

2 Simulation Tools and Models

Several simulation and analysis codes have been developed or modified to handle the addition of the floating degrees of freedom (DOFs) and the interaction with the incident waves. In phase IV of the offshore code comparison collaboration (OC3) project[1] [32] within the International Energy Agency (IEA) Wind Task 23 subtask 2, the major codes (listed in Table 1) are compared against each other for a floating

[1] The OC3 project started in 2005 and the last phase, phase IV, was finished and results published in 2010.

Table 1 Participating floating wind turbine simulation codes in the OC3 project (Adapted from [18])

	Code developer	OC3 participant	Aerodynamics	Hydrodynamics	Control system (servo)	Structural dynamics (elastic)
FAST	NREL	NREL + POSTECH	(BEM or GDW) + DS	$Airy^{+}$ + ME, Airy + PF+	DLL, UD, SM	Turbine: FEM^{P} + (Modal/MBS), moorings: QSCE
Bladed	GH	GH	(BEM or GDW) + DS	(Airy + or Stream) + ME	DLL	Turbine: FEM^{P} + (modal/MBS), moorings: UDFD
ADAMS	MSC + NREL + LUH	NREL + LUH	(BEM or GDW) + DS	$Airy^{+}$ + ME, Airy + PF + ME	DLL, UD	Turbine: MBS moorings: QSCE, UDFD
HAWC2	Risø-DTU	Risø-DTU	(BEM or GDW) + DS	Airy + ME	DLL, UD, SM	Turbine: MBS/FEM, moorings: UDFD
3Dfloat	IFE-UMB	IFE-UMB	(BEM or GDW)	Airy + ME	UD	Turbine: FEM, moorings: FEM, UDFD
Simo	MARINTEK	MARINTEK	BEM	Airy + PF + ME	DLL	Turbine: MBS, moorings: QSCE, MSB
SESAM/DeepC	DNV	Acciona energia + NTNU	None	$Airy^{+}$ + ME, Airy + PF + ME	None	Turbine: MBS, moorings: QSCE, FEM

$Airy^{+}$ airy wave theory, $^{+}$ with free surface corrections
BEM blade-element/momentum
DLL external dynamic link library
DNV det norsk veritas
DS dynamic stall
DTU technical university of Denmark
FEM^{P} finite-element method ^{P} for mode processing only
GDW generalised dynamic wake
GH Garrad Hassan and partners limited
IFE institute for energy technology
LUH Leibniz university of Hannover
MBS multibody-dynamics formulation
ME Morison's equation
MSC MSC software corporation
NREL National renewable energy laboratory
NTNU the Norwegian university of science and technology
PF linear potential flow with radiation and diffraction
POSTECH Pohang university of science and technology
QSCE quasi-static catenary equations
SM interface to Simulink® with MATLAB®
UD implementation through user-defined subroutines available
UDFD implementation through user-defined force–displacement relationships
UMB the Norwegian university of life sciences

Table 2 NREL 5 MW turbine model details

Power rating	5 MW
Rotor orientation	Upwind
Control	Variable speed, variable pitch, active yaw
Rotor, hub diameter	126, 3 m
Hub height	90 m
Rated rotor, generator speed	12.1, 1173.7 rpm
Blade operation	Pitch to feather
Maximum blade pith rate	8 deg/s
Rated generator torque	43,093 Nm
Maximum generator torque	47,402 Nm

configuration in a multinational team spanning multiple countries, universities, research institutes and industry.

In phase IV of the OC3 project, the design codes are compared against each other under a specified set of load cases designed to test different model features of the floating system. Results show that most design codes agree across the simulation load cases. Some differences were identified which lead to "greater understanding of offshore floating wind turbine dynamics and modelling techniques, and better knowledge of the validity of various approximations" used by different design codes [18].

2.1 The 5 MW Wind Turbine

For feasible offshore deployment of wind turbines, the minimum power rating of wind turbines is 5 MW [23]. Table 2 lists the properties of a generic 5 MW wind turbine created by Jonkman et al. [17] and used by most of the offshore wind turbine control community; the turbine is commonly known as the "NREL Offshore 5 MW Baseline Wind Turbine". The wind turbine is a fictitious 5 MW machine with its properties based on a collection of existing wind turbines of similar rating.

2.2 Floating Platform Models

The main properties of the three floating platforms (shown in Fig. 1) are listed in Table 3. The ITI Energy barge is a rectangular platform designed to be cost effective and easy to install on site. However, the barge is very sensitive to incident waves since most of the structure is above the water; i.e. it rides the waves rather than pass through them. The result is large platform translations and rotations which directly affect the turbine tower loads and power fluctuations. Conversely,

Table 3 Main properties of the floating platforms

Barge platform		Tension leg platform		Spar-buoy platform	
Width	40 m	Diameter	18 m	Diameter above taper	6.5 m
Length	40 m			Diameter below taper	9.4 m
Height	10 m			Freeboard	10 m
Draft	4 m	Draft	47.89 m	Draft	120 m
Water depth	150 m	Water depth	200 m	Water depth	320 m
Platform mass	5.5×10^6 kg	Platform mass	8.6×10^6 kg	Platform mass	7.5×10^6 kg

the MIT/NREL TLP experiences very little motion compared to the barge platform but the cost of the anchors for the tensioned mooring lines cannot always be justified. Finally, the OC3-Hywind spar-buoy has a cylindrical deep drafted hull with catenary mooring lines. Like the TLP, most of the platform structure is underwater and therefore will not be as sensitive to the incident waves as the barge platform. Having catenary mooring lines means that the spar-buoy experiences a larger motion envelope than the TLP; however, the spar-buoy is generally more cost effective than the TLP due to lower anchoring system costs.

3 Collective and Individual Blade Pitching

There are several actuators available for use by a control system on wind turbines in general. These include: blade pitch, generator torque, turbine yaw drive, passive devices such as tuned mass dampers. For floating systems, passive devices such as oscillating water column can also be used to absorb energy from the waves. For active control, the most commonly used actuators are the blade pitch angle (operated either collectively or individually) and the applied generator torque.

In this section, the physical mechanism behind collective blade pitching (CBP) and individual blade pitching (IBP) are described. The objective is to illustrate the differences between the two pitching schemes in terms of implementation, effectiveness at regulating multiple control objectives, and limitations with an emphasis on their impact on floating wind turbines.

3.1 Collective Blade Pitching

Collective blade pitch (CBP) control is widely used in wind turbine control simply because it provides the necessary actuation required for rotor speed control. It works by changing the symmetric thrust and torque loads on the rotor. It is also easy to implement since all three blades are commanded the same pitch angle and hence combining the actuation of the three blades into one single rotor actuator; useful for single-input single-output (SISO) control. However, when multiple objectives are to be regulated, collective blade pitching may not always provide the necessary actuation without sacrificing the regulation of other objectives or affecting other uncontrolled and/or un-modelled turbine DOFs.

With regards to floating offshore wind turbines, a major limitation of CBP is conflicting blade pitch commands issued by the control system with multiple objectives [whether implemented as separate SISO loops or a single multi-input multi-output (MIMO) controller] [28]. The objectives that conflict with each other in terms of actuator demand are rotor speed regulation and platform pitch motion regulation. Below is a description of how these conflicting blade pitch commands are issued based on physical insight into the floating system.

For simplicity, assume there are two separate control loops: rotor speed controller and platform pitch controller. For restoring platform pitch, the most useful force generated by the blades is rotor thrust. Since the blades cannot influence the platform pitch angle directly but only through the thrust force, we will consider the effects to a change in platform pitching velocity on the turbine in equilibrium. Consider a forward pitching velocity shown in Fig. 2a (negative pitch velocity according to the coordinate system). To keep the turbine in its equilibrium position (not necessarily upright due to the effect of constant wind), the platform pitch controller must generate a positive restoring pitch moment; this can be achieved by increasing the aerodynamic thrust of each blade collectively where it will create a positive pitching moment about the pitch axis as shown in Fig. 2a. Therefore, the blade pitch angles will have to be *decreased*.

Now, consider the effect of the same pitching velocity on the rotor speed. As the platform pitch controller decreases the blade pitch to restore platform pitch, it is also accelerating the rotor by generating additional aerodynamic torque. This is coupled by the fact that as the turbine pitches forward, the wind speed relative to the blade increases thus further accelerating the rotor. Observing an increase in rotor speed, the rotor speed controller commands the blades to reduce the generated torque by *increasing* the collective blade pitch angle to reduce aerodynamic efficiency of the blades. Therefore, the two objectives of speed and platform pitch regulation are competing for commanded blade pitch.

3.2 Individual Blade Pitching

Individual blade pitch (IBP) control where each blade is commanded independently uses a different mechanism from CBP control. It creates asymmetric aerodynamic loads in addition to the symmetric loads created by collective pitching thus enhancing the platform pitch restoring moment [28]. Due to the periodic nature of wind turbines, implementing IBP control results in time variant gains that vary periodically with the rotor azimuth. That is, the controller “knows”

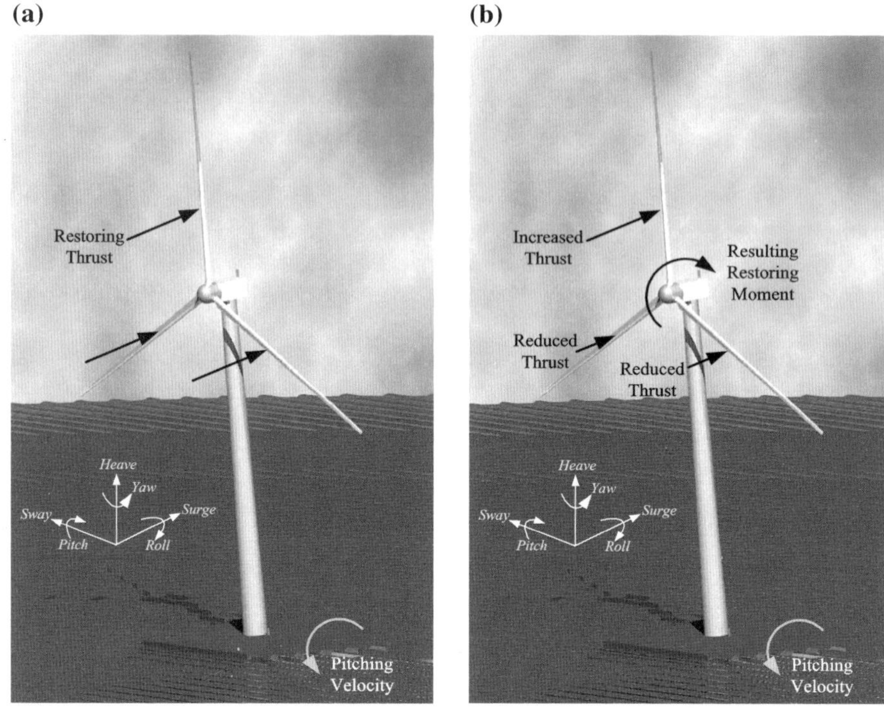

Fig. 2 Platform pitch restoring forces with different blade pitch operation: **a** collective, **b** individual

how the effectiveness of the blade changes as it rotates and acts accordingly to create the necessary restoring forces/moments to achieve its control objectives.

There are many ways to achieve individual blade pitch control depending on the control objectives. Multi-blade coordinate (MBC) transformation (also known as Coleman transformation) captures periodic properties of a system in a linear time-invariant (LTI) model—useful for control design—by transforming the rotating DOFs into a non-rotating frame of Ref. [6]. Bossanyi [8] implemented IBP control using PI controllers to mitigate blade loads. He used a direct and quadrature (d–q) axis representation (a form of MBC) to be able to allow for MIMO control using PI controllers. Direct periodic control allows the controller gains to change depending on the rotor azimuth position and control objectives based on a periodic state space model [36, 38]. Periodic control captures all the periodicity of the system. Although MBC transformation does not capture all the periodic effects, it has been shown that the residual periodic effects are negligible for analysis and control for three bladed wind turbines [37]. Wright [40] achieved IBP control using a disturbance accommodating controller. He used an internal model for the wind disturbance and the effect of wind shear to drive the individual blade pitching.

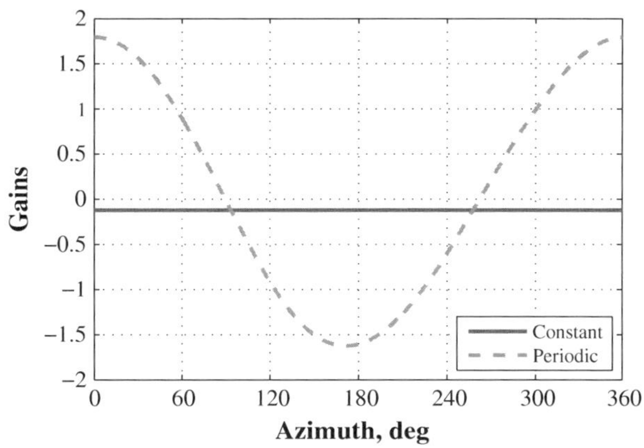

Fig. 3 Blade 1 pitch angle controller gains for the platform pitch velocity for collective and individual blade pitch controllers. Individual blade pitching results in a gain that is periodic with the rotor azimuth

As an example to illustrate the difference between the CBP and IBP controllers, blade 1 control gain for the IBP controller corresponding to platform pitch velocity is shown in Fig. 3 along with the equivalent constant CBP controller gain; both were calculated using the same weighting for the control objectives [28]. The zero azimuth position is at the 12 o'clock position on the rotor with angle increasing clockwise when looking downwind; initially blade 1 is at the 0 azimuth position. The periodic gain changes sign twice as the blade goes through a full rotor revolution. The figure does not show the gains for the other two blades as the gains are the same but shifted 120° and 240° out of phase for blades 2 and 3 respectively. The significance of the sign change will be explained next.

The mechanism can be explained by looking at the periodic gain matrix (Fig. 3). The gain for blade 1 is negative for azimuth angles approximately between 90° and 270°; these angles correspond to the blade lying in the lower half section of the rotor. Therefore, given a negative platform pitch velocity, shown in Fig. 2b, blades at the top with a positive controller gain will be commanded to reduce blade pitch thus increasing thrust. Blades at the bottom with a negative controller gain will be commanded to increase blade pitch and hence reducing thrust. This asymmetric aerodynamic loading will generate a positive restoring pitching moment in addition to the restoring moment generated by the mean thrust load (as with collective pitch) as illustrated in Fig. 2b.

Looking further at the periodic gain shown in Fig. 3, it can be seen that the maximum positive peak is when the blade is at the top (12 o'clock) position. A more positive gain value means increased blade pitching, which indicates that the controller is making use of the combined effect of increased moment arm and wind shear.

The use of IBP increases the number of available actuators to an under-actuated system where the number of degrees of freedom is more than the number of actuators. Furthermore, having the wind turbine mounted on a floating platform makes it even more difficult to regulate the control objectives than turbines with fixed foundations by introducing 6 additional DOFs (surge, sway, heave, roll,

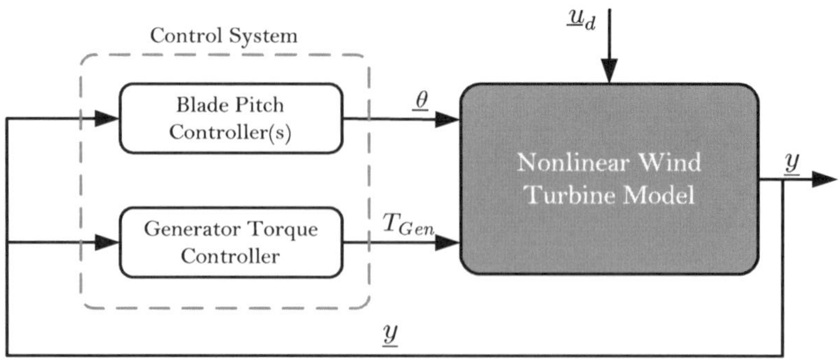

Fig. 4 Basic wind turbine control block diagram

pitch, and yaw). However, there are some issues that arise when implementing IBP that one must be aware of; these include:

- Increased blade pitch actuation which may result in blade pitch saturation and/or increased blade loads (depending on the control objectives).
- Increased computational requirements by the control system.
- The possibility of exciting or destabilising other turbine modes due to coupling with un-modelled and/or unregulated DOFs.

4 Floating Wind Turbine Controllers

Wind turbine control objectives are region dependent. In below rated wind speed region (region 2), the main objective is to maximise power capture while in above rated wind speed region (region 3) the objective is to limit power capture to the rated power of the generator. In addition to these basic control objectives, additional control objectives for wind turbine mainly focus on load reduction for the blades, tower, and drive-train. Having a floating wind turbine, allows the introduction of additional control objectives that focus on mainly reducing the platform rotational motions (roll, pitch, and yaw angles and velocities). Regulating the platform's linear displacements and velocities (surge, sway, and heave) is possible but may require more actuators than just using the blades and generator torque.

In its simplest form, a wind turbine's control system can be represented by the block diagram shown in Fig. 4 where the controller commands the actuators ($\underline{\theta}$ and T_{Gen}) based on some measurements $\underline{y}$ to achieve their control objectives in the presence of disturbances $\underline{u}_d$ such as wind and waves. The implemented controller can be a single centralised MIMO controller or consisting of several SISO control loops. Furthermore, inside each controller block there exist region transition logic and a control scheme for each operating region.

In this section, a summary of the most important controllers specifically designed for floating wind turbines found in the literature is given. These controllers range in complexity, description detail, simulation results and testing styles, and floating platform structure. Furthermore, they have not yet been compared against each other quantitatively using the same simulation conditions. However, the attributes of each control approach are documented in the literature and are listed in this section. The main attributes of these controllers are given in Table 4. In the table, the platform types barge, TLP, and spar-buoy refer to the ITI Energy barge, MIT/NREL TLP, and OC3-Hywind spar-buoy respectively whose properties are listed in Table 3; The Hywind spar-buoy is a slightly different model than the OC3 spar-buoy where it has a different wind turbine mounted from the NREL 5 MW baseline wind turbine. Descriptions of each controller follows.

4.1 Gain Scheduled PI Control

A collective blade pitch gain scheduled proportional-integral (GSPI) controller was implemented on a floating wind turbine by Jonkman to assess the dynamic performance of the barge floating platform [16]. The GSPI control loop only operates in the above rated wind speed region (region 3) to regulate the rotor speed to the rated speed by pitching the blades thus reducing their aerodynamic efficiency. The GSPI control law is given by Eq. 1 where $\theta(t)$ is the commanded collective blade pitch angle, $K_P(\theta)$ and $K_I(\theta)$ are the scheduled proportional and integral gains respectively. The error signal is denoted $e(t)$ with ω_{Rated} and ω_{Gen} as the rated and actual generator speed respectively. The GSPI controller is a SISO controller with a single objective of rotor speed regulation. The gains are scheduled as a function of blade pitch angle θ such that the controlled rotor DOF has the same prescribed natural frequency and damping ratio for all above rated wind speeds. Gain scheduling is a simple form of nonlinear control.

$$\theta(t) = K_P(\theta)e(t) + K_I(\theta)\int_0^t e(\tau)d\tau \quad where \quad e(t) = \omega_{Gen} - \omega_{Rated} \tag{1}$$

In addition to the blade pitch control loop, a separate generator torque control loop was used to maximise power capture in below rated wind speeds and regulate power in above rated wind speeds. The torque controller varies the applied generator torque as a function of filtered generator speed. In below rated wind speed (region 2), the relationship that maximises power capture is given by Eq. 2 where ρ, R_{Rotor}, N, ω_{Gen}, $C_{P,max}$, and λ_o are air density, rotor radius, gearbox ratio, generator rotational speed, maximum power coefficient and tip speed ratio that yields $C_{P,max}$ respectively [7]. In above rated wind speed region where the objective is to regulate power to the rated, the relationship is given by Eq. 3 where P_{Rated} is the rated generator power and η_{Gen} is the generator efficiency.

Table 4 Overview of most important controller applied on floating wind turbines

	Blade pitch control	Torque control	Additional control features	Simulation code	Number of simulations	Model fidelity	Sim. regions	Platform
GSPI	CBP GSPI	Constant power	TTF loop, pitch to stall, or detuned gains	FAST	Extensive	High	2, 3	Barge, TLP, Spar-buoy
GSPI Hywind	CBP GSPI	Constant torque	Constant speed region just below rated with PI torque control loop	HAWC2/SIMO-RIFLEX	Limited	High	2, 3	Hywind sparbouy
VPPC	CBP GSPI	Constant torque	Variable rotor speed setpoint. IBP for blade load reduction	FASF054	Limited	High	3	Barge
EBC	Unknown	Unknown	Wind turbine estimator to hide tower dynamics	HywindSim/SIMO-RIFLEX	Limited	Low	3	Hywind sparbouy
ASC	Unknown	Unknown	Tuned mass damper with active control	FAST-SC	Moderate	High	3	Barge
CBP SS	CBP SS	Constant torque		FAST	Limited	High	3	Barge
IBP SS	IBP SS	Constant power + SS		FAST	Extensive	High	3	Barge, TLP
DAC	IBP SS	Constant power + SS	Wind speed disturbance rejection only	FAST	Extensive	High	3	Barge, TLP

ASC active structural control
CBP collective blade pitch
DAC disturbance accommodating control
EBC estimator based control
GSPI gain scheduled proportional-integral control
IBP individual blade pitch
SS state-space
TTF tower-top feedback
VPPC variable power pitch control
Number of simulations guide:
Limited <10
Moderate <100
Extensive >100

$$T_{Gen} = \frac{\pi \rho R_{Rotor}^5 C_{P,max}}{2\lambda_o^2 N^3} \omega_{Gen}^2 = K\omega_{Gen}^2 \tag{2}$$

$$T_{Gen} = \frac{P_{Rated}}{\eta_{Gen}\omega_{Gen}} \tag{3}$$

The generator speed, used by both CBP and generator torque controllers, was filtered using a low pass filter with a cut-off frequency of 0.25 Hz. This frequency was chosen to be a quarter of the blades edgewise natural frequency to prevent the controller from exciting these modes [16]; this behaviour was also observed by Wright [40] where he found that a PI controller with a fast actuator can destabilise certain modes due to the actuation frequency.

Since the GSPI controller was one of the first well documented controllers implemented on floating wind turbines, it became the reference or baseline controller for other researchers to compare more advanced control strategies. In this chapter, this controller with detuned gains (see next section) will be referred to as the Baseline controller.

4.1.1 Barge Platform

Jonkman, using his newly developed FAST simulator with the HydroDyn module [16], conducted an extensive set of simulations on the ITI Energy barge floating platform in accordance with the IEC-61400-3 standard for ultimate loads [2]. The IEC-61400-3 is a standard for offshore wind turbines with fixed foundations. To date, no standards for floating wind turbines have been released.

The barge platform exhibited large platform oscillations especially in the pitch direction (fore-aft rocking motion) resulting in large tower loads and power fluctuations. To mitigate these effect, Jonkman implemented three modifications to the controller; these were:

Tower-Top Feedback Loop

Additional proportional control loop that aims to reduce tower fore-aft motion based on the measurement of the tower top acceleration was implemented. However, the addition of a tower top feedback control loop did not improve platform damping due to conflicting blade pitch commands issued by the separate control loops as discussed in Sect. 3.1.

Active Pitch to Stall Control

The objective of using pitch to stall was to get extra restoring thrust force once the blades are stalled when the turbine pitches forward. The active pitch to stall controller had excellent power regulation, however, platform motions were not reduced but were increased instead. This contradictory result can be explained by looking at the open and ideal closed loop damping ratios (not shown here). Jonkman concluded that the pitch to stall damping was actually positive. And, since the pitch to feather controller had better damping, it was concluded that the actual closed loop damping of the pitch to feather controller was greater than that of the pitch to stall; i.e. the system has positive damping.

Detuned Controller Gains

Detuning the controller gains reduced the use of the blades and possibly reduced the negative damping effect. The detuned gains produced the best results out of the four controllers. Reducing the controller gains made the response of the system closer to that of the open loop response hence increased damping. The reduction in controller gains produced reasonable power regulation and slightly reduced the platform oscillations.

Conclusions

The increase in the main turbine loads is staggering. For example, the tower base loads are 6 times that of an onshore wind turbine. Jonkman concludes in his PhD dissertation that further reductions in platform motions is necessary and suggests many possibilities to improve platform pitch damping such as the use of constant torque algorithm, use of MIMO state space controllers to avoid conflicting blade pitch commands, and implementing individual blade pitching control strategies.

4.1.2 TLP and Spar Buoy

Larsen and Hanson [21], working on the Hywind spar-bouy (slightly different from the OC3-Hywind model with a different mounted wind turbine), also implemented a GSPI pitch control strategy in above rated wind speed region. However, to avoid the platform pitch damping issue, they used a constant torque algorithm in the above rate wind speed region and introduced a constant speed region just below rated wind speed. A variable speed below rated region also exists to maximise power capture. In the constant speed region, a PI torque control was used to regulate the rotor speed. Switching between regions was handled by a set of variable min/max operations in addition to low pass filters to ensure smooth transition. Simulation results using HAWC2 coupled with SIMO-RIFLEX from MARINTEK showed a 30 % increase in rotor speed and power fluctuations. The motion response of the floating system was acceptable and tower loads were "reasonable" but were not compared to an onshore system. Larsen and Hanson accidentally simulated the floating system with active pitch to stall controller and achieved surprisingly good results and recommended further investigations. However, this contradicts with what Jonkman [16] found when he investigated pitch to stall on the ITI Energy barge. The difference in behaviour is most likely due to different platform dynamics and/or model complexity used.

Matha [22], continuing on Jonkman's work, extended the simulations on the Barge platform to include fatigue loads analysis. He also carried out a series of extensive simulations in accordance with the IEC-61400-3 standard on the TLP and Spar-Buoy platforms. Comparisons were made relative to an onshore wind turbine with the same controller. The Baseline controller was used on all three platforms, however, some modifications were made to the controller on the Spar-Buoy platform. A constant torque instead of constant power algorithm was used in the above rated wind speed region to improve platform pitch damping. Furthermore, the controller bandwidth was limited to avoid resonance issues due to lower natural frequencies of the system than the other platforms. Below is a brief summary of results obtained by Matha [22] on each platform.

ITI Energy Barge

The barge concept has some advantages such as cost effective construction, ability to be assembled in any port due to its shallow draft, and relatively inexpensive slack anchoring system. However, fatigue loads analysis results are no different to the ultimate results found by Jonkman [16] in terms of their relative increase to an onshore system; e.g. tower fatigue loads are 6 times that of an onshore wind turbine. Matha concludes that the barge may not be suitable for sites with "severe sea states" but could provide a cost effective solution to more sheltered sites such as the Great Lakes in the USA.

OC3-Hywind Spar-Buoy

Results from the extensive simulations show that the spar-buoy experiences significantly less loading on the turbine structure when compared to the barge. However, fatigue loads relative to an onshore wind turbine are still approximately 1.5–2.5 times more than an onshore system. This requires some degree of strengthening the turbine tower and blades to withstand these loads. Furthermore, the deep draft of the spar-buoy limits the number of ports where the hull can be fabricated and assembled. However, the first floating wind turbine prototype—the Hywind project—is a spar-buoy floating system. To date, the outcome of this project, which the offshore wind turbine community is very interested in, are yet to be analysed and published.

MIT/NREL TLP

The performance of the TLP relative to the onshore wind turbine is excellent. Almost all of the turbine loads are close to unity with the exception of tower fore-aft (FA) loads which are on average 1.5 times that of an onshore system. Therefore, the TLP probably has the best potential, with more advanced controllers, to achieve similar loading to that of an onshore wind turbine. However, the TLP may not be the most economical choice due to the relatively high cost of the taut anchoring system.

Platform Instabilities

Jonkman [16] and Matha [22] have identified certain instabilities for each platform. These arise from a unique design load case (DLC) specified by the IEC-61400-3 standard such as operating with a faulty blade or in extreme conditions. For example, all three platforms experience yaw instability in a certain DLC where the turbine is idling with two of its blades fully feathered and one stuck at the maximum lift position (blade pitch angle of 0°). Most of these instabilities are due to the design of the system rather than being caused by the controller.

4.2 Variable Power Pitch Control

A simple empirical yet effective solution to avoid the reduced or negative damping in the pitch motion was developed by Lackner [20]. As discussed in Sect. 3.1, a standard speed controller exacerbates platform pitching motion as it attempts to regulate rotor speed to the rated speed in above rated wind speed region. Lackner's solution was to change the speed set-point for the collective blade pitch GSPI controller from a constant to a linear function of platform pitching velocity and implement a constant torque algorithm in above rated wind speed region. The idea is to increase the rotor speed as the turbine is pitching forward (negative pitching velocity according to the coordinate system) forcing the rotor to increase torque to accelerate the rotor and hence create a pitch restoring moment due to increased rotor thrust force. Also, when the wind turbine is pitching backwards, the desired rotor speed is less than the nominal rated speed forcing the speed controller to feather the blades to reduce the rotor speed and thereby reducing the rotor thrust allowing the turbine to pitch forward. By holding the generator torque constant, turbine power becomes a linear function of platform pitch velocity. Furthermore, rotor speed fluctuations are reduced resulting in reduced blade pitching which limit the changes in rotor thrust that causes the reduction in platform pitch damping. This simple control strategy essentially trades power fluctuations for reduced pitching motion.

The rated rotor speed is given by Eq. 4 where ω_R and ω_O are the rotor speed reference/set-point and the nominal rated rotor speed respectively, $\dot{\varphi}$ is the platform pitch velocity and k is the gradient of the linear relationship—a design parameter. It can be seen from Eq. 4 that if the platform pitch velocity is zero, the desired rotor speed is the nominal rated speed. When the turbine is pitching forward (negative $\dot{\varphi}$), the desired rotor speed is increased and vice versa.

$$\omega_R = \omega_o(1 - k\dot{\varphi}) \tag{4}$$

Lackner applied this control strategy on the ITI Energy barge using FAST. Results were obtained after running two 600 s simulations with turbulent wind and irregular waves for three different slopes (k) of Eq. 4. Results show that platform pitch and pitch rates were, on average, reduced by 8–20 % under different k values but power and rotor speed fluctuations, on average, only increased by 3–11 and 1.5–5 % respectively. Of course, the higher the slope, the bigger the reductions in platform motions and increased power fluctuations.

In addition to the variable power pitch control method, Lackner also implemented individual blade pitch control to reduce blade loads in a similar way to Bossanyi's method [9] of using MBC (or Coleman) transformation with PID control for the cosine-cyclic and sine-cyclic components. However, Lackner found that the individual blade pitch controller was not as effective at reducing blade loads when compared to its onshore counterpart. Blade load reductions were small (0.6–1.6 % reduction) but blade pitch usage had increased significantly. Furthermore, the IBP controller increased platform rolling and hence tower side–side fatigue loads. This effect was also observed by Namik et al. [30] and later resolved by Namik and Stol [28] by including the platform roll and first tower side–side bending mode DOFs into the state space controller design as explicit objectives (see Sect. 4.5.2).

4.3 Estimator Based Control

Because the Hywind concept was designed with much lower tower resonance frequencies to avoid the wave energy spectrum [21, 35], Skaare et al. [35] implemented an estimator based control strategy in the above rated wind speed region to "hide the tower motions for the wind turbine control system and thereby avoid the negative damping effect of the tower motion". The wind turbine estimator contains a simplified SISO model of the wind turbine with the wind speed as an input and the rotor to the rotor speed as an output. The wind speed is assumed to be either directly measured or estimated via a wind speed estimator.

In below rated wind speed the control system takes the *measured* rotor speed and commands the generator torque to maximise power capture. In above rated wind speed region, the controller commands the blade pitch based on the *estimated* rotor speed from the wind turbine estimator whose output is based on the estimated wind speed. Seven simulations were carried out using HAWC2 coupled with SIMO-RIFLEX using the same wind and wave conditions but different turbulence intensity. Results show noticeable reductions in nacelle motions and hence large reductions in tower and rotor loads were observed along with increased rotor speed fluctuations and a maximum of 3.81 % reduction in the average power output.

Under these limited cases, the controller seems to be regulating the floating turbine well with promising results. However, since the controller in above rated wind speed region depends on the estimated rotor speed and not the actual which is calculated/estimated based on wind speed estimates, the robustness of such controller will heavily depend on the quality of estimates.

4.4 Active Structural Control

An active structural control of floating wind turbines using a tuned mass damper (TMD) was developed by Rotea et al. [33]. The idea is to add a tuned mass damper with an active component in the nacelle to influence the platform pitching motion of the floating wind turbine. Rotea et al. derived the general equations of motion of the newly added TMDs (one along each of the pitching and rolling directions) and incorporated these changes into FAST resulting in a special version called FAST-SC (for structural control).

In their work in [33], only the TMD along the platform pitching direction is activated and analysed. The TMD parameters (mass, spring stiffness, and damping) were chosen such that they reduce the platform pitching motion on the ITI Energy barge passively (i.e. without a controller input). This is followed by designing a family of H_∞ controllers to actively introduce additional damping by driving the TMD.

Simulation results show that the passive TMD system achieves 10 % reduction in the tower fore-aft (FA) fatigue damage equivalent load (DEL)[2] relative to a "baseline system" without structural control. However, no details were provided about the baseline system/controller. With active structural control, the tower loads are reduced by an average of 15–20 % relative to the passive TMD system at the cost of 3–4 % energy consumption of the rated power of the 5 MW wind turbine. However, the stroke of the TMD (passive or active) is large when the dimensions of the nacelle are considered. Rotea et al. recommend the use to nonlinear control to implement active TMD with stroke stop limits.

A major benefit from using a TMD (active or passive) is achieved when the turbine is operating in extreme conditions. In these conditions the blade pitch controller is normally switched off and the blades are feathered. Having a passive TMD will help improve the pitching (and rolling when implemented) motion(s) in these situations. The performance in these conditions are yet to be evaluated.

[2] Fatigue DELs are used as a metric to replace the stochastic loads on a component by a periodic load with a calculated magnitude at a known frequency.

4.5 Linear State Space Control

Linear state space (SS) control is one of the preferred types of controllers when dealing with multi-input multi-output (MIMO) systems with multiple objectives. This approach requires a linearised state space model of the *nonlinear* floating system. As a result, the system states $\underline{x}$ become perturbations, $\Delta\underline{x}$, about an operating point $\underline{x}^{op}$ such that $\underline{x} = \Delta\underline{x} + \underline{x}^{op}$. This also applies for the measurements vector $\underline{y}$, actuators vector $\underline{u}$ and disturbance inputs vector $\underline{u}_d$. A generic linearised state space model is given by Eq. 5 where A is the state matrix, B is the actuator gain matrix, B_d is the disturbance gain matrix, C relates the measurements to the states, D relates the measurements to the control inputs, and D_d relates the measurements to the disturbance inputs.

$$\begin{aligned} \Delta\dot{\underline{x}} &= A\Delta\underline{x} + B\Delta\underline{u} + B_d\Delta\underline{u}_d \\ \Delta\underline{y} &= C\Delta\underline{x} + D\Delta\underline{u} + D_d\Delta\underline{u}_d \end{aligned} \tag{5}$$

Due to the periodic nature of the wind turbine, the linearised state space model is periodic. That is, the system matrices are periodic function of the rotor azimuth angle. There are three ways to address the periodicity of the system:

1. Averaging the periodicity of each matrix across the azimuth angle results in a time invariant SS model which allows linear time-invariant (LTI) control techniques to be applied. However, some information is lost by averaging and individual blade pitching cannot be used for most control objectives.
2. MBC transformation of the periodic SS model will result in a weekly periodic system. But it was shown that little or no information is lost if it was followed by averaging the weakly periodic matrices [37].
3. Direct periodic control techniques capture all the periodicity of the wind turbine. However, it is more complicated and computationally more demanding than the previous two methods.

The latter two methods facilitate the use of IBP. However, averaging after MBC transformation method allows for the utilisation of IBP control via LTI control design without the computational complexity of direct periodic control method.

The state space control law is given by Eq. 6 where K is the controller gain matrix. In all the SS controllers described next, full state feedback (FSFB) is used to implement the control law. FSFB is where all the states are directly measured and fed to the controller. In cases where direct measurement of some of the states is not possible or desirable, a state estimator is required to estimate the missing states from the available measurements based on a linear state space model of the system.

$$\Delta\underline{u} = -K\Delta\underline{x} \tag{6}$$

4.5.1 Collective Blade Pitch State Space Control

A multi-objective collective blade pitch SS controller to regulate rotor speed and platform pitch (angle and velocity) was implemented by Namik et al. [30] on the ITI Energy barge. The torque controller in the above rated region was set to constant torque to improve platform pitch damping. Limited simulation results, using FAST, show that the SS controller was able to reduce rotor speed error, platform pitching and pitching velocity by an average of 19.5, 14.5, and 17 % respectively relative to the detuned GSPI developed by Jonkman (see Sect. 4.1).

The improvement in performance due to the additional control objective was obvious, however, the magnitude of the improvement was not expected since the CBP SS controller has the same objectives as Jonkman's controller with tower-top feedback loop. The difference was attributed to better controller tuning due to a large emphasis on regulating the platform pitch DOF since both controllers (CBP SS and GSPI with tower-top feedback loop) use collective blade pitching to regulate their objectives.

Interestingly, tower side–side loads were increased by an average of 18 %. The increased side–side tower loading is due to the controller trying to regulate the platform pitch through increased collective blade pitching; as the turbine pitches forward, the controller commands the blades to increase aerodynamic thrust and consequently the aerodynamic torque. For a three bladed wind turbine, the generated aerodynamic torque is balanced given a uniform incident wind; but, due to wind shear combined with the effects of shaft tilt and rotor pre-cone, a blade at the top generates more torque and thus creates a net sideways force resulting in tower side–side excitation. This issue was resolved by using individual blade pitching.

4.5.2 Individual Blade Pitch State Space Control

Individual blade pitching was implemented on the ITI Energy barge and the MIT/NREL TLP by Namik and Stol [25–29]. IBP was initially implemented using direct periodic techniques but then was changed to using MBC transformation for simplicity; no impact on performance was identified. What follows is a summary of all the work done by Namik and Stol regarding IBP on floating platforms.

The minimum order of the SS model for the controller design was established to be 6 DOFs to ensure stability and limit coupling between controller and uncontrolled turbine DOFs. These DOFs (also serve as control objectives) include platform roll, pitch, and yaw, first tower side–side bending mode, generator DOF, and drive-train twist DOF. The reason for these instabilities is simply due to the coupling between using the asymmetric loads created by the IBP controller and the uncontrolled DOFs. For example, when Namik et al. [30] first implemented IBP, a large increase in tower side–side fatigue load was noticed. By looking at the interaction between the blades, rolling and pitching DOFs, it was found that the turbine blades induce rolling and pitching moments on the floating system that are almost in phase. This means when the IBP controller commands the blades to create a restoring pitching moment the blades are also inducing a rolling moment. By including the roll DOF into the controller design model the issue was resolved.

The generator torque was included in the SS control design to influence rotor speed regulation, tower side–side loading, and platform roll. Since the SS controller commands torque perturbations about an operating point, the generator torque operating point was made to change as a function of rotor speed such that constant power algorithm is implemented (Eq. 3). This configuration allows for explicit power regulation as power is not a design state/DOF in the IBP SS controller model.

An extensive number of simulations (presented in [29]) using FAST were carried out in accordance with the IEC-61400-3 standard on the barge platform and TLP for design load case (DLC) 1.2; normal operation for fatigue load testing. Overall averaged results relative to an onshore wind turbine with a Baseline controller (GSPI CBP controller with constant power algorithm for torque control) for the barge and TLP are shown in Fig. 5 where RMS and LSS stand for root mean square and low speed shaft respectively. The figure also shows the performance results for the disturbance accommodating controller (DAC) which is discussed in the next section. The performance of the IBP SS controller on each platform is discussed next. For these results, the controllers are designed based on a 7 DOFs SS model; the additional DOF, first tower fore-aft bending mode, was included to further reduce tower FA fatigue DEL. The increase in blade pitch usage is definitely noticeable, but this increase does not result in prolonged blade pitch saturation.

ITI Energy Barge

Figure 5 clearly shows that IBP makes a big impact at reducing tower loads relative to the Baseline controllers. However, tower loads among others remain too high relative to an onshore wind turbine (e.g. tower FA fatigue DEL after reduction is still more than 5 times that of an onshore wind turbine). These large loads are due to the barge "riding" the waves resulting in large platform motions and hence increased tower and blade loads. Namik and Stol also conclude that the barge, in its current design, is not suitable for open sea deployment.

MIT/NREL TLP

Although Fig. 5 does not show the results for the IBP SS controller on the TLP, the performance of the IBP SS controller is very similar to the DAC; the DAC has better power and speed regulation. It is worthy to note that the performance of the TLP Baseline controller relative to the onshore system is almost unity across most metrics with the exception of RMS power and speed errors, RMS blade pitch rate, and tower FA fatigue DEL. The almost unity performance of the TLP controllers demonstrate the effectiveness of the TLP design. That is, because platform motions

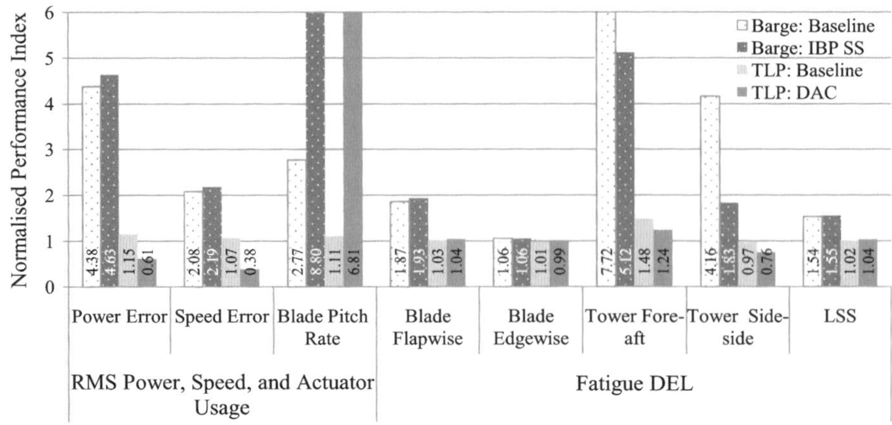

Fig. 5 Overall averaged and normalised DLC results for controllers on the *offshore* platforms relative to the Baseline controller on an *onshore* system

and rotations are kept small due to the mooring lines, turbine fatigue loads were of a comparable level to an onshore wind turbine with the same controller.

Since the IBP SS controller performance is very similar to that of the DAC (reasons will be given in the next section), the DAC performance will be described since it uses IBP and to avoid repetition. The DAC manages to have better power and speed regulation and reduce tower side–side fatigue DEL relative to the onshore system. Tower side–side fatigue DEL are reduced by 24 % relative to the onshore system. Tower fore-aft fatigue DEL are reduced from an increase of 48 % by the GSPI controller to an increase of 24 % relative to an onshore wind turbine. Namik and Stol conclude that with minor strengthening of the tower, the TLP is a good candidate for open sea deployment subject to commercial feasibility.

4.5.3 Disturbance Accommodating Control

Disturbance accommodating control is used to minimise or cancel the effects of persistent disturbances $\underline{u}_d$ that affect a dynamic system given by Eq. 5. The DAC is simply the IBP SS controller discussed in the previous section but with an additional module to reject disturbances. A realisable DAC, where direct measurement of the disturbance is not possible, requires a disturbance estimator to estimate the applied disturbance to the system and correct for that [3, 40]. The disturbance estimator requires a disturbance waveform model given by Eq. 7 where $\underline{z}$ is the disturbance states vector. The choice of matrices F, Θ, and initial conditions $\underline{z}(0)$ determines the nature of the assumed waveform (step, ramp, periodic, etc.).

$$\begin{aligned} \dot{\underline{z}} &= F\underline{z} \\ \underline{u}_d &= \Theta\underline{z} \end{aligned} \tag{7}$$

The realisable disturbance rejection control law is given by Eq. 8 where K is the state regulation gain matrix and G_d is the disturbance minimisation gain matrix. The ^ symbol indicates an estimate. G_d is calculated by evaluating Eq. 9 where the resulting gain will minimise or completely cancel the persistent disturbance. The $^+$ symbol indicates the application of the Moore–Penrose pseudo inverse.

$$\underline{u} = -K\hat{\underline{x}} + G_d\hat{\underline{z}} \tag{8}$$

$$G_d = -B^+B_d\Theta \tag{9}$$

Namik and Stol implemented DAC on the ITI Energy barge platform and MIT/NREL TLP to reject uniform wind speed perturbations [25–27, 29]. The DAC is designed to improve rotor speed regulation since it is minimising the effect of wind speed perturbations about an operating point. Simulation results using the same conditions for their IBP SS controller show that the DAC on the barge platform had limited impact. The improvement in rotor speed regulation and hence power regulation was minimal. The reason for such minimal improvement was determined to be due to the barge's motion being dominated by incident waves rather than wind. Hence, reducing the effects of wind speed perturbations will have little or no impact on performance. Since the TLP is not as dominated by incident waves as the barge, the DAC has a noticeable impact on rotor speed regulation and hence power regulation. The results are shown in Fig. 5 and discussed in the previous section.

4.6 Other Controllers

In this section, the remaining works to the authors' best knowledge that have been carried out on control of floating wind turbines are briefly discussed. These works either have limited information available and/or limited conclusions can be drawn.

In 2006, Nielsen et al. [31] implemented an additional blade pitch control algorithm for active damping of the platform pitching motion in above rated wind speed region on the Hywind floating concept. Simulation results using HywindSim (a tool to model a simple floating wind turbine) coupled with SIMO-RIFLEX in addition to scale model testing agree to some degree. The additional pitch control algorithm seems to improve the platform pitching response. Scale model testing was carried out at the Ocean Basin Laboratory at Marintek with a scale of 1:47 [4].

Henriksen [15] implemented a model predictive controller primarily on an onshore wind turbine but a simple model for a floating wind turbine was also used. However, little insight into the performance of the floating system can be offered due to limited floating model fidelity.

5 Conclusions

Offshore wind energy is part of the solution to meet the ever increasing demand for energy. To fully utilise the offshore potential, deployment of wind farms further offshore is necessary. Floating wind turbines offer a feasible solution for water deeper than 60 m. There are three main floating platform concepts; these include a buoyancy stabilised barge platform, a mooring line stabilised tension leg platform, and a ballast stabilised spar-buoy platform.

A recurring issue with floating wind turbines in terms of control is reduced or negative platform pitch damping in above rated wind speed region (region 3). Several ways of dealing with this issue were proposed and below is a summary of the most important findings (in no particular order):

- Using constant torque instead of constant power control in above rated wind speed region helps to improve the platform pitching response by reducing the use of the blade pitch actuator responsible for the reduction in damping. This is not a complete solution but it can help if used other controllers.
- The use of individual blade pitching can greatly improve power and speed regulation, reduce platform motions, and reduce tower bending loads. This comes at the cost of increased blade pitch actuation which may lead to blade actuator saturation and possibly destabilising the system.
- The addition of tuned mass dampers (TMDs) can noticeably improve the pitching motion on certain floating platform designs. Active TMDs further improve performance, however, based on current studies to date, the stroke length required to achieve these reductions cannot be used inside the nacelle of the wind turbine. A major benefit of using TMDs (active or passive) is that it has the potential to reduce pitching motion in extreme conditions where the blade pitch controller is not active.

Below is a list of topics that are yet to be explored that have the potential to improve the response of floating systems when coupled with advancements in other wind turbine fields (such as blade design and materials). These topics are listed below in no particular order of importance.

- Verification of simulation and controller responses against full scale prototypes. This is a critical issue. The verification process may take some time, but when successful verification of simulation tools is achieved, it will increase the confidence in the predicted controllers' performance.
- A detailed look at region transition to and from above and below rated for floating wind turbines is yet to be investigated. There is a potential for limit cycle oscillations (continuously switching between regions regardless of wind conditions) due to the lack of rigid foundations.
- The effectiveness of TMDs in extreme conditions. It is expected that TMDs will reduce platform motions in extreme conditions, however, the extent of this reduction needs to be established.

- Wave disturbance rejection, if possible, may significantly improve the response of floating platforms.
- Including additional actuators on the platform may help improve the response but may require too much power.
- Applying more advanced controllers such as model predictive control and nonlinear control on such systems. The benefits of implementing such complex controllers is yet to be evaluated on floating wind turbines.

References

1. (2006) Largest and deepest offshore wind turbine installed. Refocus 7(5):14. http://www.sciencedirect.com/science/article/pii/S1471084606706783
2. (2009) Wind turbines—part 3: design requirements for offshore wind turbines. http://webstore.iec.ch/webstore/webstore.nsf/mysearchajax?Openform\&key=61400-3\&sorting=\&start=1\&onglet=1
3. Balas MJ, Lee YJ, Kendall L (1998) Disturbance tracking control theory with application to horizontal axis wind turbines. In: Proceedings of the 1998 ASME wind energy symposium, pp 95–99. Reno, Nevada
4. Berthelsen AP (2006) Model testing of floating wind turbine facility. Review 1(1):6
5. Biester D (2009) Hywind: Siemens and StatoilHydro install first floating wind turbine. Retrieved 3 July 2009 from: http://www.siemens.com/press/pool/de/pressemitteilungen/2009/renewable_energy/ERE200906064e.pdf
6. Bir G (2008) Multi-blade coordinate transformation and its application to wind turbine analysis. In: 46th AIAA aerospace sciences meeting and exhibit. Reno, NV, pp CD–ROM
7. Bossanyi EA (2000) The design of closed loop controllers for wind turbines. Wind Energy 3(3):149–163. doi:10.1002/we.34
8. Bossanyi EA (2003) Individual blade pitch control for load reduction. Wind Energy 6(2):119–128. doi:10.1002/we.76
9. Bossanyi EA (2003) Wind turbine control for load reduction. Wind Energy 6(3):229–244. doi:10.1002/we.95
10. Brennan S (2008) Offshore wind energy resources. Workshop on deep water offshore wind energy systems. http://www.nrel.gov/wind_meetings/offshore_wind/presentations.html
11. Bulder B, Peeringa J, Pierik J, Henderson AR, Huijsmans R, Snijders E, Hees Mv, Wijnants G, Wolf M (2003) Floating offshore wind turbines for shallow waters. In: European wind energy conference. ECN Wind Energy, Madrid, Spain
12. Butterfield S, Musial W, Jonkman J, Sclavounos P, Wayman L (2005) Engineering challenges for floating offshore wind turbines. In: Copenhagen offshore wind 2005 conference and expedition proceedings. Danish Wind Energy Association, Copenhagen, Denmark
13. Goldman P (2003) Offshore wind energy. Workshop on deep water offshore wind energy systems
14. Henderson AR, Morgan C, Smith B, Sørensen HC, Barthelmie RJ, Boesmans B (2003) Offshore wind energy in Europe: a review of the state-of-the-art. Wind Energy 6(1):35–52
15. Henriksen LC (2007) Model predictive control of a wind turbine. Master's thesis, Technical University of Denmark
16. Jonkman JM (2007) Dynamics modeling and loads analysis of an offshore floating wind turbine. Ph.D. thesis, University of Colorado

17. Jonkman JM, Butterfield S, Musial W, Scott G (2007) Definition of a 5-MW reference wind turbine for offshore system development. Technical Report TP-500-38060, National Renewable Energy Laboratory
18. Jonkman JM, Lasren T, Hansen A, Nygaard T, Maus K, Karimirad M, Gao Z, Moan T, Fylling I, Nichols J, Kohlmeier M, Pascual Vergara J, Merino D, Shi W, Park H (2010) Offshore code comparison collaboration within IEA wind task 23: phase IV results regarding floating wind turbine modeling. In: European wind energy conference 2010. Warsaw
19. Jonkman JM, Sclavounos PD (2006) Development of fully coupled aeroelastic and hydrodynamic models for offshore wind turbines. In: Proceedings of the 44th AIAA aerospace sciences meeting and exhibit. Reno, NV, pp CD–ROM
20. Lackner MA (2009) Controlling platform motions and reducing blade loads for floating wind turbines. Wind Energy 33(6):541–553
21. Larsen TJ, Hanson TD (2007) A method to avoid negative damped low frequency tower vibrations for a floating pitch controlled wind turbine. In: Conference on Journal of Physics, series 75. doi:10.1088/1742-6596/75/1/012073
22. Matha D (2009) Modelling and loads and stability analysis of a floating offshore tension leg platform wind turbine. Master's thesis, National Renewable Energy Lab's National Wind Turbine Center and University of Stuttgart
23. Musial W, Butterfield S, Boone A (2004) Feasibility of floating platform systems for wind turbines. In: 23rd ASME wind energy symposium. NREL, Reno, NV
24. Musial W, Butterfield S, Ram B (2006) Energy from offshore wind. In: Offshore technology conference. Houston, Texas, USA, pp 1888–1898. doi:10.4043/1855-MS
25. Namik H, Stol K (2009) Control methods for reducing platform pitching motion of floating wind turbines. In: European offshore wind 2009. Stockholm, pp CD–ROM
26. Namik H, Stol K (2009) Disturbance accommodating control of floating offshore wind turbines. In: 47th AIAA aerospace sciences meeting and exhibit. Orlando, FL, pp CD–ROM
27. Namik H, Stol K (2010) Individual blade pitch control of a floating offshore wind turbine on a tension leg platform. In: 48th AIAA aerospace sciences meeting and exhibit. Orlando, FL, pp CD–ROM
28. Namik H, Stol K (2010) Individual blade pitch control of floating offshore wind turbines. Wind Energy 13(1):74–85. doi:10.1002/we.332
29. Namik H, Stol K (2011) Performance analysis of individual blade pitch control of offshore wind turbines on two floating platforms. Mechatronics 21(4):691–703, Elsevier
30. Namik H, Stol K, Jonkman J (2008) State-space control of tower motion for deepwater floating offshore wind turbines. In: 46th AIAA aerospace sciences meeting and exhibit. Reno, NV, pp CD–ROM
31. Nielsen FG, Hanson TD, Skaare B (2006) Integrated dynamic analysis of floating offshore wind turbines. In: Proceedings of the 25th international conference on offshore mechanics and arctic engineering. Hamburg, pp 671–679
32. Passon P, Kühn M, Butterfield S, Jonkman JM, Camp T, Larsen TJ (2007) OC3—benchmark exercise of aero-elastic offshore wind turbine codes. In: Conference on Journal of Physics, series 75(1):012,071. doi:10.1088/1742-6596/75/1/012071. url:http://stacks.iop.org/1742-6596/75/i=1/a=012071
33. Rotea M, Lackner MA, Saheba R (2010) Active structural control of offshore wind turbines. In: 48th AIAA aerospace sciences meeting and exhibit. Orlando, FL, pp CD–ROM
34. Shikha Bhatti TS, Kothari DP (2003) Aspects of technological development of wind turbines. J Energy Eng 129(3):81–95
35. Skaare B, Hanson TD, Nielsen FG (2007) Importance of control strategies on fatigue life of floating wind turbines. In: Proceedings of the 26th international conference on offshore mechanics and arctic engineering. San Diego, CA, pp 493–500
36. Stol K (2001) Dynamics modeling and periodic control of horizontal-axis wind turbines. Ph.D. thesis, University of Colorado

37. Stol K, Moll HG, Bir G, Namik H (2009) A comparison of multi-blade coordinate transformation and direct periodic techniques for wind turbine control design. In: 47th AIAA aerospace sciences meeting and exhibit. Orlando, FL, pp CD–ROM
38. Stol K, Zhao W, Wright AD (2006) Individual blade pitch control for the controls advanced research turbine (CART). J Sol Energy Eng, Trans ASME 22:16,478–16,488
39. Ushiyama I, Seki K, Miura H (2004) A feasibility study for floating offshore windfarms in Japanese waters. Wind Energy 28(4):383–397
40. Wright AD (2004) Modern control design for flexible wind turbines. Technical Report NREL/TP-500-35816, National Renewable Energy Lab

Modelling and Control of Wind Turbines

Luis M. Fernández, Carlos Andrés García and Francisco Jurado

Abstract This chapter provides a basic understanding of modelling of wind turbines, including both the mechanical and electrical systems, and control schemes that enable a suitable operation of the wind turbines. An overview of the most widely used wind turbine concepts is performed and their models for integration in large power system dynamic simulations are described. The wind turbine concepts studied are: fixed-speed squirrel cage induction generator (FS-SCIG), wound rotor induction generator (WRIG) with variable rotor resistance (VRR), doubly fed induction generator (DFIG), and direct drive synchronous generator (DDSG).

Nomenclature

A	Swept area of rotor
C_p	Power coefficient
D_{mec}, K_{mec}	Damping and stiffness of mechanical shaft
e_d', e_q'	Voltage behind a transient reactance in d and q axis
E_g	Excitation voltage
f_e, f_m	Electrical and mechanical frequency
H_r, H_g	Turbine rotor and generator inertia
i_{dc}, i_{qc}	Grid side converter currents in d and q axis
i_{dg}, i_{qg}	Grid currents in d and q axis

L. M. Fernández (✉) · C. A. García
Department of Electrical Engineering, University of Cadiz, EPS Algeciras, Avda. Ramon Puyol s/n 11202 Algeciras (Cádiz), Spain
e-mail: luis.fernandez@uca.es

C. A. García
e-mail: carlosandres.garcia@uca.es

F. Jurado
Department of Electrical Engineering, University of Jaen, EPS Linares, Alfonso X, Linares 23700 Jaén, Spain
e-mail: fjurado@ujaen.es

P. M. Pardalos et al. (eds.), *Handbook of Wind Power Systems*, Energy Systems, DOI: 10.1007/978-3-642-41080-2_14, © Springer-Verlag Berlin Heidelberg 2013

i_{dr}, i_{qr}	Rotor currents in d and q axis
i_{ds}, i_{qs}	Stator currents in d and q axis
I_s, I_r	Stator and rotor currents
L_c	Grid connection inductance
L_{dm}, $L_{\sigma m}$	Mutual inductances in d and q axis
L_{ldk}, L_{lqk}	Damping winding inductances in d and q axis
L_{lf}	Field inductance
L_m	Magnetizing inductance
L_s, L_r	Stator and rotor inductances
$L_{\sigma s}$, $L_{\sigma r}$	Stator and rotor leakages inductances
n_p	Number of generator poles
P_g	Active power delivered to grid
P_{gsc}	Grid side converter active power
P_s, P_r	Stator and rotor active powers
P_{wt}	Mechanical power extracted from the wind
Q_g	Reactive power delivered to grid
Q_{gsc}	Grid side converter reactive power
Q_s, Q_r	Stator and rotor reactive powers
R_c	Grid connection resistance
R_{df}	Field winding resistance
R_{dk}, R_{qk}	Resistance of damping winding in d and q axis
R_{ext}	External resistance
R_s, R_r	Stator and rotor resistances
s	Slip
T_e	Electromagnetic torque
T_{mec}	Mechanical torque from the generator shaft
T_o'	Transient open circuit time constant of induction generator
T_{wt}	Mechanical torque extracted from the wind
u	Wind speed
v_{dc}, v_{qc}	Grid side converter voltages in d and q axis
$V_{dc\text{-}link}$	DC link voltage
v_{dg}, v_{qg}	Grid voltages in d and q axis
v_{dr}, v_{qr}	Rotor voltages in d and q axis
v_{ds}, v_{qs}	Stator voltages in d and q axis
V_g	Grid voltage
V_{gsc}	Grid side converter voltage
V_s, V_r	Stator and rotor voltages
X_s'	Transient reactance of induction generator
ρ	Air density
λ	Tip-speed ratio
ω_b	Base synchronous speed
ω_g	Generator speed
ω_r	Rotor speed
ω_s	Synchronous speed

β	Pitch angle
δ_r	Load angle
ψ_{df}	Field winding flux linkage
ψ_{dk}, ψ_{qk}	Damping winding flux linkage in d and q axis
ψ_{dr}, ψ_{qr}	Rotor flux linkage in d and q axis
ψ_{ds}, ψ_{qs}	Stator flux linkage in d and q axis
ψ_{pm}	Permanent magnet flux linkage
*	Reference value
$\overline{Variable}$	Per unit variable
Variable	Phasor
AC	Alternating current
DC	Direct current
DDSG	Direct drive synchronous generator
DFIG	Doubly fed induction generator
EESG	Electrically excited synchronous generator
FS	Fixed speed
GSC	Grid side converter
IGBT	Insulated gate bipolar transistor
MSC	Machine side converter
PMSG	Permanent magnet synchronous generator
SCIG	Squirrel cage induction generator
VRR	Variable rotor resistance
WRIG	Wound rotor induction generator

1 Introduction

Because of the increasing wind power penetration on power systems and the rapid development of the wind turbine technology, the wind turbines and wind farms begin to influence power system. This justifies the development of adequate models to represent the behaviour of wind turbines in large power system dynamic simulations.

This chapter aims to provide a basic understanding of modelling of wind turbines, including both the mechanical and electrical systems, and control schemes that enable a suitable operation of the wind turbines. An overview of the most widely used wind turbine concepts is performed and their models for integration in large power system dynamic simulations are described.

The chapter starts with a brief review of the power system simulation in Sect. 2. This is followed in Sect. 3 by the description of main systems that integrate a generic dynamic model of a wind turbine. Section 4 describes the modelling and control of the mechanical system of a wind turbine, which is composed of the following systems: aerodynamic rotor, drive train, and blade pitch angle control.

Because of the electrical and control system depends on the wind turbine concept, the modelling of electrical and control system of the most important current wind turbine concepts is presented in Sect. 5. The wind turbine concepts studied are: (1) fixed-speed squirrel cage induction generator (FS-SCIG); (2) wound rotor induction generator (WRIG) with variable rotor resistance (VRR); (3) doubly fed induction generator (DFIG); and (4) direct drive synchronous generator (DDSG). They are the most commonly applied wind turbine concepts [13].

The first concept is a fixed-speed wind turbine equipped with a squirrel cage induction generator (SCIG) connected directly to the grid and coupled to the wind turbine rotor through a gearbox.

The second concept is a variable speed wind turbine that uses a wound rotor induction generator (WRIG) with variable rotor resistance (VRR) by means of a power electronic converter, mounted on the rotor shaft. In this case, the stator winding is directly connected to the grid, whereas the rotor winding is connected in series with an external resistance controlled by power converter.

The third concept is also a variable speed wind turbine with a gearbox and a doubly fed induction generator (DFIG). It presents the stator winding directly connected to grid and a power electronic converter feeding the rotor winding. This converter has a power rating of 25–30 % of the generator capacity.

The fourth concept is also a variable speed wind turbine equipped with a synchronous generator whose rotor presents a large number of poles and whose stator windings are connected to the grid through a full-scale power converter. This wind turbine presents the generator directly coupled to the wind turbine rotor, and thus the gearbox is omitted. The rotor excitation can be obtained by means of either a current-carrying winding (electrically excited synchronous generator, EESG) or permanent magnets (permanent magnet synchronous generator, PMSG).

The modelling of electrical generator, power converter (if any), and control system of each wind turbine concept is described in this chapter.

2 Power System Dynamic Simulation

A power system consists of a large number of components, such as overhead lines and underground cables, transformers, generators and loads, whose behaviour can be described by one or more differential equations. In case of a large power system, the system of equations can easily contain hundreds or even thousands of differential equations. A system of equations of this size cannot be solved analytically, so numerical integration remains the only practical possibility to analyse power system behaviour.

On the other hand, the time scale to be used in the analysis of power system depends on phenomena of interest. Thus, there are phenomena in the power systems that take micro to milliseconds (lightning overvoltage, line switching voltage, switching power electronic converters, …) or several seconds, minutes or hours (wind turbine production, daily load following, ...). The model of the power system and its components must be then selected depending on the phenomena to be studied. In fact, various simulation approaches are distinguished [27]:

- *Electromagnetic transient/Subtransient simulation*: In this approach, all differential terms in the equations of the power system are taken into account. This type of simulation can, in principle, be used to study many different types of problems. However, because the simulations are rather time consuming, they are normally only used to study problems for which a detailed representation of the power system is essential. Examples are the calculation of fault currents and the study of the fault behaviour and the controller tuning of power electronic converters.
- *Electromechanical transient/Transient/Dynamic simulations*: In this approach, only the fundamental frequency component of voltages and currents is taken into account and higher harmonics are neglected. This allows the use of a load flow representation of the network and thus cancels the differential equations associated with the network. Further, also some of the differential equations associated with generators are cancelled as well as short time constants, enabling the use of a larger time step. In combination, this increases the simulation speed substantially. This type of simulation can be used to study aspects of power system behaviour that have characteristic frequencies in the order of 0.1 or 1–10 Hz. Examples are generator rotor speeds and node voltages.
- *Power balance*: In this approach, the aim is to study the balance between power generated and power consumed, which must be maintained for a power system to function properly. The typical time frame studied is from tens of seconds to days and the time step used for the simulations is in the order of seconds.

Therefore, a suitable election of the simulation approach is necessary to avoid unreliable or incorrect results and an unnecessarily long simulation time. This chapter is focused on the modelling of wind turbines for power system dynamic simulations.

3 Wind Turbine Dynamic Model

The wind turbine dynamic model in general is composed of the following subsystems (see Fig. 1):

- *Rotor aerodynamic system*. This system represents the turbine aerodynamics and computes the mechanical torque or power extracted from the wind, depending on the incoming wind speed, turbine rotor speed and blade pitch angle.
- *Drive train system*. It represents the mechanical system of the wind turbine, composed of turbine rotor, shafts, gearbox and generator. The drive train model computes the turbine and generator speeds by using the mechanical torque extracted from the wind and the generator torque as input variables.

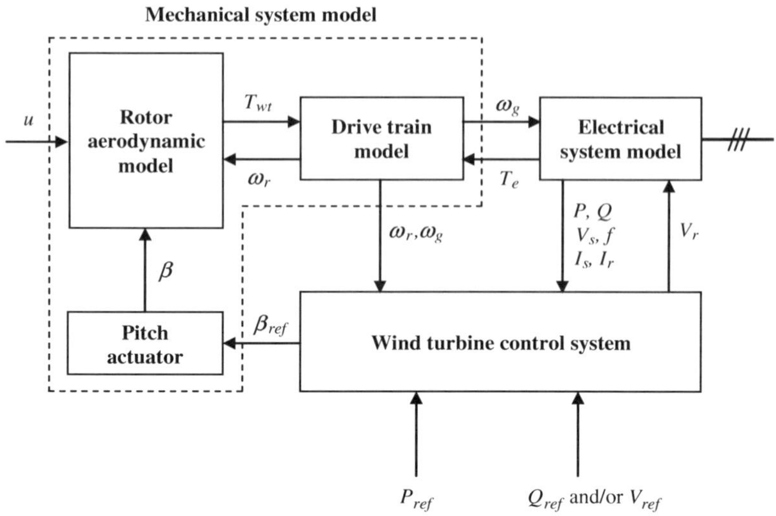

Fig. 1 Wind turbine dynamic model

- *Blade pitch angle control system.* This system is responsible of controlling the movement of the blade pitch angle.
- *Electrical system.* In the electrical system, the mechanical power is converted in electrical power and delivered to grid. It consists of the electrical generator and power converters, if any, depending on the wind turbine concept.
- *Wind turbine control system.* This system models the wind turbine control in order to achieve the operation desired (active power reference, reactive power reference or terminal voltage reference).

The modelling of these subsystems is described in the following sections of this chapter.

4 Wind Turbine: Modelling and Control of Mechanical System

The mechanical system of a wind turbine is composed of the following models: (1) the aerodynamic rotor; (2) the drive train; and (3) the blade pitch angle control. The inputs for the mechanical system model are the wind speed u, the blade reference pitch angle β_{ref} and the generator torque T_e, whereas the model output is the generator speed ω_g. The wind speed is the only independent input quantity, while the generator torque and the pitch angle are achieved from the generator model, and the wind turbine control system, respectively.

4.1 Rotor Aerodynamic Model

The rotor aerodynamic model expresses the mechanical torque or power extracted from the wind. It can be derived from the blade element momentum (BEM) method [14]. This method is based on a separation of the blades into a number of sections along the length of each blade. Each blade section is characterised by the blade geometry, and the aerodynamic properties are given for each section from the hub to the blade tip as functions of the local radius. The static forces on the blade element, and consequently the corresponding shaft torque are calculated in this method for a given wind speed, a given rotor speed and a given blade pitch angle. However, modelling the rotor using BEM method has a number of drawbacks [29]:

- Instead of only one wind speed signal, an array of wind speed signals has to be applied.
- Detailed information about the rotor geometry should be available.
- Computations become complicated and lengthy.

To solve these problems, a simplified model of the wind turbine rotor derived from the disk actuator theory is normally used when the electrical behaviour of the system is the main point of interest [14].

In the disk actuator theory, the mechanical power extracted from the wind is calculated from an algebraic equation, where the power is expressed as function of the wind speed u, the blade tip speed ratio λ, and the blade pitch angle β.

$$P_{wt} = \frac{1}{2}\rho A u^3 C_p(\lambda, \beta) \tag{1}$$

where P_{wt} is the mechanical power extracted by the wind turbine rotor, ρ is air density, A is the area of rotor disk, and C_p is the power coefficient.

The mechanical torque extracted from the wind is calculated as follows

$$T_{wt} = \frac{1}{2}\rho A R u^2 \frac{C_p(\lambda, \beta)}{\lambda} \tag{2}$$

where R is the radius of the wind turbine rotor.

The power coefficient represents the fraction of the wind power that is extracted by the rotor. It expresses the rotor aerodynamics as a function of both tip speed ratio λ and the pitch angle of the rotor blades β, as shown in Fig. 2. The tip speed ratio is defined as the ratio between the blade tip speed and wind speed, expressed as

$$\lambda = \frac{\omega_r \cdot R}{u} \tag{3}$$

where ω_r is the rotor speed.

The power extracted from the wind is maximized when the rotor speed is such that the power coefficient is the maximum value obtained for a determined tip

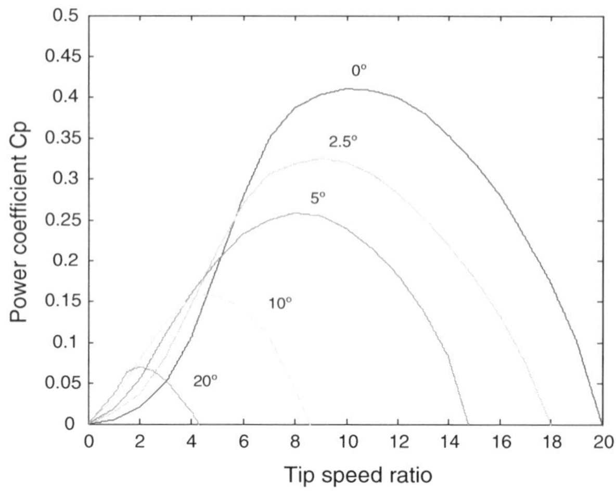

Fig. 2 Power coefficient $C_p(\lambda, \beta)$

speed ratio and blade pitch angle. The control system of a variable speed wind turbine assures variable speed operation that maximizes the output power for a wide range of wind speeds, according to the optimum power extraction curve, given by

$$P_{opt} = \frac{1}{2} \cdot \rho \cdot c_{p\,\max} \cdot \pi \cdot R^2 \cdot \left(\frac{\omega_r \cdot R}{\lambda_{opt}}\right)^3 = \left(\frac{1}{2} \cdot \rho \cdot c_{p\,\max} \cdot \frac{\pi \cdot R^5}{\lambda_{opt}^3}\right) \cdot \omega_r^3 = K_{opt} \cdot \omega_r^3. \tag{4}$$

Because the wind turbine limits the output power to the rated power of the generator for high winds, the power-speed curve is truncated to rated power. This power-speed curve serves as a dynamic reference for the control system of variable speed wind turbine. It assures that the wind turbine operation has an optimum power efficiency for below nominal winds and an output power limited to rated value for above nominal winds. Figure 3 illustrates the aerodynamic mechanical power [expressed by Eq. (1)], the optimum power extraction curve [Eq. (4)], and the power-speed control curve for a variable speed wind turbine.

4.2 Drive Train Model

The drive train of a wind turbine is composed of the rotating masses and the connecting shafts, including a possible gearbox. To represent the dynamic of the wind turbine in power system transient analysis, several models of the drive train are reported in literature.

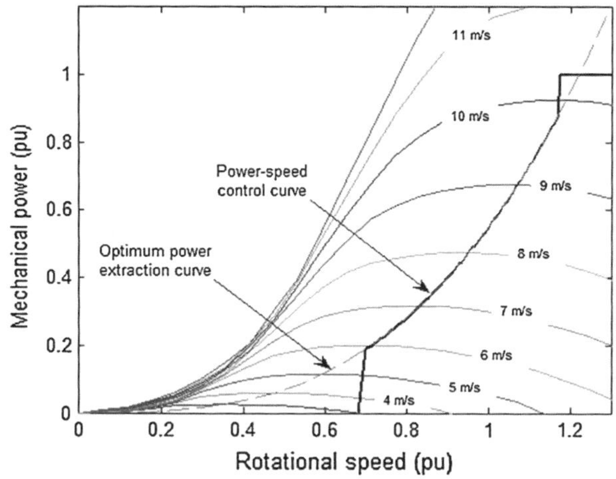

Fig. 3 Aerodynamic mechanical power, optimum power extraction *curve*, and power-speed control *curve*

Six rotating masses (three blades, hub, gearbox and generator) are considered in the six-mass drive train model [8, 21, 23]. Each rotor blade is modelled by a separate inertia, elastically connected to the hub. In addition, hub, gearbox and generator are represented by its inertia, which are considered to be elastically connected via springs.

In the three-mass drive train model [8, 21], the three rotating masses are turbine, gearbox and generator, elastically coupled each other. In this case, the turbine inertia is calculated from the combined weight of three blades and hub, and the mutual-damping between hub and blades is ignored.

However, these models are not commonly used for large power system simulation studies because the model order is high. The models mostly used are the two-mass model and the one-mass or lumped-mass model. These models are described next.

4.2.1 Two-Mass Model

In the two-mass model, the inertia of the gearbox is neglected and only the transformation ratio of the gearbox is included in the model when the wind turbine has gearbox. Thus, the model is composed of two masses, turbine (three blades and hub) and generator, elastically connected via springs. Figure 4 shows this model.

This model is described by the following equations:

$$T_{wt} - T_{mec} = 2\,H_r \frac{d\omega_r}{dt} \tag{5}$$

$$T_{mec} = D_{mec}(\omega_r - \omega_g) + K_{mec} \int (\omega_r - \omega_g)dt \tag{6}$$

Fig. 4 Two-mass drive train model

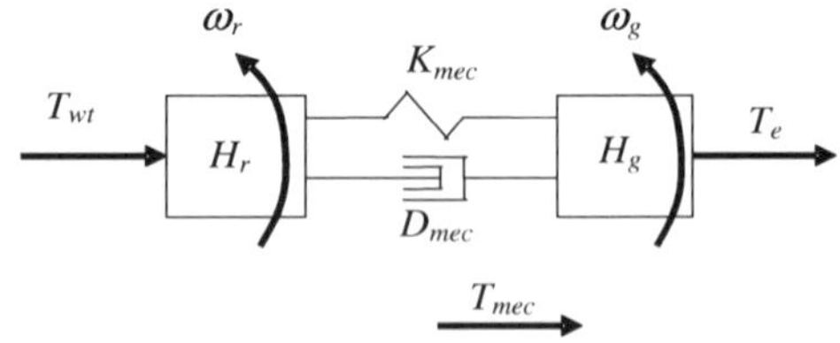

$$T_{mec} - T_e = 2H_g \frac{d\omega_g}{dt} \tag{7}$$

where T_{wt} is the mechanical torque from the wind turbine rotor shaft, T_{mec} is the mechanical torque from the generator shaft, T_e is the generator electrical torque, H_r is the turbine rotor inertia, H_g is the generator inertia, and K_{mec} and D_{mec} are the stiffness and damping of mechanical coupling.

4.2.2 Lumped-Mass Model

In the lumped-mass model, one lumped mass accounts for all the rotating parts of the wind turbine.

$$T_{wt} - T_e = 2H_{wt} \frac{d\omega_g}{dt} \tag{8}$$

where H_{wt} represents the inertia constant of all the rotating masses.

In case of fixed speed wind turbines, the two-mass model is necessary to accurately represent the dynamic of the wind turbine for the transient stability analysis. The one-mass lumped model is too simple, and it does not represent the behaviour of the wind turbine accurately enough [3, 18, 21].

Both models could be used in case of variable speed wind turbines, because the drive train properties are hardly reflected at the grid connection due to the decoupling effect of the power converter [3, 29]. However, the two-mass model is preferred, because it represents more accurately the dynamic of the wind turbine when the power converter is blocked during grid faults.

4.3 Blade Angle Control

Wind turbines are designed to produce so much electric energy as possible at wind speed below rated wind. However, at wind speed above rated wind, the mechanical power extracted from the wind must be limited in order to prevent overloads in the mechanical construction of the wind turbine and generator.

In passive-stall-controlled wind turbines, the rotor presents fixed blade angle. The geometry of the rotor blade profile is aerodynamically designed to ensure that it creates turbulence on the side of the rotor blade when the wind speed becomes too high. By using this passive method, the power extracted from the wind is reduced for high winds.

Both pitch-controlled and active-stall-controlled wind turbines present variable blade pitch angle, enabling active control of the mechanical extracted from the wind. The blade pitch angle control allows: (1) preventing overloads in the wind turbine with high wind; and (2) an active control of the power production, and thus, reducing the power production when regulation power is required by the power system operator [3].

In pitch-controlled wind turbines, the mechanical power extracted from the wind is reduced by rotating the blades into the feathered position (blade is moved against the incoming wind). It supposes that the blade pitch angle is increased and the angle of attack is decreased. This control principle can be applied in fixed speed and variable speed wind turbines, although it is mostly used in variable speed wind turbines.

In case of active-stall-controlled wind turbines, the reduction of mechanical power is achieved by rotating the blades into the stall position. In this case, the blade is moved across the incoming wind, the blade pitch angle is decreased, and thus, the angle of attack is increased. This principle is commonly applied in fixed speed wind turbines.

The blade angle is commonly controlled by the control scheme shown in Fig. 5. It is composed of the pitch angle controller and the pitch actuator.

The pitch angle controller commonly uses the power or rotational speed as control variable (X in Fig. 5) in order to define the blade reference angle.

The pitch actuator turns the blades to the angle ordered by the pitch controller. It can be modeled by a first order model, subject to physical limitations on both pitch angle and speed (Fig. 5a) or by only these physical limitations (Fig. 5b). For pitch-controlled wind turbines, the permissible range of pitch angle is between 0° and +90° (or even a few degrees to the negative side), whereas for active-stall-controlled wind turbines, it lies between −90° and 0° (or even a few degrees to the positive side). The pitch speed limit is likely to be higher for pitch-controlled wind turbines than for active-stall-controlled wind turbines, which have a higher angular sensitivity [2]. The pitch speed is normally less than 5° per second, although it may exceed 10° per second during emergencies.

5 Wind Turbine: Modelling of Electrical and Control System

The electrical and control system depends on the wind turbine concept. In this section, the modelling of electrical and control system of each wind turbine concept is presented. It is composed of the following models: (1) electrical generator; (2) power converter; and (3) control system.

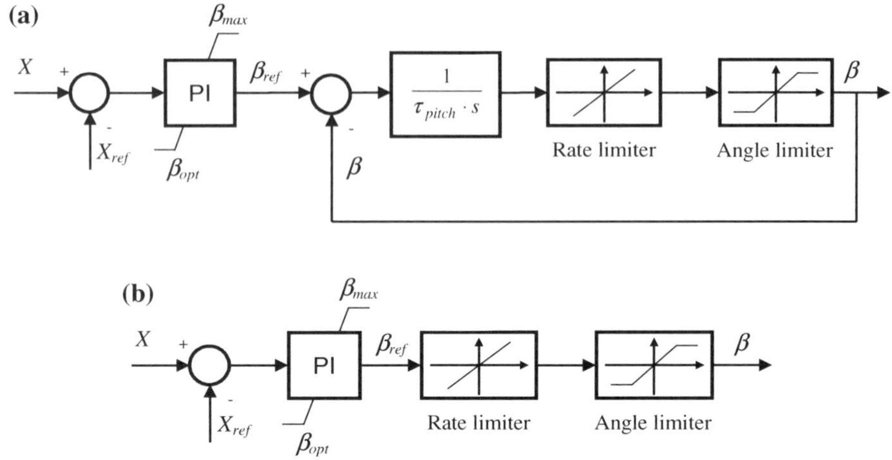

Fig. 5 Control schemes of the blade pitch angle control for case of pitch-controlled wind turbine. **a** Detailed model of pitch actuator. **b** Reduced model of pitch actuator. In case of active-stall wind turbine, the upper limit of the PI controller must be changed to β_{opt} and the lower limit to β_{min}

In power system dynamic simulation, the following assumptions are commonly applied to the electrical generator of the wind turbine:

- Magnetic saturation is neglected.
- Flux distribution is sinusoidal.
- Any losses apart from copper losses are neglected.
- Stator voltages and currents are sinusoidal at the fundamental frequency.

As usual for power system dynamic simulation where the internal dynamics of power converter are not of interest, the power converter is considered ideal. Therefore, the converters are modelled as voltage/current source.

5.1 Fixed-Speed Squirrel Cage Induction Generator (FS-SCIG)

This is the conventional concept applied by many Danish wind turbine manufacturers during the 1980s and 1990s. It presents a squirrel cage induction generator (SCIG) connected directly to the grid and coupled to the wind turbine rotor through a gearbox. This generator presents very small rotational speed variations, because the only speed variations that can occur are small changes in the rotor slip, so that these wind turbines are considered to operate at fixed speed. In order to limit the power extracted from the wind for high winds, the wind turbine rotor

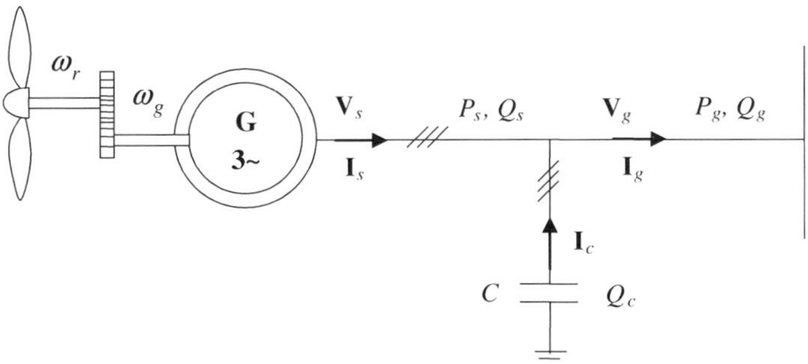

Fig. 6 FS-SCIG wind turbine configuration

limits the power extracted from the wind using passive stall effect (passive-stall-controlled wind turbine), active stall effect (active-stall-controlled wind turbine) or controlling the blade pitch angle (pitch-controlled wind turbine). A SCIG consumes reactive power, so that capacitors are added to generate the induction generator magnetizing current, thus improving the power factor of the system. Figure 6 shows FS-SCIG wind turbine configuration.

5.1.1 FS-SCIG Generation System

The dynamic behaviour of an induction generator in power systems can be represented by the fifth order model of the induction machine. This model is composed of four electrical differential equations (two equations for both the stator and rotor voltages) and one mechanical differential equation. The electrical equations, expressed in direct (d)-quadrature (q) coordinate reference frame rotating at synchronous speed ω_s and in per unit, are given by [15].

Stator voltage equations:

$$\bar{v}_{ds} = -\bar{R}_s \cdot \bar{i}_{ds} - \bar{\omega}_s \cdot \bar{\psi}_{qs} + \frac{1}{\omega_b} \cdot \frac{d\bar{\psi}_{ds}}{dt} \tag{9}$$

$$\bar{v}_{qs} = -\bar{R}_s \cdot \bar{i}_{qs} + \bar{\omega}_s \cdot \bar{\psi}_{ds} + \frac{1}{\omega_b} \cdot \frac{d\bar{\psi}_{qs}}{dt}. \tag{10}$$

Rotor voltage equations:

$$\bar{v}_{dr} = \bar{R}_r \cdot \bar{i}_{dr} - s \cdot \bar{\omega}_s \cdot \bar{\psi}_{qr} + \frac{1}{\omega_b} \cdot \frac{d\bar{\psi}_{dr}}{dt} \tag{11}$$

$$\bar{v}_{qr} = \bar{R}_r \cdot \bar{i}_{qr} + s \cdot \bar{\omega}_s \cdot \bar{\psi}_{dr} + \frac{1}{\omega_b} \cdot \frac{d\bar{\psi}_{qr}}{dt} \tag{12}$$

where $\bar{v}$ denotes voltage, $\bar{i}$ denotes current, $\bar{\psi}$ denotes flux, $\bar{R}$ denotes resistance, s is the slip speed and ω_b is the base speed; indexes d and q the direct and quadrature components, and indexes s and r refers to stator and rotor.

The stator and rotor fluxes are expressed by:

Stator flux equations:

$$\bar{\psi}_{ds} = -\bar{L}_s \cdot \bar{i}_{ds} - \bar{L}_m \cdot \bar{i}_{dr} \tag{13}$$

$$\bar{\psi}_{qs} = -\bar{L}_s \cdot \bar{i}_{qs} + \bar{L}_m \cdot \bar{i}_{qr}. \tag{14}$$

Rotor flux equations:

$$\bar{\psi}_{dr} = \bar{L}_r \cdot \bar{i}_{dr} - \bar{L}_m \cdot \bar{i}_{ds} \tag{15}$$

$$\bar{\psi}_{qr} = \bar{L}_r \cdot \bar{i}_{qr} - \bar{L}_m \cdot \bar{i}_{qs} \tag{16}$$

with the stator and rotor inductances defined as follows:

$$\bar{L}_s = \bar{L}_{\sigma s} + \bar{L}_m \tag{17}$$

$$\bar{L}_r = \bar{L}_{\sigma r} + \bar{L}_m \tag{18}$$

where $\bar{L}_{\sigma s}$ and $\bar{L}_{\sigma r}$ are the stator and rotor leakages inductances, and $\bar{L}_m$ is the magnetizing inductance.

The fifth order model is completed with the mechanical equation of the machine, expressed by Eq. (7).

$$\bar{T}_{mec} - \bar{T}_e = 2\,\bar{H}_g \frac{d\bar{\omega}_g}{dt}. \tag{19}$$

The electromagnetic torque, and the active and reactive powers can be expressed as follows:

$$\bar{T}_e = \bar{\psi}_{qr} \cdot \bar{i}_{dr} - \bar{\psi}_{dr} \cdot \bar{i}_{qr} \tag{20}$$

$$\bar{P}_e = \bar{P}_s + \bar{P}_r = (\bar{v}_{ds} \cdot \bar{i}_{ds} + \bar{v}_{qs} \cdot \bar{i}_{qs}) + (\bar{v}_{dr} \cdot \bar{i}_{dr} + \bar{v}_{qr} \cdot \bar{i}_{qr}) \tag{21}$$

$$\bar{Q}_e = \bar{Q}_s + \bar{Q}_r = (\bar{v}_{qs} \cdot \bar{i}_{ds} - \bar{v}_{ds} \cdot \bar{i}_{qs}) + (\bar{v}_{qr} \cdot \bar{i}_{dr} - \bar{v}_{dr} \cdot \bar{i}_{qr}). \tag{22}$$

In case of SCIG, the rotor is short circuited, and therefore the rotor voltage is zero ($\bar{v}_{dr} = 0\ \bar{v}_{qr} = 0$).

SCIG uses local power-factor-correction capacitor banks to provide the induction generator magnetizing current. The traditional SCIG wind turbines are equipped with standard capacitor banks using mechanical contactors. To provide a faster control possibility, new wind turbines use thyristor switches instead of mechanical contactors, which can reduce the switching transients significantly and thus make it possible to switch the capacitor much more often without reducing the lifetime significantly [30].

When a SCIG wind turbine presents capacitor banks connected in parallel to the generator, the current injected by the wind turbine at the connection point to grid is given by

$$\bar{i}_{dg} = \bar{i}_{ds} + \bar{i}_{dc} = \bar{i}_{ds} + \frac{1}{\bar{X}_c} \cdot \bar{v}_{qg} \tag{23}$$

$$\bar{i}_{qg} = \bar{i}_{qs} + \bar{i}_{qc} = \bar{i}_{qs} - \frac{1}{\bar{X}_c} \cdot \bar{v}_{dg} \tag{24}$$

where $\bar{X}_c$ is the reactance of the power-factor-correction capacitors, $\bar{i}_{dc}$ and $\bar{i}_{qc}$ are the current components of the capacitors, $\bar{v}_{dg}$ and $\bar{v}_{qg}$ are the terminal voltage components.

The active and reactive powers delivered by the wind turbine at the connection point to grid can be expressed as

$$\bar{P}_g = \bar{v}_{dg} \cdot \bar{i}_{dg} + \bar{v}_{qg} \cdot \bar{i}_{qg} \tag{25}$$

$$\bar{Q}_g = \bar{v}_{qg} \cdot \bar{i}_{dg} - \bar{v}_{dg} \cdot \bar{i}_{qg}. \tag{26}$$

For representation of an induction generator in power systems transient stability studies, it is commonly used a third order model [15]. This model is obtained from neglecting the stator transients in Eqs. (9) and (10), which corresponds to ignoring the DC component in the stator transient current. With these terms neglected, the stator voltage appears as algebraic equations. This simplification is essential to ensure compatibility with the models used for representing other system components, particularly the transmission network [15]. In this case, an induction generator can be represented by Thévenin equivalent circuit like a synchronous generator. This equivalent circuit is composed of an equivalent voltage behind a transient reactance. The components of this equivalent voltage are defined as

$$\bar{e}'_d = -\frac{\bar{\omega}_s \cdot \bar{L}_m}{\bar{L}_r} \cdot \bar{\psi}_{qr} \tag{27}$$

$$\bar{e}'_q = \frac{\bar{\omega}_s \cdot \bar{L}_m}{\bar{L}_r} \cdot \bar{\psi}_{dr}. \tag{28}$$

Substituting Eqs. (27) and (28) in Eqs. (9)–(12), it is obtained the third order model of an induction generator, given by the following electrical equations

$$\frac{d\bar{e}'_d}{dt} = -\frac{1}{T'_o} \cdot (\bar{e}'_d - (\bar{X}_s - \bar{X}'_s) \cdot \bar{i}_{qs}) + s \cdot \bar{\omega}_s \cdot \bar{e}'_q - \bar{\omega}_s \cdot \frac{\bar{L}_m}{\bar{L}_r} \cdot \bar{v}_{qr} \tag{29}$$

$$\frac{d\bar{e}'_q}{dt} = -\frac{1}{T'_o} \cdot (\bar{e}'_q + (\bar{X}_s - \bar{X}'_s) \cdot \bar{i}_{ds}) - s \cdot \bar{\omega}_s \cdot \bar{e}'_d + \bar{\omega}_s \cdot \frac{\bar{L}_m}{\bar{L}_r} \cdot \bar{v}_{dr} \tag{30}$$

$$\bar{v}_{ds} = -\bar{R}_s \cdot \bar{i}_{ds} + \bar{X}'_s \cdot \bar{i}_{qs} + \bar{e}'_d \tag{31}$$

$$\bar{v}_{qs} = -\bar{R}_s \cdot \bar{i}_{qs} - \bar{X}'_s \cdot \bar{i}_{ds} + \bar{e}'_q \tag{32}$$

with $\bar{T}'_o$ the transient open circuit time constant, $\bar{X}'_s$ the transient reactance, expressed as

$$\bar{T}'_o = \frac{\bar{L}_r}{\bar{R}_r} \tag{33}$$

$$\bar{X}'_s = \bar{\omega}_s \left(\bar{L}_s - \frac{\bar{L}_m^2}{\bar{L}_r} \right). \tag{34}$$

The third differential equation is the generator mechanical equation, Eq. (19). In this case, the electromagnetic torque is defined by

$$\bar{T}_e = \frac{1}{\bar{\omega}_s} \cdot (\bar{e}'_d \cdot \bar{i}_{ds} + \bar{e}'_q \cdot \bar{i}_{qs}). \tag{35}$$

In case of SCIG, the rotor voltage is zero, and therefore Eqs. (29) and (30) are expressed as

$$\frac{d\bar{e}'_d}{dt} = -\frac{1}{\bar{T}'_o} \cdot (\bar{e}'_d - (\bar{X}_s - \bar{X}'_s) \cdot \bar{i}_{qs}) + s \cdot \bar{\omega}_s \cdot \bar{e}'_q \tag{36}$$

$$\frac{d\bar{e}'_q}{dt} = -\frac{1}{\bar{T}'_o} \cdot (\bar{e}'_q + (\bar{X}_s - \bar{X}'_s) \cdot \bar{i}_{ds}) - s \cdot \bar{\omega}_s \cdot \bar{e}'_d. \tag{37}$$

The generation system of a SCIG, including the induction generator and compensating capacitors, can be represented by the equivalent circuit presented in Fig. 7.

If the rotor transients are also neglected, it is obtained the first order model of an induction machine. In this model, the mechanical equation, Eq. (19), is the only differential equation to be considered and the electrical equation are expressed as

$$\bar{v}_{ds} = -\bar{R}_s \cdot \bar{i}_{ds} - \bar{\omega}_s \cdot \bar{\psi}_{qs} = -\bar{R}_s \cdot \bar{i}_{ds} - \bar{\omega}_s \cdot \left(-\bar{L}_s \cdot \bar{i}_{qs} + \bar{L}_m \cdot \bar{i}_{qr}\right) \tag{38}$$

$$\bar{v}_{qs} = -\bar{R}_s \cdot \bar{i}_{qs} + \bar{\omega}_s \cdot \bar{\psi}_{ds} = -\bar{R}_s \cdot \bar{i}_{qs} + \bar{\omega}_s \cdot (-\bar{L}_s \cdot \bar{i}_{ds} - \bar{L}_m \cdot \bar{i}_{dr}) \tag{39}$$

$$\bar{v}_{dr} = \bar{R}_r \cdot \bar{i}_{dr} - s \cdot \bar{\omega}_s \cdot \bar{\psi}_{qr} = \bar{R}_r \cdot \bar{i}_{dr} - s \cdot \bar{\omega}_s \cdot \left(\bar{L}_r \cdot \bar{i}_{qr} - \bar{L}_m \cdot \bar{i}_{qs}\right) \tag{40}$$

$$\bar{v}_{dr} = \bar{R}_r \cdot \bar{i}_{dr} - s \cdot \bar{\omega}_s \cdot \bar{\psi}_{qr} = \bar{R}_r \cdot \bar{i}_{dr} - s \cdot \bar{\omega}_s \cdot \left(\bar{L}_r \cdot \bar{i}_{qr} - \bar{L}_m \cdot \bar{i}_{qs}\right). \tag{41}$$

This model can be represented by the steady state equivalent circuit shown in Fig. 8.

The steady state electromagnetic torque can be calculated from Eq. (20) or by using the following expression:

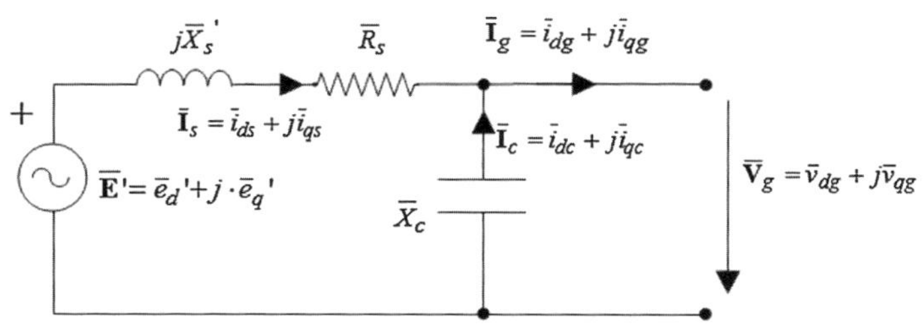

Fig. 7 Equivalent circuit of a SCIG

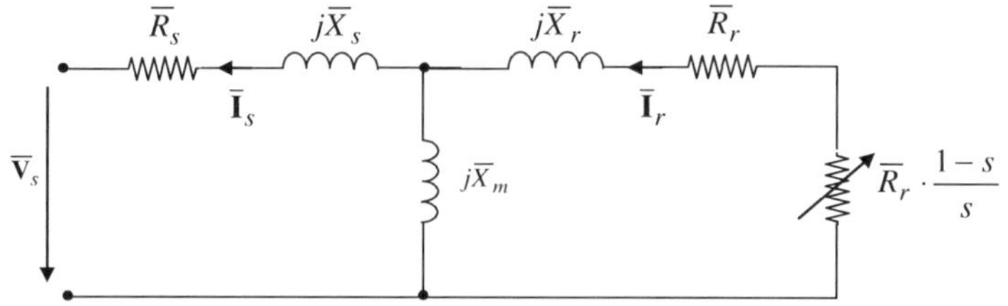

Fig. 8 Steady-state equivalent circuit of a SCIG

$$\bar{T}_e = \frac{\bar{R}_r \cdot \bar{U}_s^2 \cdot s}{\bar{\omega}_s \cdot ((\bar{R}_s \cdot s + \bar{R}_r)^2 + ((\bar{X}_s + \bar{X}_r) \cdot s)^2)}. \tag{42}$$

The active power generated by a SCIG is calculated by

$$\bar{P}_e = \bar{T}_e \cdot \bar{\omega}_g. \tag{43}$$

The reactive power consumed by a SCIG, which depends on the active power and the stator voltage, can be calculated from the equivalent circuit shown in Fig. 9. It is given by

$$\begin{aligned} \bar{Q}_e = \frac{\bar{U}_s^2}{\bar{X}_m} + (\bar{X}_s + \bar{X}_r) \cdot \frac{\bar{U}_s^2 + 2 \cdot (\bar{R}_r + \bar{R}_s) \cdot \bar{P}_e}{2 \cdot ((\bar{R}_r + \bar{R}_s)^2 + (\bar{X}_s + \bar{X}_r)^2)} - (\bar{X}_s + \bar{X}_r) \\ \cdot \frac{\sqrt{(\bar{U}_s^2 + 2 \cdot (\bar{R}_r + \bar{R}_s) \cdot \bar{P}_e)^2 - 4 \cdot \bar{P}_e^2 \cdot ((\bar{R}_r + \bar{R}_s)^2 + (\bar{X}_s + \bar{X}_r)^2)}}{2 \cdot ((\bar{R}_r + \bar{R}_s)^2 + (\bar{X}_s + \bar{X}_r)^2)}. \end{aligned} \tag{44}$$

In case of a SCIG with local power-factor-correction capacitors, the reactive power delivered to grid by SCIG wind turbine is expressed as

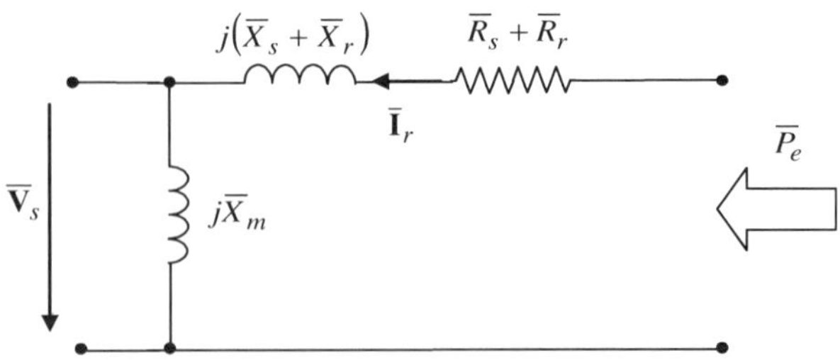

Fig. 9 Equivalent circuit used to calculate the reactive power consumed by a SCIG

$$\bar{Q}_e = \bar{U}_s^2 \frac{\bar{X}_c - \bar{X}_m}{\bar{X}_m \cdot \bar{X}_m} + (\bar{X}_{\sigma s} + \bar{X}_{\sigma r}) \cdot \frac{\bar{U}_s^2 + 2 \cdot (\bar{R}_r + \bar{R}_s) \cdot \bar{P}_e}{2 \cdot ((\bar{R}_r + \bar{R}_s)^2 + (\bar{X}_s + \bar{X}_r)^2)} - (\bar{X}_s + \bar{X}_r) \cdot \frac{\sqrt{(\bar{U}_s^2 + 2 \cdot (\bar{R}_r + \bar{R}_s) \cdot \bar{P}_e)^2 - 4 \cdot \bar{P}_e^2 \cdot ((\bar{R}_r + \bar{R}_s)^2 + (\bar{X}_s + \bar{X}_r)^2)}}{2 \cdot ((\bar{R}_r + \bar{R}_s)^2 + (\bar{X}_s + \bar{X}_r)^2)} \quad (45)$$

5.1.2 FS-SCIG Control System

In this wind turbine, the control system is responsible of controlling the blade pitch angle, if possible, and the connection and disconnection of the individual power-factor-correction capacitors.

In passive-stall-controlled wind turbines, the aerodynamic design of the rotor blade regulates the power of the wind turbine, and therefore, it does not require an active control of the blade pitch. They are the most widely used fixed speed wind turbines in wind farms.

For active-stall-controlled and pitch-controlled wind turbines, the rotor limits the power extracted from the wind by controlling the blade pitch angle in order to decrease the rotor aerodynamic efficiency and as well as the mechanical power extracted from the wind. The control scheme used in these wind turbines is shown in Fig. 5. In this case, the electrical power is commonly used as controlling variable, which is limited to rated power for high winds by acting on the blade angle.

As mentioned previously, the reactive power consumed by a SCIG is uncontrollable, because it depends on the generated active power, and it varies with the incoming wind. To minimize the reactive power absorbed by the generator from the grid, the control system controls continuously the connection and disconnection of the individual capacitors in parallel to the generator, depending on the average reactive power demand of the generator over a predefined period of time (intervals of 1–10 min).

5.1.3 Dynamic Simulation of an FS-SCIG Wind Turbine

In this section, the performance of an FS-SCIG wind turbine is illustrated by means dynamic simulations, where the response of this wind turbine is shown under wind fluctuations and a grid fault.

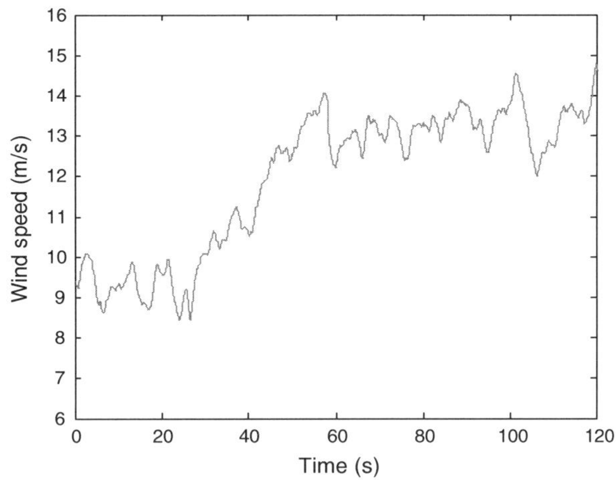

Fig. 10 Wind speed incident on the wind turbine

Figure 10 shows the wind speed incident considered in the simulations performed in this chapter for each wind turbine concept. This wind sequence corresponds to winds below nominal winds during the first seconds and above nominal winds for the rest of the simulation. This allows wind turbine to be evaluated in any operating conditions, with power optimization, power limitation or down power regulation strategies (if possible, depending on the wind turbine concept).

The response of a passive-stall-controlled FS-SCIG is depicted in Fig. 11. Figures 12 and 13 show the response of a pitch-controlled FS-SCIG wind turbine. Both wind turbines present capacitor bank and they are connected to an infinite bus. It is seen that both wind turbines produces below rated power during the first seconds. With above rated power (rated wind speed enables the wind turbine to produce the rated power), the power of a passive-stall-controlled wind turbine is limited by stall effect, because it does not present mechanical or electrical control. In pitch-controlled wind turbine, the power is limited by acting on the blade pitch angle. In both cases, it is observed that the torque, power and generator speed follows the wind speed variation.

The performance of a FS-SCIG wind turbine during a voltage dip is presented in Figs. 14 and 15. In this case, it is considered that the voltage of the infinite bus drops a 40 % during 0.1 s, and after the fault the voltage starts to recover. Because the grid disturbances are much faster than wind speed variations, the wind speed incident on the wind turbine is assumed constant. The responses obtained by using the fifth and third order model of the induction generator are compared. When the voltage dip reaches the wind turbine, it can be seen that the generated active power falls, while the mechanical power does not change and therefore the wind turbine accelerates. When the voltage starts to recover, as the rotational speed has increased, a transient period follows. The fifth order model shows an oscillatory response that is not observed in the response obtained with the third order model, because only the fundamental frequency components are represented.

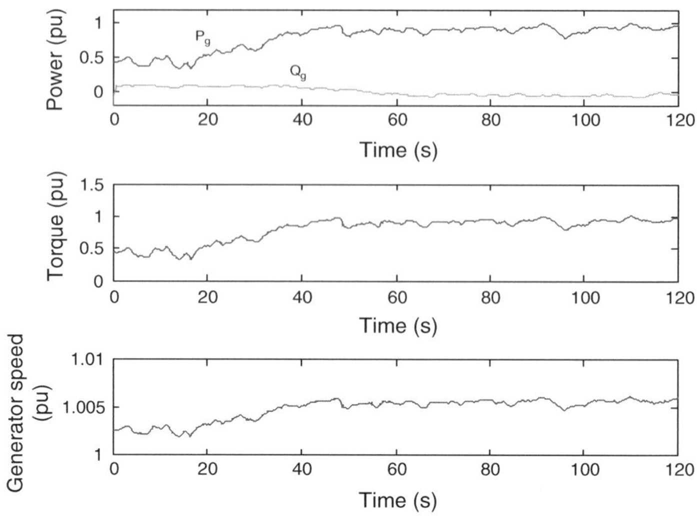

Fig. 11 Response of a stall regulated FS-SCIG: Powers, torque and generator speed

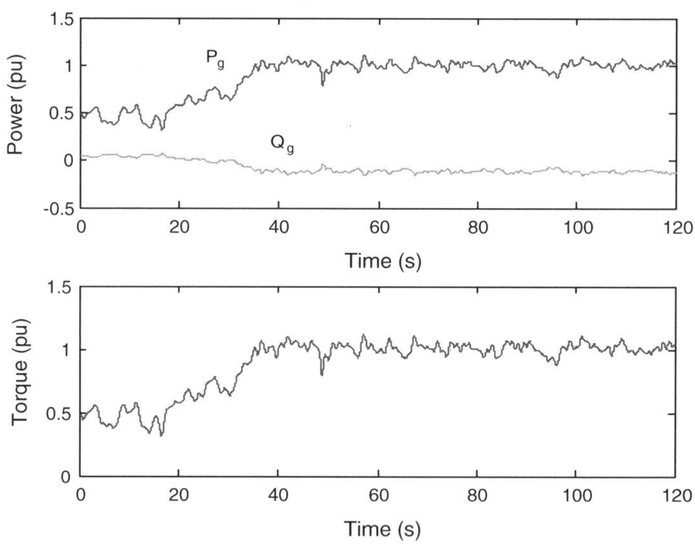

Fig. 12 Response of a pitch controlled FS-SCIG: Power and torque

5.2 Wound Rotor Induction Generator with Variable Rotor Resistance

This wind turbine uses a wound rotor induction generator (WRIG) with variable rotor resistance (VRR) by means of a power electronic converter, mounted on the

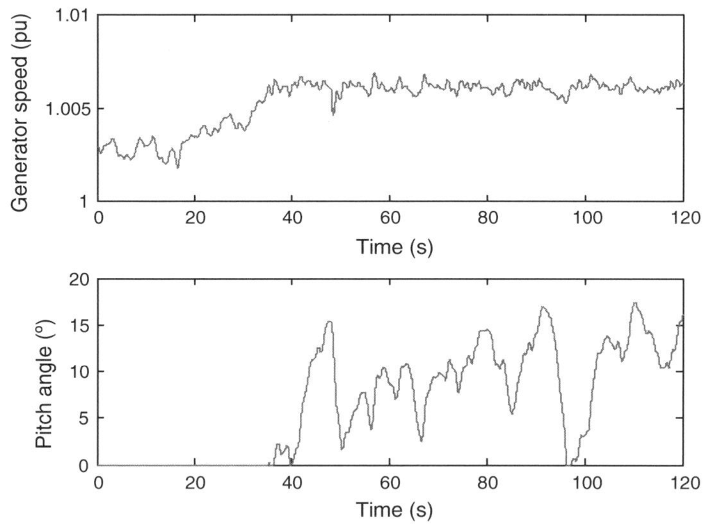

Fig. 13 Response of a pitch controlled FS-SCIG: Generator speed and blade pitch angle

rotor shaft, as shown in Fig. 16. In this case, the stator winding is directly connected to the grid, whereas the rotor winding is connected in series with an external resistance controlled by power converter. The VRR unit consists of three resistances, a diode-rectifier, an IGBT and the control system. The IGBT, as controllable switch at the DC side, is used for controlling the rotor current flow though the external rotor resistances. The converter is optically controlled, which means that no slip rings are necessary. This concept, known as the Optislip concept, was developed by the Danish manufacturer Vestas.

Variable-speed operation can be achieved by controlling the variable resistance connected to the rotor winding, and thus, the generator slip. However, this slip power is dissipated in the external resistor as losses. In this concept, the speed range is typically limited to 0–10 %, as it is dependent on the size of the variable rotor resistance. Furthermore, reactive power compensation is also required for this concept.

5.2.1 WRIG-VRR Generation System

The dynamic behaviour of a WRIG used in this wind turbine can be represented by the same models described in Sect. 5.1.1. For a WRIG, the fifth order model is defined by Eqs. (9)–(20), the third order model by Eqs. (19), (29)–(35), and the first order model by Eqs. (19), (20), (38)–(41). However, in this wind turbine, the rotor resistance $\bar{R}_r$ is dynamic, which is calculated as sum of the rotor winding resistance $\bar{R}_{rw}$ and the active external resistance $\bar{R}_{ext}$, controlled by the VRR unit.

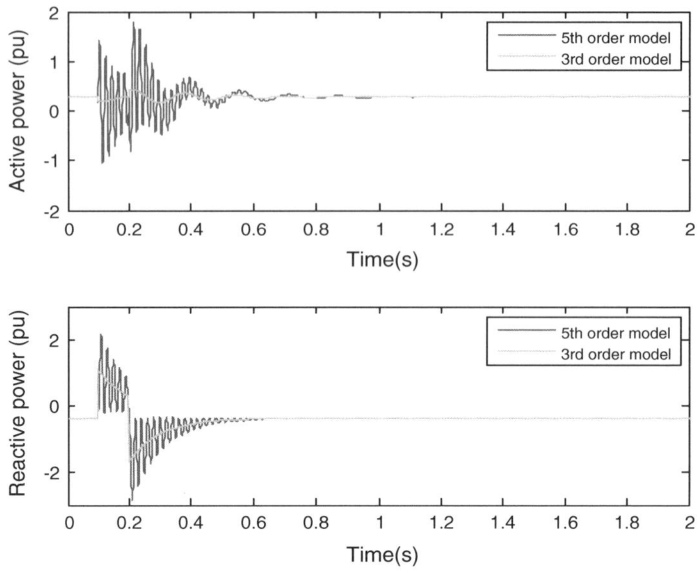

Fig. 14 FS-SCIG responses obtained by using the 5th order and 3rd order models during a voltage dip (40 % drop during 0.1 s): Active and reactive power

Thus, the dynamic rotor resistance used in the models of WRIG-VRR is expressed as follows

$$\bar{R}_r = \bar{R}_{rw} + \bar{R}_{ext}(t). \tag{46}$$

In steady-state, this wind turbine can be represented by the equivalent circuit shown in Fig. 17.

The stator and rotor active powers can be deduced from the equivalent circuit shown in Fig. 17, assuming that losses are neglected [17]:

$$\bar{P}_g = \bar{P}_s = (1 - s) \cdot \bar{P}_m \tag{47}$$

$$\bar{P}_r = s \cdot \bar{P}_m \tag{48}$$

where $\bar{P}_g$ is the total power delivered to grid, $\bar{P}_s$ is the stator power, $\bar{P}_s$ is the rotor power, $\bar{P}_m$ is the mechanical power and s is the slip speed.

The maximum generator slip depends on the active external resistance, such as deduced from calculating $d\bar{T}_e/ds = 0$ by using $\bar{T}_e$ defined by Eq. (42):

$$s_m = \frac{\bar{R}_{rw} + \bar{R}_{ext}(t)}{\sqrt{\bar{R}_s^2 + (\bar{X}_s + \bar{X}_r)^2}}. \tag{49}$$

It can be deduced from Eqs. (47)–(49) that adding external resistance to the rotor is possible to change the generator slip, and thus the electromagnetic torque and the power delivered to grid. When the mechanical power extracted from the

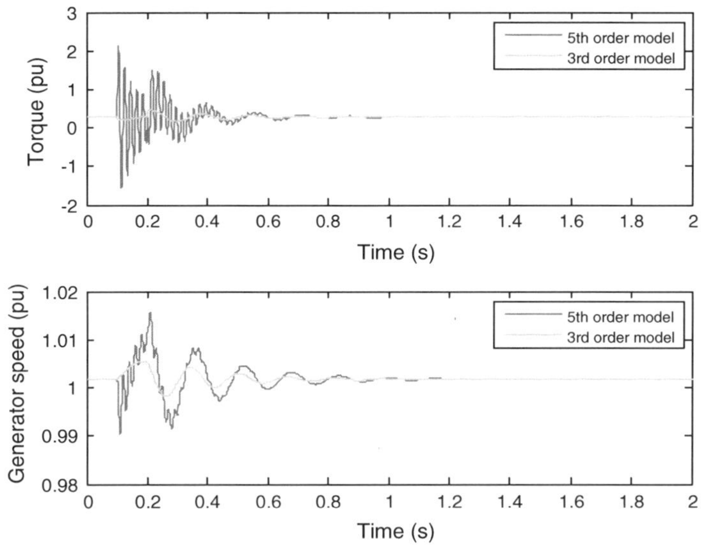

Fig. 15 FS-SCIG responses obtained by using the 5th order and 3rd order models during a voltage dip (40 % drop during 0.1 s): Torque and generator speed

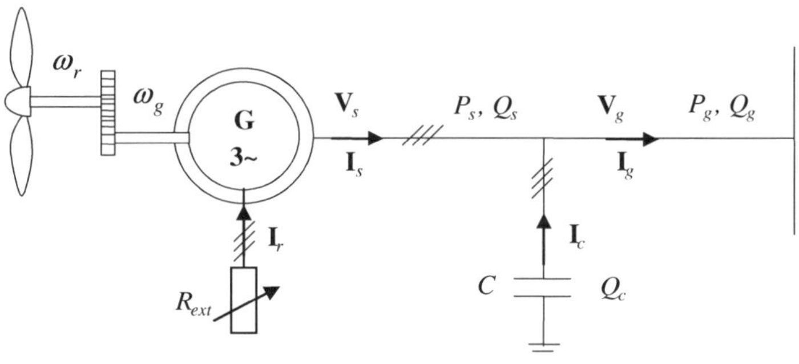

Fig. 16 WRIG-VRR wind turbine configuration

wind increases, additional resistance is added to the rotor, and the slip increases to compensate the variation and to keep the power delivered to grid constant. The excess of the energy extracted from the wind is converted into kinetic energy, causing the rotor acceleration. When the mechanical power extracted from the wind decreases, the rotor kinetic energy is converted back into electrical energy, causing the rotor deceleration. This leads to minimize the torque and power fluctuation during operation, which reduces mechanical stress and flicker level of turbine.

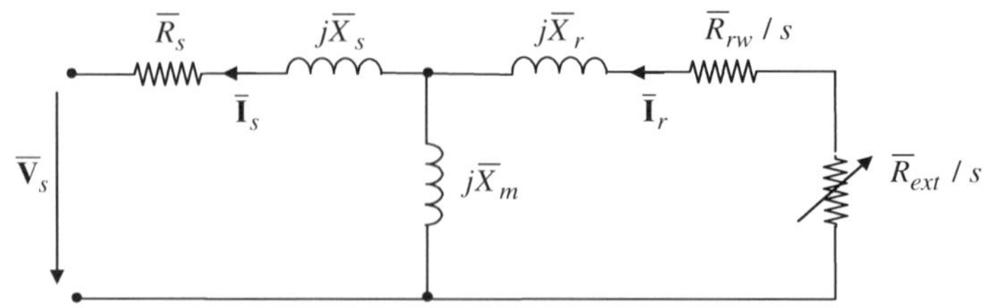

Fig. 17 Steady-state equivalent circuit of a WRIG-VRR

5.2.2 WRIG-VRR Control System

The control system of this wind turbine (Fig. 18) is composed of [6]: (1) VRR control system; and (2) blade pitch angle control system.

The external resistance connected to the rotor winding is controlled by the VRR control system. The VRR unit consists of three resistances, a diode-rectifier, an IGBT and the control system (Fig. 19). The IGBT, as controllable switch at the DC side, is used for controlling the rotor current flow though the external rotor resistances. The control system presents two control loops: a power controller as outer loop and a rotor current controller as inner loop.

The power reference is obtained from the generator slip, according to the control characteristic shown in Fig. 20. In this wind turbine, the generator slip can be changed from 1 to 10 %. The power controller uses this power reference in order to define the rotor current reference.

The rotor current controller determines the duty cycle of the power converter D (between 0 and 1), and thus, the dynamic external resistance connected to the rotor winding is calculated as follows:

$$\bar{R}_{ext}(t) = D(t) \cdot \bar{R}_{ext}. \tag{50}$$

The blade pitch angle control system is composed of the speed controller and the pitch actuator. The speed controller defines the pitch angle reference in order to adjust the rotational speed to the speed reference, which depends on the operating conditions. The pitch actuator turns the blades to the angle ordered by the speed controller.

The control strategy used in this wind turbine can be summarised as follows:

- Power optimization strategy: In this case, the wind is not high enough to produce the rated power so that the speed controller acts on the pitch angle in order to achieve the optimal rotational speed, and thus maximizing the power extracted from the wind. The power is controlled by the power controller to increase with a generator slip of 2 % as shown in Fig. 20.

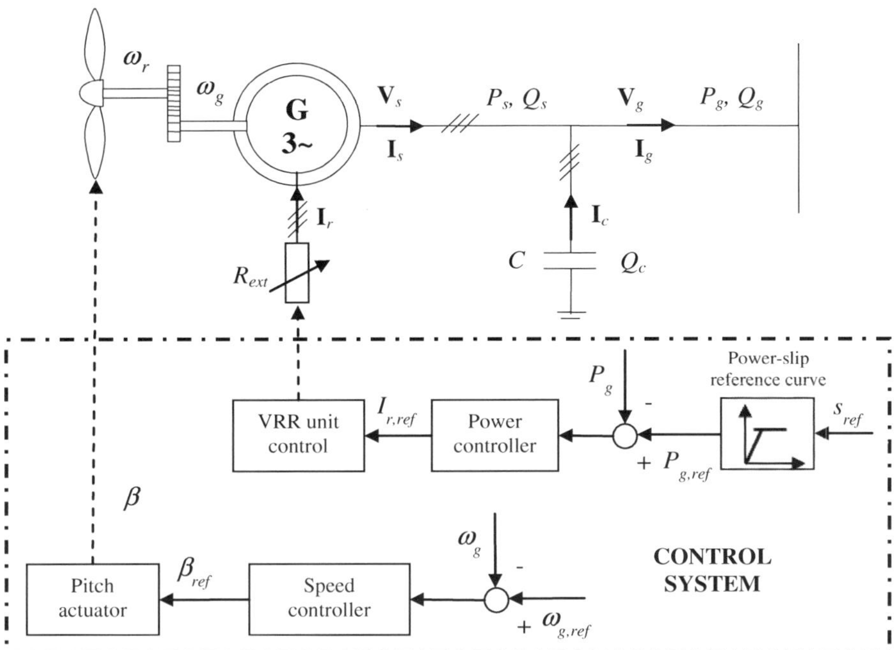

Fig. 18 Control system of a WRIG-VRR wind turbine

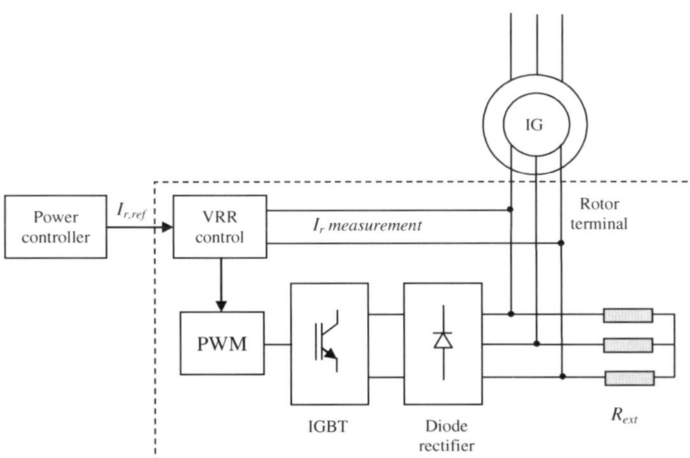

Fig. 19 VRR unit

- Power limitation strategy: When the wind is high enough to produce the rated power, the power of the wind turbine is controlled to the rated power. By adjusting the blade pitch angle, the speed controller maintains the average slip

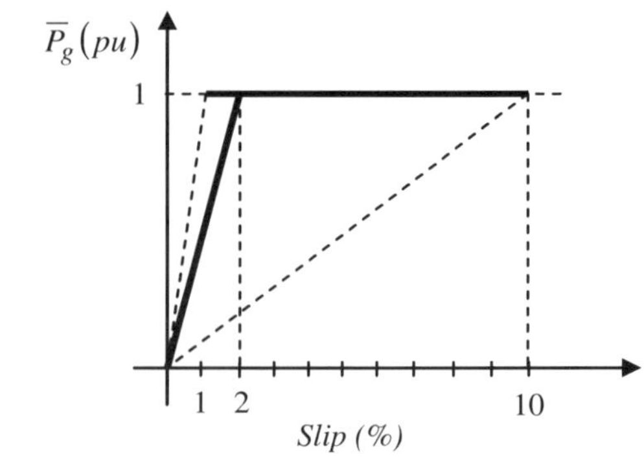

Fig. 20 Power-slip reference *curve* of the power controller of a WRIG-VRR wind turbine

equal to 4–5 %. Short time speed changes at rated power are controlled by possible generator slip changes, as shown in Fig. 20 for minimum and maximum generator slip.

5.3 Doubly Fed Induction Generator

DFIG wind turbine (Fig. 21) uses a wound rotor induction generator coupled to the wind turbine rotor through a gearbox. This generator presents the stator winding connected directly to the grid and a bidirectional power converter feeding the rotor winding. It is made up with two back-to-back IGBT bridges linked by a DC bus. This converter decouples the electrical grid frequency and the mechanical rotor frequency and thus enabling variable speed generation of the wind turbine. The wind turbine rotor presents blade pitch angle control in order to limit the power and the rotational speed for high winds. Furthermore, DFIG presents noticeable advantages such as the decoupled control of active and reactive powers, the reduction of mechanical stresses and acoustic noise, the improvement of power quality, and the use of a power converter with a rated power 25 % of total system power.

DFIG wind turbine delivers active power to the grid by both the stator and rotor winding, which can be calculated according to the following equations [20, 26]:

$$\bar{P}_s = \bar{P}_m/(1-s) \tag{51}$$

$$\bar{P}_r = -s \cdot \bar{P}_s \tag{52}$$

$$\bar{P}_g = \bar{P}_s + \bar{P}_r. \tag{53}$$

As deduced from the previous equations, the size of the converters is determined by the controllable range of the slip speed. In DFIG wind turbine, this range of slip speed commonly varies between 0.7 and 1.2 pu due to mechanical and other restrictions.

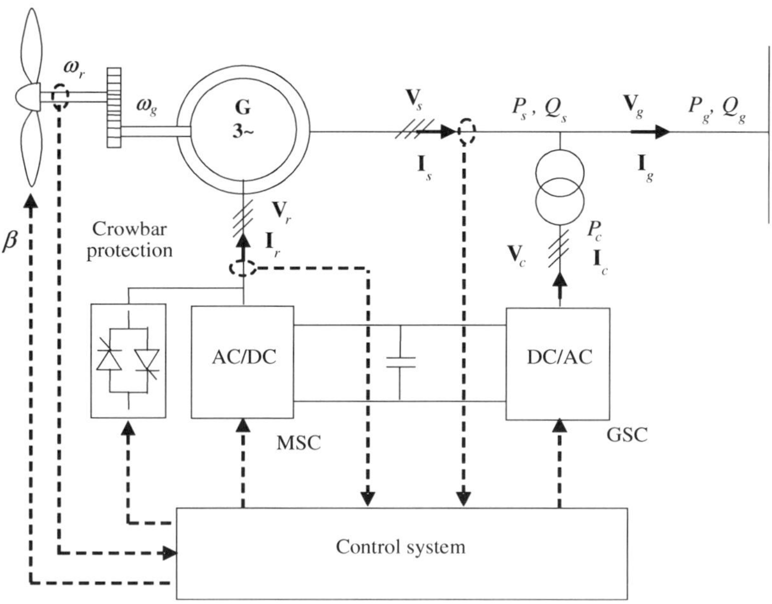

Fig. 21 DFIG wind turbine configuration

Fig. 22 Power-speed *curve* of a DFIG wind turbine

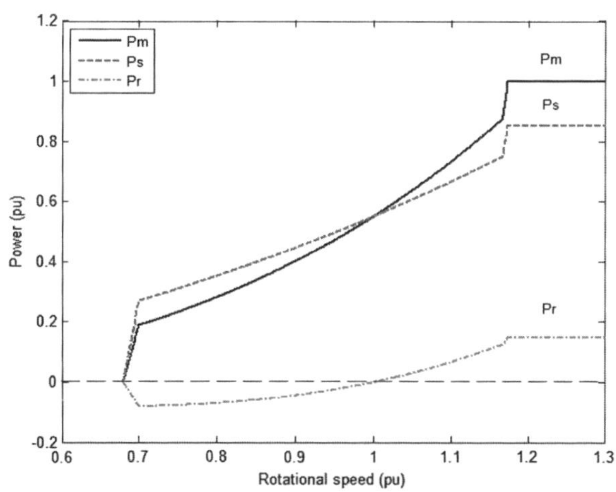

Figure 22 shows the power-speed curve of a DFIG wind turbine, where the capacity of delivering active power to the grid by both the stator and rotor winding can be observed.

However, the reactive power delivered to the grid is limited to restrictions imposed by the converter and expressed as rotor current limits to avoid an excessive heating of converters, rotor slip-rings and brushes [31]. The stator

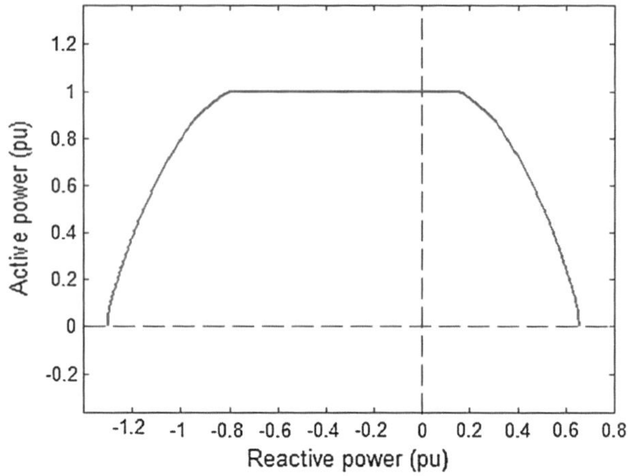

Fig. 23 Operating region of a DFIG wind turbine connected to an infinite bus

reactive power limit $\bar{Q}_{s,\lim}$ depend on the stator active power $\bar{P}_s$, the stator voltage $\bar{V}_s$ and the maximum rotor current $\bar{I}_{r,\max}$:

$$\bar{Q}_{s,\lim} = -\frac{\bar{V}_s^2}{\bar{X}_s} \pm \sqrt{\left(\frac{\bar{X}_m}{\bar{X}_s} \cdot \bar{V}_s \cdot \bar{I}_{r,\max}\right)^2 - \bar{P}_s^2}. \tag{54}$$

Equation (54) expresses the maximum reactive power that can be rendered to the grid. Figure 23 depicts the operating region Q-P for a DFIG wind turbine assuming the connection to an infinite bus.

5.3.1 DFIG Generation System

Wound Round Induction Generator

The dynamic behaviour of a WRIG used in a DFIG wind turbine can be represented by the fifth and third order models of an induction machine described in Sect. 5.1.1. For a WRIG, the fifth order model is defined by Eqs. (9)–(20), and the third order model by Eqs. (19), (29)–(35). Both models are commonly expressed in *d-q* coordinate reference frame aligned with the stator flux and rotating at synchronous speed. In this frame, the *d*-axis of the reference frame is aligned along the stator-flux position so that stator *q*-axis flux linkage is zero, and d-axis flux linkage is constant. In addition, the *d* component of the stator voltage is zero and the *q* component of the stator voltage is constant. This transformation allows the independent control of the electromagnetic torque or the power generated and the reactive power or the terminal voltage of a DFIG wind turbine, as will be shown below. Figure 24 shows the equivalent circuit corresponding to the third order model.

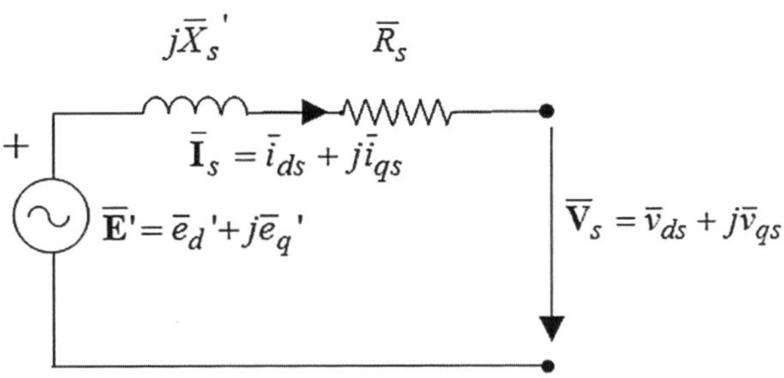

Fig. 24 Equivalent circuit of DFIG based on the third order model

For voltage and angle stability investigations, it can be also used the first order model of a WRIG [28]. This model is expressed by Eqs. (19), (20), (38)–(41). Figure 25 shows the equivalent circuit corresponding to this model. From the previous equations and the equivalent circuit, it can be deduced the rotor current.

$$\bar{\mathbf{I}}_r = \frac{\bar{\mathbf{V}}_s - \left(\frac{\bar{\mathbf{V}}_r}{s}\right)}{\left(\bar{R}_s + \frac{\bar{R}_r}{s}\right) + j(\bar{X}_s + \bar{X}_r)}. \tag{55}$$

The electromagnetic torque can be calculated from the power balance across the stator to rotor gap:

$$\bar{T}_e = \left(\bar{I}_r^2 \frac{\bar{R}_r}{s}\right) + \frac{\bar{P}_r}{s} \tag{56}$$

with the rotor power expressed as follows

$$\bar{P}_r = \frac{\bar{V}_r}{s} \bar{I}_r \cos\theta_r \tag{57}$$

where θ_r is the phase angle between the rotor current $\bar{\mathbf{I}}_r$ and the controllable rotor voltage $\bar{\mathbf{V}}_r/s$.

Power Converter

The bidirectional power converter used in a DFIG wind turbine is made up with two back-to-back IGBT bridges linked by a DC bus. The machine side converter (MSC), connected to the rotor winding, converts the low AC frequency of the generator to DC. The DC voltage is stabilised by DC link capacitor, and converted further by grid side converter (GSC) into 50 Hz AC, which is supplied to the grid.

The MSC drives the wind turbine to achieve the optimum power efficiency in winds below rated, to limit the output power to the rated value in winds above rated, or to adjust the active and reactive powers to the power references when power regulation is demanded. This converter enables the decoupled control of the

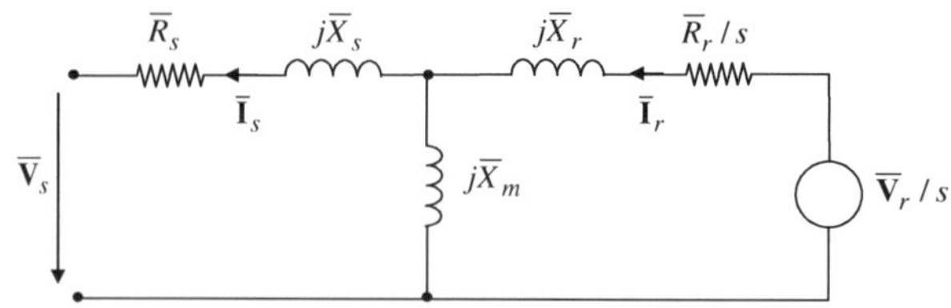

Fig. 25 Equivalent circuit of DFIG based on the first order model (steady state model)

active power, reactive power or the terminal voltage by acting on the rotor current or the rotor voltage. It is modelled as a controlled voltage source.

Through the DC link, the active power generated by the DFIG is supplied to the GSC, and from this converter, it is fed to the grid. The DC link contains a capacitor that is charged/discharged by the MSC and GSC currents, respectively. When the power losses in the DC link are ignored, the dynamic behaviour of DC voltage across capacitor, $\bar{v}_{dc-link}$, can be expressed by the following equation [3]:

$$\frac{d\bar{v}_{dc-link}}{dt} = \frac{1}{\bar{C} \cdot \bar{v}_{dc-link}} \cdot (\bar{P}_r - \bar{P}_{gc}) \tag{58}$$

where $\bar{C}$ is the value of the DC link capacitor, and $\bar{P}_{gc}$ is the power supplied to the GSC, expressed as

$$\bar{P}_{gc} = \bar{v}_{dc} \cdot \bar{i}_{dg} + \bar{v}_{qc} \cdot \bar{i}_{qg} \tag{59}$$

with $\bar{i}_{dg}$ and $\bar{i}_{qg}$ is the components of the current delivered to the grid.

When the DC link voltage is assumed constant [9, 10, 31], the power supplied to the GSC is considered equal to the power extracted from the rotor circuit.

GSC maintains the exchange power from the rotor circuit to the grid and commonly operates with the unity power factor, although it can also be used for reactive power injection. This converter is also modelled as a controlled voltage or current source, where the current or voltage are controlled in such a way to transmit the active power from the DC link to the grid and to achieve the desired power factor/voltage on the grid.

Figure 26 shows the equivalent circuits of DFIG with power converter. Different ways of controlling these converters are explained in the next section.

5.3.2 DFIG Control System

The aims of the control system of a DFIG wind turbine are:

- maximize the power extracted from the wind for a wide range of wind speeds (also known as power optimization),
- limit the output power to the rated power for high winds (power limitation),
- adjust both the active and reactive powers to a set point ordered by the wind farm control system (power regulation) when trying to adjust the wind farm production to the settings specified by the power system operator, or

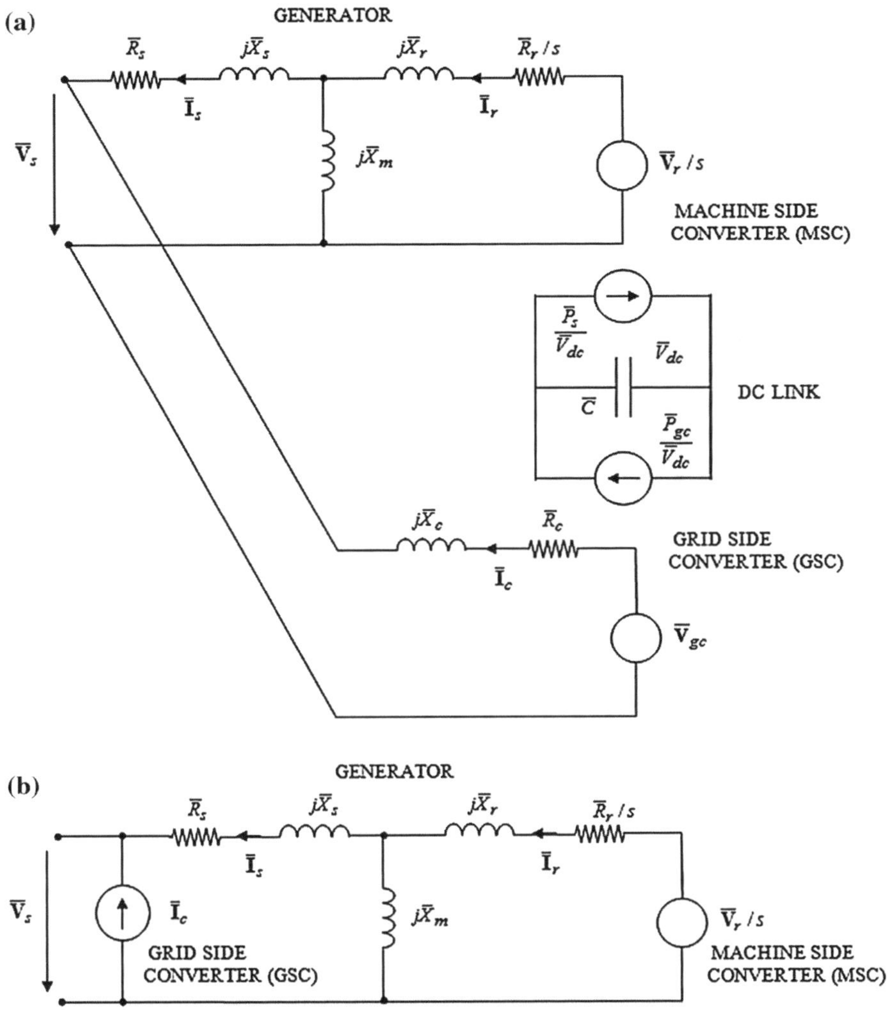

Fig. 26 Equivalent circuits of DFIG with power converter **a** with representation of DC link, **b** without representation of DC link (assuming the DC link voltage constant)

- adjust the terminal voltage to a set point (voltage regulation).

In a DFIG wind turbine, these aims are achieved by an appropriate control of the power converter and the blade pitch angle. Hence, to work effectively, the power converter is controlled in collaboration with the blade pitch angle control.

The control strategies that can be applied to the MSC, the GSC and the control of the blade pitch angle are described in this section.

MSC Control

Two control techniques can be applied to the MSC in order to achieve the independent control of the electromagnetic torque/power generated and the reactive power/terminal voltage of a DFIG wind turbine:

1. Control of the *dq* components of the rotor current/voltage.
2. Control of the magnitude and angle of the rotor voltage.

The control of the *dq* components of the rotor current/voltage is the most widely used control technique in MSC.

1. *Control of the dq components of the rotor current/voltage.*

In this control strategy, the electromagnetic torque/active power is controlled by acting on the q component of the rotor current $\bar{i}_{qr}$, and the reactive power/terminal voltage by acting on the d component of the rotor current $\bar{i}_{dr}$ [24]. This control strategy presents two controllers. The $\bar{i}_{qr}$ controller uses the optimum torque/power extraction curve (torque/power versus speed) to define the reference torque/power according to the actual rotor speed. Thus, the wind turbine can operate with variable speed below nominal winds maximizing the power extracted from the wind or limiting the output power to rated power above nominal winds. The $\bar{i}_{dr}$ controller allows the wind turbine operation with the desired power factor or terminal voltage. Figure 27 illustrates the configuration of this control strategy [4]. It is justified as follows.

When Eqs. (20), (38)–(41) are combined and the stator resistance is neglected, it is obtained the following expression relating the electromagnetic torque and the q component of the rotor current:

$$\bar{T}_e = \frac{\bar{L}_m \bar{v}_{qs}}{\bar{\omega}_s \bar{L}_s} \bar{i}_{qr}. \tag{60}$$

As the power and torque are related as Eq. (43), the q component of the rotor current can be also used to control the active power.

On the other hand, the reactive power delivered to grid by a DFIG wind turbine is the sum of the stator and the GSC reactive power. However, it is equal to the stator reactive power, because the GSC usually operates with unity power factor. When Eqs. (26) and (38)–(41) are combined and the stator resistance is neglected, the total reactive power can be expressed as directly dependent on the d component of rotor current.

$$\bar{Q}_g = \frac{\bar{L}_m \bar{v}_{qs}}{\bar{L}_s} \bar{i}_{dr} - \frac{\bar{v}_{qs}^2}{\bar{\omega}_s \bar{L}_s}. \tag{61}$$

By dividing the d component of rotor current in two components, the total reactive power can be expressed as the sum of the reactive power $\bar{Q}_{mag}$ that

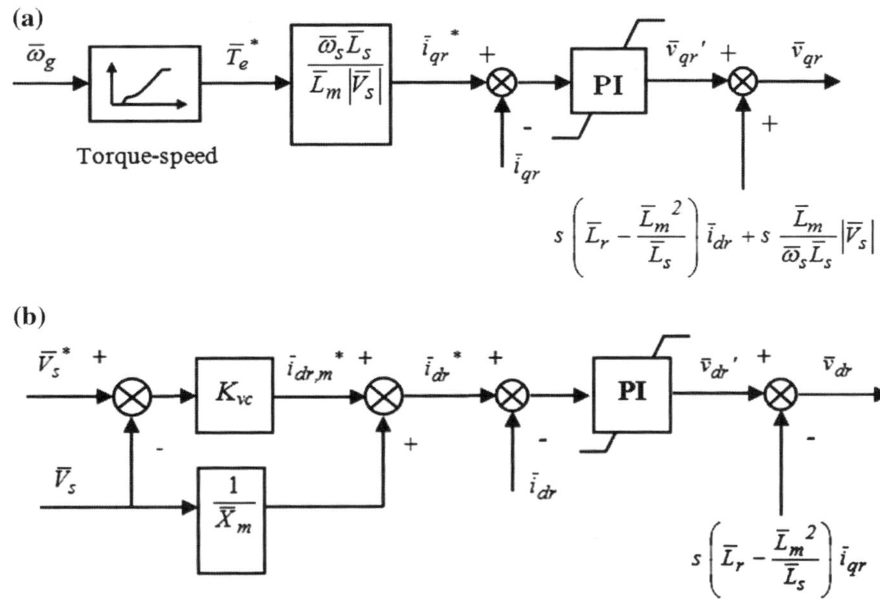

Fig. 27 MSC control scheme based on the control of dq components of the rotor current. **a** Torque controller. **b** Voltage controller

magnetizes the generator ($\bar{i}_{dr,m}$) and the reactive power delivered to grid $\bar{Q}_{gen}$ ($\bar{i}_{dr,g}$).

$$\bar{Q}_g = \bar{Q}_{mag} + \bar{Q}_{gen} = \frac{\bar{L}_m \bar{v}_{qs}}{\bar{L}_s}\left(\bar{i}_{dr,m} + \bar{i}_{dr,g}\right) - \frac{\bar{v}_{qs}^2}{\bar{\omega}_s \bar{L}_s}. \tag{62}$$

The magnetizing reactive power is compensated if $\bar{i}_{dr,m}$ presents the following value:

$$\bar{i}_{dr,m} = \frac{\bar{v}_{qs}}{\bar{\omega}_s \bar{L}_m}. \tag{63}$$

Therefore, the reactive power delivered to grid by the DFIG wind turbine results to be directly dependent on the d component of rotor current. As the terminal voltage depends on the reactive power delivered to grid, it can be also controlled by acting on this component of rotor current.

However, the rotor current components cannot be controlled independently by the rotor voltage components because of the cross-coupling effects shown in Eqs. (40) and (41). The cross-coupling effects can be cancelled by feed-forward compensation. The rotor voltage components are decided by PI controllers and the following decoupling feed-forward compensation:

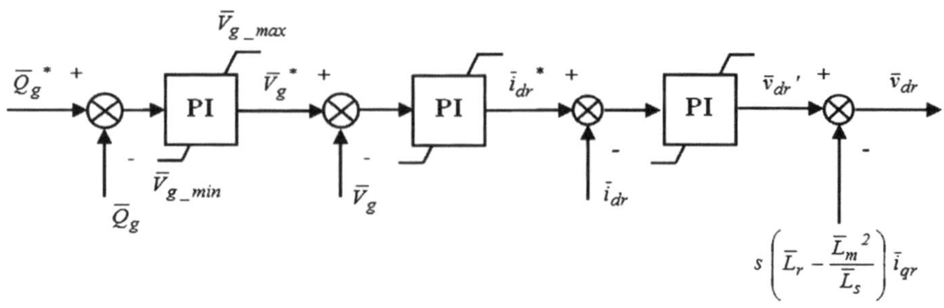

Fig. 28 $\bar{v}_{dr}$ controller

$$\bar{v}_{dr} = \bar{v}'_{dr} - s\left(\bar{L}_r - \frac{\bar{L}_m^2}{\bar{L}_s}\right)\bar{i}_{qr} \tag{64}$$

$$\bar{v}_{qr} = \bar{v}'_{qr} + s\left(\bar{L}_r - \frac{\bar{L}_m^2}{\bar{L}_s}\right)\bar{i}_{dr} + s\left(\frac{\bar{L}_m}{\bar{\omega}_s\bar{L}_s}\right)|\bar{V}_s|. \tag{65}$$

The decoupled control of the electromagnetic torque/active power and the reactive power/terminal voltage is then performed in this converter by acting on the rotor voltage components: the active power control is performed by $\bar{v}_{qr}$, and the reactive power/terminal voltage is controlled by $\bar{v}_{dr}$. In fact, this converter can be modelled as a current-controlled voltage source.

Other three strategies based on a current-controlled voltage source are described in [10]. In these three strategies, the reactive power control is performed by acting on the direct component of the rotor current $\bar{i}_{dr}$ (Fig. 28).

In the first of control strategies (Fig. 29), the speed is controlled by acting on the quadrature component of the rotor current, and the active power is controlled by acting on the pitch angle. The operating strategy of a DFIG wind turbine with this control system can be summarized as follows:

- Power optimization strategy: In this case, and as explained before, the active power reference for the wind turbine is set up at its rated value. In winds below rated, the blade pitch angle controller does not act and the pitch angle is kept at its optimal value. The $\bar{v}_{qr}$ controller controls the rotational speed, the tip speed ratio is kept at its optimal value, and thus the wind turbine operates maximizing the power extracted from the wind.
- Power limitation strategy: In this case, the active power reference for the wind turbine is set up at its rated value. In winds above rated, the pitch angle controller assures the rated power by acting on the pitch angle, and the $\bar{v}_{qr}$ controller keeps the generator speed at its rated value.
- Down power regulation strategy: When the wind turbine operates with down power regulation, the active power reference is reduced from the rated power to the reference value ordered by the wind farm control system. In this case, the blade pitch angle controller acts on the pitch blade to achieve the power set

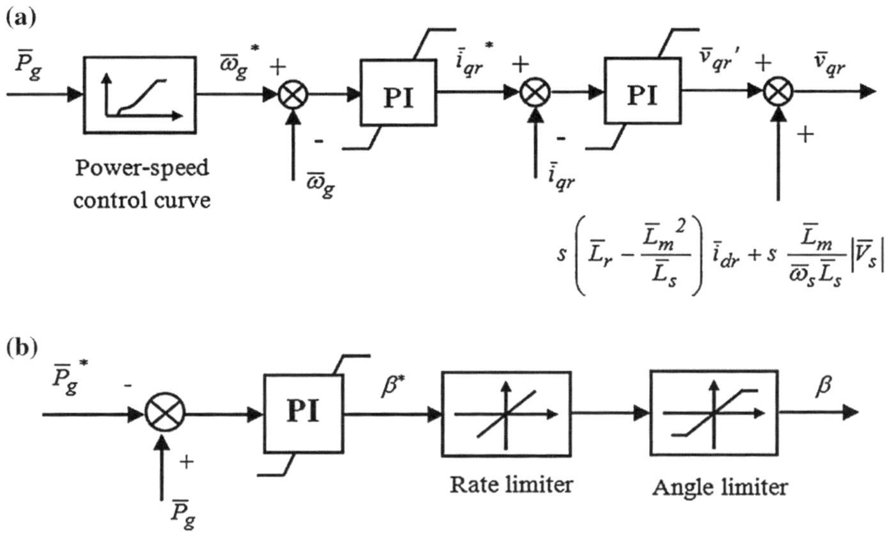

Fig. 29 MSC Control scheme 1: **a** $\bar{v}_{qr}$ controller, **b** pitch angle controller

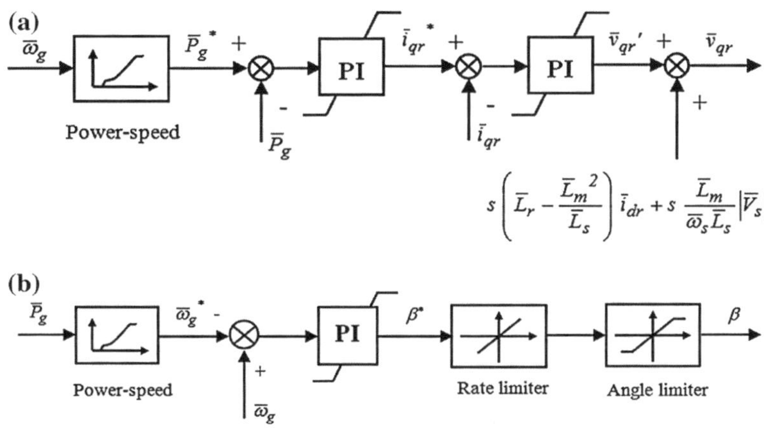

Fig. 30 MSC Control scheme 2: **a** $\bar{v}_{qr}$ controller, **b** pitch angle controller

point. The $\bar{v}_{qr}$ controller adjusts the rotational speed to the reference value derived from the power-speed curve and the reference power.

In the second control strategy (Fig. 30), the active power is controlled by the quadrature component of the rotor current, and the speed control adjusts the rotational speed to a reference speed derived from the optimum power-speed curve by acting on the blade pitch angle. The operating strategy in this case is the following:

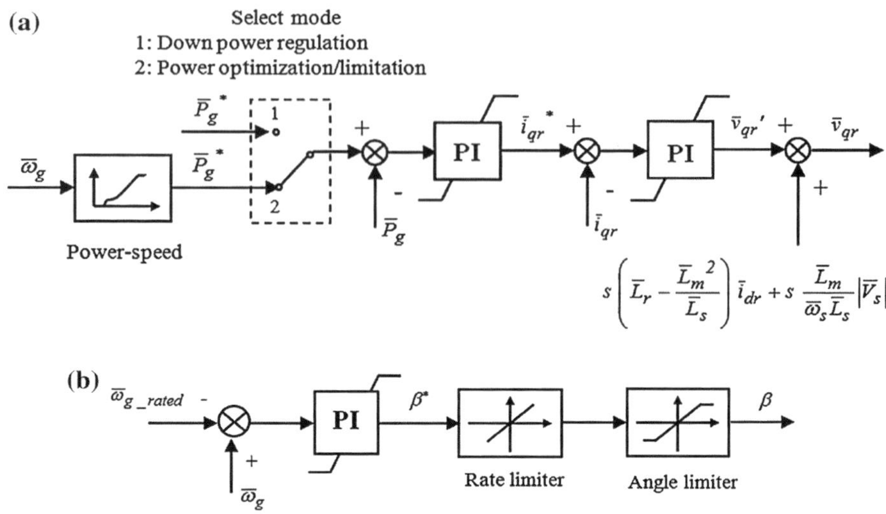

Fig. 31 MSC Control scheme 3: **a** $\overline{v}_{qr}$ controller, **b** pitch angle controller

- Power optimization strategy: In this case, the $\overline{v}_{qr}$ controller adjusts the output power to the value derived from the power-speed curve and the actual rotational speed, maximizing the power extracted from the wind. As the generator speed is less than the rated speed, the blade pitch angle controller does not act and the pitch angle is kept at its optimal value,
- Power limitation strategy: The $\overline{v}_{qr}$ controller limits the output power to the rated power. Likewise, the pitch angle controller limits the rotational speed to the rated speed.
- Down power regulation strategy: In this case, the blade pitch angle controller keeps the rotational speed to the speed reference derived from the power-speed curve and the power reference. Moreover, the $\overline{v}_{qr}$ controller adjusts the output power to the reference obtained from the power-speed curve and the actual rotational speed.

In the third strategy (Fig. 31), the active power control presents a select mode and acts on the quadrature component of the rotor voltage, and the speed control limits the rotational speed to the rated speed by acting on the blade pitch angle. Two operating modes can be selected by the select mode: power optimization/ limitation and down power regulation. In the power optimization/limitation mode, the controller uses the rotational speed to define the power reference from the power-speed curve. In down power regulation, the power is regulated to the set point demanded. This operating strategy can be summarized as follows:

- Power optimization strategy: In this case, the blade pitch angle controller keeps the pitch angle at its optimal value, whereas the $\overline{v}_{qr}$ controller controls the output power, and thus the wind turbine operates maximizing the power extracted from the wind.

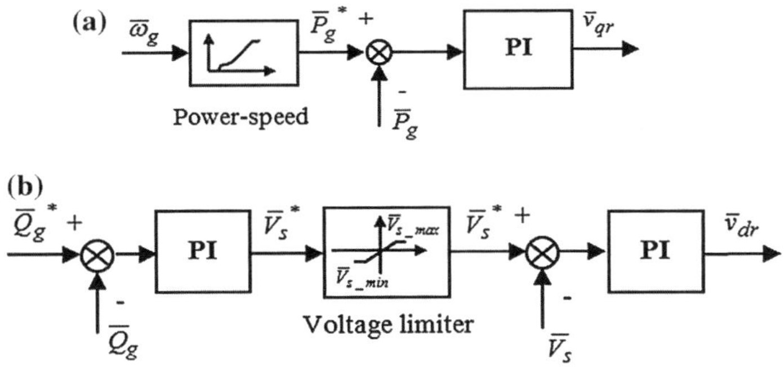

Fig. 32 MSC control scheme based on the direct control of the components of the rotor voltage. **a** Power controller. **b** Reactive power controller

- Power limitation strategy: The $\bar{v}_{qr}$ controller assures the rated power and the pitch angle controller limits the rotational speed to the rated speed.
- Down power regulation strategy: In this case, the $\bar{v}_{qr}$ controller adjusts the output power to the reference ordered by the wind farm control system, whereas the blade pitch angle controller keeps the rotational speed at the rated speed.

It is also used a more simplified control based on the direct control of the components of the rotor voltage [12, 22], where MSC is modelled as a controlled voltage source. In this case, the torque/active power is controlled by the q component of the rotor voltage $\bar{v}_{qr}$, and the reactive power/terminal voltage by the d component of the rotor voltage $\bar{v}_{dr}$. The configuration of this control scheme is shown in Fig. 32.

2. *Control of the magnitude and angle of the rotor voltage.*

In this control strategy, the power generated is controlled by acting on the magnitude of rotor voltage $\bar{V}_r$, and the terminal voltage by acting on the angle of the rotor voltage δ_r [4].

This control strategy also presents two controllers. The $\bar{V}_r$ controller allows the wind turbine operation with the desired terminal voltage or power factor. The δ_r controller allows the wind turbine to operate with variable speed below nominal winds maximizing the power extracted from the wind or limiting the output power to rated power above nominal winds. Thus, this controller uses the optimum power extraction curve to define the reference power according to the actual rotor speed. Once the rotor voltage vector is determined, it is transformed from its polar co-ordinates (magnitude $\bar{V}_r$ and angle δ_r) to rectangular dq co-ordinates ($\bar{v}_{dr}$ and $\bar{v}_{qr}$). Figure 33 illustrates the configuration of this control strategy. It is justified as follows.

The stator active and reactive power flows in the steady state can be calculated as follows

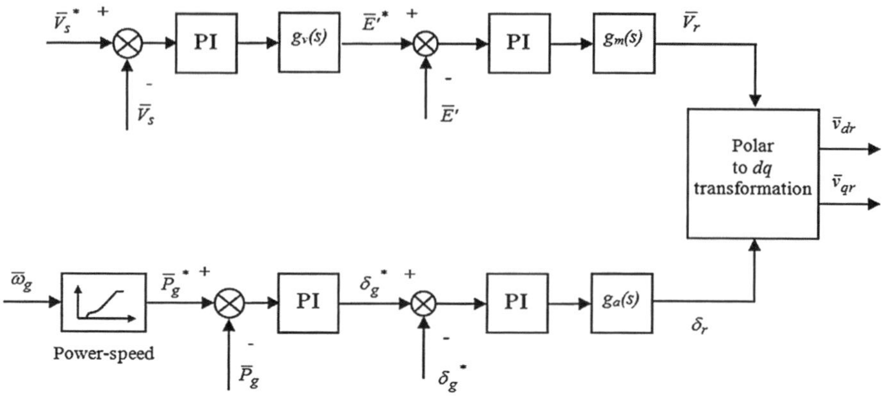

Fig. 33 MSC control scheme based on the control of the magnitude and angle of the rotor voltage

$$\bar{P}_s + j\bar{Q}_s = \bar{\mathbf{V}}_s \bar{\mathbf{I}}_s^*. \tag{66}$$

From the steady-state equivalent circuit shown in Fig. 24, neglecting the stator resistance, the stator current is expressed as

$$\bar{\mathbf{I}}_s = \frac{\bar{\mathbf{E}}' - \bar{\mathbf{V}}_s}{j\bar{X}_s'}. \tag{67}$$

Substituting Eq. (67) in Eq. (66) and separating into real and imaginary parts, it is obtained from the stator active and reactive power

$$\bar{P}_s = \frac{\bar{E}'\bar{V}_s}{\bar{X}_s'} \sin \delta_r \tag{68}$$

$$\bar{Q}_s = \frac{\bar{E}'\bar{V}_s}{\bar{X}_s'} \cos \delta_r - \frac{\bar{V}_s^2}{\bar{X}_s'} \tag{69}$$

where $\bar{V}_s$ and $\bar{E}'$ are the rms value of the stator voltage and equivalent voltage behind a transient reactance vectors, and δ_r is the load angle between both vectors.

Considering Eqs. (68) and (43), it is obtained the total active power delivered to grid as follows

$$\bar{P}_g = \bar{P}_s + \bar{P}_r = (1 - s)\frac{\bar{E}'\bar{V}_s}{\bar{X}_s'} \sin \delta_r. \tag{70}$$

And the total reactive power is equal to the stator reactive power, as mentioned previously.

$$\bar{Q}_g = \bar{Q}_s = \frac{\bar{E}'\bar{V}_s}{\bar{X}_s'} \cos \delta_r - \frac{\bar{V}_s^2}{\bar{X}_s'}. \tag{71}$$

As the load angle is generally small, sin $\delta_r \approx \delta_r$ and cos $\delta_r \approx 1$, Eqs. (70) and (71) can be simplified to

$$\bar{P}_g = (1 - s)\frac{\bar{E}'\bar{V}_s}{\bar{X}'_s}\delta_r \tag{72}$$

$$\bar{Q}_g = \bar{Q}_s = \frac{\bar{E}'\bar{V}_s}{\bar{X}'_s} - \frac{\bar{V}_s^2}{\bar{X}'_s}. \tag{73}$$

From Eqs. (72) and (73), it can be deduced that the active power $\bar{P}_g$ depends mainly on the load angle δ_r and the reactive power $\bar{Q}_g$ depends mainly on voltage magnitudes, specifically on the equivalent voltage $\bar{E}'$ when the grid voltage is assumed constant.

The magnitude of the equivalent voltage $\bar{\mathbf{E}}'$ depends on the magnitude of the rotor flux vector $\bar{\Psi}_{\mathbf{r}}$ as expressed in Eqs. (27) and (28). Although $\bar{\Psi}_{\mathbf{r}}$ is dependent on the generator stator and rotor currents, as expressed in Eqs. (15) and (16), it is derived from Eqs. (29) and (30) that $\bar{\Psi}_{\mathbf{r}}$ and therefore $\bar{\mathbf{E}}'$ can also be manipulated by adjustment of the rotor voltage vector $\bar{\mathbf{V}}_{\mathbf{r}}$ [4].

GSC Control

The GSC is normally controlled in the following way: (1) to maintain the DC-link capacitor voltage at a set value, which assures the active power exchange from the DFIG to grid; (2) to operate with unity power factor, although it could be used to the reactive power and terminal voltage control. Two control techniques can be applied to the GSC in order to achieve these objectives:

1. Vector control.
2. Load angle control.

In this case, the vector control is the most widely used control technique in GSC. On the other hand, sometimes this converter is considered very fast and often is disregarded, and the converter system of the DFIG is commonly represented by only the MSC in transient voltage stability studies (Akhmatov [3]).

1. *Vector control*

The dynamic model of the grid connection of GSC when selecting a reference frame rotating synchronously with the grid voltage space vector is

$$\bar{v}_{dg} = \bar{v}_{dc} + \bar{R}_c \cdot \bar{i}_{dc} + \bar{\omega} \cdot \bar{L}_c \cdot \bar{i}_{qc} + \bar{L}_c \cdot \frac{d\bar{i}_{dc}}{dt} \tag{74}$$

$$\bar{v}_{qg} = \bar{v}_{qc} + \bar{R}_c \cdot \bar{i}_{qc} - \bar{\omega} \cdot \bar{L}_c \cdot \bar{i}_{dc} + \bar{L}_c \cdot \frac{d\bar{i}_{qc}}{dt} \tag{75}$$

where $\bar{R}_c$ and $\bar{L}_c$ are the inductance and resistance of connection to grid, respectively, $\bar{v}_{dc}$ and $\bar{v}_{qc}$ are the voltage components of GSC, and $\bar{\omega}$ is the grid frequency.

Under steady-state condition, the derivative terms are reduced to zero. Consequently, Eqs. (74) and (75) can be expressed as follows

$$\bar{v}_{dg} = \bar{v}_{dc} + \bar{R}_c \cdot \bar{i}_{dc} + \bar{\omega} \cdot \bar{L}_c \cdot \bar{i}_{qc} \tag{76}$$

$$\bar{v}_{qg} = \bar{v}_{qc} + \bar{R}_c \cdot \bar{i}_{qc} - \bar{\omega} \cdot \bar{L}_c \cdot \bar{i}_{dc}. \tag{77}$$

The active and reactive powers delivered to grid can be calculated as

$$\bar{P}_{gsc} = \bar{v}_{dg} \cdot \bar{i}_{dc} + \bar{v}_{qg} \cdot \bar{i}_{qc} \tag{78}$$

$$\bar{Q}_{gsc} = \bar{v}_{qg} \cdot \bar{i}_{dc} - \bar{v}_{dg} \cdot \bar{i}_{qc}. \tag{79}$$

Because the d-axis of the reference frame is oriented along the grid voltage, the grid voltage vector is $\bar{\mathbf{v}}_g = \bar{v}_{dg} + j \cdot 0$. Thus, the active and reactive powers can be expressed as

$$\bar{P}_{gsc} = \bar{v}_{dg} \cdot \bar{i}_{dc} \tag{80}$$

$$\bar{Q}_{gsc} = -\bar{v}_{dg} \cdot \bar{i}_{qc}. \tag{81}$$

As deduced from Eqs. (80) and (81), the active and reactive power control can be achieved by controlling the direct and quadrature current components, respectively.

The current components cannot be controlled independently by the GSC voltage components due to the cross-coupling effects shown in Eqs. (76) and (77). The cross-coupling effects can be cancelled by the feed-forward compensation. GSC voltage components are decided by the PI controllers and decoupling feed-forward compensation:

$$\bar{v}_{dc} = -\bar{v}'_{dc} - \bar{\omega} \cdot \bar{L}_c \cdot \bar{i}_{qc} + \bar{v}_{dg} \tag{82}$$

$$\bar{v}_{qc} = -\bar{v}'_{qc} + \bar{\omega} \cdot \bar{L}_c \cdot \bar{i}_{dc} + \bar{v}_{qg}. \tag{83}$$

The decoupled control of the DC link voltage and reactive power/voltage is then performed in this converter by acting on the GSC voltage components: the DC link voltage control is performed by $\bar{v}_{dc}$, and the reactive power/voltage is controlled by $\bar{v}_{qc}$. This converter can be modelled as a current-controlled voltage source. Figure 34 illustrates the control system of the GSC. Two controllers are used to control the DC link voltage and reactive power/voltage.

The $\bar{v}_{dc}$ controller controls the DC link voltage to the rated value, which assures the active power exchange from the rotor to the grid. The $\bar{v}_{qc}$ controller allows the terminal voltage control or reactive power control. GSC usually operates with unity power factor, and therefore, the reactive power is regulated to be zero.

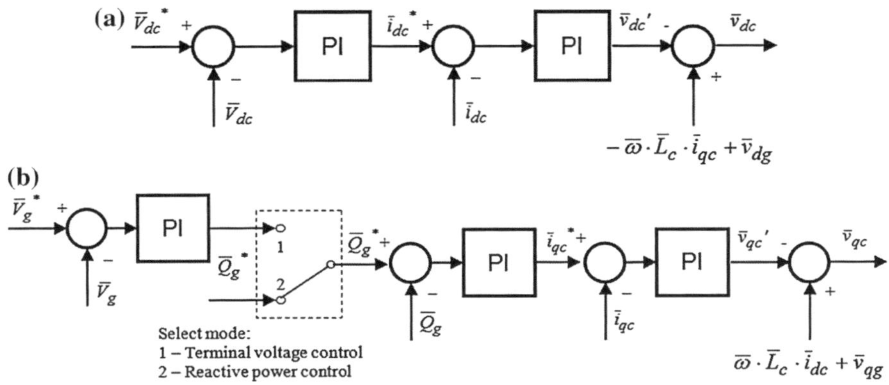

Fig. 34 Control system of the GSC. **a** DC link voltage controller ($\bar{v}_{dc}$controller). **b** Reactive power/voltage controller ($\bar{v}_{qc}$ controller)

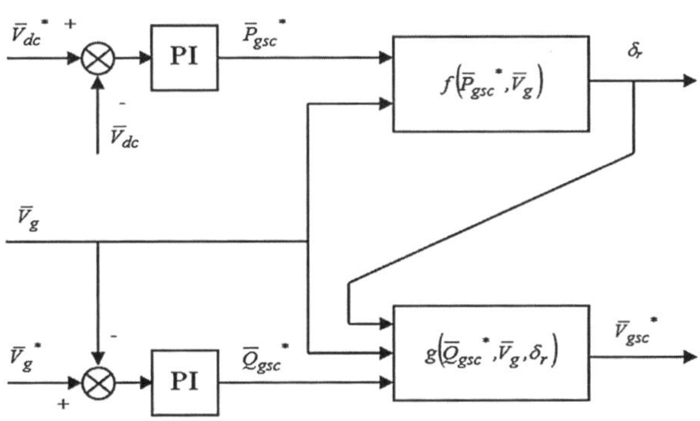

Fig. 35 Load angle control of the GSC

2. *Load angle control*

In this control strategy, the magnitude and angle of the GSC voltage vector are used to control the DC-link voltage and the terminal voltage or reactive power delivered to grid, as shown in Fig. 35 [25]. It is justified as follows.

The power transfer between the GSC and the grid, shown in Fig. 36, can be expressed as follows

$$\bar{P}_{gsc} = \frac{\bar{V}_{gsc}\,\bar{V}_g}{\bar{X}_c}\sin\delta_r \tag{84}$$

$$\bar{Q}_{gsc} = \frac{\bar{V}_{gsc}^2}{\bar{X}_c} - \frac{\bar{V}_{gsc}\,\bar{V}_g}{\bar{X}_c}\cos\delta_r \tag{85}$$

where $\bar{X}_c$ is the reactance of connection to grid.

From Eqs. (84) and (85), the magnitude and angle of the GSC voltage can be expressed as

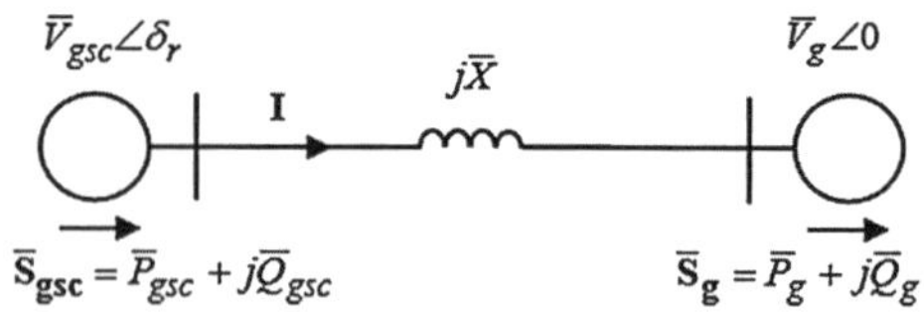

Fig. 36 Power transfer between two sources

$$\delta_r = \sin^{-1}\left(\frac{\bar{P}^*_{gsc}\bar{X}_c}{\bar{V}_{gsc}\,\bar{V}_g}\right) \tag{86}$$

$$\bar{V}_{gsc} = \frac{\bar{V}_{grid}\cos\delta_r + \sqrt{\left(\bar{V}_{grid}\cos\delta_r\right)^2 + 4\,\bar{X}_c\,\bar{Q}^*_{gsc}}}{2} \tag{87}$$

where $\bar{P}^*_{gsc}$ and $\bar{Q}^*_{gsc}$ are the power references.

5.3.3 Dynamic Simulation of DFIG Wind Turbine

To illustrate the performance of a DFIG wind turbine, the response of this wind turbine is simulated under wind fluctuations and a grid fault.

Figures 37, 38, 39 illustrate the performance of a DFIG wind turbine experiencing an incoming wind as shown in Fig. 10, when connected to an infinite bus. In the simulation, the three control systems described for the MSC (shown in Figs. 29, 30, 31) are compared, so that a DFIG wind turbine controlled by one of these control system is simulated.

To illustrate the capabilities of this wind turbine to regulate the production (active and reactive powers), in the simulation performed, the wind turbine is demanded to operate as follows:

- It produces the maximum possible output power injected to the grid during the first 60 s.
- At 60 s, the wind turbine receives a 40 % reduction of the active power reference by a slope of 0.05.
- The wind turbine operates with the unity power factor during the first 80 s, whereas it generates the maximum reactive power for the rest of the simulation, with a slope of 0.05.

As it can be observed in the results, while the wind turbine receives winds below rated (during the first 20 s), it generates below rated active power with variable rotational speed achieving optimum power efficiency. In this case, the wind turbine operates with a maximum power coefficient, and the pitch angle is kept at the optimal value. DFIG wind turbine generates rated power between the 20 and 60 s, since the incoming wind exceeds the rated wind. In the control system 1, the pitch angle controller assures the rated power by acting on the pitch angle, and the $\bar{v}_{qr}$

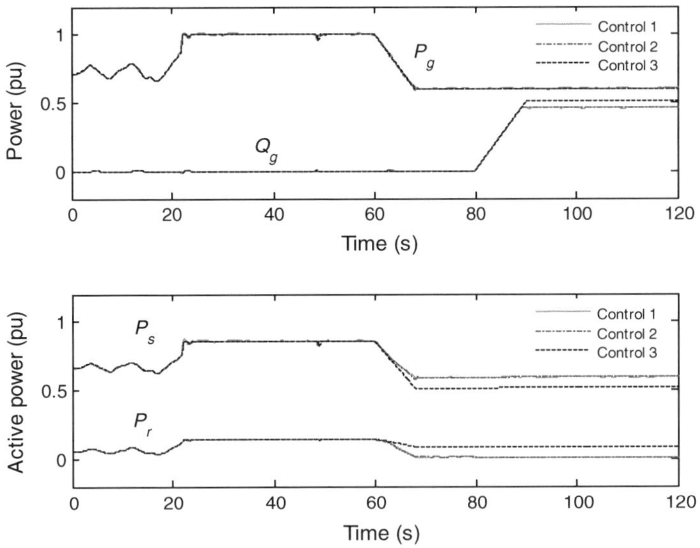

Fig. 37 DFIG response during active and reactive power regulation: Powers delivered to grid, and stator and rotor active power

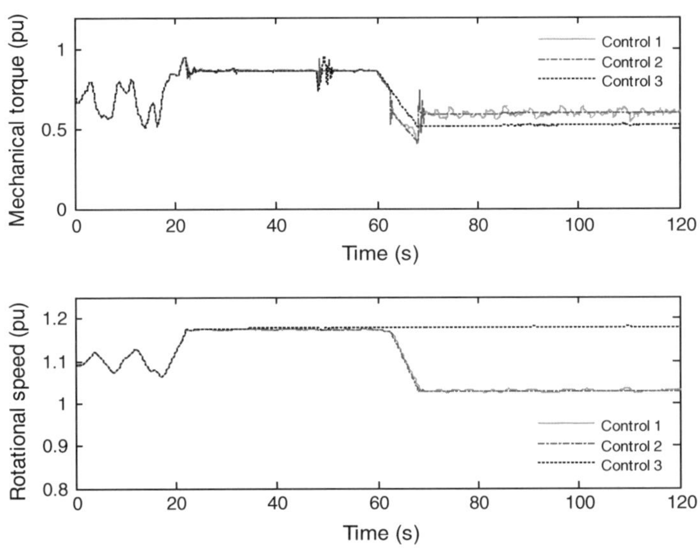

Fig. 38 DFIG response during active and reactive power regulation: Mechanical torque and rotational speed

controller keeps the rotational speed at the rated speed. Both for control systems 2 and 3, the $\bar{v}_{qr}$ controller adjusts the output power to the rated power and the pitch angle controller limits the rotational speed to the rated speed.

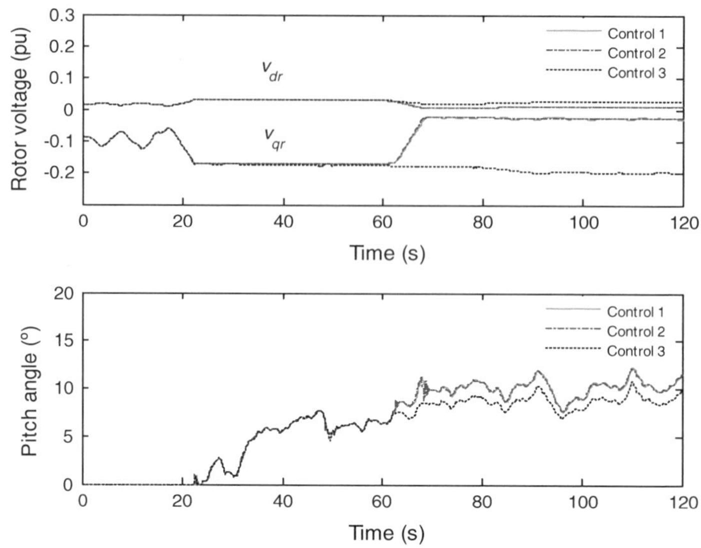

Fig. 39 DFIG response during active and reactive power regulation: Rotor voltage and pitch angle

The variability from the incoming wind and the speed limitation on the pitch blade movement cause small variability on the output power for the wind turbine control system 1 and on the rotational speed for the control systems 2 and 3. The $\bar{v}_{qr}$ control, performed through the power converter, allows a perfect adjustment of the rated speed for the control system 1 and the rated active power for the control systems 2 and 3.

For the rest of the simulation, the wind turbine is able to generate the rated power since it experiences winds above rated. However, the wind turbine is ordered to reduce its output power to 0.6 p.u. (down power regulation). Notice that, although the control systems 1 and 2 achieve the same performance of the control variables, the wind turbine with control system 1 presents a higher variability on the response. In this case, the pitch angle controller acts on the pitch angle achieving the reference power and the $\bar{v}_{qr}$ controller adjusts the rotational speed to 0.899 p.u., which is derived from the power-speed curve according to the power reference. In control system 2, the $\bar{v}_{qr}$ controller adjusts the output power to the reference derived from the power-speed curve and the rotational speed, and the pitch angle controller adjusts the rotational speed to 0.899 p.u derived from the power-speed curve and the power reference. Finally, in the control system 3, the $\bar{v}_{qr}$ controller adjusts the output power to the demanded power reference and the pitch angle controller keeps the rotational speed to a constant rated speed value.

Notice that the wind turbine with control systems 1 and 2 generate the reference power at below synchronous speed (0.899 p.u.), and therefore, the rotor winding consumes active power, as it can be deduced from Fig. 22. On the other hand, the reference power for the control system 3 is achieved at the above synchronous speed. In this case, both the stator and rotor winding generate active power, as it can be derived from Fig. 22. With down power regulation, the wind turbines with control systems 1 and 2 generate at the same rotational speed, and thus the $\bar{v}_{qr}$ variable is controlled in the same way for both control systems. As the generation speed is different for control system 3, the wind turbine control system requires a different value of $\bar{v}_{qr}$. Furthermore, the wind turbine commanded with control system 3 requires less pitch angle than the control systems 1 and 2, and as the rotational speed is higher, the tip ratio is different. This fact justifies the differences in the pitch angle, since the reduction of the power coefficient required is achieved from a different pitch angle.

During the first 80 s, the wind turbine operates with the unity power factor, whereas it generates the maximum reactive power. It is worthy of note that the reactive power and the direct component of the rotor voltage $\bar{v}_{dr}$ increase when the wind turbine changes its operating mode from the unity power factor to the maximum reactive power.

When the wind turbine operates without down power regulation (during the first 60 s of the performed simulation), the three control systems present the same performance for the $\bar{v}_{dr}$ control. Although it is worth mentioning a higher variability on the reactive power of the wind turbine commanded with the control system 1, due to the commented variability on the active power. For the remainder of the simulation, when down power regulation is required, the performance of the control systems 1 and 2 for the $\bar{v}_{dr}$ control are similar. However, they differ from the control system 3. In the operation with down power regulation, the wind turbines with control systems 1 and 2 operate at below synchronous speed, and therefore the stator winding generates power, while the rotor winding takes power from the grid. In the control system 3, the wind turbine operates at above synchronous speed, and both stator and rotor winding generate power. Hence, the wind turbine with control system 3 requires less stator power, and as the maximum reactive power depends on the stator power, this wind turbine is able to generate higher reactive power.

Figures 40 and 41 illustrate the performance of a DFIG wind turbine during a voltage dip, when connected to an infinite bus. In this case, it is considered that the voltage of the infinite bus drops a 40 % during 0.2 s, and after the fault the voltage starts to recover. During the simulation, the wind speed incident on the wind turbine is assumed constant and the wind turbine is demanded to operate with unity power factor. The responses obtained by using the fifth and third order model of the induction generator are again compared. When the voltage dip reaches the wind turbine, it can be seen that the generated active power falls, while the mechanical power does not change and therefore the wind turbine accelerates. When the voltage starts to recover, as the rotational speed has increased, a transient period follows, but it is more smooth than FS-SCIG due to the power converters. The fifth order model shows an oscillatory response that is not observed in the response obtained with the third order model, because only the fundamental frequency component are represented.

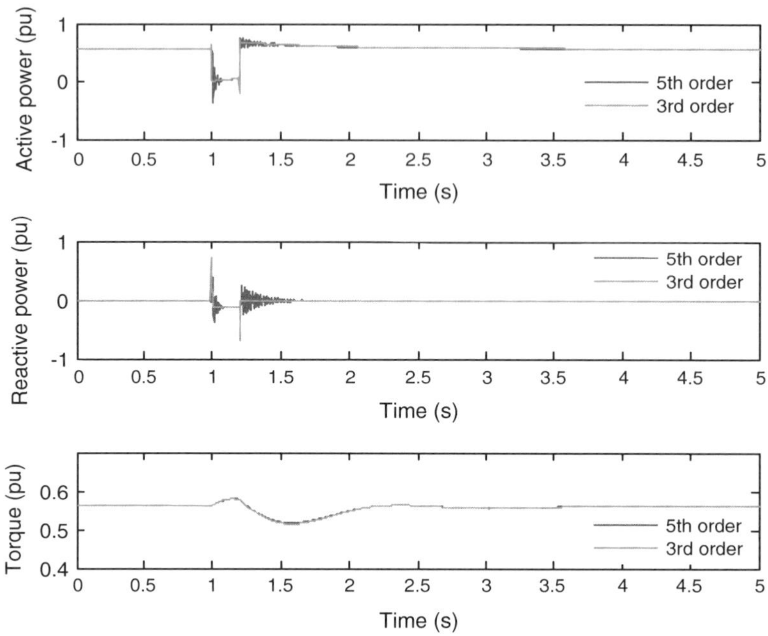

Fig. 40 DFIG responses obtained by using the 5th order and 3rd order models during a voltage dip (40 % drop during 0.2 s): Active and reactive powers, and torque

5.4 Direct Drive Synchronous Generator

A direct-drive synchronous generator uses a synchronous generator whose rotor presents a large number of poles and whose stator windings are connected to the grid through a full-scale power converter. The large number of poles mounted on the rotor allows the generator to operate at low speeds. This means that the gearbox can be omitted, and the generator is directly coupled to the wind turbine rotor. The rotor excitation can be obtained by means of either a current-carrying winding (electrically excited synchronous generator, EESG) or permanent magnets (permanent magnet synchronous generator, PMSG). Figure 42 shows the typical configuration of a direct-drive ESSG and PMSG wind turbine.

Synchronous generators combined with a full-scale power converter are an attractive concept for wind turbines. Such wind turbines have the advantage of being able to fully control the speed range from 0 to 100 % of the synchronous speed, and support both reactive power compensation and smooth grid connection. The rating of the power converter corresponds to the rated power of the generator plus losses. This power converter controls power flow to the grid. In addition, it decouples the electrical grid frequency and the mechanical rotor frequency, and controls the generator speed, thus enabling variable speed generation. Due to the variable speed operation, it presents a wide range of advantages, such as reduced

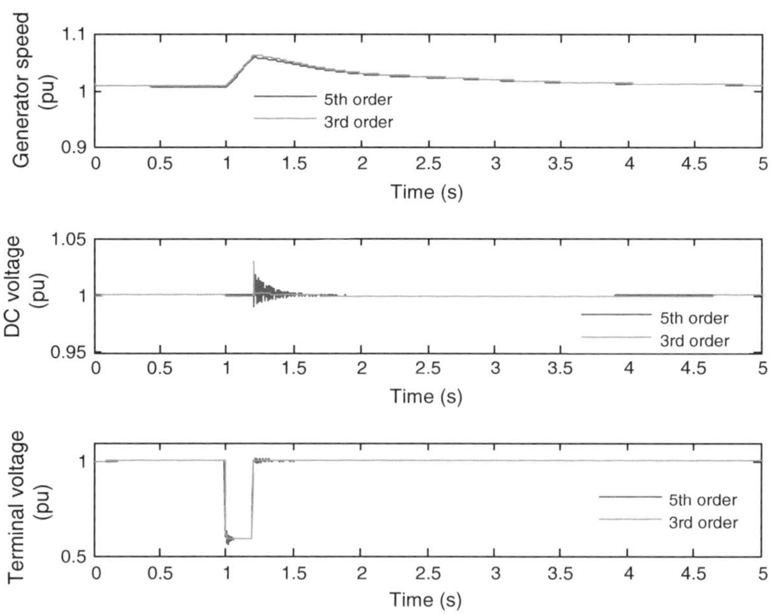

Fig. 41 DFIG responses obtained by using the 5th order and 3rd order models during a voltage dip (40 % drop during 0.2 s): Generator speed, DC voltage and terminal voltage

mechanical stress, increased power capture, reduced acoustical noise, and finally, high control capability, which is a prime concern for the grid integration of large wind farms. This wind turbine also includes blade pitch angle control in order to limit the power extracted from the wind. Compared with the DFIG, this type of wind turbine is slightly more efficient, and from an electrical engineering perspective, it is somewhat less complicated, and easier to construct. However it has the disadvantageous of being more expensive [16].

5.4.1 DDSG Generation System

Synchronous Generator: EESG and PMSG

The behaviour of a EESG can be described by means of a sixth order model, expressed in a *dq* coordinate reference frame aligned with the magnetic axis of the rotor and rotating at rotor speed ω_r. This model is composed of five electrical differential equations (two equations for the stator voltages and three for the rotor voltages) and the mechanical differential equation, described by Eq. (19). The electrical equations, expressed in per unit, are given by [5, 15]:

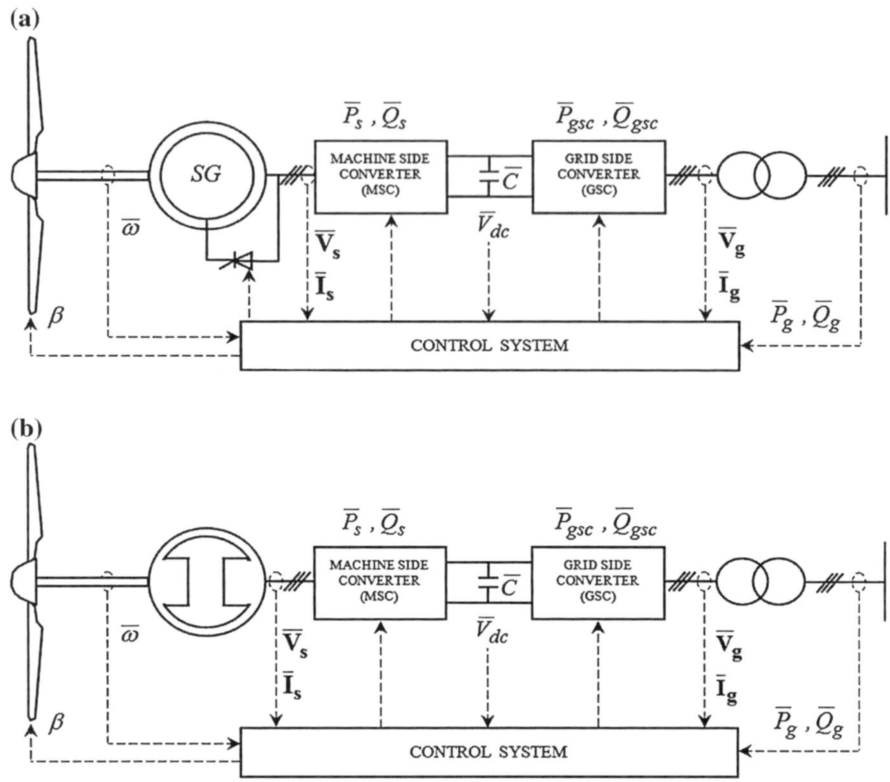

Fig. 42 Direct-drive wind turbine configuration. **a** EESG. **b** PMSG

Stator voltage equations:

$$\bar{v}_{ds} = -\bar{R}_s \cdot \bar{i}_{ds} - \bar{\omega}_e \cdot \bar{\psi}_{qs} + \frac{1}{\omega_b} \cdot \frac{d}{dt} \bar{\psi}_{ds} \tag{88}$$

$$\bar{v}_{qs} = -\bar{R}_s \cdot \bar{i}_{qs} + \bar{\omega}_e \cdot \bar{\psi}_{ds} + \frac{1}{\omega_b} \cdot \frac{d}{dt} \bar{\psi}_{qs}. \tag{89}$$

Rotor voltage equations:

$$\bar{v}_{df} = \bar{R}_{df} \cdot \bar{i}_{df} + \frac{1}{\omega_b} \cdot \frac{d}{dt} \bar{\psi}_{df} \tag{90}$$

$$0 = \bar{R}_{dk} \cdot \bar{i}_{dk} + \frac{1}{\omega_b} \cdot \frac{d}{dt} \bar{\psi}_{dk} \tag{91}$$

$$0 = \bar{R}_{qk} \cdot \bar{i}_{qk} + \frac{1}{\omega_b} \cdot \frac{d}{dt} \bar{\psi}_{qk} \tag{92}$$

where index f refers to field winding and k refers to damping winding.

The electrical speed ω_e is derived from the mechanical rotor speed ω_r and the number of generator poles n_p.

$$f_e = \frac{n_p}{2} \cdot f_m = \frac{n_p}{2} \cdot \frac{\omega_r}{2 \cdot \pi} \tag{93}$$

$$\omega_e = 2 \cdot \pi \cdot f_e = \frac{n_p}{2} \cdot \omega_r. \tag{94}$$

The stator and rotor fluxes are expressed as

Stator flux equations:

$$\bar{\psi}_{ds} = -\bar{L}_{ls} \cdot \bar{i}_{ds} + \bar{L}_{dm} \cdot \left(-\bar{i}_{ds} + \bar{i}_{df} + \bar{i}_{dk}\right) = -\bar{L}_{ds} \cdot \bar{i}_{ds} + \bar{L}_{dm} \cdot \left(\bar{i}_{df} + \bar{i}_{dk}\right) \tag{95}$$

$$\bar{\psi}_{qs} = -\bar{L}_{ls} \cdot \bar{i}_{qs} + \bar{L}_{qm} \cdot \left(-\bar{i}_{qs} + \bar{i}_{qk}\right) = -\bar{L}_{qs} \cdot \bar{i}_{qs} + \bar{L}_{qm} \cdot \bar{i}_{qk}. \tag{96}$$

Rotor flux equations:

$$\bar{\psi}_{df} = \bar{L}_{lf} \cdot \bar{i}_{df} + \bar{L}_{dm} \cdot \left(-\bar{i}_{ds} + \bar{i}_{df} + \bar{i}_{dk}\right) \tag{97}$$

$$\bar{\psi}_{dk} = \bar{L}_{ldk} \cdot \bar{i}_{dk} + \bar{L}_{dm} \cdot \left(-\bar{i}_{ds} + \bar{i}_{df} + \bar{i}_{dk}\right) \tag{98}$$

$$\bar{\psi}_{qk} = \bar{L}_{lqk} \cdot \bar{i}_{qk} + \bar{L}_{qm} \cdot \left(-\bar{i}_{qs} + \bar{i}_{qk}\right) \tag{99}$$

where $\bar{L}_{ls}$ is the stator leakage inductance, $\bar{L}_{dm}$ and $\bar{L}_{qm}$ are the mutual inductances in d- and q-axis, $\bar{L}_{lf}$ is the field inductance, $\bar{L}_{ldk}$ and $\bar{L}_{lqk}$ are the damping winding inductance in d- and q-axis, and $\bar{L}_{ds} = \bar{L}_{ls} + \bar{L}_{dm}$ and $\bar{L}_{qs} = \bar{L}_{ls} + \bar{L}_{qm}$ are the stator inductance in d- and q-axis.

The electromagnetic torque is expressed as

$$\bar{T}_e = \bar{\psi}_{ds} \cdot \bar{i}_{qs} - \bar{\psi}_{qs} \cdot \bar{i}_{ds}. \tag{100}$$

The stator active and reactive powers can be obtained as follows

$$\bar{P}_s = \bar{u}_{ds} \cdot \bar{i}_{ds} + \bar{u}_{qs} \cdot \bar{i}_{qs} \tag{101}$$

$$\bar{Q}_s = \bar{u}_{qs} \cdot \bar{i}_{ds} - \bar{u}_{ds} \cdot \bar{i}_{qs}. \tag{102}$$

In case of a PMSG, permanent magnets induces a constant flux in stator phases $\bar{\psi}_{pm}$, and therefore, its behaviour is commonly expressed by a third order model, where the electrical equations are

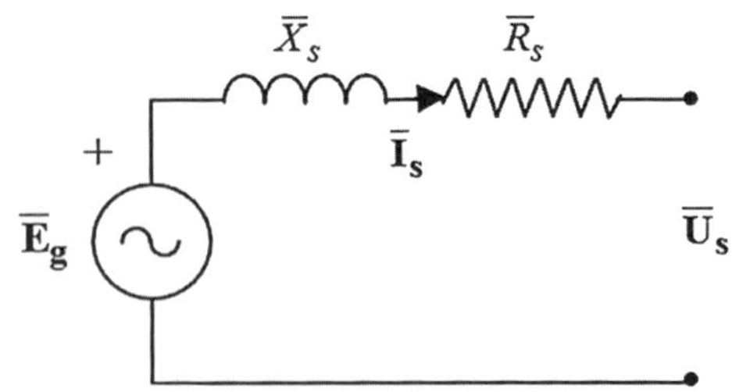

Fig. 43 Equivalent circuit of DDSG for first order model (steady state model)

$$\bar{v}_{ds} = -\bar{R}_s \cdot \bar{i}_{ds} - \bar{\omega}_e \cdot \bar{\psi}_{qs} + \frac{1}{\omega_b} \cdot \frac{d}{dt} \bar{\psi}_{ds} \tag{103}$$

$$\bar{v}_{qs} = -\bar{R}_s \cdot \bar{i}_{qs} + \bar{\omega}_e \cdot \bar{\psi}_{ds} + \frac{1}{\omega_b} \cdot \frac{d}{dt} \bar{\psi}_{qs} \tag{104}$$

$$\bar{\psi}_{ds} = -\bar{L}_{ds} \cdot \bar{i}_{ds} + \bar{\psi}_{pm} \tag{105}$$

$$\bar{\psi}_{qs} = -\bar{L}_{qs} \cdot \bar{i}_{qs}. \tag{106}$$

The excitation voltage E_g generated by the permanent magnets is given by

$$\bar{E}_g = \bar{\omega}_e \cdot \bar{\psi}_{pm}. \tag{107}$$

For voltage and angle stability investigations, it has been also used the first order model of a DDSG [5, 15]. This model is obtained by neglecting the flux transients (*d/dt* terms) in previous electrical equations and only considering the mechanical equation. Figure 43 shows the equivalent circuit corresponding to this model for both EESG and PMSG.

In case of EESG without damping winding, the electrical equations can be reduced as

Stator equations:

$$\bar{v}_{ds} = -\bar{R}_s \cdot \bar{i}_{ds} - \bar{\omega}_e \cdot \bar{\psi}_{qs} \tag{108}$$

$$\bar{v}_{qs} = -\bar{R}_s \cdot \bar{i}_{qs} + \bar{\omega}_e \cdot \bar{\psi}_{ds} \tag{109}$$

$$\bar{\psi}_{ds} = -\bar{L}_{ds} \cdot \bar{i}_{ds} + \bar{L}_{dm} \cdot \bar{i}_{df} \tag{110}$$

$$\bar{\psi}_{qs} = -\bar{L}_{qs} \cdot \bar{i}_{qs}. \tag{111}$$

Rotor equations:

$$\bar{v}_{df} = \bar{R}_{df} \cdot \bar{i}_{df}. \tag{112}$$

Substituting Eqs. (110) and (111) in the stator voltage equations, Eqs. (108) and (109), it is obtained

$$\bar{v}_{ds} = -\bar{R}_s \cdot \bar{i}_{ds} + \bar{X}_{qs} \cdot \bar{i}_{qs} \tag{113}$$

$$\bar{v}_{qs} = -\bar{R}_s \cdot \bar{i}_{qs} - \bar{X}_{ds} \cdot \bar{i}_{ds} + \bar{\omega}_e \cdot \bar{L}_{dm} \cdot \bar{i}_{df} \tag{114}$$

where $\bar{X}_{ds} = \bar{\omega}_e \cdot \bar{L}_{ds}$ and $\bar{X}_{qs} = \bar{\omega}_e \cdot \bar{L}_{qs}$

Solving $\bar{i}_{df}$ from Eq. (112) and substituting in Eq. (114) gives

$$\bar{v}_{qs} = -\bar{R}_s \cdot \bar{i}_{qs} - \bar{X}_{ds} \cdot \bar{i}_{ds} + \bar{\omega}_e \cdot \bar{L}_{dm} \cdot \frac{\bar{v}_{df}}{\bar{R}_{df}}. \tag{115}$$

Defining the variable $\bar{E}_g = \bar{\omega}_e \cdot \bar{L}_{dm} \cdot \frac{\bar{v}_{df}}{\bar{R}_{df}}$, Eq. (115) is expressed as

$$\bar{v}_{qs} = -\bar{R}_s \cdot \bar{i}_{qs} - \bar{X}_{ds} \cdot \bar{i}_{ds} + \bar{E}_g. \tag{116}$$

The terminal voltage vector $\bar{\mathbf{V}}_s = \bar{v}_{ds} + j\,\bar{v}_{qs}$ can be expressed as

$$\bar{\mathbf{V}}_s = \bar{v}_{ds} + j\,\bar{v}_{qs} = -\bar{R}_s \cdot \left(\bar{i}_{ds} + j\,\bar{i}_{qs}\right) + \left(\bar{X}_{qs} \cdot \bar{i}_{qs} - j\,\bar{X}_{ds} \cdot \bar{i}_{ds}\right) + j\,\bar{E}_g. \tag{117}$$

Neglecting the saliency ($\bar{X}_{ds} = \bar{X}_{qs} = \bar{X}_s$) and considering the current vector $\bar{\mathbf{I}}_s = \bar{i}_{ds} + j\,\bar{i}_{qs}$, the terminal voltage vector can be calculated as follows

$$\bar{\mathbf{V}}_s = \bar{v}_{ds} + j\,\bar{v}_{qs} = -(\bar{R}_s + j\,\bar{X}_s) \cdot \bar{\mathbf{I}}_s + j\,\bar{E}_g. \tag{118}$$

In case of PMSG, the electrical equations of the first order model are given by

$$\bar{v}_{ds} = -\bar{R}_s \cdot \bar{i}_{ds} - \bar{\omega}_e \cdot \bar{\psi}_{qs} \tag{119}$$

$$\bar{v}_{qs} = -\bar{R}_s \cdot \bar{i}_{qs} + \bar{\omega}_e \cdot \bar{\psi}_{ds} \tag{120}$$

$$\bar{\psi}_{ds} = -\bar{L}_{ds} \cdot \bar{i}_{ds} + \bar{\psi}_{pm} \tag{121}$$

$$\bar{\psi}_{qs} = -\bar{L}_{qs} \cdot \bar{i}_{qs}. \tag{122}$$

Substituting Eqs. (121) and (122) in Eqs. (119) and (120), and considering the lack of saliency in PMSG ($\bar{X}_{ds} = \bar{X}_{qs} = \bar{X}_s$) gives

$$\bar{v}_{ds} = -\bar{R}_s \cdot \bar{i}_{ds} + \bar{X}_s \cdot \bar{i}_{qs} \tag{123}$$

$$\bar{v}_{qs} = -\bar{R}_s \cdot \bar{i}_{qs} - \bar{X}_s \cdot \bar{i}_{ds} + \cdot\bar{\omega}_e \cdot \bar{\psi}_{pm}. \tag{124}$$

Following the same procedure as the described for EESG, it is obtained that the terminal voltage vector of a PMSG can be expressed as

$$\bar{\mathbf{V}}_s = \bar{v}_{ds} + j\,\bar{v}_{qs} = -(\bar{R}_s + j\,\bar{X}_s) \cdot \bar{\mathbf{I}}_s + j\,\bar{E}_g \tag{125}$$

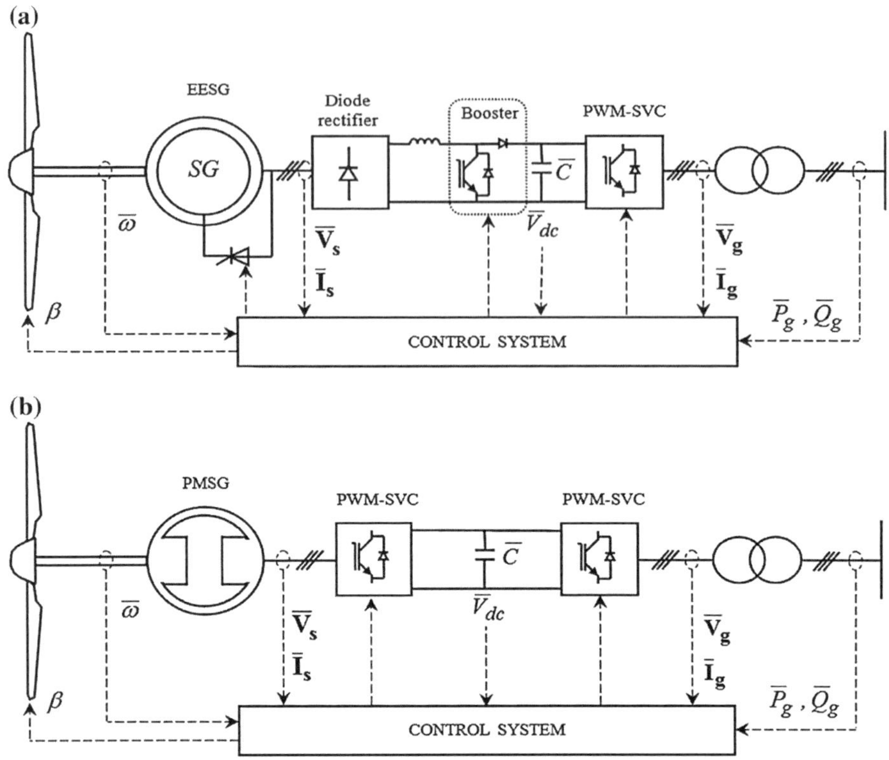

Fig. 44 Converter topology for a direct-drive wind turbine. **a** Diode-based rectifier and booster converter. **b** Two back-to-back voltage source converters (PWM-SVC)

where $\bar{E}_g$ is the excitation voltage generated by the permanent magnets, given by Eq. (107).

DDSG Power Converters

The converter topologies used for the full-scale power converter of a DDSG are:

1. An uncontrolled diode-based rectifier and a back-to-back IGBT bridge as GSC controlling the operation of the generator, linked by a DC booster controlling the DC link voltage (Fig. 44a). This configuration is very common for EESG, although it is also used for PMSG.
2. Two back-to-back IGBT bridges, the MSC and GSC, linked by a DC bus (Fig. 44b). The three-phase AC output of the generator is rectified, and the DC output from the rectifier is fed to an IGBT-based inverter. The MSC allows the generator active power control, and the control of stator current, reactive power or stator voltage, depending on the control strategy applied to the converter

[19]. The GSC has a back-to-back IGBT inverter that assures the active power exchange from the generator the grid, and generates or absorbs reactive power in order to operate at the desired power factor or terminal voltage. This configuration is very common, especially for PMSG.

5.4.2 DDSG Control System

As does DFIG wind turbine, the aims of the control system for a DDSG wind turbine are the following:

- To maximise the power extracted from the wind for a wide range of wind speeds (also known as power optimisation);
- To limit the output power to rated power for high winds (power limitation);
- To adjust both the active and reactive power to a desirable set point (power regulation);
- To adjust the voltage at the grid connection point to a desirable set point (voltage regulation).
- To work effectively, the power converter must be controlled along with the blade pitch angle control.

Control of Power Converter with Diode-Based Rectifier and Booster Converter and GSC

The control scheme of the converter topology with uncontrolled diode-based rectifier and booster converter and GSC is shown in Fig. 45 [1].

In this configuration, the DC booster converter is controlled in order to maintain the DC-link capacitor voltage at a set value, which assures the active power exchange from the DDSG to grid. An appropriate control of the fraction of time that the converter is conducting (duty cycle) allows the DC-link voltage control.

The GSC is normally controlled in the following way: (1) to control the active power delivered by DDSG to grid; (2) to operate with unity power factor, although it could be controlled by the reactive power or terminal voltage.

Similarly, vector and load angle control techniques can be applied to the GSC in order to achieve these objectives. The vector control is the most widely used control technique.

1. *Vector control*

The model of the GSC can be expressed by Eqs. (74)–(81), where $\bar{i}_{dc}$ and $\bar{i}_{qc}$ represents the total current delivered to grid by the DDSG ($\bar{i}_{dg}$ and $\bar{i}_{qg}$), because it uses a full-scale power converter.

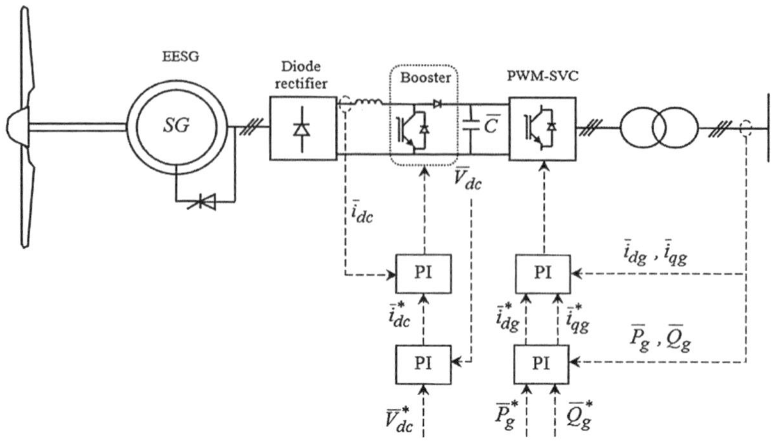

Fig. 45 Control scheme of power converter with diode-based rectifier and booster converter and GSC

As deduced from Eqs. (80) and (81), the active and reactive power control can be achieved by controlling the direct and quadrature current components, respectively.

The current components cannot be controlled independently by the GSC voltage components due to the cross-coupling effects, so that the decoupling feed-forward compensation defined in Eqs (82) and (83) must be used.

The decoupled control of the active power and reactive power/voltage is then performed in this converter by acting on the GSC voltage components: the active power control is performed by $\bar{v}_{dc}$, and the reactive power/terminal voltage is controlled by $\bar{v}_{qc}$. This converter can be modelled as a current-controlled voltage source. Figure 46 illustrates the control system of the GSC.

The $\bar{v}_{dc}$ controller controls the active power that DDSG delivers to grid. The power reference is obtained from the power-speed control curve and rotational speed. The $\bar{v}_{qc}$ controller allows the reactive power control or voltage control.

2. *Load angle control*

Similarly, the magnitude and angle of the GSC voltage vector can be used to control the active power and the terminal voltage or reactive power delivered to grid (see configuration shown in Fig. 47). It is justified as follows.

The power transfer between the GSC and the grid can be expressed by Eqs. (84) and (85). From these equations, it can be deduced that the active power depends mainly on the load angle δ_r and the reactive power depends mainly on voltage magnitudes, specifically on the GSC voltage $\bar{V}_{gsc}$ when the grid voltage is assumed constant.

In this control scheme, the active power reference is obtained from the power-speed control curve and rotational speed. The reactive power reference is derived

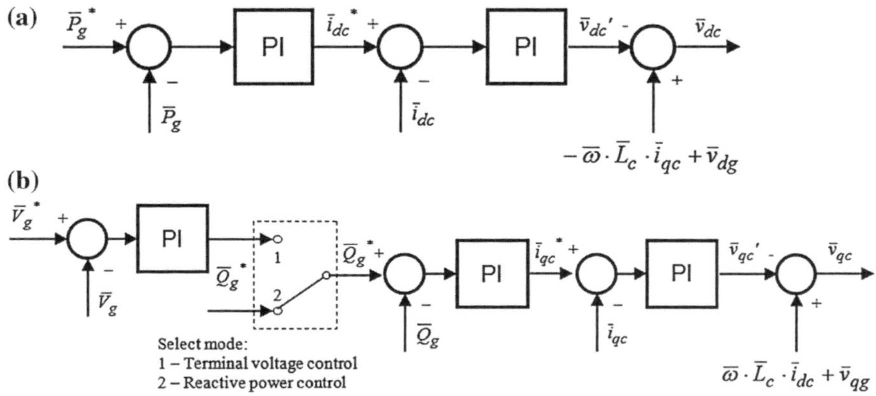

Fig. 46 Control scheme of the GSC vector control when used power converter with diode-based rectifier and booster converter. **a** Active power controller ($\bar{v}_{dc}$ controller). **b** Reactive power/voltage controller ($\bar{v}_{qc}$ controller)

Fig. 47 Load angle control of GSC

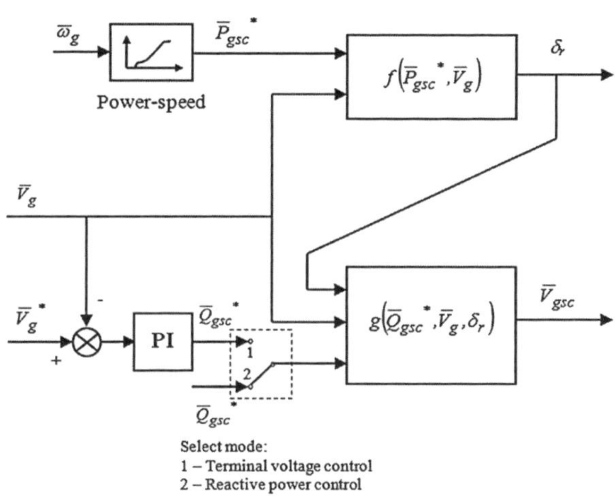

from the terminal voltage control loop, when terminal voltage control is applied to GSC, as shown in Fig. 47. When reactive power control is applied to the GSC, the demanded power is used as power reference.

Control of MSC and GSC

Figure 48 shows the control scheme of the converter topology with two back-to-back IGBT bridges, the MSC and GSC, linked by a DC bus.

The MSC commonly controls the active power generated by the DDSG. It drives the wind turbine to achieve the optimum power efficiency for below

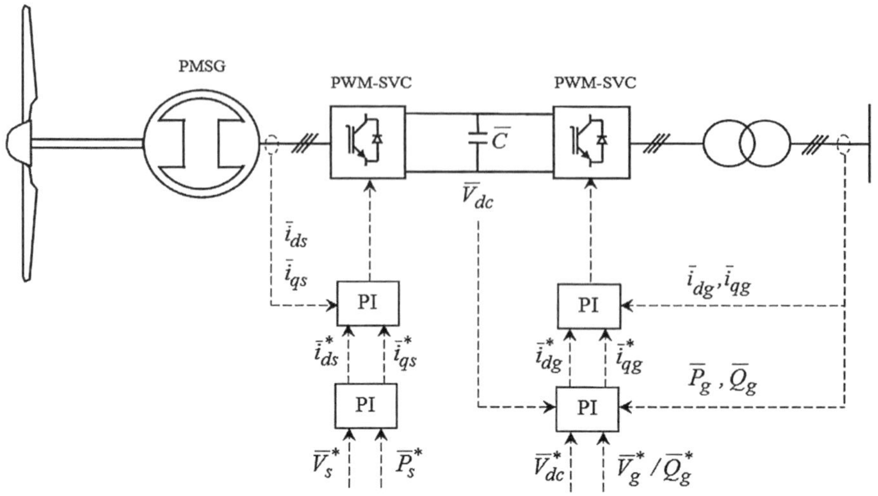

Fig. 48 Control scheme of MSC and GSC

nominal winds (power optimisation), to limit the output power to the rated value for above nominal winds (power limitation), or to adjust the active power to the power reference when power regulation is required (power regulation).

The GSC is normally controlled in the following way: (1) to maintain the DC-link capacitor voltage at a set value, which assures the active power exchange from the PMSG to grid; (2) to maintain the reactive power exchange to the grid, which guarantees a desirable power factor or terminal voltage during the wind turbine operation.

Although the control of both converters can be performed by using a different control scheme [1]. In this case, the MSC regulates DC-voltage through the d-axis current and AC-voltage through the q-axis current controller. In case of EESG with AVR, reactive power could be controlled instead of AC-voltage, because AVR allows the control of the generator voltage by acting on the excitation current. The GSC control is performed in the same way as explained for the configuration with diode-based rectifier and booster converter and GSC. Thus, the GSC allows the control of the active power delivered by DDSG to grid, and the operation with unity power factor, although it could be used to control the reactive power or terminal voltage.

The vector is the most widely used control technique for the MSC and GSC, although the load angle control can be also used. The load angle control is simpler, however it is not so effective as the vector control during transient events.

1. *MSC control*

In the *load angle control*, the magnitude and angle of the MSG voltage vector are used to control independently the active and reactive powers of the generator, as shown in Fig. 49. It is justified as follows.

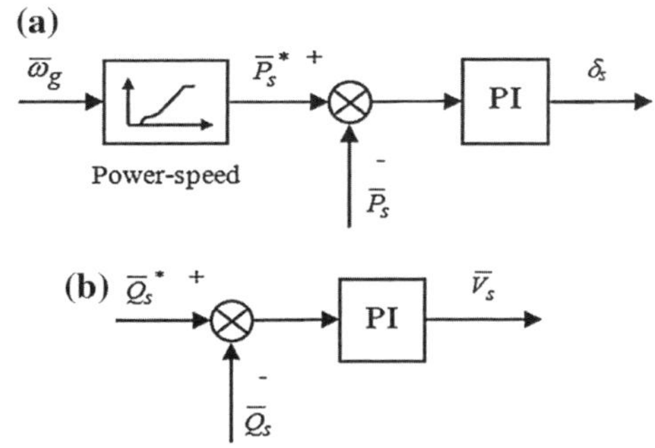

Fig. 49 Load angle control of the MSC. **a** Active power controller. **b** Reactive power controller

The active and reactive power flows between the generator internal voltage $\bar{E}_g$ and the voltage at the MSC terminal $\bar{V}_s$ depends on the synchronous reactance $\bar{X}_s$ and the phase angle between both voltages δ_s. Assuming small values of phase angle δ_s, the active and reactive power flows can be expressed as

$$\bar{P}_s = \frac{\bar{E}_g \bar{V}_s}{\bar{X}_s} \sin\delta_s \approx \frac{\bar{E}_g \bar{V}_s}{\bar{X}_s} \delta_s \tag{126}$$

$$\bar{Q}_s = \frac{\bar{E}_g^2}{\bar{X}_s} - \frac{\bar{E}_g \bar{V}_s}{\bar{X}_s} \cos\delta_s \approx \frac{\bar{E}_g^2}{\bar{X}_s} - \frac{\bar{E}_g \bar{V}_s}{\bar{X}_s}. \tag{127}$$

From Eqs. (126) and (127), it can be concluded that the active power $\bar{P}_s$ depends on the phase angle δ_s and the reactive power $\bar{Q}_s$ depends on the magnitude of the voltage at the MSC terminal $\bar{V}_s$, because the magnitude of the voltage $\bar{E}_g$ does not vary significantly during transient events of short duration. Commonly, the MSC is controlled to keep the reactive power at zero and to generate the active power derived from the power-speed control curve and rotational speed.

The *vector control* of the MSC is performed by using the *dq* components of stator current. The electromagnetic torque can be expressed as

$$\bar{T}_e = \bar{\psi}_m \cdot \bar{i}_{qs} - \left(\bar{L}_{ds} - \bar{L}_{qs}\right) \cdot \bar{i}_{ds} \cdot \bar{i}_{qs} \tag{128}$$

where

$$\bar{\psi}_m = \begin{cases} \bar{L}_{dm} \cdot \bar{i}_{df} & \text{for EESG} \\ \bar{\psi}_{pm} & \text{for PMSG} \end{cases}.$$

In case of EESG with salient pole rotor, $\bar{i}_{ds}$ is controlled so that its value is zero for the vector control. Then, the electromagnetic torque can be controlled by acting on $\bar{i}_{qs}$

$$\bar{T}_e = \bar{\psi}_m \cdot \bar{i}_{qs}. \tag{129}$$

In case of EESG with non-salient pole rotor or PMSG, the generator exhibits negligible saliency, so that the generator direct and quadrature inductances are identical ($\bar{L}_{ds} = \bar{L}_{qs}$). The electrical torque is then given by Eq. (128). In this case, different control strategies can be applied to the MSC [19]. All of these strategies are based on controlling the electrical torque and active power of the generator by acting on $\bar{i}_{qs}$: (1) minimum stator current control; (2) unity power factor control; (3) constant stator voltage control.

In minimum stator current control, $\bar{i}_{ds}$ is controlled so that its value is zero. However, this control strategy requires increasing the converter rating since the reactive power of the generator is not zero as derived from Eq. (102).

The unity power factor control of the generator can be achieved by using $\bar{i}_{ds}$ in order to compensate the reactive power demand of the generator. This control strategy minimises the converter rating. However, because the stator voltage is not directly controlled in PMSG and varies depending on speed, this control strategy could cause over-voltages for the converter and generator in the event of over-speeds.

In constant stator voltage control, the stator voltage of PMSG is controlled instead of the reactive power. The generator electrical torque and active power are controlled by acting on $\bar{i}_{qs}$. This makes it possible to control the stator voltage by acting on $\bar{i}_{ds}$, as deduced from Eq. (123). In this case, the generator and converter always operate at the rated voltage for which they were designed and optimised. Moreover, there is no risk of over-voltage and saturation of converter at high speeds. However, this control again requires increasing the converter rating, due to the reactive power demand of the generator.

The stator current components cannot be controlled independently by the stator voltage components because of the cross-coupling effects shown in Eqs. (113) and (114) for EESG and Eqs. (123) and (124) for PMSG. The cross-coupling effects can be cancelled by feed-forward compensation.

$$\bar{v}_{ds} = -\bar{v}'_{ds} + \bar{X}_{qs} \cdot \bar{i}_{qs} \tag{130}$$

$$\bar{v}_{qs} = -\bar{v}'_{qs} - \bar{X}_{ds} \cdot \bar{i}_{ds} + \bar{E}_{g}. \tag{131}$$

The decoupled control of the active power and stator voltage is then performed in this converter by acting on the stator voltage components: the active power control is performed by $\bar{v}_{qs}$, and the stator voltage is controlled by $\bar{v}_{ds}$. In fact, this converter can be modelled as a current-controlled voltage source. Figure 50 illustrates the control system of the MSC [11].

The $\bar{v}_{ds}$ controller (Fig. 50a) determines the direct component of the stator voltage in order to maintain the rated stator voltage. The $\bar{v}_{qs}$ controller (Fig. 50b) is an active power controller that controls output power by acting on the quadrature component of the stator voltage. This controller has a selector for the choice of operating mode. Two operating modes can be selected: power optimization/limitation and down power regulation. In the power optimization/limitation mode, the controller uses the rotational speed to define the reference power from the power-speed curve. The wind turbine can then be operated with variable speed, thus maximising the power extracted from the wind for below nominal winds or limiting the output power to rated power in wind conditions above nominal wind

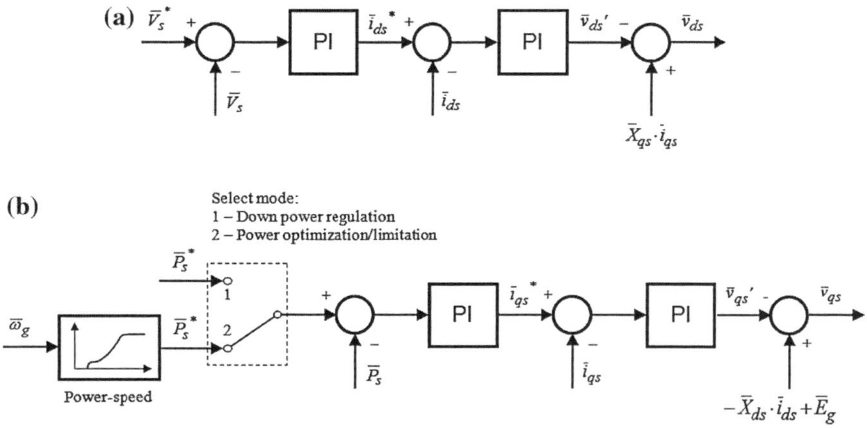

Fig. 50 Vector control of the MSC. **a** Stator voltage controller ($\bar{v}_{ds}$ controller). **b** Power controller ($\bar{v}_{qs}$ controller)

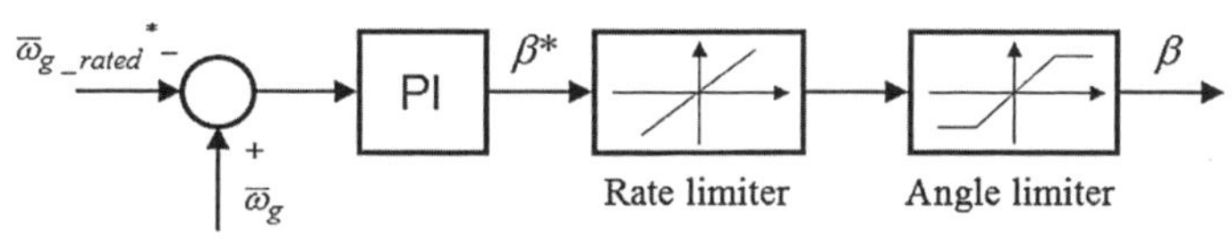

Fig. 51 Blade pitch angle controller

speed. In the down power regulation mode, the controller uses the power set point established for the wind turbine, instead of the power reference derived from the power-speed curve.

MSC is controlled in collaboration with the blade pitch angle controller, whose control scheme is shown in Fig. 51. This controller adjusts the pitch angle reducing the power coefficient, and thus, the power extracted from the wind, when the rotational speed increases up to the rated speed. It maintains the optimal pitch angle when the generator speed is less than the rated speed, and thus, the wind turbine operates with optimum power efficiency. On the other hand, when the wind turbine operates with power limitation or down power regulation strategies, this controller limits the rotational speed to the rated speed. In fact, the blade pitch angle controller acts as a rotational speed limiter in any operating conditions.

The control strategy can be summarized as follows:

- Power optimization strategy: In this case, the blade pitch angle controller keeps the pitch angle at its optimal value, whereas the power controller of the MSC acts on $\bar{v}_{qs}$ in order to maximize the power extracted from the wind.
- Power limitation strategy: The power controller of the MSC assures rated power, and the pitch angle controller limits the rotational speed to the rated speed.

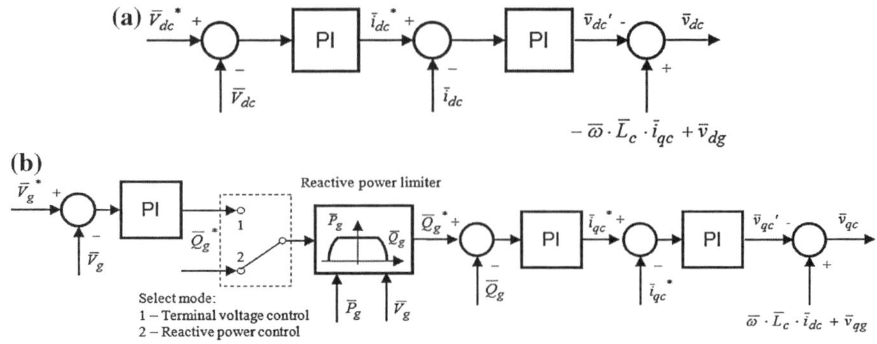

Fig. 52 Vector control scheme of the GSC. **a** DC link voltage controller ($\bar{v}_{dc}$ controller). **b** Reactive power/generation voltage controller ($\bar{v}_{qc}$ controller)

- Down power regulation strategy: In this case, the $\bar{v}_{qs}$ controller adjusts the output power to the ordered set point, whereas the blade pitch angle controller keeps the rotational speed at the rated speed.

2. *GSC control*

The GSC is controlled in the same way as explained for DFIG in section "GSC Control": (1) to maintain the DC-link capacitor voltage at a set value, which assures the active power exchange from the DFIG to grid; (2) to operate with unity power factor, although it could be used to the reactive power and terminal voltage control. Similarly, vector control and load angle control are the control techniques used.

Figure 52 illustrates the vector control of the GSC used for PMSG in [11]. In this case, the $\bar{v}_{qc}$ controller presents a selector to choose reactive power control or voltage control. When the wind turbine operates as a PV node, the voltage controller tries to adjust the voltage to the desired reference value $\bar{V}_g^*$. This controller sets the reactive power reference $\bar{Q}_g^*$. The reactive power controller adjusts the reactive power to the reference value $\bar{Q}_g^*$. A reactive power limiter is used, because the maximum reactive power in this wind turbine depends on the generated active power and the voltage as justified next.

The power capability of a PMSG wind turbine represents the maximum active and reactive power that the wind turbine can exchange to the grid. These operating limits depend on the power capability of the PSMG and its power converter capability [7]. The operating limits depend on the following constraints.

1. The rated active power that the PMSG is capable of generating.

$$\bar{P}_{g,max} = \bar{P}_{rated}. \tag{132}$$

2. The maximum apparent power that the power converter can deliver to the grid. This power depends on the rated grid voltage and the rated current of the power switches or any other limiting element in the grid interface. When the operating limits are represented at the PQ coordinate system, this constraint is represented by a circumference centred at the coordinate system origin with radius $\bar{S}_{g,\max}$.

$$\bar{P}_g^2 + \bar{Q}_g^2 = \bar{S}_{g,\max}^2 = \left(\bar{V}_g \cdot \bar{I}_{g,\max}\right)^2. \tag{133}$$

3. The maximum root mean square (rms) voltage of the fundamental component of the GSC voltage in turn depends on the DC bus voltage, on the modulation technique used, and on the maximum amplitude modulation index allowed in steady-state conditions. As the last two values are fixed, the reactive power capacity depends on the DC link voltage. From the expressions of active and reactive power delivered to grid by the power converter, and ignoring the grid resistance, it follows that the resulting limits in steady state can be expressed by the circumference:

$$\bar{P}_g^2 + \left(\bar{Q}_g + \frac{\bar{V}_g^2}{\bar{X}}\right)^2 = \left(m_a \cdot \frac{\bar{V}_g \cdot \bar{V}_{dc}}{\bar{X}}\right)^2 \tag{134}$$

where m_a is the amplitude modulation index.

It can be observed that for a given DC link voltage, the amplitude modulation index plays the role of field excitation in a synchronous generator. Moreover, the capability power depends on the grid voltage, as deduced from Eq. (134).

Figure 53 shows the active and reactive power capability that the PMSG wind turbine, simulated in this chapter, is capable of generating, in which the three constraints mentioned earlier are illustrated. As observed, for a given active power generated, the maximum reactive power that the wind turbine can generate decreases when the grid voltage increases, whereas the maximum reactive power increases when the grid voltage decreases.

5.4.3 Dynamic Simulation of PMSG Wind Turbine

Figures 54, 55, 56 illustrate the performance of a PMSG wind turbine experiencing an incoming wind as shown in Fig. 10, when connected to an infinite bus. In the simulation, vector control strategy is applied to both MSC and GSC, as performed in [11]. It is considered that MSC controls the active power of the PMSG, according to the operating conditions (optimisation power, power limitation or power regulation), and the stator voltage in order to maintain the rated value. In addition, the GSC is controlled to maintain the DC-link capacitor voltage at a rated value, which assures the active power exchange from the DFIG to grid, and to operate with unity power factor.

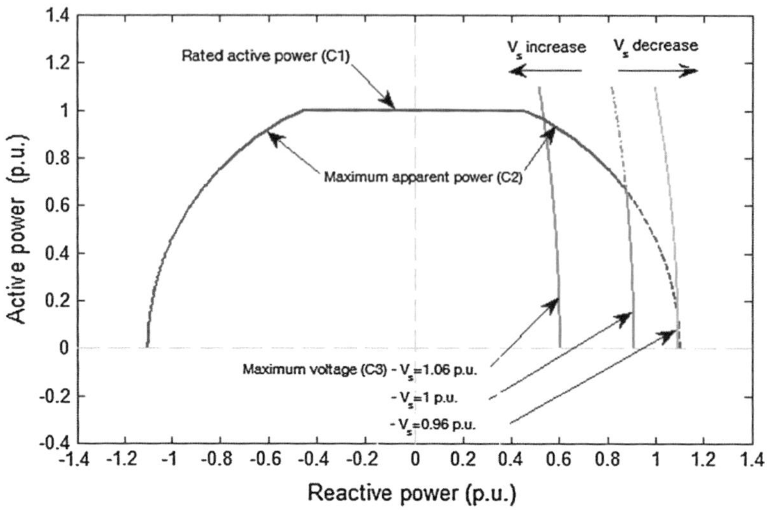

Fig. 53 Active and reactive power capability of PMSG wind turbine

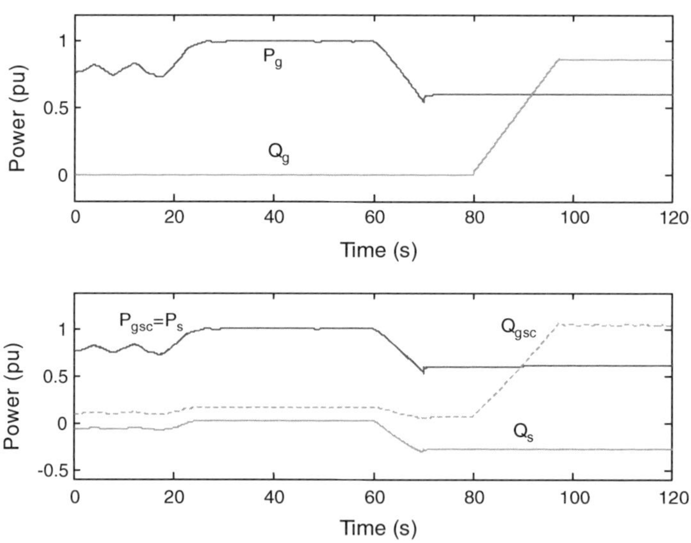

Fig. 54 PMSG response during active and reactive power regulation: Power delivered to grid, and stator powers and GSC powers

To illustrate the capabilities of this wind turbine to regulate the production (active and reactive powers), in the simulation performed, the wind turbine is demanded to operate as follows:

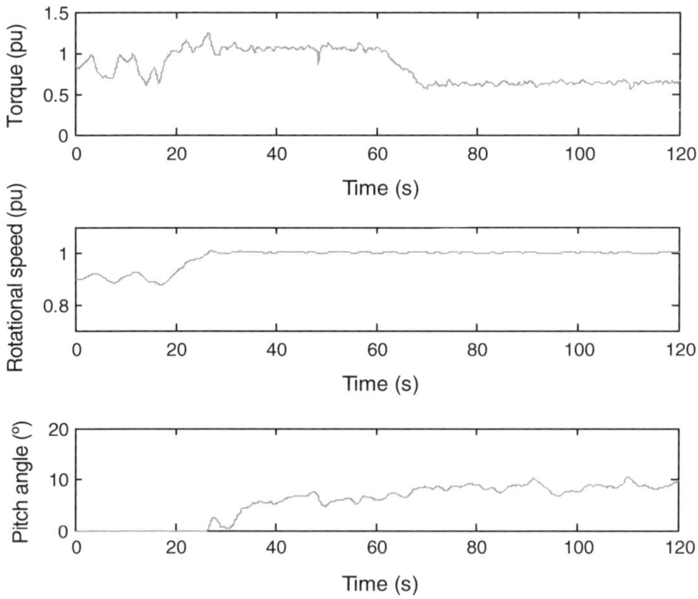

Fig. 55 PMSG response during active and reactive power regulation: Torque, rotational speed and pitch angle

- It produces the maximum possible output power injected to the grid during the first 60 s.
- At 60 s, the wind turbine receives a 40 % reduction of the active power reference by a slope of 0.05.
- The wind turbine operates with the unity power factor during the first 80 s, whereas it generates the maximum reactive power for the rest of the simulation, with a slope of 0.05.

As can be seen, the wind turbine produces the maximum possible output power injected into the grid during the first 60 s. While the wind turbine receives below rated wind (during the first 20 s), it generates below rated active power with variable rotational speed, thus achieving optimum power efficiency. In this case, the wind turbine operates with a maximum efficiency, and the pitch angle is kept at the optimal value. The wind turbine generates the rated power during 20–60 s since the incoming wind exceeds the rated wind. In this case, with power limitation, the power controller of the MSC adjusts the output power to the rated power, and the pitch angle controller limits the rotational speed to the rated speed. For the rest of the simulation, the wind turbine is able to generate the rated power since it experiences above nominal wind. However the wind turbine is ordered to reduce its output power to 0.6 p.u. (down power regulation) at 60 s by a slope of 0.05 p.u. With down power regulation, the power controller adjusts the output power to the desired power reference, and the pitch angle controller keeps the

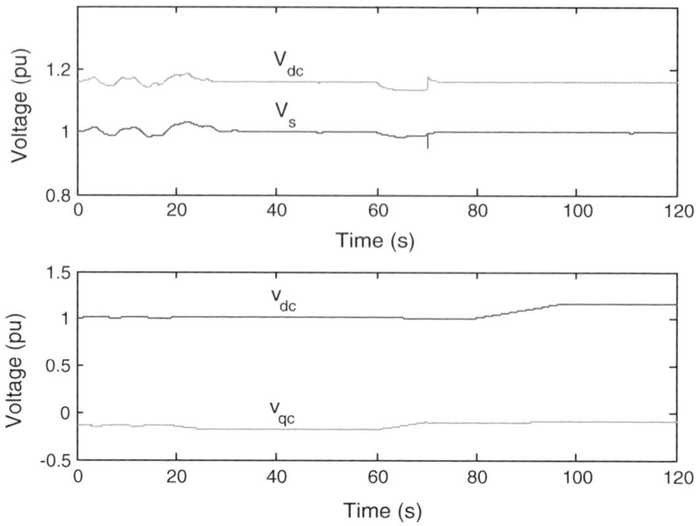

Fig. 56 PMSG response during active and reactive power regulation: Stator and DC voltage, and voltage components of the GSC

rotational speed to a constant rated speed value. During the entire simulation, the stator voltage controller of the MSC and the DC link voltage controller of the GSC try to maintain the rated stator voltage and rated DC link voltage, respectively.

On the other hand, the reactive power controller of the GSC assures that the wind turbine operated with a unity power factor during the first 80 s, and generates the maximum reactive power ordered by a slope of 0.05 p.u. for the rest of the simulation.

6 Conclusion

This chapter has presented a review of the modeling and control of the most widely used wind turbines concepts for integration in large power system dynamic simulations. The wind turbine concepts studied were: (1) fixed-speed squirrel cage induction generator (FS-SCIG); (2) wound rotor induction generator (WRIG) with variable rotor resistance (VRR); (3) doubly fed induction generator (DFIG); and (4) direct drive synchronous generator (DDSG).

In power system dynamic simulations, the wind turbine rotor is represented by a simplified model derived from the disk actuator theory, and the drive train is commonly modeled by the two-mass model.

Regarding to the electrical system, only the fundamental frequency component of voltages and currents is taken into account and higher harmonics are neglected, as usual for power system dynamic simulations. In fact, the electrical generator is represented by a fundamental frequency model, and the power converters are modelled as voltage/current source, since the internal dynamics of power converter are not of interest, and the power converter is considered ideal.

For each wind turbine concept, a review of the modelling of electrical system and control system used to achieve the operation desired (active power reference, reactive power reference or terminal voltage reference, whenever possible) was presented.

References

1. Achilles S, Pöller M (2003) Direct drive synchronous machine models for stability assessment of wind farms. In: Proceedings of 4th international workshop on large-scale integration of wind power and transmission networks for offshore wind farms, Billund, Denmark, pp 1–9
2. Ackermann T (2005) Wind power in power systems. Wiley, Chichester
3. Akhmatov V (2003) Analysis of dynamic behaviour of electric power systems with large amount of wind power. PhD thesis, Rgs. Lyngby, Denmark: Electric Power Engineering, Orsted-DTU, Technical University of Denmark
4. Anaya-Lara O, Hughes FM, Jenkins N, Strbac G (2006) Rotor flux magnitude and angle control strategy for doubly fed induction generators. Wind Energy 9(5):479–495
5. Boldea I (2005) Synchronous generators. CRC Press, Boca Raton
6. Bolik SM (2004) Modelling and analysis of variable speed wind turbines with induction generator during grid faults. Thesis, Aalborg, Denmark: Institute of Energy Technology, Aalborg University
7. Chinchilla M, Arnalte S, Burgos JC, Rodriguez JL (2006) Power limits of grid-connected modern wind energy systems. Renew Energy 31(9):1455–1470
8. CIGRE (2000) Modeling new forms of generation and storage. CIGRE Technical Brochure, TF 38.01.10
9. Ekanayake JB, Holdsworth L, Wu X, Jenkins N (2003) Dynamic modeling of doubly fed induction generator wind turbines. IEEE Trans Power Syst 18(2):803–809
10. Fernandez LM, Garcia CA, Jurado F (2008) Comparative study on the performance of control systems for doubly fed induction generator (DFIG) wind turbines operating with power regulation. Energy 33(9):1438–1452
11. Fernandez LM, Garcia CA, Jurado F (2010) Operating capability as a PQ/PV node of a direct-drive wind turbine based on a permanent magnet synchronous generator. Renew Energy 35(6):1308–1318
12. Fernandez LM, Garcia CA, Saenz JR, Jurado F (2009) Equivalent models of wind farms by using aggregated wind turbines and equivalent winds. Energy Convers Manage 50(3):691–704
13. Hansen AD, Hansen LH (2007) Wind turbine concept market penetration over 10 Years (1995–2004). Wind Energy 10:81–97
14. Heier S (1998) Grid integration of wind energy conversion systems. Wiley, Chichester
15. Kundur P (1994) Power system stability and control. McGraw-Hill, New York
16. Li H, Chen Z (2008) Overview of different wind generator systems and their comparisons. IET Renew Power Gener 2(2):123–138
17. Lubosny Z (2003) Wind turbine operation in electric power systems. Springer, Berlin Heidelberg
18. Martinsa M, Perdana A, Ledesma P, Agneholm E, Carlsona O (2007) Validation of fixed speed wind turbine dynamic models with measured data. Renew Energy 32(8):1301–1316

19. Morimoto S, Takeda Y, Hirasa T (1990) Current phase control methods for permanent magnet synchronous motors. IEEE Trans Power Electron 5(2):133–139
20. Muller S, Deicke M, De Doncker RW (2002) Doubly fed induction generator systems for wind turbines. IEEE Ind Appl Mag 8:26–33
21. Muyeen SM et al (2007) Comparative study on transient stability analysis of wind turbine generator system using different drive train models. IET Renew Power Gener 1(2):131–141
22. Nunes MVA, Pecas JA, Zürn HH, Bezerra UH, Almeida RG (2004) Influence of the variable-speed wind generators in transient stability margin of the conventional generators integrated in electrical grids. IEEE Trans Energy Convers 19(4):692–701
23. Papathanassiou SA, Papadopoulos MP (2001) Mechanical stresses in fixed-speed wind turbines due to network disturbances. IEEE Trans Energy Convers 16(4):361–367
24. Peña R, Clare JC, Asher GM (1996) Doubly fed induction generator using back-to-back PWM converters and its application to variable speed wind-energy generation. IEE Proc Electr Power Appl 143(3):231–241
25. Ramtharan G, Jenkins N, Anaya-Lara O (2007) Modelling and control of synchronous generators of wide-range variable-speed wind turbines. Wind Energy 10:231–246
26. Santos D, Arnaltes S, Rodriguez JL (2008) Reactive power capability of doubly fed asynchronous generators. Electr Power Syst Res 78(11):1837–1840
27. Slootweg JG, Kling WL (2004) Modelling wind turbines for power system dynamics simulations: an overview. Wind Eng 28(1):7–26
28. Slootweg JG, Polinder H, Kling WL (2003) Representing wind turbine electrical generating systems in fundamental frequency simulations. IEEE Trans Energy Convers 18(4):516–524
29. Slootweg JG, de Haan SWH, Polinder H, Kling WL (2003) General model for representing variable speed wind turbines in power system dynamics simulations. IEEE Trans Power Syst 18(1):144–151
30. Sørensen P, Hansen AD, Lov F, Blaabjerg F, Donovan MH (2005) Wind farm models and control strategies, report Risø-R-1464(EN). Risø National Laboratory, Roskilde
31. Tapia A, Tapia G, Ostolaza JX, Saenz JR (2003) Modeling and control of a wind driven doubly fed induction generator. IEEE Trans Energy Convers 18(2):194–204

Modelling and Control of Wind Parks

Carlos A. García, Luis M. Fernández and Francisco Jurado

Abstract Wind parks have experienced a great increase over the last years, from small wind parks with a few wind turbines connected to utility distribution systems, to large wind parks connected to transmission networks that may be considered, from the network system operators point of view, as a single wind power plant with operational capabilities similar to a conventional power plant. In this chapter, three main aspects concern to wind park grid integration are considered: the necessity of suitable wind park models for transient stability studies; the wind park control to fulfill system requirements and the application of special devices to enhance grid integration capabilities.

1 Introduction

Wind turbines can be grouped in a wind park to produce electric power as a conventional power plant. A wind park consists of a few to several hundred wind turbines over an extended area and integrated in an internal network composed by medium voltage lines and connected to grid through a common substation and feeder line.

C. A. García (✉) · L. M. Fernández
Department of Electrical Engineering, University of Cadiz, EPS Algeciras, Avda. Ramon Puyol s/n, 11202 Algeciras (Cádiz), Spain
e-mail: carlosandres.garcia@uca.es

L. M. Fernández
e-mail: luis.fernandez@uca.es

F. Jurado
Department of Electrical Engineering, University of Jaen, EPS Linares, Alfonso X 23700, Linares (Jaén), Spain
e-mail: fjurado@ujaen.es

P. M. Pardalos et al. (eds.), *Handbook of Wind Power Systems*, Energy Systems,
DOI: 10.1007/978-3-642-41080-2_15,

Wind parks have experienced a great increase over the last years, from small wind parks with a few wind turbines connected to utility distribution systems, to large wind parks connected to transmission networks that may be considered, from the network system operators point of view, as a single wind power plant with operational capabilities similar to a conventional power plant.

But, what does 'operating with the same requirements as a conventional power plant' mean? To answer this question, are at least necessary three points of view:

- For system operators and utility companies, wind parks have to be integrated in software packages for power system stability studies (dynamic studies). Thus, suitable models are needed to evaluate wind parks at the same level as conventional power plants. Nevertheless, the complexity of models is greater than in case of conventional power plant models, because those are composed of a high number of wind power generators, which require a particular study.
- System requirements are doing that wind parks operate at the same level than conventional power plants, covering production demands or ensuring the stability and reliability of the system. For that reason, the main objective of the wind park controller is to reach the power production setting in a similar way of a conventional power plant and, in some cases, to provide advanced grid support. That is, active power and voltage regulation and reactive power and frequency control. Hence, the wind park controller must to operate inside the park, fulfilling the grid demands by appropriate generation strategies at wind turbine level.
- To improve the wind park production and control capabilities, special devices can be used to enhance wind turbine capabilities and smooth the output power fluctuations caused by the wind speed variability, and thus enhance wind park performance.

Under this scenario, this chapter gives an overall description of modelling and control of wind parks from the point of view of their grid integration as a wind power plant.

Section 2 presents a wide description of the wind parks models studied in the literature, with the main focus on transitory stability studies. The wind park models accuracy relies on detailed modelling of the applied wind turbine technology, considered in a previous chapter. They allow a suitable simulation of the wind parks behaviour under wind speed fluctuations and grid faults. The clearest model (detailed model) represents all the wind turbines, with the necessary detail level of the scope of study, and the model of the internal grid of the park. Its main problem is related to the high order model if the wind park has a high number of wind turbines. Additional wind park models can reduce the complexity of detailed models based on the aggregation of the wind turbines in different ways: (1) a single equivalent model, in which the entire wind turbines are represented by a single unit; (2) a cluster equivalent model, in which the wind turbines with the same characteristics and similar incoming winds are grouped in clusters; and (3) a compound equivalent model, in which only the generation system of each wind turbine is aggregated in an equivalent one, while representing the mechanical systems of all the individual wind turbines.

As wind power plants are today required to participate actively in the power system operation, the grid connection requirements have been revised, demanding an operational behaviour similar to those of conventional power plants. Thus, in Sec 3, it is detailed some of these requirements: (1) Active power regulation and frequency control, and (2) reactive power control and voltage regulation, as previous issue to the description of the control structure of wind parks. Following, an overview of wind parks control structure is performed in this section, considering the main controller as the leading component of the structure, assuring the wind park production in a similar way as a conventional power plant. In addition, dispatch control must define the power references of the wind turbines within the wind park. Finally, an evolution of the main controller, the cluster management system, is illustrated for controlling a group of wind parks connected to a certain grid node.

One of the main issues is that wind power plants are becoming required to help the power system quality, stability and reliability, and these requirements are related to the availability of the wind and the technologies of the wind turbines. Section 4 describes some special power electronics-based system such as Flexible AC Transmission Systems (FACTS), which can be used in wind parks to improve power and voltage control capabilities and in some cases, enhancing power quality. Another way of improving wind power capabilities is to use an energy storage system integrated in wind turbine.

2 Wind Park Modelling

Dynamic studies of wind parks, in which the main focus is to study their influence on the electrical power system, are commonly called transitory stability studies. These studies use time constants between 0.1 and thousands seconds and voltages and powers as main variables. Their principal characteristic is that electrical network transient can be neglected, though network model is typically represented by fundamental frequency model. This consideration reduces the number of differential equations of the model, increases integration time step and allows using load flow algorithms.

As a result of the increase of wind capacity connected directly to transmission systems, development of wind parks models and their integration into software packages for power system stability has been stimulated. Some examples are PSS/E (PTI), PowerFactory (DigSILENT), Netomac (SIEMENS) or Simpow (ABB). Additionally, general-purpose software like MATLAB/Simulink (The Mathworks) is regularly used for transient stabilities studies when dedicated software presents restrictions in simulations [54]. The relevance of the restriction depends on the scope of the investigation and on the characteristics of the component models. Research institutes, universities, commercial entities and network operators have contributed to the development of accurate dynamic models adapted to their needs. Hence, model validation is a key issue that taken up by International Energy Agency (IEA) Wind Task 21. This international working group includes participants from nine countries, and has developed, from 2002 to 2006, a systematic approach for model benchmark testing. The rationale for the proposed benchmark testing is that dynamic wind generation models are being applied for assessing grid connection of large wind parks, even though model accuracy is not always known. This, at best, leads to uncertainty in the market, and, at worst, to an erroneous design jeopardizing power system stability. The challenge is twofold. First, the technology in modern wind parks is fairly complex, and their dynamic behaviour may differ significantly depending on the wind turbine type and manufacturer's specific technical solutions. Thus, it is not trivial to develop accurate wind generation models. Second, model validation must be transparent and adequate for providing confidence. In this respect, IEA Wind Task 21 has contributed by describing a benchmark procedure and applying this procedure for testing numerical wind generation models [26].

The wind park model has to allow the representation of the behaviour of all the wind turbines within the wind park under normal conditions (wind fluctuations) and abnormal conditions (grid faults). Accurate simulations of wind parks rely on detailed modelling of the applied wind turbines, not only the type of technology but specific variations in the same type. The benchmark test procedures suggested by IEA Wind Task 21 consider operation during normal condition and response to a voltage dip, and may include both validation against measurements and model-to-model comparisons:

- *Dynamic operation during normal conditions* Model capability to simulate wind parks characteristics power fluctuations with,
  - *Input:* Wind speed time series (optionally voltage time series).
  - *Output:* Powers and voltage (optionally) time series; power spectral density of active power and short-term flicker emission.
- Dynamic operation during abnormal conditions (*response to voltage dip*) Model capability to simulate wind parks response to voltage dip,
  - *Input* Voltage time series and constant aerodynamic torque (optionally wind speed time series).
  - *Output* Powers and voltage time series.

2.1 Detailed Wind Park Models

Wind park models may be built to various level details depending on the scope of the simulations. The clearest model is the detailed (complete) model, a one-to-one approach that consists in representing all the wind turbines and the internal grid of the park. In [49], a dynamic model of the Hagesholm wind park was implemented in power system simulation program DigSILENT, and verified against measurements to assess the ability of the simulation models to predict the influence of the park on the power quality characteristics of wind turbines specified in IEC61400-21.

In other cases, detailed models are necessary because the focus of the studies is to evaluate the internal evolution of the wind park. Akhmatov [4] presents a large offshore wind park model with eighty wind turbines to investigate how short-term voltage stability can be affected by parameters fluctuation and controllability of wind turbines in terms of dynamic stability limits. These studies include:

- In case of fixed-speed wind turbines, the necessity of dynamic compensation units, like Static Var Compensator (SVC) or STATic Synchronous COMpensator (STATCOM), adequate construction parameters or controllability of wind turbine blade angle in case of grid faults to prevent fatal over-speeding.
- In case of wind turbines with double-outage induction generator and variable rotor resistance, protection of the power electronics of the rotor converter shall be taken in account, blocking power converter and establishing power reserves.
- In case of variable speed wind turbines equipped with DFIG and partial-load frequency converters, the main challenge is to maintain interrupted operation of wind turbines without use of dynamic reactive compensation when the control feature with use of fast re-start of the rotor converters is applied.
- In case of variable speed wind turbines equipped direct-driven synchronous generators connected to the grid through frequency converters, an accurate and sufficiently detailed representation of the converter, its control and protective sequences will be absolutely recommended for investigations of transient voltage stability, because it determines the wind turbine operation during and after the transient event in the power system.

When the scope of the studies is the internal grid of the wind park, detail models are also necessary, like in [35], where a wind park with an internal DC collection grid and the same layout as the wind park in Lillgrund (Sweden), with 48 wind turbines of 2.3 MW, located in 5 radial connected to an offshore platform, is modelled in PSCAD/EMTDC to investigate dynamic operation both for normal operation and for different faults conditions. In this case, the main key is the suitable design of the DC/DC converters and DC bus, whose dynamics are much faster than those of the wind turbine, which can be neglected, and its model can be simplified with a current source connected to the DC/DC converter in the wind turbine.

Another important aspect to consider in studies with detail models is to validate them against measurements. With this consideration, project Erao-3 [42] intends to find measurement of wind turbines and wind parks in the Netherlands for model validation, under the area of IEA Annex XXI. These models can be used in wind parks and local grid studies as well as for the development of more simple, reduced order o aggregate models. In this project, Alsvik Wind Park was considered for constant speed stall wind park dynamic model validation. Comparison of measurements with simulations results showed that:

- The frequency response results of the electrical variables are good if the measured voltage is used as input instead of using a grid model;
- The frequency response results of mechanical variables are less good, which may partly be caused by grid voltage variation, which could not be used as input because it was not measured;
- The results for the voltage dip were quite good, with some mismatch in the damping.

Detailed models are also interesting when the scope of researches is the wind parks integration in the network system. In this case, the different wind power technologies establish the characteristics and control possibilities of wind parks. Gjengedal [20] evaluates the dynamic performance of three technologies, and performs a transient stability study in order to illustrate the differences between fixed speed wind turbines—Danish concept—and variable speed wind turbines (DFIG induction generator or PMSG with full converter). In order to compare the wind generator technologies, dynamic simulations were performed for a wind park with 75 wind turbines of 2 MW divided in clusters, and connected to a typical 132 kV rural grid in Norway. From the five requirements discussed (active and reactive power control, frequency control, transient voltage control and three-phase fault ride-through), fixed speed wind turbines have clearly less control possibilities than variable speed wind turbine; all the technologies have difficulties to ride-through faults that may imply additional costs; and alternative solution may be to disconnect during a fault and reconnect shortly after the fault is cleared.

In the same scope, Cartwright et al. [9] present a voltage and reactive control strategy for a variable speed wind park with 30 DFIG wind turbines of 2 MW. Dynamic models of the wind turbines and converters are developed and simulation results presented confirm the effectiveness of the control strategies within transmission and distribution system implementations.

With the same philosophy, Tapia et al. [55] developed a complete model of a wind park located in Navarra (north of Spain), composed of 33 DFIG wind turbines of 660 KW. Simulations results of the model performance were obtained and compared to the real performance of the wind park, only in case of wind fluctuations, but not in case of abnormal conditions. The wind park model developed is used for the wind park reactive power control.

The response obtained from complete wind park models is commonly used as reference for validating aggregated models. Slootweg [47] investigates an aggregation approach to be used for both fixed and variable speed wind parks, by comparing the simulation response of a detailed and an aggregated wind park models for different operating conditions. It is concluded from this study that in normal operating conditions, in which only wind speed changes occur, as well as during faults, simulation results of the detailed and the aggregated models are rather close. The same conclusion can be found in [15] and [16], where the authors present aggregated models of wind parks with SCIG and DFIG wind turbines, which are validated by means of comparison with the complete wind park models.

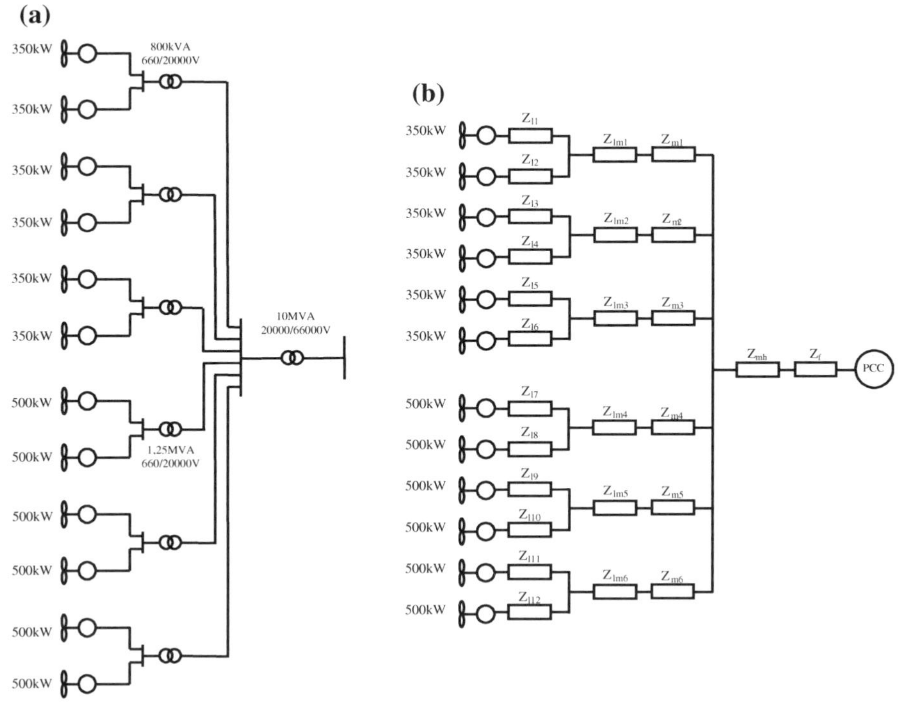

Fig. 1 Fixed speed wind park: **a** structure; **b** equivalent circuit of the internal network

Complete wind park model is always composed of all the wind turbine models of the park, with the necessary detail level of the scope of study, and the model of the internal grid of the wind park. Wind turbine models, as usual in dynamic studies, can be divided in two systems:

- The mechanical system, composed of the rotor (modelled as an ideal rotor disk) and the drive train (modelled, at least, by the two-mass model).
- The generation system, composed of the generator (SCIG, DFIG or PMSG), modelled with the first, third or fifth order model, depending on the scope of the studies. And, depending on the generation system, electronic power converters composed of two converters linked by a DC bus, modelled, in some cases, by neglecting converter dynamics.

Internal grid of the park has to be modelled including low voltage lines between wind turbines and transformers, medium voltage lines to internal substation and line feeder to the point of common couple (PCC). As usual in power systems simulation, electromagnetic transient are neglected, and all the elements of the grid are represented by constant impedances [55].

As example, a fixed-speed wind park structure is considered, as seen in Fig. 1. The park has a radial structure with twelve wind turbines in two clusters. The first one is composed of six 350 kW turbines and the other has six 500 kW wind turbines. Every pair of turbines has a common low voltage-medium voltage transformer to the medium voltage internal lines of the grid. These lines are connected to a medium voltage-high voltage transformer (substation of the park), and a line feeder connect them to the grid at the PCC. Fig. 1a shows the radial structure of the wind park, while Fig. 1b represents the impedances of the equivalent circuit of the internal network.

In this figure, Z_{li} is the low voltage line impedance of wind turbine i; Z_{lmj} is short-circuit impedance of the low/medium voltage transformer j; Z_{mj} is the medium voltage line impedance of the j line; Z_{mh} is the short-circuit impedance of the medium/high transformer and Z_f is the feeder impedance of the wind park.

In addition, the model of the wind park must include all the necessary devices to control generated power, the voltage and the frequency at the point of common coupling, or enhance generated power quality.

2.2 Equivalent Models of Wind Parks

Nevertheless, detailed model presents a main problem, they are high order models if the wind park has high number of wind turbines, and therefore the simulation time is long, because of the excessive number of equations to be computed. The complexity of the wind park model and the simulation time can be reduced by using equivalent models instead of detail models. These equivalent models result from the aggregation of the individual wind turbines into an aggregated model, whose characteristics depend on the input signals and the elements of the wind park. Thus, the models can be qualified in several groups depending on the wind speed that receive every wind turbine and the types and rated values of the machines that compose the park.

The most simply aggregation method is described in [4, 19, 41]. It consists in considering the same incoming wind to all the wind turbines of the wind park (wind speed differences are small in every location) and identical wind turbines (same mechanical and generation model). Thus, the aggregation model can be represented as one equivalent wind turbine with the same mechanical and generation models (identical parameters in per unit) as the individual ones, but with a rated power given as:

$$S_{eq} = \sum_{i=1}^{n} S_i \tag{1}$$

where S_i is rated power of the wind turbine i and n is the number of machines in the park [41].

Nevertheless, these ideal conditions are not the usual in case of large wind park with a great number of wind turbines (wind speed differences cannot be neglected), or if the wind parks are composed of different types of wind turbines or if they have different rated powers [37]. In these cases, the whole wind park is

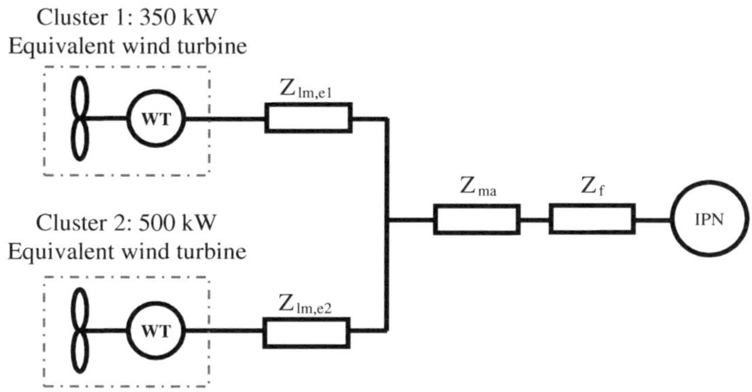

Fig. 2 Cluster representation of the aggregated model of the fixed speed wind park shown in Fig. 1a

represented by a variable number of equivalent wind turbines obtained from a simple aggregation method that consists in representing as much single equivalent wind turbines as groups (clusters) of identical wind turbines that experiencing similar incoming wind speeds are in the park.

Figure 2 shows an aggregated model of the wind park proposed in Fig. 1, considering that the six 350 kW wind turbines experience the same incoming wind and the six 500 kW wind turbines operates under similar wind speeds (with small differences).

As can be seen in Fig. 2, the first group of six 350 kW wind turbines (cluster 1) is replaced by an equivalent wind turbine that receive the same incoming wind as the individual wind turbines and has a rated power given by Eq. (1). The second group of six 500 kW wind turbines (cluster 2) experience similar incoming winds, but not identical. In this case, as illustrated in [16], an equivalent of the winds incident on the aggregated wind turbines can be used as wind incident on the equivalent wind turbine of 3 MW rated power, obtained from Eq. (1). This equivalent wind is obtained from the average of the incoming winds that experience every unit of the cluster. It is useful for wind parks located on topographically simple terrains (e.g. smooth land or off-shore) with wind turbines organized in rows at right angles to the prevailing wind direction. The wind turbines belonging to the same rows usually present similar winds, whereas the winds incident on each row are different because of shadowing between wind turbines (park effect). Akhmatov and Knudsen [5] consider this distribution pattern in the aggregated model of a large offshore wind park composed of seventy-two fixed-speed wind turbines of 2 MW rated power, whose internal network is organized in six rows with twelve wind turbines in each one. In this case, every row of wind turbines operates at the same wind conditions (same operational point). Thus, the whole wind park can be separated in six groups, composing each one of twelve wind turbines, that can be represented by one equivalent wind turbine with a rated power twelve times higher than that of the individuals.

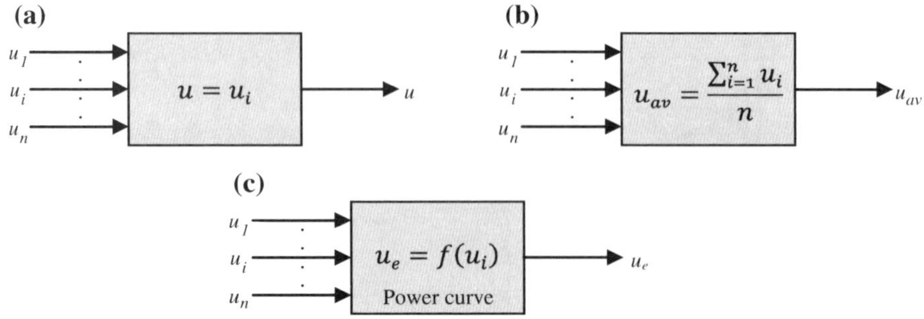

Fig. 3 Wind speed models: **a** constant wind speed model, **b** average wind speed model, **c** equivalent wind speed model

Following this assumption, a step further is presented in [14], where it is considered a new equivalent wind model in case of different incoming winds. This equivalent wind is derived from the power curve of the wind turbines in order to obtain an approximation of the power generated by the wind park as sum of the individual powers that generates each one according to its incoming wind. This equivalent wind is used as wind incident on a re-scaled wind turbine that aggregates the identical wind turbines of the wind park. The resulting model was compared with a clustering aggregated model, formed by groups of wind turbines that experience similar incoming winds, whose equivalent model receives the average of the incident winds of the group. The results analysis shows that the equivalent wind turbine with average incoming wind must be only used for units that experience similar winds (differences less than 2 m/s), while, in case of different winds, the equivalent model with the equivalent incoming wind enables the aggregation of all the wind turbines of a wind park into a single equivalent one.

Figure 3 shows the wind models considered in the literature in case of aggregated wind park models represented by a re-scaled individual wind turbine. Figure 3a presents the simplest wind model, where every wind turbine experiences the same wind speed (u). Figure 3b is useful for similar wind speeds, where the average wind speed (u_{av}) is the input to the re-scaled wind turbine. And finally, Fig. 3c represents a more suitable wind model, where an equivalent wind speed (u_e), obtained from the power curve of the individual wind turbines, allows the aggregation of all the wind turbines of the wind park, in case of different incoming winds, into a single equivalent wind turbine (without clustering). This model supposes the best way of reducing the model order and the simulation time, achieving an adequate approximation of the collective response of the wind park when carrying out dynamics simulations, e.g. in transient stability studies, when the mechanical behaviour has generally no big impact on voltages and power flows at PCC [43].

However, when simulating longer term dynamics, the aggregation of wind turbines into an equivalent one, that experience an equivalent wind speed, cannot predict the wind park's behaviour with sufficient accuracy, due to the highly

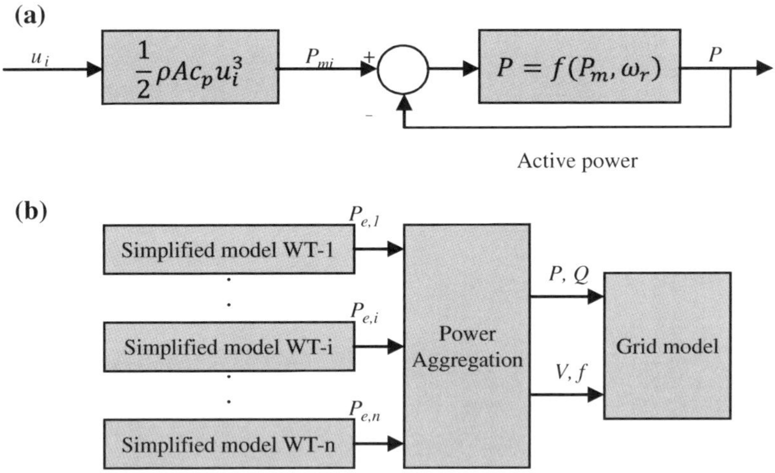

Fig. 4 **a** Simplified model of variable wind turbine. **b** Structure of the aggregated model of variable speed wind park

nonlinearity of the mechanical system. A good compromise between model accuracy and simulation time consists of aggregating just the generation systems (generators, power converters and controls) and representing the mechanical systems of all the individual wind turbines to be aggregated [43]. This 'compound' model is applied by [48] in case of variable speed wind turbines aggregation, since there is no unique relation between a wind speed value and the generated power. Therefore, the rotor speed of the individual wind turbines, combined in an aggregated model, needs to be recorded in order to get a suitable approximation of the generated power, with the following simplifications:

- The rotor model can be simplified by assuming that the wind turbine operates at the optimal value of the power coefficient over the whole operating range.
- The rotor speed controlled is simplified by approximating the rotor speed versus power control characteristic with a first order approximation.
- When the rotor speed is limited at its maximum value, the pitch angle controller can be omitted.

Figure 4 shows the simplified variable speed wind turbine model and the structure of the aggregated model of the wind park proposed in [48].

With the same philosophy, a 'compound' aggregated model is considered in [16] for fixed speed wind parks, and in [15] for a DFIG variable speed wind parks, located on topographically complex terrains (e.g. mountain ladder) or with widely separated wind turbines. The aggregated models are represented by an equivalent wind turbine with an aggregated generation system model and a simplified model of each individual wind turbine, which approximates the operational point of each one according to the incoming wind speed. These simplified models are composed of the rotor (rotor disk model) and the drive train models (two masses-model), the

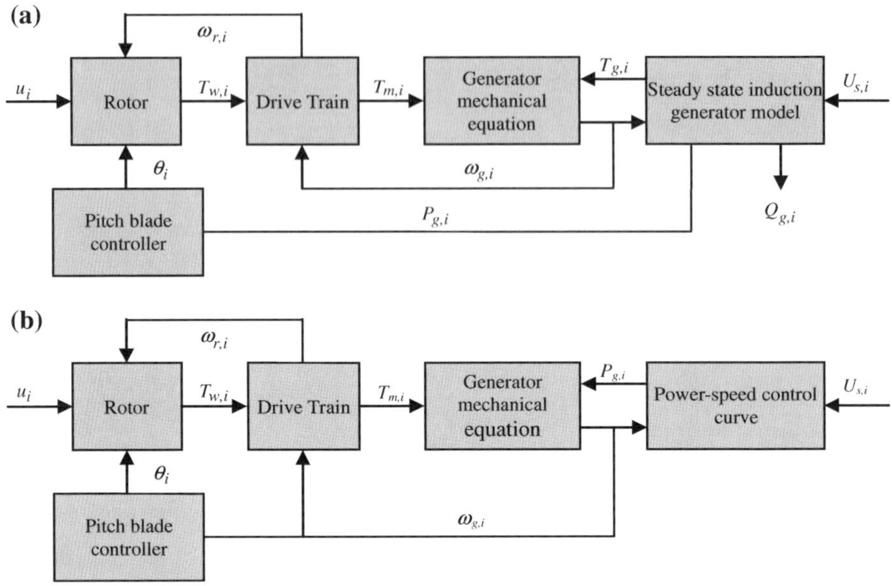

Fig. 5 Simplified models of: **a** fixed speed wind turbine, **b** variable speed wind turbine

blade pitch angle controller and the induction generator (first order model, represented by its mechanical equation). In case of fixed speed wind turbines (Fig. 5a), the simplified model is completed by the steady-state generator model to obtain the electrical torque and the calculation of the reactive power of each individual wind turbine. In case of variable speed wind turbines (Fig. 5b), the model includes the rotor speed controller represented by the power speed control curve.

The simplified models of each wind turbine approximate their operation point according to the corresponding incoming wind speed. The equivalent wind turbine presents n-times the size of the individuals, with an equivalent generation system composed of the equivalent induction generator (and power converter, in case variable speed wind parks), represented by the same model of the individual generation system, and with the same parameters at per unit. The equivalent wind turbine presents an aggregated mechanical torque, obtained from the simplified models, and used as incoming to the equivalent generation system in order to calculate the equivalent generator mechanical torque.

The structure of the equivalent wind turbine without aggregation of mechanical system is shown in Fig. 6, for both fixed speed wind parks (Fig. 6a) and variable speed wind parks (Fig. 6b). Both models have the same structure, but, the equivalent model for fixed speed wind turbines include compensating capacitors with variable reactance for a suitable approximation of reactive power, due to the great dependence of the reactive power on the active power and generation voltage. On the other hand, the equivalent model for variable speed wind turbines

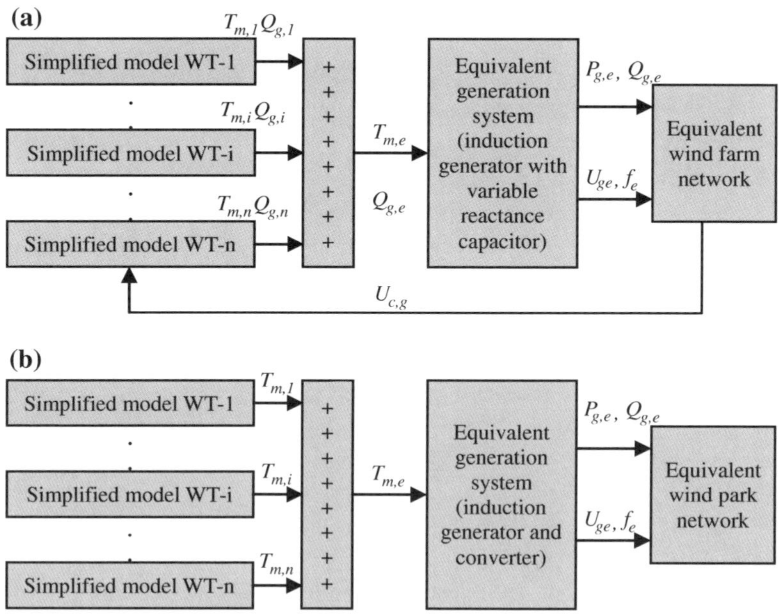

Fig. 6 Structure of the equivalent wind turbine of the: **a** fixed speed wind park, **b** variable speed wind park

needs a suitable control of active and reactive power when the wind turbines are operating with down power regulation. The active and reactive power references of the equivalent wind turbine must be equal to the sum of the active and active power references of the aggregated wind turbines.

The equivalent generation system is commonly represented by the same model of the individual wind turbines and with the same mechanical and electrical parameters at per unit [43]. Only a few authors have considered other methods of aggregating generation system and always based on aggregation methods applied to induction motors for transient stability. These aggregation methods can be seen in [46], where it was applied the method proposed by Franklin and Morelato [18], based on the steady-state theory of the induction motor and the equivalence criterion of the active power absorbed. In the same way, Trudnowski et al. [57] obtained the parameters of the equivalent generator using the method proposed by Nozari and Kankam [40]. In this method, the equivalent parameters are calculated from the rated power weighted average admittances of each branch of the induction machine equivalent circuit.

Jin and Ju [29] obtained the parameters of an equivalent induction generator by using a weighted average method based on the apparent power of the generator, although their main contribution is a suitable method of clustering aggregation in case of different kinds of wind turbines according to a slip coherency criterion. They assumed that the synchronous generators present a coherent response if they have similar rotor angles responses or the same rotor speed following a disturbance. However, they considered, with respect to the induction generator, that the slight variety of the slip causes marked changes in both active and reactive power, and the coherency analysis must be applied according to the slip response. As result of the analysis, the wind turbines of a wind park can be clustered by the value of the product of the rotor resistance by the combined inertia constants of the generator and turbine. This criterion is in agreement with the index proposed by Taleb et al. [52] as indicator for grouping induction machines.

The power converter and the control system of the equivalent wind turbine keeps the same structure as the individual, excepting the blade pitch angle controller in case of compound model, because the simplified model of each wind turbine includes this control [15].

Finally, the equivalent wind turbine operates in an aggregated wind park network, obtained from the equivalent of the lines and/or transformers of the common network of aggregated wind turbines. As usual in dynamic power systems, the internal network was commonly modelled by its static model, assuming that the short-circuit impedance of the equivalent wind park must be equal to the short-circuit impedance of the complete wind park [15]. Thus, Fig. 2 shows the equivalent internal electrical network of the wind park showed in Fig. 1, where the impedances $Z_{lm,e1}$ and $Z_{lm,e2}$ are the equivalents of the internal network of both wind parks connected to the same feeder.

When the studies focuses on developing an equivalent wind park model for power system planning studies, the equivalent internal network is modelled considering the criterion of apparent power losses, as can be seen in [39].

2.3 Wind Park Models Simulations

Some of the equivalent wind turbine models are implemented and verified their dynamic responses by comparison with the complete model. Three cases are studied in order to present results of the complete and reduced models of wind parks for dynamic power system analysis during wind fluctuations and electrical faults.

2.3.1 Case 1: Normal Operation in a Variable Speed Wind Park with Identical Wind Turbines

The wind park under consideration, Fig. 7, is composed of 6 DFIG wind turbines of 2 MW organized into a network with two sections [15].

Figure 7a shows the radial structure of the wind park, while Fig. 7b shows the impedances of the internal network. In this figure, Z_{lmij} is the short-circuit impedance of the low/medium voltage transformer of the wind turbine j of the branch i; Z_{mij} is the medium voltage line impedance of wind turbine j of the branch i; Z_{mi} is the medium voltage line impedance of the i branch; Z_{mh} is the short-circuit impedance of the medium/high transformer and Z_f is the feeder impedance of the wind park.

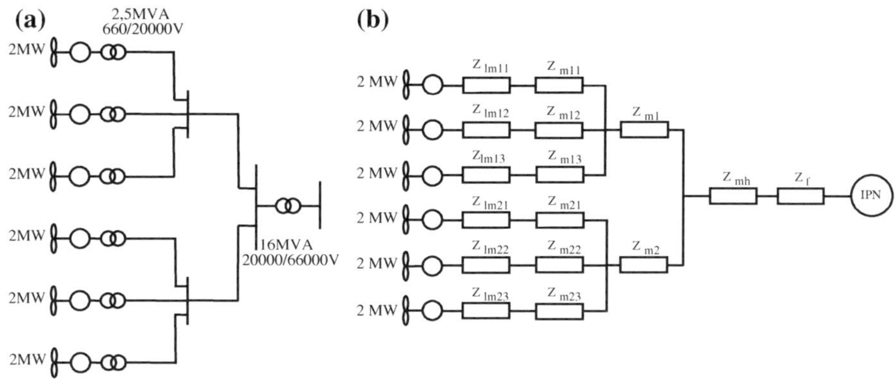

Fig. 7 Variable speed wind park: **a** structure, **b** equivalent circuit of the internal network

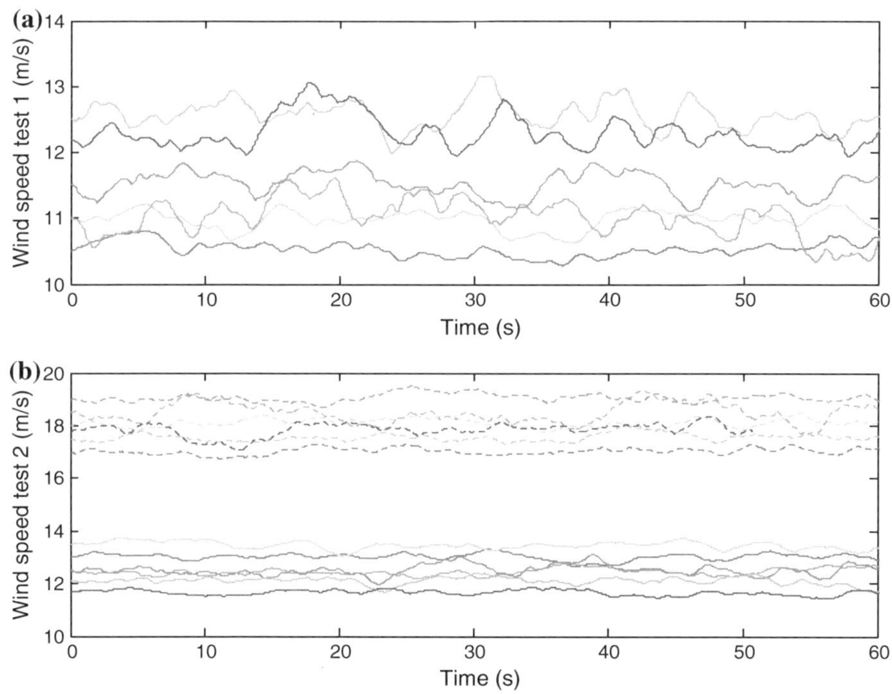

Fig. 8 Wind speeds incident on the wind turbines: **a** test 1, **b** test 2

This variable speed wind park is evaluated under wind speed fluctuations with different incoming wind between the wind turbines, as shown in test 1 of the Fig. 8, where the most of the incoming wind speeds are near to the rated wind speed.

In this case, three models of the wind park presented in Fig. 7 are implemented:

- *Complete model* Composed of all the wind turbine models of the park and the model of the internal grid of the wind park.
- A *single equivalent model*, as presented in [14] A re-scaled wind turbine that experiences an equivalent wind speed obtained from the power curve of the individuals.
- *A compound equivalent model*, as presented in [15] An equivalent wind turbine with an aggregated generation system model and a simplified model of each individual wind turbine approximating the operational points of each one according to its incoming wind speed.

The evaluation of these models is performed by comparing their active and reactive power responses for the incoming wind speeds of test 1 and considering that all the wind turbines are operating with unity power factor and rated active power reference.

Simulation results, presented in Fig. 9, show a high correspondence between the response of the complete and equivalent wind turbines, with few discrepancies in the compound equivalent model due to the approximation of the operational point of every wind turbine achieved by its simplified model and the adjust of the controller in the equivalent generation system.

2.3.2 Case 2: Normal Operation in a Fixed Speed Wind Park with Two Different Wind Turbines

In this case, the wind park under consideration is shown in Fig. 1. It presents a radial structure with six 350 kW wind turbines organized in three branch with two wind turbines for each one, and others six wind turbines of 500 kW of rated power with the same organization.

Twelve wind speed time series are used to evaluate the behaviour of the fixed-speed wind park under wind fluctuation, as shown in test 2 of Fig. 8. The 350 KW wind turbines experience different wind speeds between 11.7 and 13.5 m/s (under rated wind speed) and the 500 kW wind turbines experience wind speeds between 17 and 19 m/s (above rated wind speed).

As in the case 1, three models of this fixed speed wind park are implemented:

- *Complete model* Representing all the individual wind turbines.
- *A compound equivalent model* A simplified model of each wind turbine, as presented in [14], is used to aggregate the mechanical torque as incoming to an equivalent generation system, which is obtained from the aggregation of all the individual induction generators, applying the method proposed in [18].
- *A cluster equivalent wind model* It presents the same model than that proposed in [16], but with two clusters composed each one by a single equivalent model of six wind turbines.

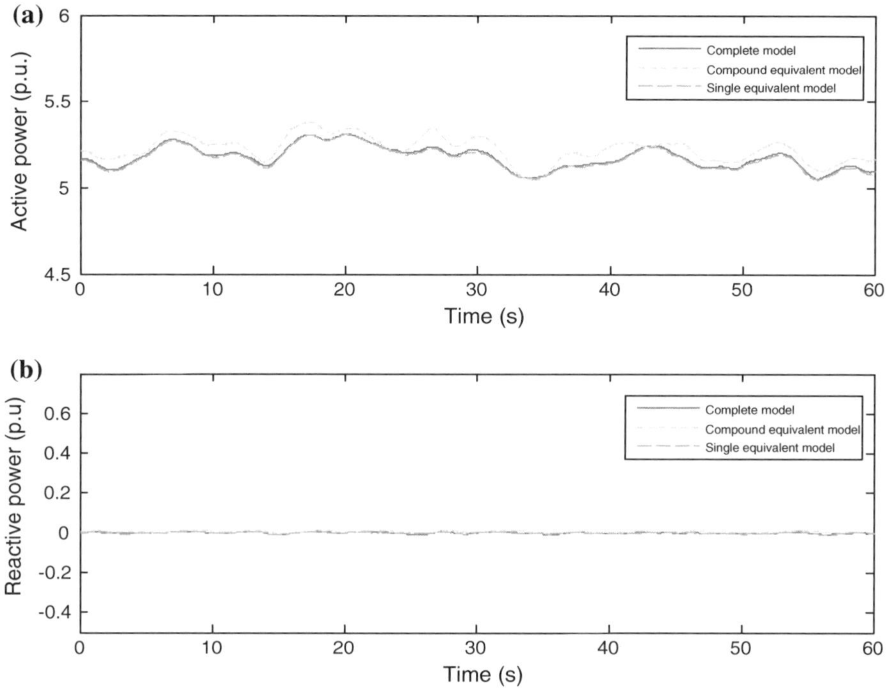

Fig. 9 Dynamic simulation results of complete and equivalent wind parks during normal operation in a variable speed wind park with identical wind turbines: **a** active power, **b** reactive power

The comparison of these models is shown in Fig. 10. It is performed assuming that the compensating capacitor banks of every wind turbine are designed to achieve unity power factor (zero reactive power) with rated incoming wind speed.

Analysing the simulation results presented in Fig. 10, it can be concluded that the equivalent models allow an accurate approximation of the dynamic response of the wind park with two different wind turbines. The best results are obtained by the equivalent wind park model without aggregation of the mechanical system (compound equivalent model). The equivalent wind model, organized in two clusters as shown in Fig. 2, presents worst results, because a worst approximation of mechanical torque obtained from the equivalent wind that experiences the single equivalent re-scaled model of the cluster 1, where all the wind turbines receive incoming wind speeds above the rated value.

As summary, regarding the simulation time, an approximate reduction of 93 % is obtained for the cluster equivalent wind model and 72 % for the compound equivalent model of the wind park shown in Fig. 1.

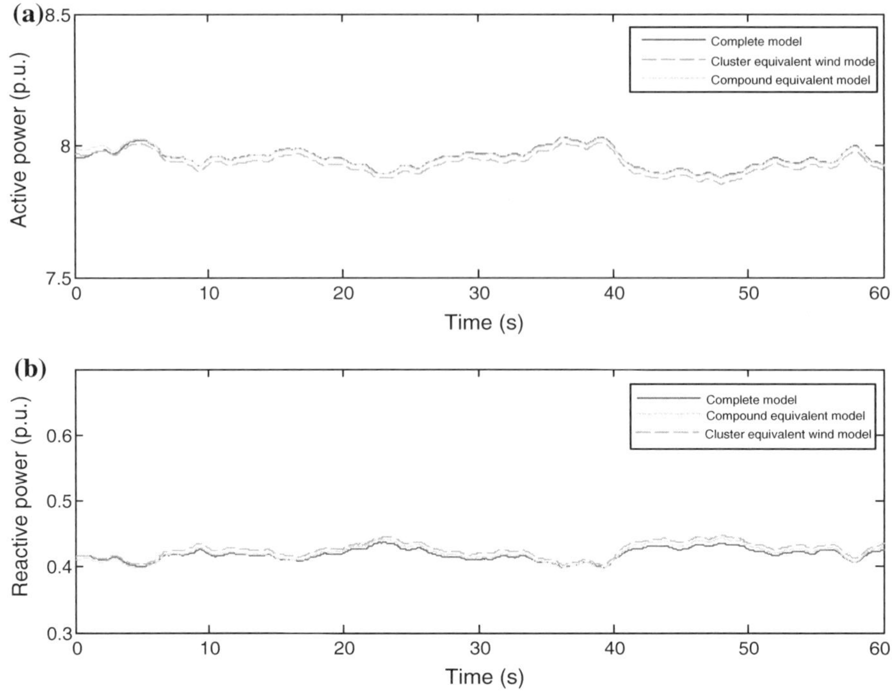

Fig. 10 Dynamic simulation results of complete and equivalent wind park during normal operation in a fixed speed wind park with identical wind turbines: **a** active power, **b** reactive power

2.3.3 Case 3: Abnormal Operation (Response to Voltage Dip) in a Variable Speed Wind Park with Two Different Wind Turbines

The equivalent models are evaluated during grid disturbances for two variable speed wind parks connected to the same feeder, the first of them with 6 wind turbines of 660 kW and the other with 6 wind turbines of 2 MW of rated power, as depicted in Fig. 11.

Figure 11a shows the radial structure of the wind parks, while Fig. 11b shows the impedances of the internal network. In this figure, Z_{lmij} is short-circuit impedance of the low/medium voltage transformer of the wind turbine j of the branch i; Z_{mij} is the medium voltage line impedance of wind turbine j of the branch i; Z_{mi} is the medium voltage line impedance of the i branch; Z_{mhk} is the short-circuit impedance of the medium/high transformer of the wind park k; and Z_f is the feeder impedance of the wind parks.

To verify adequately the robustness of the equivalent models, a voltage dip at the PCC is considered, with 0.8 p.u. of depth and a width of 1 s. As usual in short term stability simulations, constant incoming winds are assumed, with different

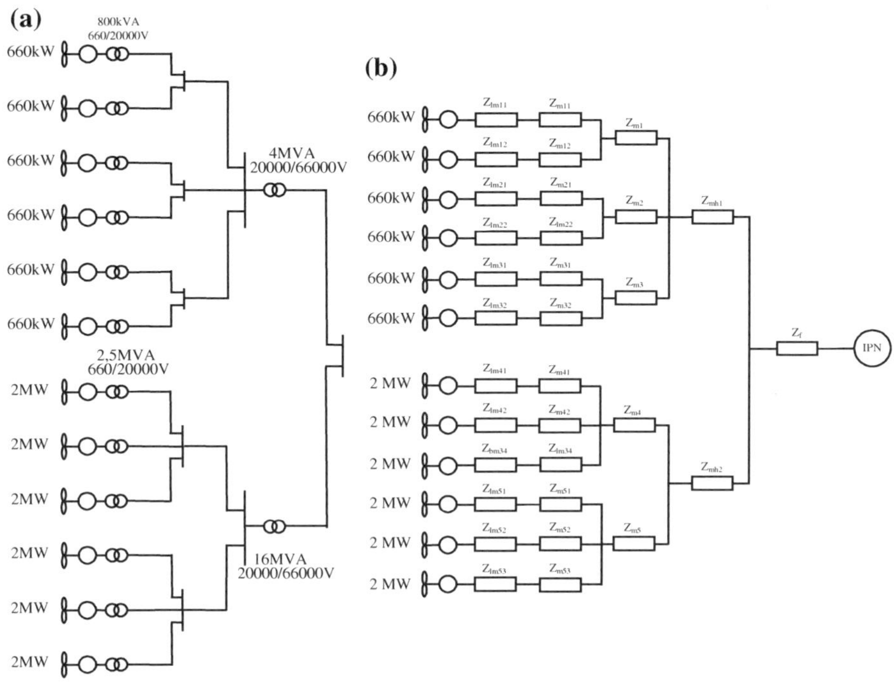

Fig. 11 Variable speed wind parks: **a** structure, **b** equivalent circuit of the internal network

values above the rated wind speed for both types of wind turbines. Figure 12 shows the power response during the slow voltage dip. The simulation results of the complete and equivalent models show a high correspondence, although a little faster and less damped in the equivalent models.

Figure 13 shows the voltage at the PCC to grid and the wind turbine voltage at every wind turbine of the complete wind park and at the aggregated wind turbine of each equivalent model. In this case, the voltage recovery in the compound model is slower than that obtained with the equivalent model, where only one re-scaled wind turbine represents the whole wind park.

3 Wind Park Control System

Until relatively recently, wind parks have commonly operated delivering the available energy to the grid. Hence, the wind turbines maximized the energy captured from the wind, without exceeding generator limits and operating with unity power factor (zero reactive power). However, the wind parks production has been voluntarily reduced by disconnecting wind turbines from the grid at special conditions (especially during low consumption periods or strong winds). During

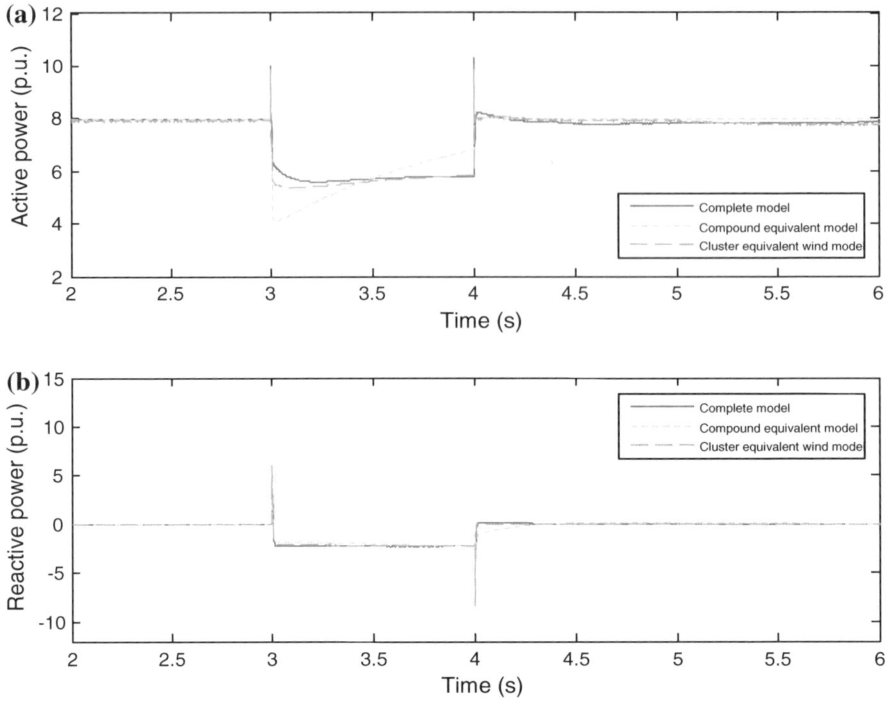

Fig. 12 Active and reactive powers of the complete and equivalent models of the wind park for a slow voltage dip

these conditions, system operators recommended the reduction of wind parks production in order to maintain the stability and reliability of the power systems with high wind power penetration.

The increase of wind power penetration on power systems has led to partial substitution of conventional power plants by current wind parks. Therefore, large wind parks are today required to participate actively in the power system operation as conventional power plants. Thus, the system operators have revised the grid connection requirements for wind parks, demanding an operational behaviour with several control tasks similar to those of conventional power plants [28]. The requirements encountered in the majority of grid codes, concerning to the wind park interconnection, including [51, 58]:

- *Fault ride-through* Wind parks connected to transmission networks must withstand voltage dips to a certain percentage of the nominal voltage and for a specified duration. Some of the grid codes prescribe that wind parks should support the grid by generating reactive power during a network fault, to support and faster restore the grid voltage.

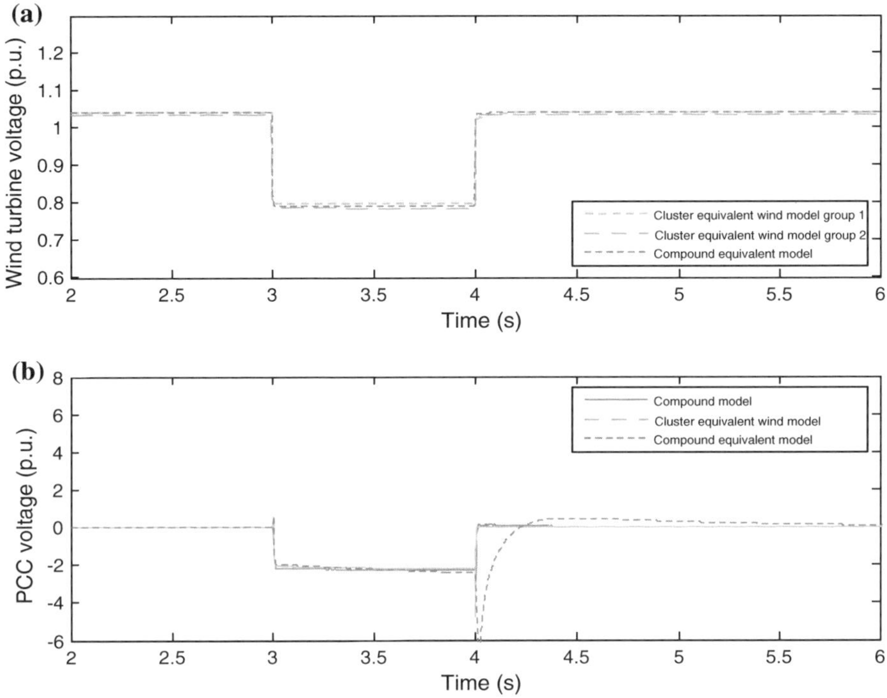

Fig. 13 Voltage at the PCC to grid and wind turbine voltage of the complete and equivalent models for a slow voltage dip

- *System voltage and frequency limits* Wind parks must be capable of operating continuously within the voltage and frequency variation limits encountered in normal operation of the system.
- *Active power regulation and frequency control* Wind parks should control their active power output to a defined level, either by disconnecting turbines or by pitch control action. In addition wind parks are required to provide frequency response, regulating their active output power according to the frequency deviations.
- *Reactive power/power factor/voltage regulation* Some grid codes demand from wind parks to provide reactive output regulation, often in response to power system voltage variations, like conventional power plants.

A main consideration to operational behaviour of large wind park is the technology of the generation system of the wind turbines, affecting the influence on the grid [20]:

- *Fixed speed wind turbines* These wind turbines are based on an induction generator directly connected to grid (Danish concept). In passive stall application, the blades are kept at a fixed angle, while in active or semi-active stall the control is used to compensate for wind variations. The induction generator draws reactive power, which might be supplied from the grid or from installed compensation equipments (capacitor banks or FACTS, Flexible AC Transmission Systems). The dynamic response and controllability are therefore poor compared with variable speed wind parks.

- *Variable speed wind turbines* The generation system is composed of a variable speed generator, that have the ability to control both the active and reactive power delivered to the grid, optimizing the grid integration with respect to steady state conditions, power quality, voltage and angular stability [34]. Two main types of generation system can be found:

– *The variable speed wind turbine with Doubly Fed Induction Generator (DFIG)* The generation system can be controlled with a back-to-back converter in the rotor. It presents some difficulties to ride-through voltage dips, and the rotor converter rating (typically 30 % of generator rated power) determines the frequency response.
– *The variable speed wind turbine with Permanent Magnetic Synchronous Generator (PMSG)* It is connected through a full back to back converter that totally decouples the generator from de grid. This provides maximum flexibility, enabling full real and reactive power control and fault ride-through capability during voltage dips.

A second consideration to their operational behaviour is how the internal network of the wind parks is connected to the grid, especially for long distance power transmission (e.g. offshore wind park), where High Voltage Direct Current (HVDC) may be an interesting option. In this case, a common or a separated AC/DC power converter system converts the AC voltage into a DC voltage and transferred to a DC/AC power converter system that are connected to the grid [10]. This power converter decouples the wind turbines from the grid and can be used for controlling the active and reactive power delivered to the grid and improving their dynamic response.

3.1 Active Power Regulation and Frequency Control

The generated power in the power system must be in balance with the demand from the loads and losses in the system [50]. The active power output of the generators is determined by the mechanical power input from their prime movers. The consequence of a mismatch between the generation and the demand for active power is a change in the rotational energy stored in the rotating mass of the generator, and hence, a drift in the system frequency [34]. Therefore, the power balance in the system can be obtained by controlling the rotational speed of the generator (primary control), and in case of conventional power plants, there is a proportionality between the speed of the generators (prime movers) and the power

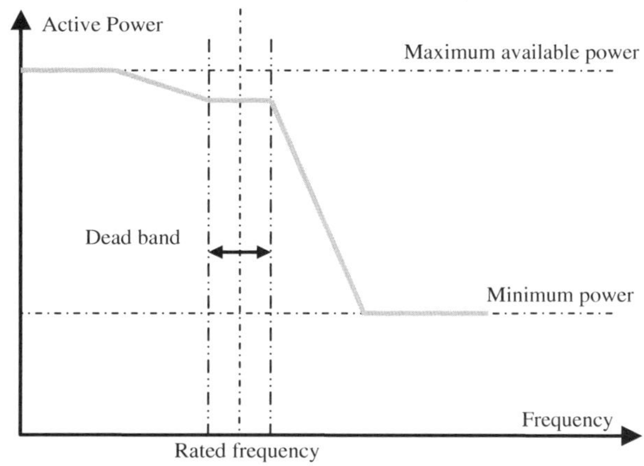

Fig. 14 Typical frequency controlled regulation of active power

system frequency. This power balance requires an operating reserve 'spinning reserve' to control those faster fluctuations and a 'supplementary reserve' to cover slower power fluctuations.

With these considerations, frequency regulation ancillary service may consider three functions:

- *Primary frequency control*, limits variations caused by sudden power unbalance in the grid. It is performed locally by the speed governor of each generation system.
- *Secondary frequency control*, restoring frequency and interchanges power to their scheduled values. It is performed by an automatic central controller.
- *Tertiary control*, supporting the secondary control and re-establish the power reserve consumed by it.

Wind power plants, as conventional power plants, must operate within strictly frequency margins. They should change their production in agree with system operators requirements to maintain frequency limits. A typical characteristic of primary frequency control is shown in Fig. 14 [34].

This typical frequency control curve shows that the frequency can be limited as function of the available power. Thus, high-frequency response can be provided from full output to a reduced output when the frequency exceeds the dead band of the rated power and the grid codes require that, when the frequency increases above the rated value, power plants should decrease their output at a given rated. On the other hand, at nominal frequency, the wind parks would be required to limit their power output below the maximum achievable power level, and so, if the frequency starts to drop, the wind park would increase the power output to the maximum achievable power, trying to sustain the frequency. As summary, a wind park should be capable of providing frequency response, and it would be used by the system operator. The active power levels of the wind park are determined by the system operator, according to the system demand.

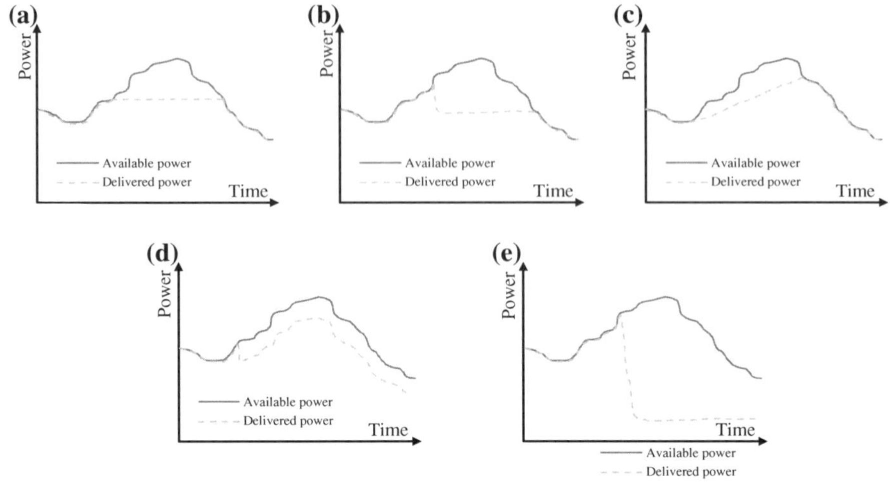

Fig. 15 Different power control possibilities: **a** absolute power limitation, **b** balance control **c** power rate limitation, **d** delta control, **e** system protection

The active power output of a wind park can be determined by the mechanical input power of the wind turbines and, thus, by the wind energy, a fluctuating source. In case of large wind parks, significant power fluctuations, caused by the uncontrollability of the wind source, may affect the grid if it is not appropriately controlled by the system operators. Hence, large wind parks may participate in grid management as a conventional power plant, including provision of regulating power, generation management and power reserve. Thus, some types of power control possibilities are necessary to use the wind park as a wind power plant supporting the power balance [32]:

- *Absolute power limitation* As seen in Fig. 15a, the delivered power output of the wind park will never exceed a pre-set maximum.
- *Balance control* In this case, the wind park must be able to change quickly its power as shown in Fig. 15b. It is helpful in order to balance the production and consumption of active power in the grid.
- *Power rate limitation* This possibility avoids fast gradients of output power in the wind park with regard to slower gradient of conventional power plants to keep the power balance. Figure 15c shows increase rate limitation.
- *Delta control* The wind park runs as a spinning reserve, with a power output below available in an adjustable margin. Figure 15d shows this type of control that allows the wind park to take part in the frequency control.
- *System protection*, such as the drastic reduction of the output power in the wind park when there are overloads in the grid, shown in Fig. 15e.

3.2 Reactive Power Control and Voltage Regulation

In power systems, the voltage is controlled by the power flow from the power plants, where the reactive power exchange controls the magnitude of the voltage, whereas the phase difference is related to the active power exchanging. Then, the active and reactive powers flow between the generation and the load in the power system must be balanced in order to avoid large voltage (and frequency) excursions.

Grids codes require that a minimum of the reactive power from a large wind park can be controlled to a specific interval, which is close to unity power factor. This requirement is usually specified with a limiting curve, as depicted in Fig. 16 [34]. If the wind power plant provides low active power, the power factor may deviate from unity because it can support additional leading or lagging currents due to the reactive power demanded. When the wind park is working under nominal conditions, the power factor must be kept close to unity or else it will be excessive currents. In addition, local reactive power generation reduces losses in the system.

The dependence between reactive power and voltage can be expressed by a voltage control, in a similar way as the frequency control shown in Fig. 14, integrating a combined droop and dead band control. But, in this case, reactive power is exchanged to the grid in order to compensate for the deviations in grid voltage.

3.3 Control Structure of Wind Parks

The basic structure of a wind park main controller is illustrated in Fig. 17. The aim of the wind park controller is to regulate, in a centralized way, the active and reactive powers injected by the whole wind park into the grid (when the it operates as a PQ node) or the active power and the voltage at the PCC (when it operates as a PV node). In addition, in some circumstances, the wind park controller has to adjust frequency margins.

As shown in Fig. 17, the Wind Park Main Controller (WPMC) receives set point commands in agree with the required operation and the wind park current status (system operators receive monitoring signals of the status of wind parks). The WPMC computes the incoming reference settings, the wind turbines status and delivers the wind park operation references that assure a correct production. Finally, a dispatch centre must distribute the operation references between the wind turbines.

The main objective of the WPMC is to assure the wind park production in a similar way as a conventional power plant. Although the WPMC performance depends on the wind turbine technology, the basic objective is the same: it controls the power production of the wind park (active and reactive power) considering

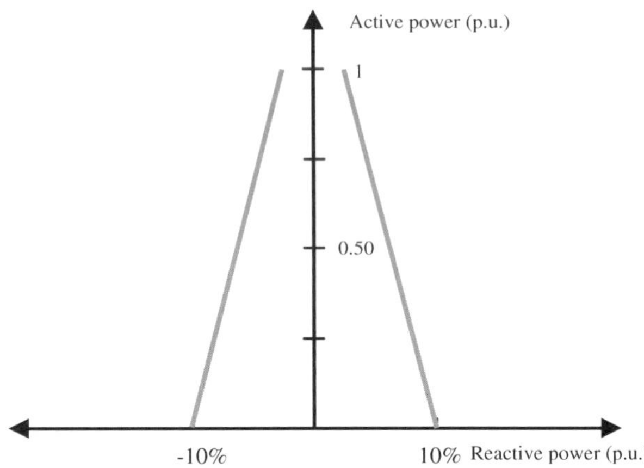

Fig. 16 Typical reactive power limiting curve

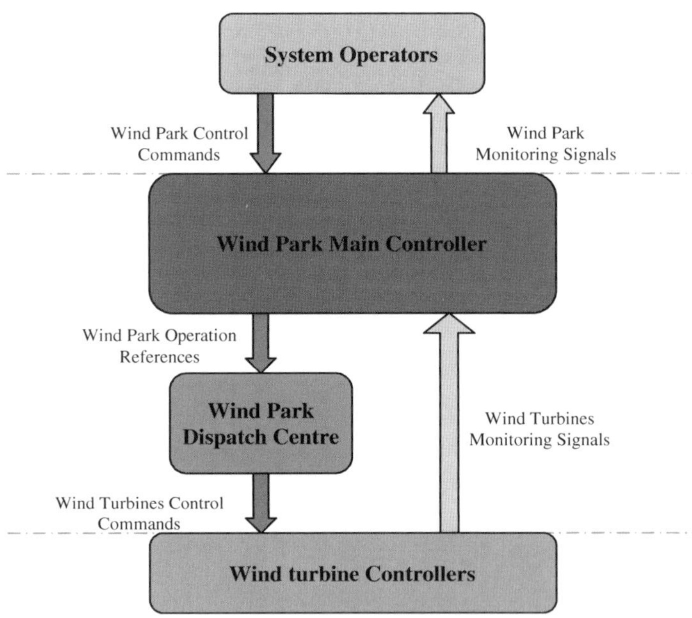

Fig. 17 Basic structure of the wind park main controller

some of the control functions described in Sect. 4.1 for the active power and for reactive power in Sect. 4.2. In addition, large wind parks must be able to provide advanced grid support (frequency and voltage control).

The most typical WPMC structure is composed of two separated PI control loops, one for the active power control (frequency control) and another for the reactive control (voltage control). This structure is used in [24] for a variable speed wind park composed of three 2 MW DFIG wind turbines, including automatic frequency control and voltage control function and focused on the wind park's capability to regulate active and reactive power production. In the actual WPMC, the active power control with balance control, delta control, ramp limitation and the reactive power control are implemented.

In a similar way, Rodríguez-Amenedo et al. [44] present two PI controllers (active power control loop and reactive power control loop) for a variable speed wind park with 37 DFIG wind turbines of 850 kW rated power. In addition, it presents the WPMC for a fixed speed wind park composed of 21 stall regulated wind turbines of 900 kW rated power. The results show that the controller performance depends on the wind turbine technology. Each fixed speed wind turbine is provided with a low voltage variable capacitor bank of a total rated of 450 kVAr at its terminals. These capacitor banks are coordinated with an under-load tap changing (ULTC) booster transformer located at the wind park substation, and an additional medium voltage capacitor bank of 6.2 MVAr.

The active power control is based in a wind turbine disconnection strategy, with a PI control loop and a hysteresis band controller to obtain the connection or disconnection order to each wind turbine. Furthermore, a PI regulator is responsible for maintaining the medium voltage level and two PI regulators control the reactive power generated by the capacitor banks at the terminals of each induction generator in order to compensate the power factor (capacitor banks of each wind turbine) and reach the wind park reactive power reference (wind park capacitor bank).

The WPMC structure presented above, with PI controllers in two separated control loops, is showed in Fig. 18 [13]. In this figure, the active power control loop is based on a PI controller that assures the wind park production according to the set point ordered by the system operator ($P_{wp\_so}^*$). This controller computes the active power error and set up the power reference (P_{wp}^*) for the whole wind park. The other control loops presents a switch to select the reactive power control or the node voltage control. Thus, the wind park can operates as a PV node, trying to adjust the voltage (V_{wp}), at the wind park node, to the voltage reference ordered by the system operator ($V_{wp\_so}^*$). This control is performed by a PI controller that sets up the reactive power reference ($Q_{wp\_vc}^*$). On the other hand, the reactive power reference ($Q_{wp\_so}^*$) is obtained from the system operator. In both cases, a PI controller sets up the reactive power reference (Q_{wp}^*) for the whole wind park.

This WPMC is implemented for the wind park shown in Fig. 19, consisting of three 2 MW rated power DFIG wind turbines. Each wind turbine is connected to the internal network through a transformer of 2.5 MVA, and the wind park connection to the grid is performed through a substation with a transformer of 8 MVA and a long feeder.

As can be seen in Fig. 17, a wind park dispatch centre is needed to compute the active and reactive power references. Different strategies for the dispatch function can be used:

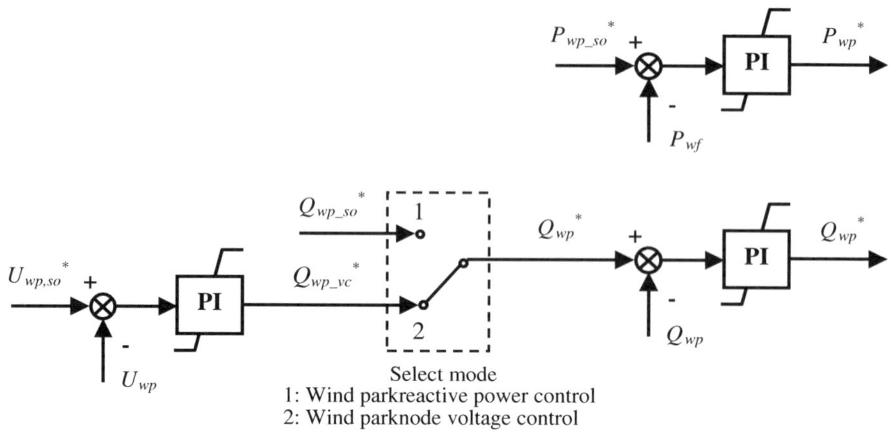

Fig. 18 Wind park main control structure

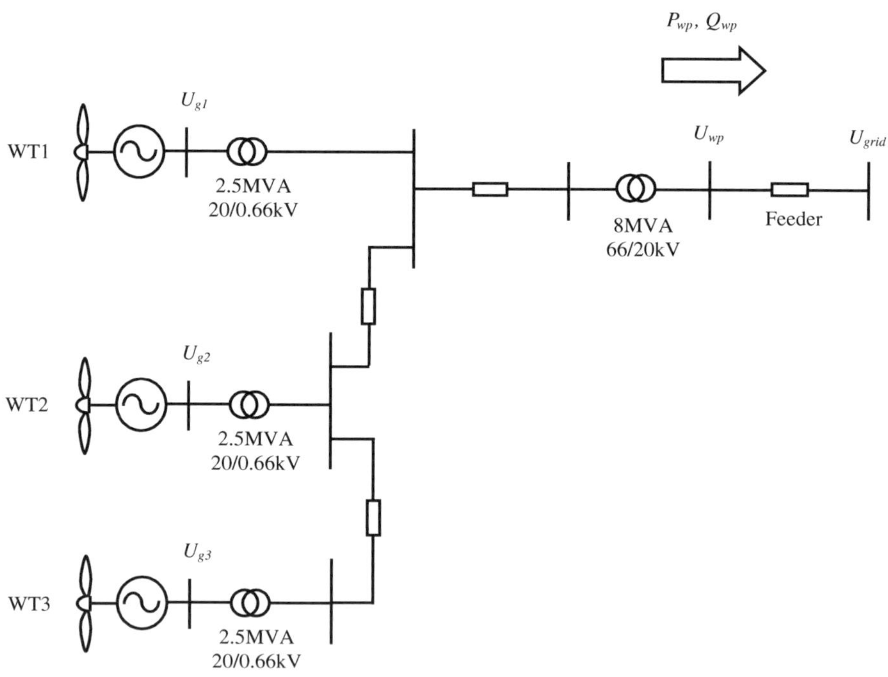

Fig. 19 Variable speed wind park layout [13]

- The simplest strategy is based on computing the same power references for each wind turbine [44]:

$$P^*_{gi} = \frac{P^*_{wp}}{n} \quad Q^*_{gi} = \frac{Q^*_{wp}}{n} \tag{2}$$

- A more efficient strategy is proposed in [24], where the power references are defined from a proportional distribution of the available active and reactive powers:

$$P^*_{gi} = \frac{P^*_{ava\_gi}}{\sum_{i=1}^{n} P^*_{ava\_gi}} P^*_{wp} \quad Q^*_{gi} = \frac{Q^*_{ava\_gi}}{\sum_{i=1}^{n} Q^*_{ava\_gi}} Q^*_{wp} \tag{3}$$

where $P^*_{ava\_gi}$ is the available active power for the *i*th wind turbine in one specific moment, calculated from the power-speed control curve of the wind turbine; and $Q^*_{ava\_gi}$ is the available reactive power for the *i*th wind turbine computed from the operating Q–P curve of the wind turbine.

- An optimized dispatch control strategy for optimizing the active and reactive power references of each wind turbine is presented in [36]. In this work, an optimal power flow algorithm is adopted at the wind park control level to define the active and reactive set points, considering the generator terminal voltage and other operational constrains, such that, a minimization of the mismatch between the total wind park generation output (active and reactive) and wind park dispatch centre request is achieved.

The dispatch control strategy applied to the WPMC of the wind park showed in Fig. 19 is based on the second option, a proportional distribution of the available active and reactive powers. In addition, the wind turbines can autonomously operate with power optimization or power limitation (the active power reference of the wind turbine is set up at its rated value), and only when down power regulation is required, the power reference is changed to the value defined by the dispatch centre control.

An adequate wind speeds time series, showed in Fig. 20, allow the evaluation of the WPMC in any operating conditions (power optimization, power limitation or down power regulation).

The performance of the described control system is assessed through two different simulations:

- *Wind park operating as a PQ node* (Fig. 21) The capabilities of the WPMC to regulate the wind park production according to the power references ordered by the system operator are tested. In this simulation, the wind park produces the maximum possible output power during the first 60 s (power optimization). At this time, the wind park receives a 60 % reduction of the active power reference by a slope of 0.1 (power down regulation and rated limitation). Finally, a change in the reactive power reference is ordered at 80 s. The wind turbine changes from unity power factor operation to maximum reactive power generation, with a slope of 0.1 (reactive power maximizing).

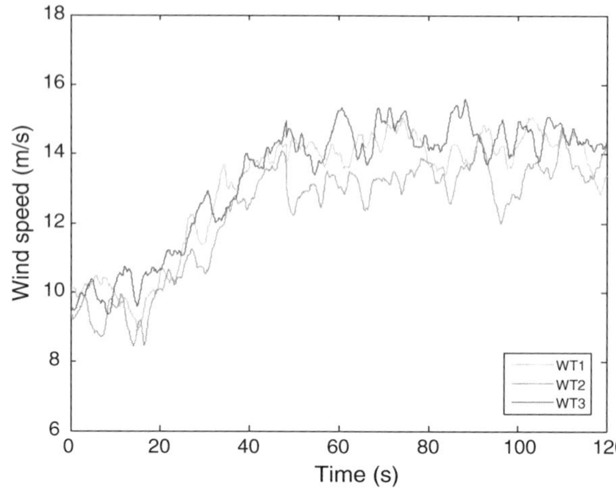

Fig. 20 Incoming wind speed time series on the wind turbines [13]

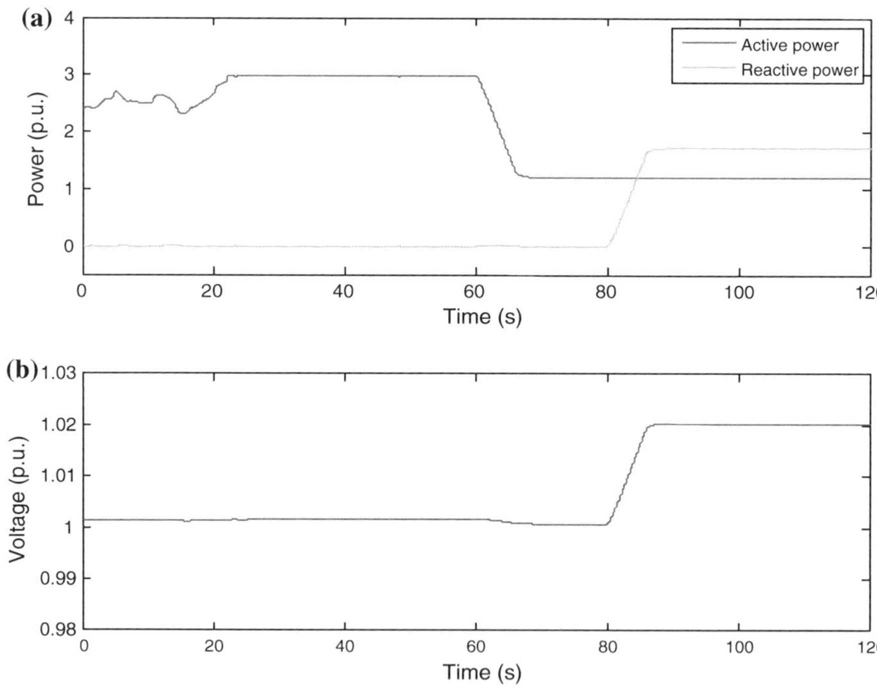

Fig. 21 Response of the wind park operating as a PQ node

- *Wind park operating as a PV node* (Fig. 22) To evaluate the WPMC when the wind park production is regulated, a second case is simulated, in which the active power and voltage at the wind park node are controlled according to

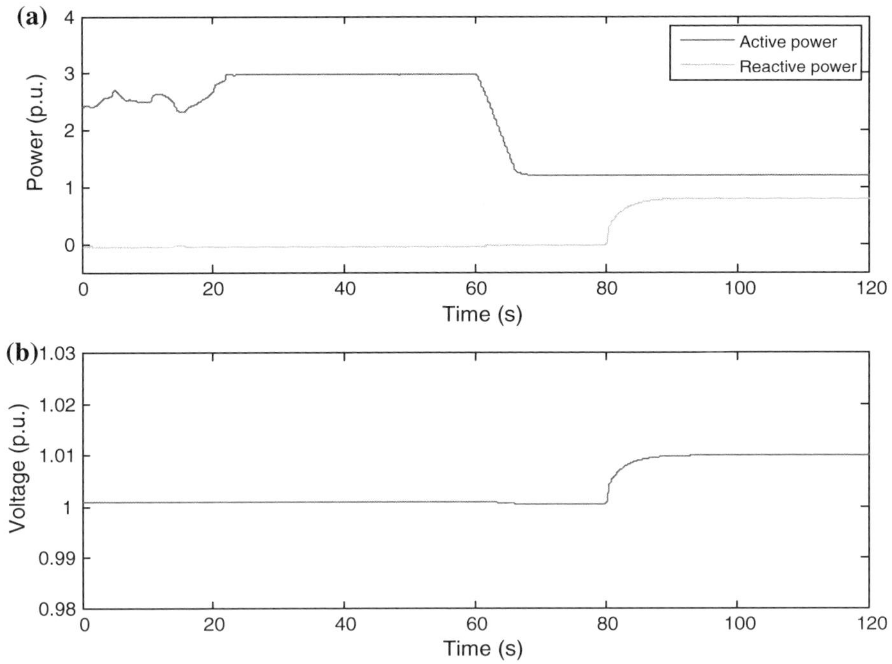

Fig. 22 Response of the wind park operating as a PV node

the references ordered by the power system operator. In this simulation, the wind park operates with the same active power references as in the first case. However, the voltage reference at the wind park node is set to 1 p.u. during the first 80 s and 1.01 p.u. for the rest of the simulation.

More complex WPMC structures can be used when the scope of the researches is more than the behaviour of a wind park as a conventional power plant, considering powers and voltage controls. In these cases, the wind parks may contribute to the electrical network stability, maintaining the network frequency or contributing to network stability, operating the wind parks as compensator devices. Fernandez et al. [17] consider this philosophy and they propose control laws based on energy function. These control laws are independent of operating points, and linearization techniques are used to assure the contribution of DFIG variable speed wind parks over a wide range of different working conditions. They proposed steady state controls (normal operating conditions) plus corrections looking for contributing to the network stability:

$$P_{wp} = P^{*}_{wp} + \Delta P_{st} \quad Q_{wp} = K_{QV} \cdot \Delta V + \Delta Q_{st} \tag{4}$$

The total active power of the wind park P_{wp} is obtained from the power reference, given by a supervisory control (P^{*}_{wp}), and the correction which contributes to the network stability (ΔP_{st}). Meanwhile, the reactive power is controlled by considering the voltage profile with the voltage gain K_{QV} for contributing to the energy quality and the correction ΔQ_{st} for collaborating with the electromechanical oscillations. These non-linear active and reactive power corrections are obtained by forcing the incremental energy function of the network to decrease.

Some authors have focused their studies on improving the voltage control at a specified location, without using additional compensating devices. Ko et al. [31] present a voltage control scheme for a DFIG variable speed wind park, using a control-design technique, known as the 'linear quadratic regulator', and a reduced-order linear approach of the wind turbines. These linear models force a reformulation of the problem in order to find a common Lyapunov function for the set of considered linear systems. These are accomplished by representing the underlying control optimization problem in terms of a system of linear-matrix-inequality constraints and matrix equation to be simultaneously solved to ensure a robust and reliable operation of the linear quadratic regulator, in case of a three-phase symmetrical fault in the transmission line of a variable speed wind park composed of 3 DFIG wind turbines of 3.6 MW rated power.

Others authors focus their researches on reactive power control generated by variable speed wind parks. Zhao et al. [64] present a reactive power control and distribution network reconfiguration to reduce power losses and improve voltage profile. They use a joint optimization algorithm to obtain the optimal reactive power output of a wind park and network reconfiguration simultaneously. To find the optimal reactive power output of a wind park, they apply an improved hybrid particle swarm optimization with wavelet mutation algorithm. Furthermore, they develop a binary particle swarm optimization algorithm to find the optimal network structure of the wind park. On the other hand, Tapia et al. [56] study the behaviour of a wind park as a continuous reactive power source. They devised a PI-based control strategy to manage the net reactive power exchanged between the grid and DFIG variable speed wind parks. They present experimental results of the proposed strategy when applied to a wind park comprising 33 wind turbines of 660 kW rated power.

Control systems for large offshore wind parks, with high voltage DC (HVDC) connection to the main onshore network, needs a special consideration. A typical offshore wind park layout is shown in Fig. 23, where each wind turbine is connected through a small transformer to a common AC grid, which finishes in a power electronic rectifier to transform the AC voltage to a DC link. At the end of this link, an output power electronic inverter controls the power flow over the AC transmission network. Another layout is also possible when the power generated by every wind turbine is rectified and connected to the DC bus using a DC/DC converter.

As considered in [61], a wind park with HVDC connection to the AC network results in a flexible installation, with the capability to control the active and reactive power in a wide range. The four-quadrant controllability of HVDC is similar to an electrical machine without inertia, but improving system dynamics. However, there are some limitations over the amount of the active and reactive power to be injected. From the AC transmission network point of view, the main limitation is the impedance, seen from the point of the common coupling where the

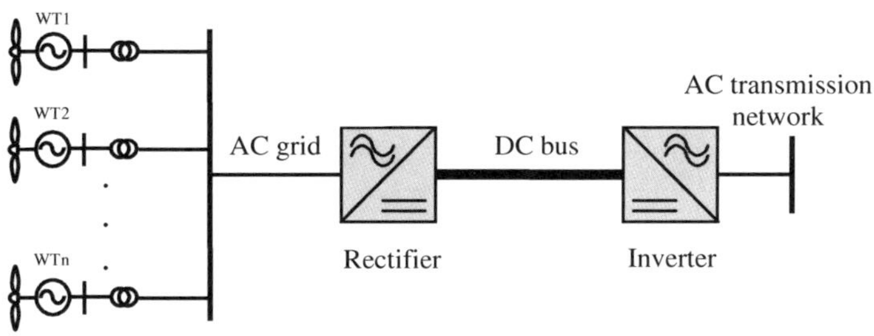

Fig. 23 HVDC wind park layout

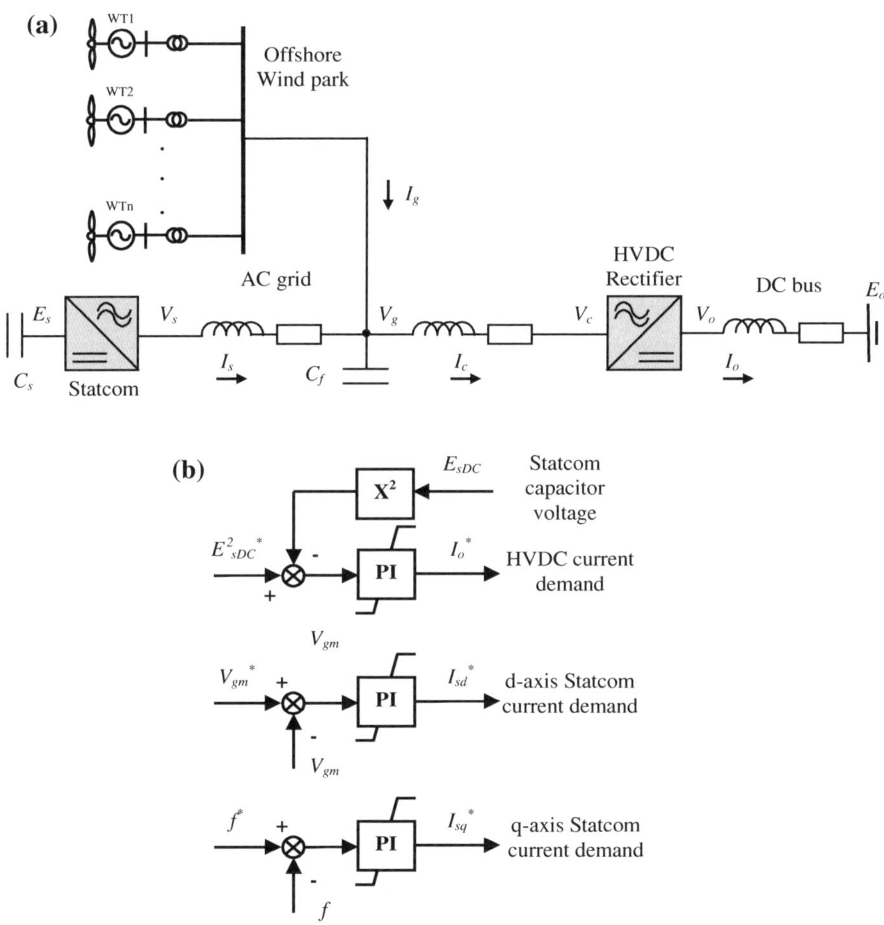

Fig. 24 **a** Simplified scheme, and **b** control structure of the system studied in [8]

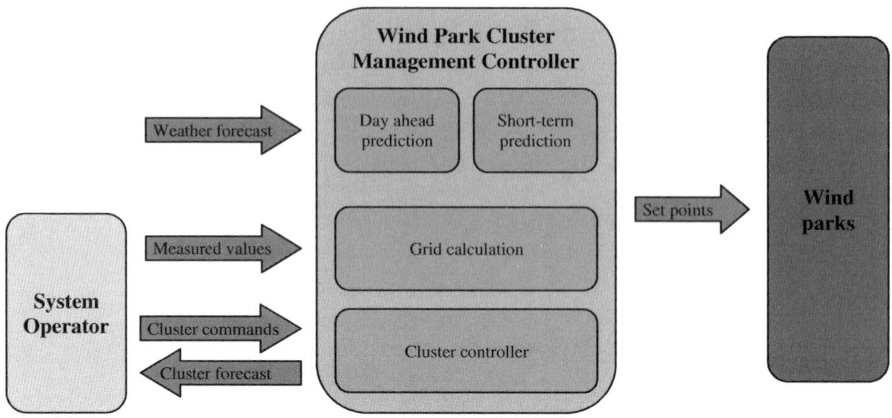

Fig. 25 Modular structures of a Wind Park Cluster Management System

wind park is installed, because, with high grid impedance, less full-regulation area of the voltage is feasible.

Bozhko et al [8] study the grid integration of large offshore DFIG wind parks with a common collection bus controlled by a STATCOM. Figure 24a shows a simplified scheme of the system to be controlled in order to regulate the voltage and frequency of the offshore AC bus, while the power flow is regulated controlling the DC bus. In this scheme, the capacitive filters compensate for the HVDC converter reactive absorption in lumped amounts, while the STATCOM provides fine reactive power control and commutation voltage during disturbances, in collaboration with the DFIG wind turbines. It can be seen in Fig. 24b the basic control structure of the system, where the DC bus current demand (I_o^*) is derived from the dc-link STATCOM voltage controller, while the *dq*-axis current demands (I_{sd}^*, I_{sq}^*) are derived from the grid voltage vector controllers (V_{gm}^*, f^*). The controlled plant is shown to be a third-order system linearized about the operating rectified firing angle, and decoupled through the use of appropriate feed forward terms.

An evolution of the WPMC can be found in [63], an initiative supported by the European Union and focused on preparation of the European electricity network for the large-scale integration of wind parks through the design, development and validation of new tools and devices for its planning, control and operation in a competitive market. This consortium has created a new concept: the Wind Park Cluster Management System (WCMS). The WCMS consists of the implementation of advanced techniques and control strategies, combined with high-tech wind energy technologies, to provide the system operators the needed tools to control and manipulate wind energy as a conventional power source by a logical aggregating (cluster) of existing wind parks, which are connected to a certain grid node.

One of the main characteristics of the WCMS architecture is to consider the existence of different companies, with different aims and business issues, relations between them and needs concerning a reliable wind energy management. In fact, some grid codes have included in their requirements that, when several wind parks are connected in a PCC of the transmission network, there must be an intermediate between the Transmission System Operator (TSO) and the wind parks owners. Thus, WCMS considers two operation layers:

- The TSO layer, who is mainly responsible for grid security issues, and
- The dispatch centres of the companies, which are in charge of the relation between the wind parks and the TSO.

In normal conditions, the TSO sends set points to all the dispatch centres, which are controlling all wind parks of a cluster. It is also required that the information data flows in the opposite way, so that the TSO receives all available power production data of the controlled wind parks into the cluster, in order to know in real-time the operational status of the cluster. Figure 25 describes the WCMS modular structure.

The WCMS includes a grid calculation module which calculates the required power flow of the cluster grid model in order to determine the capability of active and reactive powers at the PCC of the cluster. In addition, the dispatch set points for the wind parks are calculated in this module. Furthermore, the WCMS grid calculation module allows the development of a reactive power management tool to control not only the reactive power, but the power factor and voltage changes caused by wind power generation.

To achieve the controllability of wind power at a transmission cluster level, it is necessary to determine the capability of active and reactive power (P/Q availability) at this point. The active power capability is determined based upon the technical capabilities of the wind turbines within the cluster. The reactive power capability is described through the P/Q characteristic of every wind turbine. Once the P/Q profile of the wind park level is determined, it is necessary to perform grid calculations in order to determine the active losses and reactive power consumption of the wind park grid, and obtain the P/Q capability at the cluster level. Based upon this information, grid operators are able to perform internal grid calculation to set the active and reactive power requirements of the cluster for the next time interval.

Once the set points from the TSO are issued and they lies within the forecasted P/Q capabilities, the WCMS must perform a dispatch calculation procedure to obtain the clusters active and reactive power set points in agree with technical requirements and capabilities.

4 Wind Parks with Special Devices for Enhancing Their Impact into the Grid

Normally, wind power plants have less rated power than the conventional ones [7], and thus, they supply only a fraction of the total power demand, taking part of the called Distributed Generation (DG) of the power system. In some countries, penetration levels of DG have increased in the last years not only to reduce CO_2

emissions but also an interesting economic alternative in areas with appropriate wind resources [3]. Thus, the impact on the grids become more significant, that has conducted to more severe network requirements.

The problems and constrains encountered when a wind power plant is connected to the grid are similar to the one of the conventional power plants, plus those related to the stochastic nature of wind:

- *Grid Capacity Constraints* Such as steady state thermal constraints and network congestion, short-circuit powers and currents or steady state voltage profile [7].
- *Power Quality Issues* [27] Voltage fluctuations and flicker respect to impact on voltage quality, harmonics emitted by power electronic converters and disturbances of remote control signals [38].
- *Protection Issues:* Sensitivity and selectivity of the protection scheme.
- *Dynamic behaviour and stability* [11]: Capability to be able to ride through voltage dips or faults.
- *Ancillary Services Issues* [53] Voltage control-reactive power compensation and power fluctuations-frequency control.

Some of these problems are typical of weak grids, and the solutions are simple but often expensive grid reinforcement. Nevertheless some constrains require system flexibility and application of monitoring and control systems, and depends on the wind park capabilities, with the agreement of the network operators.

Wind park capabilities depend directly on wind turbines [33], so that challenges in grid requirements have forced the development trends of wind generator systems. The simply and robust fixed speed wind turbine known as the "Danish concept" is clearly in decreasing [23]. Today, variable speed concepts, where the power electronics play an important role, are widely used. DFIG wind turbines with partial-scale power converter and direct-drive PMSG wind turbines with a full-scale power converter are the most used technologies, since they enhance performance and controllability but with interaction in power quality [53]. Thus, the main characteristics of fixed speed wind turbines are [7]:

- Very simple and robust.
- Lack of power control possibilities.
- Large mechanical loads.
- Large fluctuations in output power.

On the other hand, variable speed wind turbines present power electronic converters, typically a full back-to-back converter, which increases wind turbine features:

- Higher rotor energy efficiencies due to that optimal rotor speed can be achieved for each wind speed.
- Reduction of mechanical loads.
- Controllability of active and reactive power.
- Fewer fluctuations in output power.

In spite of these advantages, the availability of the wind speed and its fluctuations are the main disadvantages of wind power generation [30].

When wind power penetration is high, wind parks must participate to maintain the short term balance, sharing frequency control participation between different generation units. Decrease output power is relatively easy switching on and off the wind turbines in the wind park and, in case of variable speed wind turbines, controlling power electronic converters. On the other hand, increase output power needs a reserve that can be used when necessary, decreasing maximum power generation to achieve controllability.

Long term balancing is very limited in wind parks due to wind speed dependence. Therefore, there must always be other means available for assuring the long term balance, such as conventional generator, electricity storage or controllable loads [3].

Voltage control capabilities of wind parks depend on the wind turbine technology. Thus, fixed speed wind parks cannot be used to control voltages nodes unless they are equipped with special devices. Variable speed wind parks can control reactive power generation by means of the power electronic converters connected to the wind turbines and taking part in voltage control at the connection points across a suitable control strategy between the different wind turbines of the park. For this reason, converter sizing is one of the priority factors when voltage control capability is desired.

4.1 Special Devices for Enhancing Reactive Control Capabilities of Wind Parks

The special devices required at the grid connection points by fixed speed wind parks or, in some circumstances, by variable speed wind parks, to satisfy voltage control capabilities are based on different compensator capacitor technologies.

4.1.1 Thyristor-Switched Capacitor

As shown in Fig. 26, Thyristor-Switched Capacitor (TSC) circuit consisting of capacitors and reactors switched on the network by solid state power electronics that are fired at the natural zero crossing of the capacitive current. As a result, capacitors are connected to the network without transients. The control is such that, only complete alternations of the current are allowed, ensuring that no harmonics are generated [1].

4.1.2 Static VAr Compensator

As can be seen in Fig. 27, basic SVC approach is based on conventional capacitor banks together with parallel thyristor controlled reactor branches, which consume the excess of reactive power generated by the capacitor bank. In the grid

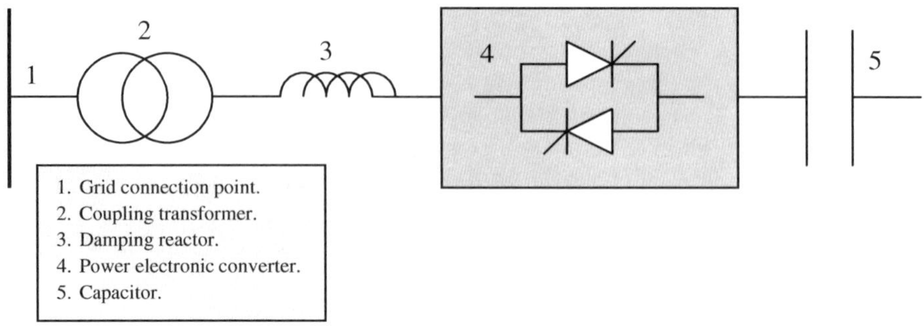

Fig. 26 TSC building blocks

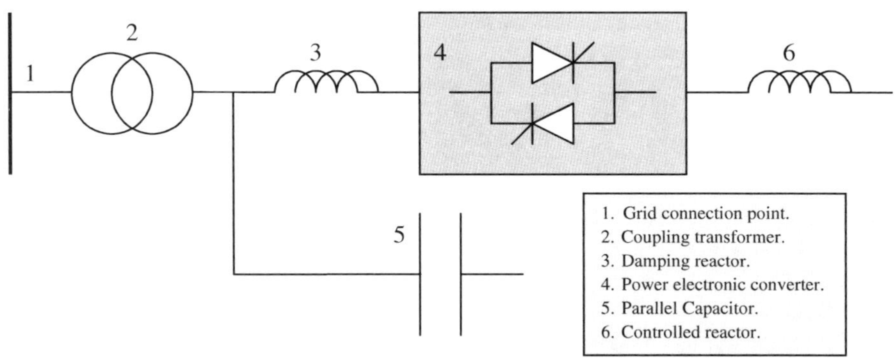

Fig. 27 SVC building blocks

connection point, a centralized SVC provides the necessary balance of reactive power and controls the voltage, making it possible to transmit the desired power levels. It is preferred to use some kind of regulating scheme continuously controlling the reactive power injection in the power grid as a function of the active power generated [12].

As much in TCS as in SVC, the power electronic converter consists of a Thyristor Switched Capacitors (TSC) or Thyristor Switched Reactances (TCR). The coordinated control of these branches varies the reactive power following the curve shown in Fig. 28.

For a sinusoidal voltage, the expression of the fundamental frequency current can be found by using Fourier analysis:

$$I_{SVC} = j\frac{V}{X_c \cdot X_L}\left\{X_L - \frac{X_c}{\pi}[2(\pi - \alpha) + \sin 2\alpha]\right\} \quad (5)$$

where X_c and X_L are de SVC reactances, and α is the firing angle of the semiconductors components.

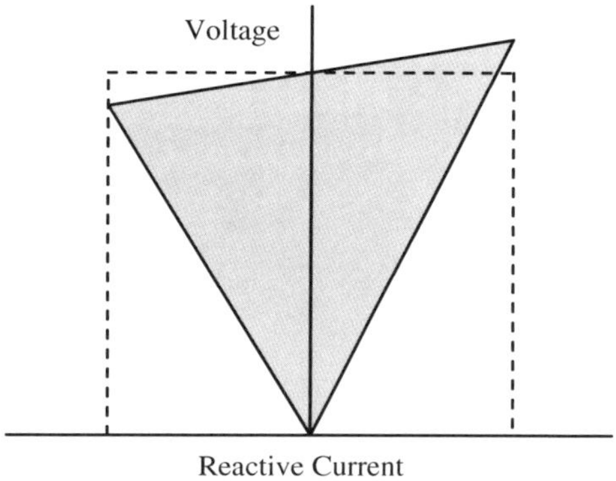

Fig. 28 Reactive current versus voltage of an SVC

4.1.3 STATic Synchronous COMpensator

A STATCOM is a voltage source converter (VSC) based device, with the voltage source behind a reactor (Fig. 29). The voltage source is created from a DC capacitor. STATCOM provides outstanding performance not only in steady state operation as in dynamic voltage control and behaviour during grid faults, adding functionality to wind parks behaviour in order to integrate them into grids with demanded connection requirements.

Figure 30 shows the reactive current versus connected voltage of a STATCOM. The performance is similar to a SVCm, but it performs smooth variation of reactive current across its operating range.

4.1.4 Simulations of a Fixed Speed Wind Park with Special Devices for Enhancing Reactive Power

A small fixed speed wind park of 9 MW is considered in this case. As can be seen in Fig. 31a, the wind park is composed of three 3 MW with an internal cluster structure connected of 25/120 kV transformer that exports the generating power to the grid through a 25 km feeder. Figure 31b presents the impedances of the equivalent circuit of the park. In this figure, Z_{lmi} is the short-circuit impedance of one of the 4 MVA low/medium voltage transformers; Z_{mi} represents the internal medium voltage line to the internal common coupling point; Z_m and Z_{mh} are the feeder and the transformer impedances. For this wind park, two FACTS devices, a 3 MVAr SVC and a 3 MVAr STATCOM with the same operating limits, are simulated.

Figure 32 depicts the behaviour of the wind park when the FACTS devices are demanded to operate at 3 MVAr (rated power) or generate only 1 MVAr. The wind speed applied to each wind turbine is set at 8 m/s, above the nominal wind speed (9 m/s) and a gust of wind is applied to the first turbine after 5 s to reach 11 m/s after 1.5 s. The same gust of wind is considered in the others wind turbines with 5 and 10 s delays.

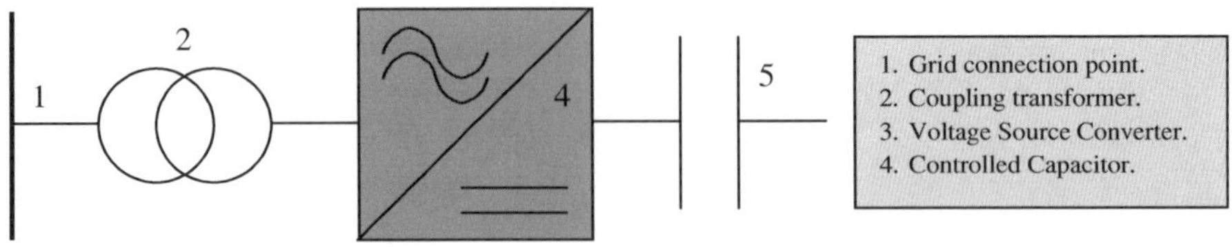

Fig. 29 Statcom building blocks

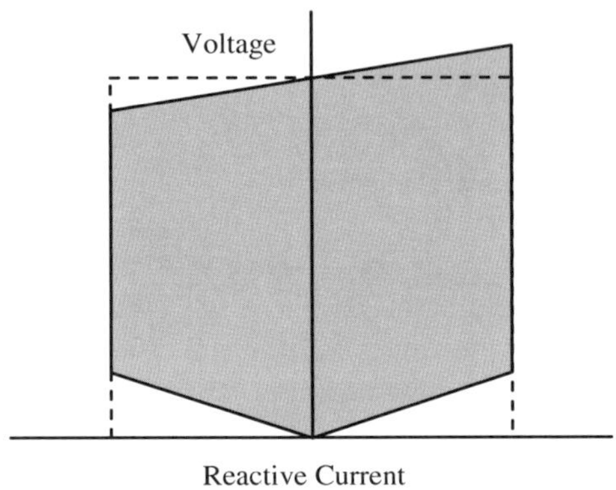

Fig. 30 Reactive current versus voltage of a STATCOM

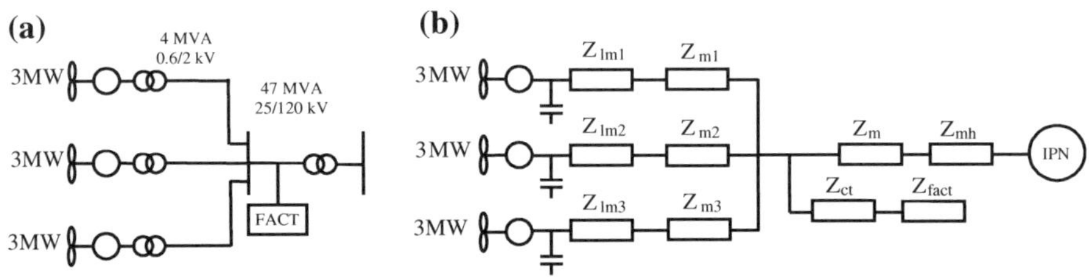

Fig. 31 A 9 MW fixed speed wind park with FACTS: **a** structure, **b** equivalent circuit of the internal network

The responses of both devices are similar because only requirement is to provide the specified reactive power. As the capacitor bank, in each wind turbine, compensate a fraction of the generator requirements (400 kVAr), the wind park consumes reactive power without FACTS device (consumer criterion). When they are working at rated power, the reactive power is less than zero (reactive power generation). The bus voltage at the wind park PCC reflects the dependence showed in Sect. 3.2, and thus, a reactive power consumption increase causes greater voltage drops. As seen in Fig. 32b, the FACTS devices do not affect active power generation.

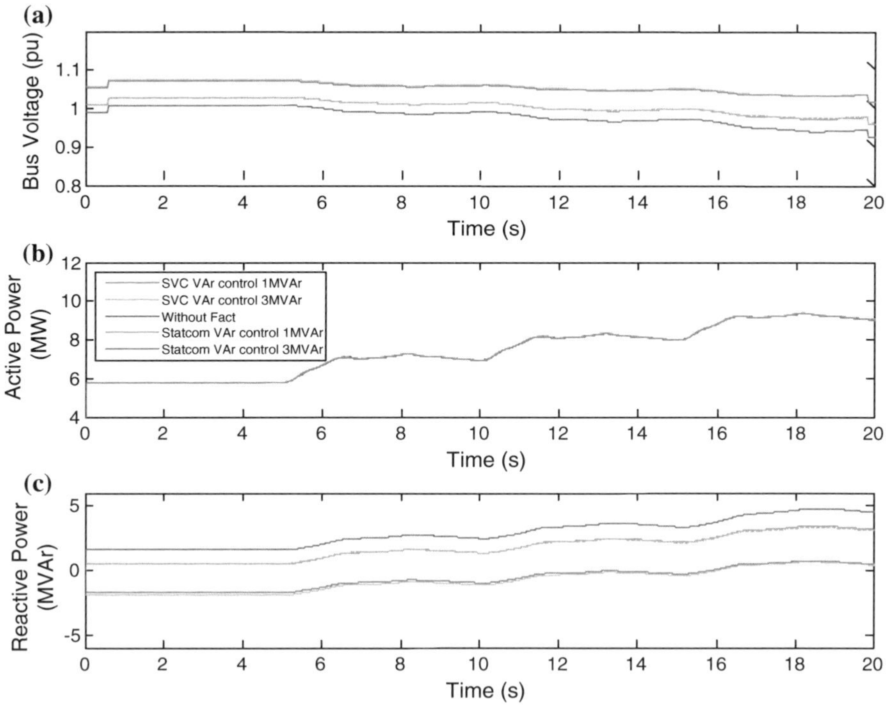

Fig. 32 Response of the fixed speed wind park with SVC and STATCOM in case of reactive power control

Figure 33 depicts the behaviour of the wind park under the same incoming wind but with a different control law in the FACTS devices. In this case, the STATCOM device is trying to maintain voltage bus level at 1 p.u., with a slope in the V–I characteristics of 0.03 p.u.

After the third wind gust at 15 s, all the wind turbines are above nominal wind speed and the wind parks needs over 4 MVAr, while the bus voltage level falls to 0.92 p.u. When the STATCOM is connected, it supplies the reactive power that the wind park demands and the bus level remains over 0.98 p.u.

4.2 Special Devices for Enhancing Power Quality of Wind Parks

Besides maintaining acceptable steady state levels and voltage profiles in all operating conditions exchanging reactive power with the grid, a fast control of this reactive compensation is required to relax possible voltage stability constrains related to the wind park [53].

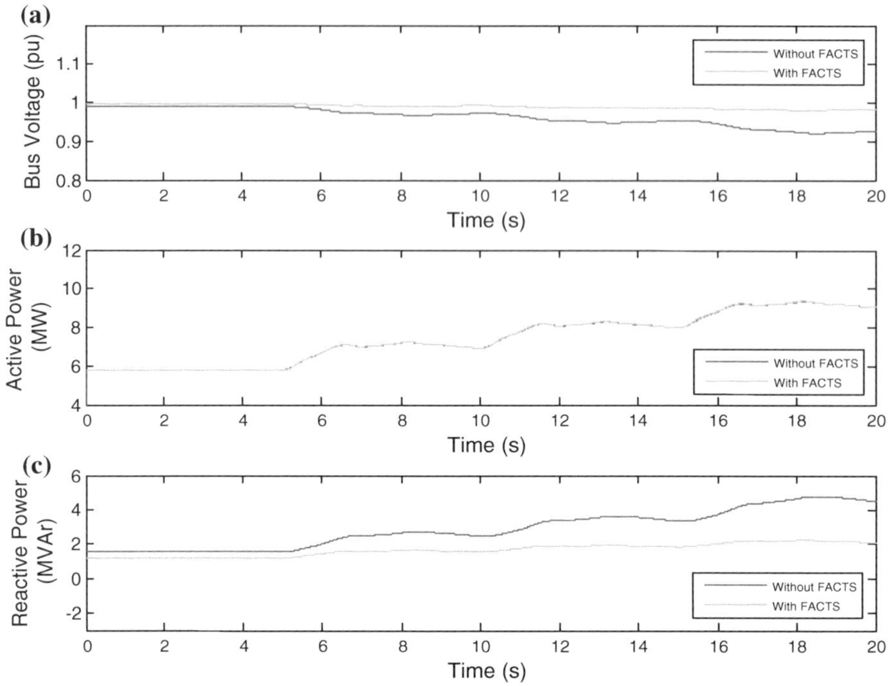

Fig. 33 Response of the fixed speed wind park with STATCOM in case of voltage control

In transient stability, grid codes have specified that wind parks have to be able to cope with grid disturbances without disconnection and they should supply active and reactive power after the fault has been cleared [6]. This implies that wind turbines must be able to 'ride through' temporary faults and contribute to the provision of short-circuit capacity. In this case, if the generation system of the wind turbine is composed of an induction generator, the risk of voltage collapse is important because this type of generator consumes large amounts of reactive power when its speed slightly deviates from the synchronous speed and fast support of reactive power is required [21]. Special devices used to address issues related to voltage control capabilities (control and stability), such as TSC, SVC and STATCOM, are not fast enough to solve satisfactory transient stability and power quality functions. In these cases, Distribution-STATic synchronous COMpensator (D-STATCOM) and Dynamic Voltage Restorer (DVR), that are composed of power electronic converters based on Voltage source converter (VSI), provide as much as power control as power quality functions [6].

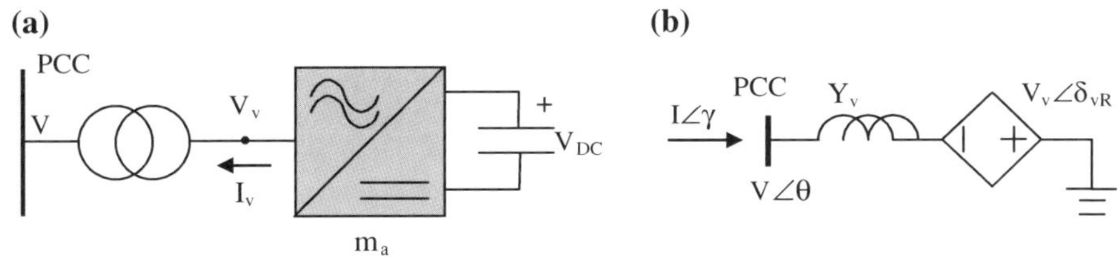

Fig. 34 D-STATCOM system: **a** VSC connected to the AC network via a shunt-connected transformer, **b** shunt solid-state voltage source

4.2.1 Distribution-STATic Synchronous COMpensator

The D-STATCOM configuration consists on a Current Source Inverter (CSI), a DC-energy storage device, a coupling transformer connected in shunt with the PCC at the wind park terminals and associated control systems [6].

A schematic representation of a D-STATCOM and its equivalent circuit are shown in Fig. 34. The equivalent circuit corresponds to the Thevenin equivalent as seen from bus, where the voltage source V_v is the fundamental frequency component of the VSC output voltage, resulting from the product of capacitor voltage (V_{DC}) and amplitude modulation ratio (m_a). Thus, the D-STATCOM is represented as a variable voltage source V_v, whose magnitude and phase angle may be adjusted, by using a suitable iterative algorithm, to satisfy a specified voltage magnitude at the point of connection with the AC network [2].

4.2.2 Dynamic Voltage Restorer

The DVR contains a three phase voltage-source inverters (VSI) with three single phase series transformer, a passive filter and an energy storage device. It can mitigate voltage dip and restore distorted voltage signal at the PCC of the wind park [59].

The DVR injects voltage in quadrature with one of the line end voltages, in order to regulate active power without drawing reactive power from the grid. It has its own reactive power provision from the capacitor, enabling the regulation of reactive power and nodal voltage magnitude. A schematic representation of the DVR and its equivalent circuit are shown in Fig. 35, where the series voltage source is a function of the capacitor rating and the phase quantities. The magnitude and phase angle of the DVR model are adjusted by using suitable iterative algorithm to satisfy a specified active and reactive power flow across the DVR.

4.2.3 Unified Power Flow Conditioner

A D-STATCOM and a DVR may be combined sharing a common capacitor on their DC side and a unified control system. A simplified representation of the Unified Power Flow Conditioner (UPFC) is shown in Fig. 36a, and its equivalent circuit in Fig. 36b, as can be seen in [2].

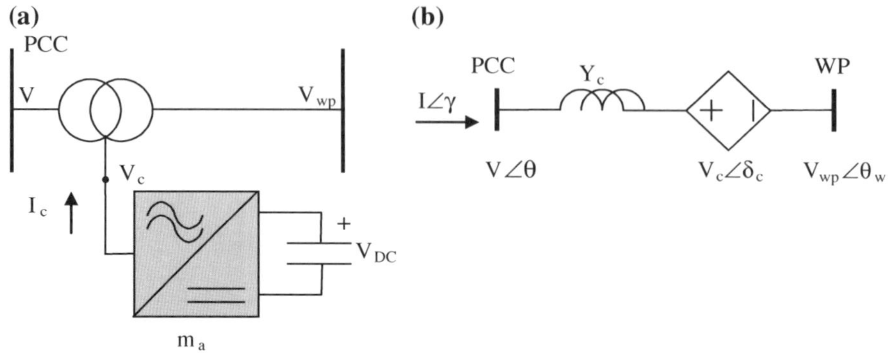

Fig. 35 DVR system: **a** VSC connected to the AC network using a series transformer, **b** series solid-state voltage source

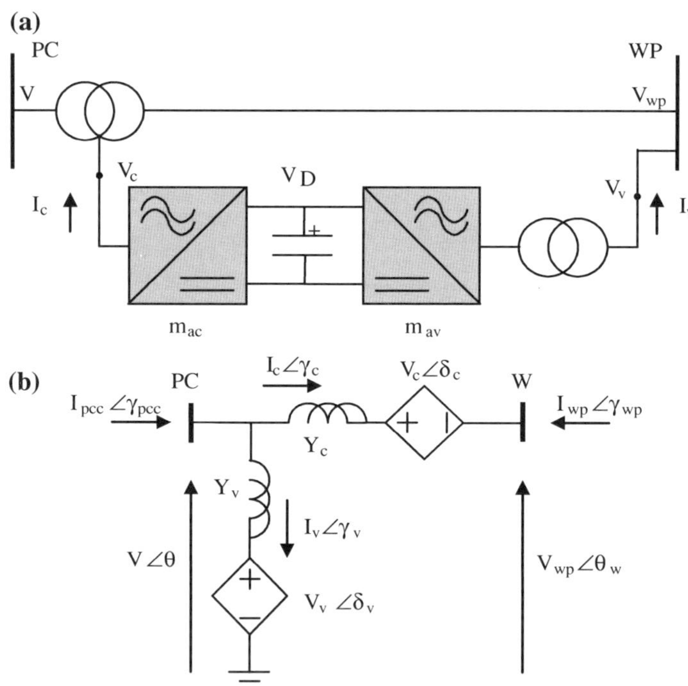

Fig. 36 UPFC system: **a** Simplified representation; **b** Equivalent circuit

The equivalent circuit shown in Fig. 36b consists of a shunt and a series VSI connected to the system through inductive reactances representing the VSC transformers. The constraint equation which links the VSI can be expressed as:

$$Re\{V_v.I_v^* + V_c.I_c^*\} = 0. \tag{6}$$

Thus, the transfer admittance equation can be written as:

$$\begin{bmatrix} I_{pcc} \\ I_{wp} \end{bmatrix} = \begin{bmatrix} (Y_c + Y_v) & -Y_c & -Y_c & -Y_v \\ -Y_c & Y_c & Y_c & 0 \end{bmatrix} \begin{bmatrix} V \\ V_{wp} \\ V_c \\ V_v \end{bmatrix}. \tag{7}$$

The UPFC allows simultaneous control of active, reactive power flow, and voltage magnitude at the UPFC terminals. Alternatively, the controller may be set to control one or more of these parameters in any combination or to control none of them.

4.2.4 High-Voltage Direct-Current Based on Voltage Source Converters

High-Voltage Direct-Current based on Voltage Source Converters (HVDC-SVC) is composed of two VSCs connected back-to back or link together by a DC cable, one operating as a rectifier and the other as an inverter. Its main function is to transmit constant DC power from one to another VSC with high controllability. A simplified representation of the HVDC-SVC is shown in Fig. 37a, and its equivalent circuit in Fig. 37b, as can be seen in [2].

The equivalent circuit showed in Fig. 37 consists of shunt-connected VSI linked together by the following active power constraint equation and transfer admittance equation:

$$Re\{V_{v1}.I_{v1}^* + V_{v2}.I_{v2}^*\} = 0 \tag{8}$$

$$\begin{bmatrix} I_{pcc} \\ I_{wp} \end{bmatrix} = \begin{bmatrix} Y_{v1} & -Y_{v1} & 0 & 0 \\ 0 & 0 & Y_{v2} & -Y_{v2} \end{bmatrix} \begin{bmatrix} V \\ V_{v1} \\ V_{wp} \\ V_{v2} \end{bmatrix}. \tag{9}$$

Besides active and reactive power flow control, voltage control capabilities and transient stabilities, these devices allow power quality enhancement in case of other different disturbances such as harmonics, interharmonics, long and short interruptions, overvoltages, etc. [25]. This group of devices, used for enhancing power quality is called Custom Power Systems (CUPS) if there are used in the distribution level (up to 60 kV) and Flexible AC Transmission Systems (FACTS) if the devices are use on transmission level [60].

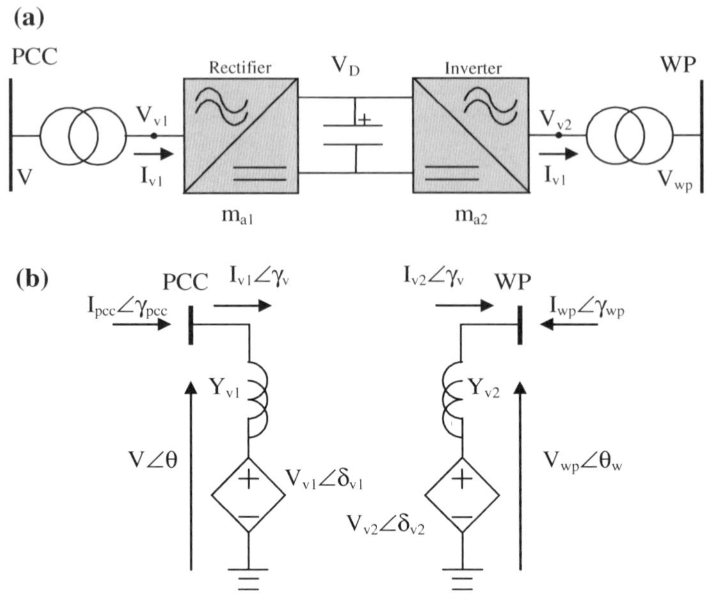

Fig. 37 HVDC-VSC system: **a** Simplified representation, **b** Equivalent circuit

4.2.5 Simulations of a Fixed Speed Wind Park with Special Device for Enhance Power Quality

In this case, a 3 MVA DVR is connected in series at the PCC of the fixed speed wind park showed in Fig. 31. The injected voltage at the wind park bus is limited to 0.3 pu and a maximum change rate of the voltage reference of 3 pu/s.

Figure 38 depicts the comparison between the responses of wind parks with and without DVR. In both cases, all the wind turbines are operating at nominal wind speed (9 m/s), so that the wind park generates about 8.5 MW and consumes over 3.8 MVAr, if the DVR is disconnected. When the VDR is connected, three different references have been applied:

- From 0 to 5 s, the quadrature voltage reference of the DVR control is set to 0, and the DVR upstream and downstream voltages are very close, since the difference is the internal drop voltage.
- From 5 to 10 s, the DVR quadrature voltage reference is increased to 0.2, and thus, the DVR upstream voltage is 1 p.u., while the downstream voltage reaches 1.05 p.u..
- From 10 to 15 s, the reference voltage is set to 0.28. In this case, the wind park achieves a voltage of 1.07 and 1.016 p.u. at the beginning of the feeder.
- At last, the reference voltage is set to 0.1, and then, both voltages decrease to 1.01 and 0.98, respectively.

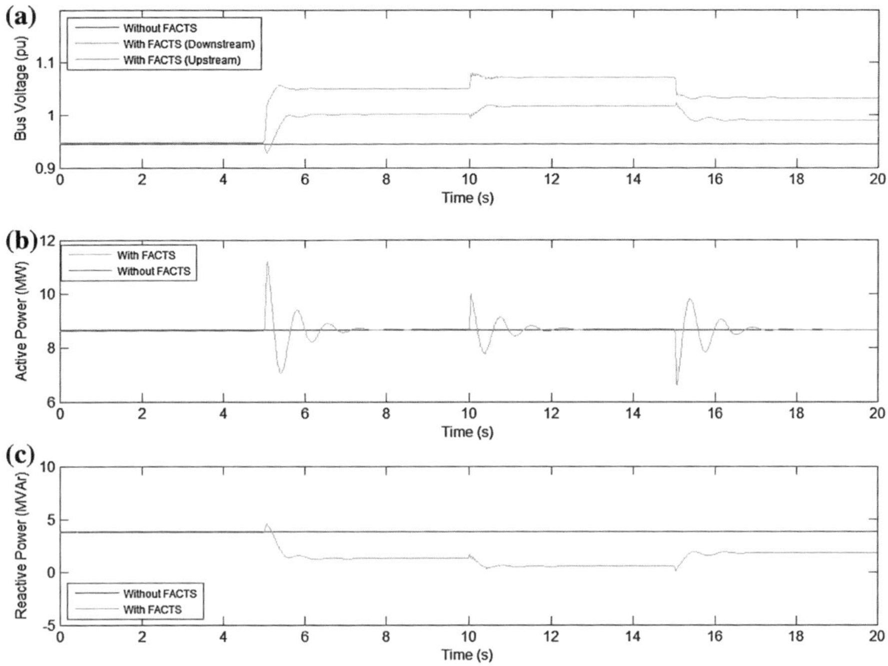

Fig. 38 Response of the fixed speed wind park with VDR in case of voltage reference changes

The DVR device is also tested during a grid disturbance. In this case, it is considered a voltage drop of 0.3 p.u. at 10 s with duration of 400 ms, and after the fault, the voltage starts to recover. As the grid disturbances are much faster than a wind speed fluctuation, the wind speed is assumed constant. Figure 39 shows the wind park response in two cases: when the DVR is bypassed and when it is controlling the injected voltage, setting its reference at 0.28 p.u. while the fault occurs.

The voltage injected by the DVR in quadrature with the line voltage (at the beginning of the feeder) allows increasing the voltage at the PCC of the wind park, while the drop voltage occurs and thus, a faster voltage recovering. During the fault, the DVR tries to regulate the active power, increasing during the first part of the drop and decreasing in the second part. Meanwhile the device provides reactive power (consumer criterion), as can be seen in Fig. 39.

Figure 40 depicts the behaviour of the DVR during the voltage drop. In Fig. 40a, it is shown the device upstream and downstream voltages. The difference between the two voltages is due to the injected voltage by the DVR, as seen in Fig. 40b. The reactive power injected by the device to the wind park is shown in Fig. 40c. In this case, only when the voltage drops occurs the device injects reactive power.

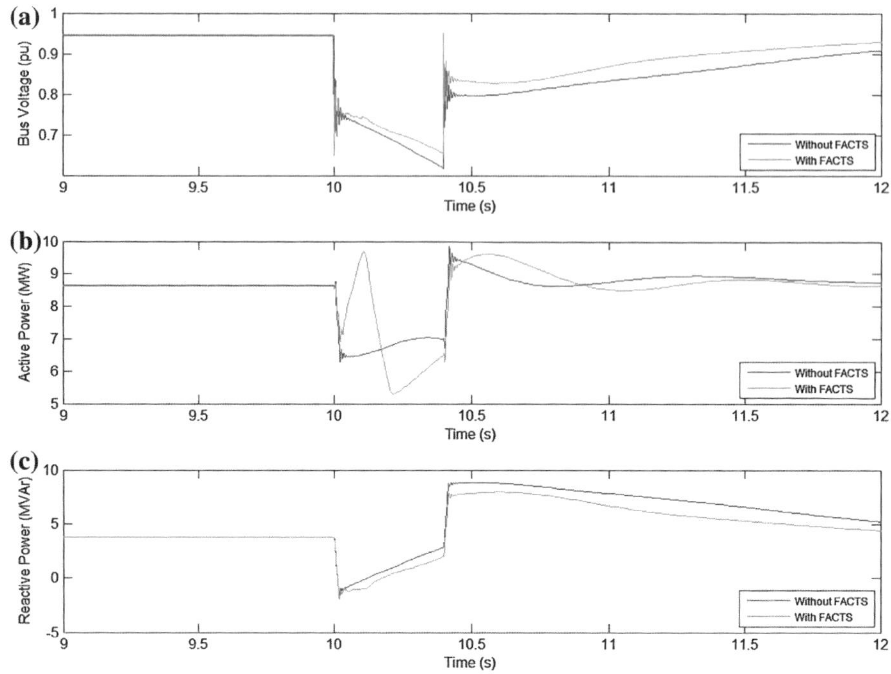

Fig. 39 Response of the fixed speed wind park with VDR in case of grid fault

4.3 Energy Storage Systems for Enhancing Power Quality of Wind Parks

Another way to improve low power quality of wind power is to use energy storage systems. They avoid short-term fluctuations and allow the perfect adjust of the demanding and generating curves, because they store energy surpluses when produced, and then release when the demanding level exceeds production. Voltage stability is also affected, because they help to minimize the peaks and sags that can appear in the grid. Furthermore, the existence of a proper storage system reduces the necessity of power plants in reserve, covering the demand peaks with the stored energy.

The technology in the field of energy storage systems is wide and varied. A huge range of devices has been developed in order to improve the capacity and efficiency of the already existing. Of all the numerous types of energy storage systems, the most important devices in the wind power generation field are [22]:

- *Batteries* The most important feature is the independence between the power and energy values. There are different types of batteries, from the classic lead acid to flow batteries, whose commercial application has not been achieved yet.

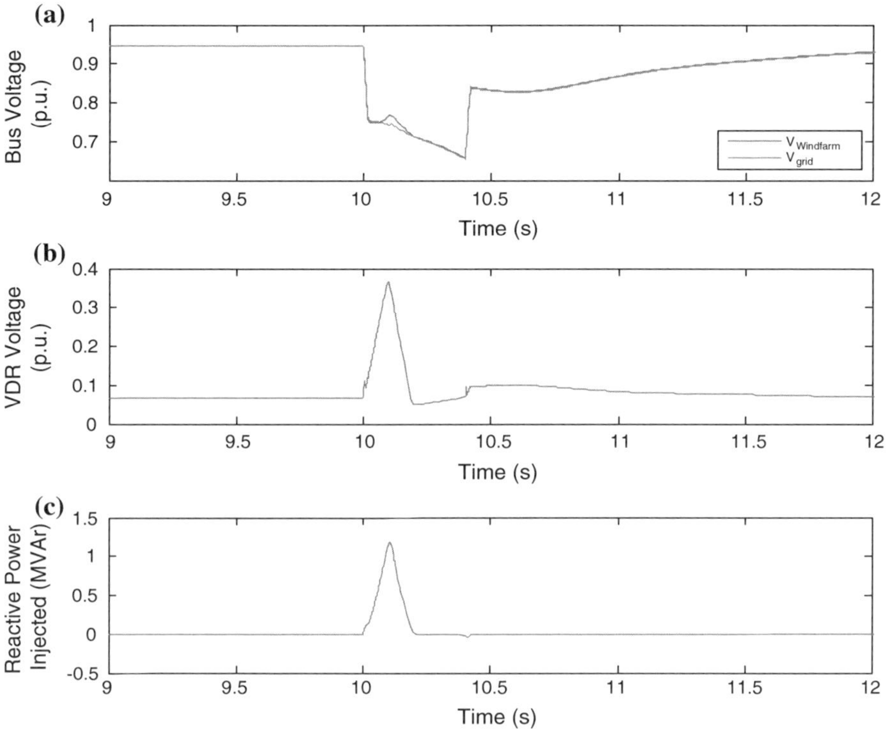

Fig. 40 Response of the VDR in case of grid fault

- *Supercapacitors* They store energy through the conservation of an electric field between two electrically conducting plates. They have several great advantages, however they are relatively new and expensive, but they are very promising devices for application in a near future.
- *Flywheels* They store electric energy as kinetic energy in a rotating mass. They have found interest as load levelling devices acting together with wind energy systems.
- *Superconducting magnetic energy storage systems (SMES)* These devices store electric energy in a magnetic field generated by a DC current flowing through a superconducting magnetic coil. They present high efficiency and no moving part. However, the main drawback is related to their high costs.
- *Fuel cells (FC)* They can act as an energy barrier and adjust power output effectively [62].
- *Hydrogen production* It is considered one of the most attractive energetic alternatives in the near future. As a result, a number of researchers are working on the development of new producing and storing possibilities. At the same time, the currently existing hydrogen storage technologies are progressing in their efficiency and capabilities.

4.3.1 Simulations of a DFIG Wind Turbine with Energy Storage System

A variable speed wind turbine integrating a battery is considered to evaluate the behaviour of a wind park with energy storage system. Figure 41a depicts the proposed structure: a 1.5 MW DFIG wind turbine connected to a battery with DC/AC inverter as energy storage system. The whole system is connected to a 30 km feeder and a 25/120 kV transformer that evacuate the generated power. Figure 41 b shows the impedances of the equivalent circuit, where Z_{lm} is the short-circuit impedance of the 1.75 MVA low/medium voltage transformer; Z_m represents the internal medium voltage line to the internal common coupling point; while Z_f and Z_{mh} are the feeder and the medium/high voltage transformer impedances.

The hybrid generation system (DFIG wind turbine with battery) and its control system are modeled and simulated as performed in [45]. The only difference is that the battery is located outside the wind turbine and connected to the 25 kV line by means of a DC/AC inverter. A supervisory control system based on a state machine control strategy is responsible for setting the battery power reference depending on the battery State of Charge (SOC) and the operating conditions.

In the performed simulation, a constant wind speed of 14 m/s is considered. The active power demanded by the grid is set to 0.8 p.u. during the first 60 s, and it is changed to 1.2 p.u. for the rest of the simulation. At the same time, the reactive power is controlled to unity power factor until 90 s, and then, it is increased up to 0.1 p.u. (consumer criterion) with a rising slope of 0.1 p.u./s. In this case, the control system maintains the reactive power consumed by the wind turbine at 0 p.u., while the battery inverter is controlled to achieve the desired grid demand.

Figure 42 shows the evolution of active and reactive powers and battery SOC during the simulation. As seen in Fig. 42a, during the first 90 s, the battery stores the surplus energy between the active power reference and the available wind power, increasing the battery SOC from 60 to 61 %. In the rest of the simulation, the active power reference exceeds the available power in the wind turbine, so that the battery supplements that generation to provide the demanded power, which causes the battery discharge (Fig. 42b). The suitable reactive power control is shown in Fig. 42b.

The energy stored in the battery allows maximum wind energy capture even when the power demand is low. It enables the possibility to decouple generation from demand, and optimize the energy. Other variables, such as reactive power and voltage, can be regulated by applying an adequate control strategy. Hence, the inclusion of an energy storage system enhances the wind turbine capabilities, allowing a higher and less fluctuating power output to the grid.

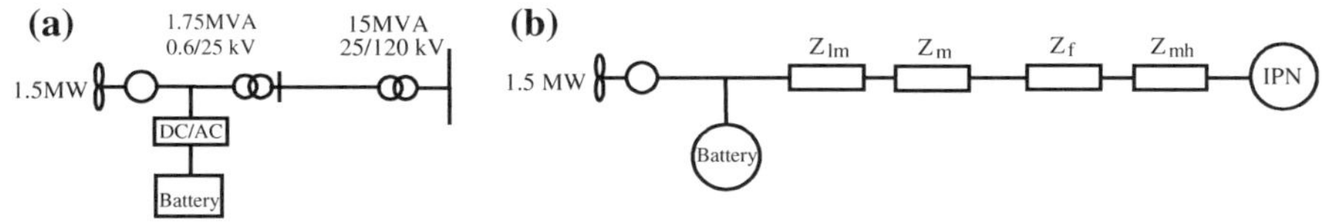

Fig. 41 A 1.5 MW DFIG wind park with battery as energy storage system: **a** structure, **b** equivalent circuit of the internal network

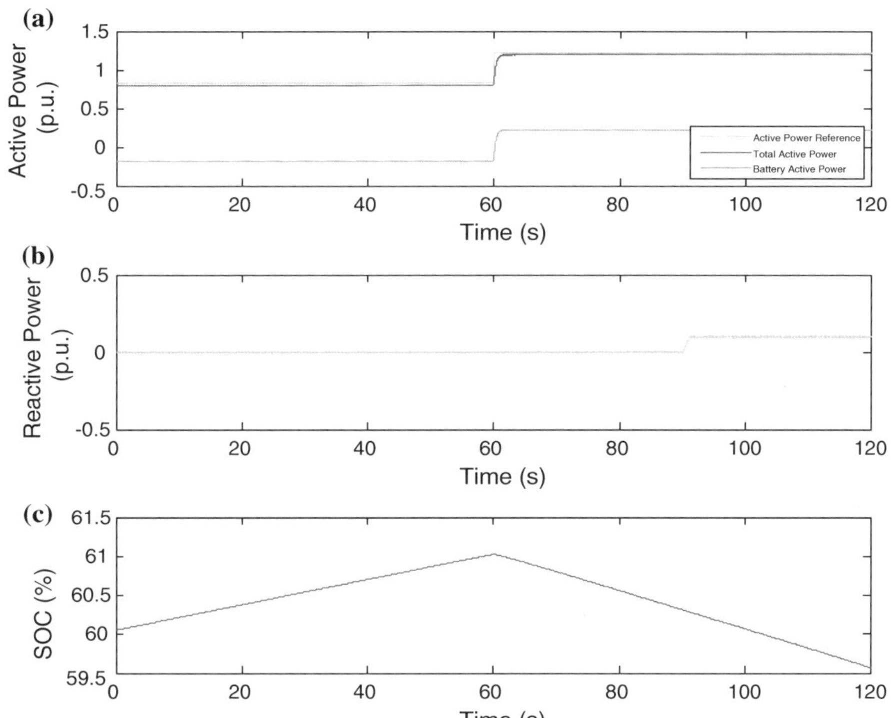

Fig. 42 Response of a 1.5 MW DFIG wind park with battery energy storage

5 Conclusion

This chapter has dealt with some of the main issues related to wind energy penetration and its grid integration, mainly focused on transient stabilities studies. First of all, it is considered the necessity of suitable models to evaluate wind parks dynamic behavior. Thus, when the scope of study is the wind park as a whole power plant, the complexity of the models is reduced by aggregating wind turbines into equivalent models, as usual in transient stability studies in case of wind fluctuations and grid disturbances. These equivalent models are qualified in several groups depending on the wind speed that receives every wind turbine and the types and rated values of the machines that compose the wind park. Secondly, the wind park control is considered, since wind parks are today required to participate in the power system operation like conventional power plants do, assuring stability and quality of the system. The main requirements to assure adequate grid integration are active power and voltage regulations and reactive power and frequency controls. Furthermore, the main controller and the dispatch center are analyzed as the leading components of the control structure. Finally, additional devices are described, such as FACTS or energy storage systems, which can be used in wind parks for improving the power quality, the control capabilities, and thus, the grid integration.

As a result, wind park capabilities are near to the ones of conventional power plants, assuring a great adaptation to grid code requirements and making feasible to integrate wind power to a significant extent without major system changes.

Appendix: Wind Park Models Parameters

(a) **350 kW fixed speed wind turbine**

Rated power: 350 kW, rated voltage: 660 V, $R = 15.2$ m, $H_r = 5$ p.u., gear box ratio: 1:44.5, $K_{mec} = 100$ p.u., $D_{mec} = 10$ p.u., $H_g = 0.5$ p.u., $R_s = 0.006$ p.u., $R'_r = 0.006$ p.u., $X_{\sigma s} = 0.007$ p.u., $X'_{\sigma r} = 0.19$ p.u., $X_m = 2.78$ p.u., $X_c =$ 2.5 p.u (Fig. 43).

(b) **500 kW fixed speed wind turbine**

Rated power: 500 kW, rated voltage: 660 V, $R = 28$ m, $H_r = 5$ p.u., gear box ratio: 1:89, $K_{mec} = 200$ p.u., $D_{mec} = 15$ p.u., $H_g = 1$ p.u., $R_s = 0.01$ p.u., $R'_r = 0.01$ p.u., $X_{\sigma s} = 0.01$ p.u., $X'_{\sigma r} = 0.08$ p.u., $X_m = 3$ p.u., $X_c = 2.3$ p.u (Fig. 44).

(c) **660 kW DFIG wind turbine**

Rated power: 660 kW, rated voltage: 660 V, $R = 23.5$ m, $H_r = 0.5$ p.u., gear box ratio: 1:52.5, $K_{mec} = 90$ p.u., $D_{mec} = 15$ p.u., $H_g = 3$ p.u., $R_s = 0.01$ p.u., $R'_r = 0.01$ p.u., $X_{\sigma s} = 0.04$ p.u., $X_{\sigma r}\prime = 0.05$ p.u., $X_m = 2.9$ p.u (Fig. 45).

(d) **1.5 MW DFIG wind turbine**

Rated power: 1.5 MW, rated voltage: 600 V, $R = 41$ m, $H = 4.64$ p.u., $R_s = 0.005$ p.u., $R'_r = 0.004$ p.u., $X_{\sigma s} = 0.125$ p.u., $X'_{\sigma r} = 0.179$ p.u., $X_m = 6.77$ p.u (Fig. 46).

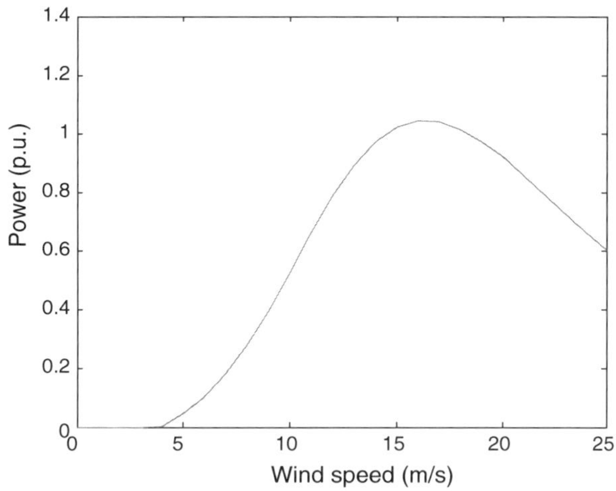

Fig. 43 Power curve of a 350 kW fixed speed wind turbine

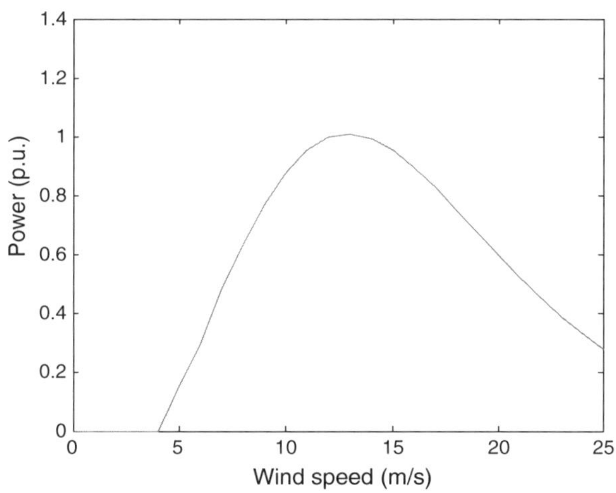

Fig. 44 Power curve of a 500 kW fixed speed wind turbine

(e) **2 MW DFIG wind turbine**

Rated power: 2 MW, rated voltage: 690 V, R = 38 m, $H_r = 0.5$ p.u., gear box ratio: 1:89, $K_{mec} = 95$ p.u., $D_{mec} = 40$ p.u., $H_g = 2.5$ p.u., $R_s = 0.01$ p.u., $R'_r = 0.01$ p.u., $X_{\sigma s} = 0.1$ p.u., $X'_{\sigma r} = 0.08$ p.u., $X_m = 3$ p.u (Fig. 47).

(f) **3 MW fixed speed wind turbine**

Rated power: 3 MW, rated voltage: 600 V, R = 45 m, $H_r = 4.29$ p.u., gear box ratio: 1:89, $K_{mec} = 296$ p.u., $D_{mec} = 15$ p.u., $H_g = 0.90$ p.u., $R_s = 0.003$ p.u., $R'_r = 0.002$ p.u., $X_{\sigma s} = 0.063$ p.u., $X'_{\sigma r} = 0.089$ p.u., $X_m = 3.38$ p.u (Fig. 48).

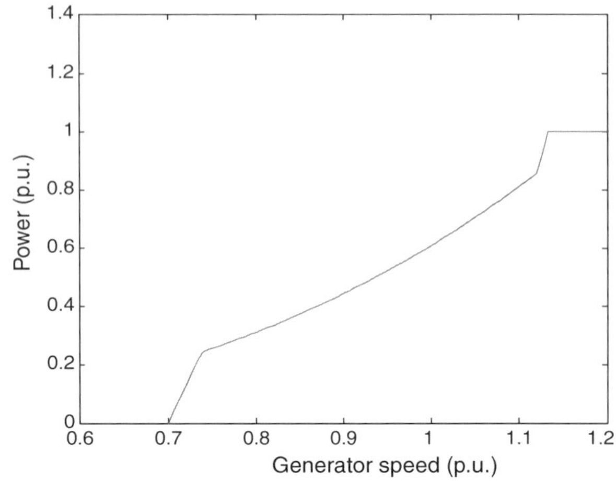

Fig. 45 Power curve of a 660 kW DFIG wind turbine

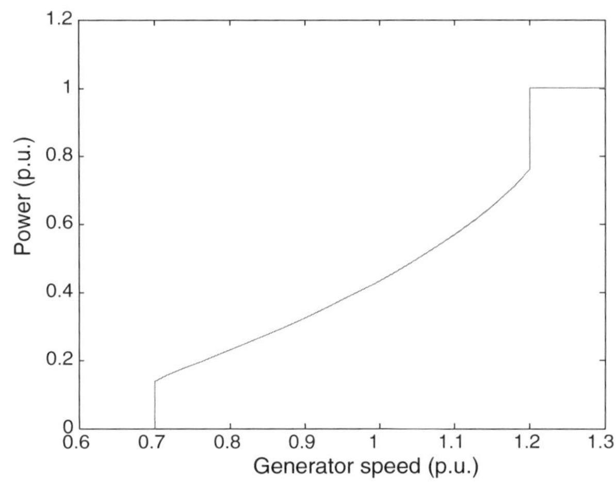

Fig. 46 Power curve of a 1.5 MW DFIG wind turbine

(g) **Electrical network of the fixed speed wind park with 6 wind turbines of 350 kW and 6 wind turbines of 500 kW**

LV lines Cluster 1 (r = 0.4 Ω/km, x = 0.1 Ω/km, length = 200 m); Cluster 2 (r = 0.4 Ω/km, x = 0.1 Ω/km, length = 300 m).
LV/MV transformers Cluster 1 (800 kVA, 20/0.66 kV, ε_{cc} = 6 %); Cluster 2 (1,250 kV, 20/0.66 kV, ε_{cc} = 5 %).
MV lines Cluster 1 (r = 0.15 Ω/km, x = 0.1 Ω/km, length = 500 m); Cluster 2 (r = 0.15 Ω/km, x = 0.1 Ω/km, length = 600 m).
MV/HV transformers (10 MVA, 20/66 kV, ε_{cc} = 8 %).
Feeder (r = 0.2 Ω/km, x = 0.4 Ω/km, length = 10 km).
Grid Short circuit power at PCC = 500 MVA, X/R ratio = 20.

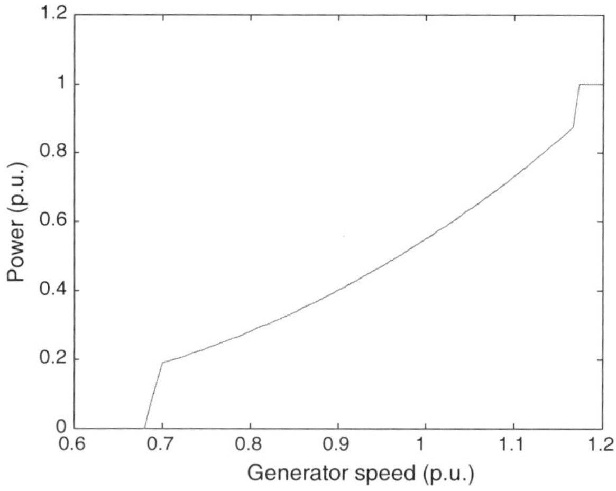

Fig. 47 Power curve of a 2 MW DFIG wind turbine

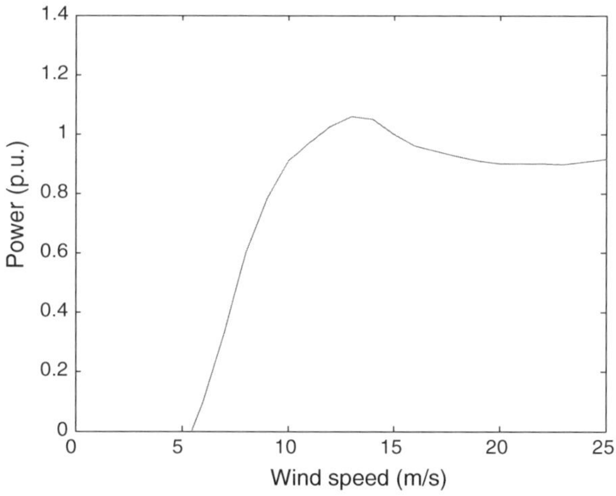

Fig. 48 Power curve of a 3 MW fixed speed wind turbine

(h) Electrical network of the DFIG wind park with 6 wind turbines of 2 MW

LV/MV transformers (2.5 MVA, 20/0.66 kV, $\varepsilon_{cc} = 6$ %).
MV lines (r = 0.3 Ω/km, x = 0.1 Ω/km, length = 200 m).
MV cluster lines Cluster 1 (r = 0.15 Ω/km, x = 0.05 Ω/km, length = 1 km); Cluster 2 (r = 0.15 Ω/km, x = 0.1 Ω/km, length = 2 km).
MV/HV transformers (15 MVA, 20/66 kV, $\varepsilon_{cc} = 8.5$ %).
Feeder (r = 0.16 Ω/km, x = 0.35 Ω/km, length = 20 km).
Grid Short circuit power at PCC = 500 MVA, X/R ratio = 20.

(i) **Electrical network of the DFIG wind park with 6 wind turbines of 660 kW and 6 wind turbines of 2 MW**

LV/MV transformers Cluster 1 (800 kVA, 20/0.66 kV, $\varepsilon_{cc} = 6$ %); Cluster 2 (2.5 MV, 20/0.66 kV, $\varepsilon_{cc} = 6$ %).
MV lines Cluster 1 (r = 0.3 Ω/km, x = 0.1 Ω/km, length = 200 m); Cluster 2 (r = 0.4 Ω/km, x = 0.1 Ω/km, length = 200 m).
MV cluster lines: Cluster 1 (r = 0.15 Ω/km, x = 0.05 Ω/km, length = 500 km); Cluster 2 (r = 0.15 Ω/km, x = 0.1 Ω/km, length = 2 km).
MV/HV transformers Cluster 1 (4 MVA, 20/66 kV, $\varepsilon_{cc} = 8$ %); Cluster 2 (15 MVA, 20/66 kV, εcc = 8.5 %).
Feeder (r = 0.2 Ω/km, x = 0.4 Ω/km, length = 10 km).
Grid Short circuit power at PCC = 500 MVA, X/R ratio = 20.

(j) **Electrical network of the fixed speed wind park with 3 wind turbines of 3 MW**

LV/MV transformers (4 MVA, 25/0.60 kV, $\varepsilon_{cc} = 7.7$ %).
MV lines (r = 0.115 Ω/km, x = 0.33 Ω/km, length = 1 km).
MV cluster line (r = 0.115 Ω/km, x = 0.33 Ω/km, length = 25 km).
MV/HV transformers (47 MVA, 25/120 kV, $\varepsilon_{cc} = 3.3$ %).
Grid Short circuit power at PCC = 2,500 MVA, X/R ratio = 10.
SVC (3 MVA, 25 kV, $t_{delay} = 4$ ms).
STATCOM (3 MVA, 25 kV, R = 0.007 pu, X = 0.22 pu, $C_{eq} = 1{,}125$ μF).

(k) **Electrical network of the DFIG wind park with 1 wind turbine of 1.5 MW**

LV/MV transformers (1.75 MVA, 25/0.60 kV, $\varepsilon_{cc} = 7.7$ %).
MV line (r = 0.115 Ω/km, x = 0.33 Ω/km, length = 1 km).
MV cluster line (r = 0.115 Ω/km, x = 0.33 Ω/km, length = 30 km).
MV/HV transformers (15 MVA, 25/120 kV, $\varepsilon_{cc} = 6.3$ %).
Grid Short circuit power at PCC = 2,500 MVA, X/R ratio = 10.
Battery (585 Ah, 624 V).

References

1. ABB (2009) DYNACOMP® The top-class reactive power compensator
2. Acha E, Fuerte-Esquivel C, Ambriz-Pérez H et al (2004) Facts, modelling and simulation in power networks. Wiley, London
3. Ackermann Te (2005) Wind power in power systems. Wiley, Stockholm
4. Akhmatov V (2003) Analysis of dynamic behaviour of electric power systems with large amount of wind power. Electric Power Engineering. Ph.D. Thesis, Ørsted-DTU, Technical University of Denmark, Kongens Lyngby
5. Akhmatov V, Knudsen H (2002) An aggregated model of a grid-connected, large-scale, offshore wind farm for power stability investigations-importance of windmill mechanical system. Int J Electr Power Energy Syst 25(9):709–717

6. Alvarez C, Amarís H, Samuelsson O (2007) Voltage dip mitigation at wind farms. In: Proceedings of European wind energy conference EWEC 2007. Milan
7. Bousseau P, Fesquet F, Belhomme R, Nguefeu S, Thai T (2006) Solutions for the grid integration of wind farms-a survey. Wind Energy 9:13–25
8. Bozhko S, Asher G, Li R, Clare J, Yao L (2008) Large offshore DFIG-based wind farm with line-commutated HVDC connection to the main grid: engineering studies. IEEE Trans Energy Convers 23(1):119–127
9. Cartwright P, Holdsworth L, Ekanayake JB, Jenkins N (2004) Co-ordinated voltage control strategy for a (DFIG)-based wind farm. IEE Proc Gener Transm Distrib 151(4):492–502
10. Chen Z, Blaabjerg F (2009) Wind farm—a power source in future power systems. Renew Sustain Energy 13:1288–1300
11. EN-50160 S (2007) Voltage characteristics of electricity supplied by public distribution. Commission of the European communities, EC
12. Eriksson K, Halvarsson P, Wensky D et al (2003) System approach on designing an offshore wind power grid connection. In: 4th international workshop on large-scale integration of wind power and transmission networks for offshore wind farms
13. Fernández L, García C, Jurado F (2008) Comparative study on the performance of control system for doubly fed induction generator (DFIG) wind turbines operating with power regulation. Energy 33:1438–1452
14. Fernández L, García C, Saenz J, Jurado F (2009) Equivalent models of wind farms by using aggregated wind turbines and equivalent winds. Energy Convers Manage 50:691–704
15. Fernández L, Jurado F, Saenz J (2008) Aggregated dynamic model for wind farms with doubly fed induction generator wind turbines. Renew Energy 33:129–140
16. Fernández L, Saenz J, Jurado F (2006) Dynamic models of wind farms with fixed speed wind turbines. Renew Energy 31:1203–1230
17. Fernandez R, Battaiotto P, Mantz R (2008) Wind farm non-linear control for damping electromechanical oscillations of power systems. Renew Energy 33:2258–2265
18. Franklin D, Morelato A (1994) Improving dynamic aggregation of induction motor models. IEEE Trans Power Syst 9(4):1934–1941
19. García-Gracia M, Comech M, Sallán J, Llombart A (2007) Modelling wind farms for grid disturbance studies. Renew Energy 33:2109–2121
20. Gjengedal T (2005) Large-scale wind power farm as power plants. Wind Energy 8:361–373
21. Grünbaum R (2001) Voltage and power quality control in wind power. In: Proceedings of powergen Europe 2001 conference, Brussels
22. Hadjipaschalis I, Poullikkas A, Efthimiou V (2009) Overview of current and future energy storage technologies for electric power applications. Renew Sustain 13:1513–1522
23. Hansen A, Hansen L (2007) Market penetration of wind turbine concepts over the years. In: Proceedings of European wind energy conference EWEC 2007, Milan EWEC
24. Hansen A, Sorensen P, Iov F, Blaabjerg F (2006) Centralised power control of wind farm with doubly fed induction generators. Renew Energy 31:935–951
25. Hansen L, Helle L, Blaabjerg F et al (2001) Risø-R-1205(EN). Conceptual survey of generators and power electronics for wind turbines. Risø National Laboratory, Roskilde
26. IEA WA (2007) Final Technical Report 2007. Dynamic models of wind farms for power system studies. IEA
27. IEC 61400-21 (2001) Wind turbine generator systems-part 21: Measurement and assessment of power quality characteristics of grid connected wind turbines, IEC 61400-21. International Electrotechnical Commission
28. Jauch C, Matevosyan J, Ackermann T, Bolik S (2005) International comparison of requirements for connection of wind turbines to power systems. Wind Energy 8:295–306
29. Jin Y, Ju P (2009) Dynamic equivalent modeling of FSIG based wind farm according to slip coherency. In: Proceedings of IEEE international conference on sustainable power generation and supply SUPERGEN'09. IEEE, Nanjing, pp 1–9
30. Kling W, Slootweg J (2002) Wind turbines as power plants IEEE/Cigre workshop on wind power and the impacts on power systems. IEEE, Oslo, p 7

31. Ko H, Jatskevich J, Dumont G, Yoon G (2008) An advanced LMI-based-LQR design for voltage control of grid-connected wind farm. Electr Power Syst Res 78:539–546
32. Kristoffersen J (2005) The Horns Rev wind farm and the operational experience with the wind farm main controller. Coopenhagen Offshore Wind 2005, Copenhagen pp 1–9
33. Li H, Chen Z (2008) Overview of different wind generator systems and their. IET Renew Power Gener 2(2):123–138
34. Martínez de Alegría I, Andreu J, Martín J, Ibañez P, Vilate J, Camblong H (2007) Connection requirements for wind farms: a survey on technical requirements and regulation. Renew Sustain Energy 11:1858–1872
35. Max L (2009) Design and control of a DC collection grid for a wind farm. Göteborg. Ph.D. Thesis, Department of Energy and Environment, Chalmers University of Technology, Sweden
36. Moyano C, Peças Lopes J (2007) Using an OPF like approach to define the operational strategy of a wind park under a system operator control. IEEE power tech 2007. IEEE, Lausanne, pp 651–656
37. Muljadi E, Parsons B (2006) Comparing single and multiple turbine representations in a wind farm simulation. In: Proceedings of European wind energy conference, EWEC'06, Athens, pp 1–10
38. Muljadi E, Butterfield C, Chacon J, Romanowithz H (2006) Power quality aspects in a wind power plant. IEEE power engineering society general meeting, IEEE Montreal, pp 1–8
39. Muljadi E, Pasupulati S, Ellis A, Kosterov D (2008) Method of equivalencing for a large wind power plant with multiple turbine representation. IEEE power and energy society general meeting-conversion and delivery of electrical energy in the 21st Century. IEEE, Pittsburgh, pp 1–9
40. Nozari F, Kankam MD (1987) Aggregation of induction motors for transient stability load modeling. IEEE Trans Power Syst PWRS-2:1096–1102
41. Perdana A, Uski S, Carlson O, Lemström B (2006) Validation of aggregate model of wind farm with fixed speed wind turbines against measurement. In: Proceedings of nordic wind power conference, NWPC'06. Future Energy, Espoo, pp 1–9
42. Pierik J, Morren J (2007) Validation of Dynamic models of wind farms: Erao 3. Delf University of Technology ECN-E-07-006
43. Pöller M, Achilles S (2004) Aggregated wind park models for analyzing power systems dynamics. In: 4th international workshop on large-scale integration of wind power and transmission networks for off-shore wind farms, Billund, pp 1–10
44. Rodríguez-Amenedo J, Arnaltes S, Rodríguez M (2008) Operation and coordinated control of fixed and variable speed wind farms. Renew Energy 33:406–414
45. Sarrias R, Fernandez L, Garcia C, Jurado F (2012) Coordinate operation of power sources in a doubly-fed induction generator wind turbine/battery hybrid power system. J Power Sour 205: 354–366
46. Shafiu A, Anaya-Lara O, Bathurst G, Jenkins N (2006) Aggregated wind turbine models for power system dynamic studies. Wind Eng 30(3):171–186
47. Slootweg J (2003) Wind power. Modelling and impact on power system dynamics. PhD. Thesis, Technische Universiteit Delft, Ridderkerk
48. Slootweg J, Kling W (2002) Modeling of large wind farms in power system simulations. IEEE power engineering society summer meeting. IEEE, Chicago, pp 503–508
49. Sørensen P, Hansen A, Iov F et al (2005) Risø-R-1464(EN). Wind farm models and control strategies. Risø National Laboratory, Roskilde
50. Sørensen P, Hansen A, Janosi L et al (2001) RisØ-R-1281. Simulation of interaction between wind farm and power. Risø National Laboratory, Roskilde
51. Sudrià A, Chindris M, Sumper A et al (2005) Wind turbine operation in power systems and grid connection requirements. In: Proceedings of ICREPQ'05, Zaragoza, pp 1–5
52. Taleb M, Akbaba M, Abdullah E (1994) Aggregation of induction machines for power systems dynamic studies. IEEE Trans Power Syst 9(4):2042–2048
53. Tande J (2003) Grid integration of wind farms. Wind Energy 6:281–295

54. Tande J, Muljadi E, Carlson O et al (2004) Dynamic model of wind farms for power system studies-status by IEA Wind R&W Annex 21. European wind energy conference, EWEC'04, London, pp 22–25
55. Tapia A, Tapia G, Ostolaza J (2004) Reactive power control of wind farms for voltage control applications. Renew Energy 29:377–392
56. Tapia G, Tapia A, Ostolaza J (2007) Proportional–integral regulator-based approach to wind farm reactive power management for secondary voltage control. IEEE Trans Energy Convers 22(2):488–498
57. Trudnowski D, Gentile A, Khan J, Petritz E (2004) Fixed-speed wind-generator and wind-park modelling for transient stability studies. IEEE Trans Power Syst 19(4):1911–1917
58. Tsili M, Patsiouras C, Papathanassiou S (2008) Grid code requirements for large wind farms: a review of technical regulations and available wind turbine technologies. European wind energy conference EWEC'08, Brussels, pp 1–11
59. Visiers M, Mendoza J, Búnez J, González F et al (2007) Windfact®, a solution for the grid code compliance of the wind farm in operation. European conference on power electronics and applications. IEEE, Aalborg, pp 1–9
60. Wachtel S, Adloff S, Marques J, Schellschmidt M (2008) Certification of wind energy converters with FACTS capabilities. European wind energy conference EWEC 2008. Brussels
61. Wämundson M, Hassan F (2009) HVDC wind park. actively interfaced to the grid. Elforsk, Stockholm
62. Wang C, Wang L, Shi L (2007) A survey on wind power technologies in power systems. In: IEEE power engineering society general meeting, IEEE, Florida, pp 1–6
63. Wind on the Grid (2008) Concepts and design of the wind farm cluster management system. Wind on the grid consortium
64. Zhao J, Li X, Hao J, Lu J (2010) Reactive power control of wind farm made up with doubly fed induction generators in distribution system. Electr Power Syst Res 80:698–706

Grid Support Capabilities of Wind Turbines

Gabriele Michalke and Anca Daniela Hansen

Abstract Wind power has gained a significant penetration level in several power systems all over the world. Due to this reason modern wind turbines are requested to contribute to power system support. Power system operators have thus introduced grid codes, which specify a set of requirements for wind turbines, such as fault ride-through and reactive power supply during voltage sags. To date different wind turbine concepts exist on the market comprising different control features in order to provide ancillary services to the power system. In the first place the present chapter emphasizes the most important issues related to wind power grid integration. Then different wind turbine concepts are characterized and their grid support capabilities are analysed and compared. Simulation cases are presented in which the respective wind turbine concepts are subjected to a voltage dip specified in a grid code.

1 Introduction

Wind power is naturally fluctuating. It possesses thus a different role in the power system compared to conventional power generation units. As long as only small and dispersed wind power units are installed in the power system, wind power does not influence power system operation and can therefore easily be integrated. However, when wind power penetration reaches a significant high level in a

G. Michalke (✉)
Robert Bosch GmbH, Corporate Sector Research and Advance Engineering,
Postfach 30 02 40 70442 Stuttgart, Germany
e-mail: Gabriele.Michalke@de.bosch.com

A. D. Hansen
Department of Wind Energy, Risø National Laboratory for Sustainable Energy,
Technical University of Denmark, Frederiksborgvej 399, P.O. Box 49 Roskilde, Denmark
e-mail: anca@dtu.dk

P. M. Pardalos et al. (eds.), *Handbook of Wind Power Systems*, Energy Systems,
DOI: 10.1007/978-3-642-41080-2_16,

respective control area and conventional power production units are substituted, the impact of wind power on the power system may become critical from the power system stability point of view and must be handled.

The increased penetration of wind power in the power system generates thus new challenges for power system operators, who have to ensure reliable and stable grid operation. In countries with large amount of wind power e.g. in Denmark, Germany, Spain, the UK or Ireland, this has resulted in the power system operators introducing grid codes [1–5]. These grid codes specify technical grid connection requirements for wind turbines connected to the power system. Wind turbines are requested to act as active components in the grid and have to take over control tasks, which are traditionally aligned to conventional power plants. Among others, these grid code specifications require fault ride-through capability of wind turbines, which means that the wind turbines must be able to remain connected to the network during grid faults. This addresses primarily the turbine's protection system and the design of the wind turbine's controller. Moreover, modern wind turbines are demanded to assist the power system by supplying ancillary services. These represent a number of services required by the power system operators such as active and reactive power supply to support the system's frequency and voltage level respectively in order to secure safe and reliable grid operation.

In the last twenty years there has been a process of maturation of wind turbine concepts. The first wind turbines were fixed speed stall controlled turbines, which means they were designed simple and robust. Today, most wind turbines are pitch controlled and operate with variable speed due to the frequency converter they are connected to. The frequency converter in turn facilitates better controllability of the turbine especially in terms of power system support. Grid support capabilities of different wind turbine concepts depend thus on their particular technology.

2 Wind Power Integration Issues

As already mentioned, wind power has reached a significant level in electrical power production in some countries such as Germany and Denmark [6]. This development is to be awaited in more and more countries focusing on renewable energies e.g. in Spain, USA, Great Britain and especially in countries like China and India with strongly rising energy demand. According to the Global Wind Energy Council [7] 38,265 MW of wind power have been installed in 2010, which results in a cumulative installed wind power capacity worldwide of 197,039 MW. The leading countries were China with 18,928 MW, the USA with 5,115 MW and India with 2,138 MW installed wind power capacity in 2010 [7]. Due to the increased penetration of wind power and its growing impact on the power system the challenge of wind power integration becomes extremely important in those countries.

Moreover, the role of wind power in the power system has changed in the last years. Traditionally wind turbines were connected to the distribution system, while today large wind farms are directly connected to the transmission system, replacing capacity, which was originally provided by conventional centralized power plants. Thus, power system operation and its system stability are significantly influenced by this development. Large wind farms must adopt the tasks of conventional power generation units, such as the control of active and reactive power flow. Moreover, as wind power is generally connected to the grid in remote areas, such as coastal areas, where the network is weaker, the power system impact of wind power is of special importance. These aspects are discussed in the following subsection.

2.1 Impact on Power System Stability

As stated in Tande [8] basic experience has been gained concerning wind power in distribution systems. Up to now, small wind power units were connected to the distribution system without any concern about the impact on power system stability and voltage quality, as their influence was little. However today, as wind farms are of larger size (several 100 MW) and therefore connected to the transmission system, they are causing a significant impact on the power system. This is due to the following reasons [9]:

- The contribution of conventional power plants to power system control is decreasing due to unbundling, decentralisation and replacement of conventional power plants by renewable energy sources, especially wind power.
- Wind power is installed in larger and concentrated units, e.g. offshore wind farms, and therefore the impact of wind power becomes significant.
- Wind power is often installed at remote places and in weak grids, e.g. at coastal areas far away from the consumption centres.
- Wind power is a fluctuating power source depending on the prevailing wind and must be balanced.
- Larger variation in generation and consumption causes larger current fluctuations and node voltage variations, which must be balanced.

Power system stability is thus a matter of interest in terms of large wind power penetration or wind power in very weak grids and needs to be investigated. Power system stability addresses primarily two tasks:

1. Voltage control—reactive power compensation.
2. Frequency control—active power dispatch

1. In order to keep the system voltage constant at each busbar in the grid, the reactive power production and consumption must be in equilibrium at each point in the system. This task was traditionally performed by large centralized power plants placed in vicinity to the grid nodes at which reactive power compensation was necessary. A problem occurs if this task is today assigned to

wind turbines located at remote places in distance to the consumption centres. Furthermore, some wind turbine concepts have reactive power demand by themselves, which varies depending on the wind speed and which must be compensated, too. The ability of wind turbines to supply reactive power is also a very important issue in respect of short term voltage stability, e.g. after voltage dips due to grid faults. Thus, voltage control capability of wind turbines is an increasingly important aspect regarding wind power grid integration and the wind turbine's market potential [9].

2. As wind energy is naturally a fluctuating power source, frequency regulation and active power dispatch become a challenging factor when the available power production is uncertain [8]. Another aspect in this context is, that wind turbines until now were allowed to disconnect in case of grid faults. In areas with high wind power this could cause a significant loss of electrical power production due to disconnection of large wind turbines, which could negatively influence system stability and lead to major blackouts. Thus, transmission system operators require of wind turbines to be able to ride-through temporary faults, so that large losses of active power supply can be avoided. It must furthermore be investigated to which extent wind power can contribute to primary and secondary control. A proposed solution to this problem is e.g. to curtail wind power and to release the power reserves when needed.

An overview over present grid code requirements is presented in the following.

2.2 Grid Codes

Before 2003, most grid codes did not require wind turbines to support the power system during a grid disturbance—wind turbines were only required to be disconnected from the grid when an abnormal grid voltage was detected. Nowadays since power systems have higher wind power capacity, the disconnections of wind turbines due to grid faults results in a significant loss of power production. This could generate frequency and voltage control problems in the system and could lead to system collapse in a worst case. The increased penetration of wind energy in the power system over the last decade has therefore lead to a serious concern about its influence on the dynamic behaviour of the power system. It has resulted in the power system operators revising the grid codes in several countries, e.g. [1–5]. Basically, for wind power these grid codes require an operational behaviour more similar to that of conventional generation units. An overview of existing grid connection codes is given in Iov [10–12]. These regulations are generally determined for wind farms connected to the transmission grid but some power system operators also specify grid codes for distribution systems. The attention in these requirements is drawn on both the wind turbines fault ride-through capability and on wind turbine grid support capability, i.e. their capability to assist the power system, by supplying ancillary services [13, 14]. These ancillary services concern today generally the following issues [15].

2.2.1 Fault Ride-Through

The fault ride-through demand requests wind turbines to stay connected to the grid during faults in order to avoid significant losses of active power production in systems with high amount of wind power. Before the fault ride-through requirement was introduced wind turbines had to disconnect in case of a fault. As most of the wind turbines were predominantly conventional fixed speed wind turbines with direct grid connected generator they could cause large inrush currents at voltage recovery [16]. With installation of the anticipated wind power (e.g. 20 % in 2020 [17]) such disconnection could however cause a sudden loss of several thousand megawatts. Thus, wind turbines, which are connected to the power system today, are requested to ride-through grid faults.

The fault ride-through requirement addresses both an advanced design of the wind turbine controller [13] and the development of new technologies/equipment in wind turbines [18] to enable the wind turbine to remain connected to the network and to continue normal operation and power production directly after fault clearing. The fault ride-through requirement defines rules of how long and how deep a voltage drop must be at which the wind turbines are allowed to disconnect [15].

2.2.2 Reactive Power Supply

During grid faults the system voltage drops in vicinity to the location of the fault. In order to support and re-establish the voltage level voltage control and reactive power supply are required from wind turbines by transmission system operators. The ability to deliver reactive power to the grid is strongly dependent on the wind turbine technology. Grid codes require not only compensation of the wind turbine's own reactive power demand but also additional reactive power supply in dependency of the voltage dip. The requirement can however be met by means of capacitor banks and power electronics connected close to the wind turbines. Moreover, modern variable speed wind turbines with frequency converters can accomplish the demand by means of advanced control.

2.2.3 Frequency Stability

Another issue related to system stability is the operation of the generation units under deviating frequency. First, it must be assured that the generation unit stays connected to the system even if the frequency varies. Second, its ability to contribute to primary and secondary control is important. Grid codes specify rules to adapt the wind turbine's active power production according to frequency changes in the system. However, wind is naturally fluctuating so that active power is not necessarily available on demand. Nevertheless, some grid codes, e.g. the Danish grid code [2], precisely specify to which extent wind turbines must contribute to adjust their power output. On demand of the transmission system operator the wind power production must be reduced below its available output (delta control) and moreover the gradient of the power production may be specified and limited (gradient constraints).

Other aspects are the contribution to active power supply on demand, the requirement to provide balancing power and to curtail wind power production in order to provide spinning reserve. In this respect wind power forecast and the introduction of incentive market mechanisms for wind turbine operators are important issues.

Additionally to the requirements mentioned above, transmission system operators demand modelling, simulation and verification of the wind turbine system. Communication and external control of the wind turbine especially of large wind farms are also required but represent anyway standard equipment in modern wind turbines. Other issues as e.g. the contribution of wind farms to perform a "black start", which means to start and operate in an islanding system, are discussed but no yet required [15].

Grid codes are generally different for wind turbines connected to medium voltage grids or high/extra high voltage grids [1, 2] and are more stringent for the high voltage network. Today, most of the wind turbines are still connected to the medium voltage grid. However, the installation of larger units and especially of large offshore wind farms will always require connection to the high voltage grid [18]. As stated in Ackermann [9], in contrast to distribution systems the R/X value of transmission systems is lower, which intensifies the need of reactive power control. Thus, the most rigorous grid code requirements for high voltage grids refer to fault ride-through capability and reactive power control especially during voltage sags [19]. In the following the focus will therefore be on fault ride-through and grid support capability of wind turbine connected to the transmission system with voltages above 100 kV.

As an example the fault ride-through requirement of one of the harshest grid codes [1] is representatively illustrated in the following.

2.2.4 Grid Code Example

The German transmission system operator E.ON Netz GmbH was the first power system operator, who introduced grid codes for wind turbines and is followed now by many other network operators in several countries. Up to now in the north of Germany power system operators have to face the highest amount of wind power in their control areas world wide.

In this grid code a voltage profile shown in Fig. 1 defines a limiting voltage, above which fault ride-through of wind turbines is required in case of a voltage drop. The voltage profile is divided into 4 areas. Above the dotted line corresponding to area 1 no tripping of the generation unit is permitted. The solid line, defining area 2 requires also fault ride-through but allows short term interruption of the generation unit in case of occurring instability. The voltage profile defines a voltage drop down to 0 % for 150 ms continuing with linear voltage recovery to

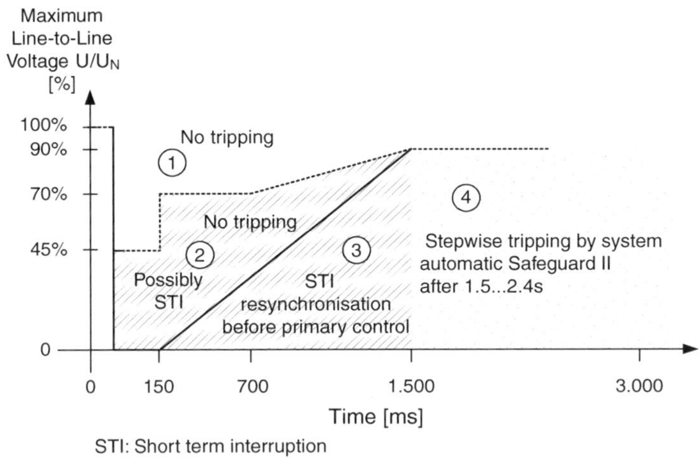

Fig. 1 Voltage profile for fault ride-through requirement of E.ON Netz GmbH [1]

90 % at 1.5 s after the fault incident. For area three and four in Fig. 1 no fault ride-through is required.

Similar voltage profiles for fault ride-through of wind turbines are defined in other grid codes such as in Denmark [2], Spain [5], UK [4], Ireland [3], USA [20] etc.

Besides the fault ride-through requirements the E.ON grid code preferentially requires reactive power supply during voltage sags and reactive current injection is prioritized against active current. In contrast, other grid codes define more precisely in which way active power can be reduced and reactive power must be increased. The fault ride-through of wind turbines exposed to the afore shown voltage profile will exemplarily be presented in Sects. 5.2 and 6.2.

3 Wind Turbine Concepts

Since the beginnings of wind turbine development ($\sim$1980) up to now (2010), wind energy has become a mature technology and has gained a significant role in the power system. During the last decades various wind turbine concepts and designs have been developed. The marketable wind turbine concepts can be distinguished by different electrical design and control and can be classified by their speed range (variable speed, fixed speed) and power controllability (stall, pitch control) [21, 14]. In the following, a general overview about wind turbine components and topology is given. Then the characteristics of different wind turbine concepts are presented and are assessed in respect to their controllability, grid code accomplishment. Finally, their market penetration is evaluated.

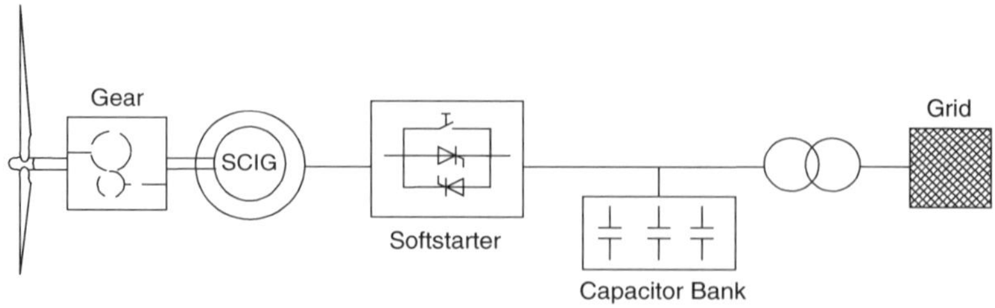

Fig. 2 Squirrel cage induction generator concept with gearbox and direct grid connection—Type A

3.1 Characteristics of Different Wind Turbine Concepts

Four different wind turbine concepts have predominated the global market in the last decade [21]. For the sake of uniformity the same classification of concepts as presented in Hansen [21] and Ackermann [9] is made in the following.

Type A: Fixed speed wind turbine concept—Danish concept

Figure 2 shows the widely used and most conventional turbine concept. A stall or active stall controlled aerodynamic rotor is coupled via gearbox to a squirrel cage induction generator (SCIG), which in turn is directly (by transformer) connected to the power grid. A capacitor bank provides reactive power compensation and improved grid compatibility during grid connection is facilitated by means of a softstarter.

Due to its direct grid connection the generator operates at fixed speed. However, the slip of the generator allows very small speed variations and slightly softens the torque-speed characteristic. Frequently, the generator is equipped with a pole changeable stator (e.g. 4-poles and 6-poles), so that the generator can operate at two speeds, which leads to a better aerodynamic utilization of the turbine at higher wind speeds. The turbine concept excels in its cheap and simple design and robustness, while however its controllability is relatively poor.

A more advanced design of this turbine concept is to use adjustable rotor blades. The turbines are then active stall controlled, which means that the stall effect, limiting the absorbed aerodynamic power, can actively be influenced. This can be utilized to achieve a higher energy capture at high wind speeds but also to curtail active power production of the turbine.

Type B: Variable speed wind turbine with variable rotor resistance

A slightly more advanced wind turbine concept is sketched in Fig. 3. The turbine setup is in principle the same as for Type A; however, now a wound rotor induction generator (WRIG) with external rotor resistance is used, which allows variable speed operation in a limited range of about 10 % above synchronous speed. The external rotor resistance enforces a higher slip, which on the one side enables variable speed operation but on the other hand increases ohmic losses, which are dumped in the resistance. As this also increases the reactive power

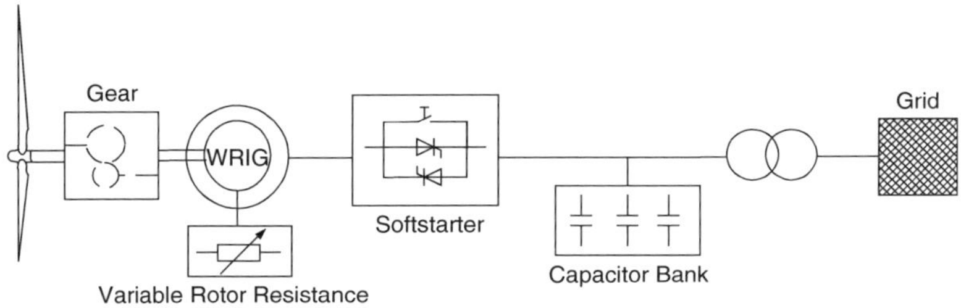

Fig. 3 Wound rotor induction generator concept with variable rotor resistance—Type B

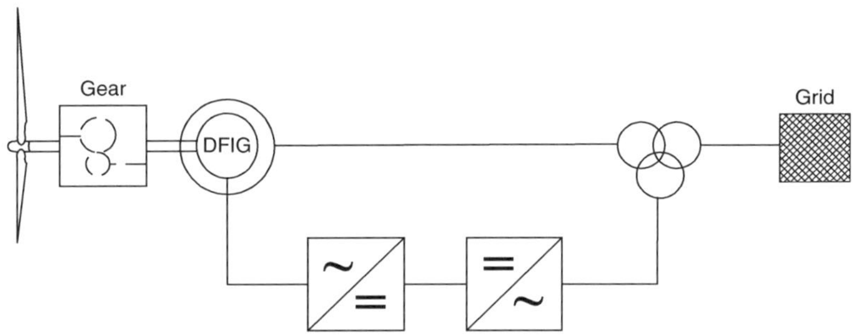

Fig. 4 Doubly-fed induction generator concept with partial-scale frequency converter—Type C

demand of the generator a capacitor bank is used for this type as well. As the turbine operates with variable speed it is advisable to use pitch control instead of stall control for active limitation of aerodynamic power above rated wind speed.

Type C: Doubly-fed induction generator wind turbine

Figure 4 shows the doubly-fed induction generator wind turbine concept, the most popular generator concept for wind turbines at the present time. The pitch-controlled aerodynamic rotor is coupled via gearbox to the generator. The doubly-fed induction generator (DFIG) provides variable speed operation by means of a partial-scale frequency converter in the rotor circuit. A wound rotor induction generator is used, in order to couple the converter via slip rings to the rotor. Depending on the converter size this concept allows a wider range of variable speed of approximately ±30 % around synchronous speed [21]. Moreover, the converter system provides reactive power compensation and smooth grid connection. As the frequency converter only transmits the rotor power it can be designed for typically 25–30 % of the total turbine power. This makes the turbine concept very attractive from an economic point of view compared to turbines with full-scale converter.

Type D: Variable speed wind turbine with full-scale frequency converter

Type D (Fig. 5) represents the variable speed, pitch controlled wind turbine concept with the generator connected to a full-scale frequency converter. The

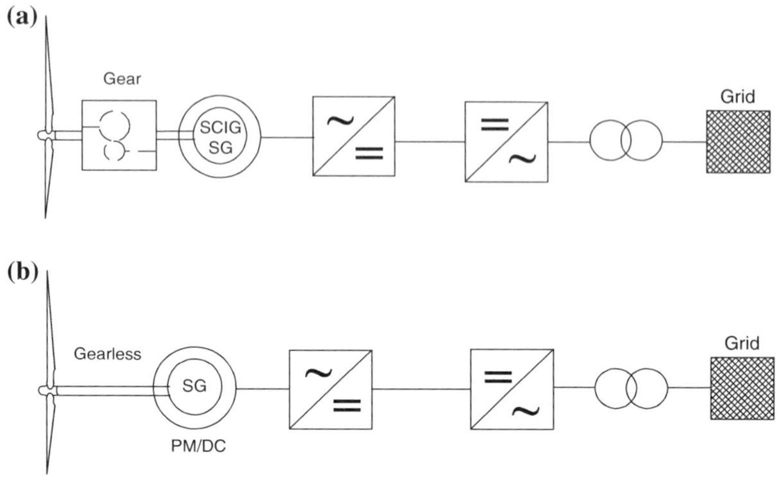

Fig. 5 Variable speed wind turbine with full-scale frequency converter—Type D: **a** Induction or synchronous generator concept in combination with gearbox and full-scale frequency converter. **b** Gearless synchronous generator concept with full-scale converter

full-scale frequency converter provides variable speed over the entire speed range of the generator. At the same time the converter guarantees reactive power compensation and smooth grid connection.

The generator can optionally be an asynchronous or synchronous generator, as shown in Fig. 5a. If a multipole synchronous generator is used instead of an induction generator, the generator can be built to operate at low speeds, so that a gearbox can be omitted. This reduces losses and maintenance requirements. The generator can be excited electrically by a DC system (DCSG) or by means of permanent magnets (PMSG). This configuration is shown in Fig. 5b.

3.2 *Market Penetration of Different Wind Turbine Concepts*

Figure 6 shows the results of a wind turbine market assessment from Hansen [21], which is based on data provided by BMT Consults.

During the years 1995–1997 the fixed speed wind turbine concept (Type A) was the predominating wind turbine technology, which was due to the simplicity and robustness of the system. However, from 1997 the DFIG wind turbine (Type C) gained an increasing market penetration. The market penetration of the variable speed wind turbine concept with full-scale frequency converter—Type D—shows no considerable changes during these considered 10 years. However, for the coming years a strongly increasing trend of Type D wind turbines is expected, due to their good control and grid support capabilities. However, recent studies such as

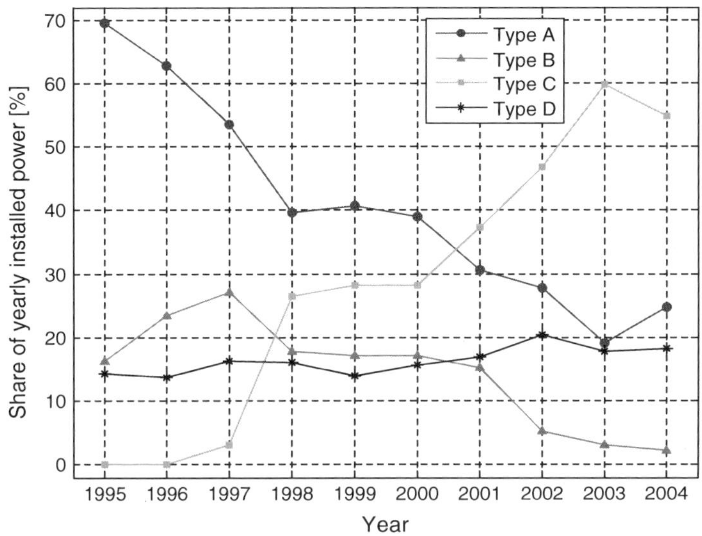

Fig. 6 World market share of yearly installed wind power during 1995–2004 of the four different wind turbine concepts Type A–D (*Source* [21])

[22] and [23] show an increasing trend of Type D wind turbines up to now (2010) and confirm the market dominance of Type C and D due to their good control and grid support capabilities.

4 Grid Support Capabilities of Direct Grid Connected Wind Turbines

In the present subsection the grid support capabilities of the above mentioned Type A and B wind turbines will be assessed, both types representing wind turbine concepts with direct grid connected asynchronous generators.

In the previous section it has been shown that fixed speed and direct grid connected wind turbines were the most installed wind turbines in the early 1990s. The major advantages of this concept are its simplicity and robustness, as standard asynchronous generators can be used and power electronics can be omitted causing economical benefits and reducing maintenance requirements. However, today this benefit turns into a major drawback, as power electronics enable wind turbines to contribute to power system support and to comply with grid code requirements.

4.1 Active Power Dispatch

Fixed speed stall controlled wind turbines (Type A) have a passive control of the absorbed aerodynamic power. Thus, active power production cannot be adjusted according to requirements in the power system and do not contribute to active power dispatch. However, due to the direct connection of the generators to the grid, these kinds of wind turbines add rotating mass and thus spinning reserve to the power system.

If in contrast the fixed speed turbine is active stall controlled, this feature can be utilized to reduce the turbine power by demand. Power can thus also be curtailed in order to provide reserve power if required. As for all wind turbines, this feature is however only available if there is sufficient wind speed.

In case of the Type B wind turbine (the asynchronous generator is equipped with variable rotor resistant) the slip and thus the speed of the generator can be actively influenced. Moreover, as the wind turbine is then pitch controlled it is able to reduce the incoming aerodynamic power. The Type B wind turbine can thus contribute to active power dispatch in the same way as the active stall wind turbine. Since its pitch drive acts generally faster than in active stall turbines, an even faster power reduction can be achieved.

4.2 Fault Ride-Through

In case of a fault the wind turbine concepts Type A and B are exposed to a severe voltage drop at the generator terminals due to their direct grid connection. This in turn causes a demagnetization of the generator during the fault. As the electrical torque is also drastically reduced, while the aerodynamic torque stays constant this causes heavy mechanical stresses for the drive train. Moreover, the turbine accelerates towards higher speeds. When the fault is cleared the generator has suddenly an increased reactive power demand, which negatively influences voltage re-establishment [24]. Fault ride-through means thus a negative impact on both, the wind turbine and the grid so that tripping of the turbines is general practice.

However, as mentioned before, Type A wind turbines with active stall control and Type B wind turbines with pitch control are able to reduce their turbine power. In this case the fault ride-through capability and the stabilization of the power system close to the turbines at a short circuit can be achieved by reducing the wind turbine power for duration of a few seconds after fault occurrence. In addition to that mechanical stress and increased reactive power demand are mitigated.

4.3 Reactive Power Supply

As the considered wind turbine concepts do not use any power electronic interface their reactive power operational point cannot be adjusted. This means, that both the Type A and Type B wind turbine concept cannot contribute to reactive power supply. As shown in Figs. 2 and 3, these concepts are normally equipped with capacitor banks, in order to compensate for the own reactive power demand. Thus a limited adjustment of reactive power is possible. However, these capacitor banks can normally not be used for dynamic voltage control in case of a fault.

Type A and B wind turbine concepts can generally not comply with modern grid code requirements and their grid support capability is relatively poor. This fact has already influenced their market penetration negatively as shown in Fig. 6. However, in distributions systems or power systems, where grid code requirements are less severe there is still a market for them. Moreover, if large wind farms of this type will be coupled to the grid connected via an HVDC connection, grid support can then be provided by the grid side converter of the HVDC station.

5 Grid Support Capabilities of DFIG Wind Turbines

5.1 DFIG Control for Grid Support

The variable speed doubly-fed induction generator (DFIG) wind turbine is today the most commonly used wind turbine concept [21]. The doubly-fed induction generator wind turbine concept uses a partial-scale frequency converter in the rotor circuit. This converter controls the rotor voltage and thus performs independent control of active and reactive power at a speed range of approx. ± 30 % around synchronous speed [25]. Compared to wind turbine concepts with full-scale frequency converter, the partial-scale converter of the DFIG has the advantage of reduced converter size, costs and losses [26]. However, this configuration also has a major drawback in terms of fault ride-through as the converter has to be protected against high transient currents.

A representative control strategy of the DFIG wind turbine's electrical system is exemplarily shown in Fig. 7. The rotor side converter controls the active and reactive power production, which is fed via the stator into the grid. The grid side converter maintains the DC-link voltage and controls the reactive power of the grid side converter, which is set to zero during normal operation. In case of a grid fault this control is extended by means of several control features such as overspeed protection, converter protection, damping and voltage control, i.e. reactive power supply by the converters, indicated by the red blocks in Fig. 7 [27]. The details of the control and their ability to enhance fault ride-through and grid support are discussed in the following.

5.1.1 Active Power Dispatch

The control of the rotor side converter allows independent control of active and reactive power. Moreover, as the turbine is pitch controlled the aerodynamic power absorbed by the turbine can be adjusted. Thus, the DFIG wind turbine is able to reduce its output power by demand and can then contribute to active power

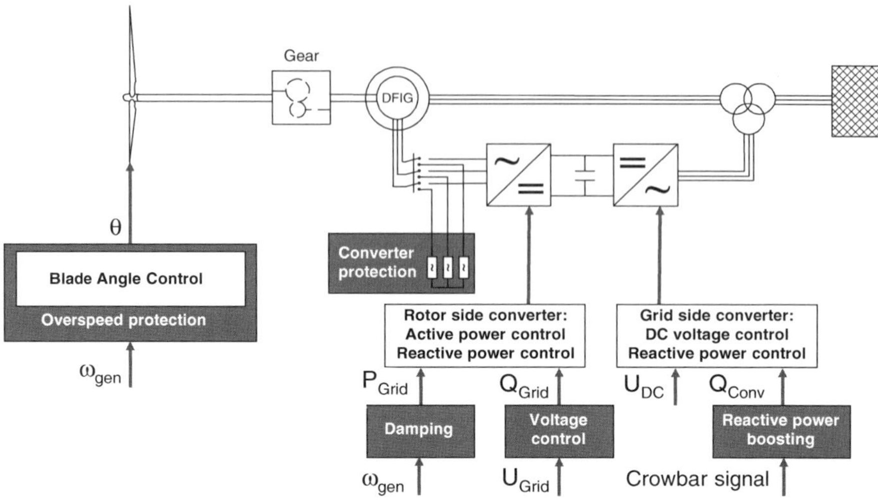

Fig. 7 Overview over the control and system configuration of a DFIG wind turbine [27]

dispatch. However, as wind is a fluctuating power source, the capability of active power dispatch is limited or dependent of the prevailing wind speed. Nevertheless, if the wind speed is large enough the DFIG wind turbine can curtail the power production and provide reserve power. Moreover, the turbine can also adjust its reactive power operational point at its terminals and contribute to power factor control. However, as the DFIG rotor current is used in inner control loops, the limitation of the rotor current size also limits the control capability of active and reactive power supply to the grid.

As the power control is realized by means of power electronics, the control has a very fast system response. Compared to conventional power plants such wind turbines can much faster contribute to power dispatch. Wind turbines with frequency converter are thus predestined to contribute to power system support requiring fast reaction time.

5.1.2 Fault Ride-Through

The fault ride-through capability of the DFIG wind turbine is realized by three units: an over-speed protection, a damping controller and the converter protection indicated in Fig. 7 [27]. During a grid fault voltage and power at the wind turbine terminal drop and thus the power in the DFIG drops, too. This results in an acceleration of turbine and generator. In this case the blade angle control serves as an overspeed protection and the absorbed aerodynamic power is reduced. Moreover, when the electrical power drops the drive train will start to oscillate. Due to this reason a damping controller is implemented, which actively damps any torsional oscillations of the drive train and prevents instability of the system and substantially reduces the mechanical stresses of the turbine.

In addition to that a converter protection needs to be implemented. Due to the DFIG's direct grid connection via the stator, high transient currents and voltages can arise in the generator. As these high transient currents and voltages are also induced in the generator's rotor the rotor side converter must be protected. A common protection system applied in many DFIG wind turbines is the crowbar protection, e.g. an external rotor resistant. In case of a severe grid fault, the crowbar is triggered and the rotor side converter is subsequently blocked. In this case the controllability of the rotor side converter is temporarily lost. Nevertheless, the crowbar guarantees, that the wind turbine can stay connected to the grid and can thus directly contribute to power production, when the fault is cleared.

5.1.3 Voltage Control

In addition to fault ride-through capability the wind turbine must furthermore be enabled to support the grid and to contribute to voltage control by means of reactive power supply. This is realized by a voltage controller, which is added in cascade to the power controller (Fig. 7). The voltage controller demands the rotor side converter control to supply reactive power to the grid in order to re-establish the voltage level. However, when the rotor side converter is blocked by the crowbar, the task of reactive power supply is then temporarily assigned to the grid side converter, which can stay active during a fault. As the grid side converter can control only a fraction of the turbine power (~ 30 %), it always contributes with its maximum reactive power to the grid during a grid fault and is thus called "reactive power boosting" in Fig. 7 and is activated by the crowbar signal.

If the DFIG wind turbine is equipped with a control system as explained above, it can accomplish fault ride-through and contribute to grid support as required in the grid codes.

5.2 Simulation Example: DFIG Wind Turbine Exposed to Voltage Drop

In order to verify the ability of wind turbines to comply with grid codes, power system operators require simulation models of the manufacturers. In the following example such a simulation model of a 2 MW DFIG wind turbine is subjected to the voltage profile defined by the E.ON grid code [1] shown in Fig. 1. The grid between the wind turbine and the occurrence of the voltage drop is simulated by means of a Thevenin equivalent with short circuit power of ten times the wind turbine power and an R/X ration of 0.1 [2, 14]. Simulation results are shown in Fig. 8.

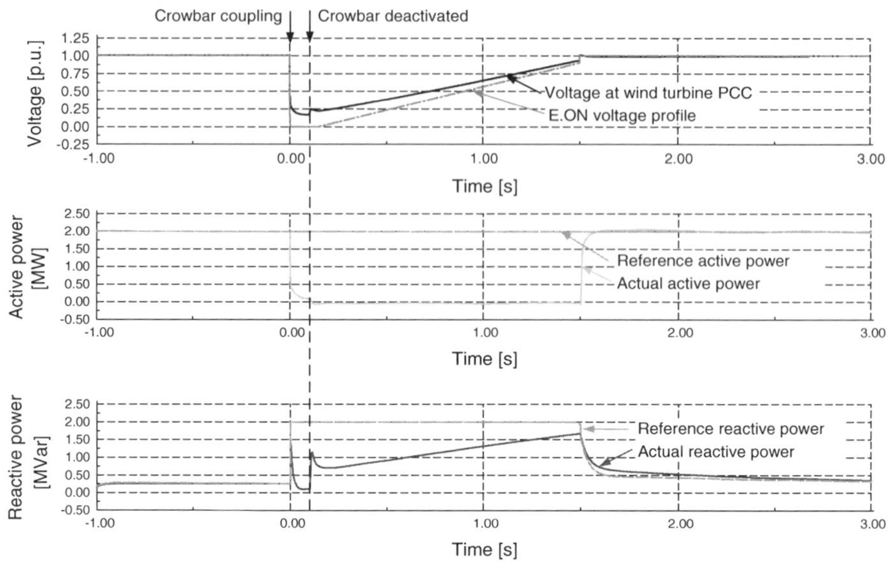

Fig. 8 Voltage profile for fault ride-through applied to a DFIG wind turbine: Voltage profile and voltage at PCC, active and reactive power production of the DFIG wind turbine

In the uppermost plot, the voltage profile is plotted together with the voltage characteristic at the wind turbine connection point (PCC). In addition to that, the active and reactive power production of the DFIG wind turbine is shown together with their corresponding reference values of the controller.

According to the voltage profile of the grid code, a voltage drop down to zero for 150 ms is followed by a slow recovery of the voltage within 1.5 s. Due to the severe voltage drop the crowbar of the DFIG is coupled immediately at the fault incident in the current example. During this time the rotor side converter is blocked but the grid side converter provides a small amount of reactive power to the grid. After the crowbar is disconnected at 100 ms the rotor side converter contributes with reactive power. In this moment the active power decreases to zero as reactive power production is prioritized by the rotor side converter control. In the simulated case the DFIG contributes with its maximum reactive power.

Due to the crowbar protection fault ride-through can be accomplished and the turbine can stay connected. The turbine is thus able to contribute to power production immediately after voltage recovering. Moreover, reactive power supply supports the grid and improves the voltage level at the wind turbine terminal. However, during crowbar coupling the controllability of the DFIG and thus the reactive power contribution is temporarily compromised. Due to this reason manufacturers try to avoid crowbar coupling or aim to provide more advanced solutions for DFIG wind turbines to comply with grid code requirements.

6 Grid Support Capabilities of Full-Scale Converter Wind Turbines

For wind turbines using grid connection via a full-scale frequency converter it can be presumed, that such wind turbine concepts have very good grid support capability compared to any other wind turbine concept. Due to the larger converter rating the amount of reactive power, which can be provided to support the grid voltage, is larger than for any other wind turbine concept [9]. Furthermore, the converter system decouples generator and turbine from the grid so that both are less subjected to the grid fault impact than turbines with direct grid connected generator [28].

Wind turbines with full-scale frequency converter can have different generator types, which are coupled to the turbine via gearbox or operate direct driven so that the gearbox can be omitted (Fig. 5). Nevertheless, because all concepts use a full-scale frequency converter their grid support capability is equal [14]. Due to this reason a multipole PMSG wind turbine is representatively chosen for the following considerations.

6.1 PMSG Control for Grid Support

The control of the PMSG's electrical system is exemplarily illustrated in Fig. 9. The control of the frequency converter is realized as follows: The generator side converter controls the DC-link voltage and the stator voltage of the generator. On the other side, the grid side converter is used to control the active and reactive power supply to the network. As stated in Michalke [14] this control strategy is very beneficial for fault ride-through and grid support but other control strategies can be used as well. In case of a grid fault this control is extended by means of several control features such as an overspeed protection, damping and chopper module. Furthermore a voltage controller is added in cascade to the converter control facilitating reactive power supply [29]. The turbine's grid support capability is assessed in the following.

6.1.1 Active Power Dispatch

Due to the coordination of converter control and pitch control, variable speed wind turbines with full-scale frequency converter are able to curtail their output power by demand. This means, that reserve power is provided and if the wind speed is large enough the turbine can increase and decrease power production to reference values imposed by the power system operator. However, as stated before power dispatch depends on the wind, a fluctuating power source. But due to the fast control of the frequency converter, active power dispatch can immediately be provided.

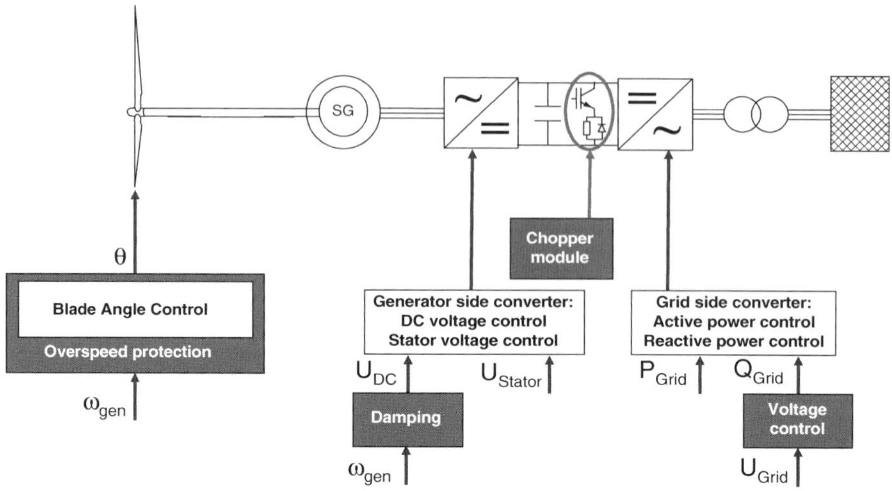

Fig. 9 Overview over the control and system configuration of a PMSG wind turbine [29]

Moreover, due to the converter the turbine can also adjust its reactive power operational point at its terminals and contribute thus to power factor control. Since this wind turbine type has a full-scale converter compared to the DFIG wind turbine its controllers have a larger range before they reach their limitations. This means, active and reactive power can be adjusted in a wider range than with the DFIG.

6.1.2 Fault Ride-Through

Similar as for the DFIG wind turbine, fault ride-through capability is facilitated and enhanced by three units (Fig. 9). Again, the pitch control serves as an over-speed protection and a damping controller is added in order to damp torsional oscillations in the drive train. However, in contrast to the DFIG wind turbine, no converter protection is necessary. As the generator is decoupled from the grid by the converter no transient currents and thus no damaging risk for the converter arise. Nevertheless, in order to enhance fault ride-through capability a chopper module is implemented in the DC-link of the PMSG wind turbine. When the terminal voltage drops in case of a fault, the output power, which is fed into the grid, is significantly reduced. However, turbine and generator will continue to deliver their power into the DC-link. In this case the chopper is triggered to dump the surplus power of the turbine during a grid fault into an additional resistance.

6.1.3 Voltage Control

The PMSG wind turbine must also be enabled to provide grid support by reactive power supply during faults. Due to this reason a voltage controller is added to the grid side converter control as shown in Fig. 9. In case of a voltage drop the grid side converter is then asked to provide reactive power in order to re-establish the grid voltage. In comparison to the DFIG the full-scale converter used in a PMSG wind turbine can provide a larger amount of reactive power. This yields that the voltage level is supported to a higher level and recovers faster.

The PMSG wind turbine itself is furthermore fully decoupled from the grid by the converter and is thus less subjected to the impact of a grid fault. Due to appropriate control both fault ride-though as well as active and reactive power supply can easily be accomplished by the PMSG wind turbine and all other wind turbines using a full-scale frequency converter.

6.2 Simulation Example: PMSG Wind Turbine Exposed to Voltage Drop

In order to exemplify the PMSG wind turbine's grid support capability, a simulation case is presented, in which a PMSG wind turbine is exposed to a voltage drop as specified in the E.ON grid code [1]. The grid between the wind turbine and the occurrence of the voltage drop is modelled by means of a Thevenin equivalent with short circuit power of ten times the wind turbine power and an R/X ration of 0.1 [2, 14]. Simulation results are shown in Fig. 10: The voltage profile is plotted together with the voltage characteristic at the wind turbine connection point (PCC). In addition to that, the active and reactive power production of the DFIG wind turbine is shown together with their corresponding reference values of the controllers.

As the PMSG wind turbine is connected via a full-scale frequency converter the control is able to provide maximum reactive current to the grid. Reactive power supply is then prioritized over active power production. However, reactive power production is of course limited due to the reduced voltage level. The active power is reduced to zero and the surplus power of the turbine is burned in the chopper. After the voltage has totally recovered the pre-fault steady state is regained.

It can be concluded that the PMSG wind turbine with full-scale frequency converter can ride-through grid faults and also comply with the reactive power supply requirements defined in the grid codes. In comparison to the DFIG wind turbine full-scale converter connected wind turbines show a slightly better performance in terms of grid code accomplishment as they can provide a higher amount of reactive power and do not need extra converter protection such as a crowbar. The only drawback of full-scale converter connected units is, that there is no direct dependency between active power feed-in, turbine speed and grid frequency. This means that these turbines do not add inertia to the power system as in contrast direct grid connected generators do.

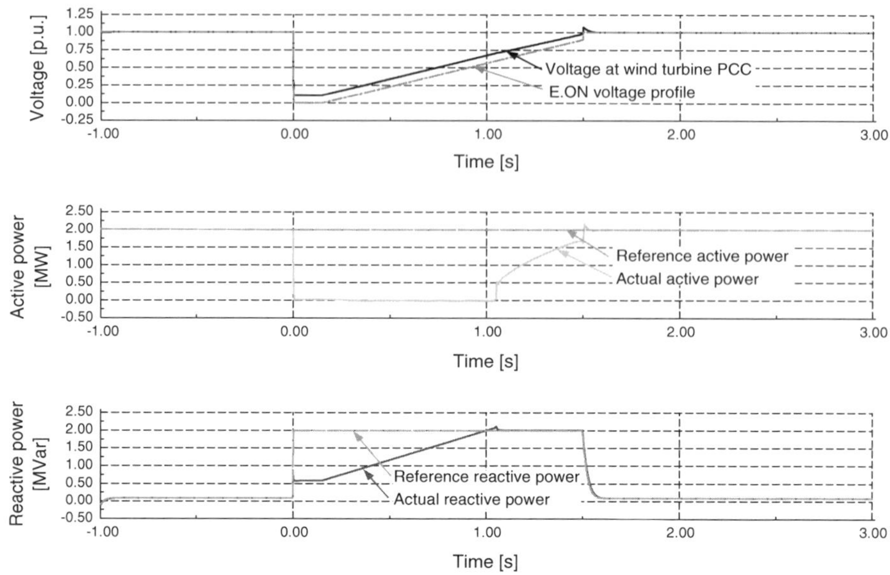

Fig. 10 Voltage profile for fault ride-through applied to a PMSG wind turbine: Voltage profile and voltage at PCC, active and reactive power production of the PMSG wind turbine

7 Discussions

Different wind turbine concepts exist on the market and have different capabilities to support the power system and to comply with modern grid code requirements.

The first generation of wind turbines used direct grid connected generators without any power electronics. Today these kinds of wind turbines are not able to provide ancillary services such as active power dispatch or reactive power supply during grid faults and are disconnected from the grid in case of a fault. However, if they are equipped with pitch or active stall control they can contribute to active power dispatch in a limited way and improve their fault ride-through capability.

In contrast to this, DFIG wind turbines use a partial-scale frequency converter and can thus control active and reactive power to imposed reference values if the wind speed is sufficient. However, the reactive power supply capability of DFIG wind turbines is smaller compared to full-scale converter connected wind turbines. Moreover, in case of a fault high transient currents arise in the DFIG. Thus the rotor side converter must be protected and might even be temporarily blocked, which reduces the ability to support the grid during faults. Nevertheless, the DFIG wind turbine can accomplish fault ride-through and complies with modern grid code requirements.

As mentioned before wind turbines with full-scale frequency converter have a better grid support capability than all other wind turbine concepts. Depending on the wind speed these wind turbines can contribute to active power dispatch. Due to the full-scale converter system they can provide a higher amount of reactive power to the grid than DFIG wind turbines. In case of a fault this yields that the voltage level is supported to a higher level and recovers faster. A full-scale converter connected wind turbine itself is furthermore fully decoupled from the grid by the converter and is thus less subjected to the impact of a grid fault.

The above summarized facts show that modern wind turbines using power electronic interface are able to support the power system by providing ancillary services. Since power electronics facilitate flexible control and fast system response, wind turbines are rather predestined to contribute to fast power system control tasks. However, up to now wind turbines were installed decentralized and in small units compared to large centralized power plants, which made power system control more challenging. But as the trend is towards larger centralized wind farms, e.g. offshore, this drawback may be mitigated. Moreover, new approaches as virtual power plants or smart grids are suitable to integrate renewable energy production units and will enhance power system operation in the future.

References

1. E.ON. Netz GmbH (2006) Netzanschlussregeln, Hoch—und Höchstspannung. www.eon-netz.com
2. Energinet dk (2004) Wind turbines connected to grids with voltage above 100 kV Technical regulation TF 3.2.5. www.energinet.dk
3. ESB National Grid Ireland (2007) Grid code version 3.0. 28th Sep 2007
4. National Grid Electricity transmission plc (2007) The grid code. Issues 3, Revision 24, UK
5. REE (2005) Requisitos de respuesta frente a huecos de tension des las instalaciones de produccion de regimen especial. PO 12.3, Spain
6. American Wind Energy Association (2004) Global wind energy market report 2004. www.awea.org. Accessed Nov 2007
7. Global Wind Energy Council (2010) Statistics. http://www.gwec.net
8. Tande JOG (2003) Grid integration of wind farms. Wind Energy 6(3):281–295
9. Ackermann T (2005) Wind power in power systems. John Wiley and Sons Ltd, New Jersey
10. Ciupuliga AR, Gibescu M, Fulli G, Abbate AL, Kling WL (2009) Grid connection of large wind power plants: a European overview In: 8th international workshop on large-scale integration of wind power into power systems as well as on transmission networks for offshore wind farms, Bremen
11. Iov F, Hansen A, Soerensen P, Cutululis N (2007) Mapping of grid faults and grid codes. Risoe Report Risoe-R-1617(EN). Risoe National Laboratory, Technical University of Denmark
12. Tsili M, Patsiouras C, Papathanassiou S (2008) Grid code requirements for large wind farms: a review of technical regulations and available wind turbine technologies, European wind energy conference (EWEC), Brussels
13. Hansen AD, Michalke G, Soerensen P, Lund T, Iov F (2007) Co-ordinated voltage control of DFIG wind turbines in uninterrupted operation during grid faults. Wind Energy 10(1):51–68
14. Michalke G (2008) Variable speed wind turbines—modelling, control, and impact on power systems. PhD thesis, Darmstadt University of Technology, Germany. Published online: http://tuprints.ulb.tu-darmstadt.de/1071/
15. Milborrow D (2005) Going mainstream at the grid face. Windpower Monthly pp 47–50

16. Soerensen P, Cutululis NA, Lund T, Hansen AD, Soerensen T, Hjerrild J, Donovan MH, Christensen L, Kraemer Nielsen H (2007) Power quality issues on wind power installations in Denmark. Power engineering society, General meeting, IEEE Volume, Issue 24–28, pp 1–6
17. European Commission (2007) Communication from the commission to the council and the European parliament, renewable energy road map, renewable energies in the 21st century: building a more sustainable future. www.ec.europa.eu/energy/energy_policy/doc/03_renewable_energy_roadmap_en.pdf. Accessible Nov 2007
18. Erlich I, Winter W, Dittrich A (2006) Advanced grid requirements for the integration of wind turbines into the German transmission system. Power engineering society general meeting 2006 IEEE, 7:18–22
19. Robles E, Villate JL, Ceballos S, Gabiola I, Zubia I (2007) Power electronics solutions for grid connection of wind farms, European wind energy conference (EWEC), Milan, 7–10 May 2007
20. Federal Energy Regulatory Commission (FERC) USA (2005) Interconnection of wind energy. 18, CFR Part 35, Docket No RM05-4-001, Order No. 661-A, 12 Dec 2005
21. Hansen AD, Hansen L (2007) Wind turbine concept market penetration over 10 years (1995–2004). Wind Energy 10(1):81–97
22. Global Direct Drive wind turbines Market (2011–2016) Key trends and opportunities, new Installations and product developments and lower-maintenance requirements Will Drive Market Growth. http://www.transparencymarketresearch.com/direct-drive-wind-turbines-market.html
23. Kearby J (2012) Wind Converters – World. http://imsresearch.com/report/Wind_Converters_World_2012&cat_id=115&type=LatestResearch
24. Akhmatov V (2003) Analysis of dynamic behavior of electric power systems with large amount of wind power. Ph.D thesis, Ørsted DTU, Denmark
25. Leonhard W (2001) Control of electrical drives. 3rd edn. Springer, Stuttgart, ISBN: 3-540-41820-2
26. Hansen AD, Iov F, Sørensen P, Blaabjerg F (2004) Overall control strategy of variable speed induction generator wind turbine. Nordic wind power conference 1–2 March, Chalmers University of Technology, Göteborg
27. Michalke G, Hansen A, Hartkopf T (2007) Dynamic behaviour of a DFIG wind turbine subjected to power system faults. European wind energy conference (EWEC), Milan
28. Akhmatov V (2006) Modelling and ride-through capability of variable speed wind turbines with permanent magnet generators. Wind Energy 9(4):313–326
29. Michalke G, Hansen A, Hartkopf T (2007) Fault ride-through and voltage support of permanent magnet synchronous generator wind turbines. Nordic wind power conference (NWPC), Roskilde, 1–2 Nov 2007

Coordination Between Wind Farms and Storage Devices, Technical and Economic Aspects

Edgardo D. Castronuovo and J. Usaola

Abstract The increased participation of the wind power production in the electric systems requires the analysis and adoption of new operational techniques. One of the alternatives most frequently cited in the literature is the coordination of wind farms with storage devices, with different purposes. The present chapter studies different approaches for this methodological option, presenting results from the recent literature. Also, this chapter presents an alternative for calculating the optimal operation of storage devices in coordination with wind generation. Finally, the interaction of wind farms with reserve markets is fully explained.

1 Introduction

In the last years, wind power has increased its participation in power systems worldwide, and especially in Europe, USA and China. Wind power has important advantages, including: reduction of the CO_2 emissions, local availability, reduction of the dependence to external energy sources, utilization of renewable energy resources, etc. One of the main problems associated to wind power generation is the impossibility to predict the time-behaviour of the actual production of the plants at a given time, due to the limited controllability of the technology and the variability of the energy resource. As wind cannot be stored, the variation in the power injected in the system with respect to the schedule (due to errors in the

E. D. Castronuovo (✉) · J. Usaola
Department of Electrical Engineering, University Carlos III de Madrid, Av. de la Universidad 30, 28911 Leganés (Madrid), Spain
e-mail: ecastron@ing.uc3m.es

J. Usaola
e-mail: jusaola@ing.uc3m.es

P. M. Pardalos et al. (eds.), *Handbook of Wind Power Systems*, Energy Systems,
DOI: 10.1007/978-3-642-41080-2_17,

production forecast) must be compensated by other generators, to maintain the balance among generation, load and losses.

The error in the forecasting production of a wind farm depends, among other factors, on the quality of the prediction methods used to estimate the probable production. Many works analyze this issue. Between them, in [1] the requirements for additional reserve in one region of the system due to the compensation of forecasting errors in the wind production are analyzed. In this study, the error is represented as a function of the prediction horizon, using real historical data of a wind farm and correlation matrices. The study [2] also considers the forecasting horizon as the only predictor variable, proposing a function to estimate the possible errors. In [3] and other, the wind prediction is obtained by using ARMA series, transforming this result in wind power generation through real wind generator curves. Other works also consider the error dependence on the predicted power level, beside the forecasting horizon. References [4, 5] consider this correlations, using non-parametric distributions.

Independently of the prediction method used to estimate the future wind production, the real production is always different (with a larger or smaller error) to the predicted value. Therefore, compensation of the production deviation must be required for the adequate operation of the power system. Several works have utilized storage devices to perform this compensation. In the present work, a review of the main approaches proposed in the literature for the cooperation between wind farms and storage devices is performed. Next, a procedure to calculate the best operation of a storage device that compensates the production deviation of a wind farm is analyzed in detail. Finally, the interaction with actual reserve markets, as the most common procedure to compensate wind farm deviations is considered.

2 Review of Alternatives for the Coordination Between Wind Power and Storage

In the following review, the main research lines to analyze the cooperation between wind power generation and storage are summarized. Several approaches have been considered, including the utilization of water pumped storage, hydropower generation, batteries, superconducting magnetic energy storage and fuel cells. Also, important aspects of the cooperation, as the coordination in isolated systems, influence of statistical analysis, market behaviors and reliability, are also included in the cited bibliography. It must be stressed that the collaboration between wind energy and storage in automotive devices is not considered in this review, due to the incipient stage of this research line.

2.1 Utilization of Water Pumped Storage

Water storage plants are intensely used in many power systems, to transfer electric generation between periods of operation. These plants consist of two reservoirs at different heights, connected by pipes and hydraulic valves. Electric pumps can raise water from the lower reservoir to the upper one, storing electric energy as water. Part of this energy can be recovered by the power system, using conventional hydraulic generators. The efficiency of the storage cycle (pumping/storage/hydraulic generation) is between 70 and 80 %, depending on the utilized equipment. When working in a daily cycle, these plants absorb electric power from the system in low-price hours (generally, at night) and sell the storage energy in the high-price periods. However, the storage capacity of these plants can be also used to compensate the deviations in the wind power production, balancing the integrated operation.

In [6], the authors propose a new approach to coordinate a wind power plant with a hydro pump storage station. The operational limitations of these two plants are taken into account in the formulation. The proposed algorithm accomplishes to identify the best operation strategy of a combined W–H power plant with little water storage ability, determining the amounts of wind and hydropower to be generated for each hourly period. Pumping power consumption and storage level profiles are also calculated for every hour. Through Monte Carlo simulations, wind-power time-series scenarios are determined. For each one of them, an optimized daily operation strategy is determined by solving a linear hourly-discretized optimization problem. The Portuguese wind-energy remuneration tariffs for wind-power generation are used in this work to illustrate the economic gains of the combined operation. Three cases are analyzed: (a) output power operational band restrictions in all periods of the day; (b) only output hourly power restrictions during the off-valley hours; and (c) calculation of the maximum lower limit in the output power band. Besides the operational advantages obtained for the coordinated operation, the predicted yearly average economic gain of the joint strategy is between 425.3 and 716.9 k€, for the analyzed test cases.

The same authors, in [7] continue the research previously described, including an extensive account of the economic profits obtained from the operational cooperation of the wind and hydro units and analyzing the optimal size of the coupled water reservoir. Special attention is given to the formulation of the optimization problem (wind park developer profit maximization), taking into consideration specific physical and technical restrictions. The formulation proposed in [8] aims to assist the owner of a wind—hydro storage generating facility in improving the quality of the estimates of the hourly energy quantities it can offer in the market for the next hours, and the intervals of confidence for these estimates. The objective of the analysis is to enclose the wind production between narrow limits, using the storage capacity. The profit in the operation is also represented in the optimization problem, configuring a multi-objective analysis. Monte-Carlo simulations are used to represent the stochastic characteristic of the wind production. The work shows that by incorporating energy storage ability in wind plants, it is possible to increase the controllability of the power to be produced, enabling its participation in the market.

In [9], the authors perform a thoughtful inspection of the main problems for a very large integration of wind generation in power systems. In the study, several techniques for improving the behavior of the wind farms are analyzed, including: collaboration of the wind generators in critical situations, participation in the primary frequency control and wind generation aggregation through Dispatch Centers of renewable production.

In [2], the authors analyze the addition of a storage device, in the same location of the intermittent generator, to compensate the power generation deviations due to forecasting errors. The algorithm estimates possible errors in the production for the next periods, calculating the requirements of the storage device. The formulation is tested in two real case studies: using wind speed measurements from northern Norway with a 2 MW wind turbine model and on measurement data from a 500 kWp photovoltaic installation in Mont Soleil, Switzerland. Different forecast horizons are analyzed, up to 24 h. From the results, the reliability of the network can be improved with an appropriate energy storage device. The characteristics of the required storage can also be calculated from the simulations in different scenarios. The advantages on using real series in the analysis are emphasized by the authors, for allowing more accurate results than by using simulations.

The paper [10] introduces an alternative formulation for the coordination of a wind farm and hydro storage facilities. In this study, the costs for unbalancing the system can be included, performing stochastic analyses. The work [11] also considers the utilization of hydro-pumped storage to cooperate with the wind power production. In the analysis, future prices and wind power generation are considered stochastic variables. Also, the pump action is represented by using discrete values, due to the connection/disconnection of no continuously controllable pumps. The stochastic characterization is performed by using the Collocation method. Realistic data is used to test the proposed algorithm. The comparison of expected profits concludes that joint operation is beneficial for both wind and pump storage units.

In [12], the authors analyze the future utilization of pumped storage plants, including the coordination of them with wind farms. The study is centered in Germany and concludes by highlighting the importance of these plants for the adequate operation of the system. Moreover, the authors suggest that more than 14,000 MW could be incorporated in the German system in the next years, if the adequate incentives are introduced in the market.

2.2 Coordination with Hydropower

Classic coordination between wind farms and hydro storage considers the utilization of pumped water storage facilities. However, water can also be stored by reducing the hydro production in same periods. The water not used for generation proposes in these periods can be utilized in following hours, resulting in a water storage operation. Some papers have followed this approach.

The authors of [13] formulate a daily planning algorithm for a multi-reservoir hydropower system coordinated with wind power. The study considers that wind power and hydropower are owned by different utilities, sharing the same transmission lines. In the analysis, hydropower production has a priority for the utilization of transmission capacity. However, when congestion restrictions are applied to the production, the hydro production is reduced to allow an increased penetration of the wind power generation. The algorithm considers two successive steps: (a) calculation of the base case hydro power planning without consideration of the wind power and (b) re-planning of hydro power production, considering wind power forecast and its uncertainty. The analysis is performed in a projected wind farm facility in northern Sweden, near the Norwegian border. In the region, the transmission capacity is limited to 350 MW, constraining the introduction of power due to the new wind farm. From the results, the coordinated operation reduces wind energy curtailment by about 50 %, allowing an adequate use of the wind farm.

The work [14] also considers the cooperation between a wind farm and a hydro generation company, offering combined bids for the portfolio. The study considers the economic remuneration conditions of Spain, aiming to calculate the optimal joint bid of the hydro+wind power arrangement. The main advantage in the cooperation results in the reduction of the imbalance costs, in around 50 %.

In [15], the author analyses different alternatives to increase the wind power integration in power systems with transmission bottlenecks. Utilization of battery storage, pumped hydro storage and coordination of wind production with hydropower generation are options considered in the study. Other alternatives formulated in the literature to increase the transmission capacity (as temporary overloading, probabilistic evaluation of operational uncertainties, wind production curtailments and others) are also cited and analyzed. The author concludes that some of these methods could be interesting alternatives to reduce the cost of transmission system reinforcements.

In [3], the authors aims to determine the optimal hourly bids for the day-ahead market (spot market) by the hydropower utility, considering the uncertainty of the wind power forecast. A long-run (1 year) evaluation of the algorithm is included, showing the advantages of the coordinated action. The results of the case study show that coordination with wind power brings additional income to the hydropower utility and leads to significant reduction of wind energy curtailments. In the paper [16], the uncertainty of the power market prices is included in the analysis. The algorithm considers both spot and regulating markets. The coordinated action allows reducing the curtailments of the wind production, only resulting in about 25 % of unutilized wind production. The paper also present the economic benefits obtained in the coordinated action.

2.3 Utilization of Storage in Batteries

In the last decade, batteries have evolved significantly. Conventional and new flux batteries have reached adequate levels of storage capacity, efficiency and allowable number-of-cycles to be used in power systems.

In [17], the authors analyze the inclusion of a large battery in a wind farm. This study considers an existing wind farm in France with an installed capacity of 12 MW and the addition of redox-flow VRB (Vanadium Redox) or classical lead-acid batteries. The economical study is performed considering the French remuneration scheme in 2005. The results show that it is difficult to economically justify the inclusion of the batteries, mainly because the low lifespan of the current batteries at this year.

The authors of [18] analyze a distributed generation grid, which include conventional thermal generation, wind farms and photovoltaic solar production. The autonomous system contains also storage capacities, by using generic batteries. Genetic algorithms are used to solve the optimization problem. The study includes a comparative study of the performance of genetic algorithms and simulated annealing in the proposed formulation. In [19], the analysis of the performance of the different generation alternatives is expanded. The results demonstrate the capacity of the algorithm to obtain profitable solutions for a complex system including storage and several types of renewable generations, emphasizing the different performances of the system components.

In [20], an industrial customer with wind turbine generators in the Taiwan Power Company (TPC) system uses a storage battery system to shift electricity use from peak load hours to light load hours, in order to reduce its electricity costs. A new algorithm (called multipass iteration particle swarm optimization) is utilized to obtain the best solution of the problem. The method is tested in the optimization of a complete 30 days horizon, showing the advantages of storing energy in weekends and in the hours with low prices for the energy. In [21], the author aims at two objectives: to reduce the thermal operating costs of a large system and, simultaneously, to maximize the wind energy utilization factor of the wind farms. For this, a representative of the TPC system was used, with a maximum load of 20,747 MW, average load of 16,348 MW, a battery storage system with a capacity of 1,050 MWh/350 MW and 10 equivalent wind farms with an added installed capacity of approximately 330 MW. The simulation of the system over 24 h of operation is shown in the paper. The results show the efficiency of the solution algorithm and the advantages obtained for the planning operation.

2.4 Utilization of Air Energy Storage

In [22], the authors analyses the utilization of compressed air energy storage to compensate the imbalances due errors in the wind power forecasts. Variable operational costs, start-up expenses, investment costs and spinning and standing reserves are considered in the formulation. The representation of the adiabatic compressed-air energy device is similar to that used for hydro pump storage facilities. The method is tested in the German electricity market context. The results show that if the option to invest in compressed air storage is available, a fraction of necessary investments in gas turbines and combined-cycle gas-fired plants is displaced in the future. The amount of investments in storage does not significantly increase with the augment of the wind penetration. The CO_2 emissions are significantly reduced by the introduction of compressed air storage, because of two reasons: (a) compressed air storage can substitute thermal generation in peak hours and (b) the adequate utilization of compressed air storage can manage large deviations of the wind power production, allowing greater wind power penetration.

In [23], the authors analyze the introduction of air storage technologies in Denmark. The compressed air energy storage is compared with similar other alternatives, including heat pumps and hydrogen storages, from an economic point-of-view. The storage plants are evaluated in relation to the system integration of wind power. The study concludes with the convenience of using air storage technologies to save investments in power plant capacities in the system, as the best option of allow a profitable inclusion for this kind of storage plants.

2.5 Superconducting Magnetic Energy Storage

A superconducting magnetic energy-storage (SMES) device uses a superconducting coil, cryogenically cooled to a temperature below its superconducting critical temperature, to store energy in its magnetic field. More details about the utilization of SMES in power systems can be found in [24]. In [25], a damping control for the SMES unit is designed, allowing to suppress the active-power variations of a large-scale wind farm subject to wind-speed fluctuations. The wind farm includes several induction generators and uses the SMES for storing a large amount of energy in the magnetic field, compensating in that way the active power fluctuations of the wind farm. The method is tested in an 80 MW wind farm, including in the paper steady-state analyses and transient simulations of the combined operation. A similar analysis is performed in [26], for a large-scale wind power generation system. In [27], the combination of a offshore wind farm, marine-current turbine generators and a SMES unit is analyzed. The damping controller for the SMES is calculated by using Modal Control theory. The paper includes steady-state analysis and comparative transient responses of the studied system (with and without the SMES and the designed PID SMES damping controller) under different transient disturbances: in the torque, by gust wind and a suddenly changed local load.

2.6 Utilization of Fuel Cell Systems

Fuel cell systems are also suitable to storage large amounts of energy for the short or large term. In [28], a storage system based in hydrogen production is analyzed. The simulations consider the Italian day-ahead market prices and a big wind farm in Sicily. The storage system includes an alkaline electrolyser for the hydrogen production, compression and storage of the hydrogen in tanks and a fuel cell for recombining hydrogen and oxygen, partially recovering the energy used in the water electrolysis. The study was performed in a yearly base, wanting to determine the optimal capacity of the various components utilized in the storage cycle. The results show the operational advantages related to the inclusion of the storage device. However, for the prices used in the simulations, the hydrogen storage cycle seems to be uneconomical.

In [29], the authors consider the cooperation of a wind farm with a hydrogen cycle. Hydrogen produced from electrolysis is used for power generation in a stationary fuel cell and as fuel for vehicles. The wind energy is traded at the day-ahead spot market, analyzing three cases: (a) spot prices of the European Energy Exchange (EEX) in 2002, (b) Norwegian market conditions and (c) isolated operation of the wind–H_2 plant (no possibility for power export). From the results, the fuel cell is used only in cases with large electricity price variations and high balancing costs, since the overall efficiency of the hydrogen storage cycle is relatively low.

The paper [30] describes the modeling and control scheme for a combined wind-fuel cell hybrid energy system. The author proposes the following operational procedure: if the wind turbine is producing enough power, the load will be supplied entirely from the wind energy; in case of low wind a share of power can be supplied from the fuel cell. Otherwise, if the output power from the wind turbine exceeds the demand, the excess power may be used to produce hydrogen for later use in the fuel cell. The fuel cell is added to a 5 kW wind turbine, in Canada. Simulation results indicate the expected transients in the system for the analyzed control strategies. The performance of the system has been found satisfactory when responding to a step change in the wind speed or a step change in the load resistance. The author suggests the convenience of using the wind turbine rotor as an additional energy storage element. In [31], a system consisting of a 400 W wind turbine, a proton exchange membrane fuel cell, ultracapacitors, an electrolyzer, and a power converter is analyzed. System dynamic modeling, simulation, and design of controller are reported in this work. Transient's duration is between 1 and 5 s. In most of the transient situations, the system behaves like an overdamped one. To regulate the variation in output voltage, the need of suitable control and power electronic mechanisms is outlined.

2.7 Coordination in Isolated systems

In isolated systems, the introduction of intermittent renewable energy (as the wind production) could be more complicated than in large systems. Therefore, storage capacities can perform an important role for balancing power and energy in the operation [32].

The inclusion of large amounts of renewable generation in isolated system could be limited by dynamic security concerns. Wind parks and other renewable sources are generally unable to assist in maintaining the system within tight margins of frequency and voltage. In [33], the authors analyses the best capacity of a pumped storage facility, considering the stochastic nature of the production and the dynamic security constraints related to frequency regulation. The goal of the study is to determine the optimal energy and power capacities, considering probable operational scenarios, fixed costs of the pump installation and operational costs of the generation plants and the pump storage device. The algorithm was tested in a real island system, with 60 MW of wind generation, 50 MW of hydropower generation and two thermal stations with a total capacity of 220 MW. The optimal storage capacity was calculated as 10.9 MW and 80.2 MWh. From the results, the main advantages of the storage inclusion were due to the ability of maintaining the security constraints.

A real case based on the "Multi-purposes Socorridos system'' located in Madeira Island, Portugal, is analyzed in [34]. This grid includes a pumping and hydropower station, with 15 MW of pump capacity and 8 MW of installed hydro generation. The isolated system is analyzed in several conditions: winter and summer profiles, with and without wind production, etc. The profits, when the wind park is included in the system, are around 5,200 Euros/day. The authors do not found a relevant difference in the profits between summer and winter wind conditions, in Madeira Island.

In [35], the authors develop a stochastic optimization approach for analyzing the characteristics of a proposed storage device in isolated grids. The formulation considers wind generation and conventional diesel-based generation, representing different levels of wind penetration in the system. In the operation of the system, the diesel plant executes both frequency and voltage controls. Three diesel operating strategies are analyzed: minimum loading (by using a dump load), new low-load diesel technologies (considering electronic fuel injectors) and diesel unit commitment (allowing shut off the diesel generator in some periods). The analysis considers possible scenarios of load and wind power generation, weighted with the probabilities of their occurrence. The economic model considers fixed and operational costs, to evaluate the advantages of the storage device. The objective of the study is to calculate the optimal size of the storage plant to minimize the dumping of energy. Interesting conclusions are obtained from the sensitivity analysis, performed in a test case. The results show that storage devices are better utilized with medium and large penetrations levels of wind power. The introduction of storage devices does not significantly influence the cost of the energy served but improves the sustainable approaches of serving the energy needs of remote communities.

In [36], the analysis of a Portuguese village with wind power production, solar generation and pumped storage capacity is considered. In the normal operation, the pumped storage capacity is used exclusively to supply the water requirements of the population. Three cases were compared in the study: (a) stand-alone operation, (b) interconnection with the national external grid and (c) interconnection with the external grid and inclusion of a water turbine (in this case, requiring reinforcements in the pump station and pipes). From the analysis in the Portuguese market, the stand-alone operation is more profitable than to build an interconnection with the national grid, if the distance to the grid is larger than 31.6 km. In case (c), the reformed pump station can be used (besides to provide the water for the village) to store and sell energy of the grid, using the incorporated water turbine and acting as a conventional energy pumped plant. However, comparing the results obtained in the present study-case, it is possible to conclude that the system without a hydro turbine is more cost-effective.

In [37], the behavior of the system constituted by a wind farm and a fuel cell is analyzed, in a isolated grid. The system has 7,500 W of rated wind power and 3,500 W of rated power in the fuel cell. The stochastic characteristic of the wind power is considered in the study. The simulation results show that the various components of the system can perform a successful and integrated operation.

The work [38] analyses the coordination of a isolated system including a diesel synchronous generator, a wind induction generator and battery storage, through dynamic simulations. Different types of loads are considered in the study. The results show that the battery system helps to reduce the harmonics in source currents, in the load balancing and load leveling. A similar approach is analyzed in [39], where the association between a wind turbine with double fed induction generator (DFIG) and a battery is analyzed in isolated and grid-connected modes. When the wind power generator operates in the grid-connected mode, the proposed control scheme preserves the characteristics of a conventional grid-connected counterpart, in addition to the capabilities augmented by the battery energy storage. In isolated operation, the battery allows the islanded operation of the DFIG wind-power unit. The proposed controller also allows an adequate behavior of the combined system facing the battery disconnection and line-to-ground faults.

2.8 Sizing of Storage Capacity Using Statistical Analysis

The requirements in storage capacity are strongly related to the quality of the prediction tool, utilized to estimate the wind power production. In particular, the dimension of the storage reservoir and the response of the hydro generator and pump station are conditioned by the prediction. In [40], a Beta distribution for the wind power forecast error is used. Real data from two sites are utilized in the study. The histograms of the Beta distribution are matched with the Gaussian distribution, showing the advantages of the approach. Once the forecast error pdf (probability density function) is known, the storage energy system designed to reduce the forecasting error can be calculated. The results show that the energy loss is proportional to the integral of the tail of the pdf. Therefore, the pdf must be especially precise in its tail region.

In [5], energy storage is used as a mean of risk hedging against penalties from the regulation market. In the analysis, market participants (including wind farms) are charged for any deviation from contracted production. The penalties are as defined as the volume of electricity in imbalance times the imbalance price. As an alternative, large-scale storage can be used as a buffer, for absorbing the fluctuations around the delivery contract proposed on the day-ahead market. The work aims to calculate the optimal size of the energy storage for any period of the future, using possible production scenarios from statistical analyses. The algorithm is applied to a Danish wind farm, considering the addition of a pumped storage facility. From the results, the influence of the initial energy in the storage facility and the losses in the storage/generation loop are critical for the efficiency of the method. Moreover, the risk of deviations that the wind farm owner is able to accept modifies the optimal size of the storage facility. In the paper, the optimal storage size for different conditions, in 1 year simulations, is calculated.

2.9 Participation in the Electricity Markets

The possibility of cooperation between wind energy generators and other technologies have been studied for some years. The aim of this collaboration is to use a flexible form of generation to compensate the imbalance between scheduled and generated wind production. This cooperation is only possible if local regulation allows it. This could also be regarded as a sort of procurement of reserves, so that it is related to the management of operating system reserves. There are different models of this management, and only in some of them the scheme could be implemented.

One model is a centralized model, where an entity (usually the Transmission System Operator, TSO) manages the total available reserves in the system. In this model, the agents that incur in imbalance should pay for it, following a single or dual price scheme.

A more decentralized model is also possible. In it, consumers and generators may cluster in a Balance Responsible Party, which aims to minimize the total imbalance for all its members. The imbalance price is paid only for the net imbalance, and it is shared among the members. The Balance Responsible Party tries to reduce this imbalance to a minimum, making use of its own means.

It is also possible to have contracts between different market participants. For instance, a producer of intermittent generation could agree with a flexible generator the use of the latter resources, in order to minimize the imbalance cost. This agreement could provide the intermittent producer a constant imbalance cost, independent of the fluctuations of the imbalance market.

While the decentralized model and the bilateral strategies provide more possibilities for the system and hedge risks of too high imbalance payments by the intermittent generator, this is made at the cost of a possible suboptimal in the management of the system reserves, as well as smaller availability of overall reserves for the TSO.

The proposals and case studies that will be studied here assess the profitability of an agreement between a wind generator and a flexible producer, with or without storage, in order to minimize the total imbalance of both. The flexible production considered may be fast thermal units, hydro generation, pumping hydro or other forms of storage. Many approaches have been made to this problem in the last years, and some of the most relevant will be mentioned below. The approaches, however, are different. Some of them try to solve the problem of *joint operation*, i.e., the flexible unit tries to minimize the imbalance of the intermittent one, when the actual production of the latter is already known, or can be figured out accurately enough. Then, the flexible unit provides the support needed to minimize the imbalance.

A greater integration could be the *joint participation* of both units in the market. In this way, the probability of imbalance due to the intermittent unit should be included in the final schedule and the profit could be bigger.

The participation of this aggregation in adjustment markets, besides daily spot markets, is also a possibility to be explored. In the case of common operation, this would imply no change in the strategy (the flexible unit would change its schedule after the market gate closure). However, the common participation in these markets would become a more complex problem, since more decision variables (trading in the adjustment markets) should be added.

In the case of common participation in markets it is also necessary to consider the accuracy and uncertainty of a short term wind power prediction program, while this is not necessary for the problem of common operation. The common participation problem is thus a stochastic optimization one, which means that different random variables should be modeled. The pertinent random variables are wind production, and prices for the day ahead market, adjustment market and imbalance prices, just as in the problem of a wind bid.

Modelling of the flexible/storage unit may be very simple (for instance, a thermal unit with no ramp constraints of start up and shut off costs) or more complex (a set of hydro plants hydraulically coupled). The following references address this problem with different degrees of detail, but the conclusion seems to show that this joint operation or participation is profitable most of the times.

Bathurst and Strbac [41] study the advantages of combining energy storage and wind in short term electricity markets. The storage device considered is generic, and the imbalance is single priced. The prediction of wind energy is made using persistence, and the imbalance prices are known in advance. Lead time, i.e. time difference between gate closure of market is just up to 4 h. The study calculates the added value of such cooperation and their conclusions are that this added value decreases with price volatility, and may even be negative with small storage capacity and small error in wind prediction, and that reduction in lead time does not imply a greater increase in added value.

Angarita and Usaola propose in [42] a joint participation of hydro and wind generation following loosely the Spanish electricity market rules. This strategy consists in the change of the hydro generation from its schedule in order to compensate the imbalance of wind generator less than 1 h before the operating time. This means that the production, and hence the imbalance of wind generation is almost perfectly known because of this short lead time. In the study, the imbalance costs are assumed to be proportional to the spot market price, and a sensitivity analysis is carried out to show the dependence of the added value of this joint participation on different parameters. A prediction program for wind energy is therefore not needed for this strategy, although the wind generation might make use of a prediction program for the updating of his position in the market in subsequent adjustment markets. The advantages of this cooperation are greater when the imbalance costs are higher, and especially if there is no possibility of participating in adjustments markets. One of the limitations of this strategy is that it is not possible to increase the production of the hydro unit when it is at its maximum, or to decrease it when it is not producing. Since the hydro and the wind units have previously presented bids separately, and the hydro generator has not considered at that stage the possibility of compensation, the capabilities for joint participation of these strategies are not fully exploited.

This limitation is addressed in [43], by using stochastic optimization techniques to produce a joint bid of both technologies for the spot market, that take into account the uncertainty of the future wind production. In this case, the updating possibility in the adjustment markets for both generators is not considered, as the imbalance price is still proportional to the daily market price. The work shows that it is in general profitable for both partners to participate together in the market. The benefits depend mainly on imbalance costs and the lead time of the forecasts. In general terms, it could be said that it is more profitable when the uncertainty of wind production is higher and to commit imbalance is more expensive.

Another study, this time about the joint participation of wind energy and a thermal unit, is given by Gibescu et al. [44]. In this paper, a method for bidding wind energy in markets, taking into account the uncertainties for its prediction, is proposed together with a bid for a mixed portfolio between wind and a thermal plant. The problem is solved using optimization methods that maximize the expected benefits for the joint operation. Start up costs and ramp constraints of the thermal plant were not considered. The study was carried out for 1 year, and real imbalance price were considered in the analysis. Imbalance prices follow a two-price system, and an estimate of their future value is made from previous values which allow defining a pattern for the typical values along a day. The results show that it is slightly profitable this joint operation and that this benefit increases when it is also possible to reduce wind production to avoid surplus penalties. This last strategy is not reasonable when subsidies to wind energy production depend on the amount of energy generated, since the cost of the imbalance is usually much lower that the premium to wind energy.

The authors of [45] compare, from a economic viewpoint, the option of either using a pumped storage hydro plant for reducing the uncertainty in the wind production or to purchase call/put options to protect against this uncertainty. In the analysis, the wind farm and the pumped storage have the same owner and the Black–Scholes options pricing model is used. Real data is used in the study, extracted from the PJM transmission organization (price data) and from a wind location in Sutherland, IA, USA (wind speed data). The results show that in some situations, purchasing options is financially competitive with the alternative of using a pumped storage plant. However, the authors highlight that at this time, a complete and competitive options purchasing market does not exist in the deregulated electric industry.

In [46], the authors perform simulations of a wind-hydro portfolio. The goal of the study is to analyze the diversification effect of the energy sources, from all-hydro to a mix of hydro and wind, to lower the risk of energy deficits. The research uses real data from Quebec, Canada, over the period 1958–2003. To analyze the dependence between the two energy sources, five classical copulas were considered, characterizing the dependence of the marginal distributions. Three periods are studied in the paper: with high inflows, with low inflows and with a combination of low and high inflows. The results show that with high inflows less risky portfolios include equal parts of wind and hydro. On the other hand, high inflows situations incentive larger penetrations of wind power, to reduce the risk of the portfolio.

Besides of the stochastic characteristics of the wind power generation, other parameters can be represented as random variables. In [47], the authors consider two stochastic variables: the wind power production and the market prices. The optimal bids for the day ahead market are calculated, also considering the probable coordination of the wind farm with a pump storage facility. Uncertainty about market prices and wind generation is introduced under a two-stage stochastic programming approach. The approach is tested by using price previsions for the Spanish daily-market hourly market. The additional utilization of a storage facility increases the wind farm profits, since the pumped-storage plant provides hedging against its production uncertainty as well as a more efficient operation, storing energy during off-peak hours to increase the energy sold during peak hours. From the results, it can be concluded that the coordination of wind farms with storage represents a profit increase for both utilities. The expected profit increase is about 2.53 %.

The paper [48] examines the operation of a system with and without a storage unit for various levels of wind power and examines the cost savings associated as the wind installed on the system increases. The analysis is performed in cases extracted from the All Island Grid study, Ireland, with a peak demand of approximately 9.6 GW. A pumped storage plant was added, with maximum generating and pumping capacity of 500 MW, round trip efficiency of 75 % and 5,000 MWh of storage capacity. Also, increased wind power was included, from 3 to 15 GW of installed capacity. The presented results show that, although storage became more viable for the system with increasing levels of wind power, it never proved to be the best option for the Irish system. An initial analysis about the capacity credit of the storage is included in the paper.

The utilization of pumped energy storage to compensate deviations in the wind power production could increase the value of these storage plants. In [49], the author proposes a valuation of the pumped storage plants considering this fact. The paper only considers the activity of the storage plant in the spot market, both day ahead and intraday operations. A daily optimization is performed to obtain the optimal schedule of the pumped energy storage plant, including binary variables to represent the pump action. The validation of the pumped storage plant is performed by using stochastic analysis of probable scenarios of operation, obtained from the daily optimization for different price profiles. The proposed validation is matched with a classic investment appraisal, which takes expected cash flows from an investment as a basis for the valuation of the storage plant. From the simulations, the conventional procedure results, due to non- consideration of future scope of actions, in substantially lower contribution margins and thus leads to a misvaluation of the investment. The proposed method can be extended to a portfolio of power plants.

In [50], the active utilization of international markets is compared with different storage alternatives [pumped energy storage, underground pumped energy storage, compressed air energy storage and combined heat and power (CHP) units] in a system including the Netherlands and their neighbor countries. Different levels of wind power production are considered in the study. The simulations show that the operating cost savings by wind power increase with the amount of wind power installed. Also, the interconnection capacities between Germany and the Netherlands and the Dutch wind power mainly decrease the full-load hours for base-load coal and lignite in Germany. Wind power furthermore reduces the exports of base-load coal power from Belgium and to a lesser extent from France during periods of low load (nights and weekends). Germany reduces its imports from France at times of high wind in the Netherlands. The results show the ability of the wind power to reducing the CO2 emissions in the integrated system. The authors highlight that international exchange is the key for wind power integration, especially at high penetration levels.

In a following section, the interaction of wind generation with reserve electricity markets is fully analyzed.

2.10 Reliability Studies for the Combined Operation of Wind Power Plants and Storage

In [51], a method to evaluate the reliability of a system with wind energy and storage is proposed, using Monte Carlo simulations. The study is performed in the Roy Billinton Test System, including an energy storage facility. The model considers random generating unit failures. Different reliability indexes [as loss of load expectation (LOLE), loss of energy expectation (LOEE) and Expected Wind

Energy stored in the Battery] are shown in the work, for different operational strategies. From the results, the reliability benefits from energy storage in the analyzed scenarios are highly dependent of the restrictions in the wind energy production. In some operational strategies, the energy storage has the ability of improving the system reliability, and large amounts of additional wind energy can be stored.

3 Optimal Schedule of a Storage Plant that Compensate the Production Deviation of a Wind Farm

In the present section, a method to calculate the optimal schedule of a large storage plant, cooperating with a wind farm, is presented. In the analysis, a daily cycle of operation for the storage plant is assumed. Moreover, it is considered that the storage plant is at now in operation, built to economically operate in the market (as current pump water stations or flux batteries in substations).

Storage plants in normal operation buy energy from the system in low-price periods, store this energy for a while and sell it at high-price periods (acting as generators). This operation constitutes the operational cycle of the storage plant. The profit in the conventional operation for one storage plant is a function of the differences between prices in the operational cycle, the efficiencies of the storage plant and the constraints in the operation (storage capacity, generation rate, etc.). In the present analysis, the storage plant wants to enlarge its operation, also compensating the deviation of one (or more than one) wind farm, partially using its storage capacity. For this objective, the storage plant acts as a contracted reserve of the wind farm. The wind farm reduces in this way the uncertainties in the expected profit, due to the high volatility present in most of the reserve markets worldwide.

Four stages in the cycle operation of the storage plant can be recognized:

(a) **Pause, waiting for low price periods**: at the beginning of the cycle, the storage plant has an energy reserve defined by E_1^{esp}. The optimal operation of the storage plant may wait for the lowest prices of the day (generally, at low load periods) to initiate the charge of the storage reservoir.

(b) **Filling up the reservoir**: at low-price periods, the storage plant connects the equipments to fill up the reservoir, acting as a load of the electric system. The initial and final intervals for the load action depend on many factors: the reservoir capacity, the maximum rate for filling up, the efficiencies of the storage plant, the forecasting prices in the day, etc. As shown in [6], the optimal action for the filling up equipment is performed by generally connecting them at their maximum capacity in the lowest price periods, complementing with partial charges in not so convenient periods.

(c) **Pause, waiting for high-price periods**: when the reservoir is full, or is not economically convenient to store more energy in the reservoir, the storage plant may wait for the most profitable periods to sell the stocked energy.

(d) **Depleting the reservoir**: at high-price periods, the storage plant may deliver the stored energy to the system, depleting the reservoir. In these periods, the storage plant acts as a generator. As in (b), the best periods to sell the energy to the system can be calculated, taking into account the predictions for the energy prices in the market, the efficiencies and the requested minimum energy level in the reservoir at the end of the cycle, E^{esp}_{n+1}. In general, the optimal action for the depleting case is to fully connect the equipment in the highest price periods, to assure the best prices for the storage plant production. After depleting the reservoir, the storage plant may return to stage (a), to perform another operational cycle.

At stage (a) (when waiting for the low-price periods), the storage plant has little energy stored at the reservoir. Therefore, it is feasible to behave as a load, absorbing part of the wind power generations, when it is above the expected value. On the other hand, when the wind farm production is below the predicted value, the storage plant may act as a generator, complementing the joint production. To complete this action, the storage plant must guarantee enough stored energy to compensate the maximum deviation, in all the intervals of the present stage. Similar analysis can be made when the storage plant is in stage (d). Furthermore, the stored energy remaining in the reservoir at the final interval of the operational cycle (E^{esp}_{n+1}) must guarantee the probable compensation of these lower wind farm generations in the next programming cycle. The value for the remaining energy at the end of the simulation (E^{esp}_{n+1}) is obtained in the present case through heuristic analyses, derived from the experience in the operation of the system.

When the storage plant is at stage (b) (filling up the reservoir), the optimal conventional operation requires the action of the feeding equipments at their maximum capacity, to fulfill the energy reservoir in the lowest-price periods. In these periods, the storage plant operates as a load of the electric system. Therefore, the storage plant can easily compensate wind generations lower than the expected value when required, by reducing the power absorption. On the other hand, to compensate probable upper deviations in the wind production, the filling up equipments must maintain a margin to increase the load action. Similar analyses can be performed by period (c).

In Eqs. (1)–(8), the optimal conventional+reserve problem for the storage plant is formulated, simultaneously considering the objectives of improving the profit in the operational cycle and maintaining a reserve for compensating wind power forecasting errors.

$$\text{Max.} \sum_{i=1}^{n} (c_i Ph_i - cp_i Pp_i) \tag{1}$$

$$\text{s.t.} \quad E_{i+1} = E_i + t\left(\eta_p Pp_i - \frac{Ph_i}{\eta_h}\right) \tag{2}$$

$$E_1 = E_1^{esp} \quad (3)$$

$$E_{n+1} = E_{n+1}^{esp} \quad (4)$$

$$Ph^L \leq Ph_i \leq (Ph^U - Pw_i^m) \quad (5)$$

$$Ph_i \leq \eta_h \frac{E_i}{t} \quad (6)$$

$$Pp^L \leq Pp_i \leq (Pp^U - Pw_i^M) \quad (7)$$

$$0 \leq E_i \leq \left(E^U - E^R\right) \quad i = 1, \ldots, n \quad (8)$$

where the variables are vectors describing: Ph, hourly active powers produced by the storage plant; Pp, hourly average active power consumed storage plant; and E, energy storage levels in the reservoir in each hour. The following parameters are also defined: c, vector of hourly active power prices; cp, vector of operation costs, for storing energy in the storage plant; E^U, reservoir storage capacity; η_p, efficiency of storing energy; η_h, efficiency of the generator in the storage plant; E_1^{esp} and E_{n+1}^{esp}, initial and final levels of the reservoir, respectively; Ph^L and Ph^U, lower and upper production power limits of the generator in the storageplant, respectively; Pp^L and Pp^U, lower and upper physical power limits for the feeding operation of the storage plant, respectively; Pw_i^m and Pw_i^M, minimum and maximum values for the forecast power production in the i-interval of the next future, required to be compensated by the storage plant; E^R energy margin required for compensation issues; t, duration of each interval (1 h, in this case); and n, number of intervals.

In this analysis, the following cases are represented, using typical data of the Spanish electricity market for the year 2008: (a) Typical Day (Wednesday) Operation; (b) Saturday Operation; and (c) Sunday Operation. The optimization algorithm is evaluated in a test case, considering a storage plant with: 2,000 MWh of storage capacity, 273 MW of fill-in rate, 336 MW of installed generation in the storage plant, $\eta_p = 92\ \%$ and $\eta_h = 88\ \%$. The minimum and maximum values for the forecast power production of a 250 MW wind farm, required to be compensated by the storage plant, are between 10 and 35 % (at 4 and 12 h ahead, respectively) [52]. From the execution of the optimization algorithm (1)–(8), the optimal operational schedule is calculated (Fig. 1).

As observed in Fig. 1, the requirements associated with the reserve activity result in lower energy quantities stored in the reservoir. As it is expected, large forecasting errors in the wind power generation result in lower amounts of energy stored in the reservoir, and subsequently in lower capacities to obtain profits in the storage operational cycle (as evaluated in the conventional operation, with null E^R, Pw_i^m and Pw_i^M).

One of the difficulties associated with the action of the storage plant as reserve of wind farms is the determination of the real costs for this activity. Opportunity

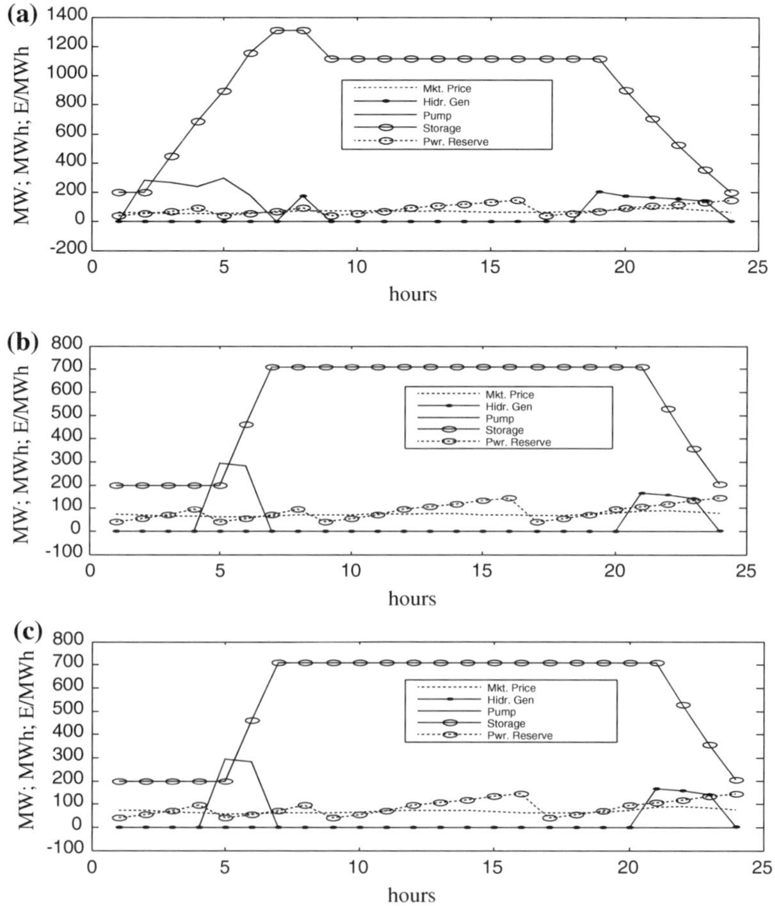

Fig. 1 Conventional+reserve operation of the storage plant. **a** Wednesday simulation. **b** Saturday simulation. **c** Sunday simulation

cost could be defined as the evaluation placed on the most highly valued of the rejected alternatives or opportunities [53]. When the storage plant chooses setting aside part of its capacity to compensate probable wind power production errors, this plant has opportunity cost that must be compensated by wind farms owners' payments. The opportunity cost can be calculated as a difference between the profits obtained in the conventional operation and in the conventional+reserve procedures. In Tables 1 and 2, the profits and opportunity costs of the simulated operation alternatives are summarized.

Table 1 Profits in the simulations

Scenario	No compensation revenue (Euros)	Compensation revenue (Euros)
Wednesday	17,368.52	13,849.66
Saturday	7,271.88	5,678.11
Sunday	12,254.59	9,558.94

Table 2 Opportunity costs

Scenario	Opportunity costs	
	Euros	(%)
Wednesday	3,518.86	20.26
Saturday	1,593.77	21.92
Sunday	2,695.65	22.00

4 Wind Energy and its Relation with Reserve Electricity Markets

An electricity market is actually a set of different submarkets where the different requirements of generators and customers are matched. Examples of these markets are the day-ahead markets, the adjustment markets, or the reserve markets. Wind energy must integrate in at least some of them and this implies challenges both for the electricity markets and for the wind energy.

In fact, wind energy is an intermittent source of energy that, besides, is difficult to predict. However, the participation of a generator in electricity markets requires delivering the settled amount of energy, at the agreed price at a particular time. This means that the production in a moment should be known more than 24 h in advance, in the case of day-ahead markets. In the case of wind energy, this is not possible and this impossibility implies that there will always be a difference between the forecasted and the actual productions, incurring in imbalances. In order to estimate this production, wind energy systems make use of short term wind power prediction programs. Even with them, imbalances are usually produced, and this means that wind energy systems make use of operating reserves continuously. As an example, it could be said that in the Spanish system (2009) for a 13 % of the demand coming from wind energy, a 34 % of the up imbalances and a 19 % of the down imbalances were produced by wind energy. As a general rule, until more accurate wind power prediction systems are available, there will be always a difference between the forecasted and the actual production of wind farms. Hence, a more intensive use of the operating reserves will be made in power systems with greater wind penetration. This inaccuracy in prediction represents also economic losses for wind producers, since the imbalance between scheduled and delivered energy have a cost for any market participant.

There are, however, means of reducing this imbalance, such as participation of several wind farms together, making use of the possibilities given in adjustment markets, or strategic bidding. This reduction will have effect mainly on the revenues of wind producer, but also may bring benefits for the system. In fact, the use of adjustment instead of imbalance markets reduces the needs of reserve.

It is also possible to reduce the imbalance due to wind energy lack of predictability by means of the coordinated participation of wind and other technology in electricity markets. This technology should be able to change its production to fit the joint production to a common schedule rapidly enough. This is even better when this technology has the possibility of storing the primary resource, like hydro generation and pumping hydro. Other means of storage, such as thermal storage are being incorporated in combination with other renewable energy such as solar thermal and could be of help in a close future. However, this flexible generation is the same that provides traditionally the running reserves in the power systems. Hence, it is necessary to ponder if the use of these resources is more efficient in this decentralized way (combinated participation in electricity markets) or through a centralized management system.

In the following sections these issues will be addressed. Firstly, the existing reserve management mechanisms will be defined. Then, the effects that wind generation will have on these systems, as well as the experience in some grids, where wind energy penetration is already considerable, will be described. Next, the economic losses for wind generation participating in electricity markets, due to imbalances will be assessed from studies already available in scientific literature. Then, an assessment of the advantages of coordinated participation of wind energy and another technology will be presented. The last section will present the conclusions.

4.1 Reserves Definitions and Uses

As already mentioned, wind energy is difficult to forecast and hence there is an imbalance between the scheduled and the actual energy delivered by a wind generator. The importance of wind energy for reserves is, therefore, very high. But not all kind of reserves are affected in the same way, and first an accurate definition of reserves is necessary.

Reserves in a power system may be classified [54] into *operating reserves*, which are required to maintain system security by handling short term disturbances to the system, and *planning reserves*, required to maintain system adequacy by meeting annual demand peaks. This work is concerned with operating reserves.

Operating reserves actually consists of different kind of reserves, and since sometimes they are given different names, the definitions given in [55] will be given. System operating reserves may be classified into frequency containment reserves (FCR), frequency restoration reserves (FRR) and replacement reserves (RR). FCR are also called *primary reserves*, and they change the power given by generators, in order to maintain the power balance in a synchronous system. They are activated in less than 30 s when there is a frequency deviation. FRR (or *secondary reserves*) have a typical activation time between 30 s and 15 min, and

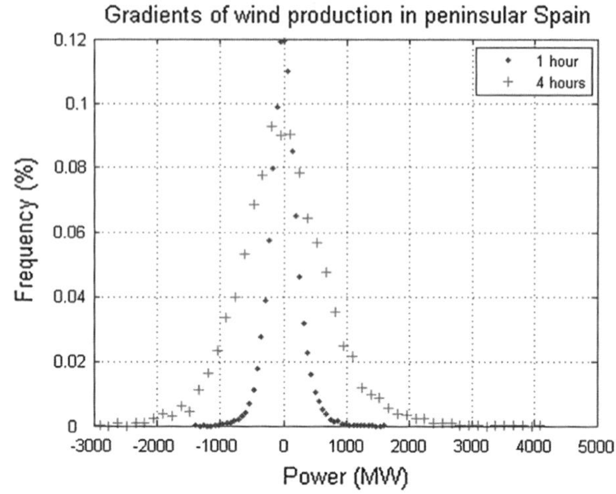

Fig. 2 Gradients of wind production in peninsular Spain (January 2007–November 2008). *Data source* REE

they aim to restore frequency and tielines power to their nominal value. They are typically managed centrally and can be activated automatically or manually. RR, also called *tertiary reserves*, restore the required level of operating reserve, both FCR and FRR, used before. Their activation time goes from several minutes up to hours.

Balancing in Europe is usually made through Balance Responsible Party (BRP), which [55] keeps "the net balance on all the connections within its control and faces the liability consequences if this is not achieved. The liability in case of imbalance involves the payment of an imbalance charge to the operator of the market area who is responsible for keeping the balance in the area."

To know which reserves are more affected by wind energy forecasting uncertainty, it is also necessary to know the range of variation of wind energy production. The change of wind production between two consecutive hours sets an upper limit to the unexpected changes in wind production (a prediction system should, in principle, reduce this figure), and gives insight into the amount and kind of reserves that must be used to compensate these changes. As an example, the changes of average hourly production for the whole Spanish peninsular power system, between January 2007 and November 2008 are given in Fig. 2. In it, the distribution of changes in wind production with a difference of 4 h is also given.

In this figure, it can be seen that hourly changes in the hourly average production of wind energy are below 1,620 MW. Although this is a significant figure, these changes are smaller than the morning rise of the demand; around 4,000 MW/h. However, changes of wind production in 4 h may be grater than 4,000 MW. The size of these changes depends primarily of the level of penetration in the system, but also on the country size and the actual siting of wind generation. At the end of 2007, the installed power in Spain was around 14 GW, for a yearly peak load of 43 GW.

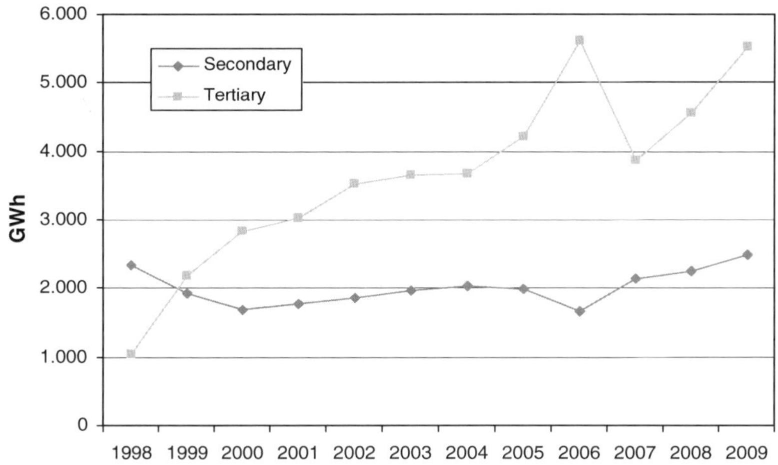

Fig. 3 Secondary and tertiary reserves used in Spain in the last years. *Data source* REE

The interaction of wind energy is with tertiary reserves may be greater. This may be seen in Fig. 3, where the use of secondary and tertiary reserves[1] in Spain in the last years is shown. This change is due mainly to demand increase in this period. Greater levels of wind penetration in the Spanish mix could lead to a higher effect on this level of reserves.

An estimate of the amount of reserves needed to compensate wind power imbalance is given in references [56, 57]. Both references are based on the expected joint standard deviation of the system due to the different system uncertainties (load, wind prediction and plant unavailability), assumed as independent. In both, the growth of the reserves is expected to be linear with the penetration level, at least beyond a certain amount. Other study [58] reach a different conclusion—the level of reserves needed depends on the Value of Lost Load (VOLL) and may be even lower for larger penetration level, depending on the case.

4.2 Reserve Pricing, Especially in Systems with Wind Penetration

Secondary and tertiary reserves can be only provided by suitable plants which comply with technical requirements. The allocation of reserves to these plants is made following market mechanisms that may be long term contracts with the

[1] It must be said that year 2006 some peculiarities due to regulatory measures taken in Spain and hence the data are atypical.

System Operator, or a marginal market. This service includes both the availability, i.e., the power margin that should be always available, and the energy, or the use of this availability by the system.

The payment of these services is 2-fold. On the one hand, availability (power) must be paid by all that do not provide this service, since it must be always available. On the other hand, the energy used is to be charged to those that have make use of it, namely, those users who incur in imbalance between the scheduled and the actual delivered power.

There are two types of imbalance price mechanisms [55]:

- Dual imbalance pricing, where a different price is applied to positive imbalance volumes and negative imbalance volumes; or,
- Single imbalance pricing, where a single imbalance price is used for all imbalance volumes.

In a dual imbalance price, there is a "main price", which is the one that applies to imbalance volumes in the same direction as the overall market, whereas the "reverse price" is that applied to imbalance volumes in opposite direction to the overall market e.g. "short" when the market is "long" or vice versa.

Where a dual imbalance pricing regime is followed, it is the "main" price that is derived from energy balancing actions. The "reverse price" can be determined by reference to a power exchange or be based on the prices of the balancing actions in the reverse direction during the settlement period.

4.3 Imbalance Costs for Wind Producers Study Cases

As mentioned before, participation in electricity markets is possible for wind energy systems in some countries. Mostly, this participation is in a day-ahead spot market, such as Nordel or MIBEL. Some adjustment markets, such as MIBEL, may be marginal, others, as Nordel, may be pay-as-bid. In this last case, a price must be put to the bid.

Since there will always be an imbalance between the actual production and the last schedule, the wind farm should pay for this imbalance: as the spilled power is paid at a price smaller than the day-ahead price, and the deficit is paid higher than it, there will be always a loss. Thus, participation of wind energy in electricity markets implies losses that are today unavoidable, because of the non dispatchable nature of the primary resource and the inaccuracy of short term wind power prediction programs.

For a given wind farm that uses a prediction program with a given accuracy, there are methods, however, to reduce these losses. The most effective is to participate in a big cluster of farms that present a bid for the joint production. The reduction is due to the so-called *portfolio effect*, that consists in the decrease of prediction error due to the compensation of individual predictions of each farm.

Another method is to participate in the adjustment markets that usually have the same average prices as the day-ahead markets. However, this is more complex in a pay-as-bid market. Many of the adjustment markets, besides, are continuous, and this means that are not liquid. It has been recommended [56] that spot adjustment markets are better fitted to systems with high wind penetration.

Other possibility is to use stochastic optimization techniques to minimize the expected imbalance costs. The problem may be set for the participation in a given day-ahead market, taking or not into account future possible updates in the adjustment markets, or only for one adjustment market. The problem must model the uncertainties of the values relevant for the imbalance minimization unknown at the time of gate closure. These uncertainties are, the future spot price (of day-ahead and/or adjustment markets), the imbalance price and the actual production of the wind farm. If the problem is set for the participation in a pay-as-bid market, the price of the bid should also be defined. These unknowns are in fact random variables whose distribution is necessary to estimate. The uncertainty of the spot price could be the result of a price prediction programs, which are numerous, so that it is reasonably well defined. However, this uncertainty is not very high, and will not likely be very important for the final decision.

The uncertainty of wind power prediction may also be provided by the short term wind power prediction programs, and it is usually higher and has a larger impact on the results. This modeling has been studied and modeled in detail, for instance in [59], and many prediction programs include this output as an standard result.

However, the uncertainty of the imbalance prices is very difficult to model because of the high variation and volatility that these prices may attain in reality. Results of the strategy depend highly on imbalance prices. A common simplification is to consider that the imbalance price is proportional to the spot price. The effect of this simplification is to underestimate possible big losses due to a miscalculation. In fact, imbalance prices may be very high and a bad forecast may imply heavy losses that could not be compensated by big earnings at other moments. For this reason, strategies based on risk reduction may be added to prevent high losses [60]. These strategies may be based on limitation of risk indices such as VaR or CVaR.

An example of the formulation of an optimization problem of this kind, is now given. In the considered case, the problem has been set to obtain the optimal bid for a marginal adjusment market, given a position at a previous day-ahead market.

The revenues of the wind producer in the hour t will, R_t, be

$$R_t = P_{d,t} \cdot \pi_{d,t} + \pi_{i,t} \cdot (P_{i,t} - P_{d,t}) + IC_t \tag{9}$$

where $P_{d,t}$ and $P_{i,t}$ are the power commited to the wind farm for the time t in the day-ahead and the adjustment market, respectively. $\pi_{d,t}$ and $\pi_{i,t}$ are the marginal prices of energy in those markets, and IC_t is the imbalance cost for that time.

$$IC_t = \begin{cases} \pi_t^{sell} \cdot \left(P_{g,t} - P_{i,t}\right) P_{g,t} > P_{i,t} \\ \pi_t^{buy} \cdot \left(P_{g,t} - P_{i,t}\right) P_{g,t} < P_{i,t} \end{cases} \tag{10}$$

Being $P_{g,t}$ the power actually generated by the wind farm in the time t and π_t^{sell} and π_t^{buy} the imbalance prices for spill and buy energy.

In these equations, $\pi_{i,t}$, π_t^{sell}, π_t^{buy} and $P_{g,t}$ are random variables that should be modeled for the stochastic optimization process. This process consists in obtaining the bid that provides maximum expected revenue. Hence, it would be:

$$P_{i,t,opt} = \arg\max E\left[R(P_{gt}, \pi_{i,t}, \pi_t^{sell}, \pi_t^{buy});\ P_{i,t}\right] \tag{11}$$

Some additional remarks might be of interest:

- The formulation of this problems assumes that the market have liquidity to allow wind producers, as well as other participants, to trade. As mentioned, this is not always the case, especially in adjustment markets, where small amounts of energy are traded.
- Wind energy is subsidized. This subsidy depends most of the cases on the amount of energy produced, so there is a very strong incentive to produce as much as possible, even if this implies an imbalance, because the subsidy is much higher than the imbalance price.

This problem has been addressed by several authors in the scientific literature. The next paragraphs will discuss some of the relevant contributions in this field, with the aim of given insight into the different settings and simplifications assumed by the different authors, as well as an estimate of the expected imbalance losses for a wind producers.

The reviewed studies have been performed under different assumptions about the timeframe participation and the strategy to be followed, including the possibility of updating the results. The subsidies have not always been considered. Despite this wide variety of premises, the studies have arrived at similar results, since the cost of this prediction errors depend mainly on the accuracy of the short term wind power prediction tools, which is rather similar for the different tools available, and the imbalance price. Any of the studies considered the effect of a massive presence of wind energy into the electricity markets. In the next paragraphs, the results of the reviewed papers are summarized, and then some common conclusions are drawn.

Bathurst et al. make a study in Ref. [61] centered on the UK market before 2002. Due to the short time between gate closure and operation time in the British market, persistence was the prediction tool used, and different bidding strategies were compared, taking into account the uncertainty of wind power production. However, since no prediction tool was used, these uncertainties were very high. Results are given as a comparison between the different proposed strategies, and the improvements that follow in each case. In the conclusions the authors underline the importance of the imbalance prices relative to the contract price, and in fact, the results are a sensitivity study for different imbalance prices. The subsidies to wind energy were not considered.

Holttinen in Ref. [62] is centered on the Danish and Nordic market. It uses the results of a short term wind power prediction tool. The envisaged participation is reduced, in the base case, to the daily market (13–37 h ahead). The possibility of updating the bids 4 times a day was also studied. The wind power considered was 2,000 MW spread throughout West Denmark. The actual imbalance prices were used, both for up and down regulation. The conclusions of the paper are quite interesting, since they remark the advantages of a large ensemble of wind producers, and the uselessness of too much back and forth trading for small individual producers trying to reduce the imbalance, while the system imbalance may not be very high. The paper also mentions the clear advantage for wind producers of trading as close to the operation time as possible.

Fabbri in Ref. [63], studies the case of different sets of wind farms under the Spanish market rules. Typical prediction errors are considered and the uncertainties are modeled parametrically assuming a Beta distribution. The imbalance costs were also different for up and down regulation, and the energy prices were those of the Spanish market for the year 2003. Different times between gate closure and operation time were studied. Among the conclusions it could again be mentioned the advantage of a combined bid of a large group of wind farms in order to reduce the imbalance costs.

Usaola and Angarita in [64] compare different bidding strategies for a single wind farm (14 MW) for a period of 3 months. The paper focuses on the benefits of considering the uncertainty of short term wind power production, when preparing a bid for the market. This is especially relevant when, as in most cases, the up and down regulation costs are different. The main conclusion of the paper is that, under these conditions, to bid a power different than the most accurate prediction, leads to a better result. The advantages of updating the bids in intraday markets were also quantified.

Matevosyan and Soder propose in Ref. [65] an optimal bidding strategy for a wind farm, with a simulated short term wind power prediction program. It also concludes that a strategic bid does not provide necessarily the best prediction. The Nordic market rules and data from January 2003 were used.

Angarita et al. in Ref. [66] compares the imbalance costs for a 14 MW wind farm in the Spanish and British electricity markets, when using a short term wind power prediction tool. The portfolio effect due to a combined bid of several wind farms is also quantified in the study. In this work, the subsidies to wind energy are considered in the final results, and among its conclusions, the reduced importance of the penalties when considering the subsidies, because penalties and subsidies are not related is underlined.

Pinson in Ref. [67] also exploits the mathematical modeling uncertainty of short term wind power prediction, and proposes a strategic bidding for a 15 MW wind farm that updates the imbalance costs to obtain better results. The conclusions are again that the most profitable bid is not the most accurate, due to the asymmetries of the uncertainty distributions and the imbalance costs. The difficulty of

Table 3 Summary of the results of some of the examined papers

	Holttinen [62]	Fabbri [63]			Usaola [64]		Angarita [66]		Pinson [67]	
Year	2005	2005			2006		2007		2007	
Wind power (MW)	2000	24.6	301.7	5000	14		14		15	
Imbalance cost (p.u. MP)	0.376	0.485			0.5		0.266		0.215	
Buy	*1.27*	*1.46*			*1.5*		–		*1.07*	
Sell	*0.51*	*0.49*			*0.5*		–		*0.64*	
Losses (% max revenue)	12.11	11.5	9.7	10.8	10.39	7.32	4.5	9	13.1	7.9
Comments	(1)	(2) (5)	(3)	(4)	(6)	(7)	(8) (10)	(9)	(11)	(12)

(1) Results for the 13–37 h forecasts. Results for shorter term are given in the paper
(2) (3) (4) Results for the powers shown above
(5) Results for 48 h forecasts. Results for shorter term are given in the paper
(6) Best prediction bid in daily market, updated in four intraday markets
(7) Results for the strategic bid, taking into account the uncertainty of wind predictions uncertainty and the asymmetry of the imbalances' costs
(8) Results with subsidies
(9) Results without subsidies
(10) Results for the Spanish markets only for the daily markets. Results for the UK market and other assumptions are given in the paper. Imbalance costs are considered equal if under or overprediction
(11) Results for the best prediction bid
(12) Results for the strategic prediction

forecasting the imbalance cost is remarked, but approximate strategies seem to give good results.

Table 3 shows some of the numerical results of the reviewed papers whenever they can be compared. In this table, the imbalance cost shown is the mean of the absolute value of the differences between the sell/buy prices and the marginal price. The losses are expressed as a percentage of the maximal revenues that would take place if the wind power prediction were perfect.

From the results given in the table, some conclusions may follow.

- In spite of the different conditions, it could be said that the losses due to imbalances, when wind farms present bids to daily markets (between 13 and 37 h in advance) reduce their income about a 10 %, compared to the situation of perfect prediction. This amount depends on the accuracy of the wind power prediction tool.
- The revenue's reduction is smaller (about 2 % less) if the wind farm updates its bid in intraday markets run six times a day with bids given 3 h before. For wind farms, the best situation would be to update their production with the shortest possible delay.
- Optimal/strategic bidding that takes into account the different sell and buy prices in the imbalance market may reduce the imbalance costs about 3 %. This is an incentive for "strategic" bidding that does not provide the best power prediction as bids.

- Premiums and subsidies may account for one half of the revenues of wind farms.
- The portfolio effect reduces also the imbalance costs. In some studies, this effect is told to account for a reduction of 50 % [68].
- The bid that produces the maximum revenues for a market participant, is not the most accurate prediction.

5 Conclusion

Due to the intermittent characteristics of the wind, the real production is always (more or less) different from the predicted one. Therefore, compensation methods must be used to maintain the balance between generation, load and losses in the system. The coordination of storage devices or traditional generation (using reserve markets) with the wind production can be performed through different alternatives. The best choice will depend of the regulatory and technical characteristics of each particular electric system. In this chapter, different alternatives to compensate the deviations in the wind farm production are presented and analyzed, extracted from the most actualized technical literature.

Acknowledgments This work was performed within the research projects *IT2009-0063: Coordinated actions of renewable resources in multi-area systems, considering the market environment* and *IREMEL (ENE2010-16074).*

References

1. Doherty R, O'Malley M (2005) A new approach to quantify reserve demand in systems with significant installed wind capacity. IEEE Trans Power Syst 20:587–595
2. Koeppel G, Korpas M (2008) Improving the network infeed accuracy of non-dispatchable generators with energy storage devices. Electri Power Syst Res 78:2024–2036
3. Matevosyan J, Soder L (2007) Short-term hydropower planning coordinated with wind power in areas with congestion problems. Wind Energy 10:195–208
4. Pinson P et al (2009) From probabilistic forecasts to statistical scenarios of short-term wind power production. Wind Energy 12:51–62
5. Pinson P et al (2009) Dynamic sizing of energy storage for hedging wind power forecast uncertainty. In: Proceedings of 2009 IEEE power and energy society general meeting, vols 1–8, pp 1760–1767
6. Castronuovo ED, Lopes JAP (2004) On the optimization of the daily operation of a wind-hydro power plant. IEEE Trans Power Syst 19:1599–1606
7. Castronuovo ED, Lopes JAP (2004) Optimal operation and hydro storage sizing of a wind-hydro power plant. Int J Electr Power Energy Syst 26:771–778
8. Castronuovo ED, Lopes JAP (2004) Bounding active power generation of a wind-hydro power plant. In: Proceedings of 2004 international conference on probabilistic methods applied to power systems, pp 705–710

9. Estanqueiro AI et al (2007) Barriers (and solutions…) to very high wind penetration in power systems. In: Proceedings of 2007 IEEE power engineering society general meeting, vols 1–10, pp 2103–2109
10. Ngoc PDN et al (2009) Optimal operation for a wind-hydro power plant to participate to ancillary services
11. Gao F et al (2009) Wind generation scheduling with pump storage unit by collocation method. In: Proceedings of 2009 IEEE power and energy society general meeting, vols 1–8, pp 2634–2641
12. Vennemann P et al (2010) Pumped storage plants in the future power supply system. VGB Powertech 1:44–49
13. Matevosyan J et al (2006) Optimal daily planning for hydro power system coordinated with wind power in areas with limited export capability
14. Martínez-Crespo J et al (2009) Tools for the effective integration of large amounts of wind energy in the system. In: Castronuovo ED (ed) Optimization advances in electric power systems. Nova Science Publishers, New York, pp 113–150
15. Matevosyan J, IEEE (2007) Wind power integration in power systems with transmission bottlenecks. In: Proceedings of 2007 IEEE power engineering society general meeting, vols 1–10, pp 3045–3051
16. Matevosyan J et al (2009) Hydropower planning coordinated with wind power in areas with congestion problems for trading on the spot and the regulating market. Electr Power Syst Res 79:39–48
17. Chacra FA et al (2005) Opportunities for energy storage associated to wind farms with guaranteed feed-in tariffs in the present French law. In: Presented at the 15th PSCC, Liege
18. Hong YY, Li CT (2006) Short-term real-power scheduling considering fuzzy factors in an autonomous system using genetic algorithms. In: IEE proceedings-generation transmission and distribution, vol 153, pp 684–692
19. Hong YY et al (2007) KW scheduling in an autonomous system
20. Lee TY (2007) Operating schedule of battery energy storage system in a time-of-use rate industrial user with wind turbine generators: a multipass iteration particle swarm optimization approach. IEEE Trans Energy Convers 22:774–782
21. Lee TY (2008) Optimal wind-battery coordination in a power system using evolutionary iteration particle swarm optimisation. IET Gener Transm Distrib 2:291–300
22. Swider DJ (2007) Compressed air energy storage in an electricity system with significant wind power generation. IEEE Trans Energy Convers 22:95–102
23. Lund H, Salgi G (2009) The role of compressed air energy storage (CAES) in future sustainable energy systems. Energy Convers Manage 50:1172–1179
24. Karasik V et al (1999) SMES for power utility applications: a review of technical and cost considerations. IEEE Trans Appl Supercond 9:541–546
25. Wang L et al (2009) Design of a damping controller for a SMES unit to suppress tie-line active-power fluctuations of a large-scale wind farm
26. Chen SS et al (2008) Power-flow control and transient-stability enhancement of a large-scale wind power generation system using a superconducting magnetic energy storage (SMES) unit. In: Proceedings of 2008 IEEE power and energy society general meeting, vols 1–11, pp 5485–5490
27. Wang L et al (2009) Dynamic stability enhancement and power flow control of a hybrid wind and marine-current farm using SMES. IEEE Trans Energy Convers 24:626–639
28. Brunetto C, Tina G (2007) Optimal hydrogen storage sizing for wind power plants in day ahead electricity market. IET Renew Power Gener 1:220–226
29. Korpas M, Holen AT (2006) Operation planning of hydrogen storage connected to wind power operating in a power market. IEEE Trans Energy Convers 21:742–749
30. Iqbal MT (2003) Modeling and control of a wind fuel cell hybrid energy system. Renewable Energy 28:223–237
31. Khan MJ, Iqbal MT (2005) Dynamic modeling and simulation of a small wind-fuel cell hybrid energy system. Renewable Energy 30:421–439

32. Khatibi M et al (2008) An analysis for increasing the penetration of renewable energies by optimal sizing of pumped-storage power plants
33. Brown PD et al (2008) Optimization of pumped storage capacity in an isolated power system with large renewable penetration. IEEE Trans Power Syst 23:523–531
34. Vieira F, Ramos HM (2008) Hybrid solution and pump-storage optimization in water supply system efficiency: a case study. Energy Policy 36:4142–4148
35. Abbey C, Joos G (2009) A stochastic optimization approach to rating of energy storage systems in wind-diesel isolated grids. IEEE Trans Power Syst 24:418–426
36. Ramos JS, Ramos HM (2009) Sustainable application of renewable sources in water pumping systems: Optimized energy system configuration. Energy Policy 37:633–643
37. Khan MJ, Iqbal MT (2009) Analysis of a small wind-hydrogen stand-alone hybrid energy system. Appl Energy 86:2429–2442
38. Bhatia RS et al (2008) Battery energy storage system based power conditioner for improved performance of hybrid power generation. IEEE, New York
39. Bhuiyan FA, Yazdani A (2009) Multimode control of a DFIG-based wind-power unit for remote applications. IEEE Trans Power Delivery 24:2079–2089
40. Bludszuweit H et al (2008) Statistical analysis of wind power forecast error. IEEE Trans Power Syst 23:983–991
41. Bathurst GN, Strbac G (2003) Value of combining energy storage and wind in short-term energy and balancing markets. Electr Power Syst Res 67:1–8
42. Angarita JM, Usaola JG (2007) Combining hydro-generation and wind energy biddings and operation on electricity spot markets. Electr Power Syst Res 77:393–400
43. Angarita JL et al (2009) Combined hydro-wind generation bids in a pool-based electricity market. Electr Power Syst Res 79:1038–1046
44. Gibescu M et al (2008) Optimal bidding strategy for mixed-portfolio producers in a dual imbalance pricing system. In: Presented at the power systems computation conference (PSCC)
45. Hedman KW et al (2006) Comparing hedging methods for wind power: using pumped storage hydro units vs. options purchasing
46. Denault M et al (2009) Complementarity of hydro and wind power: improving the risk profile of energy inflows. Energy Policy 37:5376–5384
47. Garcia-Gonzalez J et al (2008) Stochastic joint optimization of wind generation and pumped-storage units in an electricity market. IEEE Trans Power Syst 23:460–468
48. Tuohy A et al (2009) Impact of pumped storage on power systems with increasing wind penetration. In: Proceedings of 2009 IEEE power and energy society general meeting, vols 1–8, pp 2642–2649
49. Muche T (2009) A real option-based simulation model to evaluate investments in pump storage plants. Energy Policy 37:4851–4862
50. Ummels BC et al (2009) Comparison of integration solutions for wind power in the Netherlands. IET Renew Power Gener 3:279–292
51. Hu P et al (2009) Reliability evaluation of generating systems containing wind power and energy storage. IET Gener Transm Distrib 3:783–791
52. Castronuovo ED (2009) Deliverable 3.7, tools for the coordination of storage and wind generation. Public report, ANEMOS.plus project
53. Biswas T (1997) Decision-making under uncertainty. McMillan Press Ltd., London
54. Stoft S (2002) Power system economics. Wiley Interscience, New York
55. ETSO (2007) Balance management harmonisation and integration. 4th report
56. Weber C (2009) Adequate intraday market design to enable the integration of wind energy into the European power systems. Energy Policy 38(7):3155–3163
57. Anderson D (2006) Power system reserves and costs with intermittent generation. Working paper, UK Energy Research Center
58. Ortega-Vazquez MA, Kirschen DS (2009) Estimating the spinning reserve requirements in systems with significant wind power generation penetration. IEEE Trans Power Syst 24:114–124

59. Pinson P (2006) Estimation de l'incertitude des prédictions de production éolienne. Doctoral thesis Energétique, CEP Centre Energétique et Procédés, ENSMP—CEP Centre Energétique et Procédés, ENSMP
60. Morales JM et al (2010) Short-term trading for a wind power producer. IEEE Trans Power Syst 25:554–564
61. Bathurst GN et al (2002) Trading wind generation in short term energy markets. IEEE Trans Power Syst 17:782–789
62. Holttinen H (2005) Optimal electricity market for wind power. Energy Policy 33:2052–2063
63. Fabbri A et al (2005) Assessment of the cost associated with wind generation prediction errors in a liberalized electricity market. IEEE Trans Power Syst 20:1440–1446
64. Usaola J, Angarita JL (2006) Bidding wind energy under uncertainty. In: Presented at the international conference on clean electric power. renewable energy resources impact
65. Matevosyan J, Soder L (2006) Minimization of imbalance cost trading wind power on the short-term power market. IEEE Trans Power Syst 21:1396–1404
66. Angarita-Marquez JL et al (2007) Analysis of a wind farm's revenue in the British and Spanish markets. Energy Policy 35:5051–5059
67. Pinson P et al (2007) Trading wind generation from short-term probabilistic forecasts of wind power. IEEE Trans Power Syst 22:1148–1156
68. Ceña A et al (1966) Forecasting exercise (final report in Spanish, English summary)

Prototype of HOTT Generation System

Mohammad Lutfur Rahman, Shunsuke Oka and Yasuyuki Shirai

Abstract The innovative renewable energy conversion system called "Hybrid Offshore wind and Tidal Turbine (HOTT) Generation System" is proposed. "Offshore-wind and Tidal Turbine" hybrid and autonomous power system research will demonstrate the feasibility of using hybrid wind and tidal current power to provide reliable electrical energy, and to create a push toward the development of a sustainable commercial market for this technology. This chapter describes the control system for a small laboratory based hybrid power system that uses two types of power generation, offshore-wind and tidal turbines, connected on the DC side. HOTT generation system energy can have numerous benefits from both the environmental and socioeconomic perspectives. An unenclosed HOTT generation system can avoid many of the detrimental environmental effects, including CO_2 emission, which is becoming a key issue, while providing significant amounts of distributed renewable energy.

1 Introduction

In the early days, a collection of wind turbines, or a wind farm, was seen as an interesting but insignificant component in the power generation system. Terms such as intermittent generation and negative load were used to describe them [1–6].

M. L. Rahman (✉) · S. Oka · Y. Shirai
Department of Energy Science and Technology, Graduate School of Energy Science, Kyoto University, Kyoto 606-8501, Japan
e-mail: lutfur@pe.energy.kyoto-u.ac.jp; lmasum2000@gmail.com

S. Oka
e-mail: oka@pe.energy.kyoto-u.ac.jp

Y. Shirai
e-mail: shirai@energy.kyoto-u.ac.jp

P. M. Pardalos et al. (eds.), *Handbook of Wind Power Systems*, Energy Systems,
DOI: 10.1007/978-3-642-41080-2_18,

If there was a disturbance on the network, the normal thing to do was to disconnect the wind turbines, wait until the network settled down, and then reconnect [5, 6].

The wind and tide are simulated by servo-motors that drive the wind-generator and tidal generator, respectively. The offshore-wind turbine load is complicated because the load of the offshore-wind turbine fluctuates during the day. These fluctuations create imbalances in the power distribution that can affect the frequency and voltage of the power system. Offshore-wind and tidal energy systems are omnipresent, freely available, environmentally friendly, and considered to be promising power generating sources because of their availability and topological advantages for local power generation. Prototype of HOTT Generation System (PHGS) energy systems, using two renewable energy sources, allow improvements in system efficiency and power reliability, and reduce the energy storage requirements for autonomous applications. Offshore-wind systems are becoming popular around the world for power generation applications because of the advances in renewable energy technologies and substantial rise in the prices of petroleum products. This experiment was conducted to analyze the current state of the prototype, as well as the optimization and control technologies for autonomous hybrid wind-tidal energy systems without battery storage.

2 Proposed PHGS Model System

2.1 Model Setup

Figure 1a shows how the HOTT system will be set up offshore, and how it will function. A wind and tidal turbine experimental model plays an important task in the hybrid turbine modeling, particularly for analyzing the interaction between the tidal and offshore-wind power systems, which are connected on the DC side [7–10].

Figure 1a–c show a photo and schematic view of the small laboratory-based hybrid power system model that designed and fabricated. The system has two types of generation, the tidal motor/generator and the offshore wind turbine generator. The tidal turbine (induction machine) can act as either a motor or generator, depending on the need. The tidal generator provides smooth output power, whereas the output power of a wind turbine depends on the wind velocity.

Figure 1a–d show how the HOTT system is set up in the laboratory and how it will function. These figures also show the conceptual schematic of the proposed HOTT system connected to the power system, and the detailed circuit configuration. The AC power generated by the wind and tidal turbine generators is converted into DC power. It is converted again into AC power through the maximum power point tracking (MPPT) inverter.

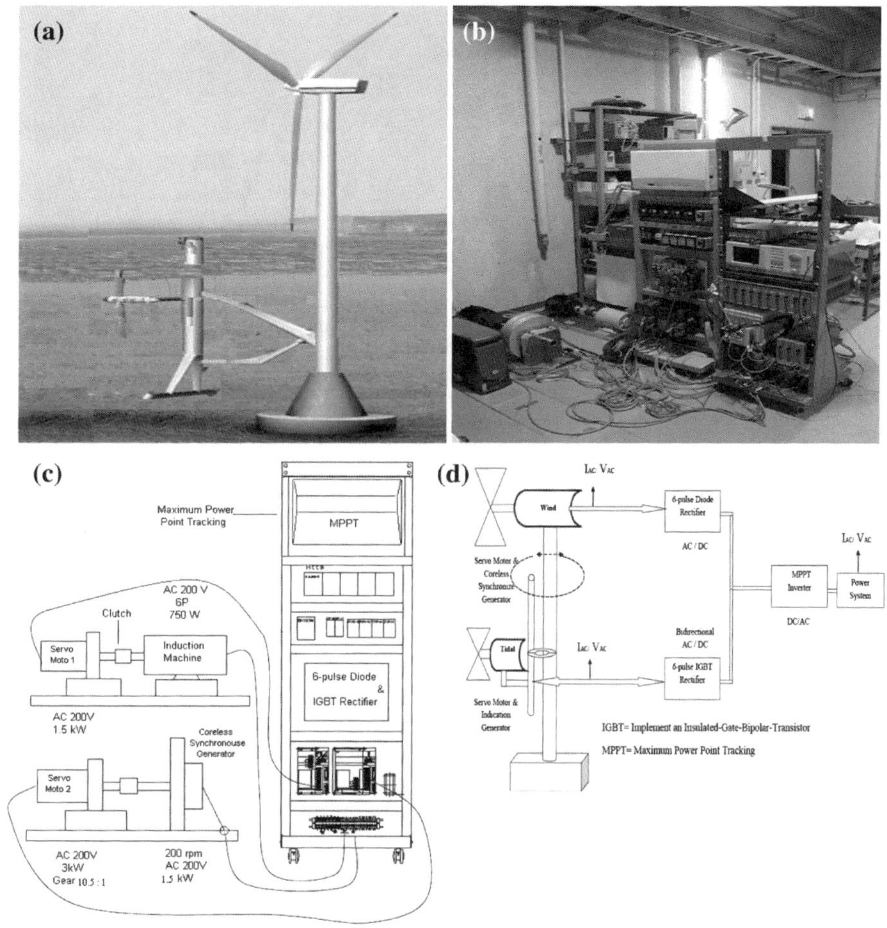

Fig. 1 **a** HOTT conceptual image ([10], ©IEEE 2010). **b** Photo of laboratory scale prototype model of hybrid offshore-wind and tidal turbine system with flywheel ([10], ©IEEE 2010). **c** Schematic of prototype model of hybrid offshore-wind and tidal turbine system with flywheel ([10], ©IEEE 2010). **d** Schematic of Hybrid Offshore-wind and Tidal Turbine (HOTT) ([10], ©IEEE 2010)

2.2 Offshore Wind Turbine

Figure 2 shows an experimental model of the offshore-wind turbine generator system. It consists of a coreless synchronous generator and a servo-motor. The offshore-wind turbine is simulated by the servo-motor. In this model system with the small servo-motor, the rated rotating speed is 2,500 rpm and the gear ratio is 10.5:1. In the real system, the wind turbine would have a slower rotating speed without the step-down gear. The rotating speed or the torque of the servo-motor is controlled by a computer. The electrical energy depends on the rpm (rotations per

Fig. 2 Offshore-wind turbine generator experimental model ([10], ©IEEE 2010)

Table 1 Rating of main components (offshore-wind servo-motor) ([10], ©IEEE 2010)

Parameter	Value
Rated output	3.0 kW
Rated voltage	200 V
Rated frequency	60 Hz
Rated speed	2,500 rpm
Gear ratio	10.5:1

Table 2 Rating of main components (coreless synchronous generator) ([10], ©IEEE 2010)

Parameter	Value
Rated output	1.5 kW
Rated voltage	200 V
Rated frequency	60 Hz
Rated speed	200 rpm

minute) of the servo-motor that rotates the coreless generator. Wind turbine generated AC power is converted to DC power with the 6-pulse diode rectifier [7–10].

The parameters of the servo-motor and the coreless synchronous generator are listed in Tables 1 and 2, respectively.

2.3 *Tidal Turbine (Flywheel)*

An induction generator produces electrical power when its shaft is rotated faster than the synchronous frequency of the equivalent induction motor. Induction machines are used in tidal system installations because of their ability to produce useful power at various rotor speeds. Induction machines are mechanically and electrically simpler than other generator types.

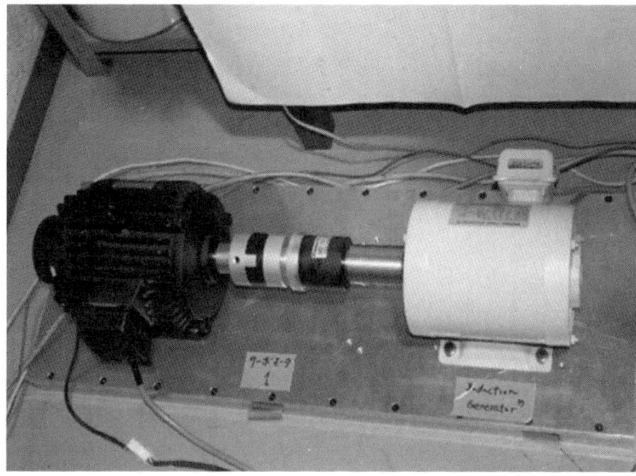

Fig. 3 Tidal turbine generator/motor experimental model. ([10], ©IEEE 2010)

The energy scenario in the world is calling for efforts toward more efficient use of electrical energy, as well as improvements in the quality of its delivery. This issue involves the use of energy storage devices, such as a tidal turbine used as a flywheel. The demand for and use of such equipment are increasing. One type of energy storage system is a Tidal Turbine Flywheel Energy System (TTFES). Due to the advancements in machines and power electronics, the flywheel is becoming more popular. Many feasible projects employing flywheel storage systems have been implemented all over the world [7–10].

Figure 3 shows the experimental model of a tidal turbine induction generator/ motor and a servo-motor. The main concept in this project is to apply and control a bi-directional (two way) energy flow scheme, so that energy is injected into the offshore wind turbine or stored as kinetic energy from/to the tidal system (induction machine).

Flywheels are one of the oldest forms of energy storage, having been used for thousands of years. The potter's wheel is one of the earliest applications of a flywheel. The kinetic energy stored in the flywheel results from spinning a disk or cylinder coupled to a machine's rotor. This energy is proportional to the flywheel mass and the square of its rotational speed:

$$E = \frac{1}{2} I \omega^2$$

where I is the moment of inertia in Kg m^2 and ω is the rotational speed in rad/s. The induction machine (tidal system) works as a motor with almost no load, and the rotational kinetic energy is stored as a function of the square of the rotational speed. The stored energy is extracted by decelerating the induction machine.

TTFES systems, in comparison with conventional batteries, present some interesting characteristics when used as an energy source to compensate for voltage sags and momentary power interruptions. The induction machine is used for bi-directional energy conversion from/to the tidal turbine. The servo-motor is

Table 3 Rating of main components (tidal servo-motor) ([10], ©IEEE 2010)

Parameter	Value
Rated output	1.5 kW
Rated voltage	200 V
Rated frequency	60 Hz
Rated speed	2,500 rpm

Table 4 Rating of main components (Induction machine) ([10], ©IEEE 2010)

Parameter	Value
Rated output	750 W
Rated voltage	200 V
Rated frequency	60 Hz
Rated speed	1,110 rpm

used as an input model of tidal energy to the induction generator, which converts the mechanical energy into electrical energy. The induction machine can work as a motor by using the bi-directional IGBT converter and a one-way clutch. When the induction machine's rotational speed is larger than that of the servo-motor, the servo-motor clutch turns to the off-state.

The design parameters of the servo-motor and the induction machine are listed in Tables 3 and 4, respectively. A speed of 1,110 rpm is selected for the induction machine to store the rotational kinetic energy. In a real system, the tidal turbine rotational speed should be much lower than that of the servo-motor, and a step-up gear will be necessary.

In this application, the TTFES supports the offshore-wind turbine, supplying power to the DC load in the case of overloads or dips. This occurs when there is a low wind speed, causing an offshore-wind turbine voltage or frequency dip or overload in the hybrid side. Thus, an offshore wind system is not capable of supply all the power needed by the DC load. The system gets help from the TTFES, which has stored kinetic energy. Therefore, the main purpose of the flywheel is to accumulate rotational kinetic energy, which can be injected into or extracted from the DC side whenever it is required.

2.4 Maximum Power Flow Control

The grid connected inverter is a current control type because the AC voltage is fixed by the grid. Therefore the DC link voltage in the HOTT is kept within a certain range for stable operation by controlling the AC output current of the grid inverter. The AC output current of the grid inverter is controlled so as to give the maximum output power with a certain DC voltage. This control is based on the Maximum Power Point Tracking (MPPT) algorithm.

In the MPPT control (Fig. 4), in order to search the DC link voltage which gives the maximum DC output power, small perturbation, ΔV (± 4 V), is given to

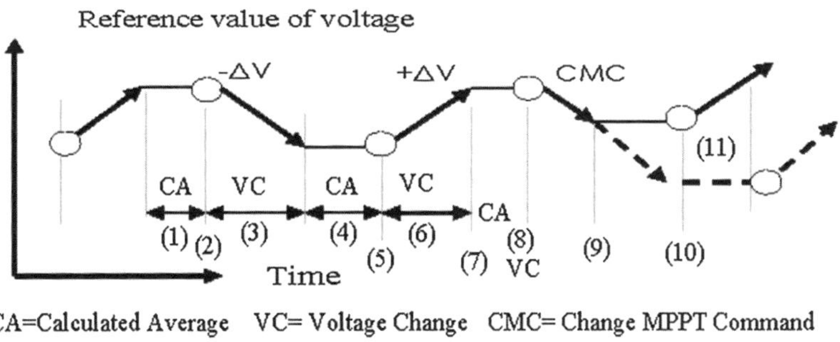

Fig. 4 MPPT electric power follow control algorithm

the DC reference voltage and check the DC output change. If the DC power increases, the perturbation is approved make a new DC reference voltage. Conversely, if it decreases, the reference voltage is changed in the opposite direction. This algorithm makes it possible to find the maximum electric power point when the characteristic of the DC side changes by wind and tide speed changes, etc. [11].

MPPT algorithm

(1)–(8). A series of processes for MPPT

- (1), (4) and (7) : Period for calculation of average value of DC voltage, current and power.
- (3) and (6): Period for DC voltage change [perturbation ΔV (±4 V)]
- (2) and (5): Check and memorize the deviation of DC output and voltage by corresponding perturbation.
- (8): New DC voltage reference is determined by the results at (2), (5) and (8).

– When the DC output power increases with DC voltage change +ΔV, the new DC voltage reference moves to (9) and start again from (1).
– Otherwise, it moves to (10) and restart from (1).

2.5 Inverter Circuit Configuration

Figure 5 shows the HOTT inverter circuit for this method. The output of the inverter is a single-phase three-wire system. A general domestic power supply is often a single-phase three-wire system. The inverter has a circuit configuration that combines two half bridges. The input can share one line with the output by using the half bridge type inverter [12]. In order to meet the grid voltage (200 V), the boost up chopper circuit is adopted in Fig. 5 to increase the voltage of the DC side.

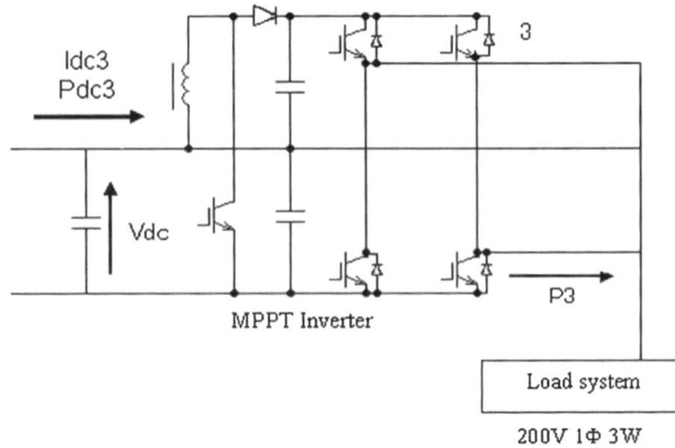

Fig. 5 Boost up chopper with a half-bridge inverter circuit

2.6 Hybrid System (Circuit Configuration)

This section describes the system and circuit configuration of the proposed HOTT with TTFES, which was designed and constructed based on the reviews of the alternatives and the components in the successive sections [7–10, 13].

The following is a block diagram of TTFES with an offshore wind system (also shown in Fig. 6). The offshore-wind coreless synchronous generator output is simply rectified by a 6-pulse diode bridge to charge a DC capacitor. The tidal turbine induction generator/motor output is connected to the DC capacitor through a 6-pulse IGBT dual converter. The DC link capacitor is connected to the commercial grid through a grid-connected, single-phase, 3-wire inverter. The grid-connected inverter is of a transformer-less half-bridge type with a boost-up chopper circuit. The voltage-source inverter output current is controlled by a PWM controller under maximum power point tracking (MPPT) control. The MPPT control monitors and maintains the DC link capacitor voltage, providing the maximum output power by controlling the output AC current. It monitors the DC voltage perturbations of 4 V up and down (2 V/s) every 4 s, calculates how to change the output power caused by them, and then determines the DC-voltage reference at the next stage to give more power. Several small controllers are implemented at both ends to provide the required performance for the system.

3 Changing Voltage Frequency 50–46–50 Hz

In order to simplify the description, an autonomous power supply with a limited capacity is substituted for the HOTT system (Fig. 7). A flywheel induction machine is connected to the offshore-wind system.

In order to compensate for wind power fluctuation, the output of the induction generator was controlled so that the total output was steady. The reference frequency for the PWM inverter signal of the bi-directional inverter was manually

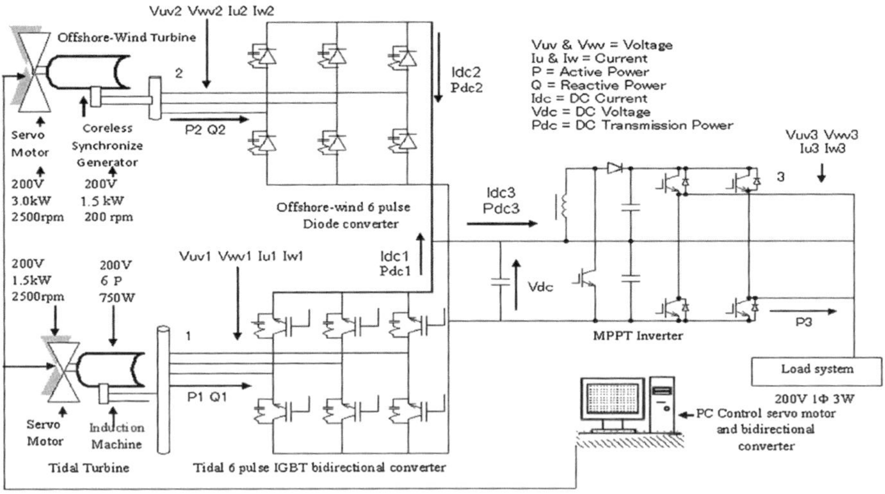

Fig. 6 PHGS system configuration ([10], ©IEEE 2010)

Fig. 7 DC Power output

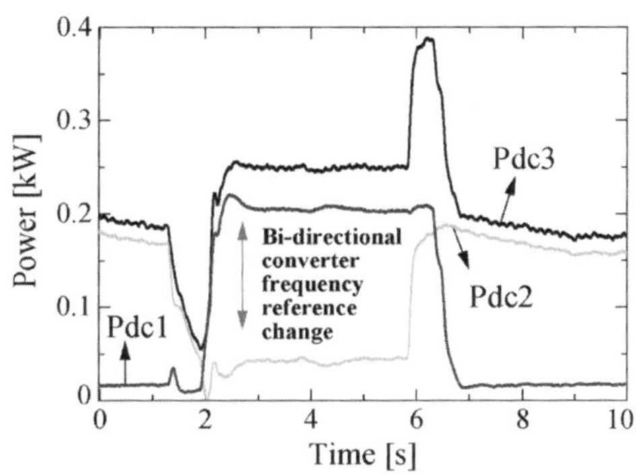

changed from 50 to 46 to 50 Hz; it is assumed that the servo-motor (tidal) was controlled to maintain a stable rotating speed (1,000 rpm), because tidal flow is more stable than wind flow. This experimental model system was able to keep the total output steady by controlling the induction generator by changing the AC voltage frequency input from the bi-directional converter.

4 Experimental Results

Figures 8, 9, 10, 11 and 12 shows some of the experimental results. From top to bottom, these figures show the generator voltages, currents, and instantaneous active and reactive powers of the tidal generation system; those of the offshore-wind

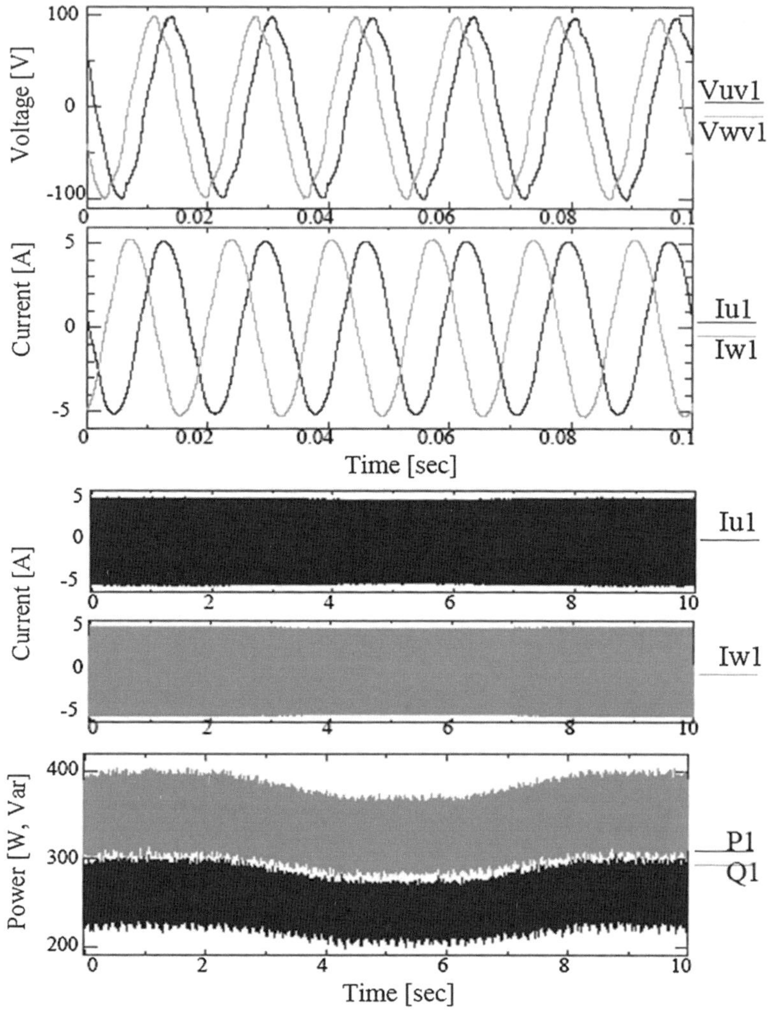

Fig. 8 Experimental results of tidal turbine generator voltages (Vuv1 and Vwv1), currents (Iu1 and Iw1), active power (P1), and reactive power (Q1) under the generator condition

generation system; the voltages, currents, and powers of the DC link circuit; those of the load (AC grid) side; and the rotating speed of the induction generator, servomotors, and coreless synchronous generator.

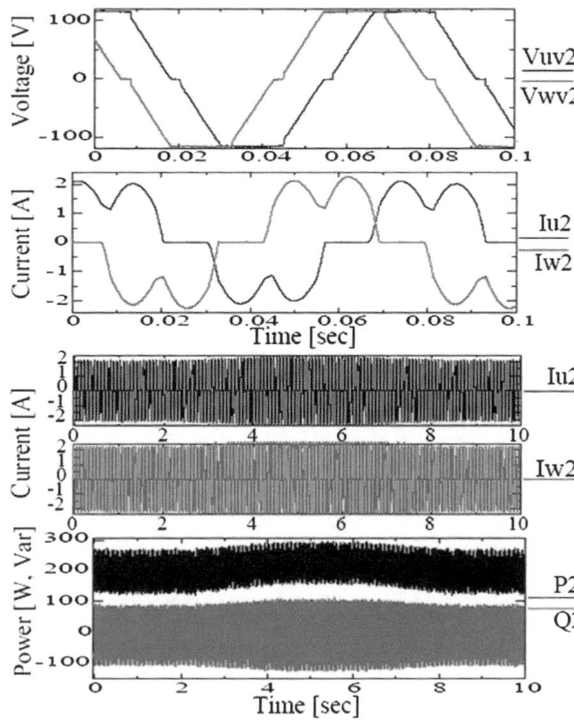

Fig. 9 Experimental results of offshore-wind turbine generator voltages (Vuv2 and Vwv2), currents (Iu2 and Iw2), active power (P2), and reactive power (Q2) under the generator condition

4.1 *Tidal Turbine (Induction Machine as Generator Mode)*

Figure 8 shows that tidal turbine servo-motor working at 1'371 rpm (shown in Fig. 12). The induction machine was driven in generator mode at 1'371 rpm (more than 1'200 rpm 60 Hz). The tidal turbine voltages (Vuv1 and Vwv1), currents (Iu1 and Iw1), and powers (P1 and Q1) were almost steady state.

4.2 *Offshore Wind Turbine (Coreless Generator)*

Figure 9 shows that the offshore wind turbine servo-motor and generator were operated at a constant speed of 82 rpm (shown in Fig. 12) throughout the test. The voltages (Vuv2 and Vwv2), currents (Iu2 and Iw2), and powers (P2 and Q2) were steady state, with a small fluctuation in the active power caused by Iu2 and Iw2.

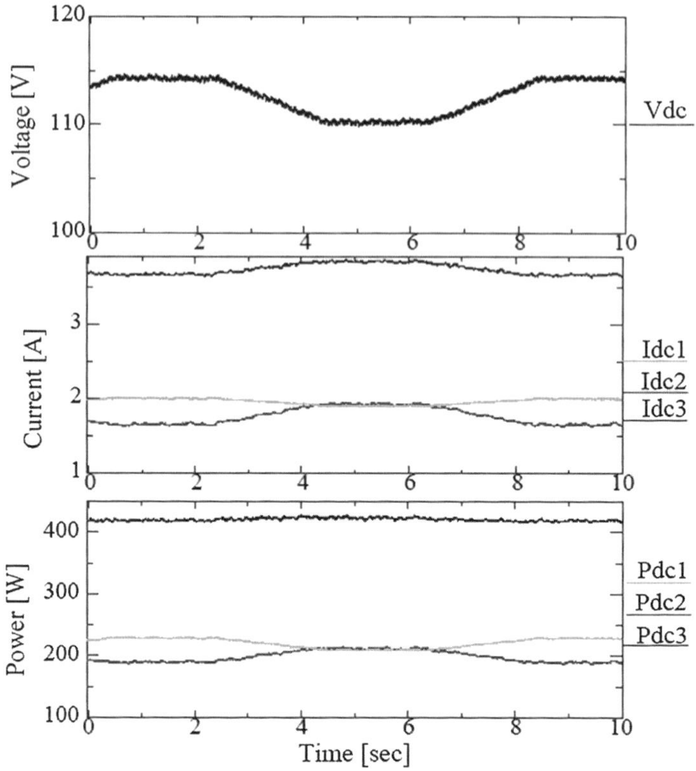

Fig. 10 Experimental results of DC (Hybrid) side voltage (Vdc), currents (Idc1, Idc2, and Idc3), and powers (Pdc1, Pdc2 and Pdc3)

4.3 DC Side

As shown in Fig. 10 of the DC link voltage, Vdc, the MPPT controller gave DC voltage perturbations of 4 V up and down (2 V/s) every 4 s. The DC side power, the offshore-wind generated power Pdc2 (offshore-wind), was almost a constant 190 W, and the tidal generated power, Pdc1 (tidal), stepped up from 230 W, with small fluctuations at 4–7 s caused by the MPPT and inverter system. The hybrid power (Pdc3) was 420 W for the PHGS steady state, although there were some fluctuations in Vdc.

4.4 Load Side

As shown in Fig. 11, the load (AC grid) side AC current and voltage were in a steady state condition, while the active power was 420 W.

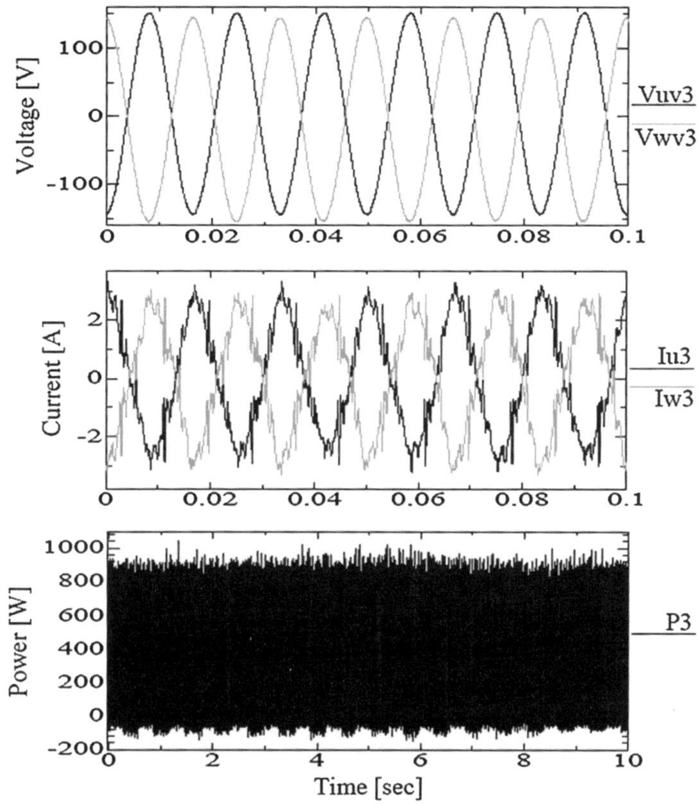

Fig. 11 Experimental results of load (AC grid) side voltages (Vuv3 and Vwv3), currents (Iu3 and Iw3), and active power (P3)

Fig. 12 Experimental results of rotating speeds (rpm) of coreless generator (offshore wind generator), induction machine (tidal generator), and servo-motor (for tidal generator)

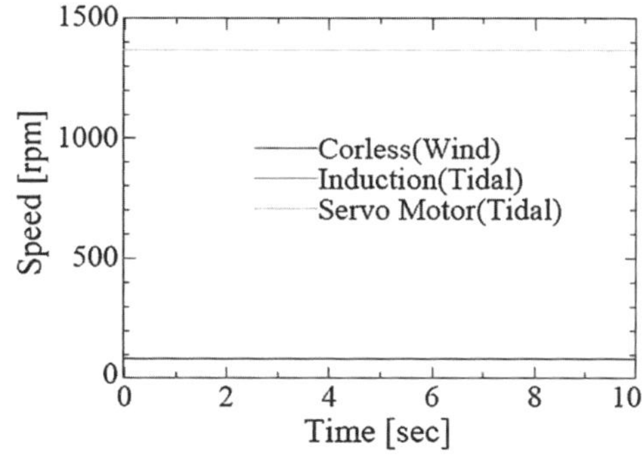

4.5 Rotating Speed (rpm)

As shown in Fig. 12, the induction machine and servo-motor rotating speeds were the same (1,371 > 1,200 rpm: generator mode), and the rotating speed of the offshore wind generator was 82 rpm. Both systems were in the generator mode and each output was summed up to the hybrid output flowing to the grid.

5 Discussion

The specific task of this research was using the prototype to evaluate the offshore wind energy and tidal technologies with respect to hybrid integration, and conveying this know-how into the ocean energy domain. The following tasks were included:

- Analyses of offshore wind and tidal energy conversion technologies and their conversion principles, control devices, and operational behaviors.
- Analyses of power quality and hybrid impact issues, as well as other peripheral technical barriers.
- Identifying possible similarities associated with hybrid integration (mixing) of offshore wind energy and tidal (induction machine) power technologies.

The PHGS conceptual demonstration prototype was constructed and successfully operated [9]. The PHGS hybrid performance was studied experimentally in stable generation ranges of the wind/tidal conditions. The proposed PHGS is more flexible than a single system, allowing the stable generation ranges of the wind/tidal conditions to be extended using an adequate system control strategy. The performance of modular hybrid energy systems can be improved through the implementation of advanced control methods in bi-directional and MPPT system controllers [11, 12, 14–16].

Optimum resource allocation, based on load demand and renewable resource forecasting, promises to significantly reduce the total operating cost of the system. The application of modern control techniques to supervise the operation of modular hybrid energy systems allows the utilization of the renewable resource to be optimized.

6 Conclusion

Executing the experimental design and analysis is a critical component in the HOTT system. The focus of this chapter is to describe the methodology and performance of a prototype experimental design hybrid system.

First, the basic performances of the conceptual demonstration prototype PHGS hybrid were studied while electric power was generated from both the offshore wind and tidal systems, and distributed to a load system. The proposed PHGS is more flexible than a single system, allowing the stable generation ranges of the wind/tidal conditions to be extended by an adequate system control strategy. The output of the tidal generator (induction machine) is controlled using an IGBT bi-directional converter system to compensate for the power fluctuation in the offshore-wind turbine generator. Additionally, the tidal induction machine rotor can be mechanically isolated from the tidal turbine shaft by a one-way clutch, allowing the induction machine to work, not only as a generator, but also as a motor (flywheel energy storage [15, 16]) using the IGBT converter control. The wind and tidal hybrid generation system circuit models linked using a DC link capacitor and MPPT grid inverter showed stable operation [13]. A boost up chopper was used to improve the power factor of the system. The grid inverter was controlled using the maximum power point tracking (MPPT) control, which kept the DC capacitor voltage at the maximum power output.

Second, two kinds of control strategies were proposed and experimentally demonstrated using the prototype test setup.

References

1. Hirachi Labrotory Technology Memo No. 20090930. http://hirachi.cocolog-nifty.com/kh. Accessed on 2010
2. Muljadi E, McKenna HE (2002) Power quality issues in a hybrid power system. IEEE Trans Ind Appl 38(3):803–809
3. Wave Energy Conversion (2008) University of Michigan College of Engineering. http://www.engin.umich.edu/dept/name/research/projects/wave_device/wave_device.html. Accessed Oct 2008
4. Polagye B, Previsic M (2006) EPRI North American tidal in stream power feasibility demonstration project, EPRI-TP-006A, 10 June 2006
5. Akagi H, Sato H (2002) Control and performance of a doubly-fed induction machine intended for a flywheel energy storage system. IEEE Trans. Power Electron 17(1):109–116
6. Wave Energy Conversion (2009) University of Michigan College of Engineering. Available: http://www.engin.umich.edu/dept/name/research/projects/wave_device/wave_device.html. Accessed 2009
7. Rahman ML, Shirai Y (2008) "Hybrid offshore-wind and tidal turbine (HOTT) energy Conversion I (6-pulse GTO rectifier and inverter)". In: IEEE International Conference on Sustainable Energy Technologies, ICSET 2008, pp.650–655
8. Rahman ML, Shirai Y DC (2010) connected hybrid offshore-wind and tidal turbine (HOTT) generation system. Green Energy Technol Springer Acad J :141–150
9. Rahman ML, Shirai Y Hybrid power system using Offshore-wind turbine and tidal turbine with flywheel (OTTF). Europe's offshore wind 2009, VIND2009 (eow2009, Stockholm) Sept 2009
10. Rahman ML, Oka S, Shirai Y (2010) Hybrid power generation system using offshore-wind turbine and tidal turbine for power fluctuation compensation (HOT-PC). IEEE Trans Sustain Energy 1(2):92–98
11. Omron Industrial Automation, Japan, model number KP40H

12. Chapmann S (2005) Electric machinery fundamentals, 4th edn. McGraw Hill International Edition, New York
13. Rahman ML, Oka S, Shirai Y (2011) Hybrid offshore-wind and tidal turbine power system for complement the fluctuation (HOTCF). Green Energy Technol Springer Acad J :177–186
14. Jeoung H Choi J (2000) High efficiency energy conversion and drives of flywheel energy storage system using high temperature superconductive magnetic bearings. Power Eng Soc Winter Meet IEEE 1:517–522
15. SilvaNeto J, Rolim G (2003) Control of a power circuit interface of a flywheel-based energy storage system; UFRJ, Cidade Universitiria, Rio de Janeiro, Brazil
16. Kim WH, Kim JS, Baek JW, Ryoo HJ, Rim GH (1998) Improving efficiency of flywheel energy storage system with a new system configuration. In: Sung-Ju Dong, Chang-Won, Kyung-Nam, Electro Technology Research Institute; Korea Institute of Machinery and Metals

Wind Energy Facility Reliability and Maintenance

Eunshin Byon, Lewis Ntaimo, Chanan Singh and Yu Ding

Abstract The global wind power industry involves operations in highly stochastic environments and thus faces challenges in enhancing reliability and reducing maintenance costs. Earlier studies related to wind energy facility reliability and maintenance focused more on qualitative aspects, discussing the unique influencing factors in wind power operations and their effects on system performance. With operational experience accumulated for more than a decade, the most recent focus has shifted to a more structured approach using analytical and/or simulation methods. In this chapter, we provide a comprehensive account of the existing research regarding wind energy facility reliability and maintenance. We group the relevant studies into three major categories. The first category addresses the degradation and failure pattern of wind turbines, aiming at optimizing the operations and maintenance. The second and third categories discuss the reliability issues in a broader sense, focusing on reliability assessment at the wind farm level and at the overall power system level, respectively.

E. Byon
University of Michigan, 1205 Beal Avenue, Ann Arbor, MI 48109-2117, USA
e-mail: ebyon@umich.edu

L. Ntaimo · Y. Ding (✉)
Texas A&M University, 4016 ETB 3131 TAMU, College Station, TX 77843-3131, USA
e-mail: yuding@iemail.tamu.edu

L. Ntaimo
e-mail: ntaimo@tamu.edu

C. Singh
Texas A&M University, 301 WERC 3131 TAMU, College Station, TX 77843-3131, USA
e-mail: singh@ece.tamu.edu

P. M. Pardalos et al. (eds.), *Handbook of Wind Power Systems*, Energy Systems,
DOI: 10.1007/978-3-642-41080-2_19,

1 Introduction

Wind power has become one of the fastest growing renewable energy sources around the world. Wind energy installed capacity increased from 31 GW in 2002 to 2082 GW in 2012 worldwide. According to the North American electric reliability corporation (NERC) [1], approximately 260 GW of new renewable nameplate capacity is projected in the US during 2009–2018. Roughly 96 % of this total is estimated to be wind energy. In fact, NERC predicts that wind power alone will account for 18 % of the U.S. total resource mix by 2018.

With operational experience accumulated for more than a decade, people have come to realize that operations and maintenance (O&M) costs constitute a substantial portion of the total costs of wind power production [2–4]. Field data from Germany [5] indicates approximately six failures per year and restoration times ranging from 60 hour to a few weeks. Overall, O&M costs account for 20–47.5 % of the wholesale market price [4]. Considering that most turbines in the U.S. were installed in the past 10–15 years, wind facilities are still operating in their relatively reliable period. In the decades to come, as turbine components are near the end of their designed life, the failure rates of wind power facilities can be expected to increase drastically, leading to higher costs of O&M.

The most common maintenance practice is to perform scheduled maintenance on a regular basis. Depending on a manufacturer's suggested guideline, scheduled maintenance is carried out usually once or twice a year for a turbine [6, 7]. Understandably, scheduled maintenance is usually incapable of addressing unexpected failures in a timely fashion, while, on the other hand, this strategy may also result in unnecessary visits. Wind farm operators have come to the consensus that more effective maintenance strategies are pressingly needed to reduce unnecessary service visits as well as turbine downtime.

Many studies have been conducted to examine effective maintenance strategies for conventional power systems, for example, [8–10] among others. However, the operating environment of wind turbines is different from that of traditional power plants in several ways. Wind turbines constantly operate under highly stochastic loading. The feasibility of both new repairs and the continuation of ongoing repairs is affected by uncertain weather conditions. The vast majority of wind turbines are installed in remote windy sites or offshore locations, hoping to harvest the maximum amount of wind energy. Due to the remote locations of the wind farms, access to turbines for maintenance can be restrictive during harsh weather seasons. For the same reason, repair actions, or even on-site observations, are very expensive. Compounded with these complexities, newly established wind farms usually house hundreds or more turbines that spread over a large geographical area. This makes O&M even more challenging and costly. We believe that wind farm operations call for new maintenance strategies that are specially tailored for handling the reliability and maintenance issues for wind energy facilities.

In this chapter, we discuss recent studies on wind energy facility reliability and maintenance. Earlier studies on wind power operations investigated unique but critical factors that have significant impact on the turbine's reliability and maintenance [6, 11–13]. Later, failure patterns and reliability trends of wind turbine were studied based on field data [2, 14, 15]. More recently, the focus has shifted towards developing a more structured model for reliability enhancement.

We group the existing research into three broad categories. The first category is degradation and failure pattern of an individual wind turbine (or, a wind turbine component) to assess its reliability over time. The major goal of these studies is to build a model for predicting the lifetime of a system, and further, to develop a cost-effective O&M decision strategy by quantifying risks and uncertainties based on the lifetime prediction. To fulfill this goal, several analytical models have been recently introduced in the wind power literature, including statistical-distribution based models [15–20], Markov process-based models [21–24] and physics-based structural load models [25–35]. Although renewal reward processes and regenerative processes have been extensively studied for general aging systems and power systems [36–38], we did not find them currently employed in the wind power literature. Concerning the question of testing, there is limited amount of literature which studies the appropriate testing procedures to certify wind turbine components [39–41].

The second category of research concerns the reliability of a wind farm with multiple turbines, and focuses on how the reliability of each wind turbine (or, turbine component) affects the performance of the whole wind farm. Many researchers have developed simulation models to represent the complex behavior of wind farm operations, as well as the interactions between wind turbines [6, 11, 16, 22, 42–46]. They investigate the effect of different maintenance strategies and evaluate the reliability of a wind farm using several performance indices [22, 45]. The effect on system performance coming from the environmental differences such as geographical terrain and wind farm sites (i.e., land-based or offshore) are also discussed in these studies [22, 46].

The third category extends the second category of studies to the overall power facilities, i.e., measure its capability to satisfy customer demands (or, loads) or system operational constraints [47]. This line of research is usually called "generation adequacy assessment". Both analytical and simulation models are used in this line of work, including multi-state models [47–52], correlation analysis [53, 54], Markov process [21, 55, 56], population-based stochastic search [57, 58], and Monte Carlo simulations [59–66]. One specific aspect these studies focus on is the effect of intermittent generation of wind energy on the power system reliability. The failure and repair rates of wind turbines are taken into account in the assessment. These studies provide practical implications about the reliability and cost issues of wind energy in the context of the entire electric power system. Table 1 provides a summary of the existing research.

The remainder of this chapter is organized as follows. We first discuss the O&M aspects of wind power operations in Sect. 2. In Sect. 3, we review the models that explore the reliability patterns of wind turbine components as well as the models that find optimal O&M strategies. Section 4 presents the reliability and maintenance studies at the wind farm level, while Sect. 5 discusses the same issue but for the entire power system (i.e., the generation adequacy assessment). Finally,

Table 1 Classification of literature at three operating levels

Level	Methodology	Literature
Wind turbine	Statistical approach	[15–20]
	Stochastic process[a]	[21–24]
	Structural load models	[25–35]
Wind farm	Simulation	[6, 11, 16, 22, 42–46]
Power system	Analytical approach	[21, 47–58]
	Simulation	[59–66]

[a] Only Markov processes are presented in this chapter because, to the best of our knowledge, other stochastic processes are not yet discussed in the wind power literature

we conclude the chapter in Sect. 6. We acknowledge that there are two other reviews on relevant literature: one is Alsyouf and El-Thalji [67], who reviewed the recent practices and studies focusing on maintenance issues, and the second is Wen et al. [47], who summarize the studies on adequacy assessment of power systems (wind power is one element of the power systems). This chapter provides a more comprehensive survey, including the literature in [67] and [47]. We also include new results from reports and papers since the last survey.

2 Understanding Current Status of Wind Turbine Operations

In this section, we examine the current O&M costs and key factors affecting wind turbine operations.

2.1 O&M Costs

Due to the relatively young history of the wind power industry, many turbines are still operating in their early age so that failure and maintenance data are not as ubiquitous as in a mature industry. Making the matter worse, modern turbines are much more diversified in terms of sizes and designs than earlier turbines. However, the current operational experience was built and accumulated based on earlier turbines. This means that evaluating O&M costs for this young industry is challenging and the results embed high uncertainty. Nevertheless, useful insights can be gained from the experience with the existing turbines.

2.1.1 Overall O&M Costs

Walford [2] evaluates how much O&M costs account for the total cost of energy (COE) by summarizing the results from Vachon [3], and Lemming et al. [68]. COE is a key metric to evaluate the marketability of wind energy. The current

COE is $0.06 - 0.08 k/kWh. According to the studies by Vachon [3] and Lemming and Morthorst [68], O&M costs are estimated to be $0.005–$0.006/kWh in the first few years of operations, but it escalates to $0.018–0.022/kWh after 20 years of operations. Based on these two studies, Walford [2] concluded that the overall O&M costs can account for 10–20 % of the total COE.

U.S. Department of Energy (DOE) [69] gives a similar estimate of $0.004–0.006/kWh in 2004 for new wind turbine projects in the Class 4 resource areas (at a Class 4 site, the wind speed is between 5.6 and 6.0 m/s at 10 m height [70]). But, according to a later report by DOE [71] in 2006, the actual O&M costs range from $0.008 to $0.018/kWh, depending on the wind farm size. Most recently, Asmus [72]'s study presents even higher O&M costs. Asmus [72] collected extensively O&M associating data from major wind turbine manufactures and wind energy operators. After compiling data from various sources, Asmus [72] concludes that the average O&M costs is $0.027/kWh (or, €0.019/kWh), which may account for half of the COE.

In summary, O&M costs are comparable to the $0.02/kWh federal production tax credit (PTC) offered in the U.S. as a subsidy. Although PTC was extended in 2012 in the U.S., the subsidy will eventually decrease or disappear in the future. Asmus [72] also notes that currently about 79 % of turbines are still under warranty, but their warranty phase will be over in the near future. The phase-out perspective of the PTC and the warranty calls for cost-effective O&M strategies for wind turbines in order to make the wind energy market competitive.

2.1.2 Cost Elements

Walford [2] identifies the cost elements associated with wind farm operations. They include the operational costs for daily activities such as scheduling site personnel, monitoring turbine operations, responding to failures, and coordinating with power utility companies. Other costs include those incurred from taking preventive maintenance actions to prevent failures in advance or taking corrective (or, unscheduled) maintenance after unanticipated failures. Krokoszinski [13] presents a mathematical model to quantify wind farm production losses in terms of the planned and unplanned downtime, and quality losses.

In general, expenses for daily operations are fairly stable over time, while repair costs fluctuate considerably, depending on the age, location, size of the turbine, and the maintenance strategy [73]. For example, offshore turbines have much higher corrective maintenance costs due to the logistic difficulties to access the turbines. Corrective maintenance cost can be more than twice the preventive maintenance cost [16, 43]. In general, corrective maintenance costs account for 30–60 % of the total O&M costs [2].

2.2 Factors Affecting O&M Costs

Many studies examine critical factors that affect the reliability and O&M costs of wind power facilities. Based on the findings, researchers suggest ways to reduce the impact of unplanned failures or minimize maintenance costs [2, 12]. Bussel [6] presents an expert system to determine the availability of wind turbines and O&M costs. The goal in this work is to find the most economical solution by striking a balance between the front-loading costs invested for reliability enhancement and the O&M costs. Pacot et al. [12] discuss key performance indicators in wind farm management, and analyze the effects of factors such as turbine age, size, and location. Here, we summarize the key factors, which we believe are the most critical in wind farm operations [22–24].

2.2.1 Stochastic Environmental Conditions

Stochastic operating conditions affect the reliability of wind turbines in several ways. First, maintenance activities can be constrained by the stochastic weather conditions. Harsh weather conditions may reduce the feasibility of maintenance. This is in contrast with traditional fuel-based power plants which operate under relatively stationary operating conditions. To maximize potential power generation, wind facilities are built at locations with high wind speeds. But climbing a turbine during wind speeds of more than 20 m/s is not allowed; when speeds are higher than 30 m/s, the site becomes inaccessible [22]. Moreover, some repair work takes days (and even weeks) to complete due to the physical difficulties of the job. The relatively long duration of a repair session increases the likelihood of disruption by adverse weather. A study using a Monte Carlo simulation [42] found that turbine availability was only 85–94 % in a 100-unit wind farm situated about 35 km off the Dutch coast. The relatively low turbine availability is in part due to the farm's poor accessibility, which is on average around 60 %. Another study [6] found that the availability of a wind farm was 76 %. The operational environments and their effect on maintenance feasibility are also discussed in our previous studies [23, 24, 45].

Weather conditions can affect the reliability of wind turbines substantially. Harsh weather events such as storms and gusts will generate non-stationary loading (or, stress) and reduce the life of key components. Tavner et al. [15] show that there is a strong relationship between the number of failures and wind speeds. Toward this claim, analysis of field data from Denmark and Germany confirms a clear 12-month periodicity of the number of failures, which coincides with annual weather seasonality.

Besides strong winds, other adverse weather conditions include temperature, icing and lightning strikes. According to Smolders et al. [74], the operating temperature for wind turbines could range from −25 to 40 °C. Wind turbines at very low temperature sites suffer from the icing problems, namely that icing decreases power generation and damages rotor blades significantly [75]. Lightning strikes can also cause serious damage to blades. Although modern turbines are equipped with lightning protection systems in blades, current technology cannot yet completely protect turbine blades from lightning damage [76].

Harsh weather conditions can cause high revenue losses during downtime. Lost productivity becomes more significant when turbine unavailability occurs during the windy periods [2]. With increasing application of wind turbines in different terrains, more research is pressingly needed to study how the adverse weather affects reliability and maintenance feasibility and to help practitioners make a better choice of farm locations [75].

It should also be noted that weather forecasting can play an important role on reliability assessment and maintenance decision-making. The benefits from weather forecasts depend on the accuracy of the forecasts. Thanks to today's sophisticated weather forecast technology, the near-future weather information is usually deemed reliable [77]. Considering the possible long repair time which can take up to several weeks, reliable forecast techniques for medium- or long-term is much needed but not yet available.

2.2.2 Logistic Difficulty

Recently established wind farms tend to house larger and larger wind turbines, in a great number and spreading over a broad area. The sheer size of wind turbine components makes it difficult to store spare parts in a warehouse waiting for repairs or replacements. Rather, they are likely ordered and shipped directly from a manufacturer when needed. Doing so often leads to long lead time in obtaining parts and can result in costly delays in performing repairs. Pacot et al. [12] point out that it may take several weeks for critical parts, such as a gearbox, to be delivered. Another aspect of the logistic difficulty is caused by the long distances of wind farms from their operation centers, and that major repairs require heavy duty equipment such as a crane or a helicopter to access the turbine. It certainly takes quite an effort to assemble the maintenance crews and prepare for a major repair. Logistic costs may escalate substantially, depending on accessibility to the turbine site, maintenance strategy, and equipment availability.

2.2.3 Different Failure Modes and Effects

A wind turbine consists of several critical components such as a drive train, gearbox, generator and electrical system, each of which has different failure frequencies and effects. Each single component may also have different failure modes. The occurring failure mode determines what type of parts/crew is required, which in turn determines the costs, lead time and repair time. Accordingly, the maintenance costs and the downtime due to the occurrence of a failure could vary significantly for different failure modes.

Arabian-Hoseynabadi et al. [78] perform a failure modes and effects analysis for the major components in various types of wind turbines and study their failure effects on the overall turbine performance. Ribrant [14] and Ribrant and Bertling [79] review the different failure modes of turbine components and the corresponding consequences. For example, a failing gearbox can lead to bearing failures, sealing problems, and oil system problems. According to Ribrant [14], it can take several weeks to fix the problems associated with bearing failures, partly because of the long lead time needed to have labor and heavy-duty equipment in place. On the other hand, oil system problems can be fixed within hours.

According to a recent industry survey [72], most serious failures are associated with the gearbox. As an alternative, direct-drive (gearbox free) turbines in which generators rotate at the same speed as the rotor were introduced. Echavarria et al. [75] show, however, that the failure rate in "generator" of direct drive turbines is higher than the failure rates in "generator" and "gearbox" combined during the first ten operational years. This implies that direct-drive may resolve the frequent failure issues of the gearbox, but cause new problems that have not been encountered in other designs. Yet, we should be cautious to conclude which design is better in terms of reliability since the reliability results of direct-drive turbines in Echavarria et al.'s study are based on a small number of direct-drive turbines. More time and efforts would be required to assess the reliability of each different design.

2.2.4 Condition-Based Maintenance

Condition-based maintenance (CBM) has become an important method to reduce the O&M costs [80]. CBM involves two processes. The first process is to diagnose the current condition through condition monitoring equipment. Sensors installed inside turbines can provide diagnostic information about the health condition of turbine components. Such data help wind farm operators estimate the deterioration level and make real-time reliability evaluation. Based on this reliability update, the next step is to establish appropriate maintenance policies before the actual occurrence of major failures to avoid consequential damages.

Comprehensive review of the recent monitoring techniques is provided by Hameed et al. [81]. In general, vibration analysis is the primary monitoring technique used for gearbox fault detection [82]. Other common monitoring systems include: measuring the temperature of bearings, analyzing the lubrication oil particulate content, and optical measurements of strains in wind turbine structures. Caselitz and Giebhardt [83] discuss the integration of a condition monitoring system into offshore wind turbines for improved operational safety.

A few studies attempt to quantify the benefits of CBM in the wind power industry [7, 22, 45]. Nilsson and Bertling [7] present an asset life-cycle cost analysis by breaking the total maintenance cost into several cost components. They analyze the benefits of CBM with case studies of two wind farms in Sweden and the UK, respectively. They conclude that CBM can benefit the maintenance management of wind power systems. The economic advantage becomes more clear when an entire wind farm rather than a single wind turbine is considered. For offshore wind farms, a condition monitoring system could provide a greater help by making maintenance planning more efficient.

McMillan and Ault [22] evaluate the cost-effectiveness of CBM via Monte Carlo simulations. They compare a six-month periodic maintenance policy with CBM with respect to the annual yield of power production, capacity factor, availability, annual revenue and failure rates. Through simulating various scenarios with different weather patterns, down-time durations, and repair costs, they show for land-based turbines that a CBM strategy could provide the operators remarkable economic benefits. For example, with an average wind speed of 6.590 m/s and cost factors obtained from [84], CBM can have operational savings of £225,000 (or, equivalently, \$350,280 with an exchange rate of 1 pound = 1.5668 dollar) per turbine, when considering 15 years as a turbine life. Sensitivity analysis with relatively large ranges in parameter values indicate that the contribution of CBM is highly dependent on replacement costs, wind regime and downtime durations. Obviously, one would expect more benefits for offshore wind turbines since the repairs of those turbines are more costly and taking maintenance actions at an offshore location faces more constraints.

McMillan and Ault [22] assume that the condition monitoring equipment reveals exactly the degradation status of each turbine component. However, condition monitoring equipment cannot in general solve the uncertainty issue [85]. Estimations rarely reveal perfectly the system's health status due to a wide variety of reasons, such as imperfect models linking measurements to specific faults, as well as noises and contaminations in sensor signals. More importantly, fault diagnosis based on sensor measurements is nontrivial, because wind turbines operate under non-steady and irregular operating conditions.

Considering these sensor uncertainties, the authors of this chapter consider two types of measurements to estimate the internal condition of each turbine component in our previous work [23, 24]: (1) inexpensive, but less reliable, real-time remote sensing and diagnosis from general condition monitoring equipment, and (2) expensive, but more certain, on-site visit/observation (OB). OB is fulfilled by either dispatching a maintenance crew or, if technologically feasible, invoking more advanced smart sensors. Both options are generally costly, but can presumably depict system conditions with a high confidence. It should be noted that the co-existence of real-time monitoring and on-site visit/observation is a unique aspect in the wind energy industry. The information uncertainty from the condition monitoring equipment must be handled with caution, and on-site observations must be integrated with planning maintenance actions. Considering the different observation options, Byon et al. propose two CBM models in [23, 24]. Their results show that OB should be taken when the estimated system conditions from condition monitoring equipment are not clear. The optimal observation actions also depend on the weather conditions. For example, OB should be taken more often under harsh weather conditions to decide the most suitable maintenance tasks than under mild weather conditions.

The authors of this chapter also develop a discrete event simulation model for wind farm operations that can evaluate different O&M strategies [45]. The current version implements two different O&M strategies: scheduled maintenance and a CBM strategy. In the CBM strategy, preventive repairs are carried out when sensors send alarm signals. The implementation results indicate that the CBM strategy provides appreciable benefits over the scheduled maintenance in terms of failure frequency, O&M costs and power generation. For example, the failure frequency is reduced by 11.7 %, when CBM instead of scheduled maintenance is used.

3 Models to Assess Reliability of a Wind Turbine and to Find Optimal O&M Strategies

In this section, we review existing studies that evaluate the reliability of a wind turbine and find a cost-effective O&M strategy. We discuss three different approaches: the first one is the statistical approach to assess the reliability level and identify the optimal repair (or, replacement) time based on the failure statistics; the second approach uses Markov models to analyze the stochastic aging behavior of wind turbine components; and the third approach is based on physical fatigue analysis.

3.1 Statistical Distribution-Based Analysis

3.1.1 Life-Time Analysis Using the Weibull Distribution

In reliability engineering, one of the most commonly used distributions to represent the life-time of a system is a Weibull distribution. This distribution has been applied in many applications to model a great variety of data and life characteristics. In wind facility modeling, several studies apply the Weibull distribution in determining the lifetime reliability of a population of wind turbines (or, turbine components).

Let T be a random variable representing the life-time of a system. The probability density function (pdf) of the Weibull distribution is given as follows:

$$f(t) = \begin{cases} \frac{\beta}{\eta}\left(\frac{t}{\eta}\right)^{\beta-1}\left(e^{-\frac{t}{\eta}}\right)^{\beta}, & t \geq 0 \\ 0, & t < 0, \end{cases} \tag{1}$$

where $\beta > 0$ is the shape parameter and $\eta > 0$ is the scale parameter. $\beta < 1$ denotes a decreasing failure rate, usually known as the "infant mortality". This happens when failures occur in an early period of operation. As the defective items are taken out of the population, failure rates decrease over time. On the other hand,

$\beta > 1$ indicates an increasing failure pattern, often found in an aging system. When $\beta = 1$, the pdf in (1) is exactly the same as the pdf of an exponential distribution. As such, $\beta = 1$ indicates a constant failure rate, or random failure pattern, implying that any maintenance activity to improve the system condition is unnecessary. For detailed discussions to assess reliability using the Weibull distribution, please refer to [86].

Andrawus et al. [16] employ the Weibull distribution to model the failure patterns of several critical components in a wind turbine such as the main shaft, main bearings, gearbox and generator. For one type of 600 kW horizontal axis turbines, they collect historical failure data from the turbine's supervisory control and data acquisition (SCADA) system over a period of 9 years. The data include the date and time of failure occurrences and the (possible) causes of the failures. They estimate the parameter values in the Weibull distribution using a maximum likelihood estimation (MLE). According to their case studies, the estimated shape parameters for the main bearing, gearbox and generator are close to "one", indicating a random failure pattern. Only the main shafts exhibit a wear-out (or, aging) failure pattern. After conducting these life-time analyses and also considering the cost factors (such as planned and unplanned maintenance costs), they made suggestions for the optimal replacement time for each component. For example, they recommend that a gearbox should be replaced every six years and a generator every three years to minimize the total maintenance costs.

Remark The results of the constant failure rates of the three components in Andrawus et al.'s study [16] are counter-intuitive, contradicting the results from other studies [19, 22]. It is commonly accepted that most wind turbine components deteriorate over time. On the other hand, electrical systems in a turbine show random failure patterns in general. We conjecture that the maintenance practices, which might have been performed while the field data were collected, may need to be considered in Andrawus et al.'s study. If the failure data used in the study are obtained from the turbines under scheduled maintenance, the random failure patterns may be because the scheduled maintenance restores the turbines back to their renewed states from time to time. If this maintenance effect is taken out, Andrawus et al.'s data may confirm the result from [19, 22], and show turbine deterioration patterns instead.

Wind turbines at different locations undergo different degradation, depending on the climate characteristics under which a turbine is operating. Vittal and Teboul [20] use a Weibull proportional hazard model to incorporate the weather climate effects on the lifetime of wind turbines. They express the scale parameter, η in (1), as a function of weather factors such as wind speed, capacity factor, and temperature. In this way, site-specific climate characteristics are combined with the general aging behavior of a population of wind turbines to derive a turbine lifetime.

Wen et al. [47] and Basumatary et al. [87] summarize different methods to estimate the parameters of the Weibull distribution. They use these methods in the wind regime analysis to represent the wind speeds model. Nonetheless, we believe that these methods can be applied to reliability analysis as well.

3.1.2 Assessing Reliability Patterns Using Homogeneous/Non-Homogeneous Poisson Process

Oftentimes, the time to failure information (which was used in Andrawus et al. [16]) may not be available in the public domain. Instead, a few survey reports and newsletters provide "grouped" data with little information on individual turbines or maintenance activities [19, 88]. In the grouped data, only is provided the number of failures occurring in a population of turbines during a specific period, whereas the actual time when a failure takes place is missing [17]. For example, Windstats [88], a quarterly newsletter issued by Haymarket Business Media Limited (Haymarket), provides a monthly or quarterly survey including the number of failures for each subsystem of wind turbines. Table 2 presents an illustrative example of the failure statistics from one Windstats newsletter [18].

To utilize the failure statistics over time, several studies use modeling tools such as homogeneous or non-homogeneous Poisson processes [15, 17–19]. Homogeneous Poisson process (HPP) is a typical Poisson process with constant occurrence rate λ per unit time, while, understandably, a non-homogeneous Poisson process (NHPP) has a time-varying failure rate, represented by an intensity function $\lambda(t)$ at time (or, age) t. NHPP was used to model the failure patterns in the repairable systems. It is assumed that a repair returns the system condition to its original state immediately prior to the failure; this repair is called *minimal repair* [89]. In NHPP, the probability of n failures occurring in the time interval (a, b] is given by [90],

$$P(N(a,b] = n) = \frac{\left[\int_a^b \lambda(t)dt\right]^n e^{-\int_a^b \lambda(t)dt}}{n!}, \tag{2}$$

for $n = 0, 1, \ldots$. For a detailed discussion of NHPP, we refer the interested reader to [90–93].

There are several techniques to estimate the intensity function $\lambda(t)$. One of the commonly used models is a power law process, or Weibull process [89, 93], where the intensity function has the following form:

$$\lambda(t) = \frac{\beta}{\eta}\left(\frac{t}{\eta}\right)^{\beta-1}. \tag{3}$$

Table 2 The number of failures in Danish wind turbines in 1994

	October	November	December
Number of total turbines reporting	2,036	2,083	2,164
Subassembly failure			
Blade	15	6	6
Gearbox	5	2	4
...		...	
Total number of failures	158	130	175

Note: The data are obtained from the study of Guo et al. [18]

Here, $\beta > 0$ is the shape parameter and $\eta > 0$ is the scale parameter. Similar to the Weibull distribution discussed above, $\beta < 1$ represents early failure patterns, whereas $\beta > 1$ indicates a deteriorating (or, aging) process, and $\beta = 1$ denotes a constant failure rate.

Tavner et al. [17] establish an HPP model using the failure statistics data obtained from Windstats over the 10-year period from 1994 to 2004. They analyze the failure data from two countries, Germany and Denmark, both of which hold a large wind energy production capacity. Tavner et al. [17] assume that failures occur randomly and their HPP model has $\beta = 1$ and $\lambda = 1/\eta$ (i.e., $\lambda(t)$ is constant) because they consider the deterioration phase has not yet arrived for the wind turbines under study due to the turbine's young age. In another study, Tavner et al. [15] quantify how the wind speeds affect turbine reliability and which components are affected the most, also using the Windstats data.

Later, Guo et al. [18] extend Tavner et al.'s study [17] by considering a NHPP model. They also use the data extracted from Windstats [88]. For handling the "grouped data", they suppose that there are $n_1, n_2, \ldots, n_k$ failures during the time intervals of $(t_0, t_1]$, $(t_1, t_2]$, ..., $(t_{k-1}, t_k]$, respectively, and assume that individual groups are independent of each other. Then, the joint probability for the k group events is:

$$\begin{aligned} P(N(t_0, t_1] = n_1, N(t_1, t_2] = n_2, \ldots, N(t_{k-1}, t_k] = n_k) \\ = \prod_{i=1}^{k} P(N(t_{i-1}, t_i] = n_i) \\ = \prod_{i=1}^{k} \frac{\left[\int_{t_{i-1}}^{t_i} \lambda(t)dt\right]^{n_i} e^{-\int_{t_{i-1}}^{t_i} \lambda(t)dt}}{n_i!}, \end{aligned} \quad (4)$$

where $\lambda(t)$ is given in (3). One additional complexity Guo et al. [18] need to handle is the uncertain running time of the turbines since their installation. This is because the data collected in Windstats does not necessarily (and usually do not) coincide with the exact date when the wind turbines started their first operations. For example, even though the Danish turbine data in Windstats start from October 1994 in Windstats, it is more likely that the reported wind turbines started their operations earlier than October 1994. For handling this problem, Guo et al. [18] introduce an additional parameter into the Weibull process and argue for using a three-parameter Weibull distribution, rather than the typical two-parameter one as discussed above. Like the two-parameter Weibull distribution, the parameter estimation part in the three-parameter model is still done through either the MLE or the least-squares (LS) methods. After analyzing the Danish and German data in Windstat, respectively, Guo et al. [18] conclude that the three-parameter model has advantages over the two-parameters model in terms of predicting the reliability trend more accurately.

The studies by Tavner et al. [17] and Guo et al. [18] make an initial attempt in utilizing grouped failure data for reliability assessment. However, their models do not consider the changing population of turbines. Apparently, individual turbines with different sizes and structural designs are added at different times. This is evident in Table 2, where the number of turbines reported in Windstats increases over time, possibly with variations in turbine sizes and designs. Consequently, turbines may be at different operating ages and have different intensity function $\lambda(t)$ in (3) during a given time interval $(t_{i-1}, t_i]$. But because of the lack of detailed information in Windstats, Guo et al. [18] do not consider the difference in turbines coming from an evolving turbine population. Rather, Guo et al. [18]'s NHHP model treats the operation time of all turbines the same. Future work will need to account for the evolving nature of turbine population for more accurate reliability assessment of the wind power facilities.

Coolen et al. [19] study the grouped failure statistics data published in LWK [94] regarding turbines in Germany. Unlike the Windstats data, LWK provides the models of wind turbines associated with their data. Coolen et al. [19] attempt to build a NHPP model for wind turbine Model *V*39/500. Unfortunately, the Chi squared test rejects the hypothesis that the NHPP model fits the data for the whole wind turbine system and the gearbox. Coolen et al. [19] argue that the rejection is due to the lack of the actual turbine operation time, which in turn makes the probability calculation in (4) inconsistent with data. This explanation appears to be in line with our concerns expressed in the previous paragraph.

3.2 Markov Process-Based Analysis

3.2.1 Reliability Assessment Using a Markov Process

Markov process-based models are widely used in reliability assessment for the conventional power systems [8–10, 95, 96]. Not surprisingly, it has become one of the most popular modeling tools for wind turbine reliability assessment as well.

Let us consider a discrete time Markov process with a finite number of states. Suppose that the deterioration levels of an operating system are classified into a finite number of conditions $1, \ldots, M$ and that there are L different types of failures. Then, the system condition can be categorized into a series of states, $1, \ldots, M+L$. State 1 denotes the best condition like "new", and state M denotes the most deteriorated operating condition before a system fails. State $M+l$ reflects the lth failed mode, $l = 1, \ldots, L$. When a system undergoes Markovian deterioration, the transition probability from one state to another only depends on the immediate past state rather than the entire state history.

Figure 1 illustrates the state transitions for a repairable system. In the figure, p_{ij}, $i \leq j$, $i, j = 1, \ldots, M$, above a solid line denotes a transition probability from an operating state to another operating state, and λ_{ij} is the failure rate from an operating state i to the failure state j. Because an operating system cannot improve on its own in general, only transitions to more degraded or failed states are

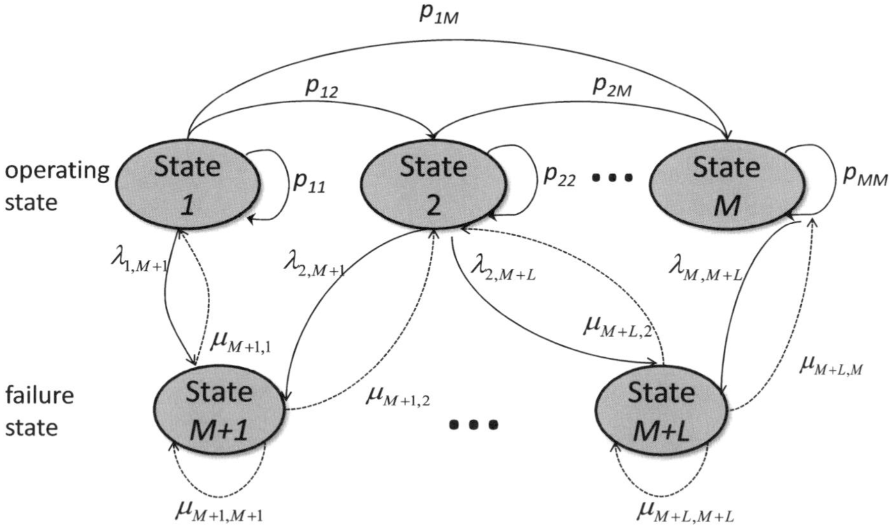

Fig. 1 State transition diagram for Markovian deterioration system

considered in the figure, denoted by the solid lines. μ_{ij} near the dashed line is the repair rate from a failure state i, improving the system to an operating state j for $j \leq i$, whereas μ_{ii}, $i = M + 1, \ldots, M + L$, is the non-repair rate. With repair, the system can be returned from a failure state back to operation.

For simplicity, suppose that $M = 1$ and $L = 1$. That is, the system condition is categorized by two simple states: up or down. Let us also assume that operators repair the system whenever it fails. Then, the transition probability matrix $P_{2\times 2}$ consists of the following four elements [24]:

$$P = \begin{bmatrix} 1 - \lambda & \lambda \\ \mu & 1 - \mu \end{bmatrix}, \tag{5}$$

where λ denotes a failure rate which is the probability that the system fails until the next period. μ is a transition probability with which the system can be restored to the operating state. People call this probability as the repair rate of the system [21]. In this simple case, the mean time to failure (MTTF) is $1/\lambda$, and the mean time to repair (MTTR) is $1/\mu$. The mean time between failure (MTBF) is simply the sum of MTTF and MTTR. By applying the limit distribution of an ergodic Markov chain [97], the limiting average availability (i.e., the limiting expected proportion of time when the system is up) becomes $\mu/\lambda + \mu$. For a more complicated system with multiple deteriorating states (i.e., $M > 1$) and/or multiple failure states (i.e., $L > 1$), similar analysis can be done.

Initial efforts using a Markov model in wind power systems can be found in Sayas and Allan's study [21]. They evaluate the generation adequacy of wind farms, in which the wind turbine condition is divided into two binary states (i.e., up and down) using the transition matrix in (5). They also model wind speeds using a Markov process with four different categories. They recognize that the failure rates and repair rates of a turbine are affected by wind regimes. The failure rate of a turbine increases under extremely high wind speeds, whereas the repair rate decreases with wind speed. Considering this weather effect, they propose a combined Markov model for a wind farm. The transition probabilities between the states are obtained using historical data. Then using the limit distribution, the average frequency and duration per year for each combined state, i.e., turbine conditions *plus* wind regimes, are calculated. The results are used for assessing a wind farm's performance, including a turbine's availability and expected generation of wind energy.

Markov processes are also employed in simulation studies [22, 45]. McMillan and Ault [22] consider an intermediate state "degraded" for each component, in addition to the "up" and "down" states. They estimate the state transition probability for four critical components, including gearbox, generator, rotor blade, and electronic device, but do so by combining the operational data and expert judgments. Then, state transitions are simulated given those transition probabilities during the simulation runs. Byon et al. [45] take a similar approach, but consider more refined component states (namely, "normal", "alert", "alarm", and "failed"). In both simulation studies, Markov processes are not used as a standalone tool, but they are embedded within simulation models to capture the evolving characteristics of component deterioration and failures.

3.2.2 Optimization Models Using Markov Decision Processes

A Markov decision process (MDP) is a sequential decision making process used to control a stochastic system based on the knowledge of system states [98–100]. This modeling method is popular in the literature of optimal preventive maintenance [99–104]. Most of the above-cited studies are not specific to wind turbine maintenance; rather they are intended for general aging systems. However, we can gain useful insights from those studies. For this reason, we first discuss the MDP studies for general aging systems and then we present the studies specialized for wind turbine maintenance.

Let S denote the state space of a Markov decision process and $V_t(s)$ be the total cost from the current period t to the terminal period (so $V_t(s)$ is also known as the total cost-to-go). When the system state is $s_t \in S$, decision makers want to find the best action to minimize $V_t(s)$ as follows:

$$V_t(s) = \min_{a_t \in A_{s_t}} \left\{ c_t(s_t, a_t) + \gamma \sum_{s_{t+1} \in S} P_t(s_{t+1}|s_t, a_t) V_{t+1}(s_{t+1}) \right\}. \qquad (6)$$

In (6), A_{s_t} is a set of possible actions at state s_t, which may include an inspection, major/minor maintenance, and taking no action. $c_t(s_t, a_t)$ is an immediate cost incurred by decision a_t. $P_t(s_{t+1}|s_t, a_t)$ is the transition probability from s_t to s_{t+1} as a result of taking the action a_t. $\gamma \leq 1$ is a discount factor. A model with $\gamma < 1$ is called a discounted cost model, whereas the model with $\gamma = 1$ is called an average cost model. Equation (6) is referred to as an "optimality equation" or "Bellman equation". Understandably, the optimal maintenance actions are found by solving the above optimality equation.

Several mathematical models have been introduced, incorporating information from condition monitoring equipment for general aging systems [99–104]. Since sensors in condition monitoring equipment provide uncertain data, most of these studies use a partially observed MDP (POMDP) to reflect the incomplete understanding about the current system state. In a POMDP setting, the system condition cannot be observed directly, so that the condition is estimated in a probabilistic sense [103, 104]. Suppose that a system condition can be categorized into $M + L$ states as explained in Sect. 3.2.1. Under a POMDP setting, a system state is defined as a probability distribution, representing one's belief over the corresponding true state. As such, the state of the system can be defined as the following probability distribution:

$$\pi = [\pi_1, \pi_2, \ldots, \pi_{M+L}], \tag{7}$$

where π_i, $i = 1, \ldots, M + L$, is the probability that the system is in deterioration level i. π is commonly known in the literature as an information state [104]. Then, the state space under the POMDP setting becomes

$$S = \left\{ [\pi_1, \pi_2, \ldots, \pi_{M+L}]; \sum_{i=1}^{M+L} \pi_i = 1, 0 \leq \pi_i \leq 1, \quad i = 1, \ldots, M + L \right\}. \tag{8}$$

Maillart [103] uses a POMDP to adaptively schedule observations and to decide the appropriate maintenance actions based on the state information. In her study, the system is assumed to undergo a multi-state Markovian deterioration process with a known and fixed transition probability matrix. Similarly, Ghasemi et al. [105] represent a system's deterioration process using the average aging behavior, provided by the manufacturer (or, from survival data), and the system utilization, diagnosed by using CBM data. They formulate the maintenance decision problem using a POMDP and derive optimal policies using dynamic programming.

Most maintenance studies in the literature to date only consider static environmental conditions. Very few quantitative studies examine systems operating under stochastic environments. Thomas et al. [106] investigate the repair strategies to maximize the expected survival time until a catastrophic event occurs in an uncertain environment. They consider the situation where a system should be stopped during inspection or maintenance action. If specific events, termed "initiating events", take place when a system is down or being replaced, they are noted as catastrophic events; examples given are military equipment or hospital systems. Kim and Thomas [107] extend the problem where the multiple environmental situations are assumed to follow a Markovian process. The criteria in both studies are to maximize the expected time until a catastrophe occurs. In other words, these studies focus on short-term availability.

There are a few relevant studies using a MDP or POMDP for wind turbine maintenances. The authors of this chapter present two optimization models and their solutions using POMDPs in [23, 24]. The presented models extend the model introduced in [103] by incorporating the unique characteristics of turbine operations. For example, to represent stochastic weather conditions, Byon et al. [23, 24] apply the initiating events idea described in [106]. Other characteristics included are long lead times after unplanned failures and the accompanying production losses.

The model presented in [23] is a static, time-independent model with homogeneous parameters in its optimality equation. The homogeneous parameters imply that the characteristics of weather conditions remain constant period by period. This relatively simple model allows us to characterize the solution structures and develop efficient solution techniques. Byon et al. [23] analytically derive the optimal control limits for each action in a set of closed-form expressions, and provide the necessary and sufficient conditions under which the preventive maintenance will be optimal and the sufficient conditions for other actions to be optimal. The second model in [24] extends the first model and is a dynamic, time-dependent model with non-homogeneous parameters. The time-varying parameters depend on prevailing weather conditions and exhibit considerable seasonal differences. Therefore, the resulting strategy is adaptive to the operating environments. The optimal policy is constructed from the evolution of the deterioration states of individual wind turbine components. A backward dynamic programming algorithm is used to solve the second problem.

In both models [23, 24], three major actions (or, polices, controls) are considered at each decision point. The first action is to continue the operation without any maintenance interruption. When this action is selected, the system would undergo Markovian deterioration with a given transition probability matrix. The second action is to perform preventive maintenance to improve the system condition and thus avoid failures. Lastly, wind farm operators can dispatch a maintenance crew for on-site observation/inspection to evaluate the exact deterioration level. Optimal action at every decision point is derived based on the system state and weather conditions.

In these studies, a season-independent degradation process is assumed. That is, the constant transition matrix is considered over the whole decision horizon. However, the degradation process can be accelerated, and consequently, more frequent failures are expected during harsh weather seasons than during mild seasons, as discussed in previous sections. Reflecting these seasonal aspects, there is a need in future research to extend the models in [23, 24] to incorporate weather impact on turbine reliability and model a season-dependent deterioration process.

3.3 Structural Load-Based Reliability Analysis

The approaches reviewed in categories of Sects. 3.1 and 3.2, namely the statistical distribution-based and Markov process-based analyses, are considered cumulative degradation modeling approaches. The influence of environmental factors on the turbine reliability and O&M costs are already examined by several lines of work [15, 22–24], falling into either category of the two previously reviewed approaches. Nevertheless, the direct relationship between structural loads (or, stresses) and physical damage is not investigated in detail in those studies. In this section, we review another category of analytical models which address the structural loads and its effect on damage.

Assuring a desired level of structural performance (in blade and tower) requires the characterization of the variability in material's resistance and stress. International Electrotechnical Commission (IEC) standards (IEC 61400-1) [108] formalized design requirements for wind turbines to ensure its structural integrity. There are two design requirements for wind turbine structures; fatigue loads and extreme loads. Fatigue load analysis focuses on the progressive structural damage that occurs when a system is subjected to cyclic loading, whereas the extreme load (or, ultimate design load) is a maximum load level that can be encountered during a turbine life [108]. In this section, we focus on the fatigue load analysis. For detailed discussions to estimate the extreme load, please refer to [109–111].

One of the initial studies on structural reliability analysis can be found in Ronold et al.'s study [25]. They derive a probabilistic model for assessing the cumulative fatigue damage of turbine blades. They first perform regression analysis to understand how weather factors, including the average wind speed and turbulence intensity, affect the bending moment at the blade root. Then, a fatigue model is developed based on an SN-curve, which is an empirical curve representing the relationship between the fatigue load stress and the number of stress cycles to failure. For example, a simple SN-curve is $N = BS^{-m}$ where N is the number of stress cycles to failure for a given S, S is the fatigue load stress, and B and m are material parameters [25]. Since there are multiple stress levels due to various wind speeds and turbulence, Ronold et al. apply a Miner's rule approach which extends the conventional SN-curve as follows:

$$D = \sum_{i=1}^{K} \frac{\Delta n(S_i)}{N(S_i)}, \tag{9}$$

where $\Delta n(S_i)$ is the number of load cycles at stress range S_i, and $N(S_i)$ is the number of cycles to failure at this stress level as determined from the SN-curve. Consequently, D denotes the weighted sum of damage from various stress ranges, $\forall S_i$, $i = 1, \ldots, K$. In Ronold et al.'s study [25], considering a 20-year design lifetime, a first-order reliability method is used to calculate the structural reliability against fatigue failure.

Manuel et al. [26] also applied parametric models to estimate wind turbine fatigue loads for turbine lifetime. In a later study, Manuel et al. [27] consider extra weather parameters including the Reynolds stresses and vertical shear exponent, in addition to the average wind speed and turbulence intensity.

Many studies on structural reliability analysis rely on a cycle counting technique over multiple stress ranges, called the "rainflow counting method" [25, 26]. However, this method is computationally extensive and requires data over the whole range of weather profiles for accurate prediction. In reality, many data sets do not include the rarely occurring (but, arguably more important) weather profiles such as the high wind speeds and turbulent inflow. To overcome these limitations, Ragan and Manuel [30] suggested using the power spectrum to estimate the stress range probability distributions. The major advantages of power spectrum analysis is that the load power spectrum may be estimated with less statistical uncertainty and a stress range probability density function can be obtained with a small number of simulations, or limited amount of field data. Ragan and Manuel [30] evaluate the advantage of this method in terms of accuracy, statistical reliability, and efficiency of calculation using field data from a utility-scale 1.5 MW turbine. Early studies making use of power spectral density can be also found in [28, 29].

Moan [31] summarizes operational experiences with respect to degradation of various types of offshore structures, and presents a reliability model focusing on fatigue failures and corrosion. Rangel-Ramírez and Sørensen [32] apply similar techniques to wind turbines, considering time-varying fatigue degradation for offshore turbines with jacket and tripod types of support structures. Especially, they analyze the "wake effect" on the performance of neighboring wind turbines. Here, wake effect implies an increased turbulence intensity, coming from the decrease of wind velocity behind a turbine. In the example of offshore wind turbines with a steel jacket support, Rangel-Ramírez and Sørensen [32] investigate the inspection schedules that are meant to achieve a prescribed reliability level. They show that earlier inspections are very beneficial and thus must be performed for wind turbines in a wind farm. This is different from the practice on those standalone, isolated wind turbines, and is primarily due to the increase of fatigue coming from the wake turbulence.

Sørensen et al. [33] take an analogous approach to model the fatigue-related reliability of main components of wind turbines, including welded details in the tower, cast steel details in the nacelle, and fiber reinforced details in the blades. Later, Sørensen [34] presents an overall decision making framework that may help decide the design for robustness, the choice of inspection methods, and scheduling and repair strategies. Sørensen [33] argues that multiple inspection techniques should be used for decision making. For instance, to assess the deterioration level of a gearbox, several techniques are available including visual inspection, oil analysis, particle counting and vibration analysis, but their abilities to detect and quantify degradation differ drastically from one another. They may provide complementary information to produce a more accurate reflection of the actual degradation. Sørensen [34] also suggests taking a Bayesian approach to update a degradation model based on inspection results. Sørensen illustrates, through an example, that inspections, scheduled for maintaining the turbine operation at an acceptable risk (e.g., in terms of the number of failures per year) can be very different, depending on inspection quality as well as the operating environment.

Bhardawaj et al. [35] illustrate a probabilistic damage model for general corrosion of the tower structure. They present a risk-based decision making methodology for undertaking the run-repair-replace actions with the ultimate goal of maximizing the net present value of the investment in maintenance.

4 Simulation Studies for Wind Farm Operations

Due to the complicated nature of wind farm operations, it is quite challenging to model wind farm operations entirely using analytical approaches. For example, wind power generation from individual turbines are not independent of each other because wind speeds at turbine sites are correlated. A failure of one component might affect the operating conditions of other components, which is known as the "cascading effect". Few analytical models have been proposed to analyze the reliability at the wind farm level. Sayas and Allan [21] extend their analysis of wind turbines (discussed in Sect. 3.2.1) to a wind farm level with multiple turbines. But the number of turbines is limited due to the curse of dimensionality.

With these reasons, simulations have been utilized to evaluate the performance of current or future wind farms with the metrics of interest such as the average number of failures per year in a wind farm, O&M costs, and availability. Most of the simulation models presented in the literature belong to the category of Monte Carlo simulation in the sense that they use random number generators to reflect the stochastic aspects of wind farm operations. To accommodate uncertainties, several probabilistic models are embedded within a simulation, and maintenance constraints under different operating environments are implemented [22, 42, 45].

Rademakers et al. [42] describe a Monte Carlo simulation model, developed by Delft University of Technology (TU-Delft), for maintaining offshore wind farms. They illustrate the features and benefits of the model using a case study of a 100 MW wind farm. The model simulates the operation aspects over a period of time, considering multiple critical factors for performing repair actions, such as turbine failures and weather conditions. The failures of turbine components are generated stochastically, based on the relevant statistics such as MTTF and reliability distributions. Weather conditions are realized using the historical summer and winter storm occurrences at specific sites. The model further categorizes different failure modes and the corresponding repair actions. For example, the first category of failure modes requires replacement of rotor and nacelle using an external crane; the second failure mode requires replacement of large components using an internal crane, and so on. The failure rates of the individual components are distributed over four maintenance categories. Please note that Rademakers et al. [42]'s model only considers corrective maintenance, and their simulation results indicate that the revenue losses during such corrective repairing account for

55 % of the total maintenance costs, largely due to the long lead time to prepare parts and the long waiting time until favorable weather conditions are met. Similar studies appear in [6, 11, 43].

Foley and Gutowski [46] investigate the energy loss associated with component failures using a simulator named TurbSim. They assume random failures and estimate the MTBF and MTTR using data obtained from [17] and [112]. In addition, they consider degraded outputs of a turbine upon a component's failure, meaning that they decrease the power output of a turbine by a certain percentage point, depending upon the impact of a broken subsystem. They use wind speed data from the national oceanic and atmospheric administration (NOAA)'s weather stations [113] and fit a Weibull distribution to simulate wind speeds. The wind speed at the ground level is adjusted to the hub height speed while considering the roughness of the surrounding terrain. Their simulation results of an artificial wind farm in Massachusetts for a 20-year period (the presumed turbine life) show that the energy loss from component failures is 1.24 % of the total generation of a wind farm with no wind turbine failures. For off-shore turbines, the simulation produces a more noticeable 2.38 % loss of power generation, still compared with the circumstance of no failure. Foley and Gutowski [46] stress that these results represent a significant loss of energy generation over the lifetime of turbines.

Negra et al. [44] describe a set of approaches to be used for wind farm reliability calculations. They especially consider unique aspects that influence the *offshore* wind farm reliability including grid connection configuration and offshore environment. Factors that are common to land-based turbines such as wake effect and correlation of power output are also considered. A Monte Carlo simulation shows how some of these factors influence the reliability evaluation. For example, the inclusion of cable and connector failures has a substantial impact on the power generation from the off-shore wind farms.

Simulations are also used for the purpose of validating various O&M approaches [16, 22, 45]. In [16], the suggested strategy resulting from the statistical model is evaluated by using Monte Carlo simulation. They assess the reliability, availability, and maintenance costs by simulating a wind farm with turbines over a period of four years. The simulation is conducted using a commercial software called ReliaSoft BlockSim-7 [114]. McMillan and Ault [22] implement a simulation model to quantify the cost-effectiveness of CBM by comparing the performance of different maintenance polices. They use an autoregressive time series analysis to generate wind speeds and consider weather constraints when performing repair actions.

The authors of this chapter develop a discrete event simulation model to characterize the dynamic operations of wind power systems, as discussed in Sect. 2.2.4 [45]. The discrete event simulation model tracks each event going from a state to another. A major difference between this model and previously discussed models is that time evolution in the latter is unimportant, and the focus is to obtain lump sum estimates for performance measures. By contrast, the discrete event model considers dynamic state changes of a wind turbine as a result of stochastic events. The model allows operators to gain a detailed knowledge of the lifetime evolution of wind power systems, in addition to gathering the performance statistics. For simulating the wind condition, the authors in [45] use the actual wind speed data measured by the west Texas Mesonet [115]. Because the simulation model considers the spatial correlation of wind speeds at wind turbine sites, power generations at different turbines naturally become correlated. This correlation phenomenon, however simple and intuitive, is often ignored in the wind energy literature, which usually assumes independent, identical turbines in a wind farm. The simulation model in [45] provides a tool for wind farm operators to select the most cost-effective O&M strategy.

5 Generation Adequacy Assessment Models

Generation adequacy assessment deals with assessing the existence of sufficient facilities within the power system to satisfy consumer demands (or, loads) or system operational constraints [44, 47]. A complete power system can be categorized into three segments: generation, transmission and distribution. And there are three hierarchical levels that provide a basic framework for the power system adequacy evaluation, as illustrated in Fig. 2 [44, 47, 116]. Most of the reliability work is focused on the hierarchical levels 1 and 2. There is very limited work on hierarchical level 3 assessment, which considers all three functional zones. One reason for this is that considering all the three zones simultaneously, the problem becomes too complex to be computationally tractable. But perhaps the more important reason is that consideration of the three zones altogether is not deemed critical or even desirable. Instead, the level 2 analysis is used to provide reliability indices (such as failure and repair rates) for major nodes to the third zone (i.e., to which distribution facilities are connected). The analysis of the third zone is then conducted independently using the reliability parameters of these nodes as the starting point. We focus on the hierarchical levels 1 and 2 in this section. We also briefly discuss the power quality issues in Sect. 5.3.

Recall that the criteria investigated in Sects. 3 and 4 focus on evaluating performance of an individual turbine, or a farm. The criteria include MTTF, MTTR,

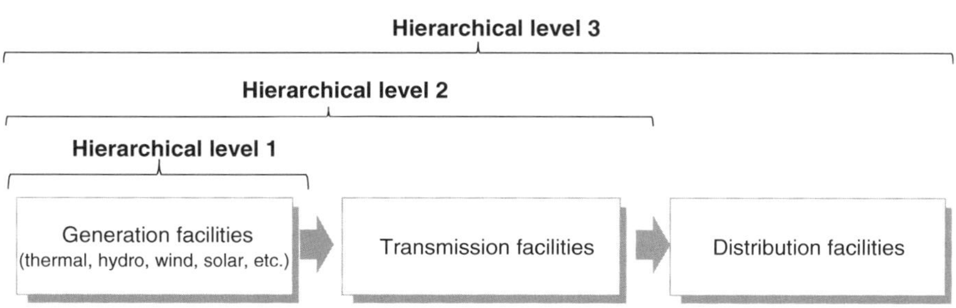

Fig. 2 Hierarchical levels for generation adequacy assessment

number of failures, O&M costs, etc. At the power system level, however, power utilities use different criteria to determine adequate generation and transmission capacity in various scenarios. We summarize the criteria commonly used in this line of literature [116]:

- Loss of load expectation (LOLE): Expected number of hours per year when demand exceeds available generating capacity.
- Loss of load probability (LOLP): Probability that loads exceed available generating capacity.
- Loss of energy expectation (LOEE), Expected unserved energy (EUE), Expected energy not supplied (EENS): Expected amount of energy not supplied by the generating system in a given time period.
- Energy index of unreliability (EIU): Ratio of the LOEE to the total energy demand.
- Capacity factor (CF): Ratio of actual energy output to energy output if generators operated at rated power outputs in a given time period.

Please note that the load in the above criteria implies customer demand, whereas the load in Sect. 3.3 denotes the stress on the facility structure. There are also newly introduced indices associated with wind energy as follows [21, 62, 117]:

- Expected wind energy supplied (EWES): Expected amount of energy offset of conventional fuel energy by wind energy application.
- Expected surplus wind energy (ESWE): Expected amount of wind energy that was available but not utilized.
- Expected available wind energy (EAWE): Expected amount of wind energy that would be generated in a year, if there were no wind turbine failures (or, outages).
- Expected generated wind energy (EGWE): Expected maximum amount of energy that would be generated in a year by wind turbines considering their failure rates.

5.1 Hierarchical Level 1 Assessment

Hierarchical level 1 (HL 1) assessment concerns evaluating the reliability of generation capacity. The goal is to determine whether the installed generating capacity satisfies the forecast system loads at an acceptable risk level, considering the uncertainty introduced by the corrective/preventive maintenance and load forecast errors [44]. In these calculations, it is assumed that the transmission system is capable of transporting all the power from generation points to the load points. We consider two main approaches: the analytical methods and Monte Carlo simulations.

5.1.1 Analytical Methods

One of the popular analytical approaches is to use a multi-state model which discretizes the power outputs into a finite number of states and assign a probability to each state based on historical data of generating units [47–49]. The number of states can be decided depending on the characteristics of wind data and accuracy requirement [47].

The initial multi-state analytical model with unconventional energy sources can be found in the study by Singh and Gonzales [50]. It is shown in [50] that under the significant penetration of variable energy sources, incorrect consideration of correlations between loads and unconventional units can give optimistic results for risk assessment. To address this issue, Singh and Kim [53] define the discretized state as a vector which contains hourly loads and output from each unconventional unit (note that the number of states here is equal to the number of hourly observations). Then, they apply a clustering algorithm to group the states into a smaller number of clusters. The clustering algorithm produces the mean value (cluster centroid) of each cluster as well as the frequency of each cluster. These pieces of information are used to calculate indices of interest such as LOLP, LOLE and EUE. Later a straightforward method [54] is introduced for computing the EUE in a fast manner when renewable sources are included.

In a multi-state model, finding the most probable states, contributing significantly to system adequacy indices, is through some optimal search algorithm. For example, Wang and Singh [57] use a population-based stochastic search algorithm to find out the set of the most probable failure states in a modified IEEE Reliability Test System (IEEE-RTS). The system adequacy indices used in [57] include LOLE, LOLF and EES. A later paper [58] compares the results obtained by various population based methods, when wind energy is included as one of the generation sources.

Dobakhshari and Fotuhi-Firuzabad [55] present an alternative procedure based on a Markov process. The Markov model is used to represent the output power of a wind farm housing multiple wind turbines. Analogously, Leite et al. [56] use a Markov model to model wind speeds. They use actual time series data of wind speeds at several Brazilian regions, and accommodate time-varying patterns of wind speeds. They cluster wind speeds into a finite number of states, and examine the consequence of having different number of states. The ideal number of clustered states depends on the characteristics of winds at a given site. The choice also depends on the computational precision desired and computational capacity. The annual power generation is estimated for the studied sites, demonstrating favorable characteristics of having wind power generation in Brazil. The resulting capacity factor for wind power generation is between 28 and 37 %, a value considered higher than the global average. The results from this study also confirm that seasonality significantly affects the reliability of wind turbines.

5.1.2 Monte Carlo Simulation

There are two approaches in generation adequacy assessment using Monte Carlo simulation [118]. The first one is a sequential Monte Carlo simulation which chronologically simulates a realization of stochastic process for a give time period. In contrast, non-sequential Monte Carlo simulation use the state sampling approach in which each state is randomly sampled without reference to the chronological system operation [118].

Sequential Monte Carlo simulations are performed in [59–62]. Billinton and Bai [59] quantify the contribution of wind energy to the reliability performance of a power system. Wind speeds are simulated using an autoregressive moving average (ARMA) time series approach. Using the two reliability indices, LOLE and LOEE, they investigate the effects of wind turbine capacity and mean wind speeds on the generation adequacy. They also compare the contribution of wind power with that of conventional generators. In doing so, they replace conventional power units with wind turbines, and investigate the capacity of wind turbines (together with mean wind speed) required to maintain the same reliability level. Based on a case study in Canada, they point out that wind power contribution is highly dependent on the wind site conditions, and that wind energy independence among multiple sites has a significant positive impact on the reliability contribution.

Wangdee and Billinton [60] further examine the impact on the system reliability indices of wind speed correlation between two wind farms, and obtain a similar conclusion. Moharil and Kulkarni [61] conduct an analogous analysis, which studies a case of three different wind-generating stations in India.

Karki and Billinton [62] use a sequential Monte Carlo simulation to help determine appropriate wind power penetration in an existing power system from both reliability and economic aspects. The generating system consists of both wind turbines and conventional generators. The time-to-failure for both types of power generators is assumed to be exponentially distributed, and then, the MTTF of each generator can be estimated from historical data. To evaluate the energy costs and utilization efficiency of wind turbines, they create a couple of new probabilistic performance indices by combining EWES, ESWE and CF as follows:

$$\text{Wind utilization factor (WUF)} = \frac{EWES}{EWES + ESWE} \times 100 \tag{10}$$

$$\text{Wind utilization efficiency (WUE)} = CF \times WUF \tag{11}$$

Note that WUF is the ratio of the wind energy supplied to the total wind energy generated. WUE implies the ratio of the actual energy utilized to the total energy generated based on the rated wind turbine capacity. Based on a case study of a small power generating system, they show that as more wind turbines are added in the power system, the reliability level increases. On the other hand, the utilization of wind power, measured by WUF and WUE, decreases with increasing wind energy capacity. Considering the fact that investment cost linearly increases with the number of wind turbines in general, they present the procedure to decide the appropriate wind penetration level in a power system.

The sequential Monte Carlo approach can provide detailed information about the system behavior over time, but it requires higher computational efforts especially when the number of generators is large. To overcome this issue, Vallée et al. [63] present a non-sequential Monte Carlo simulation to estimate the wind power capacity of a given country, and perform a case study for Belgium. They use the Weibull distribution to characterize the wind speeds at ten wind farms spreading over the country. They categorize the wind speeds into a finite number of states and assign a probability to each state. Wind power production at each site is simulated using the state probabilities, and then the country-wide wind power production in a country is computed by summing up the powers from individual wind farms. They divide the entire country into four regions (north, south, center, and off-shore) and investigate how the geographical repartition of wind turbine installations affects the country's wind production capacity. Their case study shows that spreading wind turbines over a broad geographical area help ensure a positive wind power output at any given time and avoid the situation of no wind power production for the whole country. The study also reveals that contribution of wind power production is more reliable when the offshore wind energy penetration is higher than the onshore penetration.

5.2 Hierarchical Level 2 Assessment

Hierarchical Level 2 (HL 2) assessment considers the evaluation of both generation and transmission facilities for their ability to supply adequate, dependable, and suitable electrical energy at bulk power load points [44]. This analysis is also called "composite system reliability (or, adequacy) evaluation" or "bulk power system reliability".

Generation adequacy studies accounting for wind generation have been developed for the HL 1 level (load covering with transmission system assumed always available) [64]. Relatively fewer papers and reports are found on the HL 2 assessment due to the complexity associated with modeling the generation and transmission facilities, as well as capturing the intermittent wind characteristics [65]. In the meanwhile, assessment of adequate transmission facilities to deliver wind power to the power grid becomes extremely important as many geographical sites with good wind resources are not necessarily located close to the existing power grid or load centers [66].

When a large-scale wind farm is connected to a weak transmission facility, transmission reinforcement may be necessary in order to increase the system capability to absorb more wind power at specified locations [65]. Billinton and Wangdee [65] study three different cases, and examine possible transmission reinforcement alternatives for the purpose of absorbing a significant amount of wind capacity without violating the transmission constraints. Similarly, Vallée et al. [64]

propose a tool that uses Monte Carlo simulations to assist system planners and transmission operators for the purpose of qualitatively assessing the impact of wind energy production as wind penetration increases. Karki and Patel [66] discuss the issues of determining the appropriate transmission line size and evaluating the reliability of combined wind generation and transmission systems.

A multi-state model for composite system adequacy assessment can be found in [51, 52]. Billinton and Gao [52] present a procedure that can be used in either HL 1 or HL 2 reliability evaluation. This procedure quantifies the effects of connecting multiple wind farms at different locations into a bulk power system. The benefits in doing so depend significantly on the actual transmission network. The expected energy not supplied decreases as the number of wind farms increases if the wind farms are connected at relatively strong points in the transmission system. Otherwise, the benefits are not substantial.

Another approach for the HL 2 assessment is to use a hybrid approach combining the Monte Carlo and population-based methods. One drawback of using the Monte Carlo simulation is the long computation time to achieve the satisfactory statistical convergence of adequacy index values [119]. To address this challenging problem, Wang and Singh [119] apply a classification method called an artificial immune recognition system (AIRS). They consider LOLP as the reliability criterion. Using Monte Carlo simulation, they first sample a system state consisting of a load level, generation status, and transmission line status, and then perform a power flow calculation. If the sampled state does not satisfy the sampled loads, the state is classified as a "loss-of-load" class; otherwise, it becomes a "no-loss-of-load" class. After using samples to train the AIRS algorithm, the binary classifier can predict the LOLP of a given power system. Their case studies show remarkable reductions in computation time over a pure Monte Carlo simulation. The case studies are performed with conventional generators. But the authors of this chapter believe that a similar approach can be applied to wind power systems with minimal modifications.

5.3 Power Quality

Due to the stochastic nature of wind power production, connection of wind turbine generators to the power network can lead to grid instability, wind-energy rejection (thus, the financial losses of wind farm owners), harmonics, or even failures, if these systems are not properly controlled [120–122]. Power electronic converters are used to match the characteristics of wind turbines with the requirements of grid connections, terms of including frequency, voltage, control of active and reactive power, and harmonics [123].

This chapter mainly addresses the reliability issues of wind power systems in "facility" aspects and generation adequacy of the power system. The power quality issues and those of the transients like stability are typically not included in the adequacy analysis and thus are out of the scope of this chapter. Chen et al. [123] provides a comprehensive review of the state of the art of power electronics for wind turbines. Interested readers can refer to [123] for further detailed information.

6 Summary

This chapter reviews the studies relevant to the reliability and maintenance issues of wind energy facilities, with the ultimate objective of achieving sustainable and competitive energy supply. Unlike conventional power plants largely operated under steady states, wind turbines operate in highly stochastic environments under non-stationary loading. Although more power can be harvested during the seasons with strong winds, wind turbines are prone to failures during such weather conditions and repairs are difficult to carry out and more costly. Moreover, the variable, intermittent nature of wind power generation has a significant effect on overall power system reliability.

With these unique operating characteristics for wind energy generation, traditional approaches applied in other industries may not necessarily be effective for maintaining wind energy facilities and enhancing their reliability. The wind power industry faces new challenges pertaining to developing and deploying innovative, practical operational strategies to lower maintenance cost. The exciting news is that, as evident in this chapter, there has already been a rich body of work dedicated to studies directly relevant to wind energy applications. We observe that accurate reliability predictions and cost-effective O&M strategies are pressingly needed to address the approaching challenges faced by the global wind power industry. We also believe that considering today's trend of the ever increasing scale of wind farms, a well-planned maintenance strategy will have to be coordinated with a supply chain management program that can organize repair resources in a just-in-time manner. Addressing these challenges in a comprehensive program is not trivial. However, a successful development and implementation of such a program can result in significant benefits to the wind power industry.

References

1. NERC (2009) The 2009 long-term reliability assessment. Technical report, North American Electric Reliability Corporation (NERC), Washington, DC (Online). Available: http://www.nerc.com/files/2009_LTRA.pdf
2. Walford C (2006) Wind turbine reliability: understanding and minimizing wind turbine operation and maintenance costs. Technical report, Sandia National Laboratories, Albuquerque (Online). Available: http://prod.sandia.gov/techlib/access-control.cgi/2006/061100.pdf
3. Vachon W (2002) Long-term O&M costs of wind turbines based on failure rates and repair costs. Paper presented at the WINDPOWER 2002 annual conference

4. Wiser R, Bolinger M (2008) Annual report on U.S. wind power installation, cost, performance trend: 2007. Technical report, U.S Department of Energy, Washington, DC
5. Faulstich S, Hahn B, Jung H, Rafik K, Ringhand A (2008) Appropriate failure statistics and reliability characteristics. Technical report, Fraunhofer Institute for Wind Energy, Bremerhaven
6. Bussel GV (1999) The development of an expert system for the determination of availability and O&M costs for offshore wind farms. In: Proceedings of the 1999 European wind energy conference and exhibition, Nice, pp 402–405
7. Nilsson J, Bertling L (2007) Maintenance management of wind power systems using condition monitoring systems-life cycle cost analysis for two case studies. IEEE Trans Energy Convers 22:223–229
8. Yang F, Kwan C, Chang C (2008) Multiobjective evolutionary optimization of substation maintenance using decision-varying Markov model. IEEE Trans Power Syst 23:1328–1335
9. Qian S, Jiao W, Hu H, Yan G (2007) Transformer power fault diagnosis system design based on the HMM method. In: Proceedings of the IEEE international conference on automation and logistics, Jinan, pp 1077–1082
10. Jirutitijaroen P, Singh C (2004) The effect of transformer maintenance parameters on reliability and cost: a probabilistic model. Electr Power Syst Res 72:213–234
11. Hendriks H, Bulder B, Heijdra J, Pierikl J, van Bussel G, van Rooij R, Zaaijer M, Bierbooms W, den Hoed D, de Vilder G, Goezinne F, Lindo M, van den Berg R, de Boer J (2000) DOWEC concept study; evaluation of wind turbine concepts for large scale offshore application. In: Proceedings of the offshore wind energy in Mediterranean and other European seas (OWEMES) conference, Siracusa, pp 211–219
12. Pacot C, Hasting D, Baker N (2003) Wind farm operation and maintenance management. In: Proceedings of the powergen conference Asia, Ho Chi Minh City, pp 25–27
13. Krokoszinski HJ (2003) Efficiency and effectiveness of wind farms-keys to cost optimized operation and maintenance. Renew Energy 28:2165–2178
14. Ribrant J (2006) Reliability performance and maintenance—a survey of failures in wind power systems. Master's thesis, KTH School of Electrical Engineering, Stockholm
15. Tavner PJ, Edwards C, Brinkman A, Spinato F (2006) Influence of wind speed on wind turbine reliability. Wind Eng 30:55–72
16. Andrawus JA, Watson J, Kishk M (2007) Modelling system failures to optimise wind farms. Wind Eng 31:503–522
17. Tavner PJ, Xiang J, Spinato F (2007) Reliability analysis for wind turbines. Wind Energy 10:1–8
18. Guo H, Watson S, Tavner P, Xiang J (2009) Reliability analysis for wind turbines with incomplete failure data collected from after the date of initial installation. Reliab Eng Syst Saf 94:1057–1063
19. Coolen FPA (2010) On modelling of grouped reliability data for wind turbines. IMA J Manage Math 21:363–372
20. Vittal S, Teboul M (2005) Performance and reliability analysis of wind turbines using Monte Carlo methods based on system transport theory. In: Proceedings of the 46th AIAA structures, structural dynamics and materials conference, Austin
21. Sayas FC, Allan RN (1996) Generation availability assessment of wind farms. In: Proceedings of IEEE—generation, transmission and distribution, vol 144, pp 1253–1261
22. McMillan D, Ault GW (2008) Condition monitoring benefit for onshore wind turbines: sensitivity to operational parameters. IET Renew Power Gener 2:60–72
23. Byon E, Ntaimo L, Ding Y (2010) Optimal maintenance strategies for wind power systems under stochastic weather conditions. IEEE Trans Reliab 59:393–404
24. Byon E, Ding Y (2010) Season-dependent condition-based maintenance for a wind turbine using a partially observed markov decision process. IEEE Trans Power Syst 25:1823–1834
25. Ronold KO, Wedel-Heinen J, Christensen CJ (1999) Reliability-based fatigue design of wind-turbine rotor blades. Eng Struct 21:1101–1114

26. Manuel L, Veers PS, Winterstein SR (2001) Parametric models for estimating wind turbine fatigue loads for design. J Sol Energy 123:346–355
27. Nelson LD, Manuel L, Sutherland HJ, Veers PS (2003) Statistical analysis of wind turbine inflow and structural response data from the LIST program. J Sol Energy 125:541–550
28. Veers PS (1988) Three-dimensional wind simulation. Technical report, Sandia National Laboratories, Albuquerque (Online). Available: http://prod.sandia.gov/techlib/access-control.cgi/1988/880152.pdf
29. Burton T, Sharpe D, Jenkins N, Bossanyis E (2001) Wind energy handbook. Wiley, England
30. Ragan P, Manuel L (2007) Comparing estimates of wind turbine fatigue loads using time-domain and spectral methods. Wind Eng 31:83–99
31. Moan T (2005) Reliability-based management of inspection, maintenance and repair of offshore structures. Struct Infrastruct Eng 1:33–62
32. Rangel-Ramírez JG, Sørensen JD (2008) Optimal risk-based inspection planning for offshore wind turbines. Int J Steel Struct 8:295–303
33. Sørensen JD, Frandsenb S, Tarp-Johansen N (2008) Effective turbulence models and fatigue reliability in wind farms. Probab Eng Mech 23:531–538
34. Sørensen JD (2009) Framework for risk-based planning of operation and maintenance for offshore wind turbines. Wind Energy 12:493–506
35. Bharadwaj UR, Speck JB, Ablitt CJ (2007) A practical approach to risk based assessment and maintenance optimisation of offshore wind farms. In: Proceedings of the 26th international conference on offshore mechanics and arctic engineering (OMAE), San Diego, pp 10–15
36. Aven T, Jensen U (1999) Stochastic models in reliability. Springer, New York
37. Kim H, Singh C (2010) Reliability modeling and analysis in power systems with aging characteristics. IEEE Trans Power Syst 25:21–28
38. Schilling MT, Praca JCG, de Queiroz JF, Singh C, Ascher H (1988) Detection of aging in the reliability analysis of thermal generators. IEEE Trans Power Syst 3:490–499
39. Camporeale SM, Fortunato B, Marilli G (2001) Automatic system for wind turbine testing. J Sol Energy Eng 123:333–338
40. Griffin DA, Ashwill TD (2003) Alternative composite materials for megawatt-scale wind turbine blades: design considerations and recommended testing. J Sol Energy Eng 125:515–521
41. Dutton AG (2004) Thermoelastic stress measurement and acoustic emission monitoring in wind turbine blade testing. In: Proceedings of the 2004 European wind energy conference and exhibition, London
42. Rademakers L, Braam H, Zaaijer M, van Bussel G (2003) Assessment and optimisation of operation and maintenance of offshore wind turbines. Technical report, ECN Wind Energy, Petten (Online). Available: http://www.ecn.nl/docs/library/report/2003/rx03044.pdf
43. Rademakers L, Braam H, Verbruggen T (2003) R&D needs for O&M of wind turbines. Technical report, ECN Wind Energy, Petten
44. Negra NB, Holmstrøm O, Bak-Jensen B, Sørensen P (2007) Aspects of relevance in offshore wind farm reliability assessment. IEEE Trans Energy Convers 22:159–166
45. Byon E, Pérez E, Ding Y, Ntaimo L (2011) Simulation of wind farm operations and maintenance using DEVS. Simulation - Transactions of the Society for Modeling and Simulation International, 87:1093–1117
46. Foley JT, Gutowski TG (2008) Turbsim: reliability-based wind turbine simulator. In: Proceedings of the 2008 IEEE international symposium on electronics and the environment, San Francisco, pp 1–5
47. Wen J, Zheng Y, Feng D (2009) A review on reliability assessment for wind power. Renew Sustain Energy Rev 13:2485–2494
48. Chowdhury AA (2005) Reliability model for large wind farms in generation system planning. In: Proceedings of IEEE power engineering society general meeting, pp 1–7

49. Billinton R, Gao Y (2008) Multi-state wind energy conversion system models for adequacy assessment of generating systems incorporating wind energy. IEEE Trans Energy Convers 23:163–170
50. Singh C, Lago-Gonzalez A (1985) Reliability modeling of generation system including unconventional energy sources. IEEE Trans Power Syst PAS-104:1049–1056
51. Billinton R, Li Y (2007) Incorporating multi-state unit models in composite system adequacy assessment. Eur Trans Electr Power 17:375–386
52. Billinton R, Gao Y (2008) Adequacy assessment of composite power generation and transmission systems with wind energy. Int J Reliab Saf 1(2):79–98
53. Singh C, Kim Y (1988) An efficient technique for reliability analysis of power system including time dependent sources. IEEE Trans Power Syst 3:1090–1096
54. Fockens S, van Wijk AJM, Turkenburg WC, Singh C (1992) Reliability analysis of generating systems including intermittent sources. Int J Electr Power Energy Syst 14:2–8
55. Dobakhshari A, Fotuhi-Firuzabad M (2009) A reliability model of large wind farms for power system adequacy studies. IEEE Trans Energy Convers 24:792–801
56. Leite A, Borges C, Falcao D (2006) Probabilistic wind farms generation model for reliability studies applied to Brazilian sites. IEEE Trans Power Syst 21:1493–1501
57. Wang L, Singh C (2007) Adequacy assessment of power-generating systems including wind power integration based on ant colony system algorithm. In: Proceedings of IEEE power tech conference, Lausanne, pp 1629–1634
58. Wang L, Singh C (2008) Population-based intelligent search in reliability evaluation of generation systems with wind power penetration. IEEE Trans Power Syst 23:1336–1345
59. Billinton R, Bai G (2004) Generating capacity adequacy associated with wind energy. IEEE Trans Energy Convers 19:641–646
60. Wangdee W, Billinton R (2006) Considering load-carrying capability and wind speed correlation of WECS in generation adequacy assessment. IEEE Trans Energy Convers 21:734–741
61. Ravindra M, Prakash S (2008) Generator system reliability analysis including wind generators using hourly mean wind speed. Electr Power Compon Syst 36:1–16
62. Karki R, Billinton R (2004) Cost-effective wind energy utilization for reliable power supply. IEEE Trans Energy Convers 19:435–440
63. Vallée F, Lobry J, Deblecker O (2008) System reliability assessment method for wind power integration. IEEE Trans Power Syst 23:2329–2367
64. Vallée F, Lobry J, Deblecker O (2010) Wind generation modeling for transmission system adequacy studies with economic dispatch. In: Proceedings of the 2010 European wind energy conference and exhibition, Brussels
65. Billinton R, Wangdee W (2007) Reliability-based transmission reinforcement planning associated with large-scale wind farms. IEEE Trans Power Syst 22:34–41
66. Karki R, Patel J (2009) Reliability assessment of a wind power delivery system. Proc Inst Mech Eng Part O: J Risk Reliab 223:51–58
67. Alsyouf I, El-Thalji I (2008) Maintenance practices in wind power systems: a review and analysis. In: Proceedings of the 2008 European wind energy conference and exhibition, Brussels
68. Lemming J, Morthorst PE, Hansen LH, Andersen PD, Jensen PH (1999) O&M costs and economical life-time of wind turbines. In: Proceedings of the 1999 European wind energy conference, Nice, pp. 387–390
69. U.S Department of Energy (2004) Wind power today and tomorrow. Technical report, U.S Department of Energy, Washington, DC (Online). Available: http://www.nrel.gov/docs/fy04osti/34915.pdf
70. American Wind Energy Association (2008) http://awea.org/
71. Wiser R, Bolinger M (2008) Annual report on U.S. wind power installation, cost, performance trend: 2007. Technical report, North American Electric Reliability Corporation (NERC), Washington, DC (Online). Available: http://www.nrel.gov/docs/fy08osti/43025.pdf

72. Asmus P (2010) The wind energy operations and maintenance. Technical report, Wind Energy Update, London (Online). Available: http://social.windenergyupdate.com/
73. Wind Energy—the facts (WindFacts) http://www.wind-energy-the-facts.org/
74. Smolders K, Long H, Feng Y, Tavner P (2010) Reliability analysis and prediction of wind turbine gearboxes. In: Proceedings of the 2010 European wind energy conference and exhibition, Warsaw
75. Echavarria E, Hahn B, van Bussel GJW, Tomiyama T (2008) Reliability of wind turbine technology through time. J Sol Energy Eng 130:031005 (8 pages)
76. Amirat Y, Benbouzid MEH, Bensaker B, Wamkeue R (2007) Condition monitoring and fault diagnosis in wind energy conversion systems: a review. In: IEEE international electric machines and drives conference, Antalya
77. The Weather Research and Forecasting (WRF) Model http://www.wrf-model.org/
78. Arabian-Hoseynabadi H, Oraeea H, Tavner P (2010) Failure modes and effects analysis (FMEA) for wind turbines. Int J Electr Power Energy Syst 32:817–824
79. Ribrant J, Bertling L (2007) Survey of failures in wind power systems with focus on Swedish wind power plants during 1997–2005. IEEE Trans Energy Convers 22:167–173
80. Zhang X, Zhang J, Gockenbach E (2009) Reliability centered asset management for medium-voltage deteriorating electrical equipment based on Germany failure statistics. IEEE Trans Power Syst 24:721–728
81. Hameed Z, Hong Y, Cho Y, Ahn S, Song C (2009) Condition monitoring and fault detection of wind turbines and related algorithms: a review. Renew Sustain Energy Rev 13:1–39
82. Khan M, Iqbal M, Khan F (2005) Reliability and condition monitoring of a wind turbine. In: Proceedings of the 2005 IEEE Canadian conference on electrical and computer engineering, Saskatoon, pp 1978–1981
83. Caselitz P, Giebhardt J (2005) Rotor condition monitoring for improved operational safety of offshore wind energy converters. J Sol Energy Eng 127:253–261
84. Poore R, Lettenmaier T (2006) Alternative design study report: windpact advanced wind turbine drive train design study. Technical report, National Renewable Energy Laboratories (NREL), Golden, Colorado (Online). Available: http://www.nrel.gov/wind/pdfs/33196.pdf
85. Ding Y, Byon E, Park C, Tang J, Lu Y, Wang X (2007) Dynamic data-driven fault diagnosis of wind turbine systems. Lect Notes Comput Sci 4487:1197–1204
86. Hall P, Strutt J (2007) Probabilistic physics-of-failure models for component reliabilities using Monte Carlo simulation and weibull analysis: a parametric study. Reliab Eng Syst Saf 80:233–242
87. Basumatary H, Sreevalsan E, Sasi KK (2005) Weibull parameter estimation—a comparison of different methods. Wind Eng 29:309–316
88. Windstats (2008) http://www.windstats.com/
89. Baker RD (1996) Some new tests of the power law process. Technometrics 38:256–265
90. Leemis LM (1991) Nonparametric estimation of the cumulative intensity function for a nonhomogeneous poisson process. Manage Sci 37:886–900
91. Leemis LM (1982) Sequential probability ratio tests for the shape parameter of a nonhomogeneous poisson process. IEEE Trans Reliab 31:79–83
92. Leemis LM (1987) Efficient sequential estimation in a nonhomogeneous poisson process. IEEE Trans Reliab 36:255–258
93. Rigdon SE, Basu AP (1989) The power law process: a model for the reliability of repairable systems. J Qual Technol 21:251–260
94. Landwirtschaftskammer Schleswig-Holstein Wind Energie, *Praxisergebnisse 1993–2004*. Rendsbug, Germany, 2008, Eggersglüss ed. (Note: this source appeared initially in [19] and recited here)
95. Welte T (2009) Using state diagrams for modeling maintenance of deteriorating systems. IEEE Trans Power Syst 24:58–66
96. Hoskins RP, Strbac G, Brint AT (1999) Modelling the degradation of condition indices. In: IEEE Proceedings of generation, transmission and distribution, vol 146, pp 386–392
97. Norris J (1998) Markov chains. Cambridge University Press, Cambridge

98. Puterman M (1994) Markov decision process. Wiley, New York
99. Lovejoy W (1987) Some monotonicity results for partially observed Markov decision processes. Oper Res 35:736–742
100. Rosenfield D (1976) Markovian deterioration with uncertain information. Oper Res 24:141–155
101. Ross S (1971) Quality control under Markovian deterioration. Manage Sci 19:587–596
102. Ohnishi M, Kawai H, Mine H (1986) An optimal inspection and replacement policy under incomplete state information. Eur J Oper Res 27:117–128
103. Maillart LM (2006) Maintenance policies for systems with condition monitoring and obvious failures. IIE Trans 38:463–475
104. Maillart LM, Zheltova L (2007) Structured maintenance policies in interior sample paths. Naval Res Logistics 54:645–655
105. Ghasemi A, Yacout S, Ouali MS (2007) Optimal condition based maintenance with imperfect information and the proportional hazards model. Int J Prod Res 45:989–1012
106. Thomas LC, Gaver DP, Jacobs PA (1991) Inspection models and their application. IMA J Math Appl Bus Ind 3:283–303
107. Kim YH, Thomas LC (2006) Repair strategies in an uncertain environment: Markov decision process approach. J Oper Res Soc 57:957–964
108. International Electrotechnical Commission (2005) IEC 61400-1: wind turbines, #Part 1: design requirements, 3rd edn. IEC, Geneva
109. Fitzwater LM, Winterstein SR, Cornell CA (2002) Predicting the long term distribution of extreme loads from limited duration data: comparing full integration and approximate methods. J Sol Energy Eng 124:378–386
110. Agarwal P, Manuel L (2009) Simulation of offshore wind turbine response for ultimate limit states. Eng Struct 31:2236–2246
111. Fogle J, Agarwal P, Manuel L (2008) Towards an improved understanding of statistical extrapolation for wind turbine extreme loads. Wind Energy 11:613–635
112. Hahn B, Durstewitz M, Rohrig K (2007) Reliability of wind turbines, experiences of 15 years with 1,500 WTs in wind energy. Springer, Berlin
113. NOAA—National Oceanic and Atmospheric Administration http://www.noaa.gov/
114. ReliaSoft BlocSim-7 software (2007) http://www.reliasoft.com/products.htm/
115. West Texas Mesonet (2008) http://www.mesonet.ttu.edu/
116. Billinton R, Allan RN (1996) Reliability evaluation of power systems, 2nd edn. Plenum Press, New York
117. Negra N, Holmstrøm O, Bak-Jensen B, Sørensen P (2007) Wind farm generation assessment for reliability analysis of power systems. Wind Eng 31:383–400
118. Borges CLT, Falcão DM (2001) Power system reliability by sequential Monte Carlo simulation on multicomputer platforms. Lect Notes Comput Sci 1981:242–253
119. Wang L, Singh C (2008) Adequacy assessment of power systems through hybridization of Monte Carlo simulation and artificial immune recognition system. In: Proceedings of the power systems computation conference, Glasgow
120. Blaabjerg F, Teodorescu R, Liserre M, Timbus AV (2006) Overview of control and grid synchronization for distributed power generation systems. IEEE Trans Ind Electron 53:1398–1409
121. Kaldellis JK, Kavadias KA, Filios AE, Garofallakis S (2004) Income loss due to wind energy rejected by the Crete island electrical network—the present situation. Appl Energy 79:127–144
122. Hansen AD, Cutululis N, Sørensen P, Iov F, Larsen TJ (2007) Simulation of a flexible wind turbine response to a grid fault. In: Proceedings of the 2007 European wind energy conference and exhibition, Milan
123. Chen Z, Guerrero JM, Blaabjerg F (2009) A review of the state of the art of power electronics for wind turbines. IEEE Trans Power Electron 24:1859–1875

Wind Turbine Power Performance and Application to Monitoring

Patrick Milan, Matthias Wächter and Joachim Peinke

Abstract The concept of power performance is introduced as the ability of a wind turbine to extract power from the wind. The general performance estimates such as the power coefficient or the theoretical power curve are introduced in laminar conditions. Following Betz' limit, an upper limit for the power available in the wind is derived, as well the main sources of energy loss. This laminar theory is too simple to describe operating wind turbines, and turbulent and atmospheric effects call for statistical tools. An IEC norm defines the international standard to measure and analyze power performance. The resulting IEC power curve gives a first estimation, and can be used to evaluate the annual energy production. An alternative is proposed with the Langevin power curve, which quantifies the high-frequency dynamics of a wind turbine power production to changing wind speeds. This brings further insight on the overall performance, and allows for applications such as performance monitoring or power modeling.

1 Introduction

The only purpose of a wind power system is to extract energy from the wind. From this statement, building wind turbines might sound like an outdated challenge. The first designs were successfully developed in the Antiquity to provide mechanical

P. Milan (✉) · M. Wächter · J. Peinke
University of Oldenburg, ForWind - Center for Wind Energy Research of the Universities of Oldenburg, Hannover and Bremen, Oldenburg, Germany
e-mail: patrick.milan@forwind.de

M. Wächter
e-mail: matthias.waechter@forwind.de

J. Peinke
e-mail: joachim.peinke@forwind.de

P. M. Pardalos et al. (eds.), *Handbook of Wind Power Systems*, Energy Systems,
DOI: 10.1007/978-3-642-41080-2_20, © Springer-Verlag Berlin Heidelberg 2013

energy around the globe. The integration of the first wind turbines into electrical networks goes back to the late 19th century, soon after the first power lines were erected. Wind energy is one of the oldest sources of energy. Despite a long history, it remains relatively poorly understood. While its industry recently entered a new age of its evolution, raise more questions and challenges. Before understanding fully how to harness the wind, it is necessary to understand how the wind works. The staggering level of complexity of turbulent and atmospheric effects accounts for many of the remaining challenges that encounters the wind energy industry.

This chapter is titled *Wind Turbine Power Performance and Application to Monitoring*. Most of its content is oriented towards an introduction to power performance. Power performance is a central aspect of wind energy, and somehow carries its own interest. But it is in its applicability that lies its central interest. Uncomplicated quantities like the power coefficient or power curves are the main estimates of the global *health* of a wind turbine, and can serve as simplified mathematical models. Such models as those introduced in this chapter, although very general and unspecialized, can assist more exclusive applications like monitoring, power prediction or grid integration, that remain in essence very specialized topics. This sets the direction, in between physics and wind power systems, away from dedicated applications but always in connection with them, towards a better understanding of the wind and how to harness it.

Besides the technical challenge of designing and building rotating machines over 100 m in diameter, must be considered their integration into a turbulent, ever-changing wind flow. While the wind *signal* that can feel a wind turbine displays complex statistics, its mechanical extraction and transformation into electrical energy adds to the complexity. This justifies why it is so difficult to define unambiguously power performance for wind power systems. A basic theory was developed, that sets aside the turbulent fluctuations, following Betz' developments from the 1920s. This represents the main focus of Sect. 2. Besides giving a good first estimation, this *laminar* theory remains unrealistic to solve the current challenges. More recently, statistical tools were developed in order to integrate the features of turbulence into a more realistic theory of power performance, as introduced in Sect. 3. Although these tools represent large simplifications of the actual dynamics of a wind turbine, they bring useful insights for various applications. An overview of some central applications is introduced in Sect. 4, such as prediction of annual energy production or dynamical monitoring.

2 Power Performance Theory

This section presents the concept of power performance for wind turbines, starting from momentum theory to power curves. It is meant to give an introduction to the underlying theory, before applications are presented in Sects. 3 and 4. While this chapter deals only with horizontal-axis three-bladed electrical wind turbines, there is no major limitation to its extension to other designs of wind power systems.

2.1 Momentum Theory for Wind Turbines

In this Sect. 2.1, a basic understanding of fluid mechanics will be applied to wind turbines. For a more detailed description on momentum theory, the reader is kindly referred to [6]. This theoretical approach sets ground for the further power curve analysis. The complexity of turbulence is first set aside, so as to understand the fundamental behavior of a wind turbine in a uniform flow at steady-state. More complex atmospheric effects represent active research topics, whose detailed analysis is outside the scope of this introduction, cf. [5].

As a wind turbine converts the power from the wind into available electrical power, one can assume the following relation

$$P(u) = c_p(u) \times P_{wind}(u), \tag{1}$$

where $P_{wind}(u)$ is the power contained in the wind passing with speed u through the wind turbine, and $P(u)$ is the electrical power extracted. The amount of power converted by the wind turbine is given by the so-called power coefficient $c_p(u)$, which represents the efficiency of the machine. As the input $P_{wind}(u)$ cannot be controlled, improving power performance means increasing the power coefficient $c_p(u)$. The power contained in a laminar incompressible flow of mass m and density ρ moving along the *x-axis* with constant speed u through a vertical plane of area A is

$$\begin{aligned} P_{wind}(u) &= \frac{d}{dt}E_{kin,wind} = \frac{d}{dt}\left(\frac{1}{2}mu^2\right) \\ &= \frac{1}{2}\frac{dm}{dt}u^2 = \frac{1}{2}\frac{d(\rho V)}{dt}u^2 \\ &= \frac{1}{2}\rho\frac{d(Ax)}{dt}u^2 = \frac{1}{2}\rho Au^3. \end{aligned} \tag{2}$$

Let us consider a mass of air moving towards a wind turbine, which can be represented by an actuator disc[1] of diameter D. When crossing the wind turbine, the wind is affected as parts of its energy is extracted. This extraction of kinetic energy results in a drop in the wind speed from upstream to downstream. The velocity far before the wind turbine (upstream), at the wind turbine and far behind (downstream) are labelled respectively u_1, u_2 and u_3. An illustration is given in Fig. 1, see also [6].

Mass conservation requires that the flow-rate $\dot{m} = A_i \rho u_i$ be conserved and

$$A_1\rho u_1 = A_2\rho u_2 = A_3\rho u_3, \tag{3}$$

where A_i are the respective areas perpendicular to the flow. A_2 is the area swept by the rotor blades $A_2 = A = \pi D^2/4$. As a consequence of the wind speed slowing

[1] An actuator disc is an infinitely thin disc through which the air can flow without resistance, as proposed by Froude and Rankine's momentum theory [12].

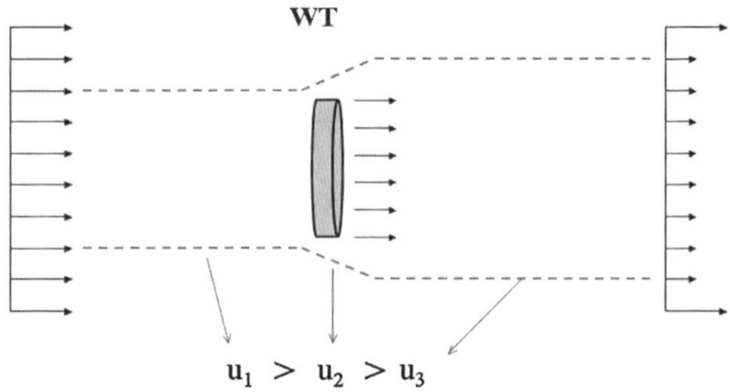

Fig. 1 Idealized flow situation around a wind turbine (*WT*) according to [20]. The wind speeds before, at and after the wind turbine are respectively u_1, u_2 and u_3

down, i.e. $u_3 < u_2 < u_1$, the area of the stream-tube[2] has to expand, and $A_3 > A_2 > A_1$. This can be observed in Fig. 1.

Also, the energy extracted by the wind turbine can be determined by the difference of kinetic energy upstream and downstream of the wind turbine

$$E_{ex} = \frac{1}{2} m (u_1^2 - u_3^2), \tag{4}$$

resulting in a power extraction

$$P_{ex} = \frac{d}{dt} E_{ex} = \frac{1}{2} \dot{m} (u_1^2 - u_3^2). \tag{5}$$

The wind power cannot be totally converted into mechanical power because the wind turbine continuously takes energy out of the wind flow, which reduces its velocity. However, the flow needs to *escape* the wind turbine downstream with a speed $u_3 > 0$. If all the power content of the wind would be extracted, the wind speed downstream would then become zero. As a consequence, the air would *accumulate* downstream and block newer air from flowing through the wind turbine, so that no more power could be extracted. This means that the wind flow must keep some energy to escape, which naturally sets a limit for the efficiency of any wind power system. The power coefficient $c_p(u)$ must be inferior to 1. An optimal ratio of wind speeds $\mu = u_3/u_1$ can be found that allows for the highest energy extraction, as introduced in Sect. 2.2.

2.2 Power Performance

In the plane of the rotor blades, an intermediate value of wind speed

$$u_2 = \frac{u_1 + u_3}{2} \tag{6}$$

[2] A stream-tube is defined here as the stream of air particles that interact with the wind turbine.

is found.[3] Knowing this value one also knows the flow-rate in the rotor plane area that is now given by

$$\dot{m} = \rho A u_2. \tag{7}$$

Inserting Eqs. (6) and (7) into Eq. (5) yields

$$\begin{aligned} P(\mu) &= \frac{1}{2}\rho A u_1^3 \times \frac{1}{2}\left(1 + \mu - \mu^2 - \mu^3\right) \\ &= P_{wind}(u_1) \times c_p(\mu). \end{aligned} \tag{8}$$

The theoretical definition of the power coefficient is then

$$c_p(\mu) = \frac{1}{2}\left(1 + \mu - \mu^2 - \mu^3\right), \tag{9}$$

where $\mu = u_3/u_1$, as shown in Fig. 2. The optimal power performance is obtained for a ratio μ such that the derivative of $c_p(\mu)$ with respect to μ is zero

$$\frac{d}{d\mu}c_p(\mu) = \left(-\frac{1}{2}\right) \times \left(3\mu^2 + 2\mu - 1\right) = 0. \tag{10}$$

This leads to $\mu_{max} = 1/3$ and a corresponding $C_p(\mu_{max}) = 16/27 \approx 0.593$, as shown in Fig. 2.

This limit is called the Betz limit, as it was found by Albert Betz in 1927 [2]. In other words, a wind turbine can extract at most 59.3 % of the power contained in the wind. This can be obtained when the wind speed downstream is one-third of the wind speed upstream.

A widely used representation of power performance is given by the relation of c_p to the tip speed ratio λ defined as

$$\lambda = \frac{\omega R}{u_1}, \tag{11}$$

where ω and R are the angular frequency and radius of the rotor. λ is the ratio of the rotational speed at the tip of the blades to the upstream wind speed. The dimensionless c_p–λ curve is introduced in Sect. 2.3.

Betz' momentum theory only considers the mechanical transfer of energy from the wind to the rotor blades. The next step of the conversion from mechanical to electrical energy has not been taken into account, as well as

[3] Following Froude-Rankine, it can be shown that this value is the optimal value.

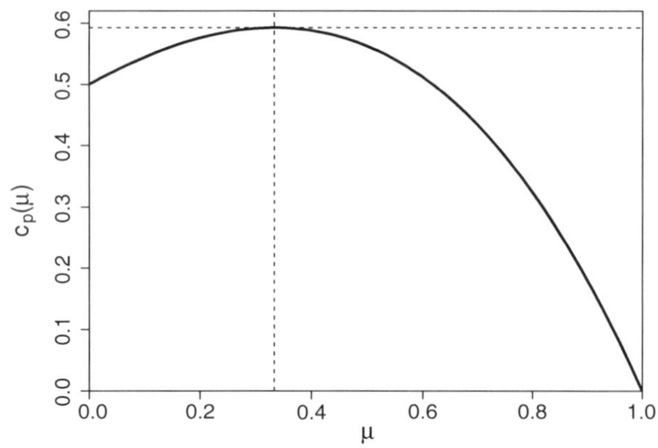

Fig. 2 Power coefficient c_p as a function of the wind speed ratio $\mu = u_3/u_1$

all energy losses. The more complex design of wind turbines causes lower values of c_p, as discussed in Sect. 2.3. The power coefficients of modern commercial wind turbines reach values of order 0.5. Also, criticism of Betz theory is given in [13, 14], leading to a less well defined upper limit of c_p.

2.3 Limitations of Betz Theory: Energy Losses

Although it is based on a simplified approach, the Betz limit is a widely used and accepted value. But more realistic considerations indicate that real wind turbine designs have even lower efficiency due to additional limitations. In this section, the three main limitations to reach the optimal value of $c_p = 16/27$ are introduced. Additional considerations, e.g. the finite number of blades and losses due to the drag and stall effects on the blades are discussed in [3, 6]. All the derivations introduced in this Sect. 2.3 are based on Appendix 1. This section only aims to give a first idea. For a more detailed understanding of the mathematical equations presented here, the reader is kindly referred to the appendix first.

2.3.1 Bouncing Losses

Betz' consideration does not take into account that there is not only a reduction of wind speed downstream, but also an additional angular momentum that is transferred to the air flow, as shown in Fig. 3.

This effect follows Newton's third law, as a reaction to the rotational motion of the rotor. The energy loss is more important for smaller tip speed ratios, following a derivation in Appendix 1. This follows Eq. (36) where a small velocity ωr

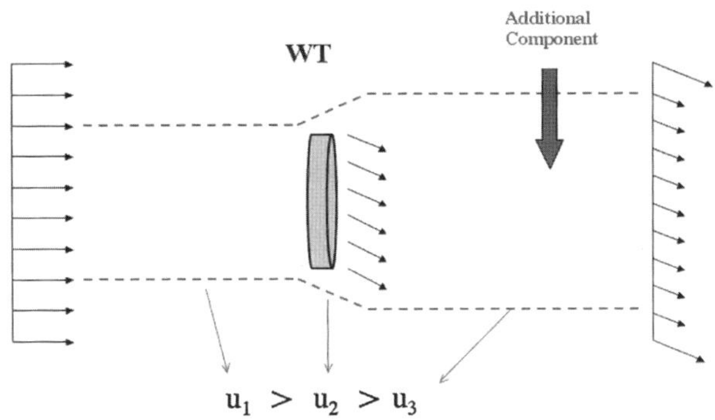

Fig. 3 Flow around a wind turbine. After passing the turbine the velocity field has a rotational component due to the rotating rotor blades

requires a larger force F_r (the subscript $_r$ indicates the rotational direction) to obtain the same power. So for slow rotating wind turbines (λ small), these losses are much more severe than for fast rotating machines. For instance, for $\lambda \approx 1$ an optimum value of c_p of only 0.42 can be reached instead of the Betz optimum of 0.59. c_p approaches the Betz optimum with increasing tip speed ratio.

2.3.2 Profile Losses

Another important source of energy loss is the quality of the airfoil profile. According to Eqs. (33) and (36) in Appendix 1, a cut dr at radius r yields a power extraction

$$dP = z\omega r \frac{\rho}{2} c^2 \times t \times dr[(C_L \cos(\beta) - C_D \sin(\beta)]. \tag{12}$$

For a perfect (ideal) airfoil the drag vanishes and

$$dP_{ideal} = z\omega r \frac{\rho}{2} c^2 \times t \times dr \times C_L \cos(\beta). \tag{13}$$

The efficiency η can now be defined as the ratio between Eqs. (12) and (13), i.e. dP/dP_{ideal}. The general definition of the efficiency is given by

$$\eta = 1 - \xi. \tag{14}$$

The profile losses ξ_{prof} follow the relation

$$\xi_{prof} \propto r\lambda. \tag{15}$$

In contrast to the bouncing losses, the profile losses mainly affect fast rotating machines. For higher tip speed ratios, the lift to drag ratio C_L/C_D must be optimized. Furthermore the losses increase with the radius, such that the manufacturing quality of the blade tips is of primary importance for power performance.

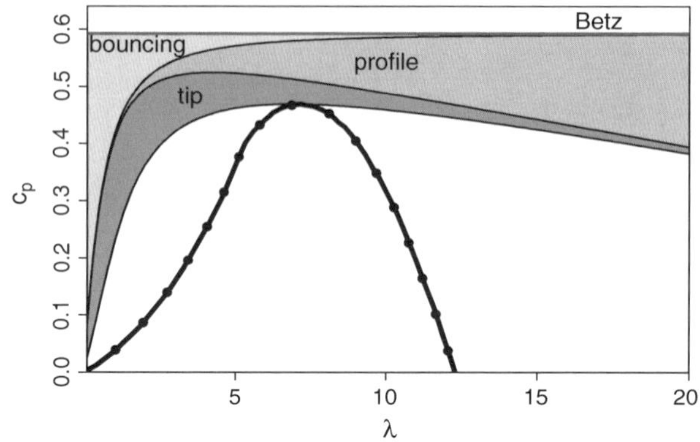

Fig. 4 Typical $c_p - \lambda$ curve (*black dotted line*). The influence on the efficiency of Betz limit, the bouncing, profile and tip losses is illustrated. These results were obtained with a lift to drag ratio $C_L/C_D = 60$ and $z = 3$ blades. Maximum performance $c_p \simeq 0.475$ is obtained at $\lambda \simeq 7$

2.3.3 Tip Losses

A good quality of the tips especially means that they should be as narrow as possible because this corresponds to an (ideal) airfoil with length infinity $(R/t \rightarrow \infty)$. For real blades there is always a flow around the end of the blade (forming an eddy that is advected by the flow) from the high pressure area to the low pressure area. This is partly levelling the pressure difference and consequently the lift force. The tip losses obey approximately the following relation

$$\xi_{tip} \propto \frac{1}{z\lambda}. \qquad (16)$$

Different to the profile losses an increasing tip speed ratio decreases the tip losses, as well as an increased number of blades.

2.3.4 Impact on Power Performance

Figure 4 shows an overview of the different kinds of losses and their influence on the value of c_p. One can see that bouncing losses cause the largest reduction in the power coefficient for small values of λ, similar to the finite number of blades. This is the opposite for the profile losses. Three-bladed wind turbines can reach optimal c_p values of order 0.50 for typical values of $\lambda \approx 6 - 8$, which naturally sets the strategy for optimal power performance in terms of rotational frequency ω.

2.4 Theoretical Power Curve

Along with the $c_p - \lambda$ curve, a standard representation of a wind turbine power performance is given by a so-called power curve. The power curve gives the relation between the simultaneous wind speed u and power output P. Following usual practice, the wind speed u will refer to the upstream horizontal wind speed u_1 from now on (such that $u = u_1$). Also, the net electrical power output P that the wind turbine actually delivers to the grid is considered, integrating all possible losses. The two quantities u and P will follow these specifications until the end of the chapter. Following Eq. (8), the theoretical power curve reads

$$\begin{aligned} P(u) &= c_p(u) \times P_{wind}(u) \\ &= c_p(u) \times \frac{1}{2}\rho A u^3. \end{aligned} \tag{17}$$

In most of the modern wind turbine designs, the regulation of the power output is performed through changes both in the rotational frequency of the generator and in the pitch angle of each blade.[4] The rotational frequency of the generator is physically linked to the wind speed, such that it cannot be changed freely. However, the pitch angle of the blades can be controlled at will, and almost independently of the wind speed, to reach the chosen control strategy, and hence represents the central mean of control for the operation. Pitching plainly consists of a rotation of the blades by a pitch angle θ in the plane of their cross-section. We will refer only to this design in this Sect. 2.4.

The power production is then controlled by changing the lift forces on the rotor blades [3, 6]. The power production can be reduced or stopped by pitching the blades towards stall.[5] In modern wind turbines, this is achieved by a so-called active pitch control. The power coefficient c_p depends strongly on this pitch angle θ and on the tip speed ratio λ, i.e. $c_p = c_p(\lambda(u), \theta)$. As λ can typically not be controlled, c_p is optimized via θ to a desired power production. In particular for high wind speeds, c_p is lowered to protect the wind turbine machinery and prevent from overshoots in the power production.

This pitch regulation is commanded by the controller of the wind turbine, which constitutes of several composite mechanical–electrical components that operate actively for the optimum power performance.[6] For the common pitch-controlled wind turbines, the control strategy gives four distinct modes of operation:

---

[4] Other wind turbine designs involve fixed rotational frequency (called fixed-speed wind turbines) or fixed pitch angle (called fixed-pitch wind turbines). A more detailed description on control strategies is given in [3].

[5] Stall effects are obtained when the angle of attack of an airfoil exceeds a critical value, resulting in a sudden reduction in the lift force generated. A detailed study on airfoil lift effects can be found in [19].

[6] Additional considerations such as mechanical loads or power stability are usually taken into account as well [3], but reach out of the scope of this chapter.

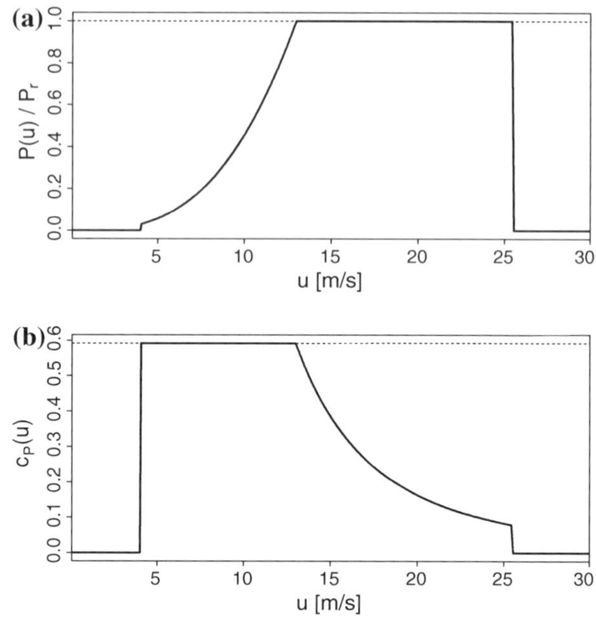

Fig. 5 **a** Theoretical power curve $P(u)$. **b** Theoretical power coefficient $c_p(u)$ for a pitch-controlled wind turbine with $u_{cut-in} = 4$ m/s, $u_r = 13$ m/s and $u_{cut-out} = 25$ m/s

- for $u \leq u_{cut-in}$,[7] the power contained in the wind is not sufficient to maintain the wind turbine into motion, and no power is produced;
- in partial load $u_{cut-in} \leq u \leq u_r$,[8] the wind turbine works at its maximum power performance, i.e. c_p is maximized, and the pitch angle θ is normally maintained constant;
- in full load $u_r \leq u \leq u_{cut-out}$,[9] the wind turbine power output is limited to the rated power P_r. In this mode of operation, the pitch angle θ is adjusted in real-time to maintain $P \approx P_r$;
- for $u > u_{cut-out}$ the pitch angle θ is maximized to the feathered position so as to eliminate the lift forces on the blades. A braking device can be used in addition to block the rotation for safety reasons. As a consequence, the power production is stopped.

An illustration of the theoretical strategy for $c_p(u)$ and $P(u)$ is given in Fig. 5.

---

[7] u_{cut-in} represents the minimum wind speed such that the wind turbine can extract power, typically in the order of 3–4 m/s.

[8] u_r represents the rated wind speed at which the wind turbine extracts the rated, maximum allowed power P_r, typically in the order of 12–15 m/s.

[9] $u_{cut-out}$ represents the maximum wind speed at which the wind turbine can safely extract power, typically in the order of 25–35 m/s.

It is important to precise that this theoretical estimation is valid for a laminar flow, which never occurs in real situations. The more complex atmospheric winds call for more complex descriptions of power performance. Following the path of turbulence research, statistical models are introduced in Sect. 3 to deal with this complexity.

3 Application to Operating Wind Turbines

The theory of wind power performance was introduced in Sect. 2. Although this theory sets a good foundation for wind energy applications, the complexity of atmospheric effects calls for a more advanced description. This Sect. 3 presents the typical complex data of wind speed u and power output P recorded on wind turbines in Sect. 3.1. It is followed by two approaches to estimate power curves in Sects. 3.2 and 3.3, respectively the IEC and the Langevin procedures.

For information, all the results presented were derived from measurements on operating multi-MW commercial wind turbines, with a sampling frequency 1 Hz (unlike stated otherwise). All power values are normalized by the rated power P_r in order not to confuse the reader. All results can straightforwardly be converted back to actual power values.

3.1 Atmospheric Turbulence: A Complex Challenge

The introduction in Sect. 2 assumed a steady, laminar wind inflow $\mathbf{u_1} = \mathbf{constant}$. Although it is a necessary assumption to derive Betz limit, atmospheric flows are turbulent. Atmospheric winds combine the complex aspects of turbulence on small scales and of climatology on larger scales. The statistics of wind measurements display complex properties like unstationarity or intermittency (such as gusts). An illustration on wind statistics is presented here, as well as their impact on power output statistics.

Two typical time series for the simultaneous measurement of wind speed u and power output P are displayed in Figs. 6 and 7. The 10-min average and standard deviation are displayed, illustrating how the IEC norm reduces the complexity of the measurement signals (see Sect. 3.2).

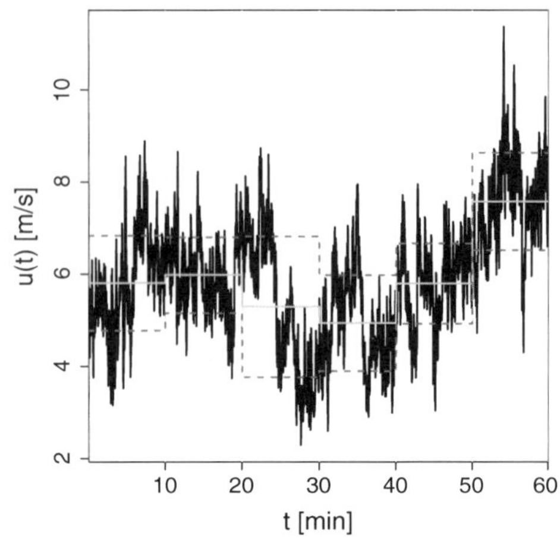

Fig. 6 Excerpt of a wind speed measurement at frequency 1 Hz for 1 h. The 10-min average and standard deviation are displayed respectively with the *solid red* and *dashed blue lines*

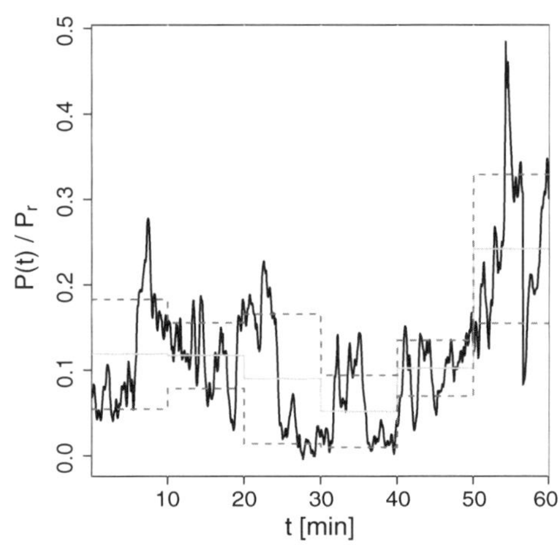

Fig. 7 Excerpt of a power output measurement at frequency 1 Hz for 1 h. The 10-min average and standard deviation are displayed respectively with the *solid red* and *dashed blue lines*

The time series can be plotted together, i.e. power output versus wind speed for the same measurement, as shown in Fig. 8.

One can see from Fig. 8 that when no time-averaging is performed, the power conversion is a highly dynamical system even on very short time scales. The power signal reacts quickly to the wind speed signal, and can be considered *turbulent* as well. An important aspect for the wind energy industry is the impact of gusts[10] on wind turbines fatigue loads, as well as on power stability. Gusts can be estimated from the statistics of the increments of wind speed

$$u_\tau(t) = u(t+\tau) - u(t), \tag{18}$$

[10] Although no unique, clear definition of gust exists, one can see a wind gust as a rapid change of wind speed (and possibly direction). Gusts are *extreme events*.

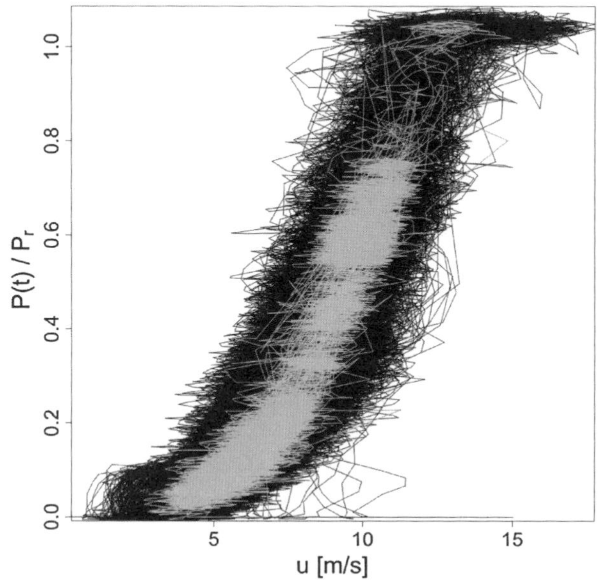

Fig. 8 1 Hz measurement of power output versus wind speed during 10^6 s. The data is plotted in *black* for the first 960.000 points in the background, then in *cyan* for the last 40.000 points in the foreground

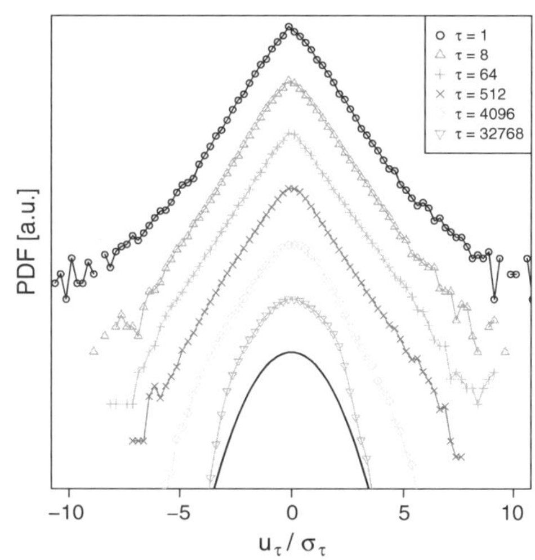

Fig. 9 PDF of normalized wind speed increments for various time increments τ (increasing τ downwards). The values of τ are given in seconds. The various PDFs are intentionally shifted vertically for clarity. A Gaussian distribution is given as a reference (*solid line*)

where τ is the time increment, or scale of interest. $u_\tau(t)$ then represents the change in wind speed between time t and time $t + \tau$. Similarly, the increments of power output can be defined as

$$P_\tau(t) = P(t + \tau) - P(t). \tag{19}$$

Following what is usually done in research on turbulence, the probability density function (PDF) of u_τ is displayed in Fig. 9, as well as the PDF of P_τ in

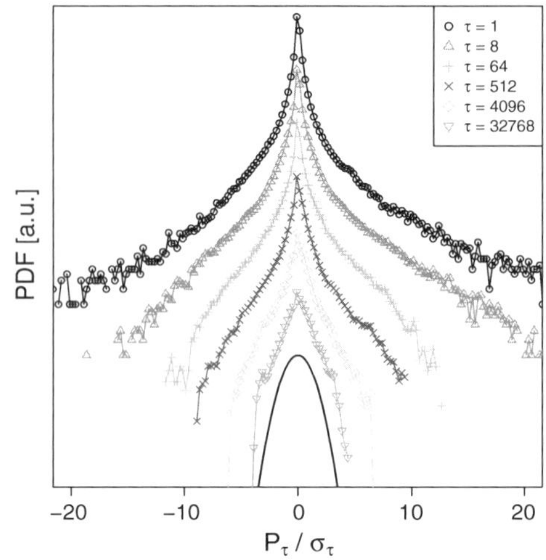

Fig. 10 PDF of normalized power output increments for various time increments τ (increasing τ downwards). The values of τ are given in seconds. The various PDFs are intentionally shifted vertically for clarity. A Gaussian distribution is given as a reference (*solid line*)

Fig. 10. For information, the increments are systematically normalized by their standard deviation σ_τ, such that only the shape of the PDF is of interest.

Compared to a normal, i.e. Gaussian distribution, the normalized PDFs of u_τ and P_τ can be seen as intermittent,[11] especially for the short time scales $\tau \approx 1-100\,\text{s}$. The PDFs are most intermittent for small values of τ, indicating that the intermittent dynamics act on short time scales. This means that the probability of sustaining a major change of wind speed or power output is higher over short time intervals. When increasing the scale τ, the increment PDFs of u and P then become less intermittent and tend towards the Gaussian distribution.

While this is a well-known result for turbulent winds [5], this aspect is seldom emphasized for wind turbines power output, see [8].

The information displayed in Fig. 10 is important to understand wind turbines behavior and for the wind energy industry in general, as it shows a non-zero probability of having extreme changes in power output. For example for $\tau = 64\,\text{s}$, events $P_\tau \approx 10\sigma_\tau \approx 0.75P_r$ were recorded, meaning that the power output can increase by about 75 % of the rated power within a minute. Also for $\tau = 8\,\text{s}$, events $P_\tau \approx 20\sigma_\tau \approx 0.6P_r$ occur, meaning that the power output can increase by about 60 % of the rated power within 8 s.

PDFs of power output increments appear to be even more intermittent than PDFs of wind speed increments. This can be justified by the cubic relation of the

[11] The notion of intermittency is related to the probability of a process to sustain extreme events. It can be identified as a large deviation from the Gaussian distribution far away from the mean value. Extreme events such as gusts yield intermittent PDFs.

power output to the wind speed. As a matter of fact, when the wind speed doubles, the power output should theoretically increase by a factor 8. This justifies why wind gusts are transferred and amplified to the power production which, as a consequence, suffers rapid changes relatively often. This certainly accounts for the shorter lifetime of wind turbines than originally designed. Also, these rapid changes in power production are fed into the grid, and raise the matter of the power stability for wind energy integration.

3.2 *The International Standard: The IEC Power Curve*

The standard power performance procedure for wind turbines was defined by the International Electrotechnical Commission in 2005 in the norm IEC 61400-12-1. For a detailed description of this norm, the reader is kindly referred to the complete proceeding [10]. This procedure provides a unique methodology to ensure accuracy, consistency and reproducibility in the measurement and in the analysis of power performance. It consists first of the minimum requirements for a power performance test, and second of a procedure to process the measured data without extensive knowledge.

3.2.1 Measurement Procedure

First, are described the necessary preparations for the performance test, such as criteria for the measurement equipments, guidance for the location and setup of the meteorological mast that will be used to measure the wind speed and other parameters like the wind direction, temperature and air pressure. The sector of the measurement is also described as the range of wind directions that are valid for a representative measurement, such that wind directions where the met mast is in the wake of the wind turbine must be excluded. A detailed assessment of the terrain at the test site is presented in the additional site calibration procedure which reports for additional obstacles (other than the wind turbine).

> The first goal of the IEC norm is to ensure that the data collection displays a sufficient quantity and quality for an accurate estimation of the power performance.

3.2.2 IEC Power Curve

Second, the measured data is processed.[12] The data processing is mainly performed in two steps.

After adequate normalization of the data, the first step consists in averaging the measured data over time intervals of 10 min. The IEC power curve is derived in a second step from the 10-min averages using the so-called method of bins, i.e. the data is separated into wind speed intervals of width 0.5 m/s.

In each interval i, bin averages of wind speed u_i and power output P_i are calculated according to

$$u_i = \frac{1}{N_i}\sum_{j=1}^{N_i} u_{norm,i,j}, \quad P_i = \frac{1}{N_i}\sum_{j=1}^{N_i} P_{norm,i,j}, \tag{20}$$

where $u_{norm,i,j}$ and $P_{norm,i,j}$ are the normalized 10-min average values of wind speed and power, and N_i is the number of 10 min data sets in the ith bin.

For the power curve to be complete and reliable, each wind speed bin must include at least 30 min of sampled data. Also, the total measurement time must cover at least a period of 180 h. The range of wind speeds must range from 1 m/s below cut-in wind speed to 1.5 times the wind speed at 85 % of the rated power of the wind turbine. The norm also provides an estimation of uncertainty as the standard error of the normalized power data, plus additional uncertainties related to the instruments, the data acquisition system and the surrounding terrain. A typical IEC power curve is presented in Fig. 11.

The IEC norm also defines the AEP (Annual Energy Production), as presented in Sect. 4.2. The AEP is a central feature for economical considerations, as it gives a first estimate of the long-time energy production of a wind turbine. As it sets a unique ground for wind power performance worldwide, the IEC norm helps building a general understanding between manufacturers, scientists and end-users. This statement comes to be ever more important as the wind energy sector grows. Hence, focusing on this standard is paramount to any study on power performance.

3.2.3 Turbulence-Induced Deviations

As a downside to its simplicity, the IEC power curve method presents a limitation. In contrast to a good definition of the requirements in Sect. 3.2.1, the definition of the power curve in Sect. 3.2.2 suffers a mathematical imperfection. In order to deal with the complexity of the wind speed and power signals, the data is systematically

[12] Additional correction of the measured data should be performed using temperature and pressure measurements.

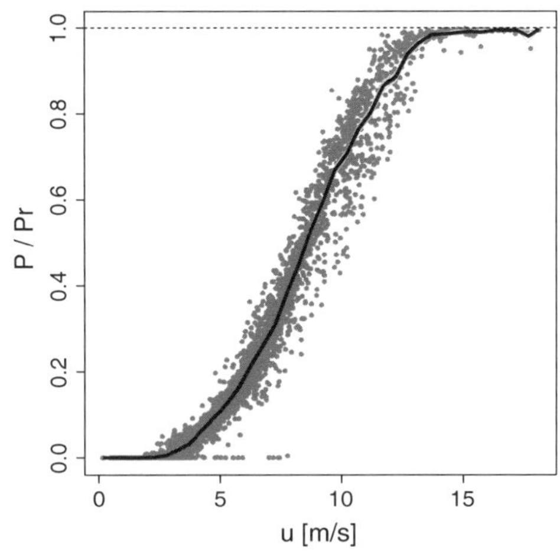

Fig. 11 Power curve (*black line*) and corresponding error bars (*thin red lines underneath*) obtained according to the IEC norm. The *blue dots* represent the 10-min average values

averaged over time. Although a statistical averaging is necessary to extract the main features from the complex processes, the averaging procedure over 10-min intervals lacks a clear physical meaning, beyond its statistical definition. As the wind[13] fluctuates on various time scales (down to seconds and less), a systematic averaging over 10 min filters out all the short-scale turbulent dynamics. Combining these turbulent fluctuations with the non-linear power curve $P(u) \propto u^3$, the resulting IEC power curve is spoiled by mathematical errors. To show this, one can first split the wind speed $u(t)$ sampled at 1 Hz into its mean value and the fluctuations around this mean value

$$u(t) = \overline{u(t)} + u'(t), \tag{21}$$

where the operation $\overline{x(t)}$ on a given signal $x(t)$ represents the average (arithmetic mean) of $x(t)$. Assuming that $u'(t) \ll \overline{u(t)}$, a Taylor expansion of $P(u(t))$ reads [4]

$$\begin{aligned} P(u(t)) = {} & P\left(\overline{u(t)}\right) \\ & + u'(t)\left(\frac{\partial P(u)}{\partial u}\right)_{u=\overline{u(t)}} \\ & + \frac{u'(t)^2}{2!}\left(\frac{\partial^2 P(u)}{\partial u^2}\right)_{u=\overline{u(t)}} \\ & + \frac{u'(t)^3}{3!}\left(\frac{\partial^3 P(u)}{\partial u^3}\right)_{u=\overline{u(t)}} \\ & + o\left(u'(t)^4\right). \end{aligned} \tag{22}$$

---

[13] To some extent the power output also fluctuates on short time scales, but its high-frequency dynamics are limited by the inertia of the wind turbine.

Averaging Eq. (22) yields

$$\begin{aligned}\overline{P(u(t))} = P\left(\overline{u(t)}\right) \\ + 0 \\ + \frac{\overline{u'(t)^2}}{2}\left(\frac{\partial^2 P(u)}{\partial u^2}\right)_{u=\overline{u(t)}} \\ + \frac{\overline{u'(t)^3}}{6}\left(\frac{\partial^3 P(u)}{\partial u^3}\right)_{u=\overline{u(t)}} \\ + o\left(u'(t)^4\right),\end{aligned} \tag{23}$$

because $\overline{u'(t)} = \overline{u(t) - \overline{u(t)}} = 0$. This means that the average of the power is not equal to the power of the average, and must be corrected by the 2nd and 3rd-order terms. As the IEC power curve directly relates the 10-min averages of wind speed and of power output, it neglects the higher-order terms in the Taylor expansion. The 2nd-order term is the product of the variance $\sigma^2 = \overline{u'(t)^2}$ of u(t)[14] and the second-order derivative of the power curve.[15] This demonstrates that the IEC power curve cannot describe in a mathematically rigorous way the nonlinear relation of power to wind speed when coupled with wind fluctuations (stemming from turbulence), at least not without higher-order corrections.

As a consequence of this mathematical over-simplification, the result depends on the turbulence intensity $I = \sigma/\overline{u}$, so on the wind condition during the measurement [4]. It is illustrated in Fig. 12, where the IEC power curve deviates from the theoretical power curve with increasing turbulence intensity, as predicted by Eq. (23). As it does not characterize the wind turbine only, but also the measurement condition, raises the question of its reproducibility and stability.

3.3 A New Alternative: The Langevin Power Curve

An alternative to the standard IEC power curve is proposed in this Sect. 3.3. As the IEC norm defines the measurement procedure with relevance (see Sect. 3.2.1), the

[14] $\sigma^2 = \overline{(u - \overline{u})^2} = \overline{u'(t)^2}$.

[15] Assuming a cubic power curve $P(u) \propto u^3$, $P(u)$ has non-zero derivatives up to 3rd-order. Moreover, the transition point to rated power may have non-zero derivatives of arbitrary order, see Fig. 12.

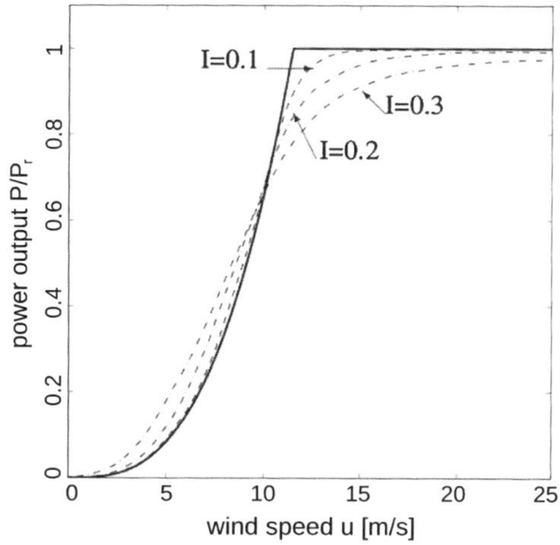

Fig. 12 IEC power curves for various turbulence intensities $I = 0.1, 0.2, 0.3$ (*dashed lines*). The *full line* represents the theoretical power curve. This result was obtained from numerical model simulations from [4]

same conditions will be considered for the Langevin analysis. The difference lies in the different approach to process the measured data.

One additional point on the sampling frequency is however important for the Langevin analysis. Because the method resolves the dynamics of a wind turbine in the order of seconds, a minimum sampling frequency in the order of 1 Hz is necessary for the measurements of wind speed and power output.

3.3.1 A Dynamical Concept

The power characteristic of a wind turbine can be derived from high-frequency measurements without using temporal averaging. One can regard the power conversion as a relaxation process which is driven by the turbulent wind fluctuations [15, 18]. More precisely, the wind turbine is seen as a dynamical system which permanently tries to adapt its power output to the fluctuating wind. For the (hypothetical) case of a laminar inflow at constant speed u, the power output would relax to a fixed value $P_L(u)$,[16] as illustrated in Fig. 13. Mathematically, these *attractive* power values $P_L(u)$ are called *stable fixed points* of the power conversion process.

[16] The subscript L stands for "Langevin" as $P_L(u)$ will be associated to the formalism of the Langevin equation.

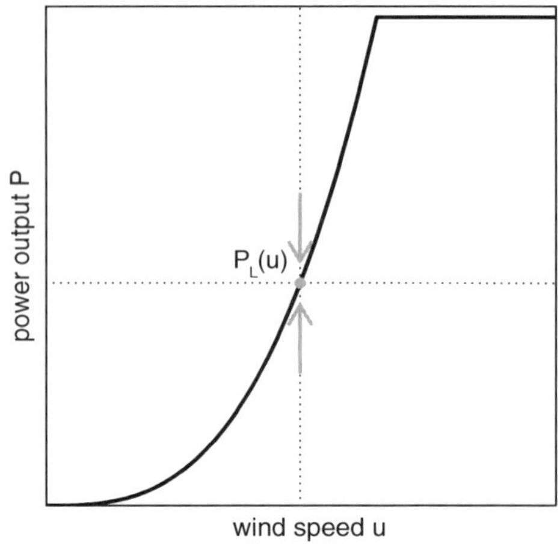

Fig. 13 Illustration of the concept of stable fixed point $P_L(u)$. For constant wind speed, the power output would relax to a stable value $P_L(u)$. This sketch is inspired from [1]

3.3.2 The Langevin Equation

The Langevin power curve[17] is derived from high-frequency measurements of wind speed $u(t)$ and power output $P(t)$. All necessary corrections and normalizations from the IEC norm [10] should be applied on the two time series.

The wind speed measurements are divided into bins u_i of 0.5 m/s width, as done in [10]. This accounts, to some degree, for the non-stationary nature of the wind, yielding quasi-stationary segments $P_i(t)$ for those times t with $u(t) \in u_i$. The following mathematical analysis will be performed on these segments $P_i(t)$. From now on, the subscript $_i$ will be omitted and the term $P(t)$ will refer to the quasi-stationary segments $P_i(t)$. The power conversion process is then modeled by a first-order stochastic differential equation called the Langevin equation[18]

$$\frac{d}{dt}P(t) = D^{(1)}(P) + \sqrt{D^{(2)}(P)} \times \Gamma(t). \tag{24}$$

In this model, the time evolution of the power output is controlled by two terms.[19]

$D^{(1)}(P)$ represents the deterministic relaxation of the wind turbine, leading the power output towards the attractive fixed point $P_L(u)$ of the system. For such, $D^{(1)}(P)$ is commonly called the *drift function*.

---

[17] In former publications on the topic, the Langevin power curve was called *dynamical power curve* or *Markovian power curve*. It is nonetheless the same approach.

[18] This equation is the reason for the name of the Langevin power curve.

[19] $D^{(1)}$ and $D^{(2)}$ are the first two Kramers-Moyal coefficients.

The second term $\sqrt{D^{(2)}(P)} \times \Gamma(t)$ represents the stochastic (random) part of the time evolution, and serves as a simplified model for the turbulent wind fluctuations that drive the system out of equilibrium. The function $\Gamma(t)$ is a Gaussian-distributed, delta-correlated noise with variance 2 and mean value 0. $D^{(2)}(P)$ is commonly called the *diffusion function*. A mathematical approach to the Langevin equation can be found in [17].

3.3.3 The Drift Function and the Langevin Power Curve

The deterministic drift function $D^{(1)}(P)$ is of interest as it quantifies the relaxation of the power output towards the stable fixed points of the system. When the system is in a stable state, no deterministic drift occurs,[20] and $D^{(1)}(P) = 0$. Following Eq. (25), $D^{(1)}(P)$ can be understood as the average time derivative of the power signal $P(t)$ in each region of wind speed u_i and power output P.

The drift and diffusion functions can be derived directly from measurement data as conditional moments [17]

$$D^{(n)}(P) = \lim_{\tau \to 0} \frac{1}{n!\tau} \langle (P(t+\tau) - P(t))^n | P(t) = P \rangle, \tag{25}$$

where $n = 1, 2$ respectively for the drift and diffusion functions. The averaging $\langle \cdot \rangle$ is performed over t, as the condition means that the calculation is only considered for those times during which $P(t) = P$.

This means that the averaging is done separately for each wind speed bin u_i and also for each level of the power P. One could speak of a *state-based* averaging on u and P, in contrast to the temporal averaging performed in the IEC norm. A typical drift function is displayed in Fig. 14.

The dynamics of the power signal can be directly related to the local sign and value of $D^{(1)}$. A positive drift indicates that the power tends to increase (arrows pointing up in Fig. 14), in regions where the wind turbine does not produce enough power for the given wind speed. On the contrary, a negative drift corresponds to a decreasing power (arrows pointing down), in regions where the wind turbine produces too much power for the given wind speed.

[20] To separate stable (attractive) from unstable (repulsive) fixed points, also the slope of $D^{(1)}(P)$ must be considered.

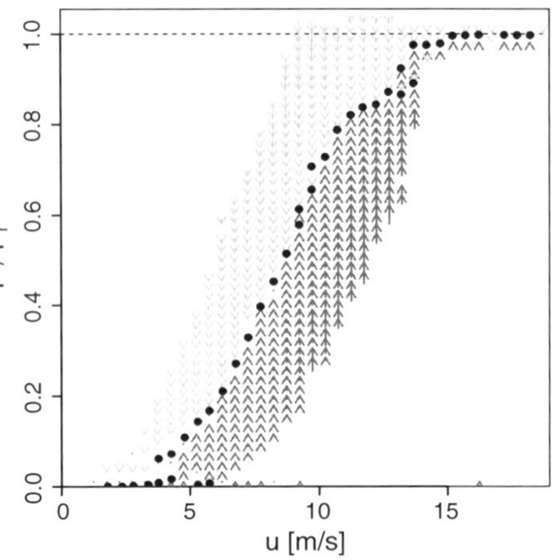

Fig. 14 Drift function $D^{(1)}(P)$. *Each arrow* represents the local value of $D^{(1)}(P)$ in magnitude (*length of the arrow*) and direction (*pointing up* for positive values). The stable fixed points where $D^{(1)}(P) = 0$ are given by the *black dots*

At the intersection are the points where $D^{(1)} = 0$, indicating that when at this value, the power output is in a stable configuration (the average time derivative is zero). The collection of all the points where the drift function is zero is defined as the Langevin power curve, and will be further labelled $P_L(u)$.

The stable fixed points $P_L(u)$ of the power conversion process can be extracted from the measurement data as solutions of

$$D^{(1)}(P_L(u)) = 0. \tag{26}$$

An illustration is given in Fig. 15.

Following the mathematical framework of Eqs. (24) and (25), an estimation of uncertainty for $P_L(u)$ can be performed [7]. One can see that for most wind speeds the power curve has very little uncertainty. Nevertheless, larger uncertainties occur in the region of transition to rated power. There the power conversion is close to stability over a wider range of power values, as a consequence of the changing control strategy from partial load to full load operation (see Fig. 5). It is a region of great interest as the controller of the wind turbine is highly solicited for the transition to rated power.

3.3.4 Advantages of the Langevin Approach

The Langevin equation (24) is a simplifying model for the power conversion process. The question of its validity for wind turbine power signals was positively answered in recent developments [11], as the power signal of a wind turbine could be successfully modelled. Predicting power signals from the Langevin equation is

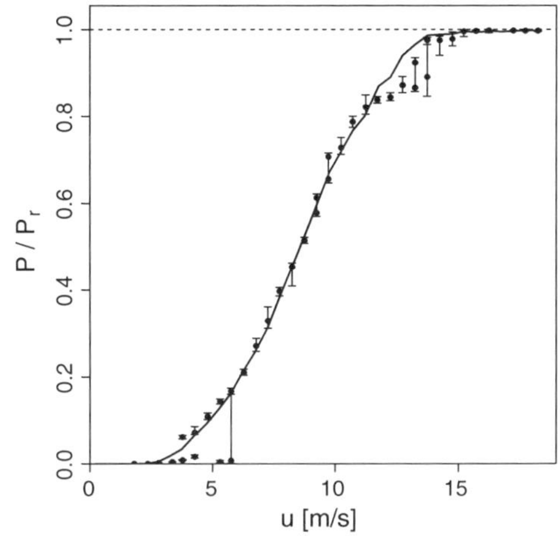

Fig. 15 Langevin power curve (*black dots with corresponding error bars*) and IEC power curve (*solid line*)

introduced in Appendix 2. Also, the drift function $D^{(1)}$ is well-defined for a large class of stochastic processes, and is not limited only to the class of the Langevin processes.

Also, the definition of the drift function does not suffer the systematic errors caused by temporal averaging. For such, the Langevin power curve characterizes the wind turbine dynamics only, regardless of the wind condition during the measurement.[21] The results are therefore machine-dependent only, and not site- or measurement-dependent, as the intensity of turbulence has no influence on the Langevin power curve.

Additionally, this approach can show complex characteristics of the investigated system, such as regions where the system is close to stability, as mentioned above, or multiple stable states, see also [1, 9]. For these various reasons, the Langevin power curve represents a promising tool for power performance monitoring, as will be introduced in Sect. 4.

3.4 Power Curve Stability Under Different Wind Measurement Technologies

Wind measurements are ordinarily performed using a cup anemometer for the speed and a wind vane for the direction. Yet new techniques appear in wind energy industry as alternatives, such as LIDAR (LIght Detection And Ranging) anemometry.[22] It represents a promising technology for wind measurements as it makes remote measurements possible. This becomes particularly interesting for

[21] Assuming that the measurement period is sufficiently long to reach statistical convergence.

[22] Another alternative is given by ultrasonic anemometers which can estimate at once the wind speed and direction.

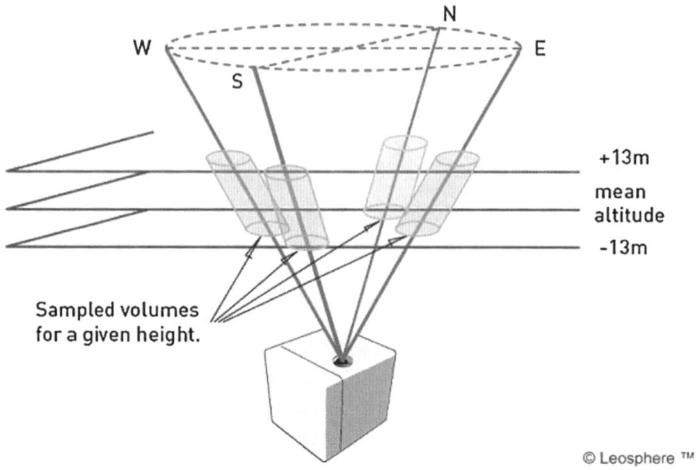

Fig. 16 Operation principle of the Leosphere WindCube. Sketch © Leosphere, Inc

larger heights, where the efforts for measurement towers reach critical levels. This holds especially for power curve measurements, considering the growing heights of recent wind turbines.

LIDAR measurements presented here are performed using a Leosphere WindCube system. It operates as a pulsed laser Doppler anemometer, see Fig. 16. An infrared laser beam is inclined by approximately 30° against the vertical direction and takes beamwise Doppler measurements of the wind velocity. These measurements are performed in the four principal directions, and a three-dimensional wind vector is then derived from the four most recent measurements. The device achieves a sampling rate of 0.67 Hz. Because of the pulsed laser operation, measurements can be obtained in up to ten height levels simultaneously, within a range between 40 and about 200 m.

Power curve measurements were performed using both LIDAR, cup and ultrasonic anemometers. The measurement was deployed in a distance of 2.5 rotor diameters away from a prototype multi-MW class offshore wind turbine. The temporal resolution of the power data, the wind speed recorded from the cup and ultrasonic anemometers was 1 Hz, while the LIDAR achieved 0.67 Hz. Further details on the measurements can be found in [21]. From these measurement data the IEC and Langevin power curves of the wind turbine could successfully be derived. Figure 17 shows the Langevin power curve from LIDAR wind measurement.

The IEC and Langevin power curves are in excellent agreement when derived from LIDAR, cup and ultrasonic anemometer measurements [21].

For the near future, a substantial increase in the use of wind LIDARs can be expected. These measurements therefore will also open the possibility of precise

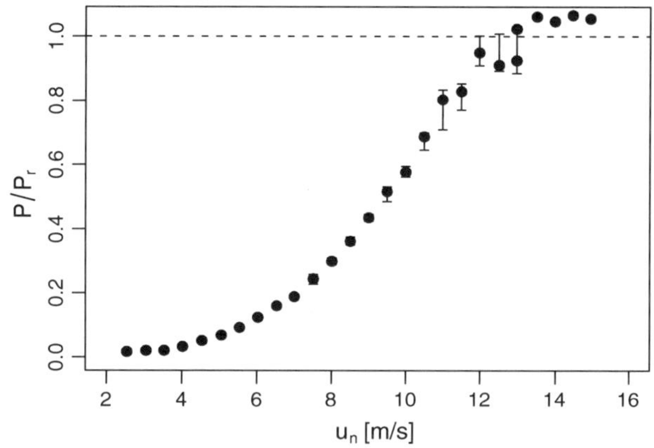

Fig. 17 Langevin power curve of a multi-MW class offshore prototype wind turbine, derived from LIDAR wind measurement [21]

and uncomplicated derivation of power curves, thanks to the portability of LIDAR devices.

4 Applications: Power Performance Monitoring

Two methods were introduced to estimate power curves in Sect. 3. Because the two methods differ by the time scales they investigate,[23] the IEC method is more suited for long-term analysis, while the Langevin method focuses more on short-time dynamics, bringing deeper insight on the inner mechanical behavior. This allows for a complementary assessment of the overall performance.

The two methods also differ by their dimension, as explained in Sect. 4.1. This aspect works constructively with their respective time scales, making the IEC power curve more suitable for an estimation of annual energy production (AEP), and making the Langevin power curve more suitable to detect dynamical anomalies. These two applications are presented respectively in Sects. 4.2 and 4.3.

4.1 One-Dimensional Versus Two-Dimensional Power Curves

The two different ways to discretize data brings the most difference between the two different approaches.

The IEC method discretizes the two-dimensional domain $\{u, P\}$ into wind speed bins of size $\delta u = 0.5$ m/s. As the domain is discretized only for the wind

[23] For reminder, the IEC method focuses on time scales of 10 min, while the Langevin approach investigates the dynamics in the order of few seconds.

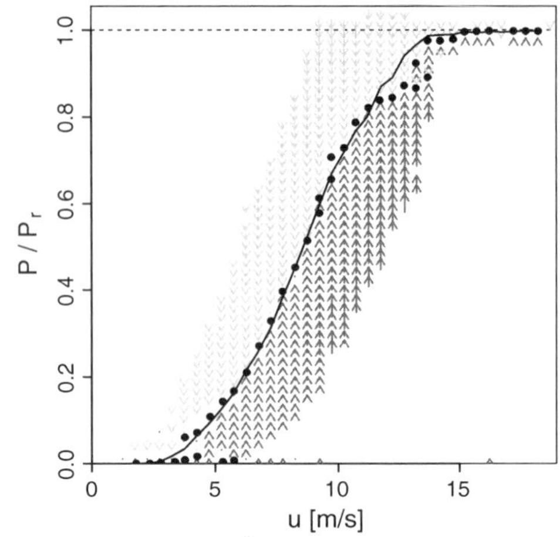

Fig. 18 Drift function $D_i^{(1)}(P)$ (*arrows*), Langevin power curve $P_L(u)$ (*dots*) and IEC power curve $P_{IEC}(u)$ (*line*)

speed, the IEC power curve depends on its unique variable u, resulting in a unique point every 0.5 m/s. The IEC power curve is hence one-dimensional, it is the line $P_{IEC}(u)$.

However, the Langevin approach discretizes the domain $\{u, P\}$ on both wind speed and power output. The drift function $D_i^{(1)}(P)$ depends on the two variables u_i and P, making $D^{(1)}$ a two-dimensional estimate. This justifies why several stable fixed points are possible for a given wind speed. An example is presented in Fig. 18.

The two-dimensional framework of the Langevin analysis allows to observe local[24] dynamics. An example can be seen in Fig. 18 for $u = 13$ m/s, as the wind turbine tends to $P = 0.9 \cdot P_r$ when in partial load operation and to $P = P_r$ when in full load operation. Limited by its unique dimension, the IEC power curve can only display the average value in between. Multi-stable behaviors, created for example by a multiple-gear gearbox, switching generator stages or changing control strategies cannot be resolved by the IEC power curve.

4.2 Annual Energy Production

The one-dimensional *limitation* of the IEC power curve becomes an advantage for long-term energy production, as $P_{IEC}(u)$ relates unambiguously a unique value of power for each wind speed. As the AEP estimates the energy produced over a year, it can be seen as a prediction estimate. A prediction of power production at high-frequency is also possible using the Langevin approach, as introduced in Appendix 2. This is a more complicated approach that is outside the scope of this section.

[24] Local in wind speed and power output.

The estimation of the AEP extrapolates the power production of a wind turbine characterized by its power curve in a given location. This Sect. 4.2 does not give an exact transcription of the AEP procedure from the IEC norm [10], but rather a comprehensive introduction on how power production can be estimated simply from a wind speed measurement. For such, the *AEP procedure* introduced here is not the official AEP procedure following IEC, but a similar version. In both cases, the availability of the wind turbine is assumed to be 100 %.

4.2.1 Estimating the Wind Resource

Any location scheduled to host a wind turbine can be categorized in advance by a characterization of its wind resource. A local measurement of wind speed from a met mast at hub height[25] of the hypothetical wind turbine must be performed, typically over 1 year.[26] From this wind speed measurement $u(t)$, a 10-min (or hourly) averaging is applied on $u(t)$. The probability density function (PDF) $f(u_i)$ of the 10-min average values u_i is established. For clarity, the values u_i will be labelled u. $f(u)$ returns the probability of occurrence of the wind speed u. For long enough measurements, $f(u)$ is known to fit a Weibull distribution [16]

$$f(u;\lambda,k) = \frac{k}{\lambda}\left(\frac{u}{\lambda}\right)^{k-1} e^{(-u/\lambda)^k}, \tag{27}$$

where k and λ[27] are called respectively the shape and scale factors.[28] Visual examples of such wind speed distributions are given in [6].

4.2.2 Estimating the AEP

A given wind site is characterized for the AEP by its wind speed PDF $f(u)$, while a given wind turbine is characterized by its IEC power curve. As P_{IEC} relates unambiguously a given wind speed u to the corresponding average power output $P_{IEC}(u)$, the power curve serves as a transfer function from wind speed to average power output. An estimation of the average power output $\overline{P}$ can be obtained following

---

[25] Typical hub heights of commercial multi-MW class wind turbines are in the order of 100 m, justifying the interest for a portable LIDAR sensor, see Sect. 3.4.

[26] A measurement of wind speed over 1 year covers the various wind situations resulting from various seasonal behaviors.

[27] One should note here that λ is not the tip speed ratio of a wind turbine, but a parameter of the Weibull distribution.

[28] The IEC norm [10] refers to the Rayleigh distribution, which is a special case of the Weibull distribution for $k = 2$.

$$\overline{P} = \int_0^\infty f(u) \times P_{IEC}(u) du, \tag{28}$$

and an estimation for the energy production over a period T reads

$$T \times \overline{P} = T \int_0^\infty f(u) \times P_{IEC}(u) du. \tag{29}$$

Over 1 year, $T = 8{,}760\,\mathrm{h}$ and

$$AEP = \overline{P} \times 8,760, \tag{30}$$

where $\overline{P}$ is given in Watt and AEP is given in Watt hour.

Thanks to its simple mathematical procedure, the AEP is commonly used to make rough predictions of energy production, as well as for financial estimations. It can predict how much energy a wind turbine will generate on a given site before installing it. This allows for an optimal choice of design for the optimal location. This result however remains a rough estimation, as it neglects e.g. wake losses generated by other surrounding wind turbines.

4.3 Detecting Dynamical Anomalies

The intention of dynamical monitoring is to detect dynamical anomalies that appear on operating wind turbines. Monitoring refers to the time evolution of the power performance here. A good monitoring procedure should be reliable,[29] as fast as available, and possibly also inform on the source of the anomaly. While monitoring procedures come to be ever more complex, the approach presented here is based only on a power curve estimation. This approach is not intended to give a full-featured method, but rather an illustration of the amount of information given by power curves. More advanced studies on the topic of power curves for monitoring are being developed, but remain outside the scope of this introduction as they represent active research topics.

The monitoring procedure simply consists in computing $P_L(u)$ at an initial time that will serve as a reference.[30] Potential changes in time of $P_L(u)$ are considered

[29] An over-sensitive procedure might indicate non-existing anomalies, while an under-sensitive procedure would fail to detect a major malfunction.

[30] The reference time is chosen when the wind turbine is believed to work with full capacity.

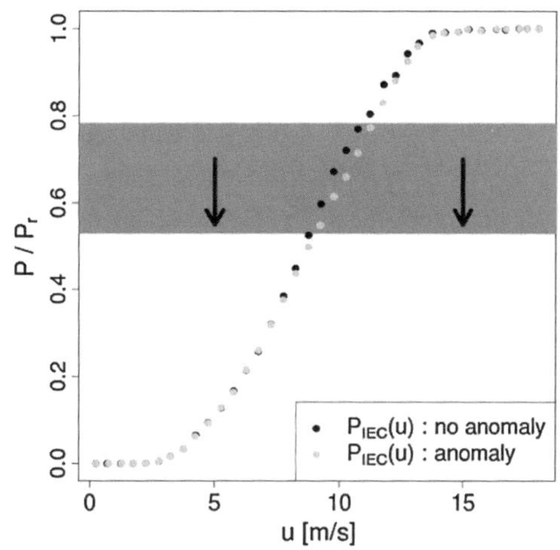

Fig. 19 Comparison of $P_{IEC}(u)$ before the anomaly (*black in background*) and after the anomaly (*grey in front*). The artificial anomaly was applied in the *grey rectangle*

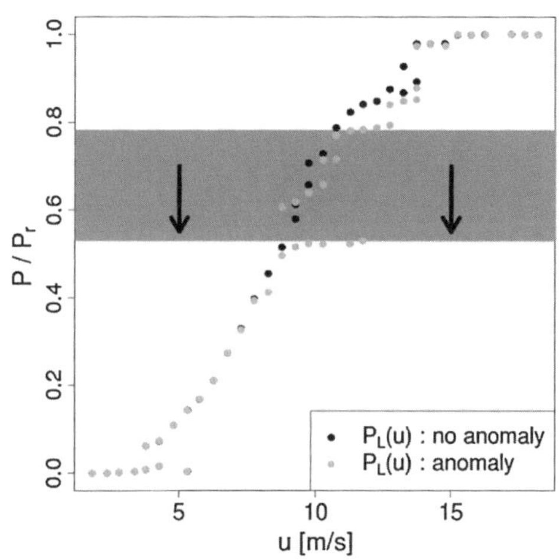

Fig. 20 Comparison of $P_L(u)$ before the anomaly (*black in background*) and after the anomaly (*grey in front*). The artificial anomaly was applied in the *grey rectangle*

anomalies, or malfunctions inside the wind turbine that spoil the conversion dynamics. While this strategy is very simple, the challenge lies in defining the right threshold for a change in $P_L(u)$ to be considered an anomaly. This threshold, along with other parameters such as the necessary measurement time or time reactivity of the method depend on the wind turbine design and location.

To illustrate the ability of the method, the monitoring procedure was applied on a numerical simulation. The simulation was applied on measurement data, where an anomaly was introduced. This artificial anomaly limits the power production to $P \approx 0.55 \cdot P_r$ for intermediate wind speeds, as represented by the grey rectangles in Figs. 19 and 20. More clearly, when in this rectangle, the power signal was sometimes forced to reduce towards $0.55P_r$. From this artificial data, $P_L(u)$ and $P_{IEC}(u)$ were then computed and compared to the original data. This is illustrated in Figs. 19 and 20.

Similar anomalies were observed on several real wind turbines. (Justifying the reason for this artificial anomaly). For information, the total energy production was reduced to 96.6 % compared to the original energy production due to the presence of the anomaly. Figure 20 illustrates the higher reactivity of $P_L(u)$. While in Fig. 19 $P_{IEC}(u)$ only shows a minor deviation in the region of the anomaly, $P_L(u)$ clearly *detaches* from the typical cubic curve to adjust to the new dynamics. $P_L(u)$ can detect changes in the dynamics of the conversion process, unlike $P_{IEC}(u)$ that is better suited for the AEP.

The Langevin power curve is more reactive to changes in the dynamics. As the IEC power curve averages over 10 min intervals, the information about high-frequency dynamics is lost. Also, the second averaging in wind speed prevents from seeing multi-stable behaviors.
In addition, the Langevin power curve does not depend on the turbulence intensity, unlike the IEC power curve, see Sect. 3.2.3. A deviation in the Langevin power curve indicates a change in the conversion dynamics, regardless of the wind situation. This makes the Langevin power curve a promising tool for dynamical monitoring.

5 Conclusion

This chapter was first dedicated to power performance for wind turbines. A general overview was introduced in Sect. 2, mainly the application of momentum theory to wind turbines, as developed by Betz. This simplified analysis sets an upper limit of $\approx 60\,\%$ for the power available in the wind, regardless of the design of the wind power system. Additional losses due to more realistic considerations bring an even lower availability for the power extraction, up to $\approx 50\,\%$ for modern commercial designs. Estimates of power performance like the power coefficient or the theoretical power curve were defined for the case of a laminar wind flow.

Facing complex turbulent and atmospheric effects, wind turbines cannot be described satisfactorily using a laminar theory. Statistical tools were introduced in Sect. 3 to integrate these complex effects into the analysis. The international procedure to estimate a reliable power curve was introduced by the IEC norm 61400-12-1. It sets good guidelines on how to perform measurements on operating wind turbines, and provides a uniform standard. It handles the measured signals of wind speed and power output through an averaging procedure both in time and on the wind speeds. It results an uncomplicated IEC power curve that accounts somewhat for the turbulent effects. Unfortunately, the result depends on the measurement condition, such that the IEC power curve does not characterize the wind turbine performance only, but also the wind condition. This procedure allows however for a gross estimation of long-time energy estimation like the AEP, as introduced in Sect. 4.2.

An alternative is proposed that is based on stochastic analysis. Instead of averaging over time, the power signal is approximated to solve a Langevin equation. A drift function is introduced, that quantifies the reaction dynamics of a wind turbine to turbulent wind fluctuations. It results a so-called Langevin power curve, which represents the stable fixed points of the conversion dynamics. The Langevin power curve and the drift function give a simplified model of how the wind turbine actively adapts the power production to the changing wind speed. Unlike the IEC power curve, the Langevin power curve does not depend on the measurement condition, and characterizes the wind turbine only. Also its two-dimensional structure allows for a more flexible result and can resolve multi-stable dynamics. This makes the Langevin approach a promising application for performance monitoring, as discussed in Sect. 4.3. Modeling of power signals is also made possible through the Langevin equation, as briefly introduced in Appendix 2.

As the wind energy industry grows rapidly, wind turbine designs might (or might not) change radically in the next decades. The freedom of change in the overall design is however limited by the physical aspects of the wind itself. The three-blade design emerged as it extracts the most power over the range of wind speeds that occur most often at hub height. Improvements in material engineering allow for larger wind turbines every year. With taller wind turbines, the impact of turbulence induced by the surface roughness of the ground is reduced. This also justifies the recent effort to build offshore wind turbines, that can benefit from smoother winds. Yet the wind remains in essence turbulent, and the wind resource is consequently intermittent. This chapter aimed to identify the response of a single wind turbine to such driving condition. On a larger scale, current challenges involve a smooth integration of a rapidly increasing amount of intermittent wind power into electrical networks. While a single wind turbine is insignificant on the scale of an entire network, a better understanding of single wind turbines, wind parks and global wind installations remains paramount for a future integration of wind energy at a global level.

Acknowledgments Parts of this work have been financially supported by the German Ministry for Environment (BMU) under grant number 0327642A. The authors would like to thank Stephan Barth, Julia Gottschall, Edgar Anahua and Michael Hölling for their pioneering work on the Langevin approach, as well as for stimulating discussions.

Appendix 1: Aerodynamics of Rotor Blades

The essential (mechanical) element of a wind turbine is the rotor, that transforms the power of the wind into a rotational or mechanical power. The ideal requirements are:

- the rotation should be steady and smooth;
- dynamical loads should be minimal;
- the regulation should be done without sudden jumps.

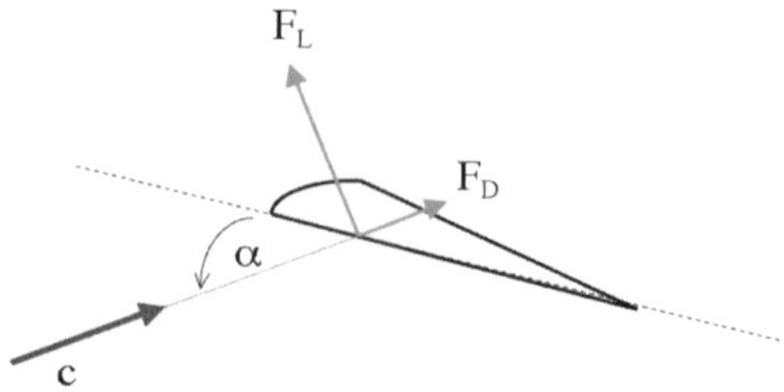

Fig. 21 Cut through an airfoil to illustrate the forces acting on it. The depth t is given by the distance between the leading and the trailing edge. The wingspan b is the length of the airfoil, here perpendicular to the illustrated plane

The number of blades, their profile and design should guarantee these features. Modern wind turbines rotate due to the lift forces F_L acting on the airfoils. For an airfoil the effective area[31] can be expressed in terms of the depth t and the wingspan b (normally equal to the rotor radius R), such that

$$\begin{aligned} F_D &= C_D(\alpha)\frac{1}{2}\rho c^2(t \cdot b) \\ F_L &= C_L(\alpha)\frac{1}{2}\rho c^2(t \cdot b), \end{aligned} \tag{31}$$

where α is the angle of attack, as displayed in Fig. 21. The lift-drag ratio F_L/F_D relates to the quality of the airfoil. The larger the ratio, the better the quality.

In Fig. 21, the velocity vector $\mathbf{c}$ gives the wind velocity in the frame of reference of the airfoil. The wind velocity is $\mathbf{u}_2$ in the frame of the ground, but the rotational motion of the rotor must be considered for the motion of the wind with respect to the blades. Hence, $\mathbf{c}$ is the superposition of the horizontal axial velocity $\mathbf{u}_2$[32] and of the rotational velocity $v = \omega r$, such that

$$c^2(r) = (2u_1/3)^2 + (\omega r)^2. \tag{32}$$

The rotor *feels* the effective wind speed $\mathbf{c}$. This is illustrated in Fig. 22.

Instead of integrating the lift and drag forces on the entire airfoil, one can estimate the local force on each infinitesimal element. Also, the total force is divided into its rotational component F_r and its axial component F_a. Considering a cut dr at r in the polar plane of the rotor, the resulting force is

$$\begin{aligned} F_r &= \frac{\rho}{2}c^2 \cdot t \cdot dr[C_L \cos(\beta) - C_D \sin(\beta)] \\ F_a &= \frac{\rho}{2}c^2 \cdot t \cdot dr[C_L \sin(\beta) + C_D \cos(\beta)]. \end{aligned} \tag{33}$$

---

[31] This effective area is the one that enters the formula to calculate drag and lift forces.

[32] For the Betz optimum, $u_2 = 2u_{1/3}$.

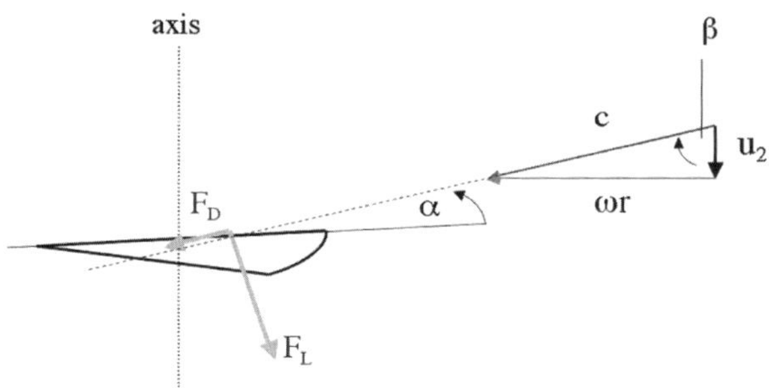

Fig. 22 Cut through a rotating airfoil. The rotational velocity ωr is perpendicular to the axial velocity vector $\mathbf{u}_2$. β denotes the angle between the resulting velocity **c** and the rotational direction

Also, on can estimate

$$\tan(\beta) = \frac{\omega r}{u_2} = \frac{\omega R}{u_1}\frac{r}{R}\frac{u_1}{u_2} = \frac{3}{2}\lambda\frac{r}{R}. \tag{34}$$

The idea is to construct the blades in such a way that for each infinitesimal radial annulus they extract the infinitesimal (Betz) optimal power out of the wind

$$dP_{r,Betz} = \frac{16}{27}\cdot\frac{\rho}{2}\cdot u_1^3\cdot(2\pi r dr). \tag{35}$$

This power also reads

$$dP = z \times F_r \times \omega r, \tag{36}$$

where z denotes the number of blades, ωr the velocity in rotational direction and F_r the force in this direction.[33] Inserting Eq. (33) and combining Eqs. (35) and (36), the optimal value of the depth t as a function of r can be determined. Assuming that $C_D \ll C_L$ and a sufficiently large tip speed ratio (for details see [20]), the profile of the airfoil $t(r)$ reads

$$\begin{aligned} t(r) &\approx \frac{16\pi}{9}\frac{R^2}{zC_L r\lambda^2} \\ &\propto z^{-1}\cdot C_L^{-1}\cdot r^{-1}\cdot\lambda^{-2}. \end{aligned} \tag{37}$$

This has an important consequence on the design of rotor blades. The depth decreases with increasing number of blades, larger lift coefficient, increasing radius and especially increasing tip speed ratios. This explains why fast rotating wind turbines tend to have only two or three narrow blades while old western-mill machines have many, rather broad blades.

[33] The force in the axial direction does not contribute to the power production of a wind turbine but to the thrust on it.

Appendix 2: A Relaxation Model for the Power Output

As introduced in Sect. 3.3.2, the power output of wind turbines is assumed to be solution of a Langevin equation

$$\frac{d}{dt}P(t) = D^{(1)}(P) + \sqrt{D^{(2)}(P)} \times \Gamma(t). \tag{38}$$

For reminder, the power value P and the functions $D^{(1)}$ and $D^{(2)}$ are conditioned on the wind speed bins, as done in the main text. The subscript $_i$ indicating the wind speed bin was dropped for simplicity. $D^{(1)}$ represents the deterministic dynamics of the conversion process, that always push the power output towards the Langevin power curve $P_L(u)$. Additional random fluctuations are superposed as a simplified model for all the microscopic degrees of freedom acting on the conversion process.[34] A simple but rather realistic ansatz for $D^{(1)}$ would be

$$D^{(1)}(P) = \alpha(P_{theo}(u(t)) - P(t)), \tag{39}$$

where $D^{(1)}$ linearly drives the power output towards the instantaneous value of the theoretical power curve $P_{theo}(u(t))$, which might read

$$P_{theo}(u) = \begin{pmatrix} P_r\left(\frac{u}{u_r}\right)^3 & amp; \text{for u} \leq \text{u}_\text{r}, \\ P_r & amp; \text{for u} \geq \text{u}_\text{r}. \end{pmatrix} \tag{40}$$

Assuming Eq. (39) and a constant diffusion function $D^{(2)}(P) = \beta$,[35] the Langevin equation becomes a relaxation model for the power output

$$\frac{d}{dt}P(t) = \alpha(P_{theo}(u(t)) - P(t)) + \sqrt{\beta} \times \Gamma(t). \tag{41}$$

Equation (41) is a phenomenological model for the power signal. This special case of the Langevin process is mathematically called an Ornstein–Uhlenbeck process [17].

Equation (41) is a simplified model for the power output, where the wind turbine design is described through the parameters α and β, as well as the power curve $P_{theo}(u)$. The parameter α is related to the reaction time of the model wind turbine,[36] while β quantifies the strength of the stochastic noise. $\Gamma(t)$ is a

---

[34] The Langevin equation relates directly the incoming wind speed and the power output. Many other variables are involved in intermediate steps of the conversion, which should be modeled by a set of mutli-dimensional deterministic differential equations. All these degrees of freedom are modeled by the one-dimensional stochastic Langevin equation instead.

[35] A constant diffusion function yields additive noise. More complex systems such as turbulence-driven systems display multiplicative noise and a non-constant diffusion function.

[36] A realistic model integrates a finite reaction time due to the inertia of the wind turbine to changing wind speeds.

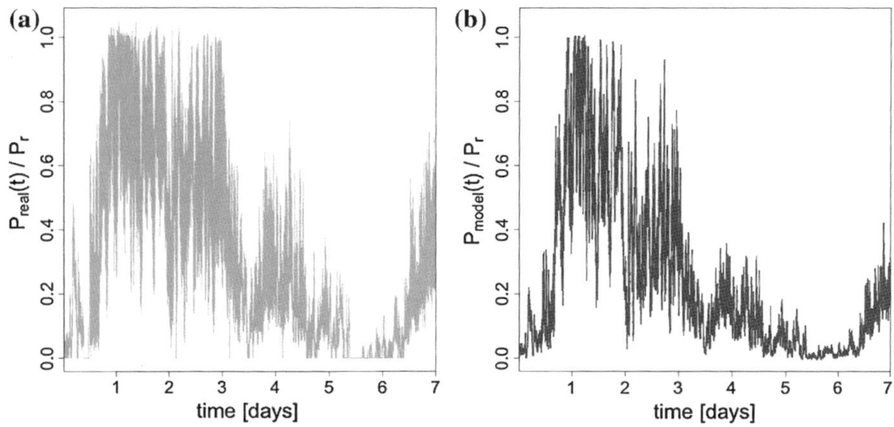

Fig. 23 **a** Power signal measured (*red line*). **b** Power signal modeled (*blue line*) following Eq. (41) with $\alpha = 0.005\,/\text{s}$, $\beta = 0.5\,\text{W/s}^2$, $u_r = 13\,\text{m/s}$ and $P_r = 1\,\text{W}$. The wind speed signal used was measured simultaneously as the measured power output

Gaussian-distributed white noise with mean value 0 and variance 2, which can be generated easily from most mathematical softwares. Using a wind speed time series $u(t)$ as an input for the model equation,[37] a time series of power output $P(t)$ can be generated at the same sampling frequency. An example is provided in Fig. 23.

From Fig. 23, it can be seen that the relaxation model manages to estimate the power output of a wind turbine to a first approximation. Fluctuations and their statistics are more difficult to reproduce than long-time behavior, which is mostly driven by the changes in wind speed rather than by the stochastic fluctuations. More advanced methods are being developed, as introduced in [11], where $D^{(1)}$ and $D^{(2)}$ are not assumed but estimated from measurement data.

References

1. Anahua E, Barth S, Peinke J (2008) Markovian power curves for wind turbines. Wind Energy 11(3):219–232
2. Betz A (1927) Die windmühlen im lichte neuerer forschung. Die Naturwissenschaften 15:46
3. Bianchi FD, De Battista H, Mantz RJ (2006) Wind turbine control systems, 2nd edn. Springer, Berlin
4. Böttcher F, Peinke J, Kleinhans D, Friedrich R (2007) Handling systems driven by different noise sources—implications for power estimations. In: Wind energy. Springer, Berlin, pp 179–182

[37] An initial condition $P(t = 0)$ for the power output is also necessary. However, the result depends only poorly on this value, as the dynamics will adjust rapidly to the given wind speed.

5. Böttcher F, Barth S, Peinke J (2007) Small and large fluctuations in atmospheric wind speeds. Stoch Environ Res Ris Assess 21:299–308
6. Burton T, Sharpe D, Jenkins N, Bossanyi E (2001) Wind energy handbook. Wiley, New York
7. Gottschall J (2009) Modelling the variability of complex systems by means of Langevin processes. PhD thesis, Carl von Ossietzky Universität Oldenburg, Germany
8. Gottschall J, Peinke J (2007) Stochastic modelling of a wind turbine's power output with special respect to turbulent dynamics. J Phys: Conf Ser 75:012045
9. Gottschall J, Peinke J (2008) How to improve the estimation of power curves for wind turbines. Environ Res Lett 3(1):015005 (7 pp)
10. IEC (2005) Wind turbine generator systems, part 12: wind turbine power performance testing. International Standard 61400-12-1, International Electrotechnical Commission
11. Milan P, Mücke T, Wächter M, Peinke J (2010) Two numerical modeling approaches for wind energy converters. In: Proceedings of computational wind engineering 2010
12. Rankine WJM (1865) On the mechanical principles of the action of propellers. Trans Inst Naval Architects 6:13–39
13. Rauh A, Seelert W (1984) The Betz optimum efficiency for windmills. Appl Energy 17:15–23
14. Rauh A (2008) On the relevance of basic hydrodynamics to wind energy technology. Nonlinear Phenom Complex Syst 11(2):158–163
15. Rauh A, Peinke J (2004) A phenomenological model for the dynamic response of wind turbines to turbulent wind. J Wind Eng Ind Aerodyn 92(2):159–183
16. Richardson LF (1922) Weather prediction by numerical process. Cambridge University Press, Cambridge
17. Risken H (1984) The Fokker-Planck equation. Springer, Berlin
18. Rosen A, Sheinman Y (1994) The average power output of a wind turbine in turbulent wind. J Wind Eng Ind Aerodyn 51:287
19. Schneemann J, Knebel P, Milan P, Peinke J (2010) Lift measurements in unsteady flow conditions. *Proceedings of the European Wind Energy Conference 2010*
20. Twele J, Gasch R (2005) Windkraftanlagen. Teubner B.G. GmbH
21. Wächter M, Gottschall J, Rettenmeier A, Peinke J (2008) Dynamical power curve estimation using different anemometer types. In: Proceedings of DEWEK, Bremen, 26–27 Dec 2008

Part IV
Innovative Wind Energy Generation

Satellite Remote Sensing in Offshore Wind Energy

Charlotte Bay Hasager, Merete Badger, Poul Astrup and Ioanna Karagali

Abstract Satellite remote sensing of ocean surface winds are presented with focus on wind energy applications. The history on operational and research-based satellite ocean wind mapping is briefly described for passive microwave, scatterometer and synthetic aperture radar (SAR). Currently 6 GW installed capacity is found in the European Seas. The European Wind Energy Association, EWEA, expects the cumulative offshore capacity in Europe will reach 150 GW in year 2030. The offshore environment is far less well-known than over land and this increases the challenge of planning, operation and maintenance offshore. Satellite-based ocean surface wind data can fill a gap in our understanding of marine winds, their temporal and spatial variations. The statistics from satellite-based ocean surface wind maps include wind resources, long-term trend analysis and daily variations in winds. Some examples using data from passive microwave radiometer, scatterometer and SAR are presented from the North Sea and Baltic Sea. These seas are home to the majority of offshore wind farms today and many new offshore wind farm projects are in progress here.

Abbreviations

ALOS	Advanced Land Observing Satellite
AMSR	Advanced Microwave Scanning Radiometer
ASAR	Advanced Synthetic Aperture Radar

C. B. Hasager (✉) · M. Badger · P. Astrup · I. Karagali
DTU Wind Energy, Department of Wind Energy, Technical University of Denmark, Risø Campus, Frederiksborgvej 399, 4000 Roskilde, Denmark
e-mail: cbha@dtu.dk

M. Badger
e-mail: mebc@dtu.dk

P. Astrup
e-mail: poas@dtu.dk

I. Karagali
e-mail: ioka@dtu.dk

P. M. Pardalos et al. (eds.), *Handbook of Wind Power Systems*, Energy Systems,
DOI: 10.1007/978-3-642-41080-2_21,

ASCAT	Advanced scatterometer
ASI	Italian Space Agency
CFOSat	Chinese French Ocean Satellite
CLS	Collecte Localisation Satellites
CNES	Centre National d'Etudes des Spatiales
CNSA	China National Space Administration
COSMO-SkyMed	COnstellation of small Satellites for the Mediterranean basin Observation
CSA	Canadian Space Agency
DLR	German Aerospace Centre
DMSP	Defense Meteorological Satellite Program
DTOC	Design Tools for Offshore wind farm Clusters
EERA	European Energy Research Alliance
EOLI-SA	Earth Observation Link-Stand Alone
ERS	European Remote Sensing satellite
ESA	European Space Agency
EUMETSAT	European Organisation for the Exploitation of Meteorological Satellites
EWEA	European Wind Energy Association
FINO	Forschungsplattformen in Nord- und Ostsee
GCOM-W2	Global Change Observation Mission, W: Water cycle
GMF	Geophysical Model Function
HH	Horizontal receive, horizontal transmit
HJ	Huan Jing (Environmental Protection and Disaster Monitoring Constellation)
HY	Chinese Ocean Color Satellite
ISRO	Indian Space Research Organisation
JAXA	Japan Aerospace Exploration Agency
JERS	Japanese Earth Remote Sensing Satellite
JHU APL	Johns Hopkins University, Applied Physics Laboratory
JPL	Jet Propulsion Laboratory
LTAN	Local Time Ascending Node
MDA	MacDonald Dettwiler and Associates
NAO	North Atlantic Oscillation
NASA	National Aeronautics and Space Administration
NASDA	National Space Agency of Japan
NESDIS	National Environmental Satellite, Data and Information Service
NOGAPS	Navy Operational Global Atmospheric Prediction System
NOAA	National Oceanic and Atmospheric Administration
NORSEWInD	Northern Seas Wind Index Database
NSCAT	NASA scatterometer
NSIDC	National Snow and Ice Data Center
NSOAS	National Satellite Ocean Application Service (in China)

PALSAR	Phased Array L-band Synthetic Aperture Radar
PO.DAAC	Physical Oceanography Distributed Active Archive
RSS	Remote Sensing Systems
SAR	Synthetic Aperture Radar
ScatSat	Scatterometer Satellite
SMMR	Scanning Multichannel Microwave Radiometer
SSM/I	Special Sensor Microwave Imager
SSMIS	Special Sensor Microwave Imager Sounder
TanDEM	TerraSAR-X add-on for Digital Elevation Measurement
TerraSAR	Terra Synthetic Aperture Radar
TMI	TRMM Microwave Imager
TRMM	Tropical Rainfall Measuring Mission
TSX-NG	TerraSAR-X Next Generation
VV	Vertical receive, vertical transmit
WRF	Weather Research and Forecasting

1 Introduction

Satellite remote sensing has various application fields in offshore wind energy. Ocean winds have been and are being observed from many different satellites. The satellite centers archive the collected data. For some satellite sensors their operational aim is to provide ocean surface wind data. For other satellites it is possible to retrieve ocean surface winds but it is not provided as part of the planned mission activity. Thus the task is to access and process the satellite data for offshore wind energy, including wind resources and long-term trend analysis.

Offshore wind farming is widespread in European Seas. The first offshore wind farm, Vindeby in the Baltic Sea in Denmark, started operation in year 1992. Since many large offshore wind farms have been constructed in Denmark, United Kingdom, the Netherlands, Sweden, among several other countries. According to EWEA the installed offshore wind power capacity in Europe was 6 GW in mid year 2013, distributed in 58 wind farms in ten countries. There are nine projects under construction and nine projects in preparation, once completed the cumulative capacity is 9 GW. Furthermore, offshore wind farms fully consented is totaling more than 18 GW. More than 100 GW of offshore wind farms are currently being planned by project developers and utilities. EWEA expects the installed capacity in the EU-27 countries to reach 150 GW in year 2030.

The planning, operation and maintenance of offshore wind farms is more challenging than on land. First of all the accessibility are limited to boat or helicopter, and weather and waves can disrupt these transports. Another general challenge is that the marine physical environment is far less well-known than for land. This is partly due to its relative inaccessible, partly due to costly operation.

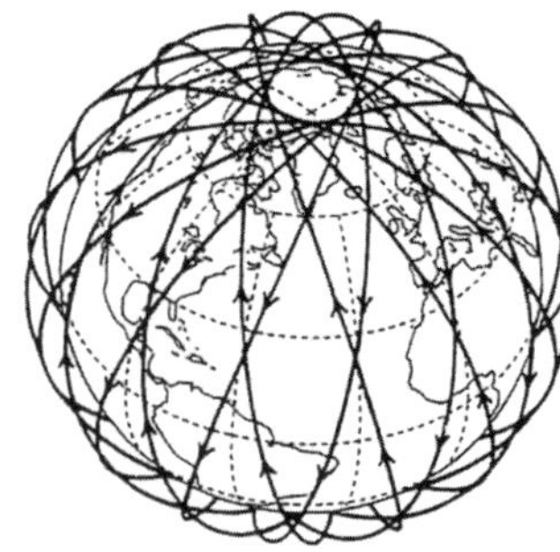

Fig. 1 Polar orbit around Earth

Ocean surface winds are mapped by satellite sensors and these data fill a gap in our knowledge on offshore wind energy potential and other statistics of relevance for offshore wind farming.

Operational satellite remote sensing for ocean surface wind observations has more than 25 year's continuous history. The Meteorological Program of the United States Department of Defense that is in charge of the Defense Meteorological Satellite Program, DMSP, launched the so-called F-8 satellite platform in year 1987. On-board this was the passive microwave radiometer, the Special Sensor Microwave Imager (SSM/I). The SSM/I instrument is dedicated to observations of ocean surface wind speed. Later more satellites in the series up to F-18 were launched and several were and are in operation in parallel, in different orbits.

Before going into further detail on the more than 30 different satellite sensors that have been or are in space today capable of observing ocean surface winds, a brief description of geostationary and polar orbiting satellites will be provided. This is to make clear the major differences in temporal and observational pattern available from these.

The geostationary satellites are in a rotation cycle identical to the Earth's rotation, i.e. the satellite is located at a fixed position above the Equator. Thus the same area is being mapped every 30 min (or if necessary every minute). From the geostationary satellites the clouds are mapped frequently enough to allow cloud tracking and from this provide wind vectors high in the atmosphere. However, no ocean surface wind sensors are carried on the geostationary platforms.

Satellites in polar orbit are recording along the track. In the ascending node (northbound) one swath will be recorded and in the subsequent descending node another swath on the other side of Earth will be recorded (Fig. 1).

Dependent upon the width of the swath and the orbit configuration it will take from one day to more than one month to complete a map of the entire globe. As example with a swath of 1,800 km 90 % global coverage can be obtained within one day. With a swath of 50 km it will take more than one month before the entire globe is mapped. At the Equator fewer observations are available for each specific local point than in areas near the poles as the swath has constant width. In general, in wide swath a lower spatial resolution will be provided than in narrow swath. There is a trade-off between high temporal and high spatial resolution. Only in

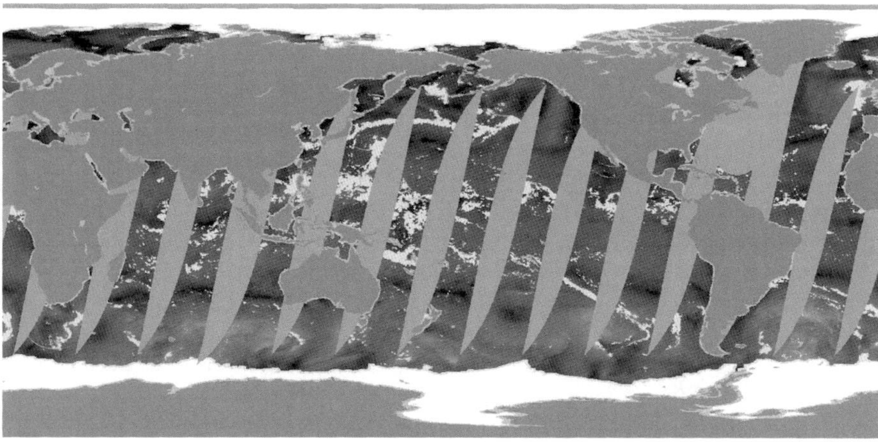

Fig. 2 SSM/I global wind speed map showing descending passes. The swath width is 1,400 km. Grey scale: *Dark* low minds, *bright* higher winds. *White* and shades of *gray* for ice, no data and land. Data from Remote Sensing Systems

cases where several similar instruments are flown on-board different platforms, such as in the F-series, global wind maps may be obtained with high temporal resolution, i.e. several times per day. A global wind speed map observed from descending passes of one of the SSM/I sensors are shown in Fig. 2. Note the gaps between passes are wider at the Equator than at higher latitudes.

A common feature of ocean surface wind instruments is that they are operated in sun-synchronous orbit which means that a specific location on Earth is mapped around the same local hour each time in the ascending and descending node, e.g. dusk/dawn, morning/evening, or afternoon/night given in Equator crossing time.

It is possible to group satellite-based ocean surface wind mapping into different categories. There are the mission specific satellites with the aim to map ocean surface wind. There are the research satellites with multi-purpose aims. Some are passive instruments observing the electromagnetic microwave radiation emitted from Earth. Others are active instruments that emit electromagnetic microwave radiation and observe the backscattered radiation from Earth to the sensor. Ocean surface wind sensors have all-weather, day and night observing capabilities. All ocean surface wind maps are valid at 10 m above sea level.

1.1 Ocean Surface Wind Missions: Passive Microwave

In year 1979 the research-based passive microwave mission, Nimbus-7 Scanning Multichannel Microwave Radiometer (SMMR) started. It lasted from 1979 to 1984 and the available data product is at 60 by 60 km spatial resolution. The *continuous* series of ocean surface wind maps starts with SSM/I in year 1987. It is the emitted microwave radiation from Earth that is observed. The swath is around 1,400 km and the spatial resolution is 25 by 25 km. The near-real-time data can be viewed at http://manati.orbit.nesdis.noaa.gov and the SSM/I data archive can be accessed at http://www.remss.com/. The archive includes data from F-8, F-10, F-11, F-13, F-14 and F-15. Only the latter is in operation at present. F-16, F-17, and F-18 with the Special Sensor Microwave Imager Sounder (SSMIS) onboard has 1,700 km swath and they operate from years 2003, 2006, and 2009, respectively, to present. Another passive microwave radiometer is the Advanced Microwave Scanning Radiometer (AMSR-E) onboard the satellite platform Aqua of National Aeronautics and Space Administration (NASA). AMSR-E provides ocean wind speed maps from 2002 to present with a swath of 1,450 km at 25 by 25 km. The Midori-2 satellite platform launched by the Japan Aerospace Exploration Agency, JAXA, carried an AMSR instrument from 2002 to 2003 but it was lost, thus the wind speed time series is only short.

Common to the passive microwave data are that wind speed only and not wind direction is mapped. Figure 3 provides an overview of the available data from passive microwave and one scatterometer, QuikSCAT. The figure details the local Equator crossing time in ascending node [Local Time Ascending Node (LTAN)]. It may be noted that the crossing time changes through time due to variations in orbit height.

In the tropics between 40° North and 40° South the TRMM, Tropical Rainfall Measuring Mission observes ocean surface winds with the TMI sensor (TRMM Microwave Imager). It provides wind speed maps from 1997 to present.

Another passive microwave radiometer in space is the fully polarimetric radiometer WindSat. From 2003 to present the Coriolis satellite with the multi-frequency polarimetric radiometer WindSat on-board has been in operation. It is developed by the Naval Research Laboratory and operated by the Air Force Research Laboratory in the USA. The instrument provides wind speed and wind direction from the microwave radiation emitted from Earth. The spatial resolution is 25 by 25 km and the data can be viewed through http://manati-test.orbit.nesdis.noaa.gov/datasets/WindSATData.php Data archiving takes place at Jet Propulsion Laboratory JPL of NASA at the Physical Oceanography Distributed Active Archive (PO.DAAC). More than ten years of data are available.

Further description of SSM/I and examples of wind analysis results based on SSM/I are provided Sect. 2.

1.2 Ocean Surface Wind Missions: Active Microwave

The first ocean surface wind scatterometer in space was on-board the SEASAT satellite in 1978. It only operated a few months. It was followed by ERS-1 SCAT in 1991 and ERS-2 SCAT in 1995 both from the European Space Agency (ESA). ERS stands for European Remote Sensing satellite. ERS-2 was taken out of service September 2011 after 16 years of operation. In 1996 NASA launched NSCAT (stands for NASA scatterometer) but unfortunately it stopped in 1997. Thereafter,

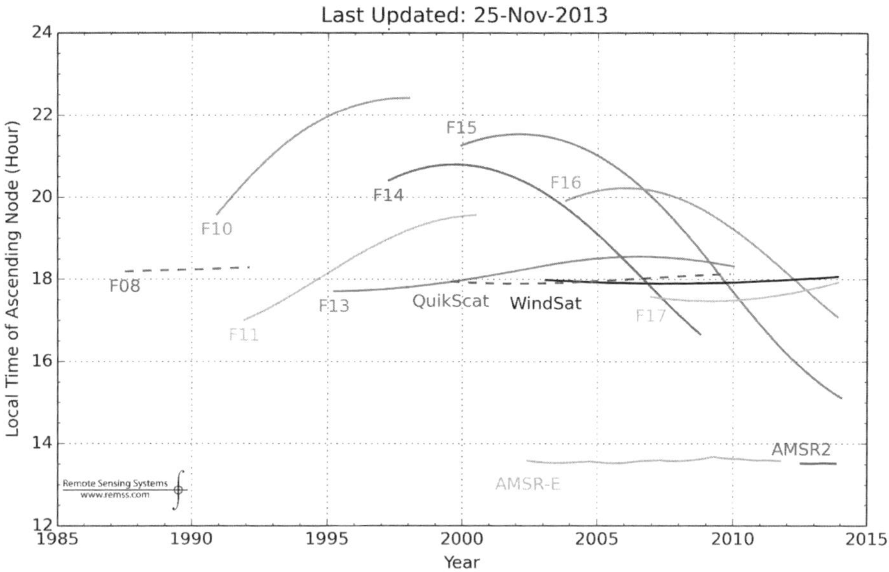

Fig. 3 Names of ocean wind mapping satellites indicating the years of operation and the ascending node, Local Time Ascending Node (LTAN), except F08 and QuikSCAT, these are at descending node (*dotted lines*). *Courtesy* Remote Sensing Systems

quickly the next mission QuikSCAT with the SeaWinds instrument on-board was launched by NASA. QuikSCAT has been a very successful mission completing more than 10 years of well-functioning operation, i.e. seven more years than scheduled. At the Midori-2 satellite another SeaWinds instrument was placed and launched by JAXA but the satellite was lost after less than one year in space, in year 2007. The NSCAT, QuikSCAT and Midori-2 missions are co-operations between NASA and JAXA. The Advanced scatterometer, ASCAT-A, was launched in 2006 and ASCAT-B in 2012 onboard meteorological satellites of EUMETSAT, the European Organisation for the Exploitation of Meteorological Satellites. The ascending node is at 21.30. The Indian Space Research Organisation, ISRO, launched Oceansat-2 in 2009. It is in operation with a scatterometer on-board and it is planned for five years operation, with Equator crossing times at 00.00 and 12.00. In China the National Satellite Ocean Application Service (NSOAS) launched the scatterometer in the HY-2 series in year 2011. HY is the Chinese Ocean Color Satellite. It has a swath of 1,700 km and orbits with Equator crossing times at 6.00 and 18.00 and map with spatial resolution at 50 by 50 km. The planned lifetime is three years. The distribution of data is through NSOAS by permission of the China National Space Administration (CNSA). The number of years in operation and the Equator crossing times are shown in Fig. 3 for QuikSCAT and Fig. 4 for ERS-1/-2. All scatterometers are dedicated to ocean surface wind observation.

Further description of QuikSCAT and examples of wind analysis results are provided in Sect. 3.

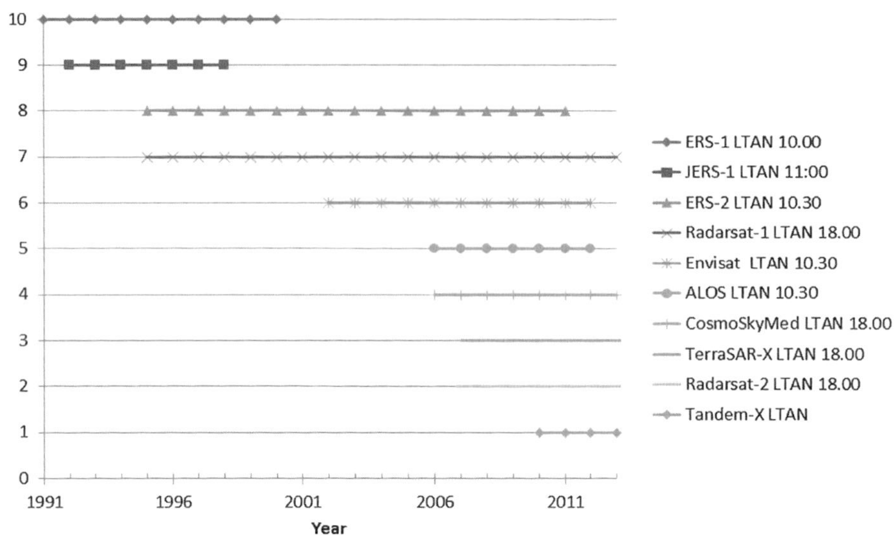

Fig. 4 Names of SAR satellites indicating the years of operation and the nominal Local Time Ascending Node (LTAN)

1.3 Multi-purpose Missions: Active Microwave

Satellite Synthetic Aperture Radar (SAR)'s are available on several satellites currently. These are in space with multi-purpose. Wind maps can be retrieved from SAR and the wind maps have one clear advantage—higher spatial resolution—compared to both passive microwave and scatterometer data. At the same time the temporal resolution is much lower than for passive microwave and scatterometer. The swath of SAR's varies greatly from a few kilometers to wideswath (also called ScanSAR) up to 400–500 km. The number of available SAR scenes can be checked for each location for several SARs through the EOLI-SA archive at ESA http://earth.esa.int/EOLi/EOLi.html the Earth Observation Link-Stand Alone archive. SAR systems typically are switched on and off during each orbit, as to map only areas where specific users have a need for data. The data downlink, on-board data storage, and ground station network limit recording. Therefore the distribution of archived SAR scenes is uneven across the globe.

The SEASAT satellite launched in 1978 carried a SAR but as mentioned earlier this satellite was soon lost. The European Space Agency launched ERS-1 in year 1991 and besides the aforementioned scatterometer it also carried an imaging SAR and an altimeter. Four years later ERS-2 was launched with similar instruments and it operated for 16 years even though it was only designed for three years nominal lifetime. Figure 4 graphs the years of operation of SAR's in space, and their approximate (nominal) Equator crossing times.

The series ERS-1/-2 with identical sensors on-board provide the longest SAR time series, around 20 years. Both have been in operation many more years than designed for. Also RADARSAT-1 launched by the Canadian Space Agency (CSA) has been in operation very long, 18 years, and now this data series is continued by RADARSAT-2 launched in year 2007 and belonging to MacDonald Dettwiler and Associates (MDA). In year 2002 ESA launched the Envisat satellite which carries a total of 12 different sensors, one of these the Advanced SAR, ASAR. The ASAR data series is 10 years long, i.e. double of the five year design for operation.

Common to the SARs on-board ERS-1/-2, Envisat and RADARSAT-1/-2 are that they operate in C-band. Other SAR's operate in L-band such as the Japanese Earth Remote Sensing Satellite, JERS-1 from year 1992 to 1998 launched by the National Space Agency of Japan, NASDA (now JAXA), SEASAT from year 1978, and ALOS PALSAR, the Advanced Land Observing Satellite (ALOS) with the Phased Array L-band SAR (PALSAR) on-board launched by JAXA in 2006.

Several satellites carry X-band sensors. These are the Italian satellite constellation of four satellites launched between year 2006 and 2010 for dual use, i.e. both civilian and non-civilian. The COnstellation of small Satellites for the Mediterranean basin Observation, COSMO-SkyMed, is funded by the Italian Ministry of Research and Ministry of Defense and conducted by the Italian Space Agency (ASI). The German TerraSAR-X1 launched in year 2007 by the German Aerospace Centre (DLR) also carry an X-band SAR as well as TanDEM-X launched in year 2010.

Currently ten SAR's are in operation. It is possible to extract wind speed from all types of SAR, however, the experience on C-band wind mapping is by far the longest. Some SARs have only one polarization, e.g. ERS-1/-2 with vertical receive, vertical transmit (VV) polarization. RADARSAT-1 on the other hand has horizontal receive, horizontal (HH) polarization. The newer SAR's have polarimetric capabilities, yet there are preferred recordings dependent upon e.g. latitude. HH is often used to map sea ice in the Arctic and Antarctica. VV is often used elsewhere for various purposes.

Altimeters provide another active microwave method for ocean surface wind mapping. Altimeters are in space with the purpose of observing the sea surface height and gravity. However, it is also possible to obtain ocean surface wind data from satellite altimeters. It is wind speed only, not wind direction. Furthermore, the sampling is rather sparse in space. There has not been much focus on using altimeter-based winds for wind energy purposes.

Further description of ERS and Envisat and examples of wind analysis results are provided in Sect. 3.

2 SSM/I

2.1 Introduction

The Special Sensor Microwave Imager (SSM/I) instrument has since July 1987 been carried on-board a number of polar orbiting satellites in the Defense Meteorological Satellite Program with up to four satellites/instruments operating simultaneously and with three operating simultaneously for eight consecutive years, 2000–2007. Three SSM/I instruments are still in operation F-15, F-16 and F-17. The SSM/I passive microwave sensor operates in seven channels, four frequency bands and observes the passive emitted radiation, i.e. the radiometric brightness temperature. An algorithm [1, 2] is used to calculate wind speed from the brightness temperature. Technical information about the SSM/I instrument can be found at National Snow and Ice Data Center [3].

SSM/I data are produced by Remote Sensing Systems and sponsored by the NASA Earth Science MEaSUREs DISCOVER Project. Data are freely available at Remote Sensing Systems (www.remss.com) from the satellites called F-8, F-10, F-11, F-13, F-14, F-15 and F-17. They are in a $0.25° \times 0.25°$ resolution raster format covering the world, i.e. in fields of $1{,}440 \times 720$ pixels. In the highest timely resolution, one file per day, each file includes a set of two times five fields, first five for the ascending paths (satellite moving north), next five for the descending paths (satellite moving south). The fields hold data for "Time of measurement (minutes of day UTC)", "10 m wind speed (m/s)", "columnar water vapor (mm)", "cloud liquid water (mm)", and "precipitation rate (mm/h)". All pixel values are given as single byte unsigned data, 0 to 255, so scaling is used to get the best resolution of the interval of interest:

Time (minutes of day, UTC) = pixel value $\times$ 6
Wind speed (m/s) = pixel value $\times$ 0.2
Columnar water vapor (mm) = pixel value $\times$ 0.3
Cloud liquid water (mm) = pixel value $\times$ 0.01
Precipitation rate (mm/h) = pixel value $\times$ 0.1

Pixel values 0–250 represent valid measurements, 251–255 is used to flag wind speed not derived, ice, no data recorded, recorded data not valid, and land.

The SSM/I instrument measures the so called brightness temperature at different wavelengths and polarizations and all four kinds of data are extracted from these measurements with an algorithm developed by Wentz and Spencer [2] as an extension of a no rain algorithm developed by Wentz [1]. The wind speed accuracy (rms) is given as 0.9 m/s. Figure 2 shows a global wind speed map from SSM/I from the descending passes.

2.2 Test of SSM/I Wind Speed Data

The algorithm of Wentz and Spencer is tested against and also based on long series of wind speed measurements at offshore buoys. In order to test the data against independent offshore wind speed measurements the SSM/I data has within the EU-NORSEWInD project [4] been compared to the German measurements at the FINO-1 offshore platform at 54°00′52″N, 06°35′16″E [5], close to the Alpha Ventus offshore wind turbine park in the North Sea. FINO stands for Forschungsplattformen in Nord- und Ostsee. These data cover with some down time

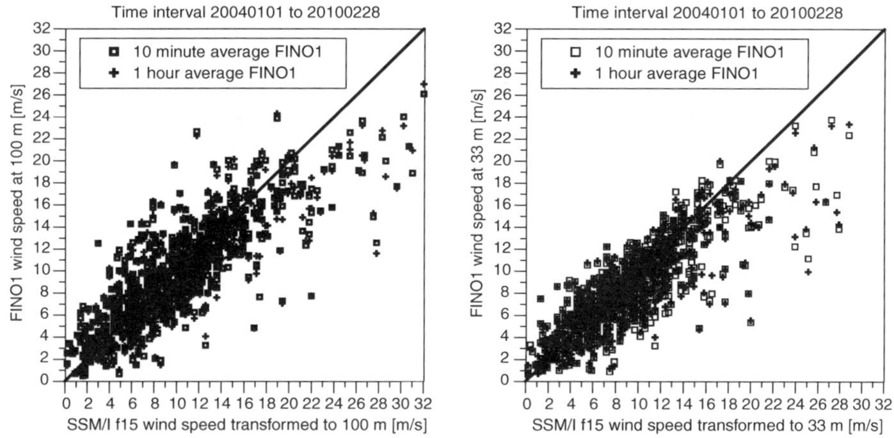

Fig. 5 Comparison of SSM/I at 6.5 E, 54 N to the FINO 1 measurements

the period 1st April 2004 to the present (pt. to 28th February 2010) and includes among many other parameters the wind speeds at eight heights between 33 and 100 m above sea.

The SSM/I data are given at 10 m above sea level but by applying Charnocks formula for sea surface roughness [6] and assuming a logarithmic wind profile, the 10 m satellite winds can be transformed to other heights.

Figure 5 shows comparisons for the 33 m wind and the 100 m wind. In both cases are the FINO data given both as the 10 min average closest to the time of the SSM/I measurement and as the 1 h average centered at that time. The trend is fine for wind speeds below around 18 m/s at 33 m, and below 20 m/s at 100 m, above these limits the SSM/I increasingly over-predicts the wind speed. There is considerable spread, however, also below the mentioned limits. For the 100 m data the average deviation SSM/I minus FINO-1 hourly averages is 0.34 m/s and the standard deviation is 3.0 m/s, appreciable more than the postulated 0.9 m/s. For the 10 min results at 100 m the average deviation is 0.35 m/s and the standard deviation is 3.1 m/s. For the 33 m results hourly and 10 min average deviations are 0.78 and 0.77 m/s, and the standard deviations 2.6 and 2.7 m/s, respectively. The FINO-1 site is at the edge of the SSM/I coverage area with only around 20 % of the satellite passages providing a validated wind speed. This may explain part of the discrepancies. Another uncertainty factor is the validity of the logarithmic profile at higher levels. The spread is higher for the 100 m than for the 30 m comparison results.

Another statistical measure is the Weibull distribution [7], as shown in Fig. 6. In correspondence with Fig. 5 the SSM/I data has a larger amount of high wind speeds than what is measured at FINO-1. The analyzed period is here 2004–2009.

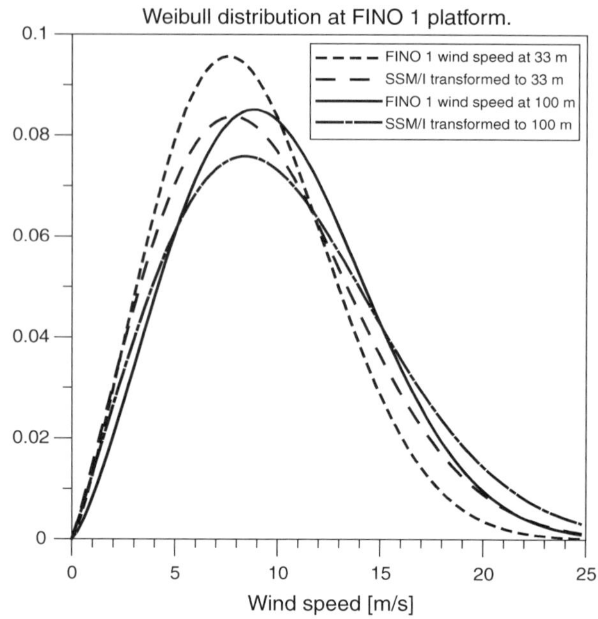

Fig. 6 Weibull distribution for SSM/I and FINO-1 data

2.3 Long-Term Investigation

The long time series makes it possible to investigate the long-term (20 year) trend in wind resources. Combining the time series from the six mentioned satellites so that the data from a higher numbered satellite takes precedence over those from a lower numbered, an almost 23 year list of two daily measurements is created. Figure 7 shows plots for the monthly wind speed averages in the period July 1987 to March 2010 for a point in the North Sea: 0°E, 59°N, and for one in the Baltic Sea: 18.75°E, 56°N. For the same points Fig. 8 shows the wind speed distributed in intervals of 5 m/s. There does not seem to be any trend in wind speed over these years, neither positive, nor negative.

2.4 Discussion

The SSM/I series of satellite data delivers the longest obtainable almost unbroken time sequence of consistent worldwide offshore wind speed data. The rms-error is given as 0.9 m/s but at the edge of the domain, i.e. relative close to the coast, where the majority of satellite passes do not result in measurement data, the rms-error is larger, and for the tested site the SSM/I data over-predicts the anemometer data at high wind speeds. The increase in number of severe storms, as predicted by some climate change investigators, is not seen in the SSM/I data so far.

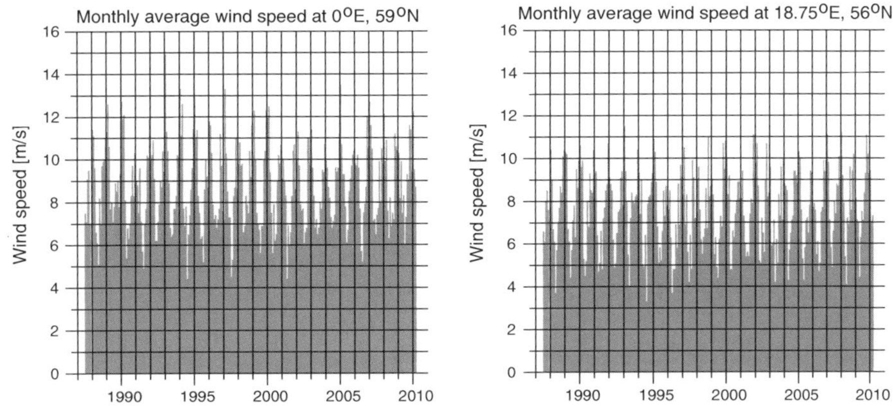

Fig. 7 Monthly average wind speeds in the North Sea and in the Baltic Sea

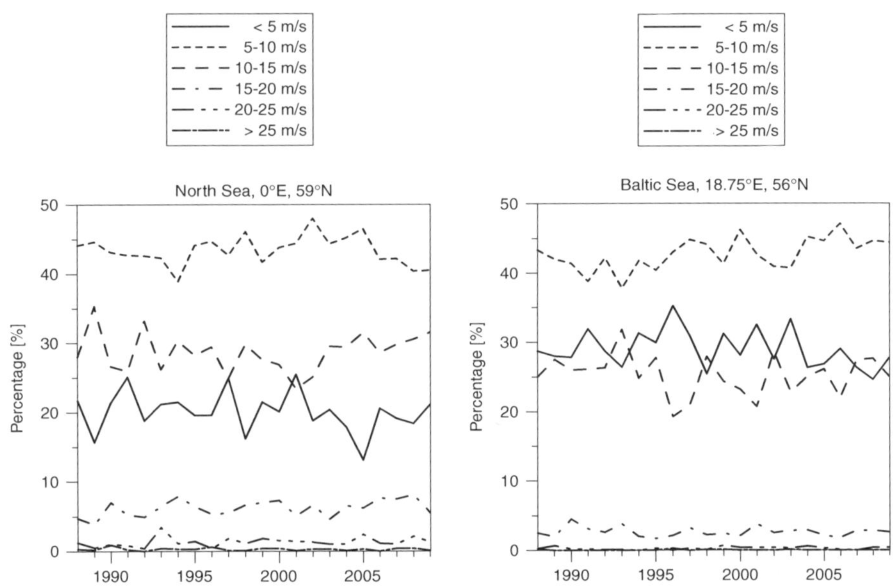

Fig. 8 Wind speeds in the North and Baltic seas binned in intervals of 5 m/s

3 QuikSCAT

3.1 Introduction

The QuikBird satellite was launched in June 1999 carrying on-board an active microwave scatterometer, SeaWinds, which was referred to as QuikSCAT. It was a sun synchronous, polar orbiting platform, with an 1,800 km wide swath. Until

November 2009, QuikSCAT recorded information about the global wind speed and direction near the ocean surface. Due to the long period of operation, it provided a valuable data set of wind vector information over the ocean. Data were available in near real time for monitoring of storms as well as data assimilation from Numerical Weather Prediction Models.

A scatterometer's principle of function is based on measuring the incoming radiation that has been backscattered from the sea surface. This backscattered radiation describes the state of the scattering surface as observed from a specific geometry, described by the azimuth and the incidence angle. For the determination of a finite set of wind vector solutions, at least two observations at different geometries are required. The existence of small scale waves on the water surface, that are assumed to be in equilibrium with the wind, applies a modification to the backscattered radiation. Based on this assumption, empirical algorithms—known as Geophysical Model Functions (GMF)—have been developed in order to relate the amount of backscattered radiation that reaches to the instrument with the near-surface wind vector.

The direction of the wind applies a modulation to the observed backscattered radiation and because of this, a correction is required. Thus, the wind vector is determined by combining several observations of the backscattered radiation from multiple viewing angles. The wind retrieval process consists of (1) wind inversion, where the GMF is inverted in order to obtain multiple maximum likelihood estimates of wind speed and direction. (2) An ambiguity removal algorithm that selects one of these estimates at each wind vector cell.

Using a geophysical model function, F, the radar backscatter σ^0 is linked to the wind vector with an expression of the form

$$\sigma^0 = F(U, \alpha, \theta, f, p)$$

where U is the neutral stability wind vector at 10 m above the sea surface, α the azimuth angle, f the frequency, p the polarization and θ the incidence angle of the observation. Modeled backscatter values are generated corresponding to the measurements. A maximum-likelihood estimator is used as the objective function, of which the multiple extrema are the wind ambiguities produced during the wind inversion process. The model function performs best for wind speeds in the range of 5–12 m/s, thus errors are larger for wind speeds outside of this range.

The ambiguity removal algorithm uses a modified median filter technique in order to select a unique wind vector out of the set of ambiguities produced during the retrieval process. Based on simulated data there are usually between two and six ambiguities for every Wind Vector Cell. The purpose of the algorithm is to choose the solution closest to the true wind [8].

An important issue is that equivalent neutral wind may differ from the actual wind due to variations in the atmospheric stability. Deviations from the well described logarithmic profile of wind speed versus height may occur, depending on the stratification of air near the water surface. Unstable atmospheric conditions, i.e. water temperature higher than atmospheric temperature, may lead to overestimated

QuikSCAT wind speeds. To the contrary, stable atmospheric conditions, dominant when the atmospheric temperature is higher than the water temperature, may lead to underestimated QuikSCAT wind speeds [9].

Since its launch, QuikSCAT has provided valuable measurements for many studies that varied from validation of the data to tropical storm monitoring. Ebuchi et al. [10] evaluated the wind vectors observed by QuikSCAT using offshore buoy data and results indicated that the rms difference for the wind speed was about 1 m/s. For the wind direction, the rms difference was about 20° when measurements higher than 3 m/s were used. Bourassa et al. [11] validated the wind vectors from QuikSCAT against observations from different research vessels, for rain free conditions. The ambiguity selection was found to be near perfect for surface wind speed higher than 8 m/s. Tang et al. [12] concluded that the accuracy of high resolution wind vectors is satisfactory in open seas while higher errors have been observed near-shore, both from the standard and the high resolution products. As presented in Boutin et al. [13], QuikSCAT operational products are accurate to 5 % or better when compared with buoy data from the South Ocean and North Atlantic.

The use of satellite data in coastal areas is a challenge due to the contamination of land. Despite that, QuikSCAT products have been used in comparison with near-shore data. Pickett et al. [14] compared QuikSCAT products with near-shore buoy wind data in order to evaluate the performance of the remotely sensed wind vectors in coastal areas. They found that satellite-buoy wind speed differences near shore were 30 % larger than those offshore. Identification of coastal winds, such as the sea breeze is of high value. The sea breeze is a phenomenon that has a diurnal cycle since it occurs because of differential heating of the land and sea masses during the day. While the air above the land becomes lighter and rises, denser air from the ocean flows in to fill the gap. There have been attempts to evaluate the sea breeze over the world ocean using the QuikSCAT swath winds [15]. An extensive list of relevant literature can be found in the review paper from Liu [16].

Radar backscatter is influenced by rain mainly due to backscatter of the signal from the rain, attenuation of the signal passing through the rain and modification of the surface due to the impact of raindrops. Large swell, rapidly changing winds and surface contaminants will also affect the water surface. Despite these known issues, QuikSCAT has been consistently providing wind vector information for the global ocean. It has been proven a most useful tool for tropical and extra-tropical storm monitoring [17], as a validation set against numerical weather prediction model outputs [18] as well as for data assimilation. It has operated far longer than originally planned, since the original mission design was three years and the instrument lifetime was five years. As of November 23, 2009 the antenna rotation rate dropped to zero due to increased friction caused by the bearings. This affected only the real time scanning equipment, while the long-term data collection system remains intact and operational.

3.2 Data

QuikSCAT was in operational polar orbit of 101 min duration, at an altitude of 803 km and travelling with a speed of approximately 7 km/s. The local Equator crossing, achieved during the ascending (morning) node, was at 06:00 ± 30 min. The scatterometer, operating in the Ku-band at a frequency of 13.4 GHz, was an active radar. It used a rotating dish antenna that radiated microwave pulses towards broad regions of the planet's surface in order to measure the power that is scattered back to the instrument. Data were collected over the ocean, land and ice in an 1,800 km wide swath, covering 90 % of the Earth's surface every day. Each geographic location, or Wind Vector Cell, has a resolution of 25 km while each radar backscatter observation corresponds to a surface of approximately 25 × 37 km [19]. Measurements included wind speed in the range of 3–20 m/s with an accuracy of 2 m/s and wind direction, with an accuracy of 20°.

There were two basic systems, the space borne observatory system (radar) and the ground based data processing one. The ground based system produced wind measurements within three days of receiving raw data from the satellite. Data are then distributed to NOAA-Suitland for operational uses and to JPL for science processing. The data used for the results included in this chapter are produced by Remote Sensing Systems—RSS—and are sponsored by the NASA Ocean Vector Winds Science Team. They are publicly available in www.remss.com. More information can be found in the JPL home page (http://winds.jpl.nasa.gov/missions/quikscat/index.cfm).

3.3 Results

The extended available data set provided by QuikSCAT can be used to derive statistic results regarding the general climatology of an area, in the present case the North Sea and the Baltic Sea. As relatively shallow seas, they provide ground for several offshore wind parks, constituted by hundreds of wind turbines. The following maps provide some relevant information about the areas which may be trusted as future sites.

The availability of QuikSCAT data for the areas of interest, as presented in Fig. 9, may be used to evaluate the statistical sufficiency of results and the validity of conclusions. Note that due to the impact of rain in the radar backscatter and thus the wind retrieval process, it is most suitable to exclude cells that are flagged with the presence of rain from analyses. The area of the North Sea shows a sufficient coverage for the very high quality pixels, i.e. no presence of rain. To the contrary, the Baltic Sea has a good coverage in the center, away from the coastal areas while inland waters, like the Danish straits, are not covered at all.

The long term data set provided for QuikSCAT is able to provide statistically relevant indications about the trends of wind speed in different areas. Mean morning wind speeds for the entire period of available data are presented in

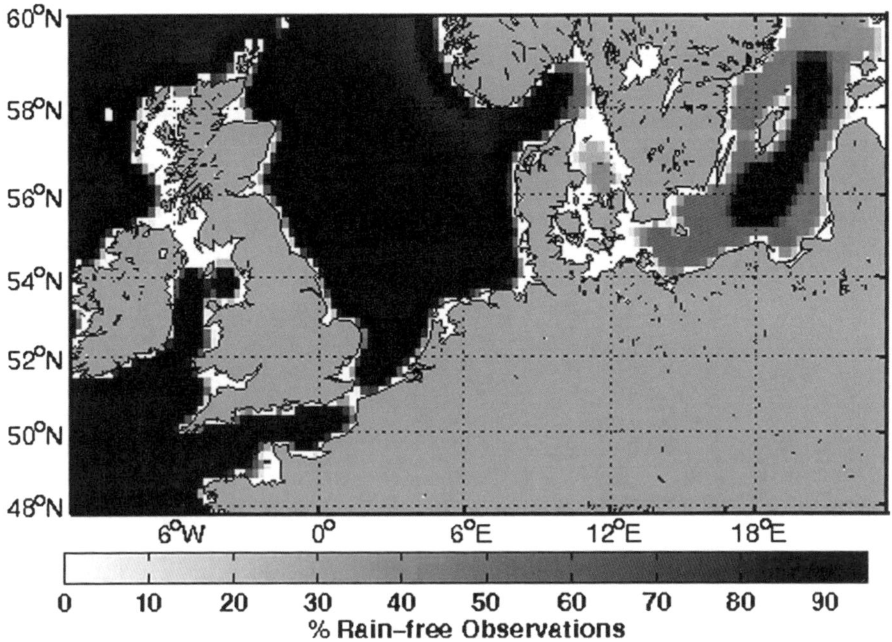

Fig. 9 Percentage of passes with no rain for every pixel of the domain, from August 1999 until October 2009 from QuikSCAT. White indicates areas with no data coverage due to the coarse resolution

Fig. 10, while the afternoon ones are shown in Fig. 11. With two passes per day above the North sea and the Baltic, QuikSCAT does not properly resolve the diurnal variability of the wind field. Despite that, a diurnal signal can be identified in Fig. 12 where the morning minus afternoon wind speed difference for the entire period of available data is shown.

Figure 13 shows the percentage of days, during the entire QuikSCAT recording period when the wind speed is larger than 10 m/s. Cases of rain have been excluded and pixels with very low data availability, i.e. less than 280 observations, have been removed. To the contrary, Fig. 14 shows the percentage of days when the wind speeds are lower than 3 m/s. Below this threshold QuikSCAT measurements are not to be trusted due to the specifications of the algorithm used during the wind retrieval process.

3.4 Discussion

Some important issues should be always kept in mind while relying on the results provided by QuikSCAT. Scatterometer winds are representative of neutral conditions hence the stability of the atmosphere may result in different real situations

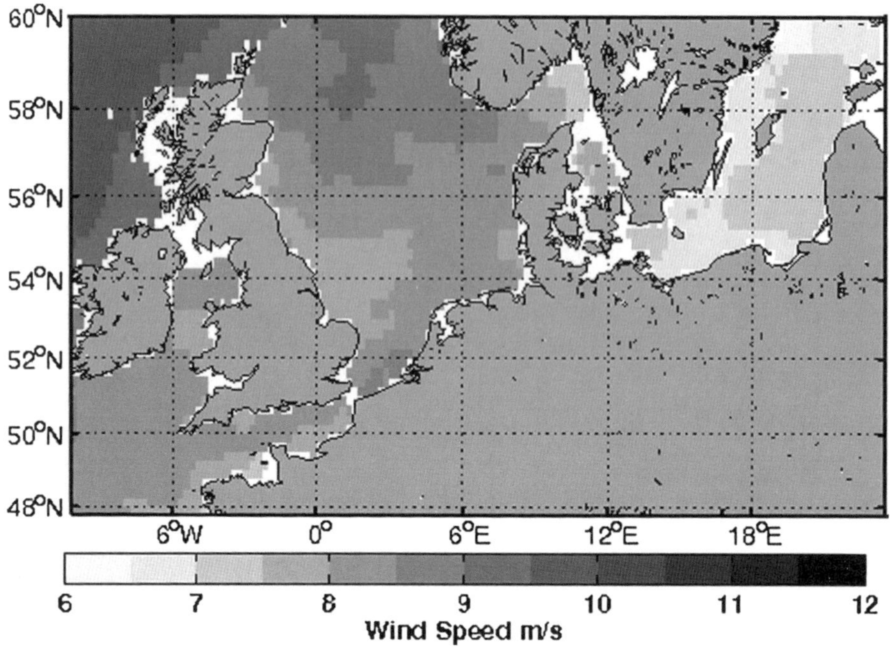

Fig. 10 Mean morning wind speeds, from August 1999 until October 2009 from QuikSCAT. Cells with presence of rain are not included in the processing. *White* areas indicate no or very few (less than 280 observations) data availability

than what the measurements imply. The wind vector cell resolution does not allow retrieval of detailed information in coastal areas, approximately 30 km from land. Despite this general feature, there are some coastal areas including the west coast of Denmark, south coast of Norway, Latvia and some parts of Sweden and Ireland for which QuikSCAT provides measurements very close to the land. Moreover, the maximum coverage of two passes per day does not properly resolve the diurnal cycle. In addition, rain alters the ocean surface, attenuates and scatters the radar signal and its overall impact is higher wind speed estimates. Therefore, rain flags are included in the gridded product which should aid for the filtering of data. Radar backscatter over ice has different properties compared to the one over the sea surface, thus no wind retrieval is obtained over ice covered areas. This is the reason for the data availability pattern seen in the Baltic Sea. The ice mask used to filter the data is obtained from SSM/I, which is a passive microwave instrument. Due to the nature of the microwave signal, no information is obtained close to land and thus, the ice mask is only available for the central part of the Baltic Sea. The coastal areas are masked out when the ice mask is active and this gives rise to the specific pattern of data availability seen in Fig. 9. This feature is inherent in the QuikSCAT product from RSS, but products from other distribution centres will be different.

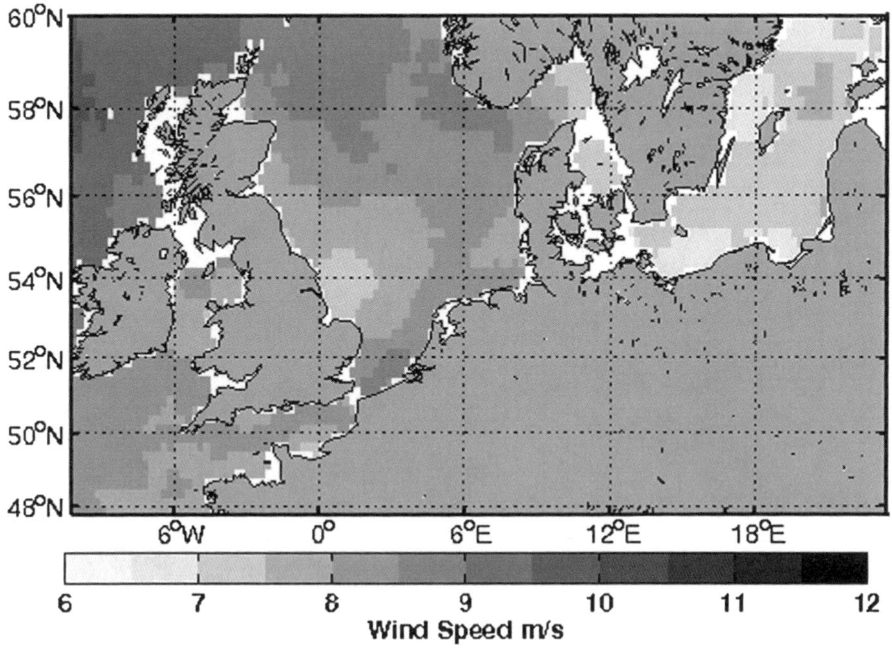

Fig. 11 Mean afternoon wind speeds, from August 1999 until October 2009 from QuikSCAT. Cells with presence of rain are not included in the processing. *White* areas indicate no or very few (less than 280 observations) data availability

As far as data availability is concerned, the largest part of the particular domain of interest is covered approximately 90 % of the time with data of trusted quality. That means that no rain is present and that the QuikSCAT measurement should be trusted. To the contrary, the Baltic Sea is only sufficiently covered in the center, away from complex coastal morphologies, while there is a buffer zone of 30–50 % coverage. Inland waters are almost never covered. The maximum percentage is a theoretical value and would correspond to a full data set, with two successful passes per day with no rain contaminated cells at all. The full data set from 10 years and three months would give 7.490 data points.

As mentioned before, the platform has two nodes, i.e. ascending during the morning and descending during the afternoon. The local time for the Danish waters is approximately around 06 and 18 h. Morning wind speeds represent post night time conditions, with maximums reaching between 10 and 11 m/s in the Atlantic. This area is complex for scatterometer wind retrievals and care has to be taken when evaluating the results. The warm surface current separating from the Gulf Stream, travels North–East towards Norway. High sea surface temperatures and currents moving in the direction of the wind speed may complicate the radar backscattered signal and give rise to overestimated winds. The North Sea has an average between 7.5 and 9 m/s while the Baltic exhibits somewhat lower speeds,

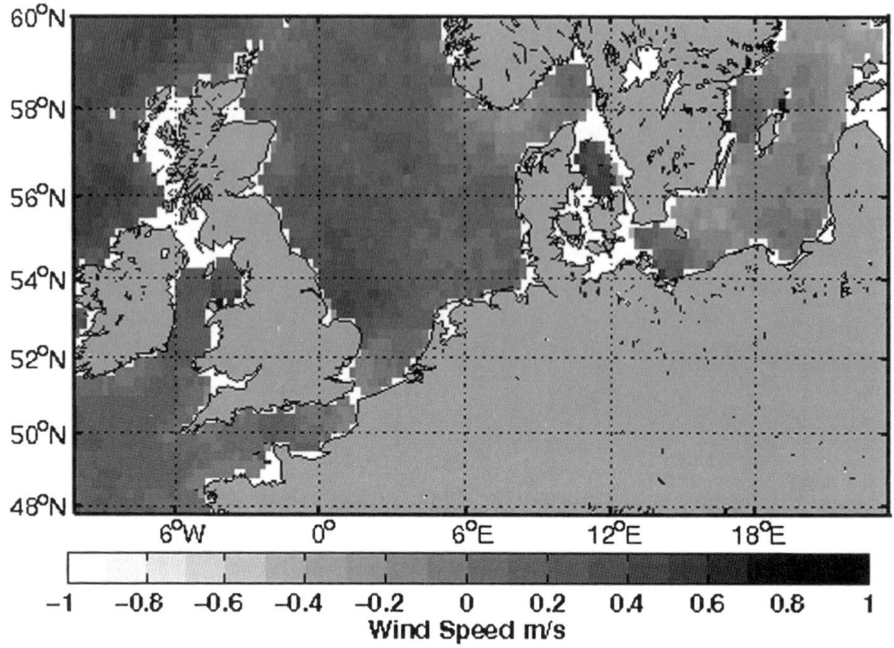

Fig. 12 Mean wind speed difference, between morning and afternoon passes, for the period between August 1999 until October 2009 from QuikSCAT. White areas indicate no or very few (less than 280 observations) data availability

in the range of 6.5 and 7.5 m/s. Protected areas show lower wind speed and the lee effects of big land masses can be observed, especially in the east coast of England. The afternoon passes, averaged through the entire period, show slightly lower wind speeds for the Atlantic, Baltic and some areas of the North Sea. What is rather striking is the amplified lee effect located in the east coast of England. This can also be an artifact of the equivalent neutral wind definition. In the case that stable conditions persist in the area due to the advection of warm air from the land over the cooler sea surface in the afternoon, the equivalent neutral wind will be underestimating the true wind and this can explain this pattern seen amplified in the afternoon mean wind speed.

Even though diurnal variability is not properly resolved, the average difference of morning minus afternoon wind speeds shows some persistent patterns. Large areas in the Baltic exhibit higher afternoon winds but the maximum difference does not exceed 0.5 m/s. The waters between Denmark and Sweden, only observed from May to November show higher morning winds. The North Sea and the North Atlantic have on average higher morning winds that reach a maximum of 0.6 m/s. Higher differences are observed for some grid cells but are spurious due to the low data availability in those grid cells.

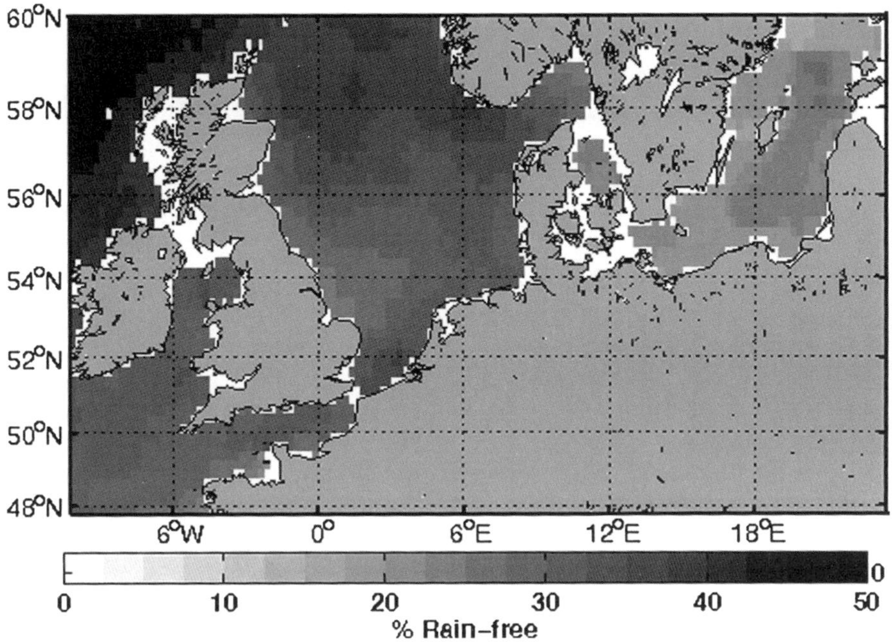

Fig. 13 Percentage of no-rain days, when wind speed is higher than 10 m/s. Period: August 1999 until October 2009 from QuikSCAT. Areas with less than 280 observations are excluded. *White* areas indicate 0 %

Since offshore wind energy is a key concept for some of the shallow areas represented in the results above, it could only make sense to use the long term data provided by QuikSCAT in order to evaluate the potential of wind availability. Wind speeds higher than 10 m/s are consistently observed 45 % of the time in the Atlantic. In the North Sea, this threshold is exceeded between 25 and 35 % of the time, depending on the area. Coastal areas and inland waters are less susceptible to high winds, but still the Irish Sea and the Wadden Sea show winds higher than 10 m/s as often as 20 % of the time. This may be an artefact of the rain flagging scheme, as the most accurate rain flag is not active in areas close to the land. Rain-contaminated observations, typically associated with higher winds, may be included in the analysis thus giving rise to higher mean wind estimates. Very low wind speeds (less than 3 m/s) are most oftenly observed in the Baltic where the percentage of time reaches a maximum of 15 %. In the most part of the coastal areas surrounding the Wadden Sea, winds lower than 3 m/s are almost never observed. Very low percentages of low winds are also recorded in the Atlantic, the English Channel and offshore from Belgium and the Netherlands.

The examples presented above demonstrate how the long temporal and large spatial coverage of QuikSCAT can be implemented, in order to evaluate various offshore wind characteristics. Results should be treated with care as the

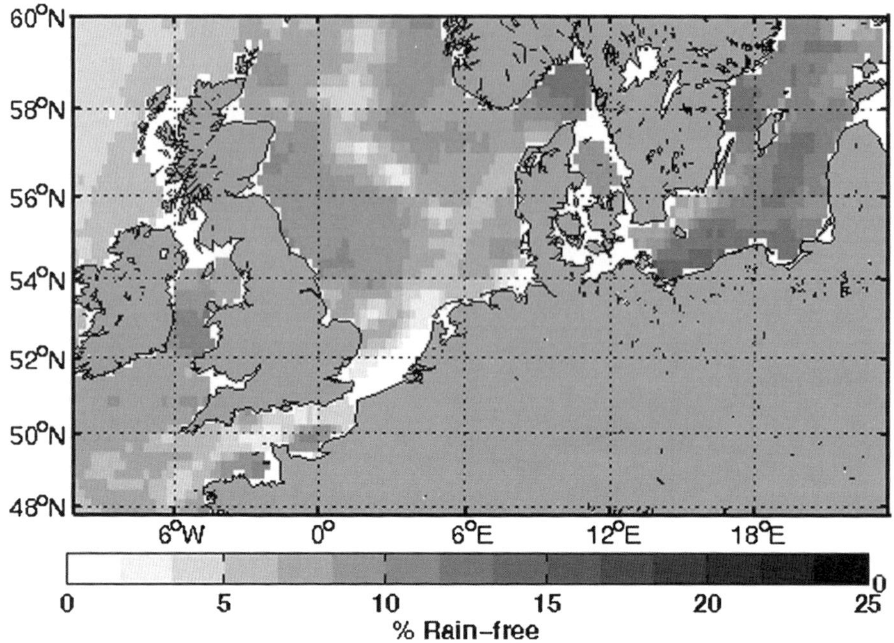

Fig. 14 Percentage of no-rain days, when wind speeds are lower than 3 m/s. Period: August 1999 until October 2009 from QuikSCAT. Areas with less than 280 observations are excluded. *White* areas indicate 0 %

instantaneous spatial series observed from the scatterometer cannot substitute the temporal series obtained from in situ observations. In addition, areas where various phenomena interact will complicate the retrieval of winds. Extended examples of the use of QuikSCAT for offshore wind resource assessment can be found in [20].

4 SAR

Synthetic aperture radar (SAR) imagery is used for a wide range of applications including land cover mapping, snow cover and sea ice monitoring, ship and oil spill detection, and wave measurements. No SAR mission has been designed specifically for wind observation but imagery from SAR has proven to be very suitable for detailed mapping of wind fields over the ocean. In contrast to the readily available wind products from passive microwave and scatterometer observations, the processing of raw SAR data to wind fields lies in the hands of individual data users. The spatial coverage of SAR data is variable over the globe and the frequency of observation is low compared to passive microwave sensors and scatterometers. The major advantage of SAR wind fields is the high spatial resolution (<1 km), which can reveal wind phenomena at the mesoscale. This resolution allows wind mapping in the coastal zone where offshore wind turbines are typically located.

SAR sensors transmit pulses in the microwave spectrum and measure the Doppler shift and time delay of the signal, as it is returned from the Earth surface. This information is used, through advanced signal processing, for generation of images showing the radar backscatter coefficient, σ^0. The amount of radar backscatter from an area unit at the surface depends on geophysical properties such as the surface roughness and moisture content and also on the radar wavelength and viewing geometry. Radar pulses interact best with elements at the surface, which are proportional in size to the radar wavelength. For example, radar pulses transmitted towards a rough sea surface interact with ocean waves of similar wavelengths. The scattering is then diffuse and known as Bragg scattering [21]. Over a smooth sea surface, specular reflection is the dominant scattering mechanism and no signal is returned to the radar. The sea surface state (i.e. rough or smooth) is determined mainly by the local wind. Radar backscatter is thus related closely to the wind speed and the wind direction relative to the radar viewing geometry.

4.1 Satellite SAR Systems

In year 1991 and 1995, ESA launched two nearly identical SAR systems, ERS-1 and ERS-2, which operated at C-band frequency ($\sim$5.3 GHz). ERS-1/2 SAR data were used, in combination with ERS-1/2 scatterometer wind fields, to develop semi-empirical geophysical model functions (GMFs) for SAR wind retrievals at C-band. The swath width of ERS-1/2 was limited to 100 km. The spatial and temporal coverage was improved significantly with the launch of RADARSAT-1 by CSA in 1995 and Envisat ASAR by ESA in 2002. Both of these sensors are operated, by default, in ScanSAR mode with 400–500 km swath widths. This facilitates operational wind mapping. The continuation of C-band SAR missions is ensured with the launch of RADARSAT-2 in 2007 and the planning of ESA's new mission Sentinel-1, which is scheduled for launch in 2014. Additional SAR sensors operating at L-band ($\sim$1.2 GHz) and X-band ($\sim$10 GHz) have been launched over the recent years and improved methods for wind retrievals at these frequencies are currently under development [22, 23].

4.2 SAR Wind Retrieval

GMF's for SAR wind retrievals are based on the functions developed for scatterometer wind retrievals [24]. Scatterometers observe a given point at the sea surface from different angles using multiple or rotating antennas. This gives a unique relationship between the radar backscatter, wind speed, and direction. In contrast, SAR sensors have a single antenna. Several wind speed and direction pairs may thus correspond to a given observation of the radar backscatter. It is therefore necessary to obtain the wind direction via other techniques before the wind speed can be retrieved from SAR.

Back in 1978, when the first spaceborne SAR sensor was launched on-board SEASAT, it became clear that signatures aligned with the wind direction are often visible in SAR imagery. These signatures are associated with atmospheric boundary layer rolls. Gerling [25] applied Fast Fourier Transformation to detect the orientation of the wind streaks. Other techniques have been developed since then to retrieve wind directions directly from SAR imagery [26–29]. The challenge is to discriminate between wind-related image streaks and other linear features, such as ice edges and coastlines, and also to resolve the 180° ambiguity problem associated with the streak detection.

Wind directions may also be obtained from alternative data sources. For local studies, site measurements can be used if available. For regional studies, it is advantageous to use spatial information about the wind direction, obtained from either atmospheric models or scatterometers. Such global, coarse-resolution data sets are very suitable for operational wind retrieval but some uncertainty is introduced through the interpolation in time and space, which is necessary to match the SAR data.

GMF's for SAR wind retrieval takes the general form:

$$\sigma^0 = A(\theta, U)\,[1 + B(\theta, U)\cos(\phi) + C(\theta, U)\cos(2\phi)]^{\gamma}$$

where σ^0 is the radar backscatter coefficient, U is wind speed at the height 10 m for a neutrally-stratified atmosphere, θ is the local incident angle, and ϕ is the wind direction with respect to the radar look direction. The coefficients A, B, C, and are functions of the wind speed and the local incident angleγ is an emperical constant . Some widely used model functions include CMOD-IFR2 [30], CMOD4 [31], and CMOD5 [32]. The last mentioned performs wind retrievals up to hurricane wind speeds whereas the valid wind speed range for previous CMOD functions was limited to approximately 2–24 ms^{-1}.

As the model functions were originally constructed for vertically polarized C-band SAR data (C_{VV}), some adjustment of σ^0 is needed when working with horizontally polarized SAR data (C_{HH}). For example, RADARSAT-1/2 operate in C_{HH} by default, as these sensors were designed mainly for ice monitoring. Some researchers have assumed that the backscatter coefficient in C_{HH} is simply equal to C_{VV} times a function that describes the ratio C_{HH}/C_{VV}; the so-called polarization ratio. Polarization ratio expressions proposed by Elfouhaily [33], Thompson et al. [34], and Vachon and Dobson [35] depend only on the radar incident angle. Mouche et al. [36] proposed a more complete model for the polarization ratio at C-band based on simultaneous measurements at C_{HH} and C_{VV} from airborne SAR.

Empirical GMF's rely on the assumption that wind speed increases logarithmically with height above the sea surface. This is normally true if the atmospheric boundary layer is neutrally stratified. Stable stratification would typically lead to an underestimation and unstable stratification to an overestimation of the 10 m wind speed. Deviations from the logarithmic wind profile are particularly strong in near-shore areas where the atmospheric boundary layer may be influenced by the land. GMFs can thus be expected to perform better over the open ocean than for coastal seas. The absolute accuracy on wind speeds retrieved from SAR data has been tested through comparisons with data from offshore masts [37–39], ocean buoys [40, 41], scatterometers [42], and atmospheric models [43, 44]. These studies all report deviations of less than 2 ms^{-1} between SAR and other wind data.

4.3 Wind Field Retrieval Over the Baltic, North and Irish Seas SAR Wind Retrieval

The DTU Wind Energy operates a system for operational SAR wind field retrieval. The system, which is developed by the Johns Hopkins University, Applied Physics Laboratory, can be used for processing of most satellite SAR data types to wind fields. Data from Envisat ASAR are obtained continuously from ESA over the Baltic, North, and Irish Seas and the wind processing is performed within a few hours from the data acquisition. Wind directions from the US Navy Operational Global Atmosperic Prediction System (NOGAPS) are used to initiate the wind speed retrieval. The six hourly model wind vectors are available at a 1° latitude and longitude grid. They are interpolated in time and space to match the satellite data.

Figure 15 shows two examples of SAR wind fields retrieved in January 2010 over the North Sea and the Baltic, respectively. In the first example, strong southeasterly winds (15–20 ms^{-1}) are blowing over the North Sea. A gradient of the wind speed is seen with increasing distance from the coastline of west Denmark. Sheltering from the land is particularly strong within the first $\sim$50 km offshore. Streaks aligned with the model wind directions are visible and there is a good agreement between the SAR and model wind speeds. In the second example, two different wind regimes are seen over the Baltic. Strong winds from the northeast are seen over the southern part of the basin whereas winds are moderate and from the northwest in the northern part. For both regimes, sheltering effects are seen downwind of the islands and coastlines and there is a good agreement between the SAR and the NOGAPS model wind speeds.

4.4 Mesoscale Wind Phenomena Observed by SAR

The high resolution of SAR wind fields facilitates description of different mesoscale wind phenomena which typically originate from the interaction of winds with the local topography. For example, a large number of wind situations over Alaska observed from RADARSAT-1 are described by Young and Winstead [45].

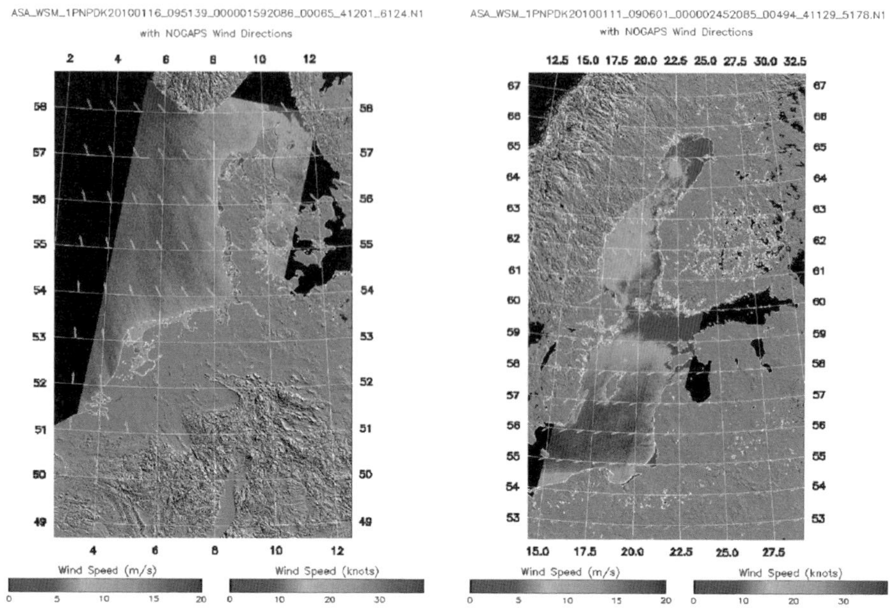

Fig. 15 Wind fields from Envisat ASAR over (*left*) the North Sea 16 January 2010 at 09:51 UTC, and (*right*) the Baltic 11 January 2010 at 09:06 UTC. *Arrows* indicate model wind vectors from NOGAPS using the same scale

Observed mesoscale wind phenomena in this high-relief and high-wind region include island and mountain wakes, barrier jets, gap flows, and lee waves. Bora events over the Adriatic and Black Seas observed from Envisat ASAR imagery are described by Alpers et al. [46]. Likewise, the SAR image archive at DTU Wind Energy contains many interesting examples of mesoscale wind phenomena of which two are presented here:

Figure 16 (left) shows a situation where the wind flow is compressed, as it is forced from easterly directions around the Norwegian coast. As a consequence, the wind speed increases significantly over a large area towards the north and west. Figure 16 (right) shows the presence of periodic wave patterns on the lee side of Scotland with winds blowing directly from the east. The waves are atmospheric gravity waves, which cause strong variations in the wind speed over scales of ~ 10 km. This SAR wind field also shows speed-up as well as sheltering effects of the rugged Scottish terrain.

4.5 Applications of SAR in Offshore Wind Energy

Wind fields from SAR are useful for several applications related to offshore wind energy. The detailed spatial information provided by SAR systems can be used to characterize offshore wind fields, as described above, and to validate model

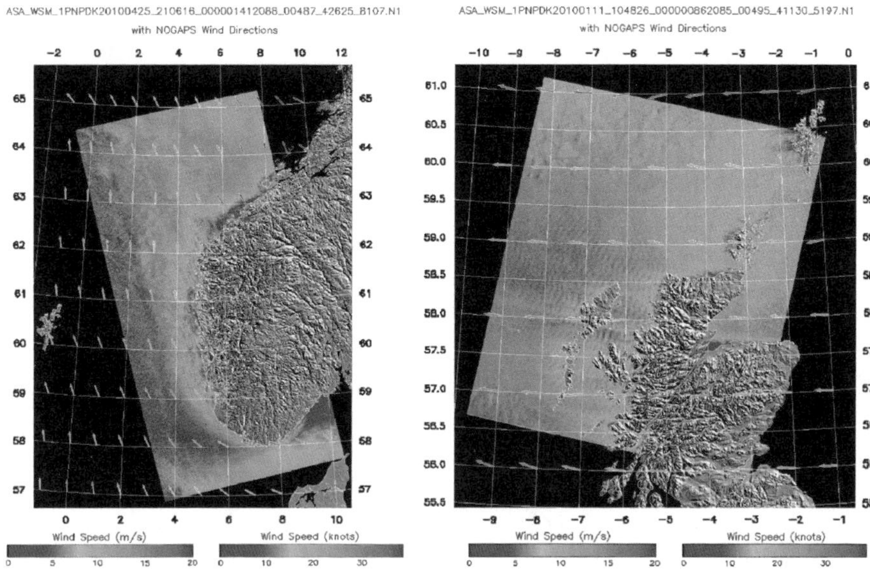

Fig. 16 Wind fields from Envisat ASAR over (*left*) parts of the North Sea 25 April 2010 at 21:06 UTC, and (*right*) the seas around Scotland 11 January 2010 at 10:48 UTC. *Arrows* indicate model wind vectors from NOGAPS using the same scale

simulations of winds where no other observations exist. The two-dimensional satellite observations may also be used to further explain one-dimensional observations from meteorological masts. In the following, different studies where SAR wind fields from DTU Wind Energy have been used for an application in wind energy are described.

Knowledge about the wind speed gradient with distance offshore is very valuable in wind farm planning as most offshore wind farms are installed within the first 50 km from the coastline where the gradient can be very strong. SAR wind fields have been used, in combination with other data sources, to determine the wind speed gradient from the coastline of western Denmark [47]. Wind speed gradients found from SAR were approximately twice as large as predicted by various models. This difference may reflect limitations in either the SAR or model data in terms of their ability to resolve coastal wind speed gradients.

Periodic wind variations caused by atmospheric gravity waves impact the wind power production and it is thus important to predict such events as accurately as possible. Larsén et al. [48] has made a case study of atmospheric gravity waves over the Danish wind farm site Nysted using spectral information from meteorological observations and the Weather Research and Forecasting (WRF) Model. A SAR wind field was used to confirm the wave patterns found from the spectral analysis.

SAR images can also be used to quantify the wake effect of large offshore wind farms. In the event that multiple wind farms are planned within a given area it is important to predict the sector-wise power loss caused by wind shadowing from one wind farm to the other. A significant reduction of the mean wind speed downwind of the 80 turbines comprising the Horns Rev I wind farm in Denmark has been found from a study of satellite and airborne SAR imagery [49, 50]. In some situations, this wake effect persisted for more than 20 km downwind and much longer than predicted from wake models.

Thanks to the growing SAR data archives, and a much easier user access to SAR data over the recent years, it is now possible to obtain a sufficient amount of SAR scenes for statistical analyses of wind resources. This type of analysis relies on fitting of a Weibull function to wind data from SAR, analogous to wind resource assessment from mast observations. A minimum of 70 overlapping SAR scenes are needed to obtain the mean wind speed or the Weibull scale parameter with ± 10 % accuracy at the 90 % confidence level, whereas thousands of samples are required for estimation of the power density and the Weibull shape parameter [51]. Wind resource maps from SAR are valuable in the pre-feasibility stage of wind farm planning but the level of uncertainty is normally too high to be bankable.

As a consequence of satellite SAR orbit dynamics and the variable coverage from site to site; wind fields from SAR are not as easily combined for statistical analyses as the gridded SSM/I and QuikSCAT wind data. Figure 17 (left) shows some SAR scenes held by DTU Wind Energy over the western part of the North Sea. The coverage varies from approximately 250–600 overlapping scenes. The mean wind speed computed from the corresponding wind fields is shown in Fig. 17 (right). The highest mean wind speed (~ 10 ms^{-1}) is seen in the northern part of the map where the topography of the Norwegian coast influences the wind fields. The remaining parts of the North Sea shows mean wind speeds of 8–9 ms^{-1} with a gradient towards the coast. This gradient varies for the different coastlines in the map and is generally strongest for the more exposed coastlines. Comparisons of the wind resource estimates from SAR with estimates from coastal and offshore meteorological masts have shown agreement within 5 % on the mean wind speed for the North Sea, which is far better than the accuracy of wind resource maps produced through mesoscale modeling [52].

5 Summary and Discussion

In broad terms, satellite remote sensing in offshore wind energy can be summarized to build upon the long-term data series of global wind speed maps from passive microwave and global wind vector maps from scatterometer. Furthermore, the long-term data series of regional wind vector maps from SAR, where the actual coverage of SAR has to be identified in the archives, gain increasing importance.

Offshore wind energy is on the rise and satellite-based ocean surface wind maps may support planning, development, maintenance and operation. In regard to planning it is of importance to understand the long-term trend of ocean winds. This is shown as examples for two locations, one in the Baltic Sea and the other in the North Sea, from SSM/I data for a 23 year period. There seems not to be any trend

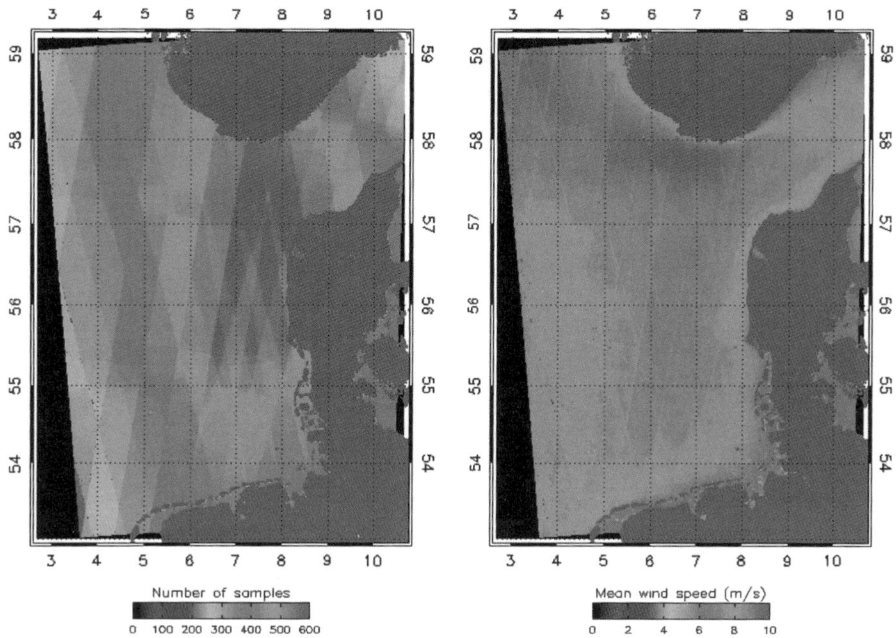

Fig. 17 Maps over the western part of the North Sea showing (*left*) the spatial coverage of SAR scenes held by DTU Wind Energy, and (*right*) the mean wind speed calculated from the corresponding SAR wind fields at a 1 × 1 km spatial resolution based on Envisat

in the data divided into different wind speed bins, i.e. the data does not exhibit more storms or calms in recent years. In a study by Hasager et al. [53] SSM/I ocean wind data were compared to the Danish wind index, i.e. the produced wind energy compared to the long-term average wind energy production. As most wind turbines in Denmark are located on land, the wind index reflects land-based winds rather than offshore winds. The study indicate that SSM/I data correlate very well with the wind index, and follow the ups and downs between windy and less wind years and also follow long-term variations in the North Atlantic Oscillation (NAO) closely. However, the ocean winds appear to be more steady with a smaller amplitude between minimum and maximum variation than the Danish wind index.

The morning and afternoon wind speed variations and their difference in the North and Baltic Seas as mapped from QuikSCAT show differences around ±0.5 m/s variation, possibly explained by major land effects. Maps of the occurrences of winds above 10 m/s and below 3 m/s show great spatial variations. The time-series span 10 years, thus the results are assumed to reflect wind statistics reasonably well. The number of available rain-free data ranges from 30 to 50 % in most of the area mapped, excluding the very near coastal parts, in particular in the Baltic Sea.

Ocean surface wind maps from SAR were at first retrieved offline in a research-based environment. Near-real-time SAR-based ocean surface wind mapping was started by JHU APL in the Alaska SAR Demonstration project in 1999 [54] for

NOAA/NESDIS, the National Oceanic and Atmospheric Administration/National Environmental Satellite Data and Information Service. Currently also Collecte Localisation Satellites (CLS) in France and DTU Wind Energy in Denmark provide near-real-time SAR ocean wind mapping. ESA has since 2008 been supporting ocean wind mapping in near-real-time from SAR through CLS www.soprano.xx as Level-2 product.

SAR-based wind mapping for wind energy applications is useful, in particular for wind resource estimation in the near coastal zone. In this zone passive microwave and scatterometer data are not available. The archive of satellite SAR images is quickly expanding and in some locations such as the North Sea and Baltic Sea several hundreds of Envisat ASAR images are recorded. In addition, SARs on-board ERS-1/-2, RADARSAT-1/-2, and other SAR sensors have observed these seas. The wind resource maps usually are provided with spatial resolution around 1 km and at this resolution many fine-scale atmospheric phenomena are resolved. This may be one reason that wind resource estimates based on SAR may have advantages to mesoscale model results for the offshore. Atmospheric fine-scale phenomena at either local or regional scale are not necessarily persistent for long but still influence long-term statistics. Thus it is stressed here that the more SAR observations become available, the better the wind resource mapping will be. Recent results from the North Sea show that SAR can compete with mesoscale model wind resource results [52]. In fact, the error as compared to three meteorological masts with high-quality time-series is smaller for SAR than the mesoscale model results.

Wind mapping from SAR is a relatively new Earth Observation discipline and further progress in GMF algorithms, polarization ratio, and calibration of sensors is expected in the future. In parallel, further analysis of the needs of the offshore wind farm developers and owners will be undertaken so as to retrieve and calculate relevant wind statistics for wind energy applications.

All satellite-based ocean surface winds are valid at 10 m above sea level and assuming neutral static stratification. It is clear that non-neutral situations occur and therefore the data should be corrected in case the necessary thermal information is available, i.e. the sea surface and air temperature. Furthermore, it has been shown from ground-based lidars that probe high in the atmosphere that the influence of the boundary layer height is important in some cases, in particular for stable conditions in the North Sea [55]. The extrapolation of winds from 10 m to wind turbine hub-height is not trivial. Thus in the EU-NORSEWInD project this topic will be investigated using ground-based wind lidars.

6 Future

The trend in offshore wind farming is on a steeply rising curve. In the European Seas the expectation is that the current 6 GW will be supplemented with another 146 GW within the next 20 years. The first 6 GW as well as the 9 GW in construction and in preparation and also the 18 GW fully consented offshore wind parks are and will be located in shallow seas. The Baltic and North Sea are on the continental shelf. In contrast, deep waters are surrounding most countries worldwide, thus new types of wind turbine installation such as floating platforms will open an immense opportunity. In parallel, the need for offshore ocean surface wind climate statistics will increase.

6.1 Passive Microwave and Scatterometer in Offshore Wind Energy

In the foreseeable future the four scatterometers in space ASCAT-A and ASCAT-B, Oceansat-2 and HY-2A will continue operation. ASCAT has a five year design lifetime, i.e. until 2011 but hopefully it will continue several more years. Oceansat-2 has a five year design lifetime until 2014. HY-2A has a three year design lifetime until 2014. The METOP-B (EUMETSAT) with a new ASCAT was launched in 2012 with a design lifetime of five years.

In Russia the Roshydromet/Roscosmos plan to launch the Meteor-M#3 with scatterometer payload in year 2014 with a planned lifetime of seven years.

China (NSAOS) and France (CNES, Centre National d'Etudes des Spatiales) plan to jointly launch a scatterometer in year 2013 on the platform CFOSat, the Chinese French Ocean Satellite. India has proposed ScatSat for launch in 2014 and China (NSAOS) has the HY-2B proposed for launch in 2014.

NOAA is proposing a scatterometer flight for JAXA's GCOM-W2 in 2016. GCOM is part of the Global Change Observation Mission. W is for Water cycle (as opposed to GCOM-C Carbon cycle). The payload will include a scatterometer as SeaWinds and a passive microwave radiometer as AMRS. Thus the AMSR-series will be continued as well as the QuikSCAT (SeaWinds) mission, the latter however with a gap in data from year 2009 to 2016. AMSR-E has been in operation for ten years.f

The F-series with F-15, F-16, F-17 and F-18 currently in operation with ocean surface wind passive microwave sensors and will be continued by two more satellites in the F-series, F-19 for launch late in year 2014 and F-20 in year 2020 by DMSP.

In summary, there will be several scatterometers and passive microwave radiometers in space in the next decade. For offshore wind energy mapping this is expected to prove very useful. The high temporal mapping of global ocean surface winds are secured. Ocean surface winds from the F-series is of particular benefit for long-term trend analysis for global wind energy applications.

6.2 SAR in Offshore Wind Energy

The present fleet of Earth Observing satellites capable of observing ocean surface winds at high spatial resolution will be significantly expanded according to present plans. There are eight active SAR instruments in space currently for civilian operation and even though three of these are far beyond their planned lifetime, there are plans to continue these missions.

For the oldest active SAR in space, the European ERS-2, the plan to continue until mid-2011 was done, i.e. completing 16 years of observations.

Also RADARSAT-1 in Canada was active after 18 years mission and the SAR archive is being continued through RADARSAT-2 that is now in space with a seven year design lifetime, i.e. until year 2014. Furthermore, the RADARSAT Constellation plan consists of three new satellites with SAR's on-board to be launched in year 2018. With RADARSAT-2 and the three next satellites in space 95 % of the world will be visited daily by these three.

Envisat, the European flagship, has operated double of its five year design lifetime. Envisat had a modification of orbital parameters in October 2010 to be able to operate with minimum hydrazine. The changes influenced overpass times. Unfortunately Envisat stopped in April 2012.

The Japanese ALOS PALSAR with design life of three years operated five years.

The four Italiean CosmoSkyMed SAR's launched between year 2006 and 2010 all have a five year design lifetime.

The German TerraSAR-X1 launched in year 2007 and TanDEM-X launched in year 2010 both have five year design lifetime. TanDEM is TerraSAR-X add-on for Digital Elevation Measurement. The TSX-NG (TerraSAR-X Next Generation) is scheduled for launch in year 2016 with seven year design lifetime. The continuation of the ERS/Envisat heritage will be the Sentinel series. ERS and Envisat are research satellites whereas the Sentinel series will be operational, thus continuation is planned. Sentinel-1 is to be launched in year 2014 with SAR payload. The second Sentinel-1 will be launched a few years later and the Sentinel-1 pair will operate in tandem and will provide coverage of Europa and Canada in less than two days. Sentinel-1 has a seven year design lifetime with consumables for 12 years and it will be in orbit in ascending time at 18.00 LTAN. ESA is planning to produce among other products, level two ocean wind maps from Sentinel-1. This will be the first SAR-based ocean wind product from a space agency.

In China the Huan Jing (HJ)-1C satellite with SAR payload was launched in year 2012 with a S-band SAR with HH polarization and 100 km swath. It is part of the Huan Jing (Environmental Protection and Disaster Monitoring Constellation) series.

Thus from the above description on future satellite SARs it is evident that available SAR imagery will increase significantly. Several SARs will be used such as Envisat, RADARSAT, TerraSAR-X, TanDEM-X and CosmoSkyMed. At the same time CLS (Collecte Localisation Satellites) in France provide SAR Level-2 product, i.e. ocean surface wind from SAR http://soprano.cls.fr/ in near real-time and so does DTU Wind Energy. DTU Wind Energy offer wind resource mapping based on satellite images since year 2006. Currently the wind maps from DTU Wind Energy and CLS are being pooled together to increase the number of data and for improvements on use for wind energy application in the EU-NORSEWInD project (2008–2012) and for wind farm wake studies in the EU EERA-DTOC project (2012–2015), the European Energy Research Alliance Design Tools for Offshore wind farm Clusters.

In summary, the future use of satellite remote sensing in offshore wind energy is by all means an obvious choice.

Acknowledgments We acknowledge the satellite remote sensing data available for analysis. This includes QuikSCAT and SSM/I. QuikSCAT data are produced by Remote Sensing Systems and sponsored by the NASA Ocean Vector Winds Science Team. Data are available at www.remss.com. SSM/I are produced by Remote Sensing Systems and sponsored by the NASA Earth Science MEaSUREs DISCOVER Project. Data are available at www.remss.com. We acknowledge satellite data provided by the European Space Agency, the EO-3644 ERS and Envisat and EO-6773 ERS, Envisat, ALOS PALSAR and RADARSAT grants. The Johns Hopkins University, Applied Physics Laboratory is thanked for use and support of the APL/NOAA SAR Wind Retrieval System. We acknowledge the meteorological data from FINO-1 the Forschungsprojekt FINO [Forschungsplattformen in Nord- und Ostsee (North and Baltic Sea)]. We acknowledge support from the EU-NORSEWInD project www.norsewind.eu TREN-FP7EN-219048, in years 2008–2012 and EERA-DTOC www.eera-dtoc.eu FP7-ENERGY-2011-1/n°282797 in years 2012–2015.

References

1. Wentz FJ (1997) A well-calibrated ocean algorithm for SSM/I. J Geophys Res 102(C4):8703–8718
2. Wentz FJ, Spencer RW (1998) SSM/I rain retrievals within a unified all-weather ocean algorithm. J Atmos Sci 55:1613–1627
3. NSIDC. http://nsidc.org/data/docs/daac/ssmi_instrument.gd.html
4. Hasager C, Mouche A, Badger M, Astrup P, Nielsen M (2009) Satellite wind in EU-Norsewind scientific proceedings of European wind energy conference and exhibition, pp 144–147, Marseille (FR) 16–19 Mar 2009
5. FINO (2002). http://www.bsh.de/en/Marine_data/Observations/Projects/FINO/index.jsp
6. Charnock H (1955) Wind stress over a water surface. Q J R Meteorol Soc 81:639–640
7. Troen I, Petersen EL (1989) European wind atlas, Risø National Laboratory ISBN 87-550-1482-8
8. NASA Quick Scatterometer (2006) QuikSCAT science data product, user's manual, overview and geophysical data products. Version 3.0, Jet Propulsion Laboratory, California Institute of Technology, D-18053-Rev A, Sep 2006
9. Liu WT, Tang W (1996) Equivalent neutral wind. National Aeronautics and Space Administration, Jet Propulsion Laboratory (US, and United States), National Aeronautics and Space Administration, Jet Propulsion Laboratory, California Institute of Technology; National Aeronautics and Space Administration; National Technical Information Service, distributor
10. Ebuchi N, Graber HC, Caruso MJ (2002) Evaluation of wind vectors observed by QuikSCAT/SeaWinds using ocean buoy data. J Atmos Oceanic Technol 19:2049–2062

11. Bourassa MA, Legler DM, O'Brien JJ, Smith SR (2003) SeaWinds validation with research vessels. J Geophys Res 108(C2):3019. doi:10.1029/2001JC001028
12. Tang W, Liu WT, Stiles BW (2004) Evaluation of high-resolution ocean surface vector winds measured by QuikSCAT scatterometer in coastal regions. IEEE Transactions on Geoscience and Remote Sensing, f42, pp 1762–1769
13. Boutin J, Quilfen Y, Merlivat L, Piolle JF (2009) Global average of air-sea CO_2 transfer velocity from QuikSCAT scatterometer wind speeds. J Geophys Res-Oceans 114:C04007
14. Pickett MH, Tang W, Rosenfeld LK, Wash CH (2003) QuikSCAT satellite comparisons with near-shore buoy wind data off the U.S. West Coast. J Atmos Oceanic Technol 20:1869–1879
15. Gille ST, Llewellyn Smith SG, Statom NM (2005) Global observations of the land breeze. Geophys Res Lett 32:1–4
16. Liu WT (2002) Progress in scatterometer application. J Oceanography 58(1)
17. Chelton DB, Freilich MH, Sienkiewicz JM, Von Ahn JM (2006) On the use of QuikSCAT scatterometer measurements of surface winds for marine weather prediction. Mon Wea Rev 134:2055–2071
18. Chelton DB, Freilich MH (2005) Scatterometer-based assessment of 10-m wind analyses from the operational ECMWF and NCEP numerical weather prediction models. Mon Wea Rev 133:409–429
19. Hoffman RN, Leidner SM (2005) An introduction to the near-real-time QuikSCAT data. Weather Forecast 20:476–493
20. Karagali I, Badger M, Hahmann A, Peña A, Hasager C, Sempreviva AM (2013) Spatical and temporal variability in winds in the Northern European Seas. Renew Energy 57: 200–210
21. Valenzuela GR (1978) Theories for the interaction of electromagnetic and ocean waves—A review. Bound-Layer Meteorol 13:61–85
22. Isoguchi O, Shimada M (2009) An L-band ocean geophysical model function derived from PALSAR. IEEE Trans Geosci Remote Sens 47:1925–1936. doi:10.1109/TGRS.2008.2010864
23. Thompson DR, Monaldo FM, Horstmann J, Christiansen MB (2008) Geophysical model functions for the retrieval of ocean surface winds. 2nd International Workshop on Advance in SAR Oceanography from ENVISAT and ERS Missions, The European Space Agency, Rome, Italy 21–25 Jan 2008
24. Stoffelen A, Anderson DLT (1993) Wind retrieval and ERS-1 scatterometer radar backscatter measurements. Adv Space Res 13:53–60
25. Gerling TW (1986) Structure of the surface wind field from the SEASAT SAR. J Geophys Res 91:2308–2320
26. Fichaux N, Ranchin T (2002) Combined extraction of high spatial resolution wind speed and direction from SAR images: a new approach using wavelet transform. Can J Remote Sens 28:510–516
27. Du Y, Vachon PW, Wolfe J (2002) Wind direction estimation from SAR images of the ocean using wavelet analysis. Can J Remote Sens 28:498–509
28. Koch W (2004) Directional analysis of SAR images aiming at wind direction. IEEE Trans Geosci Remote Sens 42:702–710
29. Horstmann J, Koch W, Lehner S (2004) Ocean wind fields retrieved from the advanced synthetic aperture radar aboard ENVISAT. Ocean Dyn 54:570–576
30. Quilfen Y, Chapron B, Elfouhaily T, Katsaros K, Tournadre J (1998) Observation of tropical cyclones by high-resolution scatterometry. J Geophys Res 103:7767–7786
31. Stoffelen A, Anderson DLT (1997) Scatterometer data interpretation: estimation and validation of the transfer function CMOD4. J Geophys Res 102:5767–5780
32. Hersbach H, Stoffelen A, de Haan S (2007) An improved C-band scatterometer ocean geophysical model function: CMOD5. J Geophys Res-Oceans, 112
33. Elfouhaily TM (1996) Modéle couple vent/vagues et son application à la télédétection par micro-onde de la surface de la mer. University of Paris 7
34. Thompson D, Elfouhaily T, Chapron B (1998) Polarization ratio for microwave backscattering from the ocean surface at low to moderate incidence angles, pp 1671–1676

35. Vachon PW, Dobson EW (2000) Wind retrieval from RADARSAT SAR images: selection of a suitable C-band HH polarization wind retrieval model. Can J Remote Sens 26:306–313
36. Mouche AA, Hauser D, Daloze JF, Guerin C (2005) Dual-polarization measurements at C-band over the ocean: Results from airborne radar observations and comparison with ENVISAT ASAR data. IEEE Trans Geosci Remote Sens 43:753–769
37. Hasager CB, Dellwik E, Nielsen M, Furevik B (2004) Validation of ERS-2 SAR offshore wind-speed maps in the North Sea. Int J Remote Sens 25:3817–3841
38. Christiansen MB, Koch W, Horstmann J, Hasager CB, Nielsen M (2006) Wind resource assessment from C-band SAR. Remote Sens Environ 105:68–81
39. Hasager CB, Badger M, Peña A, Larsén XG (2010) SAR-based wind resource statistics in the Baltic Sea. Remote Sens 3(1): 117–144. doi:10.3390/rs3010117
40. Monaldo FM, Thompson DR, Beal RC, Pichel WG, Clemente-Colón P (2001) Comparison of SAR-derived wind speed with model predictions and ocean buoy measurements. IEEE Trans Geosci Remote Sens 39:2587–2600
41. Fetterer F, Gineris D, Wackerman CC (1998) Validating a scatterometer wind algorithm for ERS-1 SAR. IEEE Trans Geosci Remote Sens 36:479–492
42. Monaldo FM, Thompson DR, Pichel WG, Clemente-Colon P (2004) A systematic comparison of QuikSCAT and SAR ocean surface wind speeds. IEEE Trans Geosci Remote Sens 42:283–291
43. Horstmann J, Schiller H, Schulz-Stellenfleth J, Lehner S (2003) Global wind speed retrieval from SAR. IEEE Trans Geosci Remote Sens 41:2277–2286
44. Furevik B, Johannessen O, Sandvik AD (2002) SAR-retrieved wind in polar regions—comparison with in situ data and atmospheric model output. IEEE Trans Geosci Remote Sens 40:1720–1732
45. Young G, Winstead N (2005) Meteorological phenomena in high resolution SAR wind imagery. High resolution wind monitoring with wide swath SAR: A user's guide. In: Beal B, Young G, Monaldo F, Thompson D, Winstead N, Scott C (eds) U.S. Department of Commerce, National Oceanic and Atmospheric Administration, pp 13–34
46. Alpers W, Ivanov A, Horstmann J (2009) Observations of Bora events over the Adriatic Sea and Black Sea by Spaceborne Synthetic Aperture Radar. Mon Weather Rev 137:1150–1161. doi:10.1175/2008MWR2563.1
47. Barthelmie RJ, Badger J, Pryor SC, Hasager CB, Christiansen MB, Jorgensen BH (2007) Offshore coastal wind speed gradients: issues for the design and development of large offshore windfarms. Wind Eng 31:369–382
48. Larsén XG, Larsen S, Badger M (2010) A case study of mesoscale spectra of wind and temperature, observed and simulated. Q J R Meteorol Soc 137(654): 264−274
49. Christiansen MB, Hasager CB (2005) Wake effects of large offshore wind farms identified from satellite SAR. Remote Sens Environ 98:251–268
50. Christiansen MB, Hasager CB (2006) Using airborne and satellite SAR for wake mapping offshore. Wind Energy 9:437–455
51. Barthelmie RJ, Pryor SC (2003) Can satellite sampling of offshore wind speeds realistically represent wind speed distributions. J Appl Meteorol 42:83–94
52. Badger M, Badger J, Nielsen M, Hasager CB, Peña A (2010) Wind class sampling of satellite SAR imagery for offshore wind resource mapping. J Appl Meteorol Climatology. doi:10.1175/2010JAMC2523.1
53. Hasager C, Peña A, Christiansen M, Astrup P, Nielsen M, Monaldo F, Thompson D, Nielsen P (2008) Remote sensing observation used in offshore wind energy. IEEE J Sel Topics Appl Earth Observations Remote Sens 1(1):67–79
54. Beal B, Young G, Monaldo F, Thompson D, Winstead N, Scott C (eds) (2005) High resolution wind monitoring with wide swath SAR: A user's guide. U.S. Department of Commerce, National Oceanic and Atmospheric Administration
55. Peña A, Hasager C, Gryning S-E (2008) Measurements and modelling the wind speed profile in the marine atmospheric boundary layer. Bound-Layer Meteorol 129:479–495

Optimization of AC Electric Power Systems of Offshore Wind Farms

Marcos Banzo and Andres Ramos

Abstract Onshore wind energy has experienced a rapid growth in recent years due mainly to the maturity achieved by its technology and the institutional support for renewable energy. Offshore wind energy also faces a very promising future as an extensive development of offshore wind farm (OWF) projects is planned in many regions of the world. The electric power system is a prominent part of an OWF since its design exceedingly affects the cost structure and operation of the entire facility. Thus the electric power system of an OWF must be optimized in order to minimize the life-cycle cost, while maintaining an adequate level of technical performance. The optimization of the electric power system of an OWF poses a complex mathematical problem since it requires considering jointly the key aspects that characterize its design: system component costs, system efficiency and system reliability. This problem may be addressed satisfactorily by using optimization models based on classical and metaheuristic optimization approaches.

1 Introduction

Wind energy has been consolidated in 2000s decade as an effective energy source in the fight against greenhouse gas emissions. The growth of wind energy in many countries has been stimulated by institutional support for renewable energy. Practically all the wind power installed in the last years has come from onshore wind farms. On the contrary, offshore wind energy has made a minor contribution

A. Ramos (✉)
Universidad Pontificia Comillas, Madrid, Spain
e-mail: andres.ramos@upcomillas.es

M. Banzo
Iberdrola Ingenieria y Construccion S.A.U, Madrid, Spain
e-mail: mbhe@iberdrola.es

P. M. Pardalos et al. (eds.), *Handbook of Wind Power Systems*, Energy Systems,
DOI: 10.1007/978-3-642-41080-2_22, © Springer-Verlag Berlin Heidelberg 2013

to growth. Nevertheless, offshore wind energy faces a promising future in many countries because of the huge offshore wind resources and the experience gained from the development of onshore wind energy. The growth perspectives of offshore wind energy in coming decades are very promising in many regions of the world such as Europe, U.S and China [1].

The high capital cost that is generally required to develop an offshore wind farm (OWF) is one of the key differences between onshore and offshore wind energy. On average, capital cost for a new OWF is in the range of 2.0–2.2 million €/MW for a near-shore shallow water facility, while for an onshore wind farm comprised of medium-sized wind turbines (1.5–2 MW) is in the range of 1.1–1.4 million €/MW [2]. This cost difference is primarily based on the higher cost required to supply and build the electric infrastructure and wind turbine foundations in offshore wind energy.

The electric power system has a significant influence on the operation and availability of an OWF. Due to the large investment required in the development of an OWF, the output energy has to be maximized to shorten the payback period. Hence the electric power system of OWFs must be optimized in order to have the lowest life-cycle cost, while maintaining an adequate level of technical performance. The main issues that characterize the design of the electric power system of an OWF have to be taken into account in the optimization problem: system component costs, system efficiency and system reliability. Likewise, specific aspects of wind energy such as wind speed conditions or geographical location of the OWF site must be considered as well. The above requirements set out a complex mathematical problem that can be tackled through classical and metaheuristic optimization approaches.

This chapter is dedicated to the optimization of AC electric power systems of OWFs. It is noteworthy that all OWFs in operation by end of year 2008 are built with AC electric power systems. This is because the installation of AC electric power systems in OWFs located near to the shore at an approximate distance no longer than the range of 80–100 km is generally more profitable than that of DC systems [3, 4].

Despite the extensive use of AC electric power systems in OWFs, there exists a lot of research to implement DC electric power systems in OWFs [5, 6] and to develop trans-national offshore transmission networks for connecting future OWFs by means of high voltage DC (HVDC) links [7].

The chapter is organized as follows. Section 2 describes the main components and the most representative layouts of the AC electric power systems of OWFs. A review of the most important issues of the reliability and efficiency evaluation of OWFs is carried out in Sect. 3. The optimization problem of the AC electric power system of OWFs and some optimization models extracted from available literature that have been used to face this problem are explained in Sect. 4. The chapter ends with an application example in which an optimization model is shown.

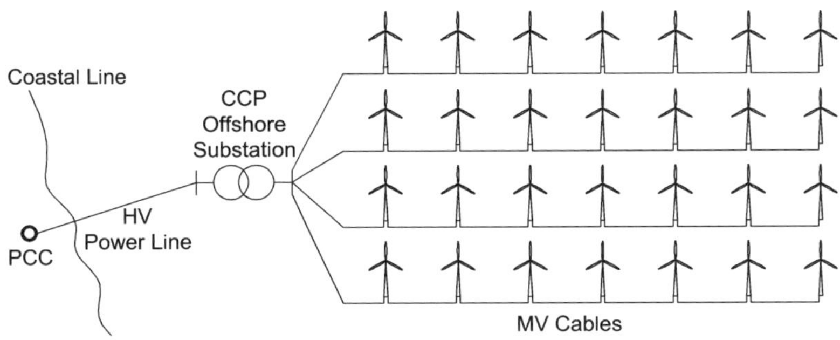

Fig. 1 General configuration with an offshore substation

2 AC Electric Power Systems of OWFs

2.1 General Configurations

The electric power system of an OWF is responsible to collect the power from the wind turbines and transmit it in an efficient, reliable and safe way to the point of common connection (PCC) of the transmission grid located onshore. It is composed by the electric power components (cables, transformers, switchgears, etc.) that are among the wind turbine generator and the PCC.

The AC electric power system of OWFs may be divided into two parts, the medium voltage (MV) collection system and the transmission system. The MV collection system collects the power produced by wind turbines and transmits it to a central collection point (CCP) through an internal grid of MV submarine cables connecting the wind turbines to each other. The power collected at the CCP is transmitted to the PCC through the transmission system in such a way that technical requirements established by the transmission system operator (TSO) at the PCC are met. The route of the transmission link consists of a submarine cable section from the CCP to the shore and an onshore section of cable from the shore to the PCC.

Two general configurations of the AC electric power system are established depending on the OWF distance to the shore and the installed power of the OWF [3]. In large OWFs far from the coast (Fig. 1), the CCP is an offshore substation located inside the OWF site where voltage level is increased from MV to high voltage (HV) by one or various step-up transformers. The power generated by the OWF is transmitted by a HV line from the CCP to the PCC. It must be noted that more than one offshore substation might be taken into account in this configuration. In small OWFs near the coast (Fig. 2), the CCP is an area at the OWF where MV submarine cables gather to follow the same route till the PCC.

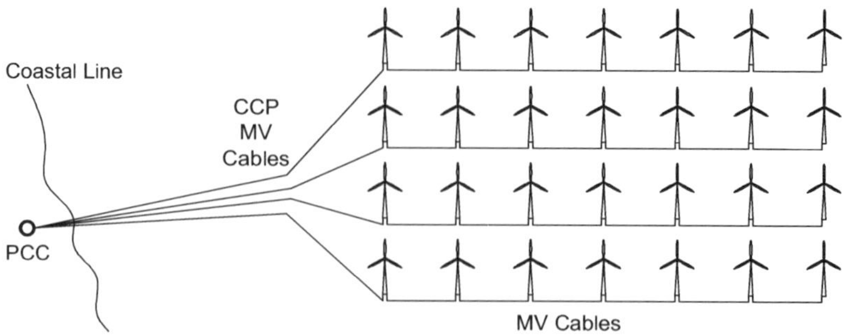

Fig. 2 General configuration with an area where MV cables gather

2.2 MV Collection System

The current trend in offshore wind energy is to install large wind turbines to exploit the better wind speed conditions at offshore sites. According to the present technology [1], the rated power of offshore wind turbines varies from 2.5 to 6 MW. The area covered by current OWF sites is extensive as the sitting of wind turbines follows the criterion of maximizing the capture of wind resource. The distance between wind turbines in each row is proportional to the rotor diameter of wind turbines (around 5–8 diameters) and the distance between rows is set to reduce wake effects of wind turbines. Besides, future and under development OWFs will have power ratings of several hundred megawatts, even on the order of one gigawatt. Thus, the extensive areas of OWF sites and the large power ratings of OWFs have an important influence on the design of MV collection systems.

As mentioned above, the MV collection system aims to collect the power generated by wind turbines and transmit it to the CCP. The principal components of the collection system are the power transformers, MV switchgears and MV submarine cables. The typical voltage level used in the collection system ranges from 30 to 36 kV. The maximum voltage level is limited by the available space inside the wind turbine to accommodate the power transformer and the MV switchgear.

The low voltage level of the wind turbine generators is transformed to MV by power transformers. On the other hand, the main functions of MV switchgears are to protect the power transformer against electrical faults and to provide connectivity to MV submarine cables. They are located inside the wind turbine tower and consist of several cubicles with the necessary electric equipment (circuit breakers, protection relays, fuses, etc.) to carry out the required function. MV switchgear operation may be manual, remote-controlled or automatically-controlled.

A grid of MV submarine cables is used to connect all wind turbines in clusters. The most common MV submarine cable adopted in OWFs is the three-core copper cable of either cross-linked polyethylene (XLPE) or ethylene propylene rubber (EPR) insulation with conductor cross-sections from 95 to 630 mm^2.

Various grid layouts for the MV collection system are available in the existing literature[8-10]. Some of these layouts have even been taken as a starting point for the electric power system design of emerging ocean energy sources [11]. Currently, most OWFs adopt the radial layout despite the variety of available grid layouts. It is foreseeable that other layouts with a higher level of redundancy than the radial one will likely have more application in future OWFs with larger power ratings and more distant from the coast. Some of the main layouts for the MV collection system of OWFs are explained in (Figs. 3, 4, 5 and 6):

- Radial layout. A MV single feeder is used to connect all wind turbines of the cluster. The maximum number of wind turbines of the cluster is limited by the rated power of feeder cable. A circuit-breaker located at the substation protects the whole MV cable of the cluster. A fault on the MV cable causes that all wind

Fig. 3 Radial layout

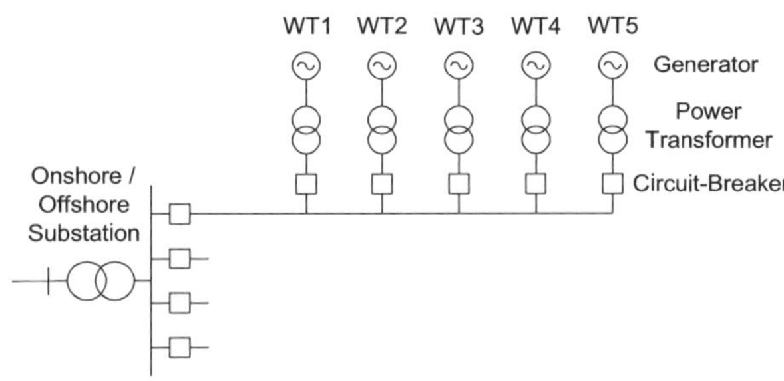

Fig. 4 Single-sided *ring* layout

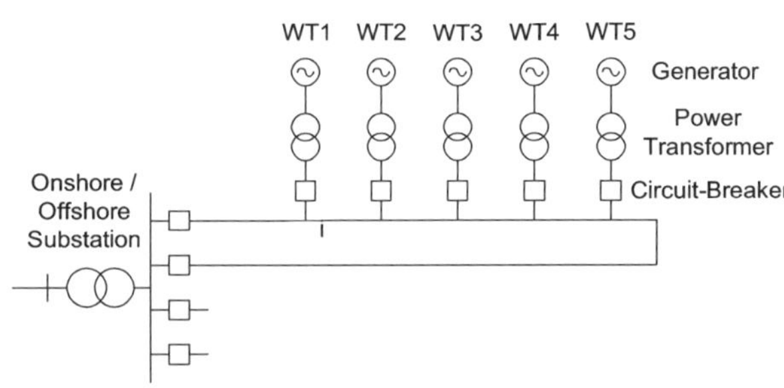

Fig. 5 Double-sided *ring* layout

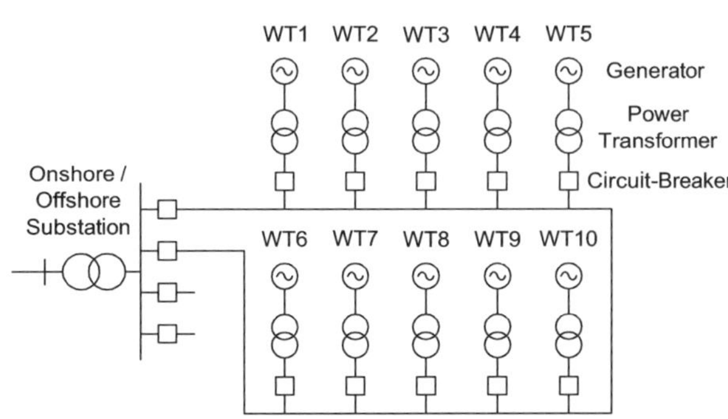

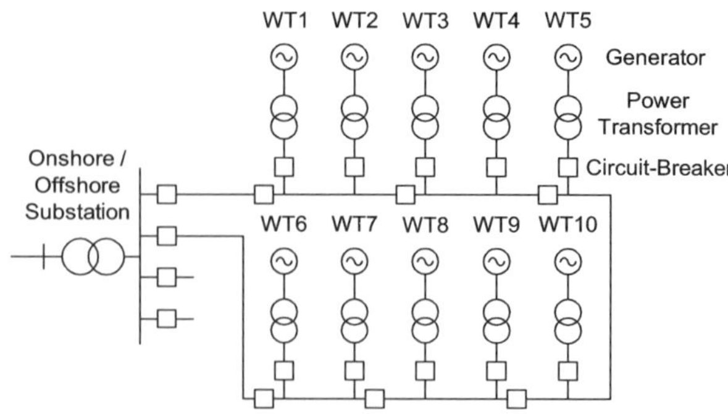

Fig. 6 Double-sided *ring* layout with circuit-breakers

turbines of the cluster are disconnected by the circuit-breaker of the substation. The path of the cluster to the substation might be partly restored by disconnecting the faulted cable from the MV switchgear of the upstream wind turbine. This is the cheapest layout but it has the disadvantage of its poor reliability.

- Single-sided ring layout. This layout is formed from the radial one by connecting and additional MV cable from the last wind turbine of the string (WT5) to a circuit-breaker of the substation. Thus, each wind turbine has two paths to the substation. The rated power of this cable has to be sufficient to transmit the entire power capacity of the cluster in the event of a fault. The availability of the cluster in this layout is greater than the radial one at the expense of installing an extra MV cable and circuit-breaker.
- Double-sided ring layout. This layout is obtained by connecting a MV cable between the ends (WT5 and WT10) of two wind turbine strings. Rated power of MV cables of both strings must be sized with an extra capacity as power might be diverted from one string to the other in the event of a fault. The definition of the maximum number of wind turbines in a string that can be diverted to another string is an important design variable in this layout [12].
- Double-sided ring layout with circuit breakers. This layout has the same MV cable connection scheme than the double-sided ring layout with the difference that a distributed control system is provided. Besides, a circuit-breaker is installed every two wind turbines. This system enables the automatic detection and location of faults with a selective isolation of them. Thus, fault analysis phase takes less time in this layout. Likewise, a greater number of wind turbines can export power during the fault analysis phase.

2.3 Transmission System

As mentioned previously, depending on the size of the OWF and its distance to the shore, power transmission in OWFs can be carried out either in MV or HV. This section is focused on HV transmission systems since MV systems are implicitly described in the last section.

The HV transmission system of an OWF basically consists of one or several offshore substations connected to the PCC through HV power lines.

The main function of an offshore substation is to increase voltage from MV to HV in order to reduce power losses in the power transmission. The main electrical components of an offshore substation are the step-up transformers, the MV switchgears and HV switchgears. All the electrical components have to be as compact as possible due to the limited space in the offshore substations.

The number and location of offshore substations to install in large OWFs is one of the main decisions that must be carefully evaluated in the design phase. By the end of year 2008 there were five OWFs in operation with a single offshore substation, namely: Horns Rev (160 MW), Nysted (165.6 MW), Lillgrund (110 MW), Barrow (90 MW) and Princess Amalia (120 MW). Moreover, the installation of more than one offshore substation is foreseen in some future larger OWFs as, for example, Krieger's Flak (640 MW) where two offshore substations are planned [13]. The installation of several offshore substations in appropriate locations of large OWFs provides some advantages such as reduction of cable length of the MV collection system or risk mitigation of the total power loss in the OWF due to a failure in a main electrical facility. By contrast, a major disadvantage of the installation of several offshore substations is the higher capital cost required.

Three-core XLPE insulated cables with copper conductor are usually adopted in HV transmission power from offshore substations to shore. Cables up to a voltage level of 150 kV have been successfully employed in HV transmission systems of OWFs, while cables of 245 kV are presently available in the market [14]. The main drawback of AC submarine cables is the high reactive power produced along the line because of the inherent large capacitive component of the cable. The resulting charging current induced in the cable reduces the current capacity available to transmit active power and increases the active power losses. In long HV lines, where this phenomenon is more pronounced, reactive power compensation devices need to be installed at ends of the lines, or even along the lines, in order to mitigate these negative effects. In terms of system reliability, it must be noted that none of the five OWFs in operation by the end of 2008 with HV lines had redundant HV cables despite the very negative consequences that a HV cable failure would cause in the availability of an OWF.

3 Assessment of AC Electric Power Systems of OWFs

3.1 System Efficiency Assessment

Efficiency of a power system is the proportion between output energy and input energy. The inequality between these terms is caused by energy losses, which in the AC electric power system of an OWF can be divided into three parts [15]:

- Fixed energy losses. These losses are independent of the power generated by the OWF, and therefore independent of wind speed too. Core losses in power transformers and compensation reactors are primarily responsible for them.
- Variable load losses. These losses vary with OWF power generation since they are function of the square of current flow. Consequently wind speed distribution has a crucial influence on them. These losses are caused by ohmic losses in cables and, to a lesser extent, by ohmic losses in windings of transformers and compensation reactors.
- Energy not generated due to constraints imposed by system unavailability. This issue is related to system reliability, which is explained in the next subsection.

Efficiency of the AC electric power system of an OWF is a significant aspect for the overall project feasibility. In fact, system efficiency assessment plays a more important role in OWFs than in onshore wind farms because of two main reasons. The first one is due to the higher wind speed regimes at offshore sites. Depending on the site, the energy production indicator for an onshore wind farm is normally around 2,000–2,500 full load hours per year, while for a typical OWF this figure reaches up to 4,000 full load hours per year [2]. The second reason is related to the increasing size of future OWFs since greater number of wind turbines and longer distances between the offshore facilities involve higher variable load losses in the electric power system.

3.2 System Reliability Assessment

3.2.1 Background of Reliability Aspects in OWFs

This section is focused on the system reliability assessment of the OWF from the perspective of the wind farm developer, whose main objective is to maximize energy produced, and thus maximize profits. Reliability assessment of OWFs has attracted more attention than that of onshore wind farms due to several aspects:

- Failure rates of components may increase because of marine environment.
- Repair times of components are longer and dependent on the season of the year when repair is carried out.

These issues greatly affect the availability of the electric power system, and therefore the overall availability of the OWF. For example, the failure of a key component, like a step-up transformer of the offshore substation or a HV cable of the transmission line, may involve a huge loss of energy if it occurs in a period of the year when accessibility to the facility is limited by adverse weather. In spite of this, the most electric power systems of OWFs in operation are designed with little or no redundancy. Moreover as reported in [16], components of the electric power system in operational OWFs have no experienced a large amount of failures. Thus, one might think that inclusion of redundancies for improving reliability in electric power systems would not be justified. However, it is also said in [16] that quantification of reliability parameters is still uncertain due to the few installations and the short operational experience. In any case, it is necessary to take system reliability assessment into account in the design of the electric power system due to the large power ratings of future OWFs.

Developing a model for system reliability assessment of an OWF requires the consideration of several factors [17]:

- Wind speed model. A probabilistic wind speed model must be developed to capture the stochastic nature of wind speed. Other significant aspects related to wind speed such as wake effects and wind speed spatial correlation can also be considered in this model.
- Wind turbine technology. The significance of wind turbine technology is primarily given by the reliability parameters that characterize its availability model and the operating parameters that define its power curve.
- Offshore environment. As mentioned above, offshore environment affects the reliability parameters of system components.
- Electric power system configuration. Several studies [4, 12, 18] have shown the influence of different electric power system configurations on the availability of the OWF. One of the main concerns is to determine the appropriate level of redundancy of the electric power system.

There are two main techniques for assessing the reliability of a generation facility, in this case an OWF: analytical and simulation techniques [19]. In the next two subsections, some methods used in wind generation that are derived from these techniques are explained.

3.2.2 Analytical Techniques in Wind Power Generation

Analytical techniques evaluate reliability indexes from a mathematical model of the system. They have been widely employed for adequacy evaluation of generating systems including wind farms. The most common way to represent these systems is to define a model for the conventional generation units and a model for the wind farms. The convolution method is generally used in the conventional generation model to obtain the capacity outage probability table of the entire system, which provides the probability of existence at each capacity level. Unfortunately, the statistical dependence of individual wind turbines on a common source, the wind, prevents the convolution method from being used in wind power generation.

This problem is overcome in [20] by merging a wind speed model and a wind turbine model. Wind speed is a stochastic process that can be represented approximately as a discrete state space (wind speed values) and continuous parameter state process (time). Wind speed is modeled by a birth-death Markov chain with a finite number of states. Some assumptions must be done in order to model the wind speed using a Markov chain:

- The wind speed model is statistically stationary, which means that the transition rates between states are constant throughout the whole process.
- The residence time of states follows an exponential distribution.
- The probability of a transition from a given wind state to another is directly proportional to the long-term average probability of existence of the new state.
- Transition between wind speeds occurs independently of transitions between wind turbines states.
- Transitions between non-adjacent states are not allowed, though it is possible to take them into account [21].

The wind speed model parameters are evaluated from a wind speed record in which wind speed measurements are usually taken at regular time intervals of 10 min. Data needed from wind speed records are the number of transitions between adjacent states and the residence time in a state before going to a different state. Wind speed parameters are sorted by wind speed states in a wind speed probability table.

The failure and repair process of a wind turbine is modeled as a two-state Markov chain since the system is stationary and the transition between the two states occurs in discrete steps. The two possible states of a wind turbine are operative or failed state. The power generated by the wind turbine in the operative state for a given wind speed is obtained through the power curve. The probabilities for the operative (p) and failed (q) states are calculated as follows:

$$\begin{aligned} p &= \frac{\mu}{(\lambda + \mu)} \\ q &= \frac{\lambda}{(\lambda + \mu)} \end{aligned} \tag{1}$$

where λ is the failure rate of the wind turbine (transition rate from operative to failed state) and μ is the repair rate of the wind turbine (transition rate from failed to operative state). It must be pointed out that this wind turbine model can be extended to the components of the electric power system of the wind farm.

The wind speed model is combined with the wind turbine model to obtain the wind farm model. The wind speed level and the status of wind turbines are known for each wind farm state. Thus the output power for each wind farm state is calculated by aggregating the output power of all wind turbines. The probabilities of the generation states of the wind farm are calculated through the solution of the stochastic system that is composed by the stochastic transitional probability matrix. Finally, a wind farm output table sorted by capacity levels is generated to provide the resulting probability states and the frequency and duration characteristics. The main drawback of this methodology is the complexity to solve the stochastic system in wind farms with a significant number of wind turbines because of the large number of states.

3.2.3 Simulation Techniques in Wind Power Generation

One important limitation of analytical techniques in wind generation applications is that chronological characteristics of wind speed can not be captured. This limitation can be overcome through the application of simulation techniques, also known as Monte Carlo simulation. Simulation techniques calculate reliability indexes by simulating the actual process and random behavior of the system. Simulation techniques in wind power generation have been used mainly to create chronological time series of wind speed and to simulate failure and repair processes in system components. The main disadvantage of simulation techniques is that they usually entail a heavy computational burden.

Auto-regressive and moving average (ARMA) time-series models [22, 23] are widely used to simulate the hourly wind speeds, and thus the available power of a wind farm. Another approach to simulate sequential wind speed series through Monte Carlo simulation is reported in [14]. This approach is based on the wind model presented in the previous subsection in which residence times of wind speed are distributed according to an exponential probability function with a constant hazard rate (transition rate). The current wind speed can reside in one of the wind speed states defined in a wind speed probability table as explained before. After a time residing in a wind speed state, the wind speed can move to one of the two adjacent wind speed states. This selection process is simulated by using the inverse transform method [19]. The new wind speed state and the residence time are determined from the smallest value of the transition times to the up and down adjacent states. These values are calculated by applying the inverse transform function of the exponential distributions (up and down transition rates) to two uniform pseudo-random numbers.

A failure and repair process for a component of the electric power system can be simulated following the same methodology used to generate the wind speed series in the previous model [19]. For a component modeled as a two-state Markov process with residence time of states distributed by an exponential density function, the duration of a state can be obtained by applying the inverse transform function:

$$\begin{aligned} TTF &= -\left(\frac{\ln U_1}{\lambda}\right) \\ TTR &= -\left(\frac{\ln U_2}{\mu}\right) \end{aligned} \tag{2}$$

where *TTF* and *TTR* are time to failure and time to repair respectively, U_1 and U_2 are pseudo-random numbers in the interval (0,1], λ is the failure rate of the component and μ is the repair rate of the component. The operating-repair cycles of a component can be simulated by generating alternatively a value of *TTF* and a value of *TTR*.

4 Optimization of AC Electric Power Systems of OWFs

4.1 Presentation of the Optimization Problem

Electric power system design of onshore wind farms is usually a trivial matter because its economic weight in the overall wind farm is relatively low and components and configurations are sufficiently proved and tested. On the contrary, electric power system design of OWFs has acquired a significant role due to its greater influence on the technical performance and economic viability of the whole OWF. The electric power system design of an OWF emerges as a complex problem with a large variable space and with various relevant issues to consider such as system efficiency, system reliability, characterization of wind speed at the site, etc. Thus, mathematical optimization techniques must lead wind farms developers and electrical engineers to facilitate decision-making by solving this optimization problem.

The optimization problem of the electric power system of OWFs is presented here from the perspective of the figure of a wind farm developer in a liberalized energy market. The objective of the wind farm developer in relation to the electric power system is mainly focused on minimizing the life-cycle costs of the project and keeping a certain level of technical performance according to grid codes and electrical standards. The most important aspects comprising the optimization problem of the AC electric power system of OWFs are described below in four parts: objective function, parameters, decision variables and constraints of the problem.

The objective function is formulated by following the objective of the wind farm developer of minimizing the total system costs for the lifespan of the project. The kinds of costs that may be taken into account in the objective function are as follows:

- Capital costs of system components. These costs are usually included through an amortization system over the lifespan of the project by taking into account the initial capital costs and the bank interest rate.
- Operation and maintenance costs of system components. They are usually included as a fixed cost per year.
- Costs due to energy losses in system components. They are composed of fixed energy losses and variable load losses as described in Sect. 3.1. Variable load losses are usually included as quadratic terms since they are proportional to the square of active power flow.
- Costs associated to energy not generated due to constraints imposed by electrical system unavailability. These costs are the result of the system reliability assessment as described in Sect. 3.2.

The parameters of the optimization problem are related to the input data. It should be noted that some parameters may be unknown depending on the progress stage of the project. The main parameters are listed below:

- Location of wind turbines. Wind turbines are located inside a predefined zone at the site by taking into account environmental conditions and following the criterion of maximizing the energy efficiency of the OWF.
- PCC. The PCC is an onshore facility of the transmission grid. The location and voltage level of the PCC are established by the TSO.
- Wind speed data. Wind speed data are collected at the site over a long enough period of time. These data are usually processed in the optimization models to obtain the power generation of wind turbines.
- Wind turbine data. The decision of which wind turbine model is installed in the OWF is taken by the wind farm developer. Basic data of the wind turbine such as rated power and power curve are usually needed in optimization models.
- System component data. Components of the electric power system that might be considered in the optimization model are MV switchgear, MV and HV cables and step-up transformers of offshore substations. Component data depend on the rated voltage of the system. Data required for system components are basically the technical characteristics, the initial capital costs, the operation and maintenance costs and the reliability data.

The selection of the decision variables determines the size and complexity of the optimization problem. The decision variable types of the problem are discrete and continuous. They are presented below:

- General configuration of the electric power system. As described in Sect. 2.1, two general configurations are possible. This choice determines whether CCPs are offshore substations or an area at the OWF site where MV cables gather on their way to the PCC.
- CCPs. The number, location and capacity of offshore substations may be decision variables in the general configuration case in which CCPs are offshore substations. Capacity of offshore substations is given by the number and rated power of step-up transformers, which are decision variables too. On the other case of general configuration, the location of an area where MV cables gather may be considered as a decision variable of the problem.
- HV power lines. HV power lines have to be installed in case of installation of offshore substations. The HV power lines can be divided into two parts, the submarine route and the onshore route. The rated HV is given by the voltage level at PCC. Variables of the HV power lines may be the number of lines and the rated power of the HV cable, which is given by the conductor cross-section and the HV cable type. It must be noted that the path of power lines are usually assumed as a straight-line between their starting and end points.
- Rated MV. The rated MV may be a variable but, in most cases, the best choice is to take the rated MV as the maximum MV voltage available on the electrical industry for MV system components of OWFs.
- Grid layout of MV collection system. As explained in Sect. 2.2, there are several layout types to form the MV collection grid. The MV switchgear arrangement, the number of clusters and the number of wind turbines per cluster may be considered as variables in addition to the type of layout.

- MV cables. MV cable type and conductor cross-section may be variables of the problem.
- Level of redundancy. Besides the inherent levels of redundancy offered by the different grid layouts of the MV collection system, the installation of redundant components may also be considered as a variable of the problem. Thus installation of redundant components may be taken into account not only for components of the MV collection system, but also for HV power lines and step-up transformers of the transmission system.
- Bus voltages. Magnitude and angle phase of system bus voltages in accordance with load flow equations.

The main constraints of the optimization problem are as follows:

- Connectivity of wind turbines. All wind turbines must be connected with power cables in order to have a path to send out the power generated to the PCC.
- Location of offshore substations. Location of offshore substations must be done in areas of the site where there is permission from relevant authorities.
- Maximum apparent power flow. The system components must withstand the maximum apparent power that can flow through them.
- Range of bus voltages. Magnitude and angle of bus voltages must remain within the permissible limits.
- Load flow equations. Balance of apparent power flow in all buses (wind turbines, CCPs and PCC) of the system must be observed. It means that the sum of all apparent power flows in a bus must be equal to zero. The load flow equations comprise a nonlinear system of equations.

This optimization problem is a mixed integer nonlinear programming (MINLP) problem because it contains both discrete variables and general nonlinear terms. The solution of the optimization model provides the optimal configuration of the electric power system of the OWF under study. This solution must be considered as a basic engineering design of the system. Therefore, this solution can be taken as a starting point to develop the detailed engineering of the system. Thus, steady-state and transient calculations of the power system should be performed for different operating conditions.

4.2 Solution Methods

According to optimization theory, the optimization problem posed in the previous subsection can be dealt with two different approaches: classical and metaheuristic optimization. The classical optimization methods consist of analytical techniques based on differential calculus. These methods search and ensure a local optimum in problems with continuous and differentiable functions. This method will be presented in the application example. By contrast, metaheuristic optimization methods have specific mechanisms to achieve the global optimum, even though this is not ensured. Metaheuristic methods are based on high level strategies to explore the solution space and search the global optimum. Metaheuristic methods can be applied to problems with non-differentiable functions. The most relevant methods used in the optimization of the electric power system of OWFs are presented below.

An optimization method that is limited to find the optimal location of offshore substations is applied in [24]. The aim of this method is to minimize the total cable length of the MV collection system. The optimal location is selected by calculating the distances from all wind turbines to all possible locations of the offshore substations that are taken into account in the problem.

Several optimization models [25–27] have been developed from metaheuristic methods. All of them adopt genetic algorithms (GAs) techniques, which are based on the Darwinian principle of natural selection of species. GAs operate a population of potential solutions (chromosomes) to find the global optimum by using evolutionary processes inspired in genetic variation and natural selection. The chromosomes are composed of genes, which represent values for a variable of the problem. These chromosomes are evaluated through an objective function called fitness function. The population of chromosomes is evolved through several iterations (generations) by three genetic operators, namely: selection, crossover and mutation. A termination criterion is defined to stop the development of generations. At the end of the process, there are generally several well-fitted chromosomes in the population.

The optimization model proposed in [27] is explained below since it is the most complete and detailed among all models based on GAs techniques. This model considers both the optimization of AC and DC electric power systems from various predefined system configurations. The wind speed distribution and location of the OWF are input data of the model. The main variables of the model are the configuration of the electric power system, the selection of key components and the voltage levels. The objective is to minimize the levelized production costs of the system with the required reliability. The capital costs, maintenance costs and the energy production are incorporated into the levelized production costs. Constraints of the model are the variation of voltage magnitude of buses within an established range, the limitation of the maximum apparent power flow in all branches, the power curve equation and the load flow equations of the system.

In a model developed by GAs techniques, the chromosomes have to be coded. The binary string method is used in this model since all variables are discrete or indexed options. Each variable is represented by several binary bits. To evaluate the objective function and check the technical feasibility of a population is also necessary to decode the binary codes of the chromosomes.

The initial population is established by using the diversity check method. The aim of this method is to ensure the diversity in the initial population of the most important genes defined by the user. Hence, all values of the key genes of the chromosomes are represented in the initial population in the same proportion.

The niching method with restricted tournament selection is used in the GA selection process to select the chromosomes of the current population (mating population). The niching methods try to search in many peaks in parallel and maintain population diversity in order to avoid stagnation in local optimum solutions.

Crossover and mutation operators are applied to the mating population to produce a new generation. Crossover operator produces a new chromosome (offspring) from the combination of two chromosomes (parents) belonging to the mating population. The uniform crossover operator is employed in this model. This method combines the parent chromosomes at gene level. Crossover operator is applied along the process with a probability defined by the user. On the other hand, the purpose of the mutation operator is to alter randomly the values of genes in the chromosomes in order to avoid stagnation in local optima. The mutation operator is applied according to a probability defined by the user, which is continuously decreased by a small step along the generations.

Various termination criteria such as generation number, fitness convergence, population convergence, best individual convergence and variation convergence are checked in each generation to assess the completion of the process.

The main problem of GA techniques applied to the optimization of the electric power system of OWFs is the premature convergence to local minima [26]. Premature convergence prevents the search of the global optimum since most part of the population gets trapped at local optima. The fast loss of diversity in the first iterations of the process has a great influence on the premature problem.

5 Application Example

5.1 Optimization Model

5.1.1 Introduction

It has been seen in the previous section that most of the optimization models available in the existing literature have been developed from metaheuristic methods based on GAs. On the other hand, an application example of an optimization model for the AC electric power system of OWFs based on classical approach is shown in this section [28] . The main aspects of this optimization model are described below, but a detailed description thereof is reported in [28].

The aim of this optimization model is to obtain the optimum electric power system design that minimizes the system costs derived from capital costs, costs of variable load losses and costs associated to energy not generated due to system unavailability. Likewise, proper technical performance in steady state (Kirchhoff's current law) must be achieved. The model takes the two general configurations of AC electric power systems into account (Figs. 1 and 2). The main components of AC electric power systems are considered in the model: MV cables, HV cables and step-up transformers of the offshore substation.

5.1.2 Wind Turbine Power Generation Model

Wind speed at the OWF site is supposed spatially uniform, that is, wind speed is the same for all wind turbines at any moment. Wind speed at the site is assumed to be distributed according to a Rayleigh probability density function [19]. This function might be approximated by a discrete function composed of a set of wind speed scenarios. The wind speed scenarios and their corresponding probabilities must be evaluated simultaneously if wind speed scenarios are not predefined in advance.

The wind speed probability density function is discretized in a predefined number of values in this application example. Each value represents a wind speed scenario (e), where the wind speed is assumed constant. The probability of each scenario is calculated by integrating the Rayleigh function between the interval bounds of each scenario. The wind power generation is obtained by transforming the values of the wind speed scenarios using the power curve of the wind turbine.

5.1.3 System Reliability Assessment

The aim of the system reliability analysis is to assess the suitability of installation of redundant components for system components subject to failure. This decision is a trade-off between the capital costs needed to have redundant components and the cost of energy not generated throughout the lifespan of the OWF due to component failures. The components subject to failure in the model are as follows:

- Step-up transformers of the offshore substation.
- HV power lines of the transmission line.
- MV cables that connect wind turbines to the CCP (CCP is an offshore substation).
- MV cables that connect wind turbines to the PCC (CCP is an area where MV cables gather on their way to the PCC).

The failure and repair process of a component subject to failure is modeled with a two-state Markov chain as explained in Sect. 3.2.2. Thus, the probabilities for the operative state and failed state of a component subject to failure are obtained through Eq. (1).

System reliability assessment is carried out through the state space method [19]. This method is based on the study of all system states. A system state (s) is defined by the availability or unavailability of each system component subject to failure. If the system states are too numerous to be enumerated, a reduction in the number of system states might be done whenever the probability of the remaining ones are small enough to be ignored. It is considered that each system state might have, at most, one component unavailable (N − 1 criterion). The probability of a system state is calculated by multiplying the state probability of each component.

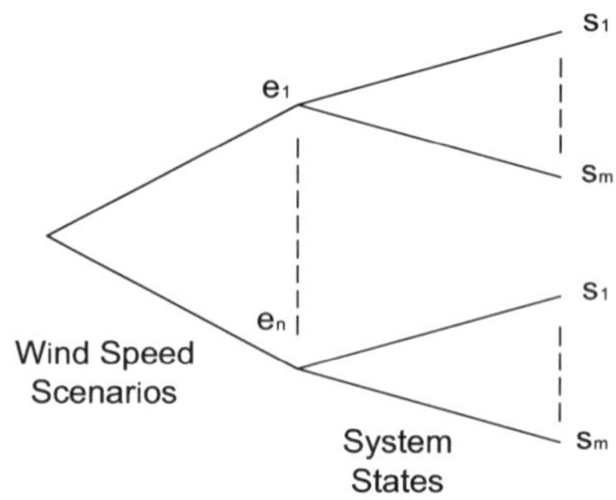

Fig. 7 Scenario tree

5.1.4 Scenario Tree

Both random variables, wind speed and system states, are implemented in the optimization model using a tree structure as shown in Fig. 7. This structure is appropriate to represent both stochastic phenomena since chronological order is not considered in this model.

The tree is composed by the branches of the wind speed scenarios and the branches of the system states. The branches of the system states sprout from the branches of the wind speed scenarios. Hence, the total number of branches of the tree is the product of wind speed scenarios by the number of system states, given that both variables are statistically independent.

5.1.5 Model Implementation

Given an OWF composed of a specific number of wind turbines, the location of a generic wind turbine is defined by its coordinates in a rectangular coordinate system. It is assumed that the possible locations of the CCP (offshore substation or area where MV cables gather) inside the OWF site are known. The location of the PCC is also known. Locations of CCP and PCC are defined by pairs of coordinates in the same coordinate system than wind turbines. The straight-line distance is taken for the calculation of cable length between the different locations of the facilities of the OWF (wind turbines, CCP and PCC).

The wind turbines power generation for each wind speed scenario is an input data that is derived from the wind speed model. The period duration of each wind speed scenario is calculated by multiplying the probability of each wind speed scenario by the lifespan of the OWF.

The rated HV is defined by the PCC voltage level, while the rated MV is selected by the user. A range of different conductor cross-sections of cable and a range of step-up transformers specified by their rated powers are defined by the user. The rated active power and initial capital costs are parameters required for the system components. Moreover, data of conductor resistance of cables are also needed to assess the variable load losses.

Failure and repair rates of system components subject to failure are also required to calculate the probability of system states through the system reliability assessment established in Sect. 5.1.3. It can be seen from the tree structure shown in Fig. 7 that the period duration of each system state can be obtained by multiplying the period duration of the corresponding wind speed scenario by the respective system state probability.

The variables of the model are described as follows:

- Binary variables for the installation or not of cables with different conductor cross-sections between all facilities of the OWF.
- Integer variables for the installation or not of a specific number of step-up transformers with different rated power in the offshore substation.
- Binary variables for the installation or not of redundant components for system components subject to failure.
- Continuous variables for the active power flow through the cables in each wind speed scenario and system state.
- Continuous variables for the active power not generated by each wind turbine in each wind speed scenario and system state due to system unavailability.
- Binary variables for the installation or not in the CCP locations of an offshore substation or an area where MV cables gather.

The objective function consists of minimizing the cost of the power system over the lifespan of the OWF. The objective function is composed of three parts:

- Capital costs of system components and redundant elements using the French system of amortization [29].
- Cost of energy not generated by wind turbines in each wind speed scenario and system state due to unavailability of system components subject to failure.
- Cost of variable load losses in MV and HV cables in each wind speed scenario and system state. Variable load losses in cables are composed of quadratic terms.

The constraints of the model are explained below:

- Balance of active power flow is required in all wind turbine buses for each wind speed scenario and system state. A set of linear equations are generated since bus voltages and reactive power flows are not contemplated in this model.
- In case of installation of offshore substation, balance of active power flow is also required in the CCP bus for each wind speed scenario and system state.
- Active power flow between any two facilities of the OWF requires the installation of a cable. The rated power of the cable has to exceed the maximum active power that can flow through it. A redundant cable can be installed to substitute a cable subject to failure in system states in which this is unavailable.
- Active power flow between the CCP (CCP is an offshore substation) and PCC requires the installation of step-up transformers in the offshore substation. The sum of the rated power of the transformers has to exceed the maximum active

power that can flow through them. A redundant transformer can be installed to substitute a principal transformer in system states in which this is unavailable.
- Only one cable with a specific conductor cross-section can be selected at any connection between two facilities of the OWF.
- One cable can leave, at most, from each wind turbine.
- There is only one location for the CCP.

This problem is a mixed integer quadratic constraint programming (MIQCP) problem since it contains both discrete variables and quadratic terms. The model is implemented in GAMS language [30], which is a high-level modeling system for mathematical programming and optimization.

The size of the optimization problem basically depends on the amount of wind turbines, wind speed scenarios and system states. For a real OWF with several tens of wind turbines, the size of the problem might reach up to tens of millions of decision variables and constraints. Thus, it is necessary to carry out some simplifications in the optimization problem in order to be processed by GAMS language. An example of such simplifications is described in subsection 5.2.2.

5.2 Case Study

5.2.1 Description of the OWF

Barrow Offshore Wind Farm (BOWF) [31], a real OWF located in the west of England at the East Irish Sea, is chosen to run the optimization model.

Location of the facilities of BOWF and layout of the electric power system is shown in Fig. 8. BOWF consists of a total of thirty wind turbines distributed in four rows, two with seven and two with eight wind turbines. The distance between wind turbines is around 500 m and between rows is about 750 m. The rated power of each wind turbine is 3 MW, so the total installed power of BOWF is 90 MW.

The MV collection system consists of four circuits that connect all wind turbines of each row to an offshore substation, which is located in the eastern part of BOWF (CCP1). Three-core cables of copper with 120 mm^2 cross-section connect the furthest wind turbines from the offshore substation in each MV circuit, whereas three-core cables of 300 mm^2 cross-section are used to connect the closest wind turbines to the offshore substation. A 120 MVA power transformer steps-up the voltage level from 33 to 132 kV at the offshore substation. The power generated by BOWF is exported to the national grid connection point (PCC) through a 27 km submarine 132 kV power line. The PCC is around 25 km to the east of BOWF. Redundant components for cable or power transformers are not installed.

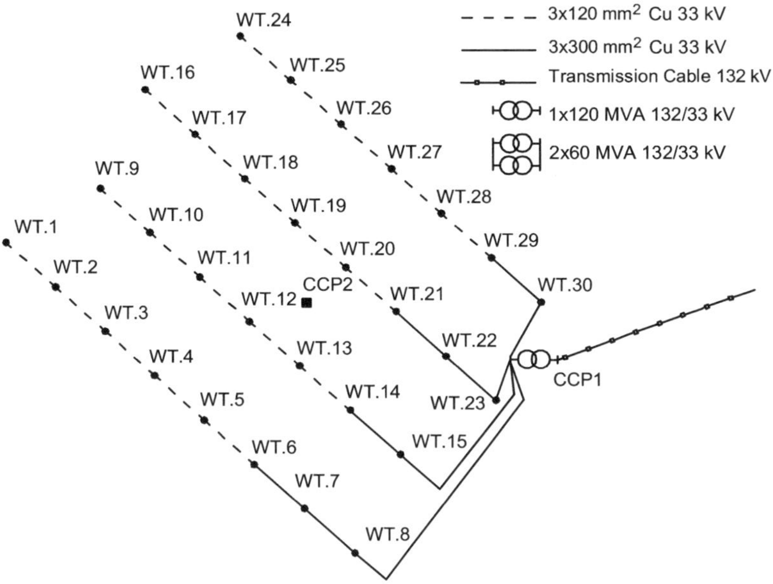

Fig. 8 Actual layout of the electric power system of BOWF

5.2.2 Input Data and Simplifications

Some simplifications are done to decrease the size of the optimization problem without losing generality in solving the problem:

- Mean wind speed at hub height in BOWF is 9 m/s [31]. The Rayleigh probability density function is discretized in five wind speed scenarios.
- The cable connections between wind turbines are limited to those pairs which are at straight-line distance of less than 700 m. This limitation is equivalent to establishing that each wind turbine can only be connected to the adjacent wind turbines of its row.
- Two possible locations of the CCP are considered, both for the offshore substation and for the area where MV cables gather. One CCP is located at the present location of the offshore substation (CCP1), while the other is located in a central point of the BOWF (CCP2).
- As many as nine wind turbines can be connected to the two possible CCP locations. The closest nine wind turbines to the CCP locations are selected.
- Two different conductor cross-sections are taken into account for MV and HV cables. Three-core cables of copper with 120 and 300 mm^2 cross-sections are chosen for MV cables, whereas 400 and 630 mm^2 cross-sections are adopted for HV cables. Data of rated power, values of conductor resistance and initial capital costs of cables are taken from [32–34], respectively.

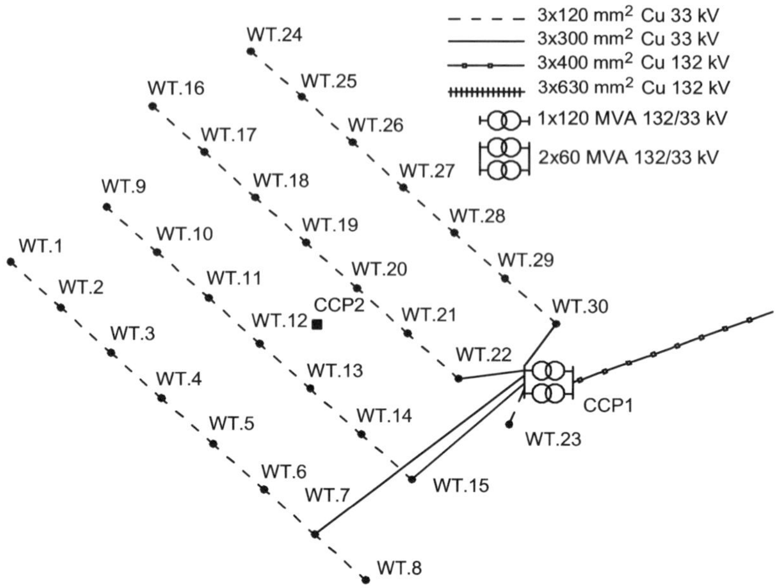

Fig. 9 Optimal solution for the electric power system of BOWF

- Two step-up transformers with different rated power are considered for the offshore substation. The rated power of the transformers is set to 60 and 120 MVA. Data cost of step-up transformers are extrapolated from [35].
- Each system state might have at most one component unavailable. Failure and repair rates of cables and transformers are taken from [36].

As a result of these simplifications, the optimization problem is composed of approximately 84,300 constraints, 83,200 continuous variables and 264 discrete variables.

5.2.3 Results

The optimal solution from the model is shown in Fig. 9. According to the optimal solution, redundancies are not required. The result is consistent with the actual layout of the electric power system of BOWF. There are some discrepancies between both layouts regarding the connections of the closest wind turbines to the offshore substation and the number of step-up transformers installed in the offshore substation.

The cost of the actual power system layout of BOWF calculated through the model is €34.9 million (a three-core cable of 400 mm^2 is considered for the HV line), while the cost of the optimal solution obtained from the model is €34.6 million. This means that by assuming the input data and hypotheses of the model, the optimal solution is more economical than the actual layout of BOWF.

6 Conclusion

The current trend of offshore wind energy is the development of large-size OWFs with power ratings of the order of several hundreds of megawatts. Thus, the design of the electric power system is a critical issue in the overall feasibility of an OWF project. Hence, the electric power system of OWFs needs to be optimized to minimize the life-cycle costs and at the same time maintaining a certain level of technical performance. The optimization of the electric power system of an OWF sets up a MINLP problem, which requires to consider jointly the key factors that determine the electric power system design (investment costs of components, system efficiency and system reliability), as well as inherent aspects of wind energy, such as the stochasticity of wind speed or the geographic location of wind turbines.

Most of the optimization models available in the existing literature that face the optimization problem of the electric power system of OWFs have been conducted from metaheuristic methods based on GAs. Utilization of appropriate GAs techniques is required in these models to ensure the diversity of the population at early stages of the process to avoid the problem of premature convergence.

An application example of an optimization model based on the classical approach of optimization has been explained in this chapter. This model poses a MIQCP problem that provides the optimal configuration of the electric power system, taking into account the three key factors that characterize its design and the main requirements of these facilities. This application example has shown that models based on the classical are also suitable to address satisfactorily the problem of optimizing the AC electric power system of OWFs.

References

1. Fichaux N, Wilkes J, Van Hulle F, Cronin A (2009) Oceans of opportunity. European Wind Energy Association. http://www.ewea.org/fileadmin/ewea_documents/documents/publications/reports/Offshore_Report_2009.pdf. Accessed 12 Dec 2009
2. Krohn S, Awerbuch S, Morthorst PE, Blanco I, Van Hulle F, Kjaer C (2009) The economics of wind energy. European Wind Energy Association. http://www.ewea.org/fileadmin/ewea_documents/documents/publications/reports/Economics_of_Wind_Main_Report_FINAL-lr.pdf. Accessed 12 Dec 2009
3. Lundberg S (2006) Evaluation of wind farm layouts. EPE J 16:14–20
4. Bresesti P, Kling WL, Hendriks RL, Vailati R (2007) HVDC connection of offshore wind farms to the transmission system. IEEE Trans Energy Convers. doi:10.1109/TEC.2006.889624
5. Robinson J, Jovcic D, Joós G (2010) Analysis and design of an offshore wind farm using MV DC grid. IEEE Transactions on Power Del. doi: 10.1109/TPWRD.2010.2053390
6. Meyer C, Höing M, Peterson A, De Doncker RW (2007) Control and design of DC grids for offshore wind farms. IEEE Transactions on Ind. Appl. doi: 10.1109/TIA.2007.908182

7. Roggenkamp MM, Hendriks RL, Ummels BC, Kling WL (2010) Market and regulatory aspects of trans-national offshore electricity networks for wind power interconnection. Wind Energ. doi: 10.1002/we.378
8. Pechey J, Taylor P, Dixon R, Lawson M, Dinning A (2004) The role of medium voltage electrical system design in risk management for offshore wind farms. Wind Eng. doi:10.1260/0309524043028154
9. Franken B, Breder H, Dahlgren M, Nielsen EK (2005) Collection grid topologies for offshore wind parks. In: The 18th international conference and exhibition on electricity distribution. Turin, Italy
10. Quinonez-Varela G, Ault GW, Anaya-Lara O, McDonald JR (2007) Electrical collector system options for large offshore wind farms. IET Renew Power Gener. doi:10.1049/iet-rpg:20060017
11. Lee MQ, Lu CN, Huang HS (2009) Reliability and cost analyses of electricity collection systems of a marine current farm—a Taiwanese case study. Renew Sustain Energy Rev. doi:10.1016/j.rser.2009.01.011
12. Liu X, Islam S (2008) Reliability issues of offshore wind farm topology. In: The 10th international conference on probabilistic methods applied to power systems. Rincon, Puerto Rico
13. Ullah NR, Larsson A, Petersson A, Karlsson D (2008) Detailed modeling for large scale wind power installations—a real project case study. In: Third international conference on electric utility deregulation and restructuring and power technologies. Nanjing, China
14. Kling WL, Hendriks RL, Den Boon JH (2008) Advanced transmission solutions for offshore wind farms. In: IEEE power and energy society general meeting. Pittsburgh, USA
15. Walling RA, Ruddy T (2005) Economic optimization of offshore windfarm substations and collection systems. In: Fifth international workshop on large-scale integration of wind power and transmission networks for offshore wind farms. Glasgow, Scotland
16. Holmstrom O, Negra NB (2007) Survey of reliability of large offshore wind farms. Part 1: Reliability of state-of-the-art wind farms. Project upwind. http://www.upwind.eu/Shared%20Documents/WP9%20-%20Publications/D9.1%20-%20Survey%20of%20reliability.pdf. Accessed 12 Dec 2009
17. Negra NB, Holmstrom O, Bak-Jensen B, Sorensen P (2007) Aspects of relevance in offshore wind farm reliability assessment. IEEE Trans Energy Convers. doi:10.1109/TEC.2006.889610
18. Sannino A, Breder H, Nielsen EK (2006) Reliability of collection grids for large offshore wind parks. In: Ninth international conference on probabilistic methods applied to power systems. Stockholm, Sweden
19. Billinton R, Allan RN (1992) Reliability evaluation of engineering systems. Plenum Press, New York
20. Castro-Sayas F, Allan RN (1996) Generation availability assessment of wind farms. IEE Proc Gener Transm Distrib. doi:10.1049/ip-gtd:19960488
21. Leite AP, Borges CLT, Falcao DM (2006) Probabilistic wind farms generation model for reliability studies applied to Brazilian sites. IEEE Trans Power Syst. doi:10.1109/TPWRS.2006.881160
22. Billinton R, Chen H, Ghajar R (1996) A sequential simulation technique for adequacy evaluation of generating systems including wind energy. IEEE Trans Energy Convers. doi:10.1109/60.556371
23. Karki R, Hu P, Billinton R (2006) A simplified wind power generation model for reliability evaluation. IEEE Trans Energy Convers. doi:10.1109/TEC.2006.874233
24. Hopewell PD, Castro-Sayas F, Bailey DI (2006) Optimising the design of offshore wind farm collection networks. In: 41st international universities power engineering conference. Newcastle upon Tyne, UK
25. Li DD, He C, Fu Y (2008) Optimization of internal electric connection system of large offshore wind farm with hybrid genetic and immune algorithm. In: Third international

conference on electric utility deregulation and restructuring and power technologies. Nanjing, China
26. Zhao M, Chen Z, Hjerrild J (2006) Analysis of the behaviour of genetic algorithm applied in optimization of electrical system design for offshore wind farms. In: 32nd annual conference on IEEE industrial electronics. Paris, France
27. Zhao M, Chen Z, Blaabjerg F (2009) Optimisation of electrical system for offshore wind farms via genetic algorithm. IET Renew Power Gener. doi:10.1049/iet-rpg:20070112
28. Banzo M, Ramos A (2011) Stochastic optimization model for electric power system planning of offshore wind farms. IEEE Transactions on Power Syst. doi:10.1109/TPWRS.2010.2075944
29. Córdoba M (2006) Fundaments and practice of financial mathematics. Dykinson, Madrid
30. Brooke A, Kendrick D, Meeraus A, Raman R, Rosenthal RE (2008) GAMS—a user's guide. GAMS Development Corporation, Washington, DC
31. BOWind. http://www.bowind.co.uk/
32. ABB: XLPE submarine cable systems, Attachment to XLPE cable systems—user's guide. Available E-mail: sehvc@se.abb.com
33. General Cable Corporation: Tables on conductors (in Spanish). http://www.generalcable.es/Productos/AyudasTécnicas/tabid/378/Default.aspx. Accessed 12 Dec 2009
34. Green J, Bowen A, Fingersh LJ, Wan Y (2007) Electrical collection and transmission systems for offshore wind power. In: Offshore technology conference. Houston, USA
35. Lazaridis LP (2005) Economic comparison of HVAC and HVDC solutions for large offshore wind farms under special consideration of reliability. Master's Thesis, Royal Institute of Technology, Stockholm, Sweden. http://citeseerx.ist.psu.edu/viewdoc/download?doi=10.1.166.4595&rep=rep1&type=pdf. Accessed 12 Dec 2009
36. Bozelie J, Pierik JTG, Bauer P, Pavlovsky M (2002) Dowec grid failure and availability calculation. http://www.ecn.nl/docs/dowec/10077_001.pdf. Accessed 12 Dec 2009

Low-Power Wind Energy Conversion Systems: Generation Configurations and Control Objectives

Iulian Munteanu, Antoneta Iuliana Bratcu and Emil Ceangă

Abstract This chapter focuses on the formulation and solving of the main control problems specific to low-power wind energy conversion systems (WECS) depending on their use either for grid-connected or for stand-alone application systems. This chapter approaches three classes of problems—namely, the low-power WECS specificity, operation of low-power WECS within grid-connected applications and use of low-power WECS as parts of stand-alone multi-source systems supplying isolated loads—with special focus on the associated control problems. A comprehensive synthesis of main results from the literature is aimed at, also giving the opportunity of presenting some results obtained by the authors.

Abbreviations and notations

WECS	Wind energy conversion system
LSS/HSS	Low-speed/high-speed shaft
MPPT	Maximum power point tracking
ORC	Optimal regimes characteristic
PMSG	Permanent-magnet synchronous generator
SCIG	Squirrel-cage induction generator
DFIG	Doubly-fed induction generator
v	Instantaneous wind speed

E. Ceangă
"Dunărea de Jos" University of Galaţi, Domnească 47, 800008 Galaţi, Romania
e-mail: emil.ceanga@ugal.ro

I. Munteanu
Grenoble Electrical Engineering Laboratory (G2ELab), 11 Rue des Mathématiques, BP 46 38402 Saint-Martin d'Hères, France
e-mail: iulian.munteanu@g2elab.grenoble-inp.fr; iulian.munt@gmail.com

A. I. Bratcu (✉)
Grenoble Image, Speech, Signal and Control Systems Laboratory (Gipsa-Lab), Control Systems Department, Grenoble Institute of Technology, 11 Rue des Mathématiques, BP 46 38402 Saint-Martin d'Hères, France
e-mail: antoneta.bratcu@gipsa-lab.grenoble-inp.fr

P. M. Pardalos et al. (eds.), *Handbook of Wind Power Systems*, Energy Systems, DOI: 10.1007/978-3-642-41080-2_23, © Springer-Verlag Berlin Heidelberg 2013

Ω_l/Ω_h	Low-speed/high-speed shaft rotational speed
λ	Tip speed ratio of a wind turbine
β	Blades' pitch angle of a wind turbine
C_p/C_Γ	Power coefficient/torque coefficient of a wind turbine
Γ_{wt}/Γ_G	Wind turbine/electrical generator torque
P_{wt}	(mechanical) power of a wind turbine
$x_{a,b,c}$	Phase components of a three-phase variable x
$x_{d,q}$	d and q components of a three-phase variable x, obtained by Park transform

1 Introduction: Basic Concepts

The last years have witnessed an exponential growth of large-power wind turbines. However, the low-power wind energy conversion systems (WECS) have not lost their importance, being nowadays of great interest in islanded generation, microgrids, distributed energy production, etc. The field specialists agree on defining these systems as the ones having power ratings below 100 kW [11]. Because of the erratic availability of the primary resource, the wind, the most encountered off-grid applications are related to battery charging or other kind of energy storage. Also, these WECS often operates in conjunction with other types of generation systems, such as photovoltaic or Diesel generators, being combined in so-called hybrid power generation systems. Research on control of low-power WECS is justified from at least two viewpoints, namely:

(a) in the case of grid-connected applications, this type of WECS implements the concept of integrated production and use of electrical energy. In such systems the energy is produced directly in the low-voltage user grid and the consumer/producer-grid transfer becomes bidirectional in order to ensure the power flow balance to the load irrespective of the wind conditions;
(b) in the case of supplying isolated (local) grids, the low-power WECS must be accompanied by back-up energy sources able to cover the wind deficit when this is the case. In this case a stand-alone multi-source system is obtained, which can eventually contain other renewable energy sources, as well as classical sources and energy storage devices (kinetic, electrochemical, etc.).

Unlike the high-power WECS, the low-power WECS are characterized by prior use of permanent-magnet synchronous generators coupled directly with the wind turbine shaft. In the case of isolated grid applications based upon such conversion configuration, a great variety of control structures are used for accomplishing different control goals, like, for example, ensuring admissible frequency deviation when no energy storage device is present.

This chapter is organized as follows. The first section is dedicated to introducing the basic concepts related to low-power wind turbines operation and modelling. The second section presents the control principles applied for this kind of WECS, namely power optimization in partial load and power limitation in full load, for both grid-connected and autonomous systems. The third and fourth section are dedicated to specific issues related to the grid-connected WECS and WECS supplying insulated loads, respectively. Whereas the grid-connected systems are mainly concerned with the power optimization as control objective, the stand-alone systems also include different configurations of hybrid systems, with or without energy storage devices—e.g., wind-battery, wind-battery-Diesel, wind-Diesel—therefore, the control goals are various. The chapter ends by a concluding section, the fifth, where open issues are also identified.

1.1 Small Wind Turbines

The prime movers used in such low-power systems employ quite diversified aerodynamic designs. While the majority still uses the classical 2- and 3-blade horizontal-axis wind turbines [11], some other rotor types can also be encountered such as the vertical-axis turbines based on Savonius and Darrieus concepts [12]. However, in the very low-power domain (under 2 kW) are emerging new designs of wind turbine rotors—see for example the multi-blade-based turbines inspired from jet engine technology [10].

The main aerodynamic features that differentiate the low-power wind turbines in relation to the medium- or high-power ones find their origin in some economical reasons that govern their designs. All the associated mechanisms are simple and their solid construction is suitable for reliable operation. However, the primary protection devices, such as the ones ensuring the power and rotational speed limitations, are still existent.

One must note that in the majority of cases the blades are not pitchable, therefore their position is fixed in relation to the hub. However, in some of the cases the blades have variable pitch capability, but their collective pitch control is achieved by using self-operated (direct-action) controllers. The energy required for control action is entirely supplied by the mechanical transducer (rotating masses) which directly (mechanically) actuates the final control element, without amplifying the control signal [11]. This system can be used at both assisting the start-up and at limiting the captured power from the wind. Anyway, these kinds of rotors, as well as the vertical-axis ones, rotate in most of their operating regimes at fixed pitch; therefore they can be modelled and controlled accordingly.

Also, the yaw mechanisms driving the rotor turbine position with respect to the air flow direction are often reduced to a tail vane. Due to the relatively low rotor inertia, this system is sufficiently efficacious to keep the turbine headed into the wind. However, this changes the normal method of yawing the rotor out of wind in order to reduce the captured power or to park the turbine [31].

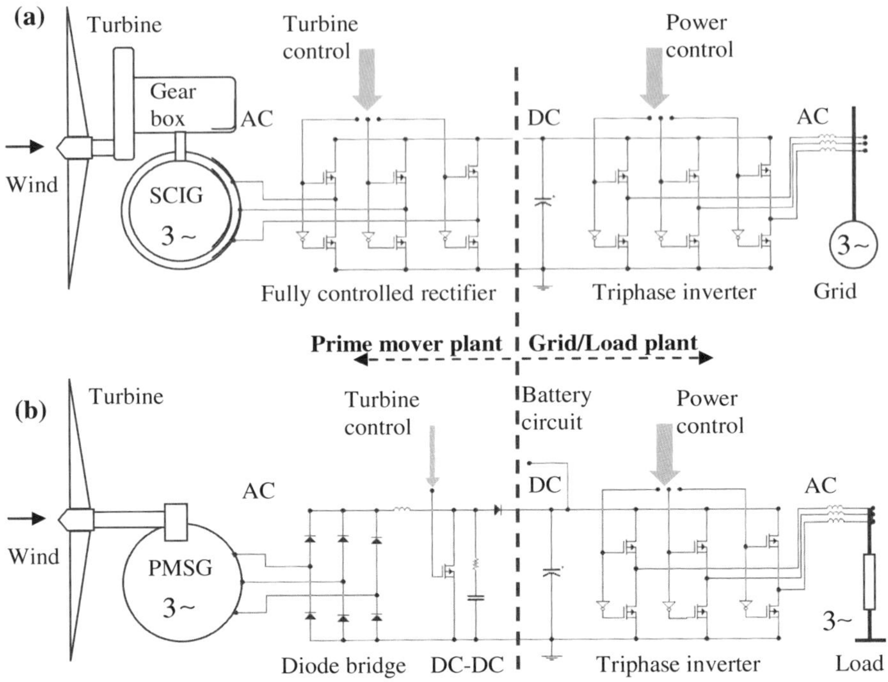

Fig. 1 Some basic low-power WECS structures. **a** Grid-connected SCIG-based. **b** PMSG-based WECS feeding a 3-phase insulated load

1.2 Electrical Generation Systems

The generator type used in low-power WECS depend not only of the prime mover but mostly on its mechanical transmission. At low-power ratings, 3-phase permanent-magnet synchronous generators (PMSGs) and squirrel-cage induction generators (SCIGs) are typically used. The SCIG is used in conjunction with a geared drive train, as the prime mover and the generator rated speeds are definitely unmatched. PMSG are preferred on insulated/remote applications. If the PMSG has a sufficiently large number of poles, the WECS may have a gearless (or direct-drive) transmission.

The interface with the power grid or with the AC insulated load has different configurations, depending on the electrical generator and on the existence of energy storage devices. Mainly, there is a generator-side converter and a grid/load-side converter that interacts via a DC-link. The former converter operates normally as an AC-DC converter and the latter as a voltage-source inverter (DC-AC—see Fig. 1). The generator-side converter may have one of two main configurations: a fully controlled rectifier or a diode rectifier linked to a DC–DC converter.

Figure 1a shows a grid-connected WECS. Its induction generator is driven by the wind turbine via a gearbox. The power electronics interface is a back-to-back converter. The generator-side converter allows both generation and motoring regimes. Also, the generator torque and the rotor magnetizing flux can be separately imposed using a highly-effective (although complex) vector control structure.

Figure 1b presents a WECS feeding an insulated load. The low-speed PMSG is directly driven; its power electronics converter is quite simple, allowing only the generation regime. The electromagnetic torque output by PMSG is not independent of the machine flux and presents a quite accentuated ripple due to the diode rectifier operation. The accumulator battery is connected to the DC-link by means of a charge controller that supervises the battery operation.

One may note that the configurations presented in Fig. 1 are not the only ones to be found in the associated literature. One may use a current-source inverter [33] or may feed a single-phase load. The generator-side converter may contain a buck converter [37], the circuit feeding the battery may contain a bidirectional DC–DC converter and so on.

These power electronics structures decouple the generator from the power grid, allowing the variable-speed operation. They also enable feeding the grid with sinusoidal currents, and varying the output reactive power. The main issue is that they enable the generator currents control—and therefore its electromagnetic torque control—in order to change the turbine rotational speed. Thus, by modifying the turbine operating point one controls its aerodynamic efficiency and therefore imposes its output power.

Therefore, from a systemic point of view, due to the DC-link intermediary stage being used, the conversion chain is split into two plants. The first, namely the *prime mover plant*, typically contains the prime mover, electrical generator and its power electronics converter. The other one, namely the *grid/load plant*, contains the power grid elements, the grid-side converter and the DC-link. The two plants are controlled separately by means of the associated converters (see Fig. 1); their control inputs are updated in order to fulfil distinct control objectives. The DC-link voltage is a variable common to both of plants and the DC-link may be considered within one plant or within the other one depending on the particular WECS structure and operating conditions [1].

1.3 Some WECS Modelling Aspects

The performance of a wind turbine is primarily characterized by the manner in which its power varies with the wind speed and is expressed by means of non-dimensional characteristic performance curves [6]. The *tip speed ratio* of a wind turbine is a variable expressed as $\lambda = R \cdot \Omega_l / v$, i.e., the ratio between the peripheral blade speed and the wind speed, where R is the blade length, Ω_l is the rotor speed (the low-speed shaft rotational speed) and v is the wind speed. The non-dimensional *power coefficient*, C_p, is an average model that describes the power extraction efficiency of a wind turbine. The variation of C_p versus λ curve usually characterizes the aerodynamic performance of a wind turbine and offers key information for the wind turbine control. The power harvested by a wind turbine is

$$P_{wt} = 0.5 \cdot \rho \cdot \pi R^2 \cdot v^3 \cdot C_p(\lambda), \qquad (1)$$

where ρ is the air density. The *torque coefficient*, denoted by C_Γ, characterizes the rotor output (wind) torque, Γ_{wt}, and is expressed by $C_\Gamma(\lambda) = C_p(\lambda)/\lambda$.

For a fixed-pitch wind turbine, the wind torque depends on the low-speed shaft rotational speed, Ω_l, the wind speed, v, and $C_\Gamma(\lambda)$:

$$\Gamma_{wt} = P_{wt}/\Omega_l = 0.5 \cdot \pi \cdot \rho \cdot v^2 \cdot R^3 \cdot C_\Gamma(\lambda) \qquad (2)$$

$C_\Gamma(\lambda)$ gives the rotor mechanical characteristic allure, $\Gamma_{wt}(\Omega_l)$, for a given wind velocity. More detailed aerodynamic models can be developed, emphasizing effects like the rotational sampling, spatial filtering or induction lag [21, 29]. For low-power wind turbines, these effects, together with the structural dynamics, can be neglected, and simplified models are preferred [38].

Typical steady-state C_Γ and C_p variations with respect to the tip speed ratio for 2-blade HAWT are given in Fig. 2. Note that the vertical-axis wind turbine average aerodynamic characteristics are not very different.

Curves belonging to a passive-stall regulated rotor (dotted line) and to a rotor type that is suitable for active-stall regulation (continuous line) have been here chosen as an example. The C_p curve is lower than the theoretical Betz limit (0.59) that assumes perfect blade design [6]. The power conversion efficiency has a well determined maximum for a specific tip speed ratio, denoted by λ_{opt}.

The turbine rotor and the electrical generator are coupled mechanically by means of the drive train, whose structure depends on each particular WECS technology. Most of the systems employ speed multipliers as drive trains. The electrical machine thus experiences an increased rotational speed and a reduced electromagnetic torque.

The mechanical transmission is dissociated into two parts: the *low-speed shaft* (LSS) on which the turbine rotor is coupled and the *high-speed shaft* (HSS) relied on by the electrical generator. The coupling between the two shafts can be either rigid or flexible. Flexible drive trains are used for damping the mechanical efforts generated by exogenous variations (wind speed or electromagnetic torque). For low-power WECS the drive train is rigid and comprises in general a single stage. The speed multiplier affects the WECS's weight, reliability and overall efficiency.

Electrical generators are systems which output the electromagnetic torque, Γ_G, as a result of the interaction between stator and rotor fluxes, which depends upon each particular generator. In WECS the generator experiences a mechanical interaction with the drive train. Generators are usually modelled in the (*d*, *q*) frame, as required by implementing the vector control [5]. Figure 3a illustrates the modelling principle for the PMSG case.

The generator modelling captures the evolution of the electromagnetic variables into a mathematical form. Thus, a set of equations involving voltages, fluxes and

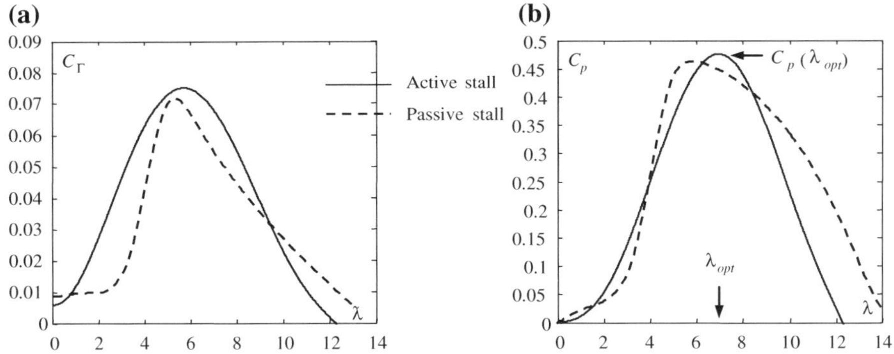

Fig. 2 Typical HAWT $C_\Gamma - \lambda$ (**a**) and $C_p - \lambda$ (**b**) curves [22]

currents results depending on each particular generator type [23]. In order to complete the prime mover plant modelling, to this set the HSS motion equation is added as being written in the form

$$J\frac{d\Omega_h}{dt} = \Gamma_{wt} - \Gamma_G, \tag{3}$$

where the static and viscous frictions have been neglected, J is the equivalent inertia rendered to the HSS, Γ_{wt} is the wind torque referred to the HSS [22], Ω_h is the HSS rotational speed and Γ_G is the electromagnetic torque.

The grid/load plant includes the DC-link and the inverter implementing the interface with either grid or load. The modelling focuses on the accumulation of energy in the DC-link capacitor and on the (d, q) model of the inverter. Figure 3b illustrates the modelling principles in the grid case. For the load case the requirements are different, therefore the control goals and modelling equations are slightly different too.

2 Control Principles of Low-Power Wind Turbines

2.1 *Issues*

The control problem is how imposing the operating point of wind turbines in order to fulfil the specific control goals. Among these, the most important are listed as follows:

- controlling the captured power for wind speeds larger than the rated (full load);
- maximising the wind harvested power in partial-load zone as long as constraints on speed and captured power are met;

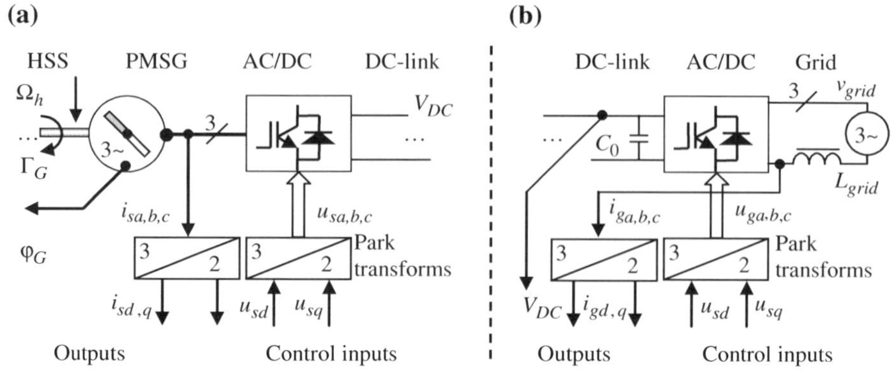

Fig. 3 **a** Generator modelling: identifying inputs, outputs and states—PMSG case. **b** Grid plant modelling: identifying inputs, outputs and states

- alleviating the variable loads, in order to guarantee a certain level of resilience of the mechanical parts;
- start and stop of the wind turbine.

In the case of variable-speed fixed-pitch wind turbines—which are of interest in this chapter—the variable-speed regime is implemented by means of the generator control. The control structures result from defining one or more of the above goals stated in relation to the mathematical model of the aerodynamic subsystem [23]. The controller determines the desired global dynamic behaviour such that to ensure power regulation, energy maximization in partial load and mechanical loads alleviation.

When the wind velocity exceeds the rated, the turbine is operating in full-load regime and the captured power—which potentially can vary with the wind speed cubed—should be limited (controlled) by aerodynamic means. Several techniques habitually used in order to satisfy this goal are briefed next [6].

Reviewing the blade element theory [6] one must note that the key variable determining the aerodynamics behaviour is the incidence angle. This one increases with wind velocity and decreases with increase of rotational speed and pitch angle, β. Figure 4 presents how the aerodynamic efficiency, $C_z(i)/C_x(i)$, is affected by the incidence angle evolution [23].

In general, when the turbine experiences high winds, the aerodynamic power can be reduced by controlling the incidence angle through the tip speed ratio and/or pitch angle. As the pitch angle is fixed, the aerodynamic power can be adjusted for constant wind speed by generator control at variable speed (varying the rotor speed). Remember that the variable-speed WECS have power electronics converter coupled with the electrical generator. This power structure allows the generator torque control and therefore imposing the turbine rotational speed.

Figure 4 shows that the turbine output power can be decreased to a suitable level either by increasing (stall effect) or decreasing (feathering) the incidence

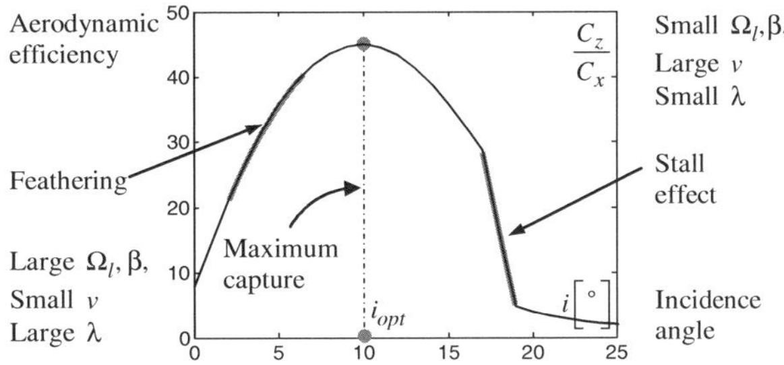

Fig. 4 Aerodynamic efficiency curve of a variable-speed variable-pitch wind turbine: emphasizing feathering and stall effects [23]

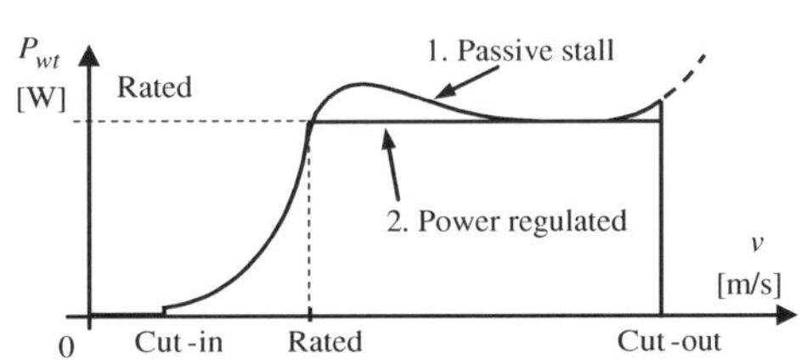

Fig. 5 Output power versus wind speed characteristics

angle. Either of these effects can be used in order to limit the turbine output power in the full-load regime. Also, an optimal regime can be obtained for a certain value of the incidence angle if output power maximization is aimed at. For fixed blade pitch, this value is obtained at an optimal value of tip speed ratio, λ_{opt}, and corresponds to the maximum aerodynamic efficiency, that is, the maximum value of power coefficient, C_p (see Fig. 2b).

The primary control goal includes two limitation constraints, in order to ensure turbine safety. One requires the captured power limitation to the maximum value, $P_{wt}(t) \leq P_{max}$, while the other one concerns the low-speed shaft rotational speed limitation: $\Omega_l(t) \leq \Omega_{max}$. While in partial load the power limitation constraint is implicitly fulfilled, at high wind speed, in full load, it must be ensured by control action. In the case of low-power WECS, this is achieved by adjusting the low-speed shaft rotational speed towards stall regime. The power limitation is fulfilled by bringing the turbine into a regime where the power coefficient can be controlled to diminish, as suggested in Fig. 5.

Note that the speed limitation necessity may also appear in partial load, mainly due to reasons related to noise. The turbine power-wind speed characteristic showing the turbine operation according to the above-stated principles is depicted in Fig. 5.

2.2 Partial-Load Operation

Control of variable-speed fixed-pitch wind turbines in the partial-load regime aims at maximizing the power harvested from wind by modifying the electrical generator speed. For each wind speed, there is a certain rotational speed at which the power curve of a given wind turbine has a maximum (C_p reaches its maximum value).

The locus of these maxima is known in the literature as the *optimal regimes characteristic*, ORC (see Fig. 6 [25]). In the $\Omega_l - \Gamma_{wt}$ plane, the ORC is placed at the right of the torque maxima locus (Fig. 6b). The maximization of the captured power is ensured by keeping the static operating point of the turbine around the ORC. In other words, this means to maintain the tip speed ratio at its optimal value, λ_{opt} (Fig. 2b) and can be achieved by operating the turbine at variable speed, corresponding to the wind speed [7].

The assumptions adopted in what concerns the known models/parameters, the measurable variables, the control method used and the version of wind turbine model play a key role in choosing the optimal control approach. Depending on how rich the information about the turbine model is, especially about its torque characteristic, the optimal control of variable-speed fixed-pitch wind turbines can be implemented by means of one of the following approaches.

2.2.1 Maximum Power Point Tracking

The Maximum Power Point Tracking (MPPT) refers to an entire class of searching algorithms; in the case of WECS, they are suitable when parameters λ_{opt} and $C_{p\,max} = C_p(\lambda_{opt})$ are not known. The reference of the rotational speed control loop is adjusted such that the turbine operates around maximum power for the current wind speed value. The speed reference must be either increased or decreased, depending on the current position of the operating point in relation to the maximum of $P_{wt}(\Omega_l)$ curve. Two main ways of estimating this position exist, namely:

- the speed reference is modified by a variation $\Delta\Omega_l$ and the corresponding change of active power, ΔP, is determined in order to estimate the value $\partial P_{wt}/\partial \Omega_l$, whose sign indicates the position of the operating point in relation to the maximum of characteristic $P_{wt}(\Omega_l)$ [22];
- a probing signal—e.g., a slowly-variable low-amplitude sinusoid—is added to the current speed reference, which produces a detectable response in the active power evolution. Further, the position of the operating point in relation to the maximum is obtained by comparing the phase lag of the probing sinusoid and that of the sinusoidal component of active power [22].

In this simplified presentation of MPPT techniques, factors like the influence of wind turbulence and system dynamics that distort the information concerning the operating point position have been neglected.

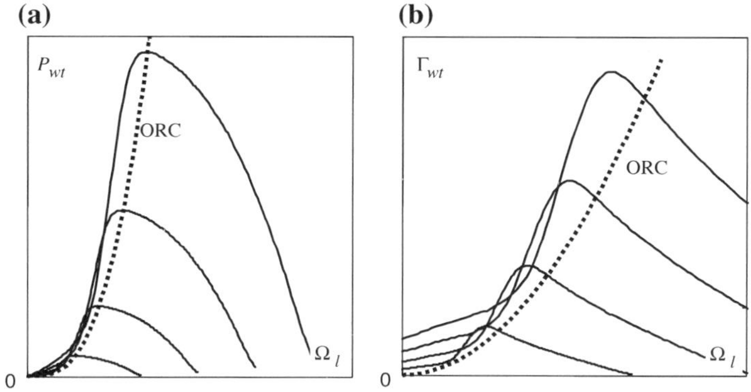

Fig. 6 Optimal regimes characteristic, ORC. **a** in the $\Omega_l - P_{wt}$ plane. **b** In the $\Omega_l - \Gamma_{wt}$ plane [23]

2.2.2 Shaft Rotational Speed Optimal Control Using a Setpoint from the Wind Speed Information

This solution can be applied if the optimal value of the tip speed ratio, λ_{opt}, is known. The shaft rotational speed is closed-loop controlled such that to ensure the operation on the ORC:

$$\Omega_{l_{opt}}(t) = \frac{\lambda_{opt}}{R} \cdot v(t). \tag{4}$$

One drawback of this approach is the necessity of measuring or estimating the wind speed. An anemometer mounted on the nacelle offers information on the fixed-point wind speed, which is quite different from the wind speed experienced by the blade [23].

2.2.3 Active Power Optimal Control Using a Setpoint from the Shaft Rotational Speed Information

This method is used when both λ_{opt} and $C_{p_{max}} = C_p(\lambda_{opt})$ are known. This approach supposes an active power control loop being used, whose reference is determined based upon the following relation:

$$P_{wt_{opt}} = P_{ref} = \frac{1}{2} \cdot \frac{C_p(\lambda_{opt})}{\lambda_{opt}^3} \rho\pi R^5 \cdot \Omega_{l_{opt}}^3, \tag{5}$$

so it is proportional with the rotational speed cubed.

In both methods an ORC tracking control loop is used, whose dynamic performances are influenced by the wind turbulence. In most of cases, classical PI or

PID control are preferred, but advanced control techniques can also be used in order to ensure better performance and robustness to modelling uncertainties [23].

The wind turbulence induces deviations of the operating point around ORC. These excursions contribute in a non-symmetrical manner to the decrease of the conversion efficiency [4]. Hence, it is appropriate to define the optimal regime as being not exactly the ORC, but the ORC slightly slipped to the right.

Another optimization strategy is to impose the electromagnetic torque corresponding to the ORC, which is proportional with the rotational speed squared [6].

2.3 Full-Load Operation

As stated before, in full load the turbine operation must ensure a limited value of captured power. In the case of fixed-pitch wind turbines this is achieved through appropriate control of the rotational speed. Therefore, the full-load regime should be detected and the outer control loop should be switched to a regulatory power control loop that imposes the rotational speed setpoint in order to reduce the output power to safe levels.

The associated strategy depends on the turbine blade profile. If the blade is designed to induce an accentuated stall effect in high winds at sufficiently large rotational speed (i.e., dotted line in Fig. 2b), the output power can be maintained within safe limits by limiting the turbine rotational speed to the rated value. In this way, behaviour similar to the stall-regulated wind turbines can be achieved (curve 1 in Fig. 5). If the turbine blades do not exhibit accentuated stall effect (see the graph represented with solid line in Fig. 2b), then the turbine can be brought in deeper stall by further reducing the rotational speed. In this case the power is strictly limited to the rated (curve 2 in Fig. 5).

3 Grid-Connected Low-Power WECS

Without loss of generality, a direct-driven PMSG generator ($\Omega_l \equiv \Omega_h$) is considered where the prime mover is a 3-blade wind turbine. The associated generation structure is also depicted in Fig. 1a. Figure 7 shows the control loops of this structure. As the WECS is grid-connected, it behaves as a current generator synchronized to the grid voltage [5]. The power transfer from the wind to the grid takes place as follows. The power converted by PMSG yields a DC current injected into the DC-link which will try to increase the DC-link voltage. The grid-side converter is operated in order to extract approximately the same current from the DC-link, therefore re-establishing the power equilibrium [32]. To this end, it will inject into the grid the alternative currents that are established by the DC-link voltage controller. In this structure, the generator-side inverter (which controls the prime mover plant) is free to be operated in MPPT mode, and it will provide

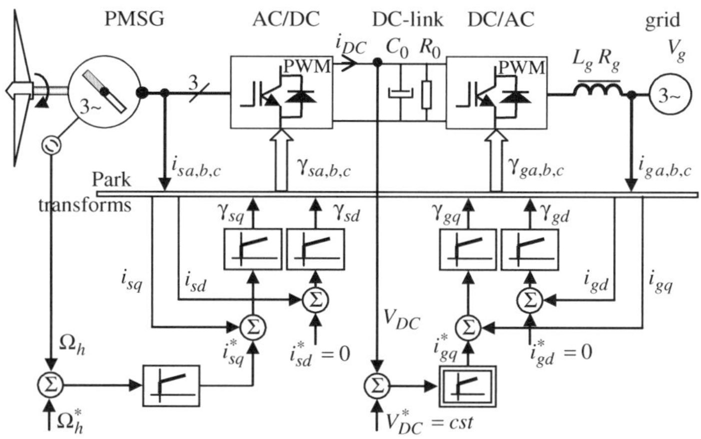

Fig. 7 Control systems of a grid-connected WECS

whichever power available in the wind within safety limits. In short, the generator-side converter provides servo-control of the turbine rotational speed and the grid-side inverter achieves DC-link voltage regulatory control.

In Fig. 7, Ω_h^* may result from maximum power (see Sect. 2.2) or power limiting (see Sect. 2.3) requirements. In order to lay out the main insights concerning the control of such structure, some modelling elements should be given.

For control purposes, the turbine mechanical characteristic for a given wind speed (see Fig. 6b) is linearized around a typical operating point [23]. The turbine torque is written in relation to the rotational speed, Ω_h:

$$\Gamma_{wt} = \Gamma_0 + D \cdot \Omega_h,$$

where Γ_0 is the zero-speed torque and the negative parameter D is the mechanical characteristic slope. Both of them depend on the steady-state operating point (and also on the current wind speed). The drive train transfer function results from the single-mass model in Eq. 3 [23].

In order to achieve a smoother generator behaviour and flexible drive system, the PMSG can be vector-controlled. This control structure is designed based upon the PMSG Park model [5]. Under simplifying assumptions (sinusoidal distribution of stator winding, electric and magnetic symmetry, negligible iron losses and unsaturated magnetic circuit, etc.) the (d,q) PMSG model can be written as

$$\begin{cases} v_{sd} = Ri_{sd} + L_d \dot{i}_{sd} - L_q i_{sq} \omega_S \\ v_{sq} = Ri_{sq} + L_q \dot{i}_{sq} + (L_d i_{sd} + \Phi_m)\omega_S \end{cases}, \tag{6}$$

where R is the stator resistance, v_{sd}, v_{sq} are d and q stator voltages, i_{sd}, i_{sq} are d and q stator currents, L_d, L_q are d and q inductances, $\omega_S = p \cdot \Omega_h$ is the stator (or else electric) pulsation, Φ_m is the rotor flux linkage. The electromagnetic torque is obtained as

$$\Gamma_G = p\left[\Phi_m i_{sq} + \left(L_d - L_q\right) i_{sd} i_{sq}\right], \tag{7}$$

where p is the number of pole pairs. If the rotor has nonsalient magnetic poles, then $L_d = L_q$ and the electromagnetic torque is expressed as

$$\Gamma_G = p\Phi_m \cdot i_{sq}. \tag{8}$$

By imposing the PMSG stator voltages, the output current and hence the machine torque is controlled by separately controlling the Park current components, i_{sd} and i_{sq}. Anti-windup PI controllers are used on each d and q channels; a decoupling structure is used for suppressing the d–q interaction [15]. The PI controllers are tuned by the well-known modulus criterion, which ensures settling time and overshoot of the closed-loop step response being optimally traded-off [2].

For maximum torque sensitivity with the stator current (i.e., maximum efficiency), the armature flux position is imposed at $\pi/2$, thus yielding reference $i^*_{sd} = 0$. The PMSG speed control relies upon employing an outer loop over the q channel control loop. The speed PI controller employed in the outer loop is designed by using the symmetrical optimum criterion, thus allowing an optimally-sized closed-loop ramp response [2]. This controller outputs the electromagnetic torque reference, hence the i^*_{sq} value.

One must note that in real-world applications this structure is more complex, as it includes the Park transformations $(d, q) \leftrightarrow (a, b, c)$, sinusoidal PWM [5] and a voltage-source inverter; the rotor absolute position is used for computing these transformations.

The generator-side converter 3-phase switching function, $\gamma_{sa,b,c}$, is also a rotating vector having components on the synchronously rotating (d, q) frame: γ_{sd} and γ_{sq}. The lossless inverter average modelling gives $v_{sd} = \gamma_{sd} V_{DC}/2$, $v_{sq} = \gamma_{sq} V_{DC}/2$ and $i_{DC} = \left(\gamma_{sd} i_{sd} + \gamma_{sq} i_{sq}\right)/2$, where i_{DC} is the DC current injected by the generator-side converter into the DC-link.

According to the above-stated modelling, the simplified plant to be controlled on the q axis can be depicted as in Fig. 8. Their control structure is then developed in order to achieve appropriate dynamics.

According to the modelling, classical PI control laws are suitable for such a task. The prime mover plant structure in Fig. 8 has as perturbation input the wind speed, whereas the average duty cycle component, β_{sq}, acts as control input. It outputs the DC-link current, i_{DC}. Provided that the rotational speed has significantly slower variations with respect to i_{sq} current, the control of this latter variable may be designed by considering that the back electromotive force, E, is constant. Note that the DC-link voltage is considered constant, as maintained by the grid-side inverter. Then, the current controller is tuned in order to compensate the stator electrical inertia on the q axis (see the second equation of set 6). For sake of stability, its reference should never overpass the value corresponding to the maximum aerodynamic torque for the current wind speed.

As the aerodynamic system is time variant (i.e., D varies with wind speed and operating point position), the speed controller design is a slighter difficult task.

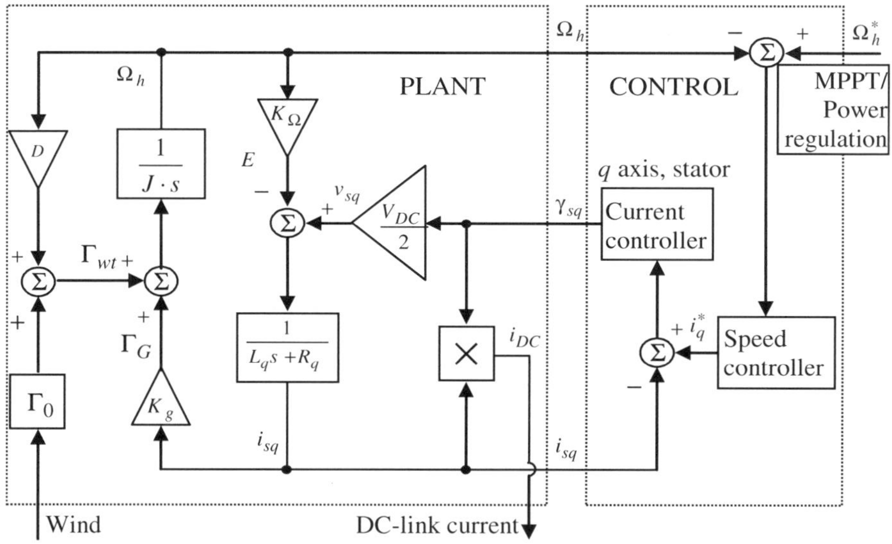

Fig. 8 Turbine-generator linear modelling and control on the q axis—the sign convention corresponds to motoring regime

First, taking into account its speed, the current loop should be considered ideal, without dynamic. Note that the slope D may take positive or negative values depending on the operating point position in relation to the maximum turbine torque point. Therefore, if a PI controller is used, it should be tuned for the worst case, when D has the largest positive value. The closed-loop system has two poles and exhibits variable dynamic behaviour that depends on the operating point. By solving the associated algebra and by imposing complex poles, one can verify that the system is stable if

$$K_g \cdot K_P + D > 0, \quad \forall D,$$

where K_g is the PMSG's torque coefficient and K_P is the PI controller gain, chosen sufficiently large. Therefore, for the above-mentioned worst case the controller yields the slowest dynamic.

The rotational speed reference can further be provided by the MPPT or by the power regulation loops. Anti-windup structure and reference gradient variations may be required for a "smooth" operation of the controlled structure.

Besides its relative simplicity, this control structure has many advantages. The machine currents and therefore its load are fully controlled (i.e., overloading risk is avoided), the turbine rotational over-speed regime can be avoided, the output power reduction can be rapidly obtained when necessary, etc. To conclude, the PMSG-inverter device converts the turbine mechanical power into DC electrical power by injecting the output current i_{DC} into the DC-link. This current will act in the sense of increasing the DC-link voltage, V_{DC}, by charging its capacitor, C_0.

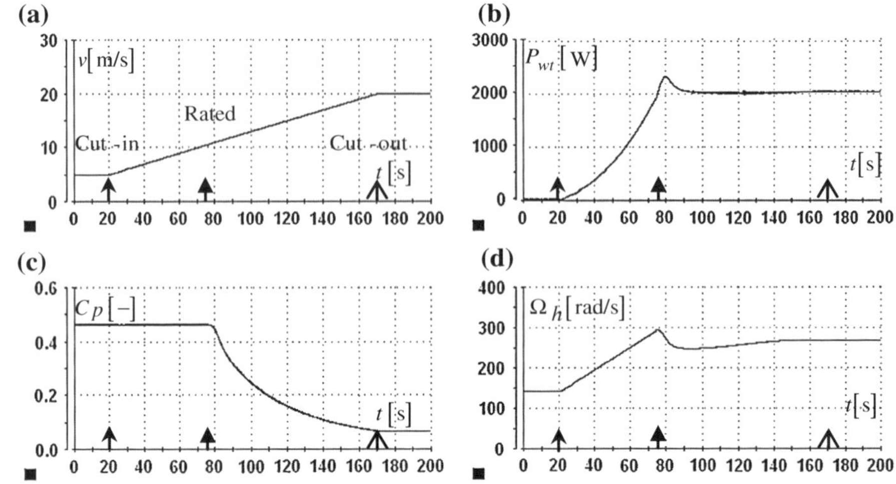

Fig. 9 Low-power WECS typical steady-state behaviour [24]

If the generator-side converter is like the one in Fig. 1b (i.e., composed of an uncontrolled rectifier and a chopper), then the generator torque is imposed by controlling the chopper inductor current. This solution yet allows the variable-speed WECS operation even if the torque ripple is quite significant.

Figure 9 displays the WECS steady-state characteristics obtained by using the above described control solution on a 1:1 WECS real-time simulator [24]. A slowly varying ramp of wind velocity, v (between 5 and 20 m/s) is applied as input signal, as Fig. 9a shows. Its variation speed is slow enough such that to allow the steady-state regime being reached. The case of a passive-stall regulated 3-blade horizontal-axis 2-kW-rated wind turbine (having the power coefficient curve like in Fig. 2a) has been chosen. The steady-state regimes, in terms of output power, P_{wt}, power coefficient, C_p, and high-speed shaft rotational speed, Ω_h, can be seen in Fig. 9b, c and d, respectively. In partial load the rotational speed reference is imposed by an MPPT algorithm (i.e., speed increases with the wind). In full load the outermost power regulation loop reduces the rotational speed reference, inducing turbine deeper stall as the wind speed increases.

Next, the grid-side converter operation in transferring the DC-link power to the grid will be described. The associated plant is composed of the DC-link bus, the inverter and the electrical grid. In order to transfer the available electrical active power, the 3-phase currents are fed into the mains while controlling the DC-link voltage at a constant value, V_{DC}. The wind power system must also be able to supply reactive power to the grid in order to compensate voltage variations in the point of common coupling.

Note that the (d, q) representation also applies on the AC system. The control structure is based upon two current control loops for the q and d output current i_g Park components respectively, and an outer voltage control loop, which regulates the DC-link voltage, V_{DC}, at an imposed value [15]. All these loops employ anti-windup PI controller which are tuned using modulus optimum criterion. The DC-link voltage controller provides the i_{gq}^* reference; i_{gd}^* is nonzero if reactive power is to be supplied to the grid. The inner loops output the grid voltage references in the (d, q) frame, to be transformed into a 3-phase system and applied to the inverter. Park transform requires the grid electrical angle being used. To this end a 3-phase phase-locked loop must be employed [19, 28].

The grid-side converter lossless (d, q) average modelling yields that [23]:

$$\begin{cases} L_g \cdot \dfrac{\mathrm{d}i_{gd}}{\mathrm{d}t} = V_g - R_g \cdot i_{gd} + L_g\omega \cdot i_{gq} - \dfrac{V_{DC}}{2} \cdot \gamma_{gd} \\ L_g \cdot \dfrac{\mathrm{d}i_{gq}}{\mathrm{d}t} = -R_g \cdot i_{gq} - L_g\omega \cdot i_{gd} - \dfrac{V_{DC}}{2} \cdot \gamma_{gq} \\ C_0 \cdot \dfrac{\mathrm{d}V_{DC}}{\mathrm{d}t} = \dfrac{3}{2} \cdot \left[i_{gd} \cdot \gamma_{gd} + i_{gq} \cdot \gamma_{gq}\right] - \dfrac{V_{DC}}{R_0} - i_{DC} \end{cases}, \tag{9}$$

where L_g, R_g are the output current filter parameters, V_g is the grid voltage amplitude, ω is the grid pulsation, C_0 and R_0 are the DC-link capacitor and resistor respectively. γ_{gd} and γ_{gq} are the (d, q) components of the converter duty cycle.

Note that the first two equations of (9) are weakly coupled (typically, the coupling term $L_g\omega/V_{DC}$ takes values much smaller than the unity), so the two current components i_{gd} and i_{gq} can be individually controlled and the associated plant is $1/(L_g s + R_g)$. Note also that the currents' closed-loop dynamics are far faster than the V_{DC} dynamic. Then, by replacing the duty cycles γ_{gd} and γ_{gq} in the third equation of (9) with their steady-state values obtained from the first two equations, and neglecting the grid filter losses, one obtains:

$$C_0 \cdot \frac{\mathrm{d}V_{DC}}{\mathrm{d}t} = \frac{3}{2V_{DC}} \cdot V_g i_{gq} - \frac{V_{DC}}{R_0} - i_{DC}. \tag{10}$$

Further, one can show that the transfer function between the current component i_{gq} and the DC-link voltage squared is a first-order filter [15]:

$$\frac{V_{DC}^2}{i_{gq}} = \frac{3V_g/2}{C_0 s + 1/R_0}. \tag{11}$$

Typically, the voltage controller may either be designed as linear (e.g., proportional-integral, PI) by using the V_{DC}^2 feedback or by linearization around a typical operating point. As a conclusion, the grid plant control structure for the q channel may be synthesized as in Fig. 10.

The plant contains the DC-link, provides the measures V_{DC} and i_{gq} to the controller and interacts with the generator-side inverter via the i_{DC} current. This latter acts as disturbance. Note that this control structure is not the sole that can be implemented—e.g., an alternative solution to the current control is the use of the generalized integrators (resonant controllers) [20].

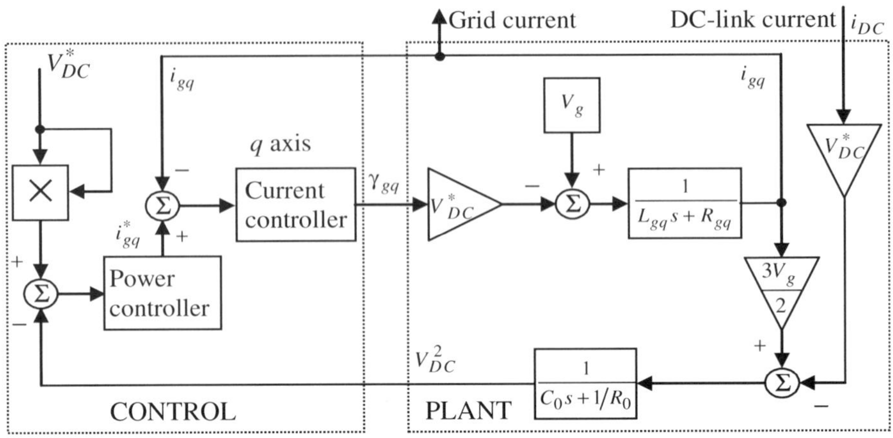

Fig. 10 Grid-side inverter control structure for the q channel

The regulatory control of the DC-link voltage has multiple consequences. Besides the electrical power evacuation into the power grid, it ensures the DC-link protection from high voltages and provides constant parameters to the plant in Fig. 8, thus ensuring an appropriate control of the electrical machine currents.

The four oscilloscope screen captures in Fig. 11 contain results—obtained using real-time simulation—illustrating the dynamic behaviour of a low-power WECS in the two main operating regimes—partial load and full load—under various wind conditions.

Thus, Fig. 11a depicts the variable-speed operation in partial load under variable wind velocity with medium turbulence intensity $I = 0.15$, obtained by using the von Karman spectrum in the IEC standard [26].

By convention, the active power that is supplied to the grid is negative, depending heavily on the wind velocity. As the WECS is operated at variable speed, the rotational speed and AC output current depend also on the wind speed. Figure 11b contains information about the WECS behaviour when the wind speed exhibits sequences of gusts having the average around 12 m/s in the full-load regime. As the power controller acts in the sense of rejecting the wind speed variations effect on the output power, the captured power variation due to the wind speed variation is not as important as in the previous case.

Figure 11c and d illustrate the power transfer to the grid, i.e., the grid-side converter operation. Figure 11c has been obtained under variable wind speed in partial load. As the active power varies with the wind speed and the DC-link voltage is maintained almost constant, then i_{DC} varies also with the wind. One can see the variations of the q-component of the PMSG stator current, proportional with the electromagnetic torque—these are due to the variable-speed operation. Figure 11d shows the behaviour of the grid electrical variables at the point of coupling, v_{ga} and i_{ga}, when the WECS injects reactive power into the grid.

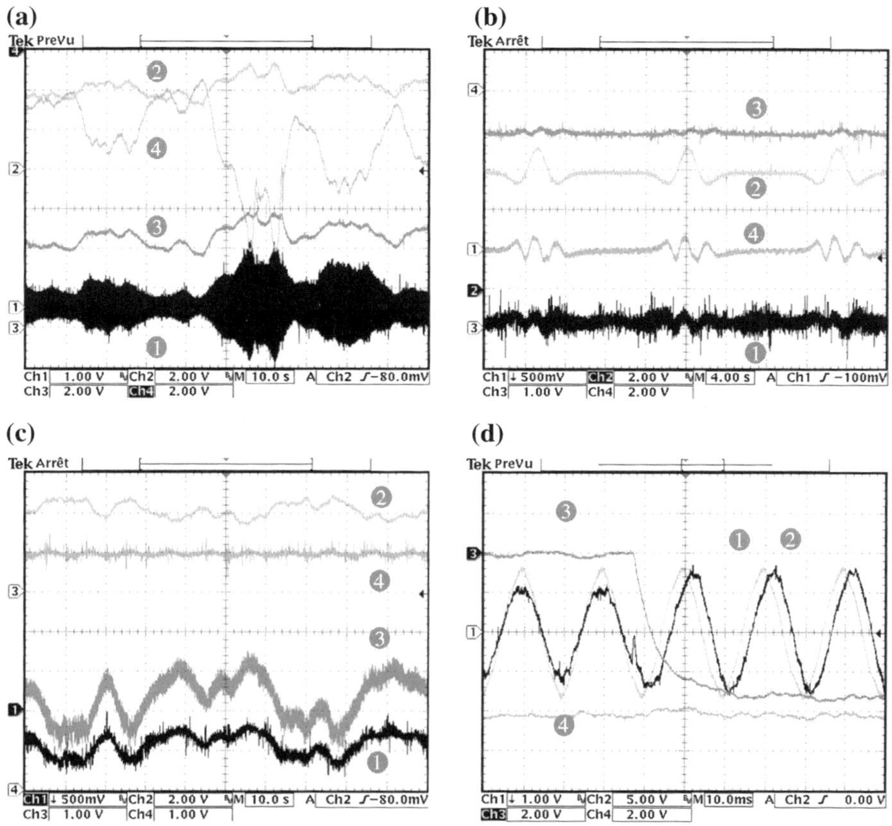

Fig. 11 Variable-speed dynamic WECS behaviour. **a** *Ch1* grid output current i_{ga}, 5A/V, *Ch2* wind speed v, 2 m/s/V, *Ch3* rotational speed Ω_h, 50 rad/s/V, *Ch4* output active power P, 200 W/V. **b** *Ch1* DC current i_{DC}, 5A/V, *Ch2* wind speed v, 2 m/s/V, *Ch3* rotational speed Ω_h, 50 rad/s/V, *Ch4* output active power P, 250 W/V. **c** *Ch1* DC current i_{DC}, 5A/V, *Ch2* wind speed v, 2 m/s/V, *Ch3* generator q component, i_{sq}, 5A/V, *Ch4* V_{DC}, 75 V/V. **d** *Ch1* line voltage v_{ga}, 25 V/V, *Ch2* grid output current i_{ga}, 5A/V, *Ch3* output reactive power Q 200VAR/V, *Ch4* output active power P, 200 W/V [24]

4 Low-Power WECS in Stand-Alone Generation Configurations

4.1 *Issues*

Because the primary resource is strongly intermittent, WECS operate in stand-alone configurations when they are used as parts of hybrid energy systems to supply isolated sites. The hybrid systems are composed of classical energy sources, such as those based on fossil fuels, energy storage devices and renewable

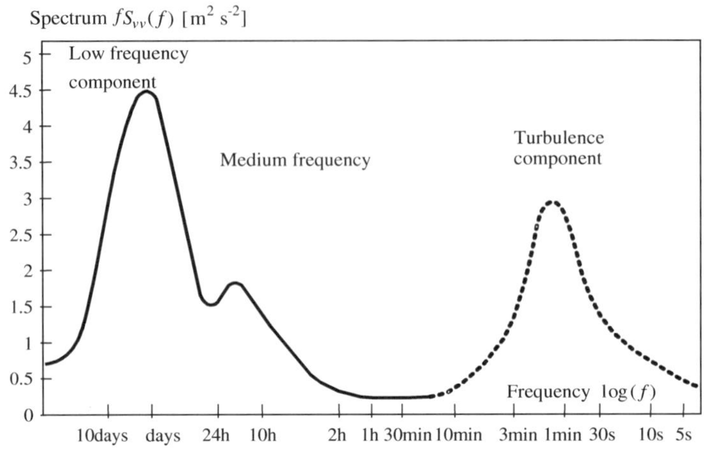

Fig. 12 Wind energy fluctuations given by Van der Hoven's wind speed spectral model

energy sources. WECS operating in stand-alone systems are subject to two types of disturbances:

- load variation in the local grid and
- variation of power provided by WECS due to the stochastic nature of the primary energy resource, the wind.

Wind speed can be modelled as a nonstationary stochastic process and described by long-, medium- and short-time spectral models. Short-time models capture the turbulence fluctuation, which depends of the average wind speed, surface roughness length and ground height [6]. The 6-decade Van der Hoven model can be seen in Fig. 12. This model gives information about the scarce availability of the resource, suggesting that a given WECS cannot guarantee an imposed level of power by itself. For example, to boost the energy level on medium term, a chemical accumulator may be used.

The issues in stand-alone WECS design basically result from the stochastic properties of the wind speed. These are:

- *necessity of uninterrupted supply of the load*, taking account of the long- and medium-term wind speed variation. To supplement the power provided by WECS one can use backup energy sources, such as electrochemical energy storage devices (accumulator batteries) on medium term or Diesel generators on long term;
- *necessity of ensuring the quality of provided energy:* frequency, voltage, harmonic content, flicker. In general, stand-alone WECS are placed near to the consumption site, where the turbulence intensity can be high. Turbulence fluctuation can produce important frequency and voltage variations in the local grid. A solution to diminish these variations consists in using kinetic energy storage devices (flywheels, for example).

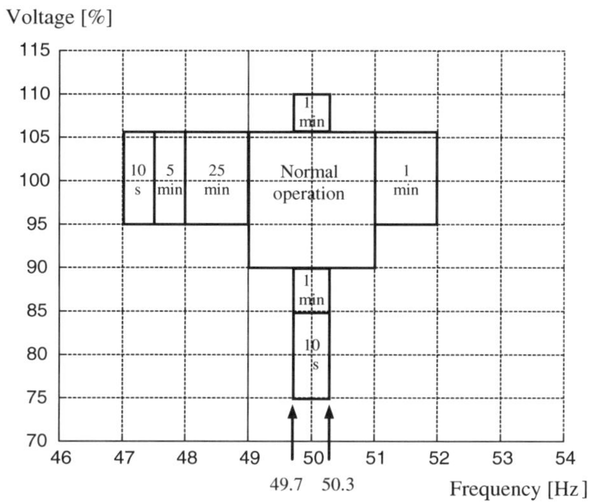

Fig. 13 Maximally admissible deviations of frequency and voltage, according to [9]

Depending on their structure, the stand-alone WECS may or may not include batteries. The latter may contain a Diesel generator and eventually a flywheel.

Performance of stand-alone WECS are reflected by the penetration rate of renewable energy, which can be defined in relation either to the power extracted from wind—instantaneous penetration—or to the energy extracted from wind (average penetration) [3]. Instantaneous penetration is defined as the ratio between the power harvested from wind and the power required by the load, whereas average penetration is a ratio of the extracted and required energy quantities on a given time horizon (hours, days, months or years).

4.2 No-Storage Wind-Diesel Systems

The simplest structure of hybrid system supplying an isolated site contains no energy storage devices [3, 13, 17, 18, 27, 36]. The elementary structure of such system includes a Diesel generator and an induction-generator-based WECS [3]. The Diesel generator provides partially or totally the required power when there is not enough power in the wind. Power limitation is achieved habitually by passive stall. Such system, in its elementary structure, ensures an average penetration below 20 %. Increase of the WECS rated power for purposes of improving the average penetration rate leads to increase of frequency and voltage fluctuations in the local grid. Requirements regarding the provided energy quality are imposed as limitations of WECS's frequency and voltage admissible deviations, as illustrated in Fig. 13 [9]. Maximally admissible deviations of frequency/voltage in relation to their respective rated values are as smaller as the voltage/frequency deviations are larger. Under these regulations, the total time of admissible deviation from normal

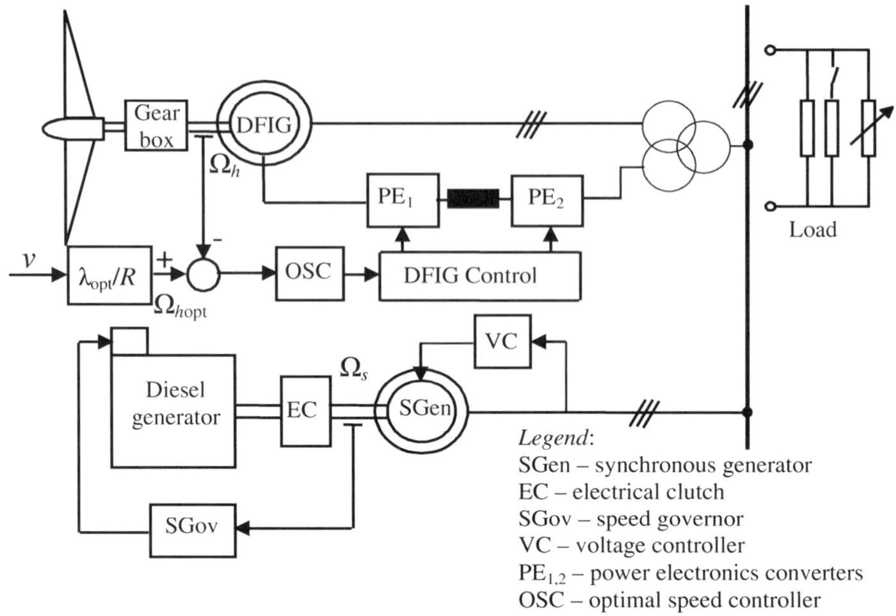

Fig. 14 Wind-Diesel system equipped with wind energy conversion optimization control law

operation must not exceed 10 h per year. This kind of demand limits severely the WECS power weight within an elementary wind-Diesel generation configuration.

The average penetration rate can be improved by controlling the WECS such as to optimize the energy conversion in partial load. Figure 14 presents the structure of a wind-Diesel system employing a doubly-fed induction generator (DFIG), where the energy optimization control loop is based on tracking the optimal rotational speed given by Relation (4).

The advantage of the structure in Fig. 14 is that the rotor converter power is typically 30 % of the generator rated power. WECS's active and reactive power control is achieved by means of converters.

Two contradictory demands must be taken into account when designing the optimal speed controller (OSC), namely maintaining the operating point around the maximum power one along with meeting the admissible frequency deviations in the local grid. Interaction between the two subsystems—WECS and Diesel generator—is illustrated in Fig. 15, where Ω_s and Ω_h are the rotational speeds of the synchronous generator and of the DFIG, respectively.

In Fig. 15 P_l denotes the load active power, P_a and T_a are the DFIG's active power and torque respectively. In response to the wind speed turbulence, the controller OSC produces variations of the DFIG's torque and power, which influence the dynamics of the synchronous generator's rotational speed loop. On the other hand, variation of the rotational speed Ω_s determines variation of the

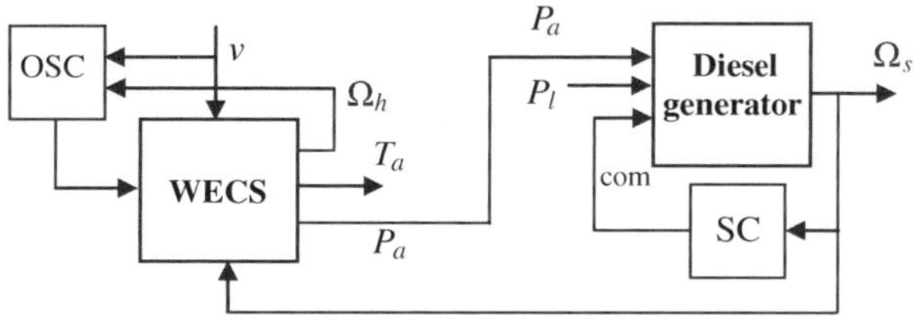

Fig. 15 Interaction between a WECS and a diesel generator

DFIG's electromagnetic torque, which is practically proportional with difference $(\Omega_s - \Omega_h)$.

In Fig. 16 one can see numerical simulation results regarding the evolution of the provided power and frequency of a wind-Diesel system for constant load and variable wind speed having the sequence in Fig. 16a. The Diesel generator power, P_{DG}, and the wind system power, P_{wt}, both expressed in per units, can be visualized for two values of the wind power instantaneous penetration rate, namely 90 %—Fig. 16b—and 50 %—Fig. 16d. Figure 16c and e contain the evolution of the frequency in the two cases respectively.

Based upon the linearized model of the wind-Diesel system, in [8] one designs the optimal control law of the hybrid system in relation with a performance index of form:

$$I = \int_0^{\infty} \left[\alpha \overline{\Delta\lambda(t)}^2 + \overline{\Delta T_a(t)}^2 \right] \mathrm{d}t, \tag{15}$$

where $\overline{\Delta\lambda(t)}$ şi $\overline{\Delta T_a(t)}$ are normalized variations of the tip speed ratio and DFIG's torque, respectively, and α is a positive coefficient weighting the influences of the energy efficiency performance and local grid frequency variations, respectively. An effective solution to compensate the dynamic frequency variations is the use of a flywheel, which achieves short-term energy storage [16]. The flywheel is habitually coupled permanently to the synchronous generator's shaft.

4.3 Hybrid Energy Conversion Systems with AC-Coupled WECS

As shown previously, WECS cannot ensure the AC parameters regulation by itself in every wind condition since this resource is intermittent. The presence of backup devices accomplishing this task is thus mandatory.

Figure 17 presents a stand-alone WECS configuration where all the sources deliver power through an AC grid. Here, the wind turbine system is equipped with a minimal control configuration: the wind power is limited by passive stall when the wind speed exceeds the rated, whereas the squirrel-cage induction generator (SCIG) is coupled directly to the AC grid.

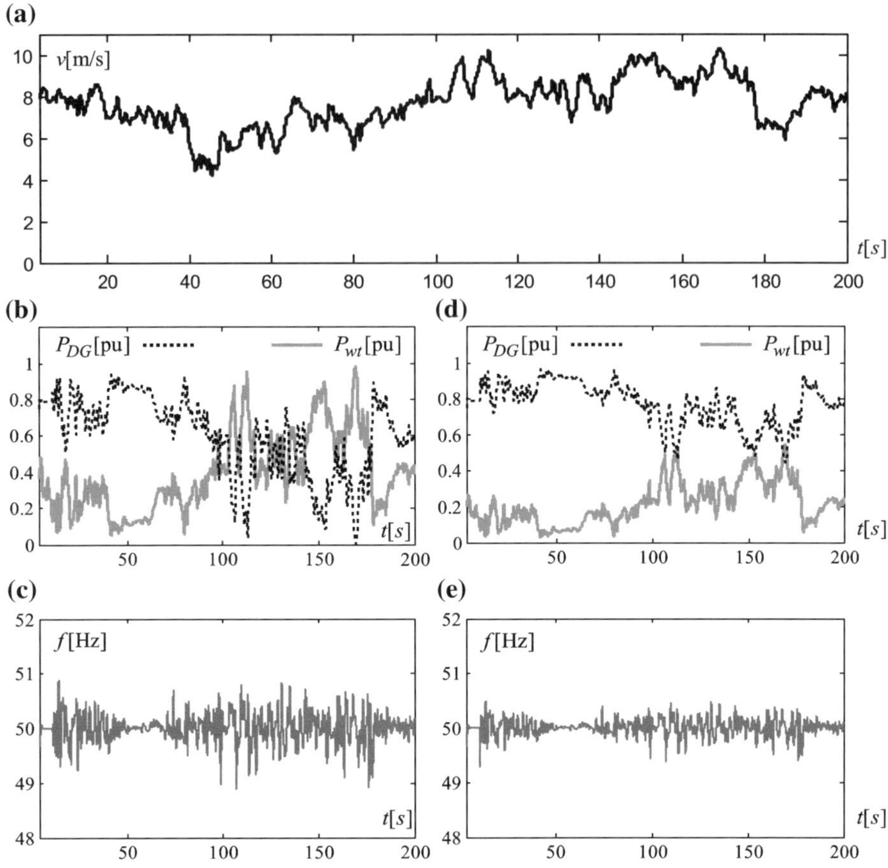

Fig. 16 Wind-Diesel behaviour under different penetration rates: **a** 90 % (**b**) and (**c**), 50 % (**d**) and (**e**)

The uncontrolled WECS induces the necessity of supplementary hardware ensuring the AC parameters regulation, as it will be described further.

The hybrid system may include a battery, achieving medium-term energy storage, as well as other renewable energy sources (such as microhydro or photovoltaic). If the hybrid system can ensure the power required by the load without using the Diesel generator, it is considered that the system has high wind penetration [30]. When the Diesel generator is not coupled and the power produced by the renewable source exceeds the power required by the load, a dump load is activated which dissipates power by means of the static converter PE2 in Fig. 17. The synchronous machine SM is in charge with ensuring the voltage and reactive power control when the Diesel generator is not coupled. An alternative solution is presented in [35], where static VAR compensators are used in order to compensate

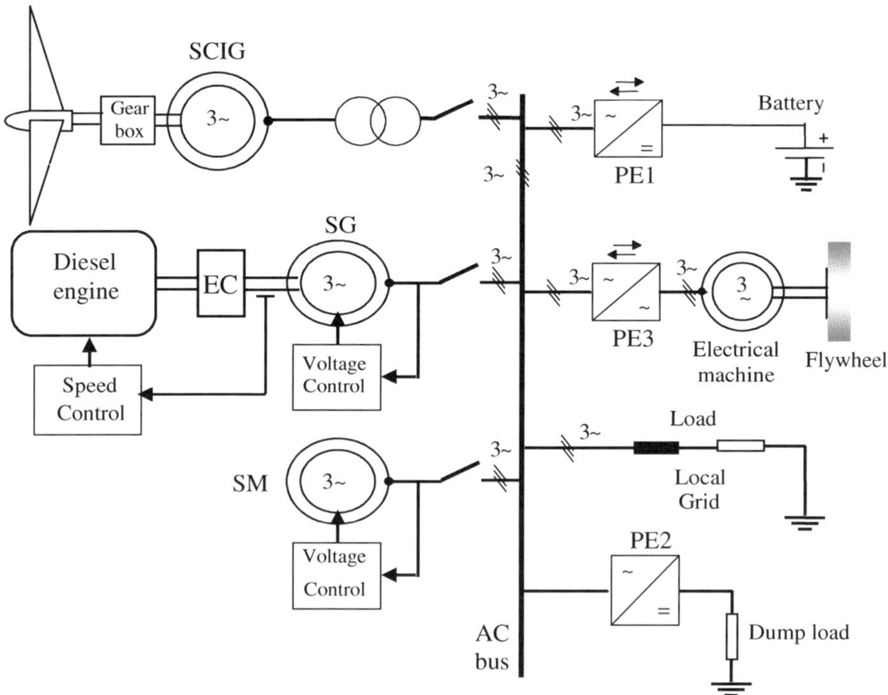

Fig. 17 Stand-alone hybrid system with battery and sources coupled through an AC link—uncontrolled WECS

the instantaneous reactive power and to diminish the voltage fluctuations in the case of a high-penetration no-storage wind-Diesel system.

Reducing the frequency variations at high wind speed—therefore, at high wind turbulence levels—is an important problem. Usually, this problem is solved by employing a kinetic energy storage device. This latter can be either a fixed-speed flywheel, coupled to the Diesel generator's shaft, between the electrical clutch and the generator, or a variable-speed high-speed flywheel. In the latter case the flywheel receives and transfers the energy through an electrical machine and an AC-AC bidirectional converter, as shown in Fig. 17.

Figure 18 describes a stand-alone hybrid configuration with reduced weight of electromechanical equipment and which allows the WECS advanced control.

The wind turbine uses passive stall to implement the power limitation; alternatively this can be achieved by varying the pitch. Here, the SCIG delivers power to the local grid by means of a back-to-back converter that controls its operation. The parameters—frequency, voltage, power factor—of the energy supplied by the WECS are controlled by the load-side converter. In order to manage the energy consumption, when the renewable energy source is in deficit, one must prioritize the local loads such that the control of the less-priority load to contribute, if necessary, at balancing the produced and required power.

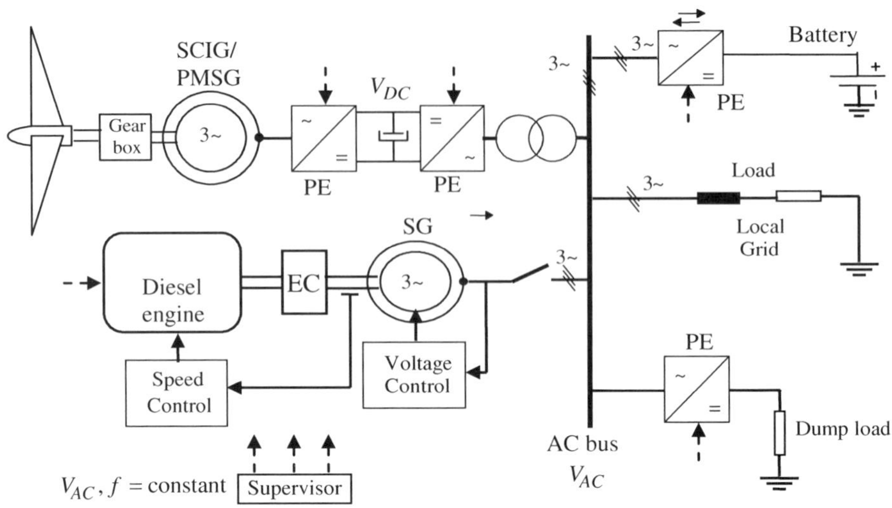

Fig. 18 Stand-alone hybrid system with battery and sources coupled through an AC link—fully controlled WECS

In the case of such hybrid system, an important issue is how to structure the control system on two hierarchical levels. The controllers of the various subsystems are placed on the low level: speed controller (SC) and voltage controller of the Diesel generator, as well as the optimal speed controller (OSC) of the WECS. The upper level (automatic dispatcher) manages the power flux provided by the various sources, taking account of the necessity of prioritizing the renewable energy sources.

Let us restate the primary objective of the power generation structure: ensuring the AC voltage parameters irrespective of the wind regime and the load power (within the structural limits). From the WECS point of view, two operation modes can be considered: when the Diesel generator is online and the WECS operates as an AC synchronized current source and when the WECS is an AC voltage source and the Diesel generator may be shut down.

The first case assumes WECS operation as in the grid-connected case, in a controlled AC environment. Note that an AC current source cannot inject current into the connection node (AC bus in Fig. 18) between an AC voltage source and its load in any conditions. The AC voltage source (i.e., the Diesel generator) should react very fast to the variable current injected by the AC current source—which is the WECS—in order to maintain the desired AC voltage. This means that the Diesel generator must always deliver current to the AC bus and avoid the voltage regulator saturation. For ensuring sufficiently fast voltage response of the Diesel generator with respect of the wind-generated AC current variations injected by the WECS, one may either use a supplementary energy accumulation in the DC-link or reduce the DC-voltage regulator bandwidth (see Fig. 7).

If the WECS operation in AC voltage-source mode is preferred, the two associated converters change their roles. While the load-side converter is dedicated to the AC voltage regulation in terms of amplitude and frequency, the generator-side converter regulates the DC-link voltage [1]. Therefore, the turbine will find the operating point corresponding to the WECS load and its rotational speed is no longer controlled. Effective measures must be taken in order to ensure the rotational speed limitation: either active stall or pitch regulation should be employed.

This operating mode is well suited for the WECS full-load regime. In this case the Diesel generator connection and synchronisation with the AC-bus is classically done; by means of the frequency-power control system, the Diesel generator equipped with a PID speed controller plays the role of regulation unit. This means that it modifies its delivered power such that this one to equal the difference between the power required by the load and the power provided by the renewable energy sources. In the case of power provided in excess by the renewable energy sources, the Diesel generator can be stopped, the battery charge is started and eventually the dump load is activated. In the contrary case, of insufficient power provided by the renewable energy sources, the battery energy is used and, depending of the predicted wind speed, one can decide to enable the Diesel generator.

The control strategy implemented on the upper control level must ensure simultaneously a set of antagonistic demands, such as:

- increase of medium penetration rate;
- reduction of start/stop order frequency for the Diesel generator, in order to keep it reliable;
- uniform charging/discharging cycles of battery, in order to ensure a long service time of this one.

4.4 Hybrid Energy Conversion Systems with DC-Coupled WECS

A distinct and widely spread kind of hybrid stand-alone systems is the one in which the different energy sources are coupled by means of a DC link. In general, these are low-power systems, where use of battery is mandatory for medium-term energy storage. When the power extracted from wind is insufficient for long time, a Diesel-generator unit is coupled to the DC link, equipped with permanent-magnet synchronous generator and electrical start/stop.

Multiple configurations of hybrid systems with DC link may be achieved, depending on how the energy sources within are controlled. One of the possibilities is where the wind source and the photovoltaic source deliver power to the battery by means of a charge controller, like presented in Fig. 19. In this case, the operating regime of the WECS is established by the charge controller, which cannot ensure the wind turbine operation on the optimal regimes characteristic

Fig. 19 Battery charge controller in a hybrid energy system including wind energy and photovoltaic sources

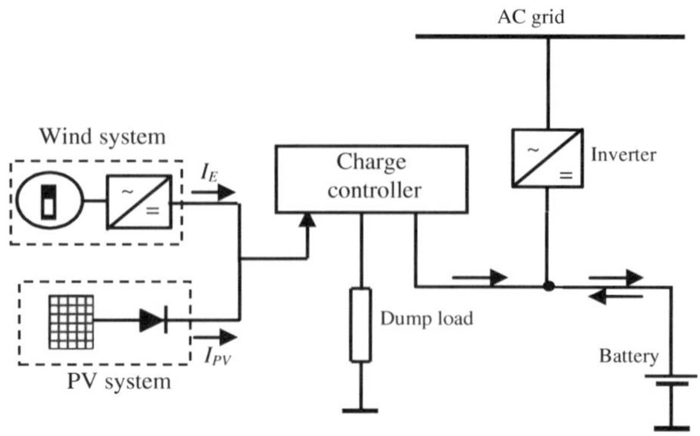

(ORC). The generator of the Diesel-generator unit delivers power directly to the local AC grid, also ensuring the battery charging by means of a rectifier.

Typically, the fixed-pitch turbine is coupled directly to the permanent-magnet multipolar synchronous generator. The charge controller uses electronic switches in order to control the battery charge process; it activates a dump load when the battery is fully charged.

For optimal and safe generation system operation, the DC-link voltage should be regulated at an imposed value.

Figure 20 shows a configuration of hybrid system with DC link that ensures better energy performance. All the sources deliver power into the DC link by means of current control loops. The wind turbine is equipped with a power control system whose setpoint is given by Eq. (5) in order to ensure the operation on the ORC.

The turbine power controller drives the inner generator current loop. When the wind power is insufficient on long term, the alternative source (the Diesel-generator unit) delivers power into the DC link by means of a rectifier and a current control loop. The DC link supplies the inverter delivering power in the local AC grid at standard frequency and voltage. The inverter draws current from DC-link according to the load necessity. In this context, the imbalance between the produced power and the power required by the load is sensed by the DC-link capacitor's voltage. The DC-link voltage control system aims at maintaining the balance between the provided power and the one required in the DC-link, when the power required by the local grid and/or the one produced by the renewable energy sources vary.

The control actions performed by this control system are:

- change of the battery current control loop reference; this loop contains a 2-quadrant DC-DC converter. The change refers to both the sign of variation—i.e., regime selection as charging/discharging—and also the value of the current injected/extracted into from the battery;

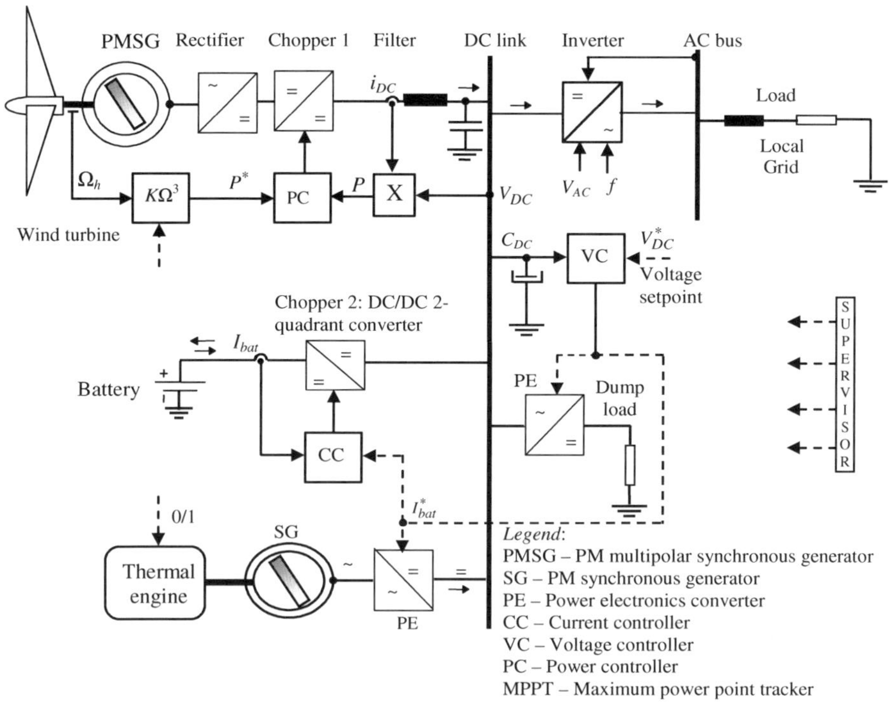

Fig. 20 Stand-alone hybrid system with sources coupled through a DC link

- the start/stop control of the alternative source (Diesel-generator unit) and establishing the reference of the current control loop of this source, in the case of long-term deficit of the wind energy source;
- control of the dump load converter when the energy production is in excess.

5 Conclusion

In the last decades the priority has been given to high-power WECS, because the energy market target was to increase the wind power penetration rate, according to actual trends in developing energy systems. Investment regarding the low-power WECS was not as well as important, since their energy weight is reduced. Low-power WECS have been used mainly in stand-alone configurations. In the last years, the energy production from distributed and decentralized sources has become increasingly successful in order to supply local grids by low-power systems, either grid connected or stand-alone. An extensive variety of generation configurations are possible: classical grid-connected WECS, stand-alone systems with either AC or DC common bus, etc. Taking account of their multitude of architectures, these latter systems exhibit control problems more various than in the case of high-power WECS.

As regards the stand-alone hybrid systems containing WECS, the main problem is to simultaneously ensure the maximum wind power extraction, increase of the battery service time (uniform charging/discharging cycles) and reduction of the coupling/decoupling frequency of the alternative source (usually, Diesel generator). Therefore, the upper-level (supervision) controls that may be capable of optimally managing various scenarios for different power generation structures, possibly in relation to the smart grids, is a fertile field of exploration.

References

1. Andreica M, Bacha S, Roye D, Munteanu I, Bratcu AI, Guiraud J (2009) Stand-alone operation of cross-flow water turbines. In: Proceedings of IEEE international conference on industrial technology-ICIT 2009, CD-ROM
2. Åström KJ, Hägglund T (1995) PID controllers: theory, design and tuning, 2nd edn. Instrument Society of America
3. Baring-Gould I, Flowers L, Lundsager P, Mott L, Shirazi M, Zimmermann J (2004) Worldwide status of wind-diesel applications. 2004 DOE/AWEA/CanWEA wind-diesel workshop http://www.eere.energy.gov/windpoweringamerica/pdfs/workshops/2004_wind_diesel
4. Bianchi FD, Mantz RJ, Christiansen CF (2005) Gain scheduling control of variable-speed wind energy conversion systems using quasi-LPV models. Control Eng Pract 13(2):247–255
5. Bose BK (2001) Modern power electronics and AC drives. Prentice–Hall, Englewood Cliffs
6. Burton T, Sharpe D, Jenkins N, Bossanyi E (2001) Wind energy handbook. Wiley, NY
7. Connor B, Leithead WE (1993) Investigation of fundamental trade-off in tracking the Cpmax curve of a variable speed wind turbine. In: Proceedings of the 12th British wind energy conference, pp 313–319
8. Cutululis NA, Bindner H, Munteanu I, Bratcu A, Ceanga E, Sørensen P (2006) LQ optimal control of wind turbines in hybrid power systems. European wind energy conference, EWEC '06, http://ewec2006proceedings.info/allfiles2/463_Ewec2006fullpaper.pdf
9. Danish Technical Regulation (2004) Wind turbines connected to grid with voltage below 100 kV. Technical regulation for the properties and the control of wind turbines, Doc. No. 177899
10. Elena Energie (2010) Elena turbowind. http://www.elena-energie.com/index.php?lang=enin. Accessed 14 June 2010
11. Heier S (2006) Grid integration of wind energy conversion systems, 2nd edn. Wiley, Chicester
12. HelixWind (2010) Savonius and Darrieus helix wind turbines. http://www.helixwind.com/en/product.php. Accessed 14 June 2010
13. Hunter R, Elliot G (1994) Wind-diesel systems. Cambridge University Press, Cambridge
14. Hur N, Jung J, Nam K (2001) A fast dynamic DC-link power-balancing scheme for a PWM converter-inverter system. IEEE Trans Ind Electron 48(4):794–803
15. Hur N, Jung J, Nam K (2001) A fast dynamic DC-link power-balancing scheme for a PWM converter-inverter system. IEEE Trans Ind Electron 48(4):794–803
16. Infield DG (1994) Wind diesel design and the role of short term flywheel energy storage. Renew Energy 5(1):618–625
17. Jeffries WQ, McGowan JG, Manwell JF (1996) Development of a dynamic model for no storage wind/diesel systems. Wind Eng 20(1):27–38
18. Kamwa I, Saulnier B, Reid R (1989) Modelling, simulation and control of a wind-diesel stand-alone grid (Modélisation, simulation et régulation d'un réseau eolien/Diesel autonome). IREQ Research Report

19. Kaura V, Blasko V (1997) Operation of a phase locked loop system under distorted utility conditions. IEEE Trans Ind Appl 33(1):58–63
20. Liserre M, Teodorescu R, Blaabjerg F (2006) Multiple harmonics control for three-phase grid converter systems with the use of PI-RES current controller in a rotating frame. IEEE Trans Power Electron 21(3):836–841
21. Molenaar D-P (2003) Cost-effective design and operation of variable speed wind turbines. Ph.D. Thesis, Technical University of Delft, The Netherlands
22. Munteanu I (2006) Contributions to the optimal control of wind energy conversion systems. Ph.D. Thesis, "Dunărea de Jos" University of Galaţi, Galaţi, Romania
23. Munteanu I, Bratcu AI, Cutululis NA, Ceangă E (2008) Optimal control of wind energy systems—towards a global approach. Springer, London
24. Munteanu I, Bratcu AI, Bacha S, Roye D, Guiraud J (2010) Hardware-in-the-loop-based simulator for a class of variable-speed wind energy conversion systems: design and performance assessment. IEEE Trans Energy Convers 25(2):564–576
25. Nichita C (1995) Study and development of structures and numerical control laws for building up of a 3 kW wind turbine simulator (Étude et développement de structures et lois de commande numériques pour la réalisation d'un simulateur de turbine éolienne de 3 kW). Ph.D. Thesis, Université du Havre, France
26. Nichita C, Luca D, Dakyo B, Ceangă E (2002) Large band simulation of the wind speed for real time wind turbine simulators. IEEE Trans Energy Convers 17(4):523–529
27. Palsson M, Uhlen K, Toftevaag T (1997) Modeling and simulation of an autonomous wind/diesel system equipped with forced commutated converter. In: Proceedings of EPE '97, pp 2646–2650
28. Rabelo B, Hofmann W (2002) DSP-based experimental rig with the doubly-fed induction generator for wind-turbines. In: Proceedings of the 10th international power electronics and motion control conference-EPE-PEMC 2002 (CD-ROM)
29. Rodriguez-Amenedo JL, Rodriguez-Garcia F, Burgos JC, Chincilla M, Arnalte S, Veganzones C (1998) Experimental rig to emulate wind turbines. In: Proceedings of the ICEM conference, vol 3, pp 2033–2038
30. Sebastián R, Quesada J (2006) Distributed control system for frequency control in a isolated wind system. Renew Energy 31(3):285–305
31. Southwest Windpower (2010) Whisper 100. http://www.windenergy.com/products/whisper_100.htm. Accessed 14 June 2010
32. Spagnuolo G et al (2010) Renewable energy operation and conversion schemes. IEEE Ind Electron Mag 4(1):38–51
33. Tan K, Islam S (2004) Optimum control strategies in energy conversion of PMSG wind turbine system without mechanical sensors. IEEE Trans Energy Convers 19(2):392–399
34. Teodorescu R, Iov F, Blaabjerg F (2004) Flexible control of small wind turbines with grid failure detection operating in stand-alone and grid-connected mode. IEEE Trans Energy Convers 19(5):1323–1332
35. Tomilson A, Quaicoe J, Gosine R, Hinchey M, Bose N (1998) Application of a static VAR compensator to an autonomous wind-diesel system. Wind Eng 22(3):131–141
36. Uhlen K (1994) Modelling and robust control of autonomous hybrid power systems. PhD Thesis, Trondheim University
37. Vlad C, Munteanu I, Bratcu AI, Ceangă E (2010) Output power maximization of low-power wind energy conversion systems revisited: possible control solutions. Energy Convers Manage 51(2):305–310
38. Wilkie J, Leithead WE, Anderson C (1990) Modelling of wind turbines by simple models. Wind Eng 4:247–274

Small Wind Driven Devices for Environment Friendly Power Generation and Consumption

N. A. Ahmed

Abstract The present chapter is written to highlight how the utilization of wind can be maximized for clean, pollution and hazard free power production and reduce consumption of electricity of households. The materials presented are built around the author's experiences while working on environmental friendly wind driven devices at the University of New South Wales in Australia. The chapter broadly considers two themes. The first theme deals with the methodologies associated with the design of rotors for small horizontal axis wind turbines to operate in low to high gust wind conditions for direct electricity generation. These are discussed in Sect. 2 of this chapter. The second theme considers wind driven devices whose operation can minimize the electricity usage of a household. The device discussed is rotating wind turbines that are used in building ventilation and can reduce the electricity bill of a household significantly with minimum carbon footprint. These are presented in Sect. 3 of this chapter. In Sect. 4, the author outlines his vision of future possibilities and foresees incorporation of flow control technologies to enhance the performance of future wind turbines and roof top ventilators. Finally in Sect. 5, overall conclusions are made about harnessing wind power for a sustainable future and living.

Symbols

A	Rotor disk area $(= \pi r^2)$
A_0	Wake cross sectional area
B	Number of rotor blades
C_P	Power co-efficient
C_T	Thrust co-efficient
C_{P0}	Power coefficient based on wake area

N. A. Ahmed (✉)
Associate Professor and Head of Aerospace Engineering, University of New South Wales, Sydney, NSW 2052, Australia
e-mail: N.Ahmed@unsw.edu.au

P. M. Pardalos et al. (eds.), *Handbook of Wind Power Systems*, Energy Systems,
DOI: 10.1007/978-3-642-41080-2_24,

C_{T0}	Thrust coefficient based on wake area
c	Local blade chord length
c_d	Local form drag coefficient
c_l	Local lift coefficient
d	Local form drag
l	Local lift drag
F	Tip loss factor
f	Sheet spacing parameter
$G(x)$	Dimensionless circulation function
P	Power
R	Blade radius (radius of rotor disk)
R_e	Reynolds number
R_O	Wake radius
r	Local blade radius
S	Slipstream distortion
T	Thrust
V	Free stream velocity
v_i	Induced velocity of the vortex sheet
W	Velocity relative to the blade section
w	Displacement velocity
$\overline{w}$	Dimensionless displacement velocity ($=\frac{w}{V}$)
χ	Dimensionless radius ($=\frac{r}{R}$)
α	Angle of attack of blade section
β	Blade angle ($=\varphi-\alpha$)
$\Gamma(x)$	Circulation at radial station, x
$\frac{\varepsilon}{\chi}$	Axial loss factor
λ	Free stream advance ratio ($=\frac{V}{\omega R}$)
λ_o	Wake advance ratio
σ	Solidity
ϕ	Relative wind angle
χ	Mass coefficient
ω	Angular velocity of the rotor

1 Introduction

The Gulf of Mexico oil spill disaster of 2010, the Exxon Valdez disaster of 1989 or the Chernobyl Disaster in the former USSR in 1986 are wake-up calls for developing clean, hazard free and renewable energy sources. The havoc such oil spills or radiation wreak on the environment will pale in comparison to the havoc that continued dependence on fossil fuel or nuclear power can wreak on the world's economy and security and even mankind's existence.

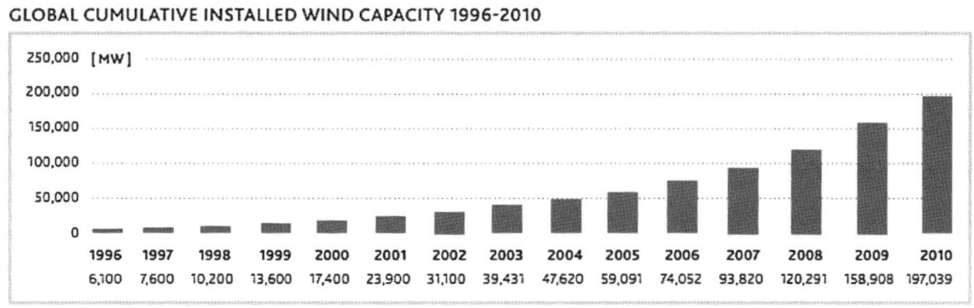

Fig. 1 Global wind power capacity (*Source* Global Wing Energy Council, Annual market update 2010)

The existence of wind turbine as a provider of useful natural source of power for the last thousand years is well documented [1–4]. The windmill, which once flourished along with the water wheel as one of the two prime movers based on the kinetic energy of natural sources, reached its peak in the 18th century. Its use then began to decline, as primary movers based on thermal energy from the combustion of fuel took precedence. In recent times, however, with increasing cost of thermal energy and a greater emphasis on sustainable environment, the wind energy is experiencing a revival. And the new millennium has commenced with the knowledge that wind represents a rapidly increasingly economically viable solution to a major problem of our time—the provision of electricity without pollution.

It is estimated that with the doubling of wind power plant capacity, which is currently happening every two and a half years, comes a 15 % drop in wind's cost, that is 6 % a year [5]. Wind power plants today contribute less than 1 % to electricity production in the world. This figure is slightly higher in the USA which is encouraging since the USA, based on the estimates published by the Department of Energy, USA [6], is by far the largest consumer of electricity in the world.

As shown in the Fig. 1, the total wind power capacity has risen dramatically [7] in the last decade. The rise in fossil fuel prices and uncertainty over resources combined with improved wind power technology and reduced costs have allowed this expansion. The number of wind farms currently in production or in planning indicates that the growth rate of the wind industry will continue to rise significantly for the foreseeable future.

Currently there is approximately 190 GW of installed wind power worldwide. China is now the largest wind power market in the world with a total installed capacity of 42 GW followed by North America [7]. Europe has approximately 85 GW of installed wind power capacity the biggest contributor being Germany [8]. One of the fastest growing markets in wind power is offshore turbines. This is due to the abundance of suitable sites and much better wind quality with less turbulence and higher speeds at lower altitude.

Australia currently has approximately 2 GW of installed wind power capacity providing 2 % of the nation's electricity needs [9]. Data on power produced by small wind power systems is limited as there is no requirement for households to report figures. The power generated by small systems would be negligible compared to utility size wind farms.

1.1 Wind Turbine Classification

Wind turbines are most often advertised according to their rated power output. This is the most important distinguishing factor between turbine models as it determines the appropriate uses of the turbine. The physical size of turbines is not always directly related to the power output as expected wind speeds vary between sites.

Despite manufactures selling turbines according to power output the international standards, applying to wind turbines, 'IEC 61400-Design Requirements for Wind Turbines' defines a small turbine as one with less than 200 m^2 swept area. This equates to approximately 16 m rotor diameter for horizontal axis turbines.

Turbines used in urban environments are often much smaller than this with a typical blade diameter of less than 2 m and an output of around 500 W. Smaller turbines of approximately 1 m rotor diameter are often used on small sailing yachts to generate a small amount of power without having to turn on engines. Larger turbines can be used on large properties where they can be mounted away from residence. Turbines up to 20 kW with approximately 10 m rotor diameter, are often used for power generation in remote areas.

Current utility size wind energy sites that supply power to the grid are using turbines in the 1–3 MW range and have a rotor diameter of 60–80 m. The largest turbines currently in used are 7.5 MW with a rotor diameter of 130 m. Large turbines are most commonly setup in groups called wind farms with output capacities often similar to small coal or nuclear power stations.

1.2 Typical Design Practices

A large proportion of horizontal axis turbines in use have a 3 blade upwind design with a generator in the nacelle behind the blades. Utility size turbines often use a gearbox to increase the rotational velocity between the rotor and the AC generator. This allows the use of smaller and cheaper generators running at higher speeds. Some manufactures such as ENERCON are moving towards using large low speed generators to remove the need for a gearbox. This reduces complexity but the generators must have hundreds of poles to produce the correct frequency, which greatly increases cost. Large turbines often use blade pitch or tip furling to control the speed of the turbine. This is critical to having a large maximum power band and preventing 'over speed' during high wind conditions.

Smaller commercially available turbines usually use DC generators as they often provide power to low voltage DC loads. To save weight and cost small turbines often use fixed blade design which gives them limited maximum power band and reduces efficiency but also reduces product and maintenance costs.

1.3 Current Issues

As wind speeds are very unpredictable the use of wind power can be very difficult to integrate into a continuous supply system or grid. If much of a grid's power is sourced by wind energy it can be very difficult to maintain a base load supply and that often leads to wasted energy due to oversupply. The output of offshore turbines is somewhat more predictable and design concepts to float wind turbines on balloons to take advantage of the fast wind currents higher up in the atmosphere could also be able to supply a base load within a grid. Developing more effective ways of integrating wind power into an energy grid is currently an area of development.

With standalone systems the unpredictable nature of the supply can also lead to additional problems. Large expensive batteries have to be used to provide power during extended periods of low wind. Turbines are, therefore, rarely used as the sole source of power in standalone systems. Improving energy storage would allow turbines to be used far more efficiently and cost effectively in off grid systems.

A major issue with smaller wind turbines is the lack of clean wind especially in urban environments. Surrounding buildings produce very turbulent and slow wind near ground level. Building turbines on tall masts can alleviate this problem but it adds significant costs and is often not welcomed by neighbours. Improving blade design in horizontal axis turbines is crucial to improving performance in poor wind conditions.

Also environmental impacts of the turbines are a big concern especially when they are located near populated areas. Some studies indicate that there are links between low frequency vibrations of large turbines and adverse health effects. Smaller turbines produce little low frequency vibration but are often located far closer to residential areas or even mounted on buildings. For these reasons understanding vibrations created by turbines and transmitted to surrounding structures is essential.

The power available from the wind has a cubic relationship with the speed of the air. Therefore it is advantageous to build turbines in high wind areas. High wind speeds though put much more stress on every component in the turbine including the electrical systems. Reliability in high speed and turbulent wind is an area of current development in turbine design. This allows turbines to be setup in areas that can yield high energy output while still maintaining a reasonable lifespan.

However, it is a fact that wind power plants are not yet as cost-effective and reliable as conventional means of electricity production.

The full exploitation of a wind turbine as a source of clean, non-polluting and renewable energy depends not only on huge multi-Mega Watt wind parks but also on small wind turbine that can generate power to meet the demand of electricity need of average households.

The present chapter deals only with small wind driven devices for power generation and power extraction, i.e., on devices that can be used to produce electricity directly and on ones that use wind directly for its operation. The materials presented are based on the author's experience on working on these devices over the past decade at the University of New South Wales in Australia. More specifically, the work relates to horizontal axis wind turbines and wind driven rotating ventilators.

2 Small Wind Turbines for Power Generation

The current state-of-the art technology and public perceptions of wind power can be aptly described in a humorous way using the following cartoon that appeared in Sunday Telegraph of Australia recently [10].

Small horizontal axis wind turbines for power generation considered here are the ones that can produce power between 1 and 10 kW. The applications sought are primarily for domestic electricity generation in residential buildings or boats or yachts. Although the ideas have been around for a long time, it appears that, it is only recently that the issue is taken up seriously by various governments around the world.

In Australia, householders in the state of New South Wales will be allowed to build wind turbines on the roofs of suburban homes to generate green electricity under a sweeping overhaul of New South of Wales planning laws. Silmalis [11] reported that the state is proposing to allow windmills with a generating capacity of 10 kW or less to be erected in residential areas, adding to solar panels as an option for domestic power generation.

A height limit of 3 m above the roof line will be imposed and turbines will have to be at least 25 m from neighbouring properties. As with solar panels, home owners will be able to sell surplus power they generate to the electricity grid, protecting them from skyrocketing power prices. Under the plan, families intending to install a wind turbine would lodge a 10-day complying development application with the local council. Strict noise and location controls would ensure neighbourhoods were not turned into turbine jungles. "Making it easier for property owners to install wind and solar systems would turn suburbs and rural areas into renewable energy harvesting areas with no, or minimal environmental and local amenity impacts", said NSW Planning minister Tony Kelly. The amendments to the law went on public exhibition on 18 April, 2010.

Under the plan wind turbines would be restricted to 10 kW in residential zones and 60 kW in rural and industrial areas. The solar energy feed-in system, introduced in January, 2010, allows households to earn 60c for each kilowatt hour of electricity produced. The government estimates that individual households could earn an average of $1,500 by selling surplus energy to the feed-in system.

In the USA, small scale residential wind turbines are allowed in parts of New York. For example, a five storey affordable housing apartment building in the South Bronx, in New York, has deployed 10 kW wind turbines to supplement the facility's conventional power usage in the building's hall ways, elevators and other common areas, the paper states.

Full exploitation of wind turbine technology will be possible when wind turbines will have the capacity to run without shut-down irrespective of wind conditions which in turn may depend on the optimum design of turbine blades that can operate efficiently under moderate to extreme variations in wind conditions. To be more effective, it may also mean identifying low wind areas and high wind areas and design optimum rotors accordingly to suit the different wind conditions.

The following section is presented to provide a simple methodology for possible optimum rotor design, or the use of the light loaded or highly loaded theories, both of them based on vortex theory. The vortex sheet inherent in the formulation itself is considered to be an infinite number of vortex filaments each of infinitesimal strength extending to the boundary of the flow. The lightly loaded theory can be applied to low free stream velocities and small wind gusts and operate with advance ratios of λ, of less than half. The point of note is that a rotor designed for light wind load applications will not be efficient in heavy wind areas.

2.1 The Lightly Loaded and Highly Loaded Theories for HAWT Rotor Design

2.1.1 Basic Consideration

The vortex theory applicable to aircraft propellers has generally been used to model the behaviour of horizontal-axis wind turbines (HAWT). The theory developed by Glauert [12], Betz and Prandtl [13] and Goldstein [14] has been adopted by Larrabee [15] in the design of manpowered aircraft propellers and wind mill rotors. Larrabee used what may be called the light loading theory, where an assumption is made that the change in the cross sectional area of the stream tube of the rotor is negligible implying thereby that the blades are widely spaced and suffer no distortion. This assumption cannot be considered appropriate for heavily loaded horizontal axis wind turbine, because the retardation of the flow through the rotor causes the vortex sheet spacing and stream tube cross section to change continuously downstream (Fig. 2).

The works [16–18] at the University of New South Wales combine the computational simplicity of the Betz and Prandtl relations [13] with the vortex theory of heavily loaded airscrews of Theodorsen [19] to yield a simple method of maximizing the power co-efficient, and consequently predicting the ideal blade

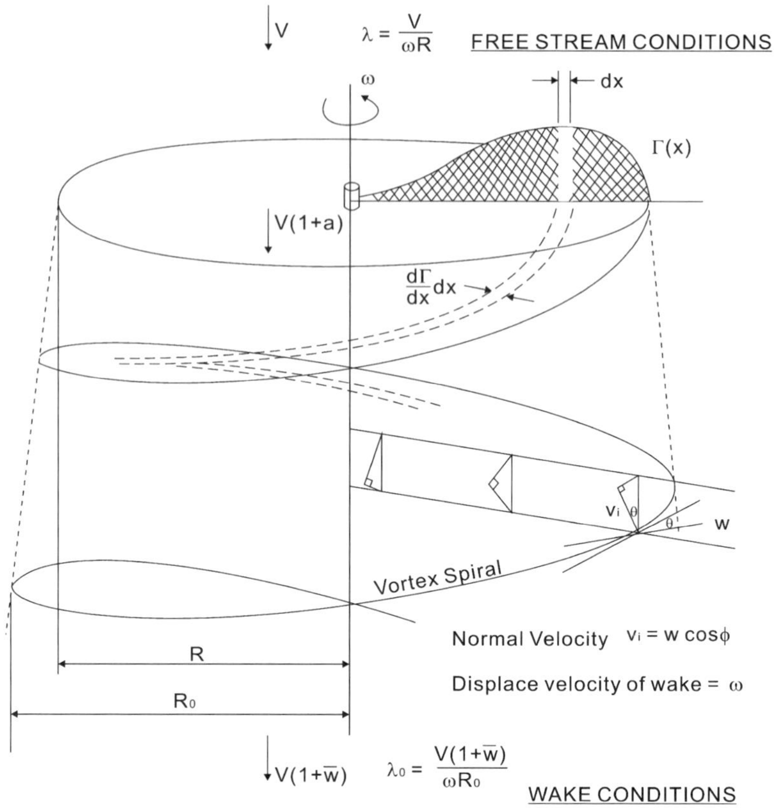

Fig. 2 Vortex sheet geometry

geometry of optimum HAWT rotors at their design advance ratio (wind speed/tip speed).

What follows next are based on the salient features of this approach.

2.1.2 Load Distribution Along the Blade

Betz and Prandtl [13] theorem of the 'rigid screw' predicts the circulation distribution along a (airscrew) rotor blade for minimum induced energy loss. The condition is analogous to the minimum induced drag (elliptic lift distribution) on a finite wing. The theorem states that the ideal circulation distribution is achieved when the helical vortex sheets, shed by the advancing blades, are displaced rearward by a radially constant displacement velocity, $\overline{w}$ (Fig. 2). The displacement velocity is an 'apparent' velocity, since the induced velocity v_i of a vortex element on the vortex sheet is directed normal to the free vortex sheet. Hence the displacement is given by:

$$\overline{w} = \frac{v_i}{v\cos\phi} = \frac{w}{V} \tag{1}$$

where ϕ is the angle between the induced velocity on the vortex sheet and a parallel to the axis of rotation (Fig. 2).

An approximate solution to the potential flow problem appropriate to this condition was provided by Betz and Prandtl [13]. Their solution offered a simple set of relationships for predicting the ideal circulation distribution along the blade. The solution was obtained by assuming the rotor to be lightly loaded and consequently, the shed vortex sheets to be regularly spaced in a constant area wake ($A = A_0$). For light loading condition, Goldstein [14] showed that the Betz and Prandtl approximations were sufficiently accurate to predict the ideal circulation distribution along the (propeller) blade at low advance ratios, $\lambda < 0.5$.

There is considerable distortion of the slipstream caused by the horizontal axis wind turbine rotor as it extracts energy from the wind. The ratio of the rotor disk to the area of the wake behind the wake, A/A_0 may be used to quantify the wake expansion. Sanderson and Archer suggests [16] that the HAWT rotor be treated as heavily loaded if $A/A_0 \leq 0.9$ or $A_0/A \geq 1.1$.

Provided that reference is made to the helix surface far behind the rotor, Theodorsen [19] has demonstrated that the predicted circulation distribution of a lightly loaded rotor was directly applicable to a heavily loaded rotor. The dimensionless circulation distribution is often referred to as the 'Goldstein function [14] and can be determined by the Betz and Prandtl relationships as expressed below:

$$G(x) = \frac{B\Gamma(x)\omega}{2\pi V^2(1+\overline{w})\overline{w}} = \frac{Fx^2}{\lambda_0^2 + x^2} \tag{2}$$

where the tip loss factor F is evaluated by:

$$F = \frac{2}{\pi}\cos^{-1}(\exp -f) \tag{3}$$

and the sheet spacing parameter, f, by:

$$f = \frac{B}{2}\frac{\sqrt{\lambda_0^2 + 1}}{\lambda_0^{2^2}}(1 - x) \tag{4}$$

The advance ratio λ_0 is defined for the wake (Fig. 1), as:

$$\lambda_0 = \frac{V(1+\overline{w})}{\omega R_0} \tag{5}$$

and is related to the free stream advance ratio by:

$$\lambda = \frac{\lambda_0}{(1+w_d)\sqrt{A/A_0}} \tag{6}$$

For the assumptions of the light-loading theory, $A/A_0 = 1$ and $\lambda = \lambda_0$. However, in the heavily loaded theory, λ_0 is adopted instead of λ as the independent variable describing the action of the rotor. Evaluation of λ is performed in the final stages of computation by Eq. (6).

2.1.3 Maximizing the Power Coefficient

The maximization of the power coefficient is expected to be carried out at the initial steps in the design of an HAWT rotor. By neglecting the form drag, the power produced by the rotor may be assumed to be dependent on the circulation distribution and displacement velocity.

Expressing the power in the form

$$P = C_{P0} 1/2\, \rho V^3 A_0 \tag{7}$$

where C_{P0} is the power co-efficient based on the wake area, consideration of the momentum changes to the flow behind the rotor shows that C_{P0} is determined from Theodorsen [19]:

$$C_{P0} = 2\chi w_d (1 + \overline{w})\left(1 + \frac{\varepsilon}{\chi}\overline{w}\right) \tag{8}$$

The mass co-efficient is the dimensionless momentum resulting from the induced velocities in the wake between successive vortex sheets and is evaluated by:

$$\chi = 2\int_0^1 G(x)\, x dx = 0(1) \tag{9}$$

It can be regarded as the mean radially weighted Goldstein function $G(x)$.

The axial loss factor relates the proportion of momentum change in the axial direction to the momentum changes in the wake. With decreasing advance ratio, the components of induced velocity, v_i in the radial and tangential directions decrease and, hence, the axial loss factor increases with decreasing advance ratio (Fig. 3). This parameter is given by Theodorsen [19]:

$$\frac{\varepsilon}{\chi} = 1 + \frac{1}{2}\frac{d(\ln \chi)}{d(\ln \lambda_o)} = 0(1) \tag{10}$$

The condition for maximum power,

$$P_{max} = C_{P0max} 1/2\, \rho\, V^3 A_0 \tag{11}$$

Is achieved when C_{P0} in Eq (8) is maximized.

If χ and ε/χ, which are functions of the wake advance ratio only are treated as constants, Eq. (8) becomes a simple cubic of $\overline{w}$. Consequently, the optimum

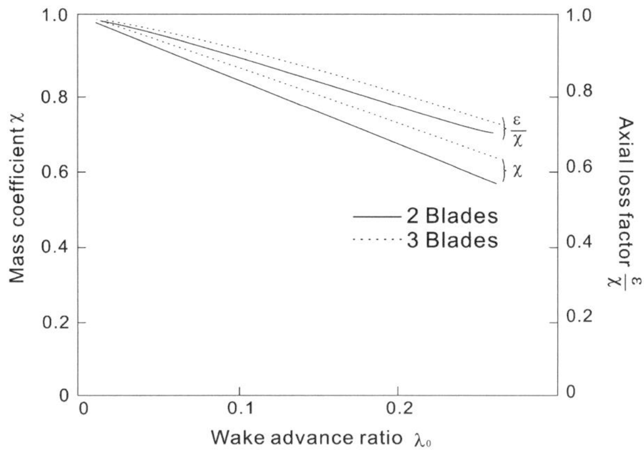

Fig. 3 Variation of mass coefficient and axial factor with wake advance ratio

condition can be derived from Eq. (8) and corresponds to the solution of the quadratic equation:

$$3\left(\frac{\varepsilon}{\chi}\right)\overline{w}^2 + 2\left(1+\frac{\varepsilon}{\chi}\right)\overline{w} + 1 = 0 \tag{12}$$

The negative root of $\overline{w}$ is applicable to horizontal axis wind turbine rotor. It is noted that that the limit as $\lambda \to 0$, of both χ and ε/χ, has a unit value and, hence, the non-trivial root of Eq. (12) is $\overline{w} = -1/3$ at $\lambda_0 = 0$. Henceforth, the subscript max will be omitted and it will be assumed that the value of C_{P0} is that obtained by Eq. (8), subject to the solution of Eq. (12).

2.1.4 Geometric Relationships

The blade geometry in its non-dimensional form, c/R, can be derived in the following manner. For an assumed value of λ_0, Eqs. (2)–(4) are used to obtain an expression for the circulation $\Gamma(\mathrm{x})$ at a radial station x:

$$\Gamma(x) = \frac{2\pi V(1+\overline{w})\, wG(x)}{Bw} \tag{13}$$

Also from airfoil theory:

$$\Gamma(x) = 1/2WC_l c \tag{14}$$

Using from velocity diagram (Fig. 4)

$$W = \frac{V}{\sin\varphi}(1 + a\cos^2\phi) \tag{15}$$

where, a is the interference factor, similar to the displacement velocity $\overline{w}$ and is negative for a windmill.

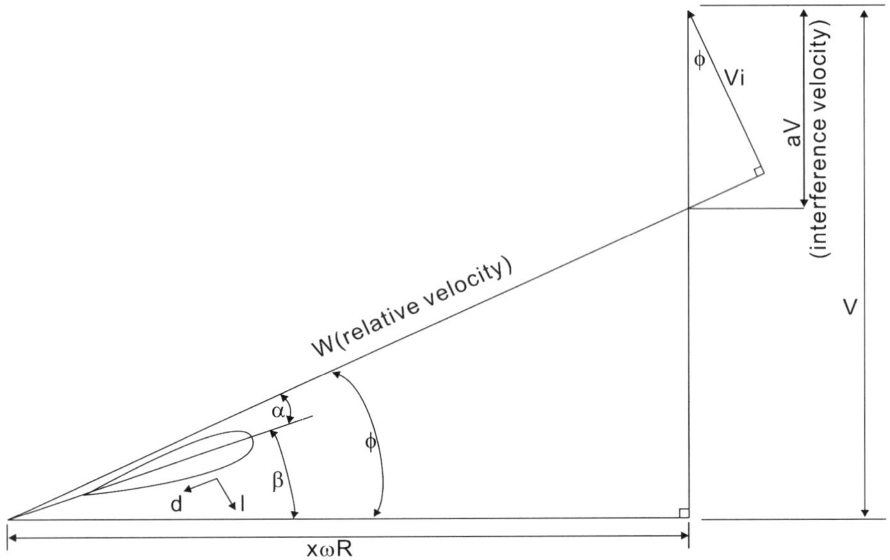

Fig. 4 Velocity diagram for radial station, x

Finally combining Eqs. (13)–(15);

$$\Gamma(x) = \frac{1}{2}\frac{V}{\sin\phi}(1 + a\cos^2\phi)C_l c \tag{16}$$

is obtained. By introducing the local solidity defined by:

$$\sigma = \frac{Bc}{2\pi r} \tag{17}$$

From Eqs. (13), (16) and (17):

$$\sigma C_l = \frac{2(1+\overline{w})wG(x)\sin\phi}{\gamma(1 + a\cos^2\phi)w} \tag{18}$$

From Eq. (5),

$$\frac{w}{\omega} = \frac{\overline{w}}{(1+\overline{w})}(\lambda_0 R_0) \tag{19}$$

Eq. (18) reduces to:

$$\sigma C_l = \frac{2\overline{w}\lambda_0 G(x)\sin\phi}{(\chi R/R_0)(1 + a\cos^2\phi)} \tag{20}$$

From Fig. 3,

$$\tan\phi = \frac{(1+a)\lambda}{\chi} \tag{21}$$

and Eq. (5)

$$\lambda = \frac{\lambda_0}{(1+\overline{w})R/R_0} \tag{22}$$

Eq. (20) reduces to:

$$\sigma C_l = \frac{2(1+\overline{w})\overline{w}G(x)\sin^2\phi}{(1+a)(1+a\cos^2\phi)\cos\phi} \tag{23}$$

Using the following expression given by Theodorsen [19] for interference factor,

$$a = \frac{0.5\overline{w} + (\varepsilon/\chi)\overline{w}^2}{1+\overline{w}(1+\varepsilon/\chi)} \tag{24}$$

And the incident angle ϕ along the blade is obtained combining Eqs. (21) and (22) to give:

$$\tan\phi = \frac{(1+a)\lambda_0}{(1+\overline{w})\chi R/R_0} \tag{25}$$

This in turn requires the evaluation of the wake expansion parameter A/A_0. Sanderson and Archer [16] proposes a method that neglects form drag and equates the thrust developed by the circulation distribution in the wake to that at the rotor. Adopting the convention that the displacement velocity is negative for HAWT rotors, Eq. (23) is rewritten as:

$$\sigma C_l = \frac{-2(1+\overline{w})\overline{w}G(x)\sin^2\phi}{(1+a)(1+a\cos^2\phi)\cos\phi} \tag{26}$$

Combining with Eq. (17), the expression of the dimensionless chord is finally obtained as:

$$\frac{c}{R} = \frac{-4\pi(1+\overline{w})\overline{w}G(x)\sin^2\phi}{C_l(1+a)(1+a\cos^2\phi)\cos\phi} \tag{27}$$

2.1.5 Determination of SlipStream Expansion, A/A_0

Neglecting form drag, the thrust developed at the rotor is expressed as (see Fig. 4):

$$T = \frac{1}{2}BR\int_0^1 W^2 C_l \cos\phi \, dx \tag{28}$$

Using Eqs. (14), (15), (25) and (28):

$$T = 2\rho V^2 A \frac{\overline{w}(1+\overline{w})}{(1+a)} \int_0^1 (1 + a\cos^2\phi) G(x)\, dx \tag{29}$$

Using the thrust co-efficient, C_{T0}, based on wake area:

$$T = C_{T0} 1/2\, \rho\, V^2 A_0 \tag{30}$$

and following the same principle s as the power coefficient C_{P0} (see Ref. [5], p. 28):

$$C_{T0} = 2\chi\overline{w}\left[1 + \overline{w}\left(\frac{1}{2} + \frac{\varepsilon}{\chi}\right)\right] \tag{31}$$

Combining Eqs. (29)–(31), the slipstream expansion expression is obtained:

$$\frac{A}{A_0} = (1+a)\chi\left[1 + \overline{w}\left(\frac{1}{2} + \frac{\varepsilon}{\chi}\right)\right] \Big/ \left[2(1+\overline{w}) \int_0^1 (1 + a\cos^2\phi) G(x) x dx\right] \tag{32}$$

Using the slipstream distortion parameter, S as:

$$S = \frac{2}{\chi} \int_0^1 \cos^2\phi G(x)\, dx \tag{33}$$

Eq. (32) is expressed as:

$$\frac{A}{A_0} = (1+a)\left[1 + \overline{w}\left(\frac{1}{2} + \frac{\varepsilon}{\chi}\right)\right] \Big/ [(1+\overline{w})(1+aS)] \tag{34}$$

The evaluation of the slipstream expansion requires the simultaneous solution of Eqs. (25), (33) and (34). These equations are found to converge rapidly, from an initial assumption of light loading $A/A_0 = 1$, using a simple convergence criterion on A/A_0. For a given A/A_0, the incident angle ϕ, along the blade can be found and the integration in Eq. (34) performed, thus providing a new value of the slip stream expansion of A/A_0 by Eq. (33).

2.1.6 Estimation of Profile Drag

Using Fig. 4, the contribution to the thrust caused by the profile drag can be accounted for by rewriting Eq. (28):

$$T = \frac{1}{2}\rho BR \int_0^1 W^2 c(C_l \cos\phi + C_d \sin\phi)dx \tag{35}$$

Substituting C_l, $\Gamma(x)$ and W from Eqs. (13)–(15),

$$T = \frac{1}{2}\rho BR \int_0^1 \frac{V(1+\overline{w})w}{B\omega} G(x) \frac{V}{\sin\phi}(1 + a\cos^2\phi)\left(\cos\phi + \frac{\sin\phi}{l/d}\right)dx \tag{36}$$

Using Eqs. (5) and (21):

$$\begin{aligned} C_T &= \frac{2T}{\rho V^2 \pi R^2} \\ &= -4\lambda \int_0^1 (1+\overline{w})\overline{w}G(x)\frac{V}{\sin\phi}(1 + a\cos^2\phi)\left(\frac{1}{\tan\phi} + \frac{1}{l/d}\right)dx \end{aligned} \tag{37}$$

Note the change of sign was made to comply with the negative convention of the displacement velocity.

Similarly, the corrected expression for power produced for form drag becomes:

$$P = \frac{1}{2}\rho R^2 \int_0^1 W^2 c(C_l \cos\phi - C_d \sin\phi)\omega x dx \tag{38}$$

In a manner similar C_T,

$$C_p = -4 \int_0^1 (1+\overline{w})\overline{w}(1 + a\cos^2\phi)\left(1 - \frac{1}{l/d\tan\phi}\right)G(x)xdx \tag{39}$$

The equivalent results from light loading theory assuming $a = \frac{\overline{w}}{2}$ are:

$$C_T = -4\lambda \int_0^1 \left(1 + \frac{\overline{w}}{2}\cos^2\phi\right)\overline{w}\left(\frac{1}{\tan\phi} + \frac{1}{l/d}\right)G(x)dx \tag{40}$$

and

$$C_p = -4 \int_0^1 \left(1 + \frac{\overline{w}}{2}\cos^2\phi\right)\overline{w}\left(1 - \frac{1}{l/d\tan\phi}\right)G(x)xdx \tag{41}$$

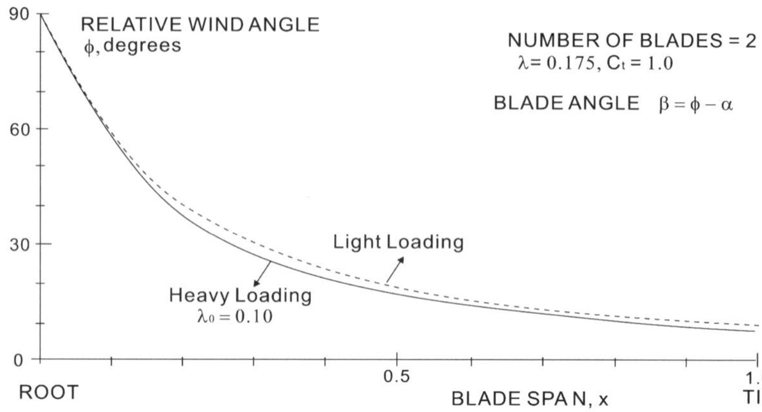

Fig. 5 Theoretical optimum rotor, for light loading and heavy loading blade angle distribution

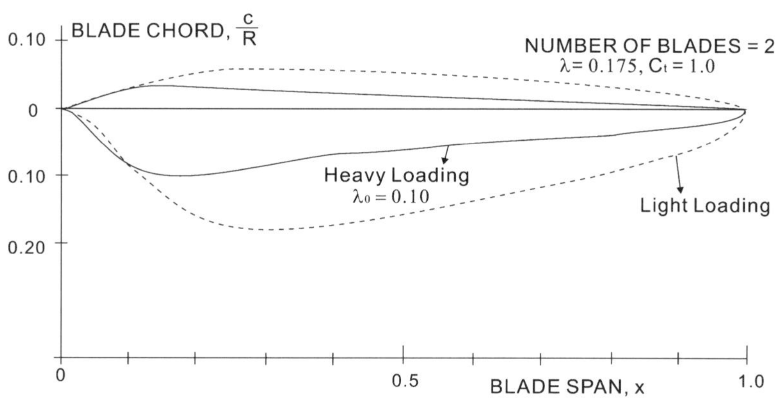

Fig. 6 Theoretical shapes of optimum rotor for light loading and heavy loading blade planform

Using the methodology described above, Sanderson and Archer [16] obtained the optimum rotor blade angle distribution and its plan form for a two rotor blade horizontal axis wind turbine for light loading and heavy loading as shown in Figs. 5 and 6 respectively.

2.2 Experimental Validation

For validation of the theories expounded earlier and to adapt a horizontal axis wind turbine for high gust wind environment, it was decided to continue with the initial work that began in the early 1980s at the University of New South Wales to produce an optimized horizontal axis wind turbine for use at the Australian

Mawson Station in the Antarctica. Meteorological conditions at the Mawson Antarctic base are characterized by strong morning winds blowing from the south east away from the high polar plateau, a high average speed of above 14 m/s, a small period of calm (5 %) per year, and hurricane strength storms with gusts exceeding 50 m/s [20]. Another impetus to this work is the scarcity of water supply at Mawson and a natural solution to this problem was to use wind energy as a heat source to melt snow. During these initial studies, however, it was realized that lightly loaded theory advanced by workers such as Glauert [12], Betz and Prandtl [13], Goldstein [14] or Larrabee [15] often used for propeller type windmill rotors, was not suitable to produce an optimum rotor design for a horizontal axis wind turbine in such circumstances.

The unusual nature of Mawson wind frequency distribution also favours a rotor design with rated speed to mean wind speed ratio approaching unity instead of the ratio of approximately 2 for the more usual Weibull distribution [20] and consequently the analysis of Sanderson and Archer [16] could be explored.

Work was, therefore, initiated to reinforce with test results the effectiveness of the highly loaded theory for the design of a Horizontal Axis Wind Turbine for optimum power production over a wide range of advance ratios. Initial tests [17] on two small rotors of 0.3 and 0.6 m diameters revealed problems with low R_e or scale effect, high g-loading, strength, stiffness and vibration. Although the theory [16] accurately predicts the blade chord and constant R_e radially, laminar separation took place at subcritical R_e with consequent loss in performance. Furthermore, the changing effect of higher R_e on l/d produced rapid changes in performance. To overcome these problems, two further 3-bladed rotors of 1.2 m blade diameter were designed and a direct means of measuring torque and speed of the rotor shaft was devised. The maximum efficiency obtained in these tests was, however, only 85 % of the theoretical value. The test facility has been modified to obtain more accurate results and further tests have been carried out. The maximum efficiency thus obtained was corrected for the mechanical power loss associated with the transmission system of the test rig and close agreement with theory was obtained.

2.2.1 Design and Manufacture of Rotor

The basis of rotor design is briefly explained here. The theoretical analysis [16], suggests that the peak performance, for example, of a two bladed rotor as expressed by the power co-efficient, C_p, is approximately 0.4 and occurs at low advance ratios, λ, of approximately 0.14. In the high average winds of Mawson, this implies a tip speed ratio, TSR, of 7.4 and a high tip speed of approximately 100 m/s. If a 1.2 m diameter rotor were designed to generate 1 kW in an average speed of 14 m/s, it would have to rotate at approximately 1,600 r.p.m. with a centrifugal g-loading of approximately 17,000. In the extreme condition, that is, at gust speeds of 50 m/s, the rotor would have to rotate at approximately 6,000 r.p.m with a centrifugal g-loading of approximately 200,000 and supersonic tip speed. If no furling or braking is to be provided then it would be necessary to adopt a higher design advance ratio or lower tip speed ratio. With this consideration, rotor design was carried out for a lower C_p than the peak optimum value of 0.4.

Because of damage to the rotors used in the study reported above, a new test rotor was produced. It was a 3-bladed, 1.2 m in diameter with a 0.4 m diameter hub, Clark Y blade section and designed for an advance ratio of 0.333. The rotor was made of wood (Queensland Maple) and manufactured using a numerically controlled machine as outlined in Ref. [17].

2.2.2 Modifications to Turbine Test Rig and Wind Tunnel Testing

Several modifications were also carried out to the turbine test rig. These included a 3:1 gearing system between the rotor and alternator to obtain peak performance results at a higher free stream speed of 10 m/s.

Also, to improve the measurement error margin, a new turbine stand was designed to ensure that the rotor was positioned centrally in the wind tunnel test cross section. A dual shaft arrangement as opposed to single shaft to minimize errors arising from misalignment during assembly and disassembly of the rig and a more accurate shaft speed tachometer were also incorporated. The total error in C_p, therefore, improved to within ± 0.0025 or approximately ± 1 % at a C_p of 0.25.

To absorb the power produced by the rotor, a DC electric motor was used. By loading the Horizontal Axis Wind Turbine test rig via an alternator, the corresponding amount of power required by the motor to achieve a predetermined speed could be measured. By comparing the value of the power produced by the alternator, the power loss was determined for a range of alternator loads and motor speed which is shown in Fig. 7. On this graph, the calibration equation which was used to correct the wind tunnel testing data is also given.

Testing was carried out using the largest available test section, 3.05×3.05 m (10×10 ft), of the large subsonic wind tunnel of the Aerodynamic laboratory [21] of the University of New South Wales. A schematic of the modified Horizontal Axis Wind Turbine test rig is given in Fig. 8.

2.2.3 Results and Discussion

Rotor performance curves obtained using the old and the new arrangements are given in Fig. 9 for comparison with highly loaded theory. For ease of viewing, the correction due to mechanical power loss is applied to the results of the new arrangement for the free stream velocity of 10 m/s only. In interpreting the results of this graph, it should be realized that the theory under consideration provides a curve which is essentially a locus of peak performance of different rotors. Thus the results for C_p can only be compared with theory at the design point. There is also a shortfall depending on the R_e and the findings of the earlier work on smaller rotors are instructive in this respect [17]. It was confirmed that the theory is expected to give an over-estimation of performance peaks for small rotors operating in the

Fig. 7 Power loss in the system as a function of alternator shaft speed

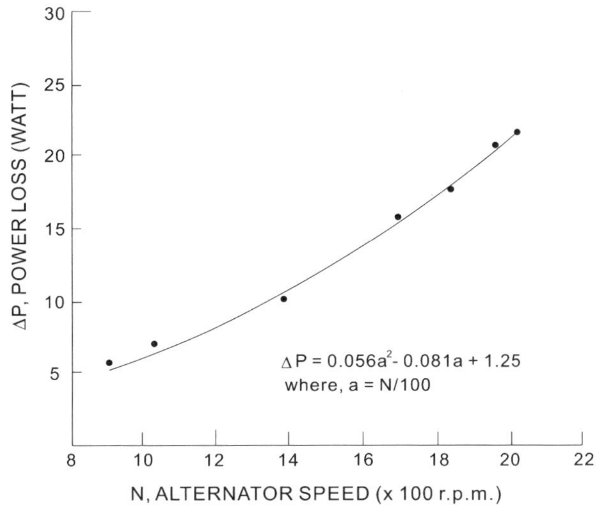

subcritical R_e regime where airfoils undergo non linear changes and that the accuracy of prediction can be expected to improve for larger rotors operating above the critical R_e range.

Results with the old rig produced a peak test value nearer to 85 % of the theoretical design value with improvements as R_e increased. Results obtained using the new test rig also shows a similar trend. The peak performance at 10 m/s was found to be around 90 % of the theoretical prediction. The power loss correction added approximately a further 3 % to the uncorrected value. It should be pointed out that further improvements to the test rig, such as streamlining, need to be carried out to obtain the performance curve for much higher free stream velocities to replicate the 14 m/s of Mawson Base.

Based on the results presented in this study, it can be concluded that the Sanderson and Archer optimum loss theory offers a rational basis for the design of propeller wind turbines for peak performance. The main attraction of the theory is its simplicity and the ease with which design curves can be set up. The design process can become rapid since required calculations can be carried out using only hand held calculators.

3 Small Roof Top Wind Driven Ventilators to Reduce Household Power Consumption

A distinction is made in this section between wind-driven devices that do not produce electricity but use wind energy directly, thereby saving electricity usage. The device considered here is wind-driven rotating roof top ventilator.

Proper ventilation in building requires that there be a movement or circulation of air within a space and that the temperature and humidity be maintained within a

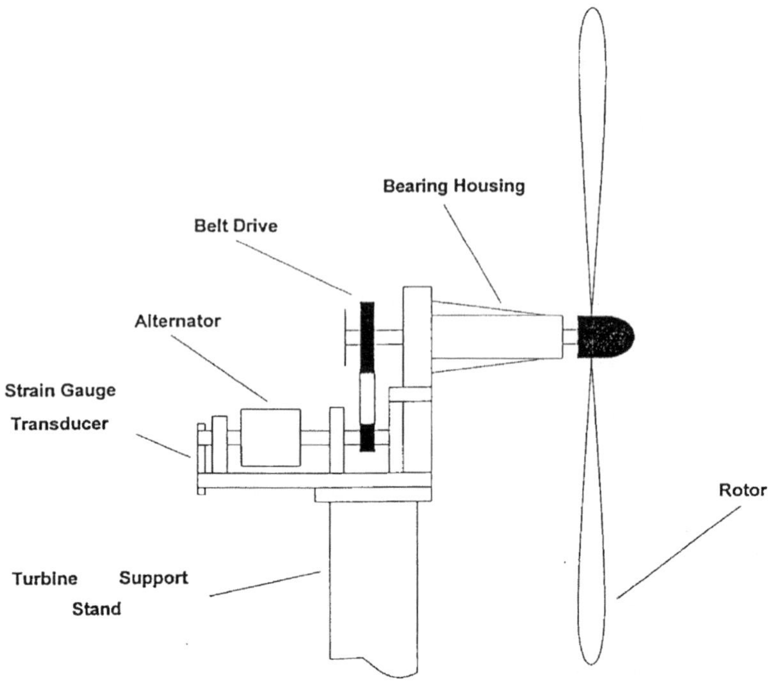

Fig. 8 Schematic of the horizontal axis wind turbine test set up

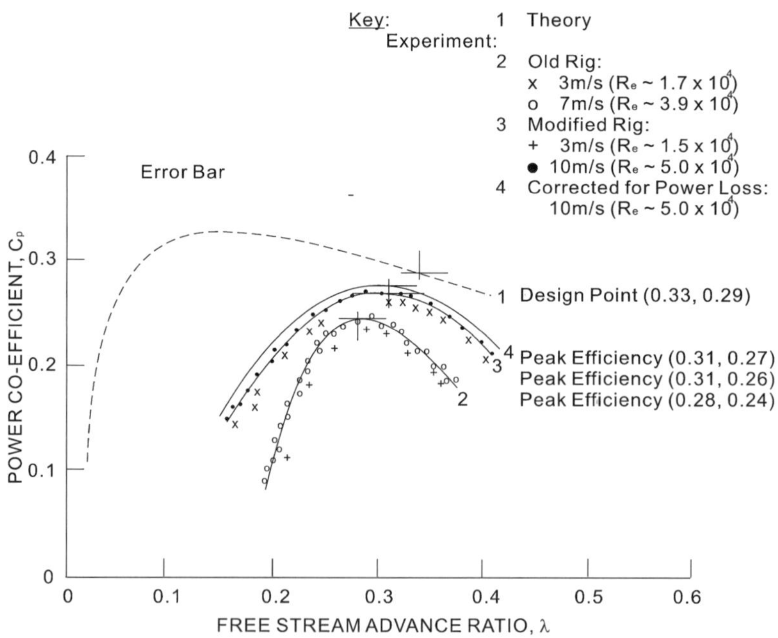

Fig. 9 Comparison of predicted power co-efficient with experiment for a 3-blade rotor

range that allows adequate evaporation of perspiration from the skin. It was formerly believed that the discomfort, headache, and lethargy were caused entirely by the increase in the amount of carbon dioxide and the decrease in the oxygen content of the air. There is now evidence to show [21] that the deleterious effects may also result largely from interference with the heat-regulating mechanism of the body. Lack of air currents and the increase in relative humidity and temperature, especially noticeable in crowded or poorly ventilated places, prevent normal evaporation of perspiration and loss of heat from the surface of the skin.

Also, despite the wide distribution of air pollutant sources, the concentration of indoor pollutants may be the dominant risk factor in relation to personal exposure, as most people spend an average of 87 and 6 % of their time within enclosed rooms and vehicles, respectively [22]. Indoor exposure may pose more harmful health effects, as the indoor concentrations of many pollutants are often higher than those typically encountered outside [23]. Under requirements to maintain a safe working environment, many dwellings and factories now need adequate fresh air exchange to remove gaseous, process emissions and/or heat build up. The high priority placed on indoor air quality from health considerations has prompted New York in the USA to pass legislation effective from December, 2008 to require landlords to notify tenants and building occupants of indoor air test results [24].

Modern day building's air conditioning system is typically responsible for around 50 % of the base building energy consumption [25]. The other 50 % typically includes other services such as common area lighting, domestic hot water, lifts, etc. As such, any reduction in air consumption or efficient use of energy utilisation will offer significant savings in total building energy consumption and carbon emissions. Under these circumstances, wind driven rotating ventilators, which use wind as a natural energy source are cheap to manufacture, install and maintain, and has, therefore, found widespread use in most parts of the world.

The rotating wind driven ventilator is environmental friendly, and costs nothing to operate. It can be installed either on the roof of a dwelling or moving vehicle or side-mounted on their windows. It is simple in structure, light in weight and cheap to install. This type of ventilator is generally manufactured from corrosion resistant aluminium. Some common forms of roof top wind driven ventilators are shown in Fig. 10. A series of cylindrical ventilators operating on the Red Centre building of the University of New South Wales can also be seen in Fig. 11.

The simplicity of the rotating wind driven ventilator belies the fact, that the actual mechanism associated with its operation is very complex. A rotating roof top ventilator can be likened in part to a vertical axis wind turbine in its rotor configuration and in the sense that it also uses wind energy directly for its rotation. However, in terms of operational details, it behaves more like a centrifugal compressor. There is, however, a major difference. A centrifugal compressor accepts air in an axial direction and expels the same air in a radial direction. A rotating ventilator, in contrast, works with air from two sources: it accepts from the free stream of the atmosphere in the axial direction, but expels air from a different source, namely the contaminated air from inside a building into the atmosphere.

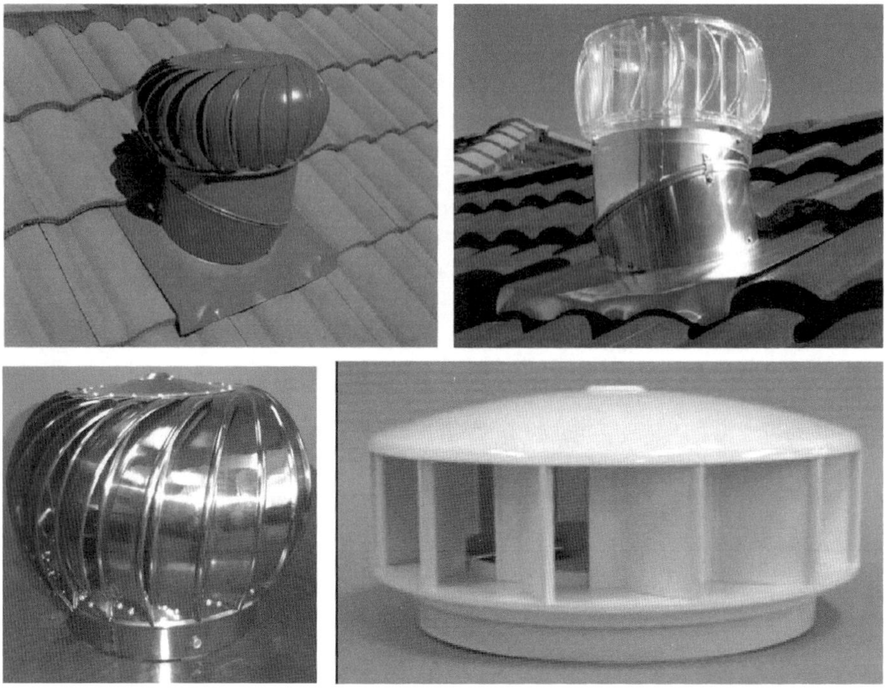

Fig. 10 Some common forms of roof top wind driven ventilator (Image www.edmonds.com.au)

Fig. 11 Cylindrical centrifugal ventilator at the University of New South Wales

Most wind riven rotating ventilators have evolved through a process of trial and error in the past as long as they met the regulatory requirements of being able to withstand wind speeds of 220 km/h without blowing off from a rooftop and not posing any hazards to the population. Works [26–30] carried out at the University of New South Wales were probably the first systematic investigations of the aerodynamic performance of these ventilators. The initial works were mainly experimental in nature using simple physical models in a wind tunnel. The models consisted of a stationary cylinder and a spinning cylinder. Such models were useful in understanding some of the aspects of aerodynamic force acting on a wind ventilator. The flow visualization experiments were limited in scope but still provided useful qualitative information about the nature of the flow, particularly, in relation to force component associated with drag that were later confirmed in force/torque transducer measurements. The size of the ventilator wake was found not to increase in size appreciably downstream of the ventilator. The rapid decay of the ventilator wake also emphasized the importance of flow mixing by the blades of the rotor.

Recent works have involved the investigation on internal flows within rotating vanes of these ventilators using computational fluid dynamics [31]. An objective this study was to explore the effect of inclined roof on the performance of these ventilators.

First some wind tunnel experiments were performed. The turbine ventilator used in this investigation is a Hurricane H100 produced by CSR Edmonds Australia Ltd. It consists of a rotating portion (Rotor) with 8 curved blades and a stationary portion in the form of a cylindrical base. The dimensions of the various components of this ventilator are shown in Fig. 12.

The physical experimentation were conducted in a 76 mm diameter open return, 0.2 % turbulence intensity open test section wind tunnel [1, 32] of the Aerodynamics Laboratory of the University of New South Wales. The experimental setup is shown in Fig. 13.

The wind tunnel measurements were obtained at three different cut planes of h = 270, 370, 420 mm, respectively (see Fig. 13) with free stream velocity kept at 10 m/s. The mean value of static pressure at each pressure ports at a measurement location were obtained from data recorded over a 10 s sampling period.

The velocity vectors and static pressure distributions thus obtained from this experiment were used to serve as the bench mark data to validate the initial aspects of the CFD simulation. The schematic diagram of the wind tunnel test set up is given in Fig. 13 (Fig. 14).

The CFD validation offered further scope to investigate both external and internal flows around and within a rotating ventilator (Figs. 15, 16). The results were encouraging as they showed trends that were in good agreement with those observed by other experimental studies [29, 33–35]. Performance studies of rooftop ventilator operating with different blade sizes were also conducted. With increases of 50 and 100 % in the ventilator blade height, the improvements of between 15 and 25 % in exhaust mass flow rate were achieved, respectively. The exhaust mass flow rate was found to have a linear relationship with the

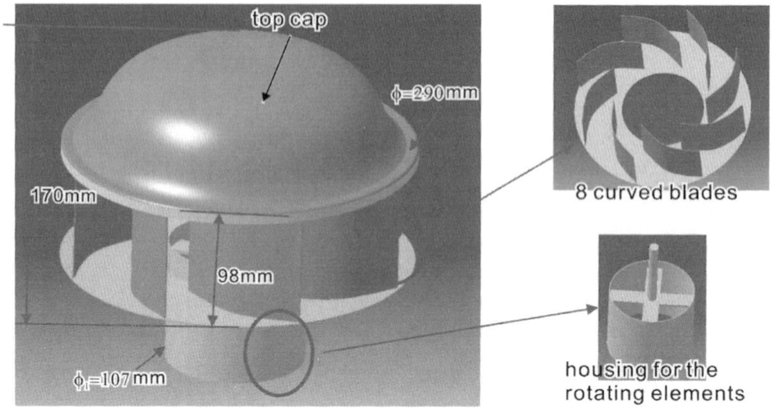

Fig. 12 Modelling of turbine ventilator

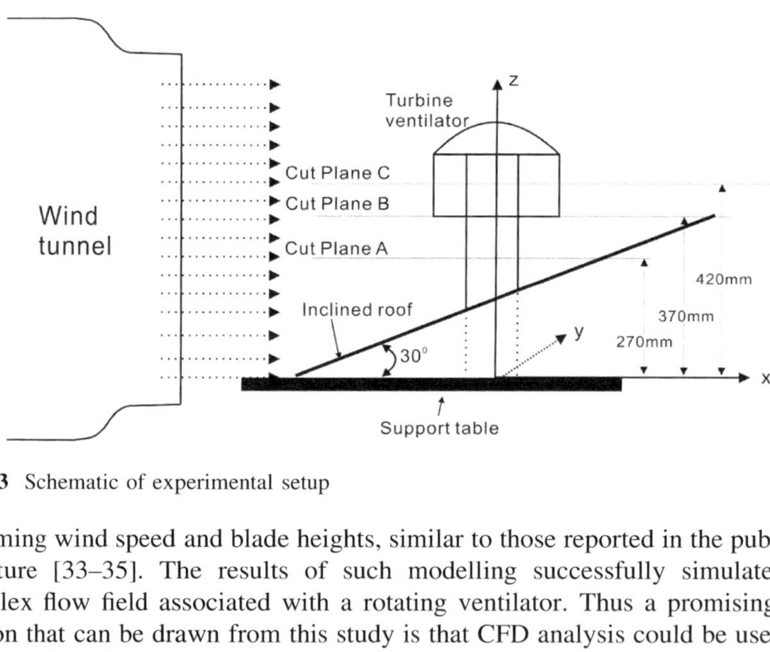

Fig. 13 Schematic of experimental setup

oncoming wind speed and blade heights, similar to those reported in the published literature [33–35]. The results of such modelling successfully simulated the complex flow field associated with a rotating ventilator. Thus a promising conclusion that can be drawn from this study is that CFD analysis could be used as a cost effective aid to future design and development of rooftop turbine ventilators with enhanced performance.

The ventilation research at the University of New South Wales with which the author is actively involved, have benefitted substantially in the recent past from two industry collaboration programs funded by the Australian Research Council that have resulted in novel aerodynamic techniques for highly three-dimensional flow measurement and analysis [36–39] and used in ventilation studies, a simple conceptual model [28] for wind driven turbine ventilator design, and a more efficient bladedesign which has been incorporated in the 'Hurricane' ventilator

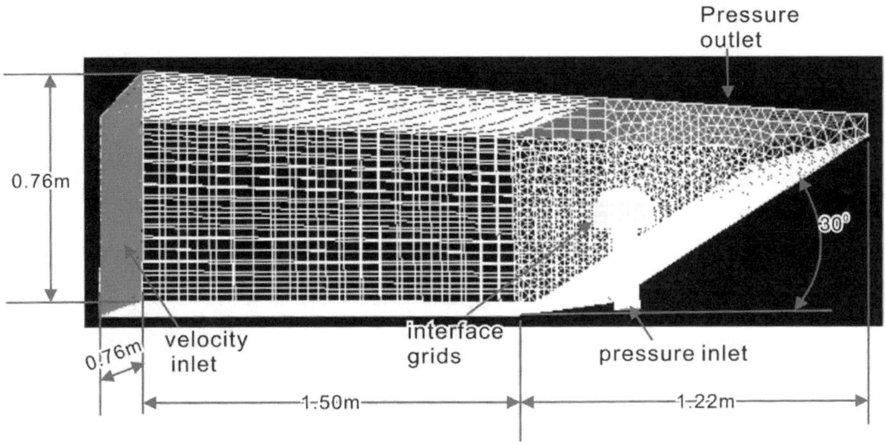

Fig. 14 Modelling of domain and dimension

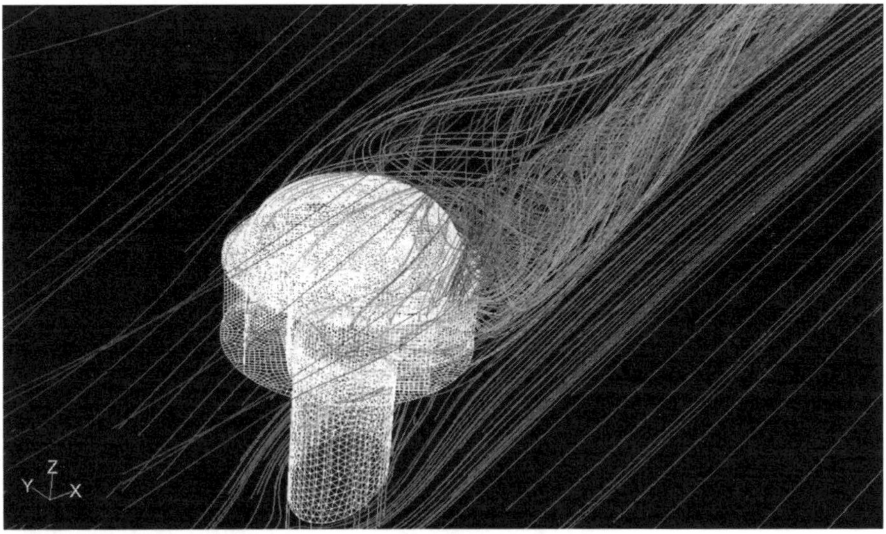

Fig. 15 Three dimensional path line of the flow associated with the rotating ventilator at 10 m/s

with nearly 15 % increase in air extraction and is currently commercially manufactured and marketed by CSR Edmonds. The works have also provided engineering solutions for enhanced safety and performance of wind driven ventilator in rain and operation at low speed and produced novel techniques for skin friction measurement [40] on roof surface, and formed the basis for the concept of a hybrid ventilator [30, 41] to overcome the dependence of conventional roof top ventilators on the availability of wind. This resulted in the 2008 AIRAH Excellence award winning hybrid ventilator, the 'ECPOWER' in the HVAC-Achiever category as a distinguished Australian product, invention or innovation in heating,

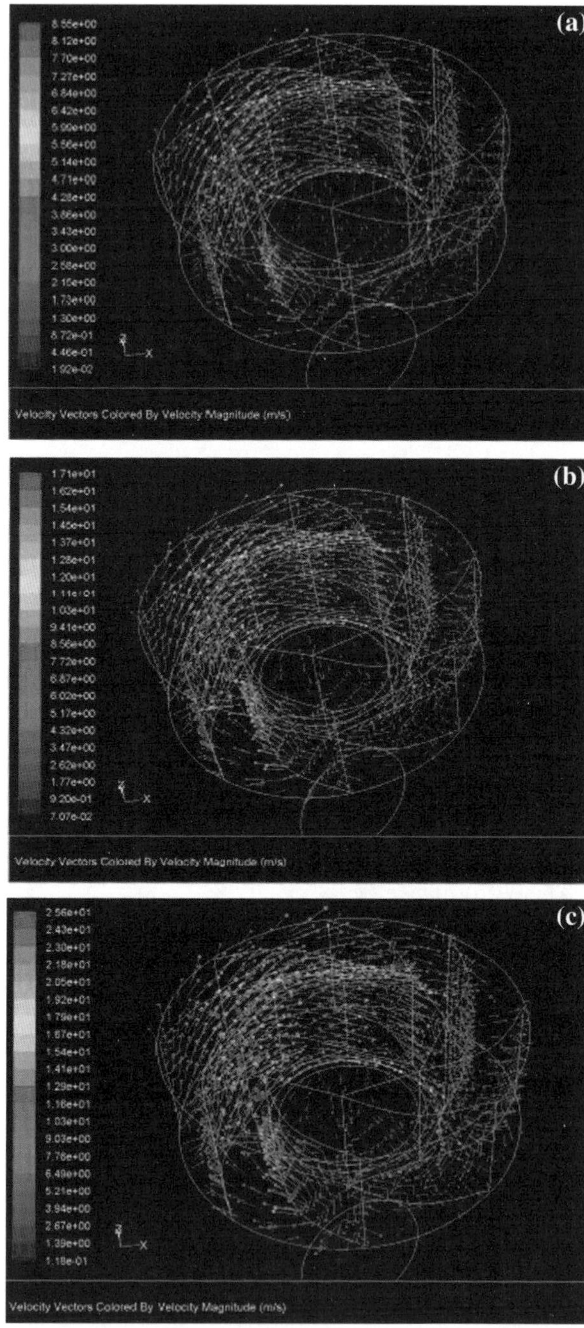

Fig. 16 **a–c** 3D velocity vector distribution inside the ventilator. **a** Free stream velocity = 5 m/s. **b** Free stream velocity = 10 m/s. **c** Free stream velocity = 15 m/s

Fig. 17 A computer aided image of wind-electric ECOPOWER

ventilation/air conditioning. The 'ECOPOWER' (see Fig. 17) is currently marketed worldwide by CSR Edmonds. Stories about these outcomes have featured in Australian Research Council report to Australian parliament [42].

4 Future Possibilities: Incorporation of Flow Control Technologies

The performance of horizontal axis wind turbines and wind driven ventilators can be substantially improved by incorporating passive or active flow control technologies. However, with the unit cost of each rotating ventilator presently ranging from \$50–\$200, incorporation of such technologies may be more appropriate to horizontal axis wind turbine purely from present economic realities.

Gyatt and Lissaman [43] carried out a theoretical and field experimental program to investigate the use of tip devices on horizontal axis wind turbine rotors. The objective was to improve performance by the reduction of tip losses. While power output can always be increased by a simple radial tip extension, such a modification also results in an increased gale load both because of the extra projected area and longer moment arm. Tip devices have the potential to increase power output without such a structural penalty. The types considered were a change in tip plan form, and a single-element and double-element non planar tip extension (winglets). Results for each of the three new tip devices, compared with the original tip, showed a small decrease (of the order of 1 kW) in power output over the measured range of wind speeds from cut-in at about 4 m/s to over 20 m/s, well into the stall limiting region. Changes in orientation and angle-of-attack of the

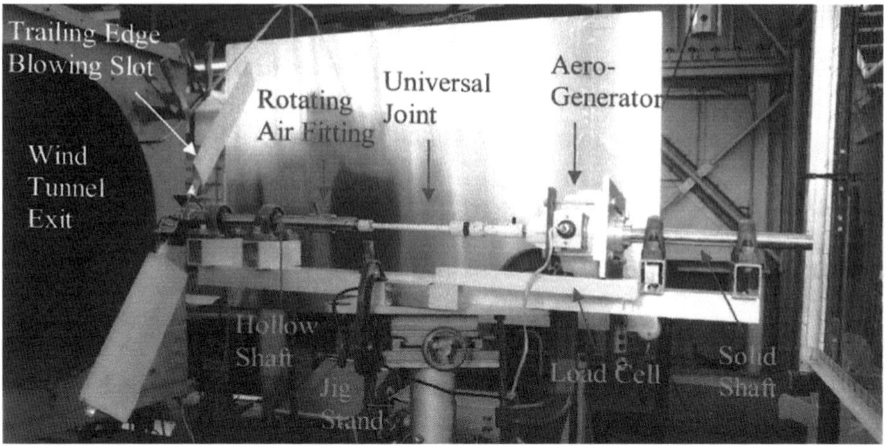

Fig. 18 Experimental set-up for active flow control application on Horizontal Axis Wind Turbine

winglets were not made. For aircraft wing tip devices, favorable tip shapes have been reported and it is likely that the tip devices tested in this program did not improve rotor performance because they were not optimally adjusted.

Gyatt [44] also tested the effectiveness of vortex generators (VGs) for a small horizontal axis wind turbine, Arrays of VGs in a counter-rotating arrangement were tested on the inbound half-span, outboard half-span, and on the entire blade. Field test data showed that VGs increased power output up to 20 at wind speeds above 10 m/s with only a small (less than 4) performance penalty at lower speeds. The VGs on the outboard span of the blade were more effective than those on inner sections. For the case of full span coverage, the energy yearly output increased almost 6 at a site with a mean wind speed of 16 mph. The VGs did reduce the performance loss caused by leading edge roughness. An increase in blade pitch angle has an effect on the power curve similar to the addition of VGs. VGs alleviate the sensitivity of wind turbine rotors to leading edge roughness caused by bugs and drift.

Some limited work conducted at the University of New South Wales suggests application of Coanda jet as a circulation control technique may be a useful means to improve the performance of Horizontal axis wind turbines. Figure 18 shows the experimental set up conducted at the aerodynamics laboratory of the University of New South Wales. The results are promising. An important lesson learnt from this experiment is that there is a need to have an efficient air delivery system. This has lead to exploring the use of an orbital pump, a concept first proposed by Day [45]. A possible adaptation of the proposed compressor to wind turbine for performance enhancement is shown in Fig. 18. However, to be effective to small horizontal axis wind turbine, will entail substantial miniaturization of the compressor. This would be a major engineering undertaking which will be highly dependent on the size of the hub of the wind turbine in adapting the air delivery system.

For effective circulation control, the flow must be prevented from completely breaking away from the surface. Since hardly any work of note exists for

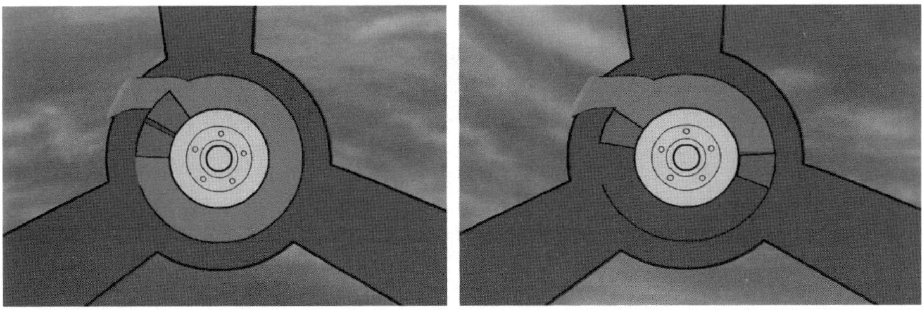

Fig. 19 A schematic diagram of possible arrangements of miniaturized compressor in the hub of a wind turbine

Fig. 20 Jessica Watson, the youngest person ever to sail around the world in 210 days (completed on 16 May, 2010, from http://www.news.com.au/)

application of circulation control on a rotating blade, information available on fixed wing or blade could be used as guide. From published works it appears, full span or tip blowing works best on fixed wings or blades. It is assumed that this will also be true for rotating blades. Moreover, it is likely that shifting the rear stagnation point and attenuating or eliminating tip vortices in wind turbines is as valid as it is for aircraft wings. Thus after all parasitic losses are subtracted rotor efficiency is likely to increase significantly resulting in higher shaft power output (Fig. 19).

An excellent example of application of small wind turbine alongside solar panels for power generation can be seen from the picture of the 'Pink Lady', the boat that Jessica Watson used for her historic round the world voyage (see Fig. 20). This may be a pointer for future with combined use of wind and solar renewable power sources, one complementing the other in the times of no-wind or no-sun conditions.

5 Conclusions

A primary concern with any wind driven device is the availability of wind, in other words, the ability to operate these devices at zero and low wind conditions. If continuous power is to be achieved, adequate means of storing of the electricity generated efficiently and economically by batteries or other means have to be ensured. Alternatively, some form of hybrid systems incorporating wind, solar and other power sources have to be devised to reduce usage of electricity produced by fossil fuel sources.

As wind power is becoming far more popular the industry is suffering from lack of regulations on both small and large turbines. Currently manufactures and installers of small wind turbines in Australia are not required to meet accreditation standards such as those set for the photovoltaic's industry. As the industry matures these standards must be adopted industry wide to avoid potential accidents or costly errors.

Finally the chapter is ended on a positive message. Greater awareness in people about environmental issues is forcing governments around the world to formulate policies towards greener power sources. This will give boost to more directed and intense research and rapid technological developments and breakthroughs. In the words of the Nobel Peace-laureate and conservation icon Al Gore, 'the time will soon come for 21st century technologies that use fuel that is free forever: the sun, the wind and the natural heat of the earth.'

Acknowledgments The author would like to acknowledge the contributions of the late Prof Archer, Terry Flynn (UNSW Laboratory Officer) and his post graduate students, Simon Shun, Jason Lien, Dewan Rashid and Anthony Pissasale. Acknowledgement is also made to Allan Ramsay (Managing Director) and Derek Nunn (Technical Director) of CSR Edmonds Australia and the Australian Research Council for providing funding to support the works that have formed the materials for this chapter.

References

1. Needham J (1965) Science and civilisation in China, vol 4, Physics and physical technology, Part II: mechanical engineering, Cambridge Univ Press, Cambridge, pp 556–560
2. Wuff HE (1966) The traditional crafts of Persia, their development, technology and influence on eastern and western civilisation. MIT Press, Cambridge, pp 284–289
3. Kealey EJ (1987) Harvesting the air. Univ of California Press, Berkeley
4. Holt R (1988) The mills of medieval England, Appendix 1. Basil Blackwell Ltd, Oxford

5. Harrison L (2001) Leader: time to turn professional. Wind Monthly, USA, January issue, 2001
6. Department of Energy, USA, 1998, as quoted in CNN.com on the topic 'Power crisis: indepth specials', 31 Jan 2001
7. Sawyer S (2010) Annual market update. http://www.gwec.net/fileadmin/images/Publications/GWEC_annual_market-_update_-_2nd_edition_April_2011.pdf
8. Wilkes J (2010) EU wind in power 2010 European statistics. http://www.ewea.org/fileadmin/ewea_documents/statistics/EWEA_Annual_Statistics_2010.pdf
9. Technologies (2010) Clean energy council. http://www.cleanenergycouncil.org.au/cec/technologies/wind.html
10. Lobbecke (2010) Home wind power, cartoon, the daily telegraph, Australia, 20 April 2010, p 109
11. Silmalis L (2010) Windmills for your roof, The Sunday Telegraph, Australia, 20 April 2010, p 9
12. Glauert H (1935) Airplane propellers, in aerodynamic theory. In: Durand WF, Div L (eds) vol IV, Springer, Berlin, also Dover Publications Inc., NY, 1963
13. Betz A, Prandtl L (1919) Screw propellers with minimum energy loss. Goettingen Nachrichten, Mathematics-Physics, vol 2, p 193
14. Goldstein S (1929) On the vortex theory of propellers. Proc R Soc Lond 123(4):440
15. Larrabee EE (1979) Design of propellers for motor soarers, science and technology of low speed and motorless flight, NASA CP-2085, part 1, pp 285–303
16. Sanderson RJ, Archer RD (1983) Optimum propeller wind turbines. J Energy AIAA 7(6):695–701
17. Cox DH, Sanderson RJ, Khoo LT, Archer RD (1986) Propeller wind turbine rotor design and testing. In: Proceedings of the 9th Australasian fluid mechanics conference, New Zealand, pp 1–4
18. Ahmed NA, Archer RD (2002) Testing of highly loaded horizontal axis wind turbines designed for optimum performance. Int J Renew Energy 25:613–618
19. Theodorsen T (1948) The theory of propellers. McGraw Hill Book Co., New York, pp 23–38
20. Bowden GJ, Adler J, Dabbs T, Walter J (1980) The potential of wind energy in Antarctica. Wind Eng 4(3):163–176
21. Yang W et al (2009) IAQ investigation according to school buildings in Korea. Envi Man 90:348–354
22. Kreichelt TE, Kern G (1976) Ventilation in hot process buildings. J Iron Steel Eng 39–46
23. Jones AP (1999) IAQ and health. Atmos Environ 33:2464–4535
24. Biblow AC (2009) NY to require landlords to notify tenants of IAQ results. Real Estate Financ 29–31
25. Air conditioning design strategy (2003) Report for Melbourne City Council (CH2) by Advanced Environmental Concepts Pty, Ltd, Report no. AESYB2000G\0\SFT3053, May 2003
26. Ahmed NA, Beck J (1996) Destructive wind tunnel tests. UNSW Unisearch Report no. 23214-10
27. Ahmed NA, Beck J (1997) Wind tunnel tests on ventilators. UNSW Unisearch Report no. 29295-01
28. Rashid DH, Ahmed NA, Archer RD (2003) Study of aerodynamic forces on a rotating wind driven ventilator. Wind Eng 27(1):63–72
29. Flynn TG, Ahmed NA (2005) Investigation of rotating ventilator using smoke flow visualisation and hot-wire anemometer. In: Proceedings of 5th pacific symposium on flow visualisation and image processing, Whitsundays, Australia, Paper No. PSFVIP-5-214, 27–29 Sept 2005
30. Shun S, Ahmed NA (2008) Utilising wind and solar energy as power sources for a hybrid building ventilation device. Renew Energy 33:1392–1397
31. Ahmed NA, Archer RD (2001) Post-stall behaviour of a wing under externally imposed sound. AIAA J Aircr 38(5):961–963

32. Khan N, Su Y, Riffat SB (2008) A review on wind driven ventilation techniques. Energy Buildings 40:1586–1604
33. Lai C (2003) Experiments on the ventilation efficiency of turbine ventilator used for building and factory ventilation. Energy Buildings 35:927–932
34. Khan N, Su Y, Riffat SB, Biggs C (2008) Performance testing and comparison of turbine ventilators. Renew Energy 33:2441–2447
35. Pissasale A, Ahmed NA (2002) Theoretical calibration of a five hole probe for highly three dimensional flow. Int J Measur Sci Technol 13(7):1100–1107
36. Pissasale A, Ahmed NA (2002) A novel method of extending the calibration range of five hole probe for highly three dimensional flows. J Flow Measur Instrum 13(1–2):23–30
37. Pissasale A, Ahmed NA (2003) Examining the effect of flow reversal on seven-hole Probe measurements. AIAA J 41(12):2460–2467
38. Pissasale A, Ahmed NA (2004) Development of a functional relationship between port pressures and flow properties for calibration and application of multi-hole probes to highly 3D flows. Exp Fl 36(3):422–436
39. Lien SJ, Ahmed NA (2006) Skin friction determination in turbulent boundary layers using multi-hole pressure probes'25th AIAA Appl Aerodyn Conference, San Francisco, USA, AIAA-2006-3659, 8–10 June 2006
40. Lien SJ, Ahmed NA (2010) Numerical simulation of rooftop ventilator flow. Build Environ 45(8):1808–1815
41. Ahmed NA (2010) Wind driven natural-solar/electric hybrid ventilators. In: Muyeen Kitami SM (ed) Wind power, Section D: the environmental issues, Chap. 23, In-Tech Organization, Austria, ISBN 978-95-7619-81-7, 04–22 Feb 2010
42. Annual Report 2008–2009 (2009) Australian research council, Canberra, ACT, p 92
43. Gyatt GW, Lissaman PBS (1985) Development and testing of tip devices for horizontal axis wind turbines. NASA-CR-174991, 1 May 1985
44. Gyatt GW (1986) Development and testing of vortex generators for small horizontal axis wind turbines. NASA-CR-179514, 1 July 1986
45. Day TD (2006) Circulation control effect and circulation control for non-aeronautical applications. Application of Circulation control Technologies. In: Joslin RD, Jones GS (eds) Progress in aeronautics and astronautics, vol 214, pp 599–613

Index

P. M. Pardalos et al. (eds.), *Handbook of Wind Power Systems*, Energy Systems,
DOI: 10.1007/978-3-642-41080-2,

Printed by Printforce, the Netherlands